KKM THEORY AND APPLICATIONS IN NONLINEAR ANALYSIS

MONOGRAPHS AND TEXTBOOKS IN PURE AND APPLIED MATHEMATICS

1. *K. Yano,* Integral Formulas in Riemannian Geometry (1970)
2. *S. Kobayashi,* Hyperbolic Manifolds and Holomorphic Mappings (1970)
3. *V. S. Vladimirov,* Equations of Mathematical Physics (A. Jeffrey, ed.; A. Littlewood, trans.) (1970)
4. *B. N. Pshenichnyi,* Necessary Conditions for an Extremum (L. Neustadt, translation ed.; K. Makowski, trans.) (1971)
5. *L. Narici et al.,* Functional Analysis and Valuation Theory (1971)
6. *S. S. Passman,* Infinite Group Rings (1971)
7. *L. Dornhoff,* Group Representation Theory. Part A: Ordinary Representation Theory. Part B: Modular Representation Theory (1971, 1972)
8. *W. Boothby and G. L. Weiss, eds.,* Symmetric Spaces (1972)
9. *Y. Matsushima,* Differentiable Manifolds (E. T. Kobayashi, trans.) (1972)
10. *L. E. Ward, Jr.,* Topology (1972)
11. *A. Babakhanian,* Cohomological Methods in Group Theory (1972)
12. *R. Gilmer,* Multiplicative Ideal Theory (1972)
13. *J. Yeh,* Stochastic Processes and the Wiener Integral (1973)
14. *J. Barros-Neto,* Introduction to the Theory of Distributions (1973)
15. *R. Larsen,* Functional Analysis (1973)
16. *K. Yano and S. Ishihara,* Tangent and Cotangent Bundles (1973)
17. *C. Procesi,* Rings with Polynomial Identities (1973)
18. *R. Hermann,* Geometry, Physics, and Systems (1973)
19. *N. R. Wallach,* Harmonic Analysis on Homogeneous Spaces (1973)
20. *J. Dieudonné,* Introduction to the Theory of Formal Groups (1973)
21. *I. Vaisman,* Cohomology and Differential Forms (1973)
22. *B.-Y. Chen,* Geometry of Submanifolds (1973)
23. *M. Marcus,* Finite Dimensional Multilinear Algebra (in two parts) (1973, 1975)
24. *R. Larsen,* Banach Algebras (1973)
25. *R. O. Kujala and A. L. Vitter, eds.,* Value Distribution Theory: Part A; Part B: Deficit and Bezout Estimates by Wilhelm Stoll (1973)
26. *K. B. Stolarsky,* Algebraic Numbers and Diophantine Approximation (1974)
27. *A. R. Magid,* The Separable Galois Theory of Commutative Rings (1974)
28. *B. R. McDonald,* Finite Rings with Identity (1974)
29. *J. Satake,* Linear Algebra (S. Koh et al., trans.) (1975)
30. *J. S. Golan,* Localization of Noncommutative Rings (1975)
31. *G. Klambauer,* Mathematical Analysis (1975)
32. *M. K. Agoston,* Algebraic Topology (1976)
33. *K. R. Goodearl,* Ring Theory (1976)
34. *L. E. Mansfield,* Linear Algebra with Geometric Applications (1976)
35. *N. J. Pullman,* Matrix Theory and Its Applications (1976)
36. *B. R. McDonald,* Geometric Algebra Over Local Rings (1976)
37. *C. W. Groetsch,* Generalized Inverses of Linear Operators (1977)
38. *J. E. Kuczkowski and J. L. Gersting,* Abstract Algebra (1977)
39. *C. O. Christenson and W. L. Voxman,* Aspects of Topology (1977)
40. *M. Nagata,* Field Theory (1977)
41. *R. L. Long,* Algebraic Number Theory (1977)
42. *W. F. Pfeffer,* Integrals and Measures (1977)
43. *R. L. Wheeden and A. Zygmund,* Measure and Integral (1977)
44. *J. H. Curtiss,* Introduction to Functions of a Complex Variable (1978)
45. *K. Hrbacek and T. Jech,* Introduction to Set Theory (1978)
46. *W. S. Massey,* Homology and Cohomology Theory (1978)
47. *M. Marcus,* Introduction to Modern Algebra (1978)
48. *E. C. Young,* Vector and Tensor Analysis (1978)
49. *S. B. Nadler, Jr.,* Hyperspaces of Sets (1978)
50. *S. K. Segal,* Topics in Group Kings (1978)
51. *A. C. M. van Rooij,* Non-Archimedean Functional Analysis (1978)
52. *L. Corwin and R. Szczarba,* Calculus in Vector Spaces (1979)
53. *C. Sadosky,* Interpolation of Operators and Singular Integrals (1979)

54. *J. Cronin,* Differential Equations (1980)
55. *C. W. Groetsch,* Elements of Applicable Functional Analysis (1980)
56. *I. Vaisman,* Foundations of Three-Dimensional Euclidean Geometry (1980)
57. *H. I. Freedan,* Deterministic Mathematical Models in Population Ecology (1980)
58. *S. B. Chae,* Lebesgue Integration (1980)
59. *C. S. Rees et al.,* Theory and Applications of Fourier Analysis (1981)
60. *L. Nachbin,* Introduction to Functional Analysis (R. M. Aron, trans.) (1981)
61. *G. Orzech and M. Orzech,* Plane Algebraic Curves (1981)
62. *R. Johnsonbaugh and W. E. Pfaffenberger,* Foundations of Mathematical Analysis (1981)
63. *W. L. Voxman and R. H. Goetschel,* Advanced Calculus (1981)
64. *L. J. Corwin and R. H. Szczarba,* Multivariable Calculus (1982)
65. *V. I. Istrăţescu,* Introduction to Linear Operator Theory (1981)
66. *R. D. Järvinen,* Finite and Infinite Dimensional Linear Spaces (1981)
67. *J. K. Beem and P. E. Ehrlich,* Global Lorentzian Geometry (1981)
68. *D. L. Armacost,* The Structure of Locally Compact Abelian Groups (1981)
69. *J. W. Brewer and M. K. Smith, eds.,* Emmy Noether: A Tribute (1981)
70. *K. H. Kim,* Boolean Matrix Theory and Applications (1982)
71. *T. W. Wieting,* The Mathematical Theory of Chromatic Plane Ornaments (1982)
72. *D. B. Gauld,* Differential Topology (1982)
73. *R. L. Faber,* Foundations of Euclidean and Non-Euclidean Geometry (1983)
74. *M. Carmeli,* Statistical Theory and Random Matrices (1983)
75. *J. H. Carruth et al.,* The Theory of Topological Semigroups (1983)
76. *R. L. Faber,* Differential Geometry and Relativity Theory (1983)
77. *S. Barnett,* Polynomials and Linear Control Systems (1983)
78. *G. Karpilovsky,* Commutative Group Algebras (1983)
79. *F. Van Oystaeyen and A. Verschoren,* Relative Invariants of Rings (1983)
80. *I. Vaisman,* A First Course in Differential Geometry (1984)
81. *G. W. Swan,* Applications of Optimal Control Theory in Biomedicine (1984)
82. *T. Petrie and J. D. Randall,* Transformation Groups on Manifolds (1984)
83. *K. Goebel and S. Reich,* Uniform Convexity, Hyperbolic Geometry, and Nonexpansive Mappings (1984)
84. *T. Albu and C. Năstăsescu,* Relative Finiteness in Module Theory (1984)
85. *K. Hrbacek and T. Jech,* Introduction to Set Theory: Second Edition (1984)
86. *F. Van Oystaeyen and A. Verschoren,* Relative Invariants of Rings (1984)
87. *B. R. McDonald,* Linear Algebra Over Commutative Rings (1984)
88. *M. Namba,* Geometry of Projective Algebraic Curves (1984)
89. *G. F. Webb,* Theory of Nonlinear Age-Dependent Population Dynamics (1985)
90. *M. R. Bremner et al.,* Tables of Dominant Weight Multiplicities for Representations of Simple Lie Algebras (1985)
91. *A. E. Fekete,* Real Linear Algebra (1985)
92. *S. B. Chae,* Holomorphy and Calculus in Normed Spaces (1985)
93. *A. J. Jerri,* Introduction to Integral Equations with Applications (1985)
94. *G. Karpilovsky,* Projective Representations of Finite Groups (1985)
95. *L. Narici and E. Beckenstein,* Topological Vector Spaces (1985)
96. *J. Weeks,* The Shape of Space (1985)
97. *P. R. Gribik and K. O. Kortanek,* Extremal Methods of Operations Research (1985)
98. *J.-A. Chao and W. A. Woyczynski, eds.,* Probability Theory and Harmonic Analysis (1986)
99. *G. D. Crown et al.,* Abstract Algebra (1986)
100. *J. H. Carruth et al.,* The Theory of Topological Semigroups, Volume 2 (1986)
101. *R. S. Doran and V. A. Belfi,* Characterizations of C*-Algebras (1986)
102. *M. W. Jeter,* Mathematical Programming (1986)
103. *M. Altman,* A Unified Theory of Nonlinear Operator and Evolution Equations with Applications (1986)
104. *A. Verschoren,* Relative Invariants of Sheaves (1987)
105. *R. A. Usmani,* Applied Linear Algebra (1987)
106. *P. Blass and J. Lang,* Zariski Surfaces and Differential Equations in Characteristic $p > 0$ (1987)
107. *J. A. Reneke et al.,* Structured Hereditary Systems (1987)
108. *H. Busemann and B. B. Phadke,* Spaces with Distinguished Geodesics (1987)
109. *R. Harte,* Invertibility and Singularity for Bounded Linear Operators (1988)

110. *G. S. Ladde et al.,* Oscillation Theory of Differential Equations with Deviating Arguments (1987)
111. *L. Dudkin et al.,* Iterative Aggregation Theory (1987)
112. *T. Okubo,* Differential Geometry (1987)
113. *D. L. Stancl and M. L. Stancl,* Real Analysis with Point-Set Topology (1987)
114. *T. C. Gard,* Introduction to Stochastic Differential Equations (1988)
115. *S. S. Abhyankar,* Enumerative Combinatorics of Young Tableaux (1988)
116. *H. Strade and R. Farnsteiner,* Modular Lie Algebras and Their Representations (1988)
117. *J. A. Huckaba,* Commutative Rings with Zero Divisors (1988)
118. *W. D. Wallis,* Combinatorial Designs (1988)
119. *W. Więsław,* Topological Fields (1988)
120. *G. Karpilovsky,* Field Theory (1988)
121. *S. Caenepeel and F. Van Oystaeyen,* Brauer Groups and the Cohomology of Graded Rings (1989)
122. *W. Kozlowski,* Modular Function Spaces (1988)
123. *E. Lowen-Colebunders,* Function Classes of Cauchy Continuous Maps (1989)
124. *M. Pavel,* Fundamentals of Pattern Recognition (1989)
125. *V. Lakshmikantham et al.,* Stability Analysis of Nonlinear Systems (1989)
126. *R. Sivaramakrishnan,* The Classical Theory of Arithmetic Functions (1989)
127. *N. A. Watson,* Parabolic Equations on an Infinite Strip (1989)
128. *K. J. Hastings,* Introduction to the Mathematics of Operations Research (1989)
129. *B. Fine,* Algebraic Theory of the Bianchi Groups (1989)
130. *D. N. Dikranjan et al.,* Topological Groups (1989)
131. *J. C. Morgan II,* Point Set Theory (1990)
132. *P. Biler and A. Witkowski,* Problems in Mathematical Analysis (1990)
133. *H. J. Sussmann,* Nonlinear Controllability and Optimal Control (1990)
134. *J.-P. Florens et al.,* Elements of Bayesian Statistics (1990)
135. *N. Shell,* Topological Fields and Near Valuations (1990)
136. *B. F. Doolin and C. F. Martin,* Introduction to Differential Geometry for Engineers (1990)
137. *S. S. Holland, Jr.,* Applied Analysis by the Hilbert Space Method (1990)
138. *J. Okniński,* Semigroup Algebras (1990)
139. *K. Zhu,* Operator Theory in Function Spaces (1990)
140. *G. B. Price,* An Introduction to Multicomplex Spaces and Functions (1991)
141. *R. B. Darst,* Introduction to Linear Programming (1991)
142. *P. L. Sachdev,* Nonlinear Ordinary Differential Equations and Their Applications (1991)
143. *T. Husain,* Orthogonal Schauder Bases (1991)
144. *J. Foran,* Fundamentals of Real Analysis (1991)
145. *W. C. Brown,* Matrices and Vector Spaces (1991)
146. *M. M. Rao and Z. D. Ren,* Theory of Orlicz Spaces (1991)
147. *J. S. Golan and T. Head,* Modules and the Structures of Rings (1991)
148. *C. Small,* Arithmetic of Finite Fields (1991)
149. *K. Yang,* Complex Algebraic Geometry (1991)
150. *D. G. Hoffman et al.,* Coding Theory (1991)
151. *M. O. González,* Classical Complex Analysis (1992)
152. *M. O. González,* Complex Analysis (1992)
153. *L. W. Baggett,* Functional Analysis (1992)
154. *M. Sniedovich,* Dynamic Programming (1992)
155. *R. P. Agarwal,* Difference Equations and Inequalities (1992)
156. *C. Brezinski,* Biorthogonality and Its Applications to Numerical Analysis (1992)
157. *C. Swartz,* An Introduction to Functional Analysis (1992)
158. *S. B. Nadler, Jr.,* Continuum Theory (1992)
159. *M. A. Al-Gwaiz,* Theory of Distributions (1992)
160. *E. Perry,* Geometry: Axiomatic Developments with Problem Solving (1992)
161. *E. Castillo and M. R. Ruiz-Cobo,* Functional Equations and Modelling in Science and Engineering (1992)
162. *A. J. Jerri,* Integral and Discrete Transforms with Applications and Error Analysis (1992)
163. *A. Charlier et al.,* Tensors and the Clifford Algebra (1992)
164. *P. Biler and T. Nadzieja,* Problems and Examples in Differential Equations (1992)

165. *E. Hansen*, Global Optimization Using Interval Analysis (1992)
166. *S. Guerre-Delabrière*, Classical Sequences in Banach Spaces (1992)
167. *Y. C. Wong*, Introductory Theory of Topological Vector Spaces (1992)
168. *S. H. Kulkarni and B. V. Limaye*, Real Function Algebras (1992)
169. *W. C. Brown*, Matrices Over Commutative Rings (1993)
170. *J. Loustau and M. Dillon*, Linear Geometry with Computer Graphics (1993)
171. *W. V. Petryshyn*, Approximation-Solvability of Nonlinear Functional and Differential Equations (1993)
172. *E. C. Young*, Vector and Tensor Analysis: Second Edition (1993)
173. *T. A. Bick*, Elementary Boundary Value Problems (1993)
174. *M. Pavel*, Fundamentals of Pattern Recognition: Second Edition (1993)
175. *S. A. Albeverio et al.*, Noncommutative Distributions (1993)
176. *W. Fulks*, Complex Variables (1993)
177. *M. M. Rao*, Conditional Measures and Applications (1993)
178. *A. Janicki and A. Weron*, Simulation and Chaotic Behavior of α-Stable Stochastic Processes (1994)
179. *P. Neittaanmäki and D. Tiba*, Optimal Control of Nonlinear Parabolic Systems (1994)
180. *J. Cronin*, Differential Equations: Introduction and Qualitative Theory, Second Edition (1994)
181. *S. Heikkilä and V. Lakshmikantham*, Monotone Iterative Techniques for Discontinuous Nonlinear Differential Equations (1994)
182. *X. Mao*, Exponential Stability of Stochastic Differential Equations (1994)
183. *B. S. Thomson*, Symmetric Properties of Real Functions (1994)
184. *J. E. Rubio*, Optimization and Nonstandard Analysis (1994)
185. *J. L. Bueso et al.*, Compatibility, Stability, and Sheaves (1995)
186. *A. N. Michel and K. Wang*, Qualitative Theory of Dynamical Systems (1995)
187. *M. R. Darnel*, Theory of Lattice-Ordered Groups (1995)
188. *Z. Naniewicz and P. D. Panagiotopoulos*, Mathematical Theory of Hemivariational Inequalities and Applications (1995)
189. *L. J. Corwin and R. H. Szczarba*, Calculus in Vector Spaces: Second Edition (1995)
190. *L. H. Erbe et al.*, Oscillation Theory for Functional Differential Equations (1995)
191. *S. Agaian et al.*, Binary Polynomial Transforms and Nonlinear Digital Filters (1995)
192. *M. I. Gil'*, Norm Estimations for Operation-Valued Functions and Applications (1995)
193. *P. A. Grillet*, Semigroups: An Introduction to the Structure Theory (1995)
194. *S. Kichenassamy*, Nonlinear Wave Equations (1996)
195. *V. F. Krotov*, Global Methods in Optimal Control Theory (1996)
196. *K. I. Beidar et al.*, Rings with Generalized Identities (1996)
197. *V. I. Arnautov et al.*, Introduction to the Theory of Topological Rings and Modules (1996)
198. *G. Sierksma*, Linear and Integer Programming (1996)
199. *R. Lasser*, Introduction to Fourier Series (1996)
200. *V. Sima*, Algorithms for Linear-Quadratic Optimization (1996)
201. *D. Redmond*, Number Theory (1996)
202. *J. K. Beem et al.*, Global Lorentzian Geometry: Second Edition (1996)
203. *M. Fontana et al.*, Prüfer Domains (1997)
204. *H. Tanabe*, Functional Analytic Methods for Partial Differential Equations (1997)
205. *C. Q. Zhang*, Integer Flows and Cycle Covers of Graphs (1997)
206. *E. Spiegel and C. J. O'Donnell*, Incidence Algebras (1997)
207. *B. Jakubczyk and W. Respondek*, Geometry of Feedback and Optimal Control (1998)
208. *T. W. Haynes et al.*, Fundamentals of Domination in Graphs (1998)
209. *T. W. Haynes et al.*, Domination in Graphs: Advanced Topics (1998)
210. *L. A. D'Alotto et al.*, A Unified Signal Algebra Approach to Two-Dimensional Parallel Digital Signal Processing (1998)
211. *F. Halter-Koch*, Ideal Systems (1998)
212. *N. K. Govil et al.*, Approximation Theory (1998)
213. *R. Cross*, Multivalued Linear Operators (1998)
214. *A. A. Martynyuk*, Stability by Liapunov's Matrix Function Method with Applications (1998)

215. *A. Favini and A. Yagi,* Degenerate Differential Equations in Banach Spaces (1999)
216. *A. Illanes and S. Nadler, Jr.,* Hyperspaces: Fundamentals and Recent Advances (1999)
217. *G. Kato and D. Struppa,* Fundamentals of Algebraic Microlocal Analysis (1999)
218. *G. X.-Z. Yuan,* KKM Theory and Applications in Nonlinear Analysis (1999)

Additional Volumes in Preparation

KKM THEORY AND APPLICATIONS IN NONLINEAR ANALYSIS

George Xian-Zhi Yuan
The University of Queensland
Brisbane, Australia

MARCEL DEKKER, INC. NEW YORK • BASEL

ISBN: 0-8247-0031-7

This book is printed on acid-free paper.

Headquarters
Marcel Dekker, Inc.
270 Madison Avenue, New York, NY 10016
tel: 212-696-9000; fax: 212-685-4540

Eastern Hemisphere Distribution
Marcel Dekker AG
Hutgasse 4, Postfach 812, CH-4001 Basel, Switzerland
tel: 44-61-261-8482; fax: 44-61-261-8896

World Wide Web
http://www.dekker.com

The publisher offers discounts on this book when ordered in bulk quantities. For more information, write to Special Sales/Professional Marketing at the headquarters address above.

Current printing (last digit)
10 9 8 7 6 5 4 3 2 1

PRINTED IN THE UNITED STATES OF AMERICA

To Mei, Susan and Steven

Preface

Since Knaster, Kuratowski and Mazurkiewicz formulated the so-called KKM principle in 1929, the study of nonlinear analysis associated with the KKM principle has been a rapidly developing area in mathematics which is emerging as an independent field. Owing to the significant contributions by Professor Ky Fan in the last few decades, a large part of nonlinear analysis related to the KKM principle and its remarkable applications in both the natural and social sciences has been established. Though this subject is still in its formative stage, its applications are so rich and broad that such an area of study definitely qualifies for a special name: **KKM theory** in nonlinear analysis.

The main objective of this book is to provide an introduction to basic and advanced aspects of the major part of KKM theory. It is our purpose both to build a bridge for readers who want to know and enter this exciting subject to learn the new ideas and methods developed in the last several decades and to provide a reference for researchers in areas of nonlinear analysis such as variational inequality theory, minimax theory, mathematical economics, optimization, game theory and related fields. A comprehensive bibliography at the end of this book gives interested readers the option of delving deeper into KKM theory and its applications. We have tried to gather together, using a uniform, consistent terminology, a number of the most important results in KKM theory which are scattered widely in the literature of the natural and social sciences of the last several decades. Our objective is to make those results accessible, in a more or less self-contained presentation, to readers with basic knowledge of mathematical analysis, economics, optimization theory, and game theory. Traditionally, the study of minimax theory is related to KKM theory. In this book, from the point of view of the author's preference and research, we include a chapter presenting some major advances in topological minimax theory by using the so-called *connectedness approach*, which was discovered by Professor Wu in the late 1950s. The importance of Wu's connectedness approach finally was widely recognized more than ten years after publication in 1959. Though the original motivation for the study of (topological) minimax theorems was, of course, von Neumann's work on the game theory of strategies in 1928, after a generalization of von Neumann's original results for matrices appeared, the study of minimax theory became progressively more remote from game theory. Minimax theory started to become an object of study in its own right, and so did the study of topological minimax theory and related topological intersection problems. As there are many questions and since more applications need to be established in this subject, we think it is worthy to be included in this book so as to provide an opportunity for more people to be aware of this very new topic which promises to play an important role in nonlinear analysis and its applications in the near future.

We believe that KKM theory is one of the few areas among the subjects from nonlinear analysis that could provide the easy and convenient forms and tools for the study of problems from social and applied sciences such as economics, optimization and game theory. We hope this book will give readers a clear outline of how KKM theory provides a unified approach to the study of general equilibrium theory in mathematical economics, variational inequalities, optimization, game theory, and many other subjects that seem to be quite diverse. Thus, we wish to convince readers that the KKM theory is not only a powerful subject in nonlinear analysis itself, but is also a very useful tool for applied and social sciences.

The second objective of the book is to present a unified treatment (the so-called *generic method*) of stability and uniqueness in nonlinear analysis from the point of view of Baire category theory. Our study shows that the generic approach works very well for the problems treated in this book, including special cases of minimax inequalities, fixed and coincidence points, quasi-variational inequalities, nonlinear complementarity problems, saddle points of multi-objective optimization problems, Nash equilibria, and Walras equilibrium price models. Though the study of generic stability from the point of view of Baire category theory was started in 1952 by Professor Fort to establish the first (essential) stability of fixed point result (later improved by Professors Jiang and Wu in the early 1960s), it was only in the 1970s that the importance of this method was rediscovered by many researchers (e.g., Professors Dierker, van Damme, Yu, and others) with emphasis on economics, optimization and game theory. In the line of mathematics, since the 1970s, Professor Kenderov and his Hungarian school established many generic results not only on the continuity of set-valued mappings and generic differentiability for convex functions in Banach spaces, but also on generic stability, generic uniqueness and generic well-posedness of optimization problems, Nash equilibrium, and Banach-Mazur games. As generic study is so important today in nonlinear analysis and related subjects such as optimization, game theory and economics, it is our goal in this book to explain this methodology as simply as possible.

As most theorems we give appear for the first time in book form, we have tried to assign all credit for results to the best of our knowledge. However, inadvertently, there might be some ideas which have been included without attribution of credit. If so, please accept our sincere apology. We try to provide readers with a picture of the significant advances in KKM theory and its applications in the last few decades, and all necessary and basic results are self-contained. But the book is not a history of KKM theory. What we intend is to present our personal view of some aspects of its development during the last three decades since Professor Fan established his famous geometric lemma in 1961 and his minimax inequality in 1972. It is our aim through this book to present recently formulated results and new methods for nonlinear analysts in both applied and social sciences, hoping that new and deeper theoretical results and interesting applications will follow. The theoretical aspects of KKM and topological minimax theory are established in Chapter 2 and Chapter 3, respectively, which include most of the new fixed point theory of the Browder-Fan, and Fan-Glicksberg types for set-valued mappings in the framework of locally convex MC-spaces, G-convex spaces, H-spaces, and hyperconvex metric spaces. Their applications to variational inequalities, mathematical economics, optimization, and game theory are extensively illustrated in Chapter 4 through Chapter 9. We have also tried to present up-to-date results on the interconnection between KKM theory and various subjects such as variational inequalities, mathematical economics, optimization, and game theory. Since the KKM theory and its applications are still in their formative stage and undoubtedly more and more new and exciting results will be found in the near future, this book can't be said to cover all aspects of KKM theory and its applications. The structure of this book is shown in the following figure. We hope this diagram helps readers with different backgrounds by giving an outline of what KKM theory covers.

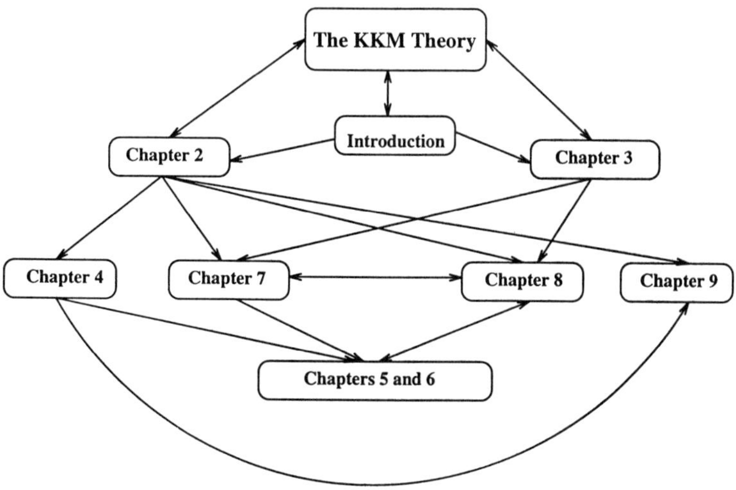

Most of the results appearing in the book are new. We hope it helps interested readers to access current research in this active field readily and easily.

I have many people to thank, in addition to those mentioned above, who developed KKM theory. But first of all, I express my sincere gratitude to Professor Ky Fan for suggesting to me the general outline of this book. This monograph is due mainly to his remarkable and fundamental contributions to the study of nonlinear analysis, and I acknowledge his valuable advice and encouragement during the preparation of the book. I owe him a lot. I have benefited from research collaboration over several years with Professors Chang, Chen, Cho, Deguire, Cubiotti, Ding, Mehta, Huang, Isac, Kim, Kirk, Lee, Luo, Sims, Smith, Takahashi, Tan, Tarafdar, Thompson, Watson, Wu, Xu, Yang, Yu, Zhu and a few others; otherwise, this book would never have been finished. In the last few years of my studies in KKM theory and applications, many people have influenced and helped my thinking. It is my pleasure to warmly thank them for their generous advice, critical suggestion, fruitful and stimulating discussions, or for their published work I read and too many to name here.

I am greatly indebted to the editorial office of Marcel Dekker, Inc., and particularly to the acquisitions editor, Maria Allegra, and staff members Tamara Nicol and Linda Ogden-Wolgemuth for their hard work; especially, the copy editor Eric Stannard deserves a special acknowledgement for so much time spent on my manuscript. I am grateful for their cooperation, patience, understanding and efficient assistance. I also wish to express my thanks and appreciation to Professors Earl Taft, Zuhair Nashed, and anonymous referees whose valuable comments and very helpful suggestions led to the present version of this book.

Finally, I owe my deepest thanks to my wife, Mei, my daughter, Susan, and my son, Steven for their understanding, support and patience through the many days it has taken to bring this work to fruition.

George Xian-Zhi Yuan

Contents

KKM THEORY AND APPLICATIONS IN NONLINEAR ANALYSIS

Chapter 1

Introduction

1.1 KKM Theory: An Outline

I: KKM Theory- An Outline. Let n be a positive integer, $N = \{0, 1, \cdots, n\}$ and Δ_N denote the unit n-simplex in $(n+1)$-dimensional Euclidean space $\mathbb{R}^{n+1}$. For each $S \subset \{0, 1, \cdots, n\}$, we denote by Δ_S the face of Δ_N spanned by the unit vectors e_i for $i \in S$. A closed covering $C = \{C_0, C_1, \cdots, C_n\}$ of Δ_N is said to be a Knaster-Kuratowski-Mazurkiewicz (KKM) covering if $\Delta_S \subset \cup_{i \in S} C_i$ for all $\emptyset \neq S \subset N$. In 1929, Knaster, Kuratowski and Mazurkiewicz [319] proved that if $\{C_0, C_1, \cdots, C_n\}$ is a KKM closed covering of Δ_N, then $\cap_{i=0}^{n} C_i$ is non-empty. This result is now called the KKM principle or KKM lemma today; and its geometric meaning is illustrated by the Figure 1.1. in next page.

In 1961, Fan generalized the classical KKM theorem above to infinite dimensional Hausdorff topological vector spaces and established an elementary but very basic *geometric lemma* for set-valued mappings which is called Fan's geometric lemma. In 1968, Browder gave a fixed point version of Fan's geometric lemma and this result is now known as the Browder-Fan fixed point theorem. Since then there have been numerous generalizations of the Fan-Browder fixed point theorem and their applications to coincidence and fixed point theory, minimax inequalities, variational inequalities in nonlinear analysis, convex analysis, game theory, mathematical economics, social sciences, and so on.

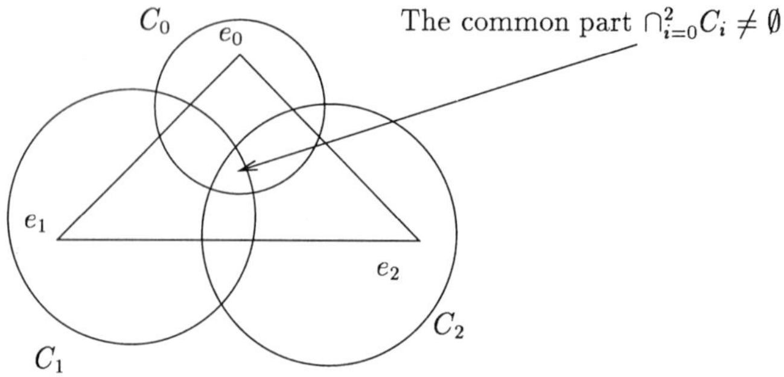

Figure 1.1: The geometric structure of the KKM principle.

By applying his geometric lemma in 1972, Fan [164] established a minimax inequality which is called the Ky Fan minimax inequality or Ky Fan minimax principle) which is regarded as one of the most important principles in nonlinear analysis as it has been successfully applied to potential theory, partial differential equations, monotone operators, variational inequalities, optimization, game theory, linear and nonlinear programming, operator theory, topological group and linear algebra. In particular, by using the Ky Fan minimax inequality, the Fan-Glicksberg fixed point theorem for upper semicontinuous set-valued mappings, which are either inward or outward in the sense of Fan's definition given in [162] (which are more general than Halpern's definitions for inward or outward mappings in 1965), could be established.

Recently, Horvath obtained some generalizations of Fan's geometric lemma and the Ky Fan minimax inequality in [236] and [237] by replacing the convexity assumption with topological properties such as pseudo-convexity and contractibility. By extending Horvath's concepts, Bardaro and Ceppitelli [24] in 1988 obtained generalizations of Ky Fan minimax inequalities to topological spaces with special properties called *H-Structures* (such spaces are also called H-spaces). Following this line, a number of generalizations of Ky Fan's minimax inequalities are given by Horvath [239], Baradaro and Ceppitelli [25], Chang et al. [66] -[70], Ding and Tan [140], Ding et al. [141]-[142], Chang and Ma [70], Lan [333], Park [422], Tarafdar [534], Tan et al. [518] in topological spaces which need not have a linear structure but have an H-Structure. Secondly, by introducing new notions called the 'G-convex space' and 'MC-space' due to Park and Kim [427] and Llinares [348], respectively, including H-spaces as a special class, a number of new results in the study of KKM theory, fixed point theorems and related applications in mathematical economics and game theory have been developed by Park et al. [427]-[416], Llinares [349]-[350], Isac and Yuan [259], Singh et al. [489], Yuan [610], and many others.

The importance of fixed point theory in mathematics is well known. By following Aubin and Frankowska [16], an example to illustrate and emphasize the close relationship between nonlinear analysis (in particular, fixed point theory) and economic science (in particular, mathematical economics) is as follows: It was Leon Walras, who was the first person to suggest using mathematics in economics. He described certain economic agents as automata seeking to optimize evaluation functions (utility, profit, etc) and posed the problem of economic equilibria. However, this area did not blossom until the birth of nonlinear analysis in 1912, with Brouwer's fixed point theorem [56], the usefulness of which was recognized by von Neumann [567] when he developed the foundation of game theory in 1928. In the wake of von Neumann came the works of Fan, Nash, Kakutani, Aumann, Shapley and many others who provided tools used by Arrow, Debreu, Gale, Nikaido, and many others to complete Walras' construction, culminating in the 1950s in the proof of the existence of economic equilibria. Debreu's book [118] entitled *Theory of Value* is a very good survey of economic equilibria at that time.

The natural extension of fixed point theory is the study of coincidence points. Let X and Y be Hausdorff topological spaces and $S, T : X \to 2^Y$ be two set-valued mappings. The *coincidence problem* for (S, T) is to find $(x_0, y_0) \in X \times Y$ such that $y_0 \in S(x_0) \cap T(x_0)$. Geometrical problems of this type in an appropriate context turn out to be intimately related to some basic problems arising in convex analysis. This important fact was also discovered by von Neumann [567] in 1928, who established a coincidence theorem in $\mathbb{R}^n$ and made a direct use of it in the proof of his well-known minimax theorem. Since then, geometrical problems of a similar kind (as well as their analytic counterparts) have attracted many people as well as finding new applications in various fields. In particular, since Eilenberg and Montgomery [152] studied coincidence theory in topological settings in 1946, this topic has been comprehensively developed by contributions due to Kakutani, Nash, Fan, Kneser, Gale, Debreu, Nikaido, Park, Sion, Gorniewicz, Granas, Liu, Chang, Ben-El-Medchaiek, Deguire, Kryszewski, Shih, Tan, Powers, Tarafdar, and many others. This topic has so many applications in mathematics and other subjects. For example, in 1988, Ichiishi [246] successfully used Fan's coincidence theorem to give another proof of Scarf's existence theorem [452] for the non-emptiness of the core of balanced n-person game without side payments. More details in this line can be found in the literature: Aubin [10]-[13], Aubin and Cellina [14], Ichiishi [245], Aliprantis et al. [3], Border [48], Yuan [609]-[610], Zeidler [613]-[614] and many others.

Owing to the significant contributions by Professor Ky Fan in the past few decades, a large part of the theory associated with the KKM principle and its remarkable applications in both natural and social sciences have been established. Though this subject is still in its formative stage, its applications are so rich and broad that such an area of study definitely qualifies to be named KKM theory [417]. This is first time the most recent results of KKM theory are presented in the book form from the point of view of author's preference and research. However, it is not a history of the development of KKM theory. What I intend to do is to present my personal view of some aspects of its development during the last 30 years or so since Professor Ky Fan established his geometric lemma in 1961 and his minimax inequality in 1972, respectively. We try to present the progres of KKM theory and some state of the art in applications. This is not a comprehensive study of the 'general theory', however all necessary and basic results are given in Chapter 2. Our main goal is to show some recently formulated results and new methods to non-linear analysts and scholars in both applied and social sciences, hoping that more interesting applications will follow. We have tried to make this monograph as self-contained as possible and a comprehensive list of literature is also included. By reading this book, we wish to show people that KKM theory is not only a powerful subject in nonlinear analysis itself, but also very useful for applied and social science.

In Chapter 2, we first establish some general theory of the KKM principle. Then the characterization of generalized KKM mapping (GHKKM) which is a natural generalization of the classical KKM mapping is established in topological spaces which, in turn gives several Ky Fan's minimax inequalities in H-spaces or in Hausdorff topological vector spaces. Moreover, several fixed point theorems and coincidence theorems for non-self set-valued mappings are derived under weaker continuity and boundary conditions. As applications, several matching theorems for closed coverings of convex sets are also derived. Furthermore, the concepts of the Ky Fan (KF) point and the essential KF point are introduced. Then the generic stability of KF points and coincidence points are established. These results improve or unify many corresponding results in the literature. For instance, our Ky Fan minimax inequalities show that the *lower semicontinuity* condition, which is always assumed in the literature by many authors, is not necessary. Furthermore, our generalization of the Fan-Glicksberg fixed point theorem shows that the condition that *the domain is paracomapct* posed in many literature is superfluous. These results are necessary when we carry out the applications of KKM theory to general equilibrium theory, variational inequality theory, multi-objective optimization, game theory, Pareto optima in Chapter 4 through Chapter 9. In particular, general fixed point theory for set-valued mappings either being upper semicontinuous or having open inverse values have been developed in the framework of locally convex H-spaces, MC-spaces, G-convex spaces and hyperconvx metric spaces. Moreover, the characterization of KKM principle and generic stability of fixed points for upper semicontinuous set-valued mappings in hyperconvex metric spaces are also developed.

II: Topological Intersection Theorems and Minimax Inequalities. It is clear that many existence problems in mathematics can be reduced to the following so called *intersection problem:* Let Y be a non-empty set, X an index set and $\{F(x) : x \in X\}$ a family of non-empty subsets of Y. Now the question is: *when does the family $\{F(x) : x \in X\}$ have a non-empty intersection*, i.e., $\cap_{x \in X} F(x) \neq \emptyset$?

It is convenient to formulate the intersection problem above in terms of correspondences. More precisely, let X and Y be two non-empty sets and $F : X \to 2^Y$ a correspondence with non-empty values. A single-valued mapping $f : X \to Y \setminus \{\emptyset\}$ is said to be a selector for F if $f(x) \in F(x)$ for all $x \in X$. Thus the intersection question is when does the correspondence F possess a constant selector?

A mathematical problem that has a very close relationship with intersection problems is the following so-called minimax problem formulated as follows: Let X and Y be non-empty sets and $f : X \times Y \to \mathbb{R}$ a function. A *minimax problem* is to find certain conditions such that the following holds:

$$\inf_{y \in Y} \sup_{x \in X} f(x, y) = \sup_{x \in X} \inf_{y \in Y} f(x, y). \tag{1.1.1}$$

The original motivation for the study of minimax theorems was, of course, von Neumann's work on game theory of strategies in 1928. After a generalization of von Nemuann's original results for matrices appeared, the study of minimax theorems became progressively more remote from game theory, and minimax theory started becoming objects of study in their own right. Of courses, there are many ways to study the minimax theory. For example, by using the Ky Fan minimax inequality, we have established some minimax theorems in Chapter 2. Also we can use either the fixed point theorem or its more general form of coincidence theorems to establish the existence of minimax theorems.

However, the importance of *connectedness* in the study of minimax theory was first recognized by Wu [579]. Then this idea was picked up by Terkerson [550]. By a refined method, it is Professor Tuy, who derived a generalized version of Sion's classical minimax theorem in [490] (see also Geraghty and Lin [194]). Independently, inspired by Joó's paper [274], the *method of level sets* was developed by Joó and his Hungarian compatriots Stachó [276] and, Komornik [328]. For example, by introducing the concept of the *interval space*, it was Stachó [499], who established an intersection theorem which was used by Komornik [328] to derive a generalization of Ha's minimax theorem [213]. All these results were unified by Kindler and Trost [310]. Following this line, many minimax theorems which only involve *connectedness* instead of convexity were obtained by Komiya [327], Horvath [237], Lin and Quan [344], Kindler [305]-[308], König [329], Simons [483]-[485], and many others. We only mention a few names here; for the historical trace of the development of minimax theory starting from von Neumann's work, we refer the reader to Simons' recent survey paper [485]. There have been various types of minimax theorems as mentioned by Professor Simon, for example, such as (i): *topological minimax theorems* in which various connectedness hypotheses are assumed for X, Y and the function f; (ii) *quantitative minimax theorems* in which no special properties are assumed for X and Y, but various quantitative properties are assumed for f; (iii) *mixed minimax theorems* in which the quantitative and the topological properties are mixed; and (iv) *unified metaminimax theorems* which include the minimax theorems of types (i)-(iii) above as special cases. Using the idea that the minimax theorems can be reduced to the equivalent existence problems of non-empty intersection of set-valued mappings (e.g., see Remark 5 of Kindler [305, p.1008]), it was Kindler who gave a number of topological (resp., abstract set theory) characterization to the existence of non-empty intersection theorems [305]-[307], by unifying ideas of Wu [579], Terkerson [550], Tuy [563], Joó [274]-[275], Joó and Stachó [276], Komyia [327], Komornik [328], König [329], Simons [483]-[485], Horvath [238] and others. As applications of his intersection theorems for set-valued mappings, many minimax theorems are derived.

In Chapter 3 of this book, by the connectedness of topological spaces, we present several sufficient conditions to guarantee the existence of the non-empty intersection for a family of sets in topological spaces. As applications, topological characterizations for the existence of a non-empty intersection of a set-valued mapping in topological spaces are given, which in turn are applied to derive topological fixed point theorem for upper semicontinuous set-valued mappings. Our results include the corresponding results of Chang et al. [67], Kindler [305], Joó and Sövegjártó [274], and Joó [275] as special cases. The idea behind this section is quite simple, i.e., based on the the following property of connectedness:

Let X be a non-empty connect topological space and both A and B are open (resp., closed) subsets X such that $X = A \cup B$ and $A \cap B = \emptyset$, then A or B must be empty.

This simple idea was first used by Wu [579] in the study of the minimax theory. Since the sufficient and necessary existence of the equality (1.1.1) can be reduced to the equivalent sufficient and necessary existence problem of the non-empty intersection of the corresponding set-valued mappings (e.g., see Kindler [305]), it is very important to study the topological characterization for the non-empty intersection of set-valued mappings. In particularly, our arguments in this chapter are motivated by the connectedness approach developed in the study of minimax theory and their applications in game theory and mathematical economics due to Wu [579], Tuy [563], Terkelsen [550], Joó [274], Kindler and Trost [310], Komornik [328], Stachó [499], König [329], Simons [485], Chang et al. [67], Cheng and Lin [85]-[87], Thompson and Yuan [551], and many others.

III: The Game Theory and Abstract Economics. A game is a situation in which each of several players has partial control over some outcome but generally conflicting preferences over the outcome: each player has a fixed range of strategies among which he selects one so as to bring about the best outcome according to his own preferences. An n-person game is a game in which the strategies of n players can't be made independently: each player must select a strategy in a subset determined by the strategies chosen by the other players. Formally, the situation can be described as follows. Let $N = \{1, 2, \cdots, n\}$ denote the set of players and for each $i \in N$, let X_i denote the set of strategies of the ith player. Each element of $X = \Pi_{i=1}^n X_i$ determines an outcome. The payoff to the ith player is a real-valued function f_i defined on X. Given $x_{-i} \in X_{-i}(= \Pi_{j \in N, j \neq i} X_j$, the strategies of all the others), the choice of the ith player is restricted to a non-empty subset $A_i(x^{\hat{i}})$ of X_i; the ith player chooses x_i in $A_i(x^{\hat{i}})$ so as to maximize $f_i([x_i, x^{\hat{i}}])$. An equilibrium point in such an n-person game is a strategy vector $x \in X$ such that for all $i \in N$, $x_i \in A_i(x^{\hat{i}})$ and $f_i(x) = \max_{y_i \in A_i(x^{\hat{i}})} f_i([y_i, x^{\hat{i}}])$. In this subject, Nash published four brilliant papers [380]-[383] in 1950s, in which he made several fundamentally important contributions to game theory. For example, he introduced the 'distinction' between *cooperative* and *non-cooperative games.* He introduced the concept of *equilibrium point*, which is now usually described as *Nash equilibria* and established their existence in all finite games as applications of Kakutani fixed point theorem which is a special case of the Fan-Glicksberg fixed point theorem; and he also proposed the *Nash bargaining solutions* and studied their existence.

The existence theorems for equilibria of an n-person game with compact strategy sets in $\mathbb{R}^n$ was proved in a seminal paper of Debreu [115] in 1952. The theorem of Debreu extended the earlier work of Nash [382] which also covers the existence theorem of equilibria of the general economic model presented by von Neumann [568] in 1937 (see also von Neumann and Morgenstern [569]) in game theory. Since then there have been many generalizations of Debreu's theorem by Arrow and Debreu [8] in 1954, Mas-Colell [362], Gale and Mas-Colell [193], Borglin and Keiding [51] in 1976, and others.

Following Debreu [115] and Shafer and Sonnenschein [51], a *generalized game* (resp., an *abstract economy*) is a family $\Gamma = (X_i; A_i; U_i)_{i \in I}$, where I is any (countable or uncountable) set of players (resp., agents) such that for each $i \in I$, X_i is the strategy set or choice set, $A_i : X = \Pi_{i \in I} X_i \to 2^{X_i}$ is the constraint correspondence and $U_i : X \to \mathbb{R}$ is the payoff or utility function, where X_i is a subset of a topological space or a topological vector space for each $i \in I$. We denote the product $\Pi_{j \in I, j \neq i} X_j$ by X_{-i} and a generic element of X_{-i} by x_{-i}. Note that a generalized game instead of being given by $(X_i; A_i; U_i)_{i \in I}$ may be given by $\Gamma = (X_i; A_i; P_i)_{i \in I}$, where for each $i \in I$, $P_i : X \to 2^{X_i}$ is the preference correspondence. The relationship between the utility function U_i and the preference correspondence P_i may be exhibited by defining for each $x \in X$,

$$P_i(x) = \{y_i \in X_i : U_i([y_i, x_{-i}]) > U_i(x)\}, \tag{1.1.2}$$

where for each $i \in I$, x_{-i} is the projection of x onto X_{-i} and $[y_i, x_{-i}]$ is the point y in X whose ith coordinate is y_i and $y_{-i} = x_{-i}$. In the case of a generalized game being given by $\Gamma = (X_i; A_i; U_i)_{i \in I}$, a point $\hat{x} \in X$ is said to be a *generalized*

Nash equilibrium point of Γ if

$$U_i(\hat{x}) = U_i([\hat{x}_i, \hat{x}_{-i}]) = \max_{z_i \in A_i(\hat{x})} U_i([z_i, \hat{x}_{-i}])$$

for each $i \in I$, where $\hat{x}$ and $\hat{x}_{-i}$ are respectively projections of $\hat{x}$ onto X_i and X_{-i}. In this case a equilibrium point is a natural extension of the equilibrium point introduced by Nash [380] in 1950. Now suppose $\Gamma = (X_i; A_i; U_i)_{i \in I}$ is a generalized game and for each $i \in I$, let P_i be defined as (1.1.2) above. Then it can be easily checked that a point $\hat{x} \in X$ is an equilibrium point of Γ if and only for each $i \in I$, $\hat{x}_i \in A_i(\hat{x})$ and $P_i(\hat{x}) \cap A_i(\hat{x}) = \emptyset$. Various generalizations of this model have been introduced and investigated by many people. For more details in this line, a number of references and related progresses have been given in Yuan [609]. Following the work of Sonnenschein [495] in 1971, Gale and Mas-Colell [191] in 1975 and Borglin and Keiding [51] in 1976 on non-ordered preference relations, many theorems on the existence of maximal elements of preference relations which may not be transitive or complete, have been proved by Aliprantis and Brown [2], Bergstrom [42], Kim [299], Mehta and Tarafdar [370], Shafer and Sonnenschein [459], Sonnenschein [495], Tan and Yuan [507], Tarafdar [535], Toussaint [560], Tulcea [562], Yannelis [586], Yannelis and Prabhakar [587], and others. These papers generalize Debreu's theorem by considering preference correspondences that are not necessarily transitive or total, by allowing externalities in consumption and by assuming that the commodity space is not necessarily finite-dimensional.

However in the most of existing literature, the domain (and or codomain) of the preference and constraint correspondences are assumed to be compact or paracompact, and the preference correspondences (respectively, payoff functions) are assumed to have open lower sections or open graphs (respectively, to be continuous). Secondly, most of these existence theorems for maximal elements and equilibrium points deal with preference correspondences which have open lower sections or are majorized by correspondences with open lower sections. Note that every correspondence with open lower sections must be lower semicontinuous, but the converse is not true in general. Moreover, in most cases, preference and constraint correspondences may be upper semicontinuous (or been majorized by upper semicontinuous correspondences) instead of being lower semicontinuous (or been majorized by lower semicontinuous), or the preference and constraint mappings are condensing. Furthermore, in the study of equilibrium theory in most economic models, the feasible sets or the budget constraints are generally not (weakly) compact in infinite dimensional commodities and are not convex in the case of the indivisibility of commodities and the underlying spaces do not have a linear structure. Thus, relaxation of convexity of choice sets and generalizations of spaces enable us to deal with the existence of maximal elements and equilibrium points even though commodities are indivisible (not necessarily convex).

Therefore it is necessary and important to study the existence of equilibria for generalized games in which the preference and constraint correspondences need not have open lower sections nor open upper sections, and also the underlying spaces need not have any linear structures and so on.

The objective of Chapter 4 is to systematically study the existence of maximal elements and equilibria for generalized games under various hypotheses, such as the preference and constrained correspondences are lower semicontinuous, upper semicontinuous or condensing, and the strategy sets may not be compact and the underlying spaces may not have a linear structure. Moreover, we also study some properties of lower semicontinuous set-valued mappings in finite dimensional spaces, which in turn give several fixed point theorems and existence theorems for equilibria of generalized games. In particular, the questions raised by Yannelis and Prabhakar [587] and Kim [299] are answered in the affirmative with weaker assumptions. The essential idea behind these existence theorems for equilibria of generalized games is to formulate them into the corresponding models of so-called qualitative games which, in turn, are reduced to the existence problem of maximal elements for preference correspondences. Since existence of maximal elements of correspondences have equivalent formulations in terms of fixed point theorems, which can be derived from Ky Fan's minimax inequalities, thus the KKM theory in Chapter 2 are applicable. The general KKM theory and its applications to the study of abstract economics has been illustrated extensively through Chapter 4.

IV: Variational Inequalities. Even though the topic of variational inequalities has a very long history, it has only been studied systematically since the 1960's (e.g., see Fichera [173] and Stampacchia [497] and others). The variational inequality theory is related to the simple fact that the minimum of the differentiable convex functional on a convex set D in a Hilbert space can be characterized by an inequality of the type $\langle I'(u), v-u\rangle \leq 0$ for all $v \in D$, where $I'(u)$ is the derivative of the functional $I(u)$. However, it is remarkable that the variational inequality theory has many diversified applications. Since its discovery three decades ago, the important developments in variational theory are formulations that can use variational inequalities to study problems of fluid flow through porous media (e.g., see Baiocchi and Capelo [22]), contact problems in elasticity (e.g., see Kikuchi and Oden [296]), transportation problems (see Bertsekas and Gafni [44] and Harker [219]), and economic equilibria (see Dafermos [113]). An additional main area of applications for variational inequalities arises in control problems with a quadratic objective functional, where the control equations are partial differential equations. A detailed discussion of this can be found in Lions [347]. The connection between control problems and quasi-variational inequalities is presented in Aubin [10] and Zeidler [613]. There also exist intimate interconnections between variational inequalities, stochastic differential equations, and stochastic optimization. One can find these in Friedman [181]-[182], Bensoussan and Lions [38], and Bensoussan [37].

In recent years, various extensions and generalizations of variational inequalities have been considered and studied. It is clear that in a variational inequalities formulation, the convex set involved does not depend on solutions. If the convex set does depend on solutions, then variational inequalities are called quasi-variational inequalities. These useful and important generalizations are mainly due to Bensoussan and Lions [39]. Applications of quasi-variational inequalities can be found in Aubin [10], Aubin and Cellina [14], and Zeidler [613].

In 1982, for the study of operation research, mathematical programming and optimization theory, Chan and Pang [64] first introduced the so-called '*generalized quasi-variational inequalities*' in finite dimensional Euclidean spaces. Chan and Pang's generalized quasi-variational inequalities can be illustrated as follows. Let $\mathbb{N}$ and $\mathbb{R}$ denote the set of all natural numbers and the set of all real numbers respectively. Let X be a non-empty subset of $\mathbb{R}^n$, where $n \in \mathbb{N}$. Let $A : X \to 2^X$ and $B : X \to 2^{\mathbb{R}^n}$. The generalized quasi-variational problem associated with A and B (denoted by GQVI (X; A; B)) is to find $(\hat{x}, \hat{u}) \in X \times \mathbb{R}^n$ such that $\hat{x} \in A(\hat{x})$, $\hat{u} \in B(\hat{x})$ and $\sup_{y \in A(\hat{x})} \langle \hat{u}, \hat{x} - y \rangle \leq 0$. The existence theorem of Chan and Pang [64] can be stated as follows.

Theorem 1.1.1. Let X be a non-empty compact convex subset of $\mathbb{R}^n$ and $A : X \to 2^X$ and $B : X \to 2^{\mathbb{R}^n} \setminus \{\emptyset\}$ are such that $A(x)$ is compact convex and $B(x)$ is contractible and compact for each $x \in X$. Moreover assume that A is continuous and B is upper semicontinuous. Then GQVI(X; A; B) has at least one solution.

In 1985, Shih and Tan [466] were the first to study the $GQVI(X; A; B)$ in infinite dimensional locally convex Hausdorff topological vector spaces and the following result has been established (e.g., see Kim [298]).

Theorem 1.1.2. Let E be a locally convex Hausdorff topological vector space, E^* be the dual space of E and X be a non-empty compact convex subset of E. Let $A : X \to 2^X$ be continuous such that for each $x \in X$, $A(x)$ is a non-empty closed convex subset of X, and $B : X \to 2^{E^*}$ be upper semicontinuous from the relative topology of X to the strong topology of E^* such that for each $x \in X$, $B(x)$ is a non-empty strongly compact subset of E^*. Then there exists a point $\hat{y} \in X$ such that

$$\begin{cases} \hat{y} \in A(\hat{y}) \text{ and} \\ \sup_{x \in A(\hat{y})} \inf_{w \in B(\hat{y})} Re\langle w, \hat{y} - x \rangle \leq 0. \end{cases}$$

Since then, there have been a number of generalizations for the existence of $GQVI(X; A; B)$, e.g., see Cubiotti [105]-[110], Ding and Tan [135], Harker and Pang [220], Kim [298], Shih and Tan [474], Tian and Zhou [556], Yao [589]-[590], Yuan [609], and references therein. These results have wide applications to problems in game theory and economics, mathematical programming (e.g., see Aubin [10], Aubin and Ekeland [15], Chan and Pang [64], Harker and Pang [220] and reference therein). Most existence theorems mentioned above, however, are obtained on compact sets in finite dimensional spaces or infinite dimensional locally convex Hausdorff topological vector spaces, and both A and B are either continuous or upper (lower) semicontinuous.

In recent years, many sources have exposited the interconnections among minimax inequalities, equilibria of generalized games and variational inequalities. For instance, Tulcea [562] gives a number of minimax inequalities which are derived by the applications of existence theorems for equilibria of generalized games. Dafermos [113] formulated the problems of finding equilibria of generalized games (in particular, equilibria of pure exchange equilibria) to the problems of finding solutions of variational inequalities. On the other hand, in economic and game applications, it is known that the choice space (or the space of feasible allocations) generally is not compact in any topology of the choice space (even though it is closed and bounded), a key situation in infinite dimensional topological vector spaces. Moreover, we note that there are essentially no existence theorem of solutions of generalized quasi-variational inequalities on non-compact sets in infinite dimensional spaces. This motivates our work in Chapter 5 to develop a series of existence theorems on generalized quasi-variational inequalities by relaxing the compactness conditions and continuity. Secondly, the general convergent theory of the corresponding algorithms for variational inequality theory is devloped. Thirdly, the generic stability property of solutions for two types of generalized quasi-variational inequalities and the existence of essentially connected components of solutions for Hartman-Stampacchia type variational inequalities have been established. These result shows that almost all of varitaional inequality problems are stable in the sense of Baire category theory.

Equally important is the area of mathematical programming known as complementarity theory, which was introduced and studied by Lemke [341] in 1965. Cottle and Dantzing [100] defined the complementarity problem and called it the fundamental problem. For recent results and applications, see Harker and Pang [220], Hyers et al. [244], Noor and Rassias [403], and references therein. However, it was Karamardian [282], who proved that if the set involved in a variational inequality and complementarity problem is a convex cone, then both problems are equivalent. After that, many generalizations have been given by Goeleven, Motreanu and Panagiotopoulos [199], Naniewicz and Panagiotopoulos [379], Panagiotopoulos [409], Shih and Tan [465], Ding [131], Isac [252]-[253], Chang and Huang [68] and references therein. For more details on the discussion between the variational inequalities and complementarity problems, we refer to Cottle, Giannessi and Lions' book [101], Hyers et al.'s book [244] and references therein.

In Chapter 6, as applications of existence results in the study of generalized games, we first develop many existence theorems of quasi-variational inequalities for various classes of mappings such as set-valued mappings without upper semicontinuity, lower semicontinuous mappings, monotone and quasi-monotone operators in the setting of locally convex topological vector spaces, Banach spaces. Then applications to fixed point theorems, minimization problems and generalized complementarity problems are given. By using the concept of a quasi-monotone operator, some fixed point theorems for set-valued mappings which are quasi-contractive (which are generalizations of pseudo-contractive mappings for set-valued mappings) have been established. In its first time, the generic stability and the existence of essentially connected components of solutions for nonlinear complementarity problems have been developed and our results show that almost all of complementarity problems are stable from viewpoint of Baire category theory.

We remark that the development of variational inequalities can be viewed as the simultaneous pursuit of two different lines of research. On the one side, it reveals fundamental facts on the qualitative behaviour of solutions (such as its existence, uniqueness, and regularity) to important classes of problems. On the other side, it also enables us to develop highly efficient and powerful new numerical methods to solve, for example, free and moving boundary value problems and the general equilibrium problems. In this book, we design some algorithms for generalized quasi-variational inclusions. Then their convergence theory has also been established. A comprehensive investigation of numerical methods for variational inequalities is contained in the book of Glowinski et al. [197]. For more details, see Cottle, et al. [101], Crank [104], Harker and Pang [220], Noor [400]-[402], Noor et al. [403], Rodrigues [448] and and many others. Among the most effective numerical techniques are projection methods and variant forms, linear approximation method, relaxation method, auxiliary principle and penalty function techniques. In addition to these methods, the finite element technique is also being applied for the approximate solution of variational inequalities, and has been used by many research workers including Falk [155], Mosco and Strang [376], Noor et al. [403], Noor [401], and references therein.

V: Interconnections. In this book, we try to present interconnections among the KKM principles, in particular, Ky Fan minimax inequalities, fixed point theorem, coincidence theorems, equilibria of generalized games, and variational inequalities in the following way: We reduce the existence problems of variational inequalities to the existence problem for equilibria of generalized games; that means, the solutions of variational inequalities are nothing else, but are exactly the equilibria of their equivalent model of generalized games. This simple fact enables us to consider the existence of solutions for non-compact variational inequalities and generalized quasi-variational inequalities in infinite dimensional Hausdorff topological vector spaces. As we mention above, the existence problems of equilibria for generalized games can be reduced to the existence problems for equilibria of qualitative games. The latter existence problems are equivalent to finding maximal elements of their preference mappings. Note that maximal elements are equivalent forms of their fixed point theorems which can be derived by Ky Fan type minimax inequalities. Therefore we establish the interconnections among minimax inequalities, fixed point theorems of set-valued mappings, generalized games in mathematical economics, and variational inequalities and generalized quasi-variational inequalities.

VI: Price Equilibria, Complementarity Problems, and Variational Inequalities. We recall that the equilibrium price problem is to find a price vector p whih claims the markets for all commodities. Suppose given a price vector p, there is a vector $f(p)$ of excess demands for each commodity. We assume that f is a continuous function of p. A very important property of market excess demand function is *Walras' law.* The mathematical statement of Walras' law can take either strong or weak forms. The *strong form* of Walras' law as $p \cdot f(p) = 0$ for all p in some price simplex. The *weak form* of Walras' law is $p \cdot f(p) \leq 0$ for all p in some price simplex. The economy's meaning of Walras' law is that in a closed economy, almost all of incomes are spent, i.e., there is no net borrowing. For example, we consider a pure exchange economy. The ith consumer comes to market with vector w^i of commodities and leaves with a vector x^i of commodities. If all consumers face the price vector p, their individual budgets require that $p \cdot x^i \leq p \cdot w^i$, that is, they can't spend more than they earn. In this case, the excess demand vector $f(p)$ is just $\Sigma_i x^i - \Sigma_i w^i$, the sum of total demands minus the total supply. Then we can get the inequality $p \cdot f(p) \leq 0$, which is the weak form of Walras' law. The strong form is obtained if each consumer spends all his income. The case of a production economy is similar. The jth supplier produces a net output vector y^j, which yields a net income of $p \cdot y^j$. In a private ownership economy this net income is redistributed to consumers. The new budget constraint from a consumer is that $p \cdot x^i \leq p \cdot w^i + \Sigma_j \alpha_j^i (p \cdot y^i)$, where α_j^i is consumer i's share of supplier j's net income. Thus we have $\Sigma_i \alpha_j^i = 1$ for each j. The excess demand value of f is $\Sigma_i x^i - \Sigma_i w^i - \Sigma_j y^i$. Again adding up the budget constraints, it implies that $p \cdot f(p) \leq 0$. Note that the derivation of Walras' law only requires that consumers satisfy their budget constraints. The weak form of Walras' law is *robust* (as pointed out by Border in [48, p.39]) to the behavioral assumptions made about consumers and supplies.

By the construction of the equilibrium price problem above, it shows us that the price equilibrium problem has a lot of structure imposed on it from economic considerations. A more general mathematical form leads to what is well-known as the '*nonlinear complementarity problem*' which is stated as follows: The function f is no longer assumed to satisfy Walras' law or homogeneity. Instead, f is assumed to be a continuous function defined in a closed convex cone C.

The problem is to find a p in C such that $f(p) \in C^*$ and $p \cdot f(p) = 0$, where C^* is the dual cone of C. If C is the nonnegative cone $\mathbb{R}^n_+$, then the condition $f(p) \in C^*$ and $p \cdot f(p) = 0$ implies that $f(p) \leq 0$. Thus, the main difference between the complementarity and the equilibrium price problems is that f is assumed to satisfy Walras' law in equilibrium problem, but f does not need to be defined for the zero price vector. In the complementarity problem f must be defined at zero, but need only satisfy Walras' law at the solution. In order to guarantee the existence of solutions for the complementarity problem, of course, additional hypotheses on f are needed. The nonlinear complementarity problem was first studied by Cottle [98]. The literature on the complementarity problem is very extensive. For references to applications see, Isac [253] and its references wherein.

From our discussion above for both the price problem and the nonlinear complementarity problem, there is a cone C and a function f defined on a subset of C and we are looking for some $p \in C$ such that $f(p) \in C^*$, or equivalent to say, $q \cdot f(p) \leq 0$ for all $q \in C$. Since in both problems (assuming strong form of Walras' is held) we have $p \cdot f(p) = 0$, we can also rewrite it as $q \cdot f(p) \leq p \cdot f(p)$ for all $q \in C$. This kind of inequalities is the so-called *variational inequality* because it compares an expression involving $f(p)$ and p with expressions involving $f(p)$ and q, where q can be viewed as a variation of p, which was studied by Hartman and Stampacchia [221] about twenty years ago. Thus the general study of variational inequalities could provide another strong tool for the investigation of both equilibrium problems in general equilibrium theory and (nonlinear) complementarity problems. These kinds of ideas will be used in Chapters 5 and 6.

VII: Minimax Theorems and Multi-Objective Optimization. It is well-known in the games theory that the existence of equilibria of scalar games for two persons is equivalent to the following minimax theorem:

$$\inf_{x \in X} \sup_{y \in Y} f(x, y) = \sup_{y \in Y} \inf_{x \in X} f(x, y)$$

where, A and B are non-empty subsets of topological vector spaces, f is a real-valued function on $A \times B$. It is clear that the real-valued function f above has a saddle point $(x^*, y^*) \in A \times B$ if and only if the minimax theorem above holds. However, if the specified game has multiple non-comparable criteria, the game is called a *multi-criteria game*. In this case, the playoff function f takes its value in a vector space (which is infinite dimensional in the most cases) instead of being in the real line. Moreover the set of all equilibrium optimal strategy pairs does not always have the interchangeability condition, which holds for scalar games. The observation above motivates us to study the general existence of saddle points and minimax theorem for vector-valued functions in topological vector spaces, and indeed this problem has been studied in the past of a few decades (e.g., see Blackwell [47]; Chen [80], Corley [97]; Luc [351]-[352]; Takahashi [502];

Tan et al. [514], Tanaka [523]-[527]; Yu [602]-[601] and others). On the other hand, minimax theorems for a vector-valued mapping and some generalizations for saddle point concepts have been explored actively (e.g., see Ferro [172]; Nieuwenhuis [386]; Simons [485]; Sion [490]; Tanaka [523] and many others).

A multi-criteria two-person zero-sum game, or say, a two-person vector-valued zero-sum game is a quadruple-tuple

$$\Gamma = (X, Y; -f, f),$$

where X and Y are non-empty sets and f is a mapping $f : X \times Y \to Z$, where Z is an ordered topological vector space with an ordering defined by a convex cone C. The set X (resp., Y) is the set of strategies of player 1 (resp., player 2) and the mapping $-f$ (resp. f) is the playoff function of this player. When player 1 and player 2 choose a strategy $x \in X$ and a strategy $y \in Y$, respectively, the payoffs with respect to player 1 and player 2 are given by the vectors $-f(x, y)$ and $f(x, y)$, respectively. Each player chooses a strategy in order to increase his/her payoff and wants to find a strategy maximizing his payoff.

Recently, the study of the existence of Pareto equilibria in game theory with vector payoffs has attracted many authors, for example, see Bergstresser and Yu [41], Chose and Prasad [91], Luc [351]-[352], Szidarovszky et al. [500], Tanaka [523], Wang [571], Yu [601], Zeleny [615] and references therein. The motivation for the study of multicriteria models can be found in Bergstresser and Yu in [41], Szidarovszky et al. [500] and Zeleny [615].

We recall that a ***finite-players game*** with multicriterica in its strategic formulation (also called a normal form) is in the form of $G := (X_i, F^i)_{i \in N}$, where $N := \{1, 2, \cdots, n\}$. For each $i \in N$, X_i is the set of strategies in $\mathbb{R}^{k_i}$ for the player i; and each F^i is a mapping from $X := \Pi_{i \in N} X_i$ into $\mathbb{R}^{k_i}$, which is called the payoff function (or say, multicriteria) of the i's player, here k_i is a positive integer. If a choice $x := (x^1, x^2, \cdots, x^n) \in X$ is played, each player i is trying to get his/her payoff function $F^i := (f^i_1(x), f^i_2(x), \cdots, f^i_{k_i}(x))$, which consists of non-commensurable outcomes. Each player i has a preference ' $\succeq$' over the outcome space $\mathbb{R}^{k_i}$. For each player $i \in N$, its preference ' $\succeq$' is given as follows:

$$z^1 \succeq_i z^2 \text{ if and only if } z^1_j \geq z^2_j$$

for each $j = 1, 2, \cdots, k_i$, where $z^1 := (z^1_1, z^1_2, \cdots, z^1_{k_i})$ and $z^2 = (z^2_1, z^2_2, \cdots, z^2_{k_i})$ are any elements in $\mathbb{R}^{k_i}$. The players' preference relations induce the preferences on X, defined for each player i; and the choices $x = (x^1, \cdots, x^n)$ and $y = (y^1, \cdots, y^n) \in X$ by

$$x \succeq_i y \text{ whenever } F^i(x) \succeq_i F^i(y).$$

Also we assume that the model of the game is a non-cooperative game, i.e., there is no replay communicating between players and so players act as free agents.

Each player is trying to maximize his or her own payoff according to his or her preferences. For the games with vector payoff functions (or say, multi-criteria), it is well known that in general there does not exists a strategy $\hat{x} \in X$ to minimize (or equivalently to say, maximize) all f_j^is for each player, for example, see the reference of Yu [602]. We know that the existence of saddle points and Pareto equilibria are fundamental problems in multi-objective optimization and game theory.

In Chapter 7, as applications of KKM theory established in Chapter 2, in particular, the Browder-Fan fixed point theorem, we first present extensive study to the existence of saddle points (resp., vector-valued saddle points) mappings (resp., vector-valued mappings) defined in underlying spaces that may be topological vector spaces, locally convex spaces, Banach spaces and even the underlying spaces are H-spaces. Then the stability of vector-valued saddle points and minimax theorems have been investigated. Our generic stability results show that the majority of C-saddle points are stable in the sense of the Baire category. Finally vector-valued minimax inequalities have been established, too.

As the existence problem of Pareto equilibria can be formulated into the model in which solutions correspond to either fixed points of certain set-valued mappings or Ky Fan type minimax inequality, thus both the *fixed point method* and the *Ky Fan minimax inequality method* provide a unified way for the study of existence of Pareto equilibria problems in multi-objective games under weak hypotheses. It is our purpose in Chapter 8 to develop such existence theory of Pareto equilibria by both the *fixed point method* and the *Ky Fan minimax inequality approach.* Our results in Chapter 8 could be regarded as a unified improvement of the corresponding existence results of Pareto equilibria given in the literature. In the last part of Chapter 8, as applications of generic stability theory for KF points developed in Chapter 2, we also establish the generic stability theory for Nash-equilibria of noncooperative games.

It is well known that the fixed point method has a wide application in the study of economics, optimization, and mathematical programming, for example, see Robinson [447] and other references cited therein. On the other hand, the Ky Fan minimax inequality plays a very important role in the study of nonlinear analysis, and is even regarded as one of the most important principles in nonlinear analysis, for instance, see Aubin [11], Aubin and Ekeland [15], Lin and Simons [345], and many others. However, to the best knowledge of the author, not enough attention has been put on the study of optimization, game theory, and mathematical programming by using the Ky Fan minimax inequality. It is our purpose to present how the Ky Fan type minimax inequality can be applied to study the existence of weight Nash-equilibria and Pareto equilibria. The Ky Fan minimax inequality method would also play another important role in the study of optimization, in particular for the investigation of the existence of Pareto equilibria in the vector-valued optimization problems and multi-objective game. Moreover it is possible to consider the numerical aspects of weight Nash-equilibrium and Pareto equilibrium points for multi-objective games by the computation method for fixed points, which has been established and used successfully in the computation of mathematical economics, and optimization problems in last three decades, e.g., see Robinson [447], and references therein.

VIII: The Debreu-Gale-Nikaido Lemma and Pareto Optima. The Debreu-Gale-Nikaido (DGN) Lemma (see Debreu [117], Gale [189] and Nikaido [387]) is fundamental to proving the existence of solutions for market equilibria of many problems in mathematical economics (e.g., see Border [48, Theorem 18.1, p.81], Aubin [13] and Debreu [120]). Let $\Delta_n = \{x = (x_1, \cdots, x_{n+1}) \in \mathbb{R}_+^{n+1} : \Sigma_{i=1}^{n+1} x_i = 1\}$. The DGN lemma claims that: if $T : \Delta_n \to 2^{\mathbb{R}^{n+1}}$ is an upper semicontinuous set-valued mapping with non-empty values such that (1) for each $x \in \Delta_n$, the set $T(x) - \mathbb{R}_+^{n+1}$ is convex closed; and (2) for each $x \in \Delta_n$, $\sup_{y \in T(x)} \langle y, x \rangle \geq 0$ (Walras's law), then there exists $\hat{x} \in \Delta_n$ such that $T(\hat{x}) \cap \mathbb{R}_+^{n+1} \neq \emptyset$, where the set $\{\hat{x} \in \Delta_n, T(\hat{x}) \cap \mathbb{R}_+^n \neq \emptyset\}$ is also called the set of *free disposal equilibrium prices.* We know that KKM theory, in particular, the KKM principle, and the Ky Fan minimax inequality [169] play essential roles in the study of mathematical economics. Recently, there are many generalization of the DGN Lemma, see Granas and Liu [209], Grandmont [211], Mehta and Tarafdar [370], Neuefeind [385] and Yannelis [585].

In Chapter 9, as applications of KKM theory developed in Chapter 2, mainly the Browder-Fan fixed point theorems and fixed point theorems, we shall first study the generalization of the famous DGN Lemma which is a new generalization of the corresponding generalizations due to Grandmont [211]. Secondly, the conccept of Pareto optima is so important for example that it has been used in almost all subjects from natural, applied and socical sciences. In this part, we study the existence of Pareto optima from the viewpoint of set-valued dynamical systems. The study of set-valued dynamic systems has received much attention in recent years. For example, Aubin and Siegel [17] established some existence and stability of endpoint for set-valued dynamic systems (see also Aubin and Ekeland [15]). Justman [273] provided some conditions that guarantee the existence of an endpoint and conditions that guarantee that each dynamic process converges and the limit of a dynamic process is an endpoint. While Maschler and Peleg [361] investigated the stability of (generalized) nucleolus sets for set-valued dynamic systems, and gave their applications to the study of game theory. In Chapter 9, we also study the existence of endpoints for set-valued dissipative dynamic systems, which are upper or lower semicontinuous in metric spaces. Then the existence, uniqueness and algorithms of endpoints for set-valued dynamic systems which are either generalized contractions in metric spaces or topological contractions in topological spaces, are given. These existence theorems of endpoints for set-valued dynamic systems are then used to derive the existence of Pareto optima for mappings that take values in ordered

Banach spaces. Finally our generic stability results established in this chapter show that '*almost all*' of production economics are stable in the sense of the Baire category.

1.2 Generic Study Method for Nonlinear Problems

In this part, we shall describe the baisc idea of the so-called *generic study method* for nonlinear problems, which will be used throughout this book.

For a given (nonlinear) problem, after we prove its existence of solutions, the equally important issue is the study of its stability of solutions subject to small perturbations in the initial values (resp., datum). Unfortunately, most nonlinear problems need not be robust with respect to small perturbations in the datum of problems. Still, we wish to show that for '*most*' nonlinear problems, their solutions sets are stable in the sense that the change of solution set depends continuously on each individual's initial data. This is very important in applications. For example, in the study of general equilibrium theory of mathematical economics, one is not only interested in the existence of price equilibria, but also in how the set of price equilibria depends on each individual's endowment and how it depends on each individual's taste. Since we do not have a measure on the set of all nonlinear problems and the set of all initial values of associated (nonlinear) problems as both of them may be sets in infinite dimensional spaces that do not have reasonable measure, the term '*most*' can only be understood in a topological sense. For example, let $\mathcal{D}^0$ be the space of all m-tuples of demand functions c^0 of which at least one satisfies Debreu's famous *assumption* (A) in the paper [121, p.180] (in the sense that each commodity is desired by at least one of its all consumers) and $\mathcal{D}^0$ is equipped with uniform topology. Assume that P is the $(l-1)$-dimensional price simplex, i.e., $P = \{x := (x_1, \cdots, x_l) \in \mathbb{R}^l_+ : \Sigma^l_{i=1} x_i = 1\}$ and $Y = \mathcal{D}^0 \times P^m$, where $m \in \mathbb{N}$. For each econmoy $e \in Y$, we do know that the price equilibria set $W(e)$ of e is not empty, and indeed W is an upper semicontinuous set-valued mapping from Y to the price simplex P. (e.g., see the Lemma 9.5.3 in this book, or Theorem of Debreu in [119, p.389] or Theorem 8.3 of Dierker [128, p.78]). However, we do not have a nontrivial measure on $\mathcal{D}^0$, in general. Note that a nice excess demand function in $\mathcal{D}^0$ with only one intersection with the x-axis (which implicitly contains one essential equilibrium point) can always be uniformly approximated by one having lots of intersections with the x-axis (which implicitly implies that there are a lots unessential equilibria). Therefore the set of economies that have only essential equilibria (in the sense of our Definition 9.5.9 in Chapter 9) is not an open set (for more details, see Dierker [128, p.82-85]). These economies are *many* in the sense of Baire category theory. For instance, in a real line, the set of all irrational numbers is *many* and the set of rational numbers is *few* in the sense that *all irrational numbers form a residual set*; however, *all rational numbers does not form a residual set.* We recall that a *residual set* (or say, *Baire set*) contains a countable intersection of dense and open sets. For example, the set Y is a Baire space, i.e., Y has at least one residual set that is dense in Y. In this book if we accept the viewpoint of '*many*' and '*few*' above, which is common and understood in differential topology, that a residual set is sufficiently large, then it is not necessary to study the continuity of solution sets depending continuously on all initial datum; it suffices to study the '*generic stability*' of a solution set in the sense a solution set is continuous on a sufficiently large set. In our specific case, we can show that the solution-set mapping W is continuous on a residual set of Y (by Lemmas 9.5.1 and 9.5.6), thus we claim that price equilibria of Y has the so-called *generic stability*.

Our second main objective in this book is to present a unified treatment for the study of *generic stability* for topics in nonlinear analysis including multi-objective optimization problems, nonlinear complementarity problems, coincidence points and fixed points, Nash equilibria, saddle points, Walras equilibrium prices, Ky Fan points and quasi-variational inequalities in locally convex spaces and so on. In order to explain clearly the way for the study of '*generic*' stability, we are going to give an exact definition of '*generic*' which has been used in the mathematical literature. As this approach depends heavily on the the Baire Theory, we first recall the following Baire theorem which is now a standard result in any analysis book, e.g., see Aubin and Ekeland [15].

The Baire Theorem. The intersection of a sequence of open dense sets in a complete metric space X is still dense in X.

Let (X, d) be a complete metric space. A subset G of X is said to be *residual* if it contains a countable intersection of open dense subsets. The Baire theorem states that in a complete metric space, any residual subset is dense.

We see that residual subsets behave much better than ordinary dense subsets in the sense that *two dense subsets may not intersect* (e.g., the set of all rational numbers Q and $\mathbb{R} \setminus Q$), whereas two residual subsets always do! In this way, we can think of residual subset G as being a *full* subset of X, as $\mathbb{R} \setminus Q$ is a *full* subset of $\mathbb{R}$. The following definition emphasizes this property and has become very popular in various areas of mathematics and its applications.

Definition 1.2.1. A '*property* $P(x)$', where x runs through a space X is said to be '*generic*' if the set of points $G(\subset X)$ where it holds true is residual.

For example, the property that a real number be irrational is generic. The interesting thing is that if two properties $P_1(x)$ and $P_2(x)$ are generic, then so is the property '$P_1(x)$ and $P_2(x)$'.

In general, a particular (linear or nonlinear) problem P may fail to have any of the following properties (i): *the existence of solutions*; (ii): *the uniqueness of solutions*, and (iii): *the stability of solutions depends continuously on the data*. But the question which may be raised and which is of some interest is to measure the 'topological bigness' of the set of data from certain spaces E which possess some or all of the properties (i)-(iii) above. The 'topological bigness' is understood here in the *Baire category sense*; . We recall that if A is a residual subset of E, then its complement is of the first Baire category in E; or equivalently, the set A is residual if it contains a dense G_δ-subset of E. Residual sets in E are regarded as a '*Big*' set in E from a topological point of view instead of measure theory as a dense subset may have zero measure!

In this book, by using nonlinear analysis, set-valued analysis and Baire theory, we shall investigate problems such as under what hypotheses (necessary and/or sufficient) on X does the set of solutions for a given problem P contains a dense G_δ subset in certain space E, or equivalently to say, when do we have the generic existence of solutions to roblems from E? Or under what hypotheses (necessary and/or sufficient) on X does the set consisting of problems with a unique solution, contains a dense G_δ subset in E? Or equivalently to say, when do we have generic uniqueness to problems from E? In Chapter 8, the stability in the traditional sense (i.e., it may be not necessarily generic) of generalized nucleus sets is first established. Then, the generic stability of Pareto optima has been obtained. Moreover, the concept of essential equilibria for production economies is given. We then prove that all equilibria are essential for almost all production economies in the sense of the Baire category.

We would like to emphasis that the way proposed above for the study of generic stability (and also generic uniqueness) of nonlinear problems can be also used for the generic and uniqueness study of other linear or nonlinear problems, which are not discussed in this book and we omit all details here as the essential idea behind them is almost the the same as we used here.

To the best knowledge of the author, Fort [179] was the first person to use the generic method above to study the stability of fixed points and he established the existence of essential fixed points. By applying Fort's theory of essential fixed points for continuous mappings, in 1962 and 1963, Wu and Jiang [580], and Jiang [268] introduced the definition of essential equilibria for n-person non-cooperative games and they established the existence of essential equilibria. By applying the concept of essential equilibria due to Wu and Jiang [580] and also the concept of strongly stable equilibria (which is stronger than the definition of essential equilibria) due to Komjima et al. [325], van Damme [566, p.35] proved that every essential equilibrium is strictly perfect. Also as early as 1974, Dierker [128] introduced the concept of essential equilibria for pure exchange economics and proved that *almost all* of pure exchange economics in the sense of Baire category all equilibria are essential. Moreover, he proposed a general topological method for the study of Walrasian economics and lots of generic stability results have been established. On the other hand, as early as in the 1970s by using Baire theory, which is the so-called generic method described in this book, Kenderov not only established several generic results on the continuity of set-valued monotone operators and generic differentiability for convex functions in Banach spaces, with his Hungarian school, but also established several generic stability, generic uniqueness, and generic well-posedness of optimization problems, Nash equilibrium, Banach-Mazur game, continuous selections of set-valued mappings and others (e.g., see Kenderov [288]-[289], Kenderov and Ribaraka [290], Kenderov and Revalski [291], Coban and Kenderov [95], Christensen and Kenderov [94], Borwein et al. [53], Lucchetti and Revalski [354], and related references wherein). In this line, some other results on generic stability and/or uniqueness have been also given by Beer [27], Damme [566], Dierker [128], Yu [594]-[596], Tan et al. [514]-[516] in recent years.

Before we conclude this introduction, it should be noted that due to historic reasons, the term '*almost all*' also has been used to means a set with Lebesgue measure zero. For example, in Harsanyi's paper [225], by introducing the concept of regular equilibria, he proved an '*almost all*' result which says that for '*almost all*' normal games all equilibria are regular (for more details, see also Theorem 2.6.1 of van Damme [566, p.42]). It should be noted that Harsanyi also uses the term '*almost all*', but his '*almost all* ' means a set whose complement is with Lebesgue measure zero and indeed his '*almost all*' result was derived as an application Sard's Theorem in finite dimensional space in the same way as initiated by Debreu in his paper [119] to prove the existence of equilibria with a finite number in 1970. Thus Harsanyi's term '*almost all*' from [225] is not the same meaning as the term '*almost all*' used in the generic study of this book, by a fact that a residual set (which is our *almost all* set in the generic study) may be with Lebesgue measure zero from the point of view of Baire category theory, but Harsanyi's '*almost all*' set would never have Lebesgue measure zero.

1.3 Characterization of Set-Valued Mappings and Strong Separation Property without Compactness

Since this book deals with most nonlinear problems associated with set-valued mappings, the aim of this section is to establish some basic properties for set-valued mappings and related strong separation properties in locally convex spaces without compactness. We first give a characterization of an upper hemicontinuous mapping with non-empty convex values being upper demicontinuous, i.e., we show that an upper hemicontinuous set-valued mapping with non-empty convex values (not necessarily compact-valued) is upper demicontinuous if and only if the set-valued mapping has no interior asymptotic plane.

Then, a characterization of the strong separation for two non-compact convex sets in locally convex spaces is also established, which, in turn, includes the corresponding results such as the celebrated Hahn-Banach theorem and related separation results for a pair of convex sets in locally convex spaces in existing literature as special cases.

1.3.1 Characterization of Upper Heimcontinuous Set-valued Mappings being Upper Demicontinuous without Compact-Values

The purpose of this section is first to give a characterization of an upper hemicontinuous mapping with non-empty convex values being upper demicontinuous, i.e., we show that an upper hemicontinuous set-valued mapping with non-empty convex values (not necessarily compact-valued) is upper demicontinuous if and only if the set-valued mapping has no interior asymptotic plane. Then we give another characterization for a set-valued mapping being upper semicontinuous if and only if the underlying space is finite dimensional.

When studying nonlinear set-valued problems such as existence of fixed points for set-valued mappings, the existence of solutions for differential inclusions, and others, we often face the problem of dealing with different kinds of continuity for set-valued mappings. In the case of the single-valued mapping, the continuity defined by neighborhood is equivalent with that definition by using language of sequences or nets. However, in the case of strict set-valued mappings, these two definitions are no longer equivalent. As a result, a few concepts about continuity of set-valued mappings, for example, upper semicontinuity, upper demicontinuity and upper hemicontinuity have been introduced and often used in the study of nonlinear set-valued analysis (e.g., see Aubin and Ekeland [15], Aubin and Frankowska [16], Fan [164], Klein and Thompson [317], and related references therein). Thus, it is necessary to illustrate the intrinsic quality among these different continuities. In order to bring readers to the problem we wish to study, we first recall three concepts of continuity for set-valued mappings.

Suppose X is any given set. Throughout this book, we denote by 2^X and $\mathcal{F}(X)$ the families of all subsets and all non-empty finite subsets of X, respectively. If S is a non-empty subset of a topological space X, we also denote by $\overline{S}$ (or clS) and $intS$ the closure and interior of the set S in X.

Let X and Y be two topological spaces and $F : X \to 2^Y$ a set-valued mapping. As usual, we say that F is upper semicontinuous (USC) on X if for each $x \in X$ and for each open set G in Y with $F(x) \subset G$, there exists an open neighborhood N_x of x in X such that $F(z) \subset G$ for all $z \in N_x$. A set-valued mapping is said to be USCO if it is an USC mapping with non-empty and compact values. Let E^* be the dual space (i.e., the vector space of all continuous linear functionals) of a real Hausdorff topological vector space E. According to Fan [164], a set-valued mapping $F : X \to 2^E$ is said to be upper demicontinuous (UDC) on X if for each $x \in X$ and any open half-space (or say, hyper-plane) H in E containing $F(x)$, i.e., $F(x) \subset H$, there exists an open neighborhood N_x of x in X such that $F(z) \subset H$ for all $z \in N_x$. We recall that an open half-plane H in E is set of the form $H := \{x \in E : \phi(x) < \lambda\}$ for some $\phi \in E^*$ and $\lambda \in R$. By following Aubin and Ekeland [2, p.122], a set-valued mapping $F : X \to 2^E$ is said to be upper hemicontinuous (UHC) on X if for each $\phi \in E^*$ and for each $\lambda \in R$, the set $\{x \in X : \sup_{u \in F(x)} \phi(u) < \lambda\}$ is open in X. From the definitions above, it is obvious that each upper semicontinuous set-valued mapping is upper demicontinuous and each upper demicontinuous set-valued mapping is upper hemicontinuous, but the inverse is not true in general (more details and examples for the discussion of relationship among three different types of semicontinuity of set-valued mappings can be found in Section 7 of Chapter 2) However, if set-valued mapping is compact-valued, the upper hemicontinuity implies the upper demicontinuity. In this case, the concepts of upper demicontinuity and upper heimcontinuity coincide. Furthermore, when the image space of a set-valued mapping with compact-valued is compact, all three definitions above are equivalent, e.g., see Proposition 2 of Shih and Tan [469, p.238] or see books of Aubin and Ekeland [15], Aubin and Frankowska [16], and Klein and Thompson [317]. But, under what kind of conditions is the upper hemicontinuity equivalent to the upper demicontinuity without assumption of compactness for set-valued mappings? To the best of our knowledge, there are not many results available in this direction. By following the study of Yang and Yuan [583] and [584], it is our aim to answer this question.

Throughout this section, let X be a Hausdorff topological space and E a real Hausdorff locally convex topological linear space. For the sake of convenience, we first introduce the definition of asymptotic plane which could give us the characterization of upper hemicontinuity being upper demicontinuity for set-valued mappings with non-compact values.

Definition 1.3.1. Let E^* be the dual space of a Hausdorff locally convex topological linear space E and A a non-empty subset of E. For each given $\phi \in E^*$ and $\lambda \in R$, the set $M_{\phi,\lambda} = \{x \in E : \phi(x) = \lambda\}$ is said to be an *asymptotic plane* of A with respect to $(\phi, \lambda) \in E^* \times R$ (in short, $M_{\phi,\lambda}$ is said to be an asymptotic plane of A if there is no confusion in the content) if $M_{\phi,\lambda} \cap A = \emptyset$ and there exists a positive number $\delta \in R$ such that $M_{\phi,\mu} \cap A \neq \emptyset$ for all $\mu \in (\lambda - \delta, \lambda)$.

Remark 1.3.2. By Definition 1.3.1 above, we know that all compact sets and weakly convex compact sets in locally topological vector space E have no asymptotic planes (for details, see the proofs of Corollaries 1.3.7 and 1.3.8

below).

Definition 1.3.3. Let X be a Hausdorff topological space, E a Hausdorff locally convex topological linear space and $F : X \to 2^E \setminus \{\emptyset\}$ a set-valued mapping with non-empty values. For any given point $x_0 \in X$, the asymptotic plane $M_{\phi,\lambda}$ of $F(x_0)$ is said to be a *boundary asymptotic plane* of F at x_0, if there exists an open neighborhood $N(x_0)$ of x_0 such that

$$F(x) \subset H_{\phi,\lambda} = \{y \in E : \phi(y) < \lambda\}$$

for all $x \in N(x_0)$. The asymptotic plane $M_{\phi,\lambda}$ of $F(x_0)$ is said to be an '*interior asymptotic plane*' of F at x_0, if for any neighborhood $N(x_0)$ of x_0, there exists $x \in N(x_0)$ such that

$$F(x) \not\subset H_{\phi,\lambda} := \{y \in E : \phi(y) < \lambda\}.$$

If for all $x \in X$, the set $F(x)$ has no interior asymptotic plane, then the mapping F is said to have no interior asymptotic plane.

Remark 1.3.4. By Definition 1.3.3, it follows that each asymptotic plane of a given set A in E is either a boundary or interior asymptotic plane. Now we have the following characterization for a set-valued upper hemicontinuous mappings being upper demicontinuous.

Theorem 1.3.5. Let X be a Hausdorff topological space and E be a Hausdorff locally convex topological linear space. Suppose $F : X \to 2^E$ is an upper hemicontinuous set-valued mapping with non-empty convex values (not necessarily compact-valued). Then F is upper demicontinuous if and only if the mapping F has no interior asymptotic plane on X.

Proof. Necessity: Suppose the conclusion is false, then there exists $x_0 \in X$ and an asymptotic plane $M_{\phi,\lambda}$ at x_0 such that $M_{\phi,\lambda}$ is an interior asymptotic plane of $F(x_0)$ for some $\phi \in E^*$ and $\lambda \in R$. By Definition 1.3.1, there exists some $\delta > 0$, such that for any $\mu \in (\lambda - \delta, \lambda)$, we have that $H_{\phi,\lambda} \cap F(x_0) \supset M_{\phi,\mu} \cap F(x_0) \neq \emptyset$. Next we want to show that $F(x_0) \subset H_{\phi,\lambda}$. In fact, if there exists $y \in F(x_0)$ such that $y \notin H_{\phi,\lambda}$, it follows that $\phi(y) \geq \lambda$. Taking $y_0 \in H_{\phi,\lambda} \cap F(x_0)$, we have that $\phi(y_0) < \lambda$. Let

$$y_1 := \frac{\phi(y) - \lambda}{\phi(y) - \phi(y_0)} y_0 + \frac{\lambda - \phi(y_0)}{\phi(y) - \phi(y_0)} y.$$

We know that $y_1 \in F(x_0)$ and $\phi(y_1) = \lambda$. Thus $M_{\phi,\lambda} \cap F(x_0) \neq \emptyset$. This contradicts our assumption that $M_{\phi,\lambda}$ is an asymptotic plane of $F(x_0)$. Therefore, we must have that $F(x_0) \subset H_{\phi,\lambda}$. Secondly by the definition of interior asymptotic planes, for neighborhood $N(x_0)$ of x_0, there exists $x \in N(x_0)$ such that $F(x) \not\subset M_{\phi,\lambda}$. This contradicts with our assumption that F is demicontinuous at $x_0 \in X$. Hence F has no interior asymptotic plane at x_0.

Sufficiency: Suppose that F is not demicontinuous on X. Then there exists $x_0 \in X$ and an open half space $H_{\phi,\lambda} = \{y \in E : \phi(y) < \lambda\}$ with $F(x_0) \subset H_{\phi,\lambda}$ for some $\phi \in E^*$ and $\lambda \in R$, such that for any neighborhood $N(x_0)$ of x_0, there exists $x \in N(x_0)$ such that $F(x) \not\subset H_{\phi,\lambda}$. Therefore, for any neighborhood $N(x_0)$ of x_0, there exist $x_{N(x_0)} \in N(x_0)$ and $y_{N(x_0)} \in F(x_{N(x_0)})$ such that $\phi(y_{N(x_0)}) \geq \lambda$. In what follows, we want to show that $M_{\phi,\lambda}$ is not a boundary asymptotic plane of $F(x_0)$. In order to do so, it suffices to show that $M_{\phi,\lambda}$ is an asymptotic plane of $F(x_0)$. Now let the net $\{x_v\}_{v\in\mathcal{V}}$ be a direct family of the set $\{N_{\mathcal{V}}(x_0)\}_{v\in\mathcal{V}}$ with that $\lim_{v\in\mathcal{V}} x_v := x_0$, where $\{N_v(x_0)\}_{v\in\mathcal{V}}$ is the family of all neighborhood $N(x_0)$ at the point x_0. As we just discussed above, for each neighborhood $N_v(x_0)$ of x_0, there exists $x_v \in N_v(x_0)$ and $y_v \in F(x_v)$ such that $\sup_{u\in F(x_v)} \phi(u) \geq \phi(y_v) \geq \lambda$. By the hemicontinuity of F, we do have that

$$\sup_{u\in F(x_0)} (u) \geq \overline{\lim_{v}} \sup_{u\in F(x_v)} \phi(u) \geq \lambda.$$

Note that $F(x_0) \subset H_{\phi,\lambda}$, it follows that that $\sup\limits_{u\in F(x_0)} \phi(u) = \lambda$. Now take $y_0 \in F(x_0)$, it follows that $\phi(y_0) < \lambda$ as $F(x_0) \subset H_{\phi,\lambda}$. Let $\delta = \lambda - \phi(y_0)$. We then have that $\delta > 0$ and $(\lambda - \delta, \lambda) = (\phi(y_0), \lambda)$. For any $\mu \in (\lambda - \delta, \lambda)$ and $\sigma \in (\mu, \lambda)$, as $\sup\limits_{u\in F(x_0)} \phi(u) = \lambda$, there exists $y_1 \in F(x_0)$ such that $\phi(y_1) > \sigma$. This implies that $\mu \in (\phi(y_0), \phi(y_1))$. By the convexity of $F(x_0)$, we know that $F(x_0)$ is a path connected set. By the intermediate-value theorem, there exists $\hat{x} \in F(x_0)$ such that $\phi(\hat{x}) = \mu$. It follows that $M_{\phi,\mu} \cap F(x_0) \neq \emptyset$ for all $\mu \in (\lambda - \delta, \lambda)$, i.e., $M_{\phi,\lambda}$ is an asymptotic plane of $F(x_0)$. Therefore the set $M_{\phi,\lambda}$ is not a boundary asymptotic plane of F at x_0 and thus $M_{\phi,\lambda}$ is an interior asymptotic plane of F at x_0. This contradicts our assumption that F has no any interior asymptotic plane on X and thus the proof is completed. □

Example 1.3.6. Let $E := R^2$, $X := [0, 1]$ and define a set-valued mapping from X to E by $F(x) = \{(u, v) : u > x, v \geq 1/(u - x)\}$ for each $x \in [0, 1]$. Then F: $X \to 2^E$ is a set-valued mapping with non-empty and convex values. For any given $\phi \in E^*$, without loss of its generality, we might assume that the functional ϕ has the form as $\phi(u, v) := au + bv$, where both a and b are two constants determined by the functional ϕ. If either a or b is positive, it is clear that $\sup\limits_{(u,v)\in F(x)} \phi(u, v) = +\infty$. In the case both a and b are negative, we can also check out that $\sup\limits_{(u,v)\in F(x)} \phi(u, v) = ax - 2\sqrt{ab}$. Thus the function $x \mapsto \sup\limits_{(u,v)\in F(x)} \phi(u, v)$ is continuous, which means that the set-valued mapping F is upper hemicontinuous. However, for any $x \in X$, let $u := x$. Then u is an interior asymptotic plane of $F(x)$. By applying Theorem 1.3.5, F is not upper demicontinuous.

Corollary 1.3.7. Let X be a Hausdorff topological space and E a Hausdorff locally convex topological vector space. If the set-valued mapping $F : X \to 2^E \setminus \{\emptyset\}$ is upper hemicontinuous with non-empty and compact convex values, then F is upper demicontinuous.

Proof. We first show that each compact set A of E has no asymptotic plane as pointed out in the previous Remark 1.3.2. Indeed, if A has an asymptotic plane $M_{\phi,\lambda}$, where $\phi \in E^*$ and $\lambda \in R$, by the Definition 1.3.1, for any $u \in A$, we have that $\phi(u) \neq \lambda$ and there exists $\delta > 0$ such that for any $\mu \in (\lambda - \delta, \lambda)$, there is $\hat{x} \in A$ such that $\phi(\hat{x}) = \mu$. Now define $B := \{u \in A : \phi(u) \leq \lambda\}$. Then B is also a compact subset of E. Therefore, there exists $u_0 \in B$ such that $\phi(u_0) = \sup_{u \in B} \phi(u)$. By the arbitrariness of $\mu \in (\lambda - \delta, \lambda)$, it follows that $\phi(u_0) = \lambda$. Note that $M_{\phi,\lambda}$ is an asymptotic plane of A, and by the definition of set B, it follows that for each $u \in B$, $\phi(u) < \lambda$. Therefore $\phi(u_0) < \lambda$. This is impossible as we just showed that $\phi(u_0) = \lambda$. Thus the compact set A in E has no asymptotic planes. By Theorem 1.3.5, F is upper demicontinuous and the proof is completed. □

As another application of Theorem 1.3.5, we have the following result which says that each upper hemicontinuous mapping with non-empty convex and weakly compact values is upper demicontinuous.

Corollary 1.3.8. Let X be a Hausdorff topological space and E a Hausdorff locally convex topological vector space. If the set-valued mapping $F : X \to 2^E \setminus \{\emptyset\}$ is hemicontinuous with non-empty weakly compact convex values, then F is demicontinuous.

Proof. If there exists $x \in X$ such that $F(x)$ has an asymptotic plane $M_{\phi,\lambda}$, then for any $u \in F(x)$, we have that $\phi(u) \neq \lambda$ and secondly, there exists $\delta > 0$ such that for each $\mu \in (\lambda - \delta, \lambda)$, there is $u \in F(x)$ such that $\phi(u) = \mu$. Now define a set $B := \{u \in F(x) : \phi(u) \leq \lambda\}$. Then for any $u \in B$, it follows that $\phi(u) < \lambda$. Let $\{\mu_n\}_{n \in N}$ be a sequence of the set $(\lambda - \delta, \lambda)$ such that $\lim_{n \to \infty} \mu_n = \lambda$. Then by the definition of the set B and our discussion above, there exists $\{u_n\} \subset B$ such that $\phi(u_n) = \mu_n$. As each weakly compact set in the Hausdorff locally convex topological vector space E is weakly closed, B is also convex weakly compact, and thus there exists a directed subfamily $\{u_v\}_{v \in D} \subset \{u_n\}$ and a point $u_0 \in B$ such that for any $\phi \in E^*$, we have that $\lim_v \phi(u_v) = \phi(u_0)$. Therefore, $\phi(u_0) = \lim_v \phi(u_v) = \lim_v \mu_v = \lambda$. This contradicts with that for each $u \in B$, $\phi(u) < \lambda$ as shown above. Therefore F has no asymptotic planes. By Theorem 1.3.5, F is demicontinuous and the proof is completed. □

Corollary 1.3.9. Let X be a Hausdorff topological space and E a reflexive Banach space. If the set-valued mapping $F : X \to 2^E \setminus \{\emptyset\}$ is hemicontinuous with non-empty bounded and closed convex set, then F is demicontinuous.

Proof. Indeed, it is a special case of Corollary 1.3.8. Here we do give a short proof by using Theorem 1.3.5 as follows. If there exists $x \in X$ such that $F(x)$ has an asymptotic plane $M_{\phi,\lambda}$, then on one hand, we know that the sets $F(x)$ and $M_{\phi,\lambda}$ can't be strongly separated by any real hyper-plane; on the other hand, note that E is reflexive Banach space, $F(x)$ is bounded closed convex and so is $M_{\phi,\lambda}$, then the sets $F(x)$ and $M_{\phi,\lambda}$ can be strongly separated by a real hyper-plane. This is impossible. Finally the conclusion is derived by Theorem 1.3.5. □

Remark 1.3.10. We would like to note that both Corollaries 1.3.7 and 1.3.8 improve the corresponding results of Proposition 2 of Shih and Tan [469]. Also all results still hold if the locally topological vector space E is replaced by a topological vector space (not necessarily having locally convex structure) with many sufficient continuous linear functionals.

Before we end this section, we will give another characterisation for the set-valued mapping F_r defined below on a normed space being upper semicontinuous.

Let X be a normed space and for any given $r > 0$, we define a set-valued mapping $F_r : X \to 2^X$ by

$$F_r(x) = \{y \in X : \|x - y\| \leq r\}$$

for each $x \in X$. For each sets A and B in X, the distance of A and B is defined by $dist(A, B) := \inf_{a\in A, b\in B} \|a - b\|$. We first need the following result.

Lemma 1.3.11. Let $(E, \|\cdot\|)$ be a (finite-dimensional) normed space and U a non-empty closed subset of E. Let $B_r(x) := \{z \in E : \|z - x\| \leq r\}$ and $d := dist(B_r(x), U)$, where $r \in \mathbb{R}$ and $x \in E$. If $U \cap B_r(x) = \emptyset$, then for each $y \in U$, we have $\|y\| \geq r + d$.

Proof. For each $y \in U$, as $y \notin B_r(x)$, so that $\|y - x\| > r$. Now we rewrite y as $y = (1 - \frac{r}{\|y-x\|})(y - x) + \frac{r}{\|y-z\|}(y - x) + x$. Let $b := (1 - \frac{r}{\|y-x\|})(y - x)$ and $c := \frac{r}{\|y-z\|}(y - x)$. As $\frac{r}{\|y-x\|} < 1$, we have that $\|y - x\| = \|b\| + \|c\|$. Note that $c \in B_r(x)$ and $b \in U - B_r(x)$. Thus $\|b\| \geq d$ and we have $\|y - x\| \geq r + d$.

We also need the following classic result from functional analysis, for example, see Wong [574, p.67].

Riesz's Lemma. Let X_0 be a proper and closed subspace of a normed space X. Then for any $r \in (0, 1)$, there exists $x_r \in X$ such that $\|x_r\| = 1$ and $\|x - x_r\| \geq r$ for all $x \in X_0$.

Now we have the following characterisation for the mapping F_r being upper semicontinuous defined on a normed space if and only if the underlying space is finite dimensional.

Theorem 1.3.12. For any given $r > 0$, the set-valued mapping F_r is upper semicontinuous if and only if X is a finite dimensional space.

Proof. Sufficiency: Suppose it is not true. As the translation of the mapping F_r preserves its upper semicontinuity at a point, without loss of generality, we may assume that F_r is not upper semicontinuus at 0. From the definition of upper semicontinuity, there exist a non-empty open subset O of E with $F_r(0) \subset O$ and a sequence $\{x_n\}_{n\in\mathbb{R}}$ with $\lim_{n\to\infty} x_n = 0$ such that for each $n \in \mathbb{N}$, there exists $y_n \in F_r(x_n)$ with $y_n \notin O$. Let U be the complement of u in E, i.e., $U = \{x \in E : x \notin O\}$. Then U is a closed subset of E. Let $d := dist(B_r(0), U)$. We have that $d > 0$ as $B_r(0) \cap U = \emptyset$ and both $B_r(0)$ and U are non-empty and closed. Moreover $\|y_n\| \geq d + r$ for all $n \in \mathbb{N}$ from Lemma 1.3.11. On the other hand, $y_n \in B_r(x_n)$, so that $\|y_n - x_n\| \leq r$ for all $n \in \mathbb{N}$. Therefore

$$\limsup_{n\to\infty} \|y_n\| \leq \limsup_{n\to\infty} \|y_n - x_n\| + \lim_{n\to\infty} \|x_n\| \leq r.$$

This conclusion contradicts with that $\|y_n\| \geq r + d$ for all $n \in \mathbb{N}$. Thus F must be upper semicontinuous at 0.

Necessity: By following Kucia's idea [330], without loss of generality, we may assume that $r = 1$. Indeed, if X is infinite dimensional, by considering the point $x = 0$, and Riesz's Lemma, there exists $b_n \in F_1(0)$ such that $\|b_n\| = 1$ and $\|b_n - b_m\| \geq \frac{1}{2}$ for all $n \neq m$. Let $a_n = (1 + \frac{1}{n})b_n$ and $x_n = \frac{1}{n}b_n$. Then

$$\frac{1}{2} \leq \|b_n - b_m\| = \|a_n - a_m - \frac{1}{n}b_n + \frac{1}{m}b_m\| \leq \|a_n - a_m\| + \frac{1}{n} + \frac{1}{m}.$$

Note that $\frac{1}{4} \leq \|a_n - a_m\|$ for all $m > n \geq 8$, it follows that the set $C = \{a_n : n = 1, 2, 3, \cdots\}$ is closed in X and $F_1(0) \subset X \backslash C = U$. It is clear that $\lim_{n\to\infty} x_n = 0$, $a_n \in F_1(x_n)$ and $F_1(V) \subset U$ as $a_n \notin U$ (i.e., for any neighbourhood V of the point 0, $F_1(V) \not\subset U$). This means that the mapping $x \mapsto F_1(x)$ is not upper semicontinuous at $x = 0$.

1.3.2 Characterization of Strong Separation Property without Compactness in Locally Convex Spaces

In this section, a characterization of the strong separation for two non-compact convex sets in locally convex spaces is established, which, in turn, include the corresponding results such as celebrated Hahn-Banach theorem and related separation results for a pair of convex sets in locally convex spaces in the existing literature as special cases.

Throughout this section, E denotes a real Hausdorff locally topological vector space unless specified and E^* is the dual space of E consisting of all continuous linear functionals defined on E. For each $x \in E$, we denote by $\mathcal{N}(x)$ the family of all open neighborhoods of x in E. For each $\psi \in E^*$ and $\lambda \in R$, the *closed hyperplane* $M_{\psi,\lambda}$ is defined as $M_{\psi,\lambda} := \{x \in E : \psi(x) = \lambda\}$. For the convenience of our discussion, we first recall some definitions and related notions as follows.

Definition 1.3.13. For any given nonzero functional $f \in E^*$ and $\alpha \in \mathbb{R}$, let $[f = \alpha]$ denote the set $\{x \in E : f(x) = \alpha\}$, and $[f > \alpha]$ denote the set $\{x \in E : f(x) > \alpha\}$, and so on. A *closed hyperplane* is a set of the form $[x \in E : f = \alpha]$, where $f \in E^*$ and $\alpha \in \mathbb{R}$. A hyperplane $[f = \alpha]$ defines two strict half spaces $[f > \alpha]$ and $[f < \alpha]$, and two closed half spaces $[f \geq \alpha]$ and $[f \leq \alpha]$. The hyperplane $[f = \alpha]$ *separates* two sets of A and B if either $A \subset [f \leq \alpha]$ and $B \subset [f \geq \alpha]$ or if $B \subset [f \leq \alpha]$ and $A \subset [f \geq \alpha]$. We say that the hyperplane $H := [f = \alpha]$ *properly separates* A and B if it separates them and $A \cup B$ is not included in H. A hyperplane $[f = \alpha]$ is said to *substrictly separates* A and B if it separates A and B, and in addition, $A \subset [f > \alpha]$ and $B \subset [f \leq \alpha]$ or vice-versa. A hyperplane $[f = \alpha]$ *strictly separates* A and B if it separates them, and in addition, $A \subset [f > \alpha]$ and $B \subset [f < \alpha]$ or vice-versa. We say that $[f = \alpha]$ *strongly separates* A and B if there is a positive ϵ with $A \subset [f \leq \alpha]$ and $B \subset [f \geq \alpha + \epsilon]$ or vice-versa. We may also say that the linear functional f itself separates the sets when some hyperplane $[f = \alpha]$ separates them. If A is a subset of a topological vector space E, a closed real hyperplane $H := [f = \alpha]$ is said to be a *supporting hyperplane* of A if $A \cap H \neq \emptyset$, and in addition A is contained in either $[f \geq \alpha]$ or $A \subset [f \leq \alpha]$.

The geometric form of the celebrated Hahn-Banach separation theorem is one of the most important tools in the study of convex analysis, linear and nonlinear analysis, and in particular for the study of fixed point theory of set-valued mappings (e.g., see Aubin and Ekeland [15], Fan [164], Jiang [269]-[271], Zeidler [613] and others). It is well-known that there are two main results for the separation of convex sets. The first one (e.g., see Schaefer [453]) says that *two non-empty convex sets A and B in E can be separated strongly by a closed hyperplane if and only if there exists a neighborhood V of zero in E such that $(A+V) \cap B = \emptyset$.* The second separation theorem (e.g., see Rudin [449]) says that *two non-empty closed and convex sets A and B in E with $A \cap B = \emptyset$, if either A or B is compact, then A and B can be separated strongly by a closed hyperplane.* The condition of the first separation theorem above is equivalent to $0 \notin \overline{A - B}$, which, in general, is not easy to verify, where $\overline{A - B}$ denotes the closure of $A - B$ in E. We wish to find some criterion to verify the strong separation for both A and B instead of $0 \notin \overline{A - B}$, if we only know that A and B are substrictly separated. The second separation result above overcomes the shortage to test the condition $0 \notin \overline{A - B}$. However the price we should pay is to assume that either A or B is compact. The requirement of compactness leads the second separation theorem above not applicable for even very simple case such as follows:

Let $A = \{(x, y) \in \mathbb{R}^2 : y \geq 1 + x^2\}$ and $B = \{(x, y) \in \mathbb{R}^2 : y \leq -x^2\}$. Note that both A and B are not compact and thus the second separation theorem is not applicable (however, we note that the first strong separation theorem is applicable for this case). Because of this reason, many coincident theorems in set-valued analysis and nonlinear analysis usually need the compactness assumption.

In this section, by following Yang and Yuan [584] to introduce the notion of the so-called *boundary compact* (for details, see Definition 1.3.18 below) to replace the traditional compactness of strong separation theorems, we establish some criteria for the strong separation property of non-compact convex sets in locally convex spaces, which, in turn, improve the corresponding results in existing literature.

In order to establish our strong separation theorem without compactness, we first introduce the following definitions.

Definition 1.3.14. Let $\psi \in E^*$, $\lambda \in R$ and let A be a non-empty subset of E. The closed hyperplane $M_{\psi,\lambda} := \{x \in E : \psi(x) = \lambda\}$ is said to be a *boundary plane* of A if for any $\epsilon > 0$, there exists $\mu \in (\lambda - \epsilon, \lambda]$ such that $M_{\psi,\mu} \cap A \neq \emptyset$ and there exists $\delta > 0$ such that $M_{\psi,\mu} \cap A = \emptyset$ for all $\mu \in (\lambda, \lambda + \delta)$. The hyperplane $M_{\psi,\lambda}$ is said to be a boundary plane of A respective to H if $M_{\psi,\lambda}$ is a boundary plane of $A \cap H$, where H is a non-empty subset of E.

Remark 1.3.15. From Definitions 1.3.1 and 1.3.13, it is clear that each supporting plane of A is its a boundary plane. Secondly, for each connected set A, its asymptotic plane is its boundary plane; and each boundary plane which is not an asymptotic plane is also its supporting plane. Moreover it is easy to verify that each compact has no any asymptotic plane.

Definition 1.3.16. Let $\psi \in E^*$, $\lambda \in R$ and let A and B be two non-empty subsets of a locally topological vector space E. Then the closed hyperplane $M_{\psi,\lambda}$ is said to be an *asymptotic boundary plane* of A and B if $M_{\psi,\lambda}$ is an asymptotic plane of A and $M_{-\psi,-\lambda}$ is a boundary plane of B, or vice-versa.

Definition 1.3.17. The asymptotic boundary plane $M_{\psi,\lambda}$ of A and B is said to be a *common asymptotic plane* of A and B if for any $\phi \in E^*$, and any $\mu \in R$, there exists a real number $\alpha \neq 0$ such that for any $\sigma \in (\min\{\mu, \mu + \alpha\}, \max\{\mu, \mu + \alpha\})$, one of the following two conditions are satisfied:

(I): The hyperplane $M_{\psi,\lambda}$ is an asymptotic boundary plane of A and B with respect respective to the closed half-space $H^{-1}_{\phi,\sigma} := \{x \in E : \phi(x) \leq \sigma\}$, or

(II): The hyperplane $M_{\psi,\lambda}$ is an asymptotic boundary plane of A and B with respect to the closed half-space $H^{+}_{\phi,\sigma} := \{x \in E : \phi(x) \geq \sigma\}$.

Definition 1.3.18. A closed subset A of a locally convex space E is said to be *boundary compact* if A has no asymptotic plane and the intersection of A with its boundary plane (if it exists) is non-empty and compact.

Remark 1.3.19. Each compact set must be boundary compact, but the converse is not true as shown by the following example.

Example 1.3.20. Let $A := \{(x,y) \in R^2 : y \geq 1 + x^2\}$. Then A has no asymptotic plane and each boundary plane of A is its tangent line. The intersection of A with its boundary plane is a singleton set and thus it is compact; however A is not compact.

To develop our main results, we first need the following lemma.

Lemma 1.3.21. (1): Let A and B be two subsets of E which is separated by the closed hyperplane $M_{\psi,\lambda}(:= \{x \in E : \psi(x) = \lambda\})$, then A and B could not be separated strongly by $M_{\psi,\lambda}$ if and only if for each $V \in \mathcal{N}(0)$, there exist $x \in A$, $y \in B$ and $z \in M_{\psi,\lambda}$ such that $x - z \in V$ and $y - z \in V$.

(2): Suppose A and B are two non-empty closed and convex sets in E which are separated substrictly by a closed hyperplane $M_{\psi,\lambda}$ (i.e., A and B are contained separately in the open and closed half-spaces $\{x \in E : \psi(x) > \lambda\}$ and $\{x \in E : \psi(x) \leq \lambda\}$ or vice-versa). If A and B could not be separated strongly by the closed hyperplane $M_{\psi,\lambda}$, then $M_{\psi,\lambda}$ is an asymptotic boundary plane of A and B.

Proof. (i) Sufficiency: Note that E is a Hausdorff locally convex space, it suffices to show that $0 \notin \overline{A - B}$. For each $V \in \mathcal{N}(0)$, there exists a balanced neighborhood V_1 of *zero* in $\mathcal{N}(0)$ such that $V_1 + V_1 \subset V$. By the assumption, there exists $x \in A$, $y \in B$ and $z \in M_{\psi,\lambda}$ such that $x - z \in V_1$ and $y - z \in V_1$. Then $x - y = (x - z) - (y - z) \in V_1 + V_1$, so that $0 \notin \overline{A - B}$.

Necessarily: Assume that A and B can't be separated strongly by any closed hyperplane. Then $0 \notin \overline{A - B}$. For each $V \in \mathcal{N}(0)$, there exists a balanced convex neighborhood $V_1 \in \mathcal{N}(0)$ such that $V_1 + V_1 \subset V$. Since $0 \notin \overline{A - B}$, there exist $x \in A$, $y \in B$ such that $x - y \in V_1$. Note that A and B are separated by some closed hyperplane $M_{\psi,\lambda}$, without loss of generality, we may assume that $\psi(x) \leq \lambda \leq \psi(y)$. If $\psi(x) = \psi(y)$, let $z := \frac{1}{2}(x + y)$. Then $z \in M_{\psi,\lambda}$ and $x - z \in V$ and $y - z \in V$. In the case $\psi(x) \neq \psi(y)$, let $z := y + \frac{\psi(y)-\lambda}{\psi(y)-\psi(x)}(x - y)$. Then $z \in M_{\psi,\lambda}$. As $|\frac{\psi(y)-\lambda}{\psi(y)-\psi(x)}| \leq 1$, so that $y - z \in V_1 \subset V$ and $x - z = (x - y) + (y - z) \in V_1 + V_1 \subset V$.

(2): Since A and B are separated substrictly, without loss of the generality, we may assume that $A \subset \{x \in E : \psi(x) < \lambda\}$ and $B = \{x \in E : \psi(x) \geq \lambda\}$. Then $M_{\psi,\lambda} \cap A = \emptyset$. As A is non-empty, there exists $x_0 \in A$ such that $\psi(x_0) < \lambda$. Let $\delta = \lambda - \psi(x_0)$. Then for any $\mu \in (\lambda - \delta, \lambda)$ and $V \in \mathcal{N}(0)$, there exists a balanced neighborhood $V_1 \in \mathcal{N}(0)$ such that $V_1 + V_1 \subset V$. As A and B are not separated strongly, by (1), there exist $x_v \in A$, $y_v \in B$ and $z_v \in M_{\psi,\lambda}$ such that $x_v - z_v \in V_1$ and $y_v - z_v \in V_1$. Thus $y_v - x_v \in V$. Note that ψ is continuous, there exists V_o such that

$$0 \le \psi(y_{v_0} - x_{v_0}) < \lambda - \mu, \tag{1.3.1}$$

that is $\psi(y_{v_0}) \ge \lambda > \psi(y_{v_0} - x_{v_0}) + \mu \ge \lambda + \mu - \psi(x_{v_0})$. Therefore $\psi(x_{v_0}) > \mu > \lambda - \delta = \psi(x_0)$. Let $y_0 = x_0 + \frac{\mu - \psi(x_0)}{\psi(x_{v_0}) - \psi(x_0)}(x_{v_0} - x_0)$. Then $y_0 \in A$ and $\psi(y_0) = \mu$. So that $M_{\psi,\lambda}$ is an asymptotic plane of A.

Next we show that $M_{-\psi,-\mu}$ is a boundary plane of B. By (1.3.1), it follows that for each $\epsilon > 0$, there exists $\hat{V} \in \mathcal{N}(0)$ such that $0 \le \psi(y_{\hat{v}} - x_{\hat{v}}) < \epsilon$. Then $-\lambda \ge -\psi(y_{\hat{v}}) > -\psi(x_{\hat{v}}) - \epsilon > -\lambda - \epsilon$. Therefore there exists $\mu \in (-\lambda - \epsilon, -\lambda]$ such that $M_{-\psi,-\mu} \cap B \ne \emptyset$ for any $\mu \in (-\lambda, -\lambda + \delta)$ and $M_{-\psi,\mu} \cap B = \emptyset$, where $\delta > 0$. Thus $M_{\psi,\lambda}$ is an asymptotic plane of A and B. $\square$.

Theorem 1.3.22. Let A and B be non-empty and convex subset of E which are separated substrictly. Then A and B are separated strongly by a closed hyperplane if and only if A and B have no common asymptotic planes.

Proof. Necessarily: Suppose A and B are separated strongly by a closed hyperplane $M_{\psi,\lambda}$. Then there exists $\lambda_1 < \lambda_2$ such that for each $\mu \in (\lambda_1, \lambda_2)$, the hyperplane $M_{\psi,\mu}$ separates A and B strongly. Let $M_{\hat{\phi},\hat{\lambda}}$ be any asymptotic plane of A and B and set $\psi := \phi$. Then for each $\mu \in (\lambda_1, \lambda_2)$ and any $\lambda \in R$ with $\lambda \ne 0$, there exists $\delta \in (\min\{\mu, \mu + \alpha\}, \max\{\mu, \mu + \alpha\})$ such that $M_{\psi,\delta}$ separates A and B strongly. Thus $M_{\hat{\psi},\hat{\lambda}}$ is not an asymptotic boundary plane respective to both $H^+_{\psi,\delta}$ and $H^-_{\psi,\delta}$. Thus A and B have no common asymptotic planes.

Sufficiency: Suppose A and B can't be separated strongly by any closed hyperplane. Let $M_{\psi,\lambda}$ substrictly separates A and B. By Lemma 1.3.21(2), $M_{\psi,\lambda}$ is an asymptotic boundary plane of A and B. We may assume that $A \subset H^-_{\phi,\lambda}$, $B \subset H^+_{\phi,\lambda}$ and $M_{\psi,\lambda}$ is an asymptotic plane of A. For any $\phi \in E^*$ and $\mu \in R$, we first claim that: (1) $M_{\psi,\lambda}$ must be an asymptotic plane of either $A \cap H^-_{\phi,\mu}$ or $B \cap H^+_{\phi,\mu}$; and (2): $M_{-\psi,-\lambda}$ must be a boundary plane of either $B \cap H^-_{\phi,\mu}$ or $B \cap H^+_{\phi,\mu}$.

Indeed for the claim (1), if $M_{\psi,\lambda}$ is not an asymptotic plane of $A \cap H^-_{\phi,\mu}$, by the fact that $M_{\psi,\lambda} \cap (A \cap H^-_{\phi,\mu}) \subset M_{\psi,\lambda} \cap A = \emptyset$, it follows that there exists a sequence $\{\delta_n\}_{n\in\mathbb{N}}$ such that $\delta_n \to 0^+$ such that $M_{\psi,\lambda-\delta_n} \cap (A \cap H^-_{\phi,\mu}) = \emptyset$. Since $A \cap H^-_{\phi,\mu}$ is convex, it is connected. Therefore the set $\psi(A \cap H^-_{\phi,\mu})$ is path-connected, there exists an integer $n \in \mathbb{N}$ such that for each $\delta \in (\lambda - \delta_n, \lambda)$, $M_{\psi,\delta} \cap (A \cap H^-_{\phi,\mu}) = \emptyset$. Note that $M_{\psi,\lambda}$ is an asymptotic plane of A, then there exists $\delta > 0$ such that for any $\sigma \in (\lambda - \delta, \lambda)$, $M_{\psi,\sigma} \cap A \ne \emptyset$. Let $\delta_0 = min\{\delta_n, \delta\}$. Then for any $\sigma \in (\lambda - \delta_0, \lambda)$, $M_{\psi,\sigma} \cap A \ne \emptyset$ and $M_{\psi,\sigma} \cap (A \cap H^-_{\phi,\mu}) = \emptyset$. Therefore $M_{\psi,\sigma} \cap (A \cap H^+_{\phi,\mu}) \ne \emptyset$. As $M_{\psi,\lambda} \cap (A \cap H^+_{\phi,\mu}) = \emptyset$, it follows that $M_{\psi,\lambda}$ is an asymptotic plane of $A \cap H^+_{\phi,\mu}$.

For claim (2), if $M_{-\psi,-\lambda}$ is not a boundary plane of $B \cap H^-_{\phi,\mu}$, as $B \cap H^-_{\phi,\mu} \subset H^+_{\phi,\lambda}$, it follows that for each $\delta > 0$, $M_{-\psi,\mu} \cap (B \cap H^-_{\phi,\mu}) = \emptyset$ for all $\mu \in$

$(-\lambda, -\lambda+\delta)$. Therefore there exists $\epsilon > 0$ such that for any $\sigma \in (-\lambda-\epsilon, -\lambda)$, $M_{-\psi,\sigma} \cap (B \cap H^-_{\phi,\mu}) = \emptyset$. As $M_{-\psi,-\lambda}$ is a boundary plane of B, for any $0 < \epsilon' < \epsilon$, there exists $\delta \in (-\lambda-\epsilon', -\lambda]$ such that $M_{\psi,\delta} \cap B \neq \emptyset$, so that $M_{-\psi,\delta} \cap (B \cap H^+_{\phi,\mu}) \neq \emptyset$. Therefore $M_{-\psi,-\mu}$ is a boundary plane of of $B \cap H^+_{\phi,\mu}$.

In order to finish the proof, it suffices to show that there exists a real number $\alpha \neq 0$ such that either the condition (i) or (ii) of Definition 1.3.17 above is satisfied. We shall prove it by considering the following 4 cases.

(a): Suppose $M_{\psi,\lambda}$ is an asymptotic plane of $A \cap H^-_{\phi,\mu}$ and $M_{-\psi,-\lambda}$ is a boundary plane of $B \cap H^-_{\phi,\mu}$, let $\alpha = 1$. Then for any $\sigma \in (\mu, \mu+\alpha)$, it follows that $M_{\psi,\lambda}$ is an asymptotic boundary plane of A and B with respect to $H^-_{\phi,\sigma}$.

(b): Suppose $M_{\psi,\lambda}$ is an asymptotic plane of $A \cap H^+_{\phi,\mu}$, and $M_{-\psi,-\lambda}$ is a boundary plane of $B \cap H^+_{\phi,\mu}$, let $\alpha = -1$. Then for any $\sigma \in (\mu+\alpha, \mu)$, the closed hyperplane $M_{\psi,\lambda}$ is an asymptotic boundary plane of A and B respective to $H^+_{\phi,\sigma}$.

(c): Suppose $M_{\psi,\lambda}$ is an asymptotic plane of $A \cap H^+_{\phi,\mu}$ and $M_{-\psi,-\lambda}$ is a boundary plane of $B \cap H^-_{\phi,\mu}$. If $M_{\psi,\lambda}$ is an asymptotic plane of $A \cap H^-_{\phi,\mu}$, then it reduces back to the case (a). Thus it suffices to discuss the case that $M_{\psi,\lambda}$ is not an asymptotic plane of $A \cap H^-_{\phi,\mu}$ and $M_{-\psi,-\lambda}$ is not a boundary plane of $B \cap H^+_{\phi,\lambda}$. Let $\alpha = 1$ and for any $\sigma \in (\mu, \mu+\alpha)$, then $\sigma > \mu$. For any $V \in \mathcal{N}(0)$, by the continuity of ϕ, there exists $x \in \{z \in E : \mu < \phi(z) < \sigma\}$ such that $\phi(x) = \frac{\mu+\sigma}{2}$. Let $\hat{U} = \{x \in E : |\phi(x)| < \frac{\sigma-\mu}{2}\}$. Then $\hat{U} \in \mathcal{N}(0)$ and such that for any $y \in \hat{U}$, $\mu < \phi(x+y) < \sigma$. Thus $x + \hat{U} \subset \{x \in E : \mu < \phi(x) < \sigma\} \subset H^-_{\phi,\sigma} \setminus H^-_{\phi,\mu}$. This implies that there exists $U \in \mathcal{N}(0)$ such that $U \subset V \cap \hat{U}$ and $U + H^-_{\phi,\mu} \subset H^-_{\phi,\sigma}$. For a given $U \in \mathcal{N}(0)$, there exists a balanced neighborhood $\hat{V}$ of zero such that $\hat{V} + \hat{V} \subset U$. Since we assume A and B can't be separated strongly, it follows by Lemma 1.3.21(1) that there exists $x_{\hat{v}} \in A$, $y_{\hat{v}} \in B$ and $z_{\hat{v}} \in M_{\psi,\lambda}$ such that $x_{\hat{v}} - z_{\hat{v}} \in \hat{V}$ and $y_{\hat{v}} - z_{\hat{v}} \in \hat{V}$. Note that $M_{\psi,\lambda}$ is an asymptotic plane of A and $M_{-\psi,-\lambda}$ is a boundary plane of A, it follows that $M_{\psi,\lambda}$ is not an asymptotic boundary plane of $B \cap H^+_{\phi,\mu}$ and $B \cap H^-_{\phi,\mu}$; and $M_{-\psi,-\lambda}$ is not a boundary plane of $A \cap H^-_{\phi,\mu}$ and $A \cap H^+_{\phi,\mu}$. Moreover by our assumption that $M_{\psi,\lambda}$ is not an asymptotic plane of $A \cap H^-_{\phi,\mu}$ and $M_{\psi,\lambda}$ is not a boundary plane of $B \cap H^+_{\phi,\mu}$, it follows that

(1): $M_{\psi,\lambda}$ is not an asymptotic boundary plane of both $A \cap H^-_{\phi,\mu}$ and $B \cap H^-_{\phi,\mu}$ (as $M_{\psi,\lambda}$ is not an asymptotical plane of both $A \cap H^-_{\phi,\mu}$ and $B \cap H^-_{\phi,\mu}$);

(2): $M_{\psi,\lambda}$ is not an asymptotic boundary plane of both $A \cap H^+_{\phi,\mu}$ and $B \cap H^+_{\phi,\mu}$ (as $M_{-\psi,-\lambda}$ is not a boundary plane of both $A \cap H^+_{\phi,\mu}$ and $B \cap H^+_{\phi,\mu}$); and

(3): $M_{\psi,\lambda}$ is not an asymptotic boundary plane of both $A \cap H^-_{\phi,\mu}$ and $B \cap H^+_{\phi,\mu}$ (as $M_{\psi,\lambda}$ is not an asymptotic plane of $A \cap H^-_{\phi,\mu}$ and $B \cap H^+_{\phi,\mu}$).

Then by Lemma 1.3.21(2), we have that $A \cap H^-_{\phi,\mu}$ and $B \cap H^-_{\phi,\mu}$, $A \cap H^+_{\phi,\mu}$ and $B \cap H^+_{\phi,\mu}$, and the sets $A \cap H^-_{\phi,\mu}$ and $B \cap H^+_{\phi,\mu}$ can be separated strongly, respectively. Note that A and B can't be separated strongly, we must have that $A \cap H^+_{\phi,\mu}$ and $B \cap H^-_{\phi,\mu}$ could not be separated strongly; otherwise, there exists $W \in \mathcal{N}(0)$ such that $(A \cap H^-_{\phi,\mu} + W) \cap (B \cap H^-_{\phi,\mu}) = \emptyset$, $(A \cap H^+_{\phi,\mu} + W) \cap (B \cap H^+_{\phi,\mu}) = \emptyset$, $(A \cap H^-_{\phi,\mu} + W) \cap (B \cap H^+_{\phi,\mu}) = \emptyset$, and $(A \cap H^+_{\phi,\mu} + W) \cap (B \cap H^-_{\phi,\mu}) = \emptyset$. Then it is easy to verify that

$$(A + W) \cap B = [(A \cap H^+_{\phi,\mu}) \cup (A \cap H^-_{\phi,\mu}) + W] \cap [(B \cap H^+_{\phi,\mu}) \cup (B \cap H^-_{\phi,\mu})] = \emptyset$$

which contradicts our assumption that A and B can't be separated strongly. Thus $A \cap H^+_{\phi,\mu}$ and $B \cap H^-_{\phi,\mu}$ can not be separated strongly. Let $\hat{W} \in \mathcal{N}(0)$ such that $\hat{W} \in W \cap \hat{V}$. As A and B are substrictly separated by $M_{\psi,\lambda}$, it follows that $M_{\psi,\lambda}$ also substrictly separates $A \cap H^+_{\phi,\mu}$ and $B \cap H^-_{\phi,\mu}$. Therefore by Lemma 1.3.21(1) again, there exist $x_{\hat{w}} \in A \cap H^+_{\phi,\mu}$, $y_{\hat{w}} \in B \cap H^-_{\phi,\mu}$ and $z_{\hat{w}} \in M_{\psi,\lambda}$ such that $x_{\hat{w}} - z_{\hat{w}} \in \hat{W}$ and $y_{\hat{w}} - z_{\hat{w}} \in \hat{W}$. Then we have that $x_{\hat{w}} - y_{\hat{w}} = (x_{\hat{w}} - z_{\hat{w}}) - (y_{\hat{w}} - z_{\hat{w}}) \in \hat{W} + \hat{W} \subset \hat{V} + \hat{V} \subset U$. This implies $x_{\hat{w}} \in U + y_{\hat{w}} \subset U + H^-_{\phi,\mu} \subset H^-_{\phi,\sigma}$ and thus $x_{\hat{w}} \in A \cap H^-_{\phi,\sigma}$. Therefore we have shown that for each $V \in \mathcal{N}(0)$, there exists $x_{\hat{w}} \in A \cap H^-_{\phi,\sigma}$ and $y_{\hat{w}} \in B \cap H^-_{\phi,\sigma}$ and $z_{\hat{w}} \in M_{\psi,\lambda}$ such that $x_{\hat{w}} - z_{\hat{w}} \in V$ and $y_{\hat{w}} - z_{\hat{w}} \in V$. Then it follows that $A \cap H^-_{\phi,\sigma}$ and $B \cap H^-_{\phi,\sigma}$ can't be separated strongly by a closed hyperplane. Thus $M_{\psi,\lambda}$ is an asymptotic boundary plane of $A \cap H^-_{\phi,\sigma}$ and $B \cap H^-_{\phi,\sigma}$.

(d): If $M_{\psi,\lambda}$ is an asymptotic plane of $A \cap H^-_{\phi,\mu}$ and $M_{-\psi,-\lambda}$ is an asymptotic plane of $B \cap H^+_{\phi,\mu}$. Let $\alpha = -1$. Then for any $\sigma \in (\mu + \alpha, \mu)$, we have that $\sigma > \mu$ and by the similar proof used in the case (c), we can show that $M_{\psi,\lambda}$ is an asymptotic boundary plane of $A \cap H^+_{\phi,\sigma}$ and $B \cap H^+_{\phi,\sigma}$.

By summarizing cases (a), (b), (c) and (d) above, it follows that $M_{\psi,\lambda}$ is a common asymptotic plane of A and B, which contradicts our assumption. Thus A and B can be separated strongly by some closed hyperplane and the proof is complete. □

As an application of Theorem 1.3.22, we now have the following strong sepapation result without compactness in locally convex spaces.

Corollary 1.3.23. Let A and B be non-empty closed and convex subsets of a locally topological vector space E which are separated substrictly. Suppose A and B either have no asymptotically boundary planes or the intersections of any asymptotic plane of A and B (if it exists) with A and with B are non-empty and compact. Then A and B can be separated strongly.

Proof. Suppose A and B can't be separated strongly. Since A and B are substrictly separated, it follows from Lemma 1.3.21(2), that A and B have an asymptotic boundary plane. Without loss of the generality, let $M_{\psi,\lambda}$ be an asymptotic plane of A and B and assume that $M_{\psi,\lambda}$ is an asymptotic plane of A. Then the intersection $M_{\psi,\lambda} \cap B$ is non-empty and compact. If A and $B \cap M_{\psi,\lambda}$ can not be separated strongly, for any $V \in \mathcal{N}(0)$, it follows by Lemma 1.3.21(1) that there exists $x_v \in A$ and $y_v \in M_{\psi,\lambda} \cap B$ such that $x_v - y_v \in V$. We define a directed order '$\geq$' on the set $\{y_v\}_{V \in \mathcal{N}(0)}$ by $y_{v_1} \geq y_{v_2}$ if and only if $V_2 \subset V_1$. Then the family $\{y_v\}_{V \in \mathcal{N}(0)}$ is a directed net. Note that $M_{\psi,\lambda} \cap B$ is compact, we may assume that $\lim_{v \in \mathcal{N}(0)} y_v = y_0$. Then $\lim_{v \in \mathcal{N}(0)} x_v = \lim_v[(x_v - y_v) + y_v] = 0 + y_0$ and A is closed, it follows that $y_0 \in A$. Thus $A \cap B \neq \emptyset$, which contradicts our assumption that A and B are substrictly separated. So that A and $M_{\psi,\lambda} \cap B$ can be separated strongly by some hyperplane $M_{\psi,\mu}$. Therefore there exists $\lambda_1 < \mu < \lambda_2$ such that $A \subset \{x \in E : \phi(x) < \lambda_1\}$ and $M_{\psi,\lambda} \cap B \subset \{x \in E : \phi(x) > \lambda_2\}$. Then $M_{\psi,\lambda}$ is not an asymptotic boundary plane of A and B with respect to either $H^-_{\phi,\mu}$ or $H^+_{\phi,\mu}$. Therefore $M_{\psi,\lambda}$ is not a common asymptotic plane of A and B. By Theorem 1.3.22, it follows that A and B can be separated strongly and completing the proof. □.

Corollary 1.3.24. Let A and B be non-empty closed and convex subsets of a locally topological vector space E which are substrictly separated. Suppose either A or B is boundary compact. Then A and B can be separated strongly.

Proof. Let $M_{\psi,\lambda}$ be any asymptotic boundary plane of A and B. By Definition 1.3.18, $M_{\psi,\lambda}$ is an asymptotic plane of A and $M_{\psi,\lambda} \cap B$ is compact and convex. It follows by Corollary 1.3.23 that A and B can be separated strongly and thus we complete the proof. □

Note that strong separation property implies the substrictly separation property, it follows by Remark 1.3.19 that Corollary 1.3.24 above include the following traditional strong separation theorem as a special case (e.g., see Schaefer [453, p.64-65], Theorem 3.4 of Rudin [449, p.58] and related references therein).

Corollary 1.3.25. Let A and B be non-empty closed and convex subsets of a locally topological vector space E such that $A \cap B = \emptyset$ and one of them is compact. Then A and B can be separated strongly.

Corollary 1.3.26. Let A and B be non-empty closed and convex subsets of a reflexive Banach space X such that A and B are substrictly separated. If B has no asymptotic planes and the interaction of B with its boundary plane is non-empty and bounded, then A and B can be separated strongly.

Proof. Suppose A and B could not be separated strongly. As A and B can be substrictly separated, it follows from Lemma 1.3.21(2) that A and B have asymptotic planes. Let $M_{\psi,\lambda}$ be any asymptotic plane of A and B. As B has no asymptotic plane, it follows that $M_{\psi,\lambda}$ must be an asymptotic plane of A and $M_{\psi,\lambda}$ must also be a boundary plane of B. Note that $M_{\psi,\lambda} \cap B$ is non-empty, closed convex and bounded, it follows that by Theorem 4.2.1 of Yu [599, p.180-181] (see also Day [114], Diestel [129], Holmes [235] and related references therein) that there exists a closed hyperplane $M_{\psi,\mu}$ which separates A and B strongly. Therefore there exist $\lambda_1 < \mu < \lambda_2$ such that $A \subset \{x \in E : \phi(x) < \lambda_1\}$ and $M_{\psi,\lambda} \cap B \subset \{x \in E : \phi(x) > \lambda_2\}$. Note that $M_{\psi,\lambda}$ is not a boundary plane of A and B with respect to either $H^-_{\phi,\mu}$ or $H^+_{\phi,\mu}$, it follows that A and B have no common asymptotic planes. By Theorem 1.3.22, it follows that A and B are separated strongly and we complete the proof. □

Example 1.3.27. Let $X := \mathbb{R}^2$ and $A = \{(x, y) \in X : xy \geq 1, x, y \geq 0\}$ and B be any non-empty closed, convex and bounded subset which intersects with x and y axes of $\mathbb{R}^2$. Suppose C is any non-empty closed and convex subset of X which contains $(0, 0)$ as its interior point and C has no asymptotic planes. If A and B, A and C are substrictly separated, respectively, then the intersection of any asymptotic plane of A and B with B is non-empty and compact. By Corollary 1.3.25, it follows that A and B can be separated strongly by a closed hyperplane. Note that all asymptotic planes of A are straight lines, and indeed both x and y axes are both asymptotic planes of A and B, which are not boundary planes of C. Thus A and C have no common asymptotic planes. It follows by Theorem 1.3.22 that A and C can be separated strongly.

Remark 1.3.28. Both Examples 1.3.20 and 1.3.27 show that our Theorem 1.3.22 and Corollaries 1.3.23-1.3.26 are the improvement of the corresponding results in the literature (e.g., see Day [114], Diestel [129], Holmes [235] and so on) for the study of strong separation property of convex sets in locally convex spaces. We also note that as applications, some new concidence theorems have been developed by Yang and Yuan [584] recently.

Chapter 2

The KKM Theory and Related Topics

This chapter is the fundamental and the most important part of this book as the general Knaster-Kuratowski-Mazurkiewicz theory (KKM theory) will be developed through topological version of the KKM principle, the Ky Fan's celeberated minimax inequality (principle) that was established in 1972 [164] as an application of his geometric lemma in 1961 from [157], which is an infinite dimensional version of the classical KKM principle in 1929 [319]; and the Browder-Fan fixed point theorem which is an equivalent form of Fan's geometric lemma in 1961 (also equivalent to Fan's minimax inequality in 1972) which was first given by Browder in 1968 [59] and related Fan's section theorem and Fan's best approximation theorem for set-valued mappings in both topological vector spaces and locally convex topological vector spaces. Thus the main contribution to KKM theory is due to Professor Ky Fan's work in the last four decades. Secondly, the generic stablity of Ky-Fan point (KF point) which are solutions of Ky Fan minimax inequalities and generic stability of coincidence and fixed point theorems established from the point of view of the Baire category. Our generic stablity results show that almost all of Ky Fan minimax inequalities and coincidences and fixed points are stable in the sense of the Baire category.Moreover various generalizations of Ky Fan minimax inequality, coincidence and fixed points in topological vector spaces and the famous Fan-Glicksberg fixed point theorem of upper semicontinuous set-valued mappings in topological vector spaces with sufficient continuous linear functionals (which is automatically satisfied by locally convex topological vector spaces) have been established. Also several topological versions of Fan's section theorems have been established which also play a very important role in the study of minimax inequalities and game theory. Then Fan's best approximation theorem for upper semicontinuous set-valued mappings have been obtained which, in turn imply several fixed point theorems for nonself set-valued mappings in locally convex topological spaces.

The fixed point theory for set-valued mappings, in particular, the Browder-Fan type and the Fan-Glicksberg type fixed point theorems have been extensively studied in the setting of locally convex spaces, H-spaces, MC-spaces, G-convex spaces and hyperconvex spaces. By using its own feature of hyperconvex metric spaces being a special class of H-spaces, we also establish the general KKM theory and then its various applications given.

Finally, we develop the general fixed point theory for set-valued analysis in the setting of the abstract convexity-called the MC-space. This theory unifies the most existing results on fixed point theorems of set-valued mappings, in particular for set-values with open-inversed values and upper semicontinuous set-valued mappings with closed acyclic values in the literature. As MC-structures inlcude H-spaces as a special class, fixed point theory establishesd in this chapter will allow us to cover more general models for the study of the existence for equilibria in mathematical economics, game theory, optimization, and nonlinear analysis itself.

The theory established in this chapter forms the main part of KKM theory which will be applied to the study of abstract economics in Chapter 4 for various situations from infinite dimensional topological vector spaces to finite dimensional spaces; multi-objective optimization problems in Chapter 7; game theory in Chapter 8; and Pareto optima in Chapter 9.

2.1 Introduction

The classical theorem of Knaster-Kuratowski-Mazurkiewicz (often called the KKM theorem, or the KKM lemma or KKM principle in [319]) has numerous applications in various fields of pure and applied mathematics. These studies and applications are called the KKM Theory today.

In 1961, Ky Fan proved the generalization of the classical KKM theorem in infinite dimensional Hausdorff topological vector spaces and established an elementary but very basic geometric lemma for set-valued mappings. In 1968, Browder gave a fixed point form of Fan's geometric lemma and it is now called the Browder-Fan fixed point theorem. Since then there have been numerous generalizations of the Browder-Fan fixed point theorem and their applications in coincidence and fixed point theory, minimax inequalities, variational inequalities, nonlinear analysis, convex analysis, game theory, mathematical economics, and so on.

By applying his geometric lemma in 1972, Ky Fan obtained a minimax inequality which plays a fundamental role in nonlinear analysis and mathematical economics and has been applied to potential theory, partial differential equations, monotone operators, variational inequalities, optimization, game theory, linear and nonlinear programming, operator theory, and topological group and linear algebra. In particular, by using Ky Fan's minimax inequality, a more general form of the Fan-Glicksberg fixed point theorem is derived for set-valued mappings which are inward (or outward) as defined by Fan in 1969 which are more general than Halpern's definitions for inward (or outward) mappings in 1965.

Recently, Horvath obtained some generalizations of Fan's geometric lemma and his minimax inequality in 1983 and 1987 by replacing convexity assumption with topological properties: pseudo-convexity and contractibilty. By extending Horvath's concepts, Bardaro and Ceppitelli [24] in 1988 obtained generalizations of Ky Fan minimax inequalities to topological spaces which have a so called H-structure (also called H-spaces).

Following this line, a number of generalizations of Ky Fan's minimax inequalities are given by Horvath [239], Baradaro and Ceppitelli [25], Ding and Tan [140], Ding et al. [141]-[142], Chang and Ma [70], Park [422], Tarafdar [534], and Tan et al. [517] in topological spaces which need not have a linear structure but with an H-structure.

On the other hand, for the need of applications, various generalizations of the classical KKM principle and Sperner's lemma [498] have been given by Fan [160]-[161], [163] and [169], Ding and Tan [140], Gale [190], Idzik and Tan [250], Shapley [460], Shih and Tan [468], [469], [470], Ichiishi [246], and Ichiishi and Idzik [248]. Recently, Horvath [239] obtained some intersection theorems for closed coverings of a topological space with a contractible structure.

In this chapter, based on the classical KKM principle and its dual form given by Shih and Tan in 1987, we first study the closed (resp., open) covering properties and intersection properties of topological spaces in Section 2. These results generalize the corresponding results of Alexandroff-Pasynkoff [1], Berge [40], Klee [315], Fan [157], Horvath [239] and Spener [498]. As applications, we give a characterization of a set-valued mapping which is a generalized HKKM mapping (which in turn, includes the classical KKM theorem in topological vector spaces as a special case.

In Section 3, by applying our generalized HKKM theorem, Ky Fan type minimax inequalities with weaker continuity conditions are given in topological spaces. Our results show that the traditional condition *lower semicontinuity* posited by many authors (e.g., see Fan [167] and references therein) is not essential for the existence of solutions for the Ky Fan type minimax inequalities.

By employing a new coercive concept called *escaping sequences* which was first introduced by Border [48] in 1985, several non-compact minimax inequalities are derived. As consequences, several equivalent fixed point theorems and maximal element theorems are given in H-spaces and topological spaces. In particular, the well-known Browder-Fan fixed point theorem has been improved.

Since the Browder-Fan is so important in the study of nonlinear analysis, in Section 4, we first present several equivalent forms of the Browder-Fan fixed point theorem and then by introducing the concept called *couple majorized mapping* , the product version of the Browder-Fan fixed point theorem has been established, which in turn enables us to obtain the system version of Ky Fan minimax inequalities

As Ky Fan minimax inequalities is so important, in Section 5, we study the stability of solutions of Ky Fan minimax inequality in both compact and non-compact settings. From the point of view of the Baire category, we prove that almost all of Ky Fan minimax inequalities are stable.

It is well-known that Fan's best approximation theorem [162] and its generalizations have many applications in fixed point theory and approximation theory (for instance, see Ding and Tan [136], Reich [435] and Sehgal et al. [455] and references therein). In Section 6, we first generalize Fan's best approximation theorem to a Hausdorff topological vector space for set-valued mappings. Then as applications, several coincidence and fixed point theorems are given for non-self set-valued mappings under weaker boundary conditions. These results improve and generalize the corresponding results of Komiya [326], Ha [213]-[214], and many others.

In Section 7, as applications of Fan's concept called inward and out mappings for non-self set-valued upper hemicontinuous mappings in topological vector spaces in [162], various generalizations of the Fan-Glicksberg fixed point theorem (which is an infinite-dmensional version of Kakutain fixed point theorem which, in turn is a set-valued version of Brouwer fixed point for a single-valued continuous mapping in finite dimensional spaces) have been obtained. These results unify most results of fixed point theorems and coincidence theorems in the literature, e.g., see Park [419] and references therein). As applications, several matching theorems for closed coverings of convex sets are given which include the well-known Shapley generalization [460] of the classical KKM theorem. These unify the corresponding results in the literature.

In Section 8, by employing the method used in section 5, we study the generic stablity of coincidence and fixed points for set-valued mappings. Our results also show that coincidence and fixed points are stable in the sense of the Baire category. These reuslts improve and cover the corresponding results given by Fort [179] and Jiang [267] in several ways.

In Section 9, corresponding Fan's famous lemma in [157], we have established section theorems of Fan's section theorem in topological spaces which may not have any linear structure. As applications several miniamx minimax inequalities and best approximation theorems have been established which in turn include the corresponding results as special cases.

As an applicaion of minimax inequalities established in Section 9; and in Section 10, the Fan's alternativr theorem has been obtained for upper semicontinuous set-valued mappings in locally convex spaces which also has various applications in the study of existence of solutions for nonlinear applications. Then the interconnection between KKM theory and related topics has been also discussed. By using notion called locally convex H-spaces, we establish fan-Glicksberg fixed point theorem for upper semicontinuous set-valued mappings with closed acyclic values, which, in turn, includes the most fixed point theorems for upper semicontinuous set-valued mappings in existing literature as special cases.

In Section 10, we develop a general KKM theory for one special class of H-spaces, so-called hyperconvex spaces and then various applications have been given.

Finally in Section 12, the general fixed point theory for set-valued analysis in the setting of the abstract convexity-called MC-spaces and G-convex spaces, have been established. This theory unifies the existing results on fixed point theorems of set-valued mappings, in particular for set-valued with open inversed values and upper semicontinuous set-valued mappings with closed acyclic values in the literature. As MC-structures and G-convex spaces are general encough to inlcude H-spaces, simplicial convexity as a special class, our fixed point theory established in this section will allow us to cover more general models for the study of the existence of equilibria in mathematical economics, game theory, optimization, and for the study of problems in nonlinear analysis itself.

Throughout this book all spaces are assumed to be *Hausdorff* unless otherwise specified. Let X and Y be non-empty sets. We shall denote by 2^Y the family of all non-empty subsets of Y, $\mathcal{F}(X)$ the family of all non-empty finite subsets of X. Let X be a topological space. For each non-empty subset A of X, we denote the closure of A in X by $cl_X A$ (in short, clA) or $\overline{A}$ if there is no confusion. A subset A of X is said to be compactly closed (resp., open) if $A \cap C$ is closed (resp., open) in each non-empty compact subset C of X.

2.2 Knaster-Kuratowski-Mazurkiewicz Principle

In this section, based on the classical KKM theorem [319] and its *dual* form given by Shih and Tan [470], we first discuss some properties of contractible subsets in topological spaces by employing Horvath's approach [237]. As applications, a characterization of generalized HKKM mappings is given. These results improve Fan's famous geometric result given in his celebrated paper [157], and the corresponding results due to Fan [167]-[169], Klee [315], Alexandroff-Pasynkoff [1], Berge [40], Horvath [237], and Chang and Zhang [76], Chang and Yan [72], and Chang and Ma [70].

A topological space X is said to be *contractible* if the identity mapping I_X of X is homotopic to a constant function. Let $\mathbb{N}$ and $\mathbb{R}$ denote the set of all natural numbers and the set of all real numbers, rspectively. Let $v_0, v_1, \cdots, v_n$ be $n+1$ points in a Euclidean space which are not contained in any linear manifold (i.e., a translation of a vector subspace) of dimension less than n. The convex hull of these $n+1$ points is called an n-dimensional simplex and is denoted by $v_0 v_1 \cdots v_n$. The points $v_0, v_1, \cdots, v_n$ are also called vertices of the simplex. For $0 \leq k \leq n$ and $0 \leq i_0 < i_1 < \cdots < i_k \leq n$, the k-simplex $v_{i_0} v_{i_1} \cdots v_{i_k}$ is a subset of the n-simplex $v_0 v_1 \cdots v_n$. It is called a k-dimensional face (or say, simply k-face) of the simplex $v_0 v_1 \cdots v_n$.

For the convenience of our discussion, for $n \in \mathbb{N}$ and $N = \{0, 1, \cdots, n\}$, we also denote by $\Delta_N = co\{e_0, \cdots, e_n\}$ the standard simplex of dimension n, where $\{e_0, \cdots e_n\}$ is the canonical basis of $\mathbb{R}^{n+1}$ and for $J \in \mathcal{F}(N)$, $\Delta_J = co\{e_j : j \in J\}$. The points $e_0, e_1, \cdots, e_n$ are called vertices of the simplex Δ_N.

Suppose that $T = \{\tau_0, \tau_1, \cdots, \tau_m\}$ is finite collection of n-simplexes in $v_0 v_1 \cdots v_n$ such that (i): $\cup_{i=1}^m \tau_i = v_0 v_1 \cdots v_n$; and (ii): $\tau_i \cap \tau_j$ is either empty or a common face of any dimension of τ_i and τ_j, where $i, j \in \{1, 2, \cdots, m\}$. Then T is called a triangulation of $v_0 v_1 \cdots v_n$. Let T be a triangulation. A vertex of a simplex in T is called a vertex of T. An $(n-1)$-face of a simplex from T is called a boundary $(n-1)$ -simplex of T if it is a face of exactly one n-simplex of T.

In this section, we first present an extremely useful result published in 1929 due to Knaster, Kuratowski, and Mazurkiewicz [319], which is regarded as one of the most important principles in nonlinear analysis, and also provides the foundation for our study in this book. Its proof is based on the celebrated Sperner's combinatorial lemma [498] published in 1928 which we will discuss below. There are several versions of Sperner's Lemma and its extensions, for more details, see Border [48, p.23-24], Cohen [96], Dugundji and Granas [149], Fan [167], Shashkin [462, p.9-14], Shih and Lee [464], Wong [574], Zeidler [613] and many other books for its numerous and different proofs.

KKM Principle. Let $C_0, \cdots, C_n$ be closed subsets of the standard n-dimensional simplex Δ_N and let $\{e_0, \cdots, e_n\}$ be the set of its vertices. If for each $J \in \mathcal{F}(N)$, $\Delta_J \subset \cup_{j \in J} C_j$. Then $\cap_{i=0}^n C_i \neq \emptyset$.

In order to prove the KKM principle above, we begin with the celebrated Sperner's combinatorial lemma appeared in 1928.

Sperner's Lemma. Let T be a triangulation of an n-simplex $\Delta = v_0 v_1 \cdots v_n$. To each vertex v of T, let $\psi(v) \in \{0, 1, \cdots, n\}$ be such that whenever $v \in$

$v_{i_0}v_{i_1}\cdots v_{i_k}$, $\psi(v) \in \{i_0, i_1, \cdots, i_k\}$ for $0 \leq i_0 < i_1 < \cdots i_k \leq n$. Then the number of n-simplexes in T whose vertices receive $n+1$ different values is odd.

Proof. By following arguments due to Cohen [96], Fan [160], and Shih in Wong [574, p.332-333], we might think of T as a house. The n-simplexes of T are the rooms of the houses. A room will be considered to be a good room if its n_1 vertices are labeled with $n+1$ different integers, otherwise it will be called a bad room. If the vertices of an $(n-1)$ face of a simplex τ of T are labeled with the integers $0, 1, \cdots, n-1$, then that face is considered to be a door to the room τ. A door which is a boundary $(n-1)$ simplex of T is called an external door. Clearly a good room has exactly one door and a bad room has either no doors or two doors. Let us trace paths through the house, going through each door exactly one time. To start a path we either enter a room through an external door or we leave a good room through its unique door. Each time we enter a room through a door and see a second door in the room, we go out of the room through the second door. This will lead us either to the outside of the house or to a new room. By this rule, our path will be stopped either when we get to the outside of the house through an external door, or when we enter a good room. After going through each door exactly once by tracing various paths, we pair off the two ends of each paths. In this way, each good room is paired either with another good room or with an external door. Also each external door is paired either with another external door or with a good room. This pairing implies that the number of good rooms and the number of external doors are congruent modulo 2, i.e., theses number have the same parity. To complete the proof of Sperner's Lemma, it suffices to show that the number of boundary $(n-1)$-simplexes in T with vertices labeled $0, 1, \cdots, n-1$ is odd. This is immediate by induction argument on n. □

As an application of Sperner's Lemma, we can prove the famous KKM principle.

The Proof of the KKM Principle: Let T be an arbitrarily triangulation of the standard simplex Δ_N. For each vertex e_i of T, we assign an integer $\psi(v) \in \{0, 1, \cdots, n\}$ such that $e_i \in C_{\psi(e_i)}$ and $\psi(e_i) \in \{i_0, i_1, \cdots, i_k\}$ whenever $e_i \in \{e_{i_0}, e_{i_1}, \cdots, e_{i_k}\}$, this is possible by virtue of the condition given in KKM principle. Thus, by Sperner's Lemma, there exists a simplex $w_0w_1\cdots w_n \in T$ such that $\psi(w_i) = i$ for $i = 0, 1, \cdots, n$. It follows that $w_i \in C_{\psi(w_i)} = C_i$ for $i = 0, 1, \cdots, n$. Consider now a sequence of triangulations $\{T^m\}$ of $e_0e_1\cdots e_n$ such that the mesh of T^m tends to zero as $m \to \infty$. From the above argument, we find that for each m, the n-simplex $w_0^m w_1^m \cdots w_n^m$ of T such that $w_i^m \in C_i$ for $i = 0, 1, \cdots, n$. By the compactness of $e_0e_1\cdots e_n$, we can choose a subsequenct $\{m_j\}$ of $\{m\}$ with $m_1 < m_2 < \cdots$ such that $\lim_{j\to\infty} w_i^{m_j} = q_i$ exists for each $i = 0, 1, \cdots, n$. As the mesh of T_j^m tends to zero when $j \to \infty$, it follows that $q_0 = q_1 = \cdots = q_n$, and hence we have that $\lim_{j\to\infty} w_i^{m_j} = q$ for all

$i = 0, 1, \cdots, n$. Note that C_i is closed for each $i = 1, 2, \ldots, n$, it follows that $q \in \cap_{i=1}^n c_i$ and hence we complete the proof. □

On the other hand, Shih and Tan [470] (see also Kim [297] or Lassonde [335]) provided the following *dual form* of KKM theorem in the sense that the word *closed* could be replaced by the word *open*.

The dual form of the KKM principle. Let $C_0, \cdots, C_n$ be open subsets of the standard n-dimensional simplex Δ_N and let $\{e_0, \cdots, e_n\}$ be the set of its vertices. If for each $J \in \mathcal{F}(N)$, $\Delta_J \subset \cup_{j\in J} C_j$. Then $\cap_{i=0}^n C_i \neq \emptyset$.

In order to prove the dual form of the KKM principle, by following Shih and Tan [470], we need to introduce the concept of the Shapley-mapping. Let $\mathcal{F}$ denote the family of all faces of the n-simplex Δ_n. Now we have the following so-called Shapley mapping from [470].

A set-valued mapping $B : \mathcal{F} \to 2^{\Delta_N}$ is said to be Shapley-mapping if for each $\tau \in \mathcal{F}$, we have

$$\tau \subset \cup_{\tau \supset \rho \in \mathcal{F}} B(\rho).$$

Let $C_0, C_1, \cdots, C_n$ be subsets of a standard n-simplex Δ_N. If we define a mapping $B : \mathcal{F} \to 2^{\Delta_N}$ by $B(i) = C_i$ for each $i \in N = \{0, 1, \cdots, n\}$ and $B(\rho) = \emptyset$ for all non 0-dimensional faces of Δ_N. Then it is easy to see that B is a Shapley mapping if and only if the following condition is satisfied:

$$\Delta_J \subset \cup_{j\in J} C_j = \cup_{j\in j} B(j)$$

for each $J \in \mathcal{F}(N)$.

Now we have the following selection theorem due to Shih and Tan in [470].

Theorem 2.2.1. If $B : \mathcal{F} \to 2^{\Delta_N}$ is a Shapley-mapping such that each $B(\rho)$ is an open subset of Δ_N for each $\rho \in \mathcal{F}$. Then B admits a Shapley-selection $A : \mathcal{F} \to 2^{\Delta_N}$ such that each $A(\rho)$ is a closed subset of Δ_N and $A(\rho) \subset B(\rho)$ for all $\rho \in \mathcal{F}$.

Proof. Since B is a Shapley-mapping, it follows that $\Delta_N = \cup_{\rho\in\mathcal{F}} B(\rho)$. For each $y \in \Delta_N$, let $H_y := \cap\{B(\rho) : y \in B(\rho)\}$. Then H_y is an open set in Δ_N containing y, and therefore there exists an open neighborhood U_y of y in Δ_N such that $y \in U_y \subset \overline{U_y} \subset H_y$. For each $\tau \in \mathcal{F}$, we have

$$\cup\{B(\rho) : \rho \subset \tau\} = \cup\{U_y : y \in B(\rho) \text{ for some } \rho \subset \tau\}.$$

As B is a Shapley-mapping, the compactness of τ ensures that there exists a finite set B_τ of $\cup\{B(\rho) : \rho \subset \tau\}$ such that $\tau \subset \{U_y : y \in B_\tau\}$. We now set $K := \cup\{B_\tau : \tau \in \mathcal{F}\}$. Then K is a finite set. Define a mapping $A : \mathcal{F} \to 2^{\Delta_N}$ by

$$A(\rho) := \cup\{\overline{U_y} : y \in K \text{ and } U_y \in B(\rho)\}$$

for each $\rho \in \mathcal{F}$. Then it is easy to check that $A(\rho)$ is a closed subset of Δ_N such that $A(\rho) \subset B(\rho)$. It remains to verify that A is also a Shapley mapping. Let $\tau \in \mathcal{F}$ be arbitrarily fixed. For each $z \in \tau$, there exists $\rho \in \mathcal{F}$ with $\rho \subset \tau$ and $y \in B(\rho)$ such that $z \in U_y$. It follows that $y \in K$, and $U_y \in \overline{U_y} \subset H_y \subset B(\rho)$, and hence $z \in A(\rho)$ for some $\rho \subset \tau$. Therefore $\tau \subset \cup_{\tau \supset \rho \in \mathcal{F}} A(\rho)$. This concludes the proof of the Theorem. □

As an application of Shih and Tan's selection theorem for Shapley mapping and the KKM principle, we can now prove the dual form of the KKM principle above.

The Proof of the dual form of the KKM principle: We define a mapping $B : \mathcal{F} \rightarrow 2^{\Delta_N}$ by $B(i) = C_i$ for each $i \in N = \{0, 1, \cdots, n\}$ and $B(\rho) = \emptyset$ for all non-0-dimensional faces of Δ_N. As we note above that B is a Shapley mapping if and only if $\Delta_J \subset \cup_{j \in J} C_j = \cup_{j \in j} B(j)$ for each $J \in \mathcal{F}(N)$. By Theorem 2.2.1, there exists another Shapley mapping $A : \mathcal{F} \rightarrow 2^{\Delta_N}$ such that $A(\rho)$ is closed and $A(\rho) \subset B(\rho)$ for each $\rho \in \mathcal{F}$. Let $D_i = A(i)$ for each $i \in N = \{0, 1, \cdots, n\}$. Then we have that D_i is closed and $D_i \subset C_i$ for all $i \in N$. Since A is also a Shapley mapping, it follows that $\Delta_J \subset \cup_{j \in J} D_j = \cup_{j \in j} A(j)$ for each $J \in \mathcal{F}(N)$. By the KKM principle, we have that $\cap_{i=0}^{n} A(i) \neq \emptyset$. Thus it implies that $\cap_{i=0}^{n} B(i) \neq \emptyset$ and the proof is completed. □.

The following notions introduced by Bardaro and Ceppitelli in [24] were motivated by the earlier work of Horvath [237].

Definition 2.2.2. A pair $(X, \{\Gamma_A\})$ (also called an *H-structure* is said to be an *H-space* (also called *C-space* according to Horvath [239]) if X is a topological space and $\{\Gamma_A\}_{A \in \mathcal{F}(X)}$ a given family of non-empty contractible subsets Γ_A of X, indexed by $A \in \mathcal{F}(X)$ such that $\Gamma_A \subset F_B$ whenever $A \subset B$. Let $(X, \{\Gamma_A\})$ be an H-space. A non-empty subset D of X is said to be: (i) H-convex (also called an *F-set* by Horvath [239]) if $\Gamma_A \subset D$ for each $A \in \mathcal{F}(D)$; (ii) weakly *H-convex* if $\Gamma_A \cap D$ is contractible for each $A \in \mathcal{F}(D)$ (or equivalently, $(D, \{\Gamma_A \cap D\})$ is an H-space); and (iii) *H-compact* in X if for each $A \in \mathcal{F}(X)$, there exists a compact, weakly H-convex subset D_A of X such that $D \cup A \subset D_A$. It is clear that the product space of a family of H-spaces is also an H-space.

The following example from Horvath [239, p.345] shows that an H-space may not be a convex subset of a topological vector space.

Example 2.2.3. Let X be a convex set in a topological vector space E and Y any topological space. Suppose that $f : X \rightarrow Y$ is a continuous bijection. For given $A \in \mathcal{F}(Y)$, let $D_A := co\{x \in X : f(x) \in A\}$. Then D_A is convex, so that D_A is contractible. Since D_A is also compact, so that $f : D_A \rightarrow f(D_A)$ is an homeomorphism. Let $\Gamma_A = f(D_A)$. Then Γ_A is contractible and $\Gamma_A \subset \Gamma_{A'}$ whenever $A \subset A' \in \mathcal{F}(Y)$. Therefore $(Y, \{\Gamma_A\})$ is an H-space. Note that the space Y itself may be a torus, the Mobius band, or the Klein bottle. This example shows that an H-space does not have to be contractible.

Definition 2.2.4. Let D be a non-empty subset of an H-space $(X, \{\Gamma_A\})$. The H-convex hull of D, denoted by $Hco(D)$, is defined by

$$Hco(D) = \cap\{B \subset X : B \text{ is } H\text{-convex and } D \subset B\}$$

(e.g., see Tarafdar [534]). Clearly, $Hco(D)$ is the smallest H-convex subset of X containing D and the intersection of any family of H-convex set is also H-convex. Let X be a topological space. By following Horvath [239], an n-dimensional singular face structure on X is a mapping $F : \mathcal{F}(N) \to 2^X$ such that (a) for each $J \in \mathcal{F}(N)$, $F(J)$ is not-empty and contractible and (b) for any $J, J' \in \mathcal{F}(N)$, $J \subset J'$ implies $F(J) \subset F(J')$.

Let $N = \{0, 1, \cdots, n\}$. We denote by $\partial\Delta_N$ and Δ_N^k the boundary and kth-skeleton of n-dimensional simplex Δ_N, respectively. By the fact that if X is a contractible topological space, each (single-valued) continuous function $f : \partial\Delta_N \to X$ can be continuously extended to Δ_N, i.e., there exists a continuous function $g : \Delta_N \to X$ such that $g(x) = f(x)$ for all $x \in \partial\Delta_N$, we have the following continuous selection result for H-structures (see also Horvath [239]).

Lemma 2.2.5. let X be a topological space. For each non-empty subset J of $N = \{0, 1, \cdots, n\}$, let F_J be a non-empty contractible subset of X with $F_J \subset F_{J'}$ whenever $\emptyset \neq J \subset J' \subset \{0, 1, \cdots, n\}$. Then there exists a continuous function $f : \Delta_N \to X$ such that $f(\Delta_J) \subset F_J$ for each non-empty subset J of $\{0, 1, \cdots, n\}$.

Proof. For each singleton $\{i\}$, take any fixed point $x_i \in F_{\{i\}}$. Then $f^0 : \Delta_N \to X$ defined by $f^0(e_i) := x_i$ is a continuous function on the 0th-skeleton Δ_N^0 of Δ_N. Now assume that a continuous function $f^k : \Delta_N^k \to X$ on the kth-skeleton of Δ_N has been constructed in a such way such that $f^k(\Delta_J) \subset F_J$ if card$J \leq k + 1$. Let Δ_J be a face of dimension $k + 1$ of Δ_N and for each $i \in J$, let $J_i := J \setminus \{i\}$. Then the boundary set $\cup_{i\in J}\Delta_{J_i}$ of Δ_J is contained in the kth-skeleton of Δ_N and $f^k(\partial\Delta_J) \subset \cup_{i\in J} f^k(\Delta_{J_i}) \subset \cup_{i\in J} F_{J_i} \subset F_J$. By the fact that F_J is contractible, f^k can be extended to a continuous function $f_J^{k+1} : \Delta_J \to F_J$. If Δ_J and $\Delta_{J'}$ are two different $k+1$ dimensional faces of Δ_N with non-empty intersection, then f_J^{k+1} and $f_{J'}^{k+1}$ coincide with the continuous function f^k on $\Delta_J \cap \Delta_{J'}$. By gluing the functions f_J^{k+1} together, we obtain a continuous function $f^k : \Delta_N^{k+1} \to X$, which has the desired property. □

Proposition 2.2.6. Let X be a topological space. Let $F : \mathcal{F}(N) \to 2^X$ be a singular face structure on X and $\{M_i : i = 0, \cdots, n\}$ be a family of closed (resp., open) subsets of X such that for any $J \in \mathcal{F}(N)$, $F(J) \subset \cup_{i\in J} M_i$. Then $\cap_{i=0}^n M_i \neq \emptyset$.

Proof. By Lemma 2.2.5, there is a continuous function $f : \Delta_N \to X$ such that for each $J \in \mathcal{F}(N)$, $f(\Delta_J) \subset F(J)$. Let $C_i = f^{-1}(M_i)$ for each

$i = 0, \cdots, n$. Then $\{C_i\}_{i=0}^n$ is a family of closed (resp., open) subsets of Δ such that for any $J \in \mathcal{F}(N)$, $\Delta_J \subset \cup_{i\in J} C_i$. By the KKM principle (resp., its dual form), $\cap_{i=0}^n C_i \neq \emptyset$. Take any $x_0 \in \cap_{i=0}^n C_i$, then $f(x_0) \in \cap_{i=0}^n M_i \neq \emptyset$. □

As an application of Proposition 2.2.6, we have the following result.

Theorem 2.2.7. Let X be a contractible topological space, $\{M_i : i = 0, \cdots, n\}$ be a closed (resp., open) covering of X and $\{F_i : i = 0, \cdots, n\}$ a family of contractible subsets of X such that (i) for any $i \in \{0, \cdots, n\}$, $F_i \cap M_i = \emptyset$ and (ii) for any $J \in \mathcal{F}(N)$ with $J \neq N$, $\cap_{i\in J} F_i$ is non-empty and contractible. Then $\cap_{i=0}^n M_i \neq \emptyset$.

Proof. Define $F : \mathcal{F}(N) \to 2^X$ by $F(N) = X$ and $F(J) = \cap_{i\notin J} F_i$ if $J \in \mathcal{F}(N)$ with $J \neq N$. Then F is a singular face structure on X. Note that for any $J \in \mathcal{F}(N)$, $F(J) \subset \cup_{i\in J} M_i$ since $F(J) \subset \cup_{i\in J} M_i \cup \cup_{i\notin J} M_i$ and $F(J) \cap M_i = \emptyset$ whenever $i \notin J$. Then Proposition 2.2.6 implies that $\cap_{i=0}^n M_i \neq \emptyset$. □

Theorem 2.2.7 is clearly a generalization of the KKM lemma. For $X = \Delta_N$ and $F_i = co(\{e_j : j \neq i\}$ for $i \in N$, Theorem 2.2.7 reduces to a theorem of Alexandroff-Pasynkoff in [1]. For $X = \Delta_N$ and $F_i = X \setminus M_i$ which is a closed convex subset for each $i \in N$, Theorem 2.2.7 is a generalization of a corresponding result due to Klee [315] and is known as Berge's intersection theorem [40].

Theorem 2.2.8. Let X be a contractible topological space and Y be a topological space, $\{M_i : i = 0, \cdots, n\}$ be an open (resp., closed) covering of Y and $\{F_i : i = 0, \cdots, n\}$ be a family of contractible subsets of X. Let $S : X \to Y$ be continuous such that

(a) for each $i \in \{0, 1, \cdots, n\}$, $F_i \subset S^{-1}(M_i)$; and

(b) for each $J \in \mathcal{F}(N)$ with $J \neq N$, $\cap_{i\in J} F_i$ is non-empty and contractible.

Then $\cap_{i=0}^n M_i \neq \emptyset$.

Proof. Suppose the contrary, so that $\cup_{i=0}^n M_i^C = Y$, where $M_i^C = Y \setminus M_i$ denotes the complement of M_i in Y for each $i = 0, 1, \cdots, n$. Thus $\{M_i^C : i = 0, \cdots, n\}$ is an open (resp., closed) covering of Y. So that $\{S^{-1}(M_i^C) : i = 0, \cdots, n\}$ is an open (resp., closed) covering of X and by condition of (a), for each $i = 0, 1, \cdots, n$, $F_i \cap S^{-1}(M_i^C) = \emptyset$. Therefore F_i and $S^{-1}(M_i^C)$ for $i = 0, \cdots, n$ satisfy all hypotheses of Theorem 2.2.7. By Theorem 2.2.7, $\cap_{i=0}^n S^{-1}(M_i^C) \neq \emptyset$ which contradicts that $\{M_i : i = 0, \cdots, n\}$ is a covering of Y. Thus $\cap_{i=0}^n M_i \neq \emptyset$. □

We remark that Horvath [239, p.343] proved Theorem 2.2.8 under the additional assumptions that $X = Y$ and X is a normal space.

Since Chang and Yang in [72] gave a generalization of the KKM theorem in which the domain need not be a subset of its range, there are several generalizations in this direction. For example, Chang and Ma [70] extended this definition into H-spaces and later Zhou [620] gave a more generalized definition and obtained a characterization of the generalized HKKM mapping which is also a generalization of the corresponding result given by Chang and Zhang [76].

Definition 2.2.9. Let X be a non-empty set and Y a topological space. A mapping $G : X \to 2^Y \cup \{\emptyset\}$ is said to be *transfer closed valued* (e.g., see Zhou and Tian [623]) if for each $x \in X$ and $y \notin G(x)$, there exist $x' \in X$ and an open neighborhood $N(y)$ of y in Y such that $y' \notin G(x')$ for each $y' \in N(y)$. It is obvious that if a mapping $G : X \to 2^Y$ is transfer closed valued, then for each $x \in X$ and $y \in Y$ with $y \notin G(x)$, there exists some $x' \in X$ such that $y \notin clG(x')$.

The following lemma was first proved by Zhou and Tian [623] for the case when the domain X is a topological space and in the present form by Zhou [620]; for completeness, we include its proof here.

Lemma 2.2.10. Let X be a non-empty set, Y a topological space and $G : X \to 2^Y$. Then $\cap_{x\in X}G(x) = \cap_{x\in X}clG(x)$ if and only if the mapping G is transfer closed valued.

Proof. Sufficiency: It is clear that $\cap_{x\in X}G(x) \subset \cap_{x\in X}cl_Y G(x)$. It is sufficient to show that $\cap_{x\in X}clG(x) \subset \cap_{x\in X}G(x)$. Suppose that $y \notin \cap_{x\in X}G(x)$. Then there exists some $x \in X$ such that $y \notin G(x)$. Since G is transfer closed valued on X, there exists some $x' \in X$ such that $y \notin cl_Y G(x')$, so that $y \notin \cap_{x\in X}clG(x)$. Therefore $\cap_{x\in X}clG(x) = \cap_{x\in X}G(x)$.

Necessity: Suppose $(x, y) \in X \times Y$ such that $y \notin G(x)$; then $y \notin \cap_{z\in X}G(z) = \cap_{z\in X}clG(z)$ so that there exists $x' \in X$ such that $y \notin clG(x')$. But then there exists an open neighborhood $N(y)$ of y in Y such that $N(y) \cap G(x') = \emptyset$ so that $y' \notin G(x')$ for all $y' \in N(y)$. Thus G is transfer closed valued. □.

Definition 2.2.11. Let D be a non-empty subset of an H-space $(X, \{\Gamma_A\})$. A mapping $F : D \to 2^X$ is called *HKKM* if $\Gamma_A \subset \cup_{x\in A}F(x)$ for each $A \in \mathcal{F}(X)$. When X is a non-empty convex subset of a topological vector space and $\Gamma_A = coA$, the convex hull of A for each $A \in \mathcal{F}(X)$, then $(X, \{\Gamma_A\})$ becomes an H-space. In this case, the notion of an HKKM mapping $F : D \to 2^X$ coincides with the notion of a KKM mapping F, i.e., $coA \subset \cup_{x\in A}F(x)$ for each $A \in \mathcal{F}(D)$.

Definition 2.2.12. Let X be a non-empty set and Y a topological space. A mapping $G : X \to 2^Y$ is said to be a *generalized HKKM* (in short, GHKKM) if for each finite subset $A = \{x_1, \cdots, x_n\}$ of X, there exists a corresponding finite subset $B = \{y_1, y_2 \cdots, y_n\}$ (y_i's need not be distinct here) in Y and a family $\{\Gamma_C\}_{C\in\mathcal{F}(B)}$ of non-empty contractible subsets of Y such that $\Gamma_C \subset \Gamma_{C'}$

whenever $C \subset C' \in \mathcal{F}(Y)$ such that

$$\Gamma_{\{y_{i_j}:j\in J\}} \subset \cup_{j=1}^{s} G(x_{i_j})$$

for $\emptyset \neq J \subset \{0, 1, \cdots, n\}$.

It is clear that each $HKKM$ mapping is $GHKK$ and the following example from Chang and Yang [72] shows that the converse does not hold.

Example 2.2.13. Let $E = (-\infty, +\infty)$ and $X = [-2, +2]$. Define a set-valued mapping $G : X \to 2^E$ by $G(x) = [-(1 + \frac{x^2}{5}), \frac{x^2}{5}]$, for each $x \in X$. Since $\cup_{x\in X} H(x) = [-9/5, 9/5]$ and $x \notin G(x)$ for each $x \in [-2, -9/5) \cup (9/5, 1]$. This shows that G is not a KKM mapping. Next we prove that G is a generalized KKM mapping. In fact, for any finite subset $\{x_1, \cdots, x_n\} \subset X$, take $\{y_1, \cdots, y_n\} \subset [-1, 1]$, then for any finite subset $\{y_{i_1}, \cdots, y_{i_k}\} \subset \{y_1, \cdots, y_n\}$, we have

$$co\{y_{i_1}, \cdots, y_{i_k}\} \subset [-1, 1] = \cap_{x\in X} G(x) \subset \cup_{j=1}^{k} G(x_{i_j})$$

Thus G is GHKKM.

Theorem 2.2.14. Let X be a non-empty set and both Y and Z two topological spaces. Suppose $S : Y \to Z$ is continuous and $G : X \to 2^Z$ is such that:

(1) the composition mapping $S^{-1} \circ G : X \to 2^Y$ defined by $(S^{-1} \circ G)(x) = \cup_{z\in G(x)}\{y \in Y : z = S(y)\}$ for each $x \in X$, is a generalized HKKM mapping; and

(2) for each $x \in X$, $G(x)$ is closed (resp., open) in Y.

Then the family $\{G(x) : x \in X\}$ has the finite intersection property, i.e., for each $A \in \mathcal{F}(X)$, $\cap_{x\in A} G(x) \neq \emptyset$.

Proof. For any finite subset $\{x_0, x_1, \cdots, x_n\}$ of X, since $S^{-1} \circ G : X \to 2^Y$ is a generalized HKKM mapping, there exists a finite subset $B = \{y_0, y_1, \cdots, y_n\}$ of Y and a family $\{\Gamma_C\}_{C\in\mathcal{F}(B)}$ of non-empty contractible subsets of Y such that $\Gamma_C \subset \Gamma_{C'}$ whenever $C \subset C'$ such that

$$\Gamma_{\{y_{i_0}, y_{i_1}, \cdots, y_{i_s}\}} \subset \cup_{j=0}^{s} (S^{-1} \circ G)(x_{i_j})$$

for each finite subset $\{y_{i_0}, y_{i_1}, \cdots, y_{i_s}\}$ of $\{y_0, y_1, y_2, \cdots, y_n\}$, where $(0 \le s \le n)$. Let $M_i = S^{-1}(G(y_i))$ for each $i = 0, 1, \cdots, n$; and define a mapping $F : \mathcal{F}(N) \to Y$ by $F(J) = \Gamma_{\{y_k:k\in J\}}$ for each $J \in \mathcal{F}(N)$. Since S is continuous, M_i is closed (resp., open) in Y for $i = 0, 1, \cdots, n$ by the assumption (2). Moreover the mapping F is a singular face structure on Y. Therefore all hypotheses of Proposition 2.2.6 are satisfied. By Proposition 2.2.6, $\cap_{i=0}^{n} M_i \neq \emptyset$. Take any $y_0 \in \cap_{i=0}^{n} M_i$, then $S(y_0) \in \cap_{i=0}^{n} G(x_i) \neq \emptyset$. □

As an application of Theorem 2.2.14, we have the following result due to Zhou [620].

Theorem 2.2.15. Let X be a non-empty set and Y a topological space. Let $G : X \to 2^Y$ be such that

(a) G is transfer closed valued on X; and

(b) there exists a non-empty finite subset X_0 of X such that the set $Y_0 = \bigcap_{x \in X_0} clG(x)$ is non-empty and compact in Y.

Then the intersection $\bigcap_{x \in X} G(x)$ is non-empty and compact if and only if the mapping clG is a generalized HKKM mapping.

Proof. Necessity: Suppose $\bigcap_{x \in X} G(x)$ is non-empty and compact. Take any $y_0 \in \bigcap_{x \in X} G(x)$. Note that the singleton set $\{y_0\}$ is contractible. For each $A = \{x_1, \cdots, x_n\} \in \mathcal{F}(X)$, take $B = \{y_1, \cdots, y_n\}$ with $y_i = y_0$ for all $i = 1, 2, \cdots, n$ and let $\Gamma_{B'} = \{y_0\}$ for all $B' \in \mathcal{F}(B)$. Since $y_0 \in clG(x)$ for all $x \in X$, it is clear that the mapping clG is generalized HKKM.

Sufficiency: Since the mapping clG is a generalized HKKM by Theorem 2.2.14 with $Y = Z$ and S being the identity mapping on Y, the family $\{clG(x) : x \in X\}$ has the finite intersection property. Now define a mapping $G'(x) = cl_Y G(x) \cap Y_0$ for each $x \in X$. Then the family of non-empty compact subsets $\{G'(x) : x \in X\}$ has the finite intersection property, so that $\bigcap_{x \in X} cl_Y G(x) = \bigcap_{x \in X} G'(x) \neq \emptyset$. Since G is transfer closed, $\bigcap_{x \in X} G(x) = \bigcap_{x \in X} clG(x)$ by Lemma 2.2.10. Therefore $\bigcap_{x \in X} G(x) \neq \emptyset$ □

An an immediate consequence of of Theorem 2.2.15, we have the following characterization of non-empty intersections for generalized HKKM mappings.

Theorem 2.2.16. Let X be a non-empty set and Y a compact topological space. Let $G : X \to 2^Y$ be transfer closed valued on X. Then the intersect $\bigcap_{x \in X} G(x)$ is non-empty if and only if the mapping clG is a generalized HKKM mapping.

Theorem 2.2.16 is a generalization of the corresponding results given by Chang and Yang [72] and Chang and Ma [70].

As special cases of Theorem 2.2.16, we have the following famous extension of the classical KKM principle (KKM Lemma) in topological vector spaces to Fan [157] which plays a very important role in the study of nonlinear analysis today.

Theorem 2.2.17. Let X be a non-empty subset of a topological vector space E and $F : X \to 2^E$ be a set-valued KKM mapping. Then the family $\{F(x) : x \in X\}$ has the finite intersection property, i.e., $\bigcap_{x \in A} F(x) \neq \emptyset$ for each $A \in \mathcal{F}(X)$.

As the finite dimensional version of Theorem 2.2.17, we have the following classical FKKM principle.

Corollary 2.2.18 (The FKKM Principle). Let X be a non-empty subset of a finite dimensional space $\mathbb{R}^n$, where $n \in \mathbb{N}$. Suppose $F : X \to 2^{\mathbb{R}^n}$ is a set-valued KKM mapping. Then the family $\{F(x) : x \in X\}$ has the finite intersection property.

We wish to remark that more study on the non-empty intersection for a family of non-empty subsets in topological spaces will be studied in Chapter 3 by a totally different approach which mainly depends on the pure topological property of '*connectedness*'.

2.3 Ky Fan Minimax Principle

The minimax inequality of Fan [164] is fundamental in proving many existence theorems in nonlinear analysis. There have been numerous generalizations of Fan's minimax inequality by weakening the compactness assumption or the convexity assumption. In [24], using Horvath's approach [237], Bardaro and Ceppitelli obtained some minimax inequalities in topological spaces which have H-*space* structure. Following this line, there are many generalizations given by Horvath [237], Tarafdar [534], Ding and Tan [135], Ding et al. [142], Chang and Ma [70], Park [422], Tan et al. [518]. These results generalize most of the corresponding results given by Fan [157] and [167], Degundji and Granas [148], Lassonde [334], Simons [482], Zhou and Chen [622] to topological spaces which have the so-called H-structure. However, most of results mentioned above require lower semicontinuity to guarantee the existence of solutions. Our results in this section shows that the *lower semicontinuity* is not essential for the existence of solutions for Ky Fan's minimax inequalities.

In this section, by weakening the compactness and continuity assumption on H-spaces, we obtain some new minimax inequalities. Then several non-compact minimax inequalities are obtained by using the concept of "*escaping of sequence*" introduced by Border [48] which is different from other non-compact minimax inequalities given by Allen [5], Aubin [10], Aubin and Ekeland [15], Lassonde [334], Fan [167], Ding and Tan [135], Chang and Zhang [76], Yen [591] and Tian and Zhou [556]. Finally, several fixed point theorems and existence theorems for maximal elements are given in H-spaces (resp., in topological vector spaces) which are equivalent to the minimax inequalities in H-spaces (resp., in topological vector spaces). These results will be needed in our study of abstract economics in Chapter 4.

Theorem 2.3.1. Let X be a non-empty set and Y a compact topological space and $\phi : X \times Y \to \mathbb{R} \cup \{-\infty, +\infty\}$ be such that:

(a) the mapping $x \mapsto \{y \in Y : \phi(x,y) \leq 0\}$ is transfer closed valued;

(b) the mapping $x \to cl_Y\{y \in Y : \phi(x,y) \leq 0\}$ is generalized HKKM on X.

Then there exists $y^* \in Y$ such that $\phi(x, y^*) \leq 0$ for all $x \in X$.

Proof. Define a mapping $G : X \to 2^Y$ by $G(x) = \{y \in X : \phi(x,y) \leq 0\}$ for each $x \in X$. Then we have: (1) the mapping G is transfer closed valued and (2) the mapping clG is generalized HKKM. By Theorem 2.2.16, $\cap_{x\in X} G(x) \neq \emptyset$. Take any $y^* \in \cap_{x\in X} G(x)$, then $\sup_{x\in X} \phi(x,y^*) \leq 0$ for all $x \in X$. □

Remark 2.3.2. It is clear that condition (a) of Theorem 2.3.1 is equivalent to the following condition which first appeared in Tan et al. [518]:

The fact $(a)'$: *for each* $y \in Y$ *with* $\{x \in X : \phi(x,y) > 0\} \neq \emptyset$, *there exists* $x' \in X$ *such that* $y \in int_Y\{y' \in Y : \phi(x',y') > 0\}$.

Theorem 2.3.2. Let X be a non-empty subset of a compact H-space $(Y, \{\Gamma_A\})$ and $\phi : X \times Y \to \mathbb{R} \cup \{-\infty, +\infty\}$ be such that:

(a) the mapping $x \mapsto \{y \in Y : \phi(x,y) \leq 0\}$ is transfer closed valued on X; and

(b) the mapping $x \mapsto cl_Y\{y \in Y : \phi(x,y) \leq 0\}$ is HKKM on X.

Then there exists $y^* \in Y$ such that $\phi(x,y^*) \leq 0$ for all $x \in X$.

Proof. Since $(Y, \{\Gamma_A\})$ is an H-space, each HKKM mapping is automatically a generalized HKKM mapping. Therefore all hypotheses of Theorem 2.3.1 are satisfied. By Theorem 2.3.1, there exists $y^* \in Y$ such that $\phi(x,y^*) \leq 0$ for all $x \in X$. □.

Corollary 2.3.4. Let X be a non-empty subset of a compact H-space $(Y, \{\Gamma_A\})$ and $\phi : X \times Y \to \mathbb{R} \cup \{-\infty, +\infty\}$ be such that:

(a) for each $x \in X$, $y \mapsto \phi(x,y)$ is lower semicontinuous on Y; and

(b) the mapping $x \mapsto cl_Y\{y \in Y : \phi(x,y) \leq 0\}$ is HKKM on X.

Then there exists $y^* \in Y$ such that $\phi(x,y^*) \leq 0$ for all $x \in X$.

Proof. Suppose $y \in Y$ is such that $\{x \in X : \phi(x,y) > 0\} \neq \emptyset$. Fix any $x' \in X$ with $\phi(x',y) > 0$. By (a), there exists an open neighborhood $N(y)$ of y such that $\phi(x',y') > 0$ for each $y' \in N(y)$. Hence $y \in \text{int}_Y\{y' \in X : \phi(x',y') > 0\}$. Now the conclusion follows from Theorem 2.3.3 and the fact $(a)'$ preceding it. □

Corollary 2.3.5. Let X be a non-empty subset of a non-empty compact convex set Y in a topological vector space and $\phi : X \times Y \to \mathbb{R} \cup \{-\infty, +\infty\}$ be such that:

(a) the mapping $x \mapsto \{y \in Y : \phi(x,y) \leq 0\}$ is transfer closed valued on X; and

(b) the mapping $x \mapsto cl_Y\{y \in Y : \phi(x,y) \leq 0\}$ is KKM on X.

Then there exists $y^* \in Y$ such that $\phi(x,y^*) \leq 0$ for all $x \in X$.

Proof. For each $A \in \mathcal{F}(Y)$, let $\Gamma_A = co(A)$. Then $(Y, \{\Gamma_A\})$ is an H-space and the mapping $x \mapsto cl_Y\{y \in Y : \phi(x,y) \leq 0\}$ is HKKM on X. Thus the conclusion follows from Theorem 2.3.3. □

Corollary 2.3.6. Let X be a non-empty subset of a non-empty compact convex set Y in a topological vector space and $\phi : X \times Y \to \mathbb{R} \cup \{-\infty, +\infty\}$ be such that:

(a) the mapping $x \mapsto \{y \in Y : \phi(x,y) \leq 0\}$ is transfer closed valued on X;
(b) for each $A \in \mathcal{F}(X)$ and each $y \in co(A)$, $min_{x \in A}\phi(x,y) \leq 0$.

Then there exists $y^* \in Y$ such that $\phi(x,y^*) \leq 0$ for all $x \in X$.

Proof. By Corollary 2.3.5, we only need to prove that the map $x \to cl_Y\{y \in Y : \phi(x,y) \leq 0\}$ is KKM on X. Suppose not, then there exist $A \in \mathcal{F}(X)$ and $y \in co(A)$ such that $y \notin \cup_{x \in A} cl_X\{y \in X : \phi(x,y) \leq 0\}$. It follows that $\phi(x,y) > 0$ for each $x \in A$, so that $min_{x \in A}\phi(x,y) > 0$ which is a contradiction. □.

As seen from the proof of Corollary 2.3.4, the condition '*for each* $x \in X$, $y \to \phi(x,y)$ *is lower semicontinuous*' implies the condition '*for each* $y \in X$ *with* $\{x \in X : \phi(x,y) > 0\} \neq \emptyset$, *there exists* $x' \in X$ *such that* $y \in int_X\{y' \in X : \phi(x',y') > 0\}$'. Thus Corollary 2.3.5 and hence Theorem 2.3.1 and Theorem 2.3.3 generalize Theorem 1 of Yen [591] (see also Theorem 2.2 of Simons [481]) and Theorem 2.11 of Zhou and Chen [622]. The following is an example for which Theorem 2.3.3 is applicable while Theorem 1 of Yen [591] and Theorem 2.11 of Zhou and Chen [622] are not.

Example 2.3.7. Let $Y = [0,1]$ and X be the set of all rational numbers in $[0,1]$. Define $\phi : X \times Y \to \mathbb{R}$ by

$$\phi(x,y) = \begin{cases} x - y, & \text{if } y \text{ is rational,} \\ 2, & \text{if } y \text{ is irrational} \end{cases}$$

for each $(x,y) \in X \times Y$. Suppose $(x,y) \in X \times Y$ and $\phi(x,y) > 0$. If y is irrational, then clearly $y < 1$. If y is rational, then since $\phi(x,y) = x - y > 0$, we also have $y < x \leq 1$. In either case, take $x' = 1$ and note that $\{y' \in Y : \phi(x',y') > 0\} = [0,1)$, so that $y \in [0,1) = \text{int}_Y\{y' \in Y : \phi(x',y') > 0\}$. Thus condition $(a)'$ and hence condition (a) of Theorem 2.3.3 is satisfied. Moreover, for each $x \in X$, $cl_X\{y \in Y : \phi(x,y) \leq 0\} = cl_X\{y \in Y : y \text{ is rational and } y \geq x\} = [x,1]$. It follows that the map $x \to cl_Y\{y \in Y : \phi(x,y) \leq 0\}$ is KKM on X. Thus condition (b) of Theorem 2.3.3 is also satisfied. Therefore Theorem 2.3.3 is applicable. However, for each $x \in X$, the map $y \to \phi(x,y)$ is not lower semicontinuous and hence Theorem 1 of Yen [591] and Theorem 2.11 of Zhou and Chen [622] are not applicable.

The Example 2.3.7 above shows that for each $x \in X$, the lower semicontinuity of the mapping $y \mapsto \phi(x,y)$ is not essential for the existence of solutions for minimax inequalities.

In order to obtain our main results on minimax inequalities, we need the concept of an escaping sequence which was first introduced by Border in [48, p.34].

Definition 2.3.8. Let X be a topological space such that $X = \cup_{n=1}^{\infty} X_n$, where $\{X_n\}_{n=1}^{\infty}$ is an increasing sequence of non-empty compact sets. A sequence $(y_n)_{n=1}^{\infty}$ in X is said to be *escaping* from X (relative to $\{X_n\}_{n=1}^{\infty}$) if for each $n = 1, 2, \cdots$, there exists a positive integer M such that $y_k \notin X_n$ for all $k \geq M$.

Now we have the following result.

Theorem 2.3.9. Let X be a non-empty set and Y a topological space such that $X = \cup_{n=1}^{\infty} X_n$ and $Y = \cup_{n=1}^{\infty} Y_n$ where $\{X_n\}_{n=1}^{\infty}$ and $\{Y_n\}_{n=1}^{\infty}$ are increasing sequences of non-empty sets and of compact spaces respectively. Let $\phi : X \times Y \to \mathbb{R} \cup \{-\infty, +\infty\}$ be such that:

(a) for each $n \in \mathbb{N}$, the mapping $x \to \{y \in Y_n : \phi(x, y) \leq 0\}$ is transfer closed valued on X_n;

(b) for each $n \in \mathbb{N}$, the mapping $x \mapsto cl_{Y_n}\{y \in Y_n : \phi(x, y) \leq 0\}$ is generalized HKKM on X_n; and

(c) for each sequence $(y_n)_{n=1}^{\infty}$ in Y with $y_n \in Y_n$ for each $n \in \mathbb{N}$ which is escaping from Y relative to $\{Y_n\}_{n=1}^{\infty}$, there exists $n_0 \in \mathbb{R}$ and $x_{n_0} \in X_{n_0}$ such that $\phi(x_{n_0}, y_{n_0}) > 0$.

Then there exists $y^* \in Y$ such that $\phi(x, y^*) \leq 0$ for all $x \in X$.

Proof. For each $n \in \mathbb{N}$, by Theorem 2.3.1, there exists $y_n \in Y_n$ such that $\phi(x, y_n) \leq 0$ for all $x \in X_n$. Suppose the sequence $(y_n)_{n=1}^{\infty}$ were escaping from Y relative to $\{Y_n\}_{n=1}^{\infty}$. By (c), there exist $n_0 \in \mathbb{N}$ and $x_{n_0} \in X_{n_0}$ such that $\phi(x_{n_0}, y_{n_0}) > 0$ which is a contradiction. Therefore the sequence $(y_n)_{n=1}^{\infty}$ is not escaping from Y relative to $\{Y_n\}_{n=1}^{\infty}$, so that some subsequenct of $(y_n)_{n=1}^{\infty}$ must lie entirely in some Y_{n_1}. Since Y_{n_1} is compact, there exists a subnet $\{z_\alpha\}_{\alpha \in \Gamma}$ of $(y_n)_{n=1}^{\infty}$ in Y_{n_1} and a point $y^* \in Y_{n_1}$ such that $z_\alpha \to y^*$. Denote $z_\alpha = y_{n(\alpha)}$ for each $\alpha \in \Gamma$. If $x \in X$ is given, there exists $n_2 \geq n_1$ such that $x \in X_{n_2}$. If $\phi(x, y^*) > 0$, then $\{x' \in X_{n_2} : \phi(x', y^*) > 0\} \neq \emptyset$, by (a) and fact $(a)'$ proceeding Theorem 2.3.3, there exists $x' \in X_{n_2}$ such that $y^* \in \text{int}_{Y_{n_2}}\{y' \in Y_{n_2} : \phi(x', y') > 0\}$. Since $z_\alpha \to y^*$, there exists $\alpha_0 \in \Gamma$ such that $n(\alpha_0) \geq n_2$ and $z_{\alpha_0} \in \text{int}_{Y_{n_2}}\{y' \in Y_{n_2} : \phi(x', y) > 0\}$; hence $\phi(x', z_{\alpha_0}) > 0$. But $x' \in X_{n_2} \subset X_{n(\alpha_0)}$ so that $\phi(x', z_{\alpha_0}) = \phi(x', y_{n(\alpha_0)}) \leq 0$ which is a contradiction. Therefore $\phi(x, y^*) \leq 0$ for all $x \in X$. $\square$.

By Theorem 2.3.9, we have the following result.

Theorem 2.3.10. Let X be a non-empty subset of a topological space Y such that $X = \cup_{n=1}^{\infty} X_n$ and $Y = \cup_{n=1}^{\infty} Y_n$ where $\{X_n\}_{n=1}^{\infty}$ and $\{Y_n\}_{n=1}^{\infty}$ are increasing sequences of non-empty sets with $X_i \subset Y_i$ and Y_i is a compact H-space for each $i = 1, 2, \cdots$. Let $\phi : X \times Y \to \mathbb{R} \cup \{-\infty, +\infty\}$ be such that:

(a) for each $n \in \mathbb{N}$, the mapping $x \mapsto \{y \in Y_n : \phi(x,y) \leq 0\}$ is transfer closed valued on X_n;

(b) for each $n \in \mathbb{N}$, the mapping $x \mapsto cl_{Y_n}\{y \in Y_n : \phi(x,y) \leq 0\}$ is HKKM on X_n; and

(c) for each sequence $(y_n)_{n=1}^{\infty}$ in Y with $y_n \in Y_n$ for each $n \in \mathbb{N}$ which is escaping from Y relative to $\{Y_n\}_{n=1}^{\infty}$, there exist $n_0 \in \mathbb{N}$ and $x_{n_0} \in X_{n_0}$ such that $\phi(x_{n_0}, y_{n_0}) > 0$.

Then there exists $y^* \in Y$ such that $\phi(x, y^*) \leq 0$ for all $x \in X$.

Proof. Since each HKKM mapping is also generalized HKKM, all hypotheses (a), (b) and (c) in Theorem 2.3.9 are satisfied. By Theorem 2.3.9, the conclusion follows. □

Similar to Corollaries 2.3.4, 2.3.5, and 2.3.6, we have the following Corollaries 3.8, 3.9, and 3.10.

Corollary 2.3.11. Let X be a non-empty subset of a topological space Y such that $X = \cup_{n=1}^{\infty} X_n$ and $Y = \cup_{n=1}^{\infty} Y_n$ where $\{X_n\}_{n=1}^{\infty}$ and $\{Y_n\}_{n=1}^{\infty}$ are increasing sequences of non-empty sets for which $X_i \subset Y_i$ and Y_i is a compact H-space for each $i = 1, 2, \cdots$. Let $\phi : X \times Y \to \mathbb{R} \cup \{-\infty, +\infty\}$ be such that:

(a) for each $n \in \mathbb{N}$ and for each $x \in X_n$, $y \mapsto \phi(x,y)$ is lower semicontinuous on X_n;

(b) for each $n \in \mathbb{N}$, the mapping $x \mapsto cl_{Y_n}\{y \in Y_n : \phi(x,y) \leq 0\}$ is HKKM on X_n; and

(c) for each sequence $(y_n)_{n=1}^{\infty}$ in Y with $y_n \in Y_n$ for each $n \in \mathbb{N}$ which is escaping from Y relative to $\{Y_n\}_{n=1}^{\infty}$, there exists $n_0 \in \mathbb{N}$ and $x_{n_0} \in X_{n_0}$ such that $\phi(x_{n_0}, y_{n_0}) > 0$.

Then there exists $y^* \in Y$ such that $\phi(x, y^*) \leq 0$ for all $x \in X$.

Corollary 2.3.12. Let E be a topological vector space. Let X be a non-empty subset of a non-empty set Y such that $X = \cup_{n=1}^{\infty} X_n$ and $Y = \cup_{n=1}^{\infty} Y_n$ where $\{X_n\}_{n=1}^{\infty}$ and $\{Y_n\}_{n=1}^{\infty}$ are increasing sequences of non-empty sets for which $X_i \subset Y_i$ and Y_i is compact convex in E for each $i = 1, 2, \cdots$. Let $\phi : X \times Y \to \mathbb{R} \cup \{-\infty, +\infty\}$ be such that:

(a) for each $n \in \mathbb{N}$, the mapping $x \mapsto \{y \in Y_n : \phi(x,y) \leq 0\}$ is transfer closed valued;

(b) for each $n \in \mathbb{N}$, the mapping $x \mapsto cl_{Y_n}\{y \in Y_n : \phi(x,y) \leq 0\}$ is KKM on X_n; and

(c) for each sequence $(y_n)_{n=1}^{\infty}$ in Y with $y_n \in Y_n$ for each $n \in \mathbb{N}$ which is escaping from Y relative to $\{Y_n\}_{n=1}^{\infty}$, there exist $n_0 \in \mathbb{N}$ and $x_{n_0} \in X_{n_0}$ such that $\phi(x_{n_0}, y_{n_0}) > 0$.

Then there exists $y^* \in Y$ such that $\phi(x, y^*) \leq 0$ for all $x \in X$.

Corollary 2.3.13. Let E be a topological vector space. Let X be a non-empty subset of a non-empty set Y such that $X = \cup_{n=1}^{\infty} X_n$ and $Y = \cup_{n=1}^{\infty} Y_n$ where $\{X_n\}_{n=1}^{\infty}$ and $\{Y_n\}_{n=1}^{\infty}$ are increasing sequence of non-empty sets for which $X_i \subset Y_i$ and Y_i is compact convex in E for each $i = 1, 2, \cdots$. Let $\phi : X \times Y \to \mathbb{R} \cup \{-\infty, +\infty\}$ be such that:

(a) for each $n \in \mathbb{N}$, the mapping $x \mapsto \{y \in Y_n : \phi(x, y) \leq 0\}$ is transfer closed valued on X_n;

(b) for each $n \in \mathbb{N}$, $A \in \mathcal{F}(X_n)$ and $y \in co(A)$, $min_{x \in A} \phi(x, y) \leq 0$; and

(c) for each sequence $(y_n)_{n=1}^{\infty}$ in Y with $y_n \in Y_n$ for each $n \in \mathbb{N}$ which is escaping from Y relative to $\{Y_n\}_{n=1}^{\infty}$, there exist $n_0 \in \mathbb{N}$ and $x_{n_0} \in X_{n_0}$ such that $\phi(x_{n_0}, y_{n_0}) > 0$.

Then there exists $y^* \in Y$ such that $\phi(x, y^*) \leq 0$ for all $x \in X$.

Corollary 2.3.13 generalizes Theorem 3.1 of Tan and Yu [505]. Now we give equivalent formulations to our minimax inequalities. We first show that Theorem 2.3.9 implies the following one.

Theorem 2.3.14. Let X be a non-empty set and Y a topological space such that $X = \cup_{n=1}^{\infty} X_n$ and $Y = \cup_{n=1}^{\infty} Y_n$, where $\{X_n\}_{n=1}^{\infty}$ and $\{Y_n\}_{n=1}^{\infty}$ are increasing sequences of non-empty sets and of compact spaces respectively. Let B be a non-empty subset of $X \times Y$ such that:

(a) for each $n \in \mathbb{N}$, the mapping $x \mapsto \{y \in Y_n : (x, y) \notin B\}$ is transfer closed valued on X_n;

(b) for each $n \in \mathbb{N}$, the mapping $x \mapsto cl_{Y_n}\{y \in Y_n : (x, y) \notin B\}$ is generalized HKKM on X_n; and

(c) for each sequence $(y_n)_{n=1}^{\infty}$ in Y with $y_n \in Y_n$ for each $n \in \mathbb{N}$ which is escaping from Y relative to $\{Y_n\}_{n=1}^{\infty}$, there exist $n_0 \in \mathbb{N}$ and $x_{n_0} \in X_{n_0}$ such that $(x_{n_0}, y_{n_0}) \in B$.

Then there exists $y^* \in Y$ such that $\{x \in X : (x, y^*) \in B\} = \emptyset$.

Proof. Let $\phi : X \times Y \to \mathbb{R}$ be defined by

$$\phi(x, y) = \begin{cases} 1, & \text{if } (x, y) \in B, \\ 0, & \text{if } (x, y) \notin B. \end{cases}$$

Then the hypotheses of Theorem 2.3.9 are all satisfied. Hence by Theorem 2.3.9, there exists $y^* \in Y$ such that $\phi(x, y^*) \leq 0$ for all $x \in X$, i.e., $(x, y^*) \notin B$ for all $x \in X$ so that $\{x \in X : (x, y^*) \in B\} = \emptyset$. □

It is clear that Theorem 2.3.14 implies the following result.

Theorem 2.3.15. Let X be a non-empty set and Y be a topological space such that $X = \cup_{n=1}^{\infty} X_n$ and $Y = \cup_{n=1}^{\infty} Y_n$, where $\{X_n\}_{n=1}^{\infty}$ and $\{Y_n\}_{n=1}^{\infty}$ are increasing sequences of non-empty sets and of compact spaces respectively. Let C be a non-empty subset of $X \times Y$ such that:

(a) for each $n \in \mathbb{N}$, the mapping $x \mapsto \{y \in Y_n : (x, y) \in C\}$ is transfer closed valued on X_n;

(b) for each $n \in \mathbb{N}$, the mapping $x \mapsto cl_{Y_n}\{y \in Y_n : (x, y) \in C\}$ is generalized HKKM on X_n; and

(c) for each sequence $(y_n)_{n=1}^{\infty}$ in Y with $y_n \in Y_n$ for each $n \in \mathbb{N}$ which is escaping from Y relative to $\{Y_n\}_{n=1}^{\infty}$, there exist $n_0 \in \mathbb{N}$ and $x_{n_0} \in X_{n_0}$ such that $(x_{n_0}, y_{n_0}) \notin C$.

Then there exists $y^* \in Y$ such that $X \times \{y^*\} \subset C$.

Definition 2.3.16. Let X and Y be two topological spaces and $F : X \to 2^Y \cup \{\emptyset\}$ be a set-valued mapping. Then we have that:

(i) F is *transfer open inverse valued* on X if for each $y \in Y$ and $x \in X$ with $x \in F^{-1}(y) = \{x \in X : y \in F(x)\}$, there exist some $y' \in Y$ and a non-empty open neighborhood $N(x)$ of x in X such that $N(x) \subset F^{-1}(y')$. It is clear that $F : X \to 2^Y \cup \{\emptyset\}$ is transfer open inverse valued on X if and only if the mapping $G : Y \to 2^X \cup \{\emptyset\}$ defined by $G(y) = X \setminus F^{-1}(y)$ for each $y \in Y$ is transfer closed valued; and

(ii) a point $x \in X$ is said to be a maximal of the mapping F provided $F(x) = \emptyset$.

The example after Theorem 2.3.22 below shows that a transfer open inverse valued mapping may be not open inverse valued.

Now we shall show that Theorem 2.3.15 implies the following maximal element theorem.

Theorem 2.3.17. Let X be a non-empty set and Y be a topological space such that $X = \cup_{n=1}^{\infty} X_n$ and $Y = \cup_{n=1}^{\infty} Y_n$, where $\{X_n\}_{n=1}^{\infty}$ and $\{Y_n\}_{n=1}^{\infty}$ are increasing sequences of non-empty sets and of compact spaces respectively. Suppose the mapping $F : Y \to 2^X \cup \{\emptyset\}$ is such that:

(a) for each $n \in \mathbb{N}$, the mapping $F : Y_n \to 2^{X_n} \cup \{\emptyset\}$ is transfer open inverse valued on X_n;

(b) for each $n \in \mathbb{N}$, the mapping $x \mapsto cl_{Y_n}\{y \in Y_n : x \notin F(y)\}$ is generalized HKKM on X_n; and

(c) for each sequence $(y_n)_{n=1}^{\infty}$ in Y with $y_n \in Y_n$ for each $n \in \mathbb{N}$ which is escaping from Y relative to $\{Y_n\}_{n=1}^{\infty}$, there exists $n_0 \in N$ and $x_{n_0} \in X_{n_0}$ such that $x_{n_0} \in F(y_{n_0})$.

Then there exists $y^* \in Y$ such that $F(y^*) = \emptyset$.

Proof. Let $C = \{(x, y) \in X \times Y : x \notin F(y)\}$, then all the conditions of Theorem 2.3.15 are satisfied. Hence by Theorem 2.3.15, there exists $y^* \in Y$ such that $X \times \{y^*\} \subset C$, i.e., $x \notin F(y^*)$ for all $x \in X$ so that $F(y^*) = \emptyset$. □

We shall now prove that Theorem 2.3.17 implies Theorem 2.3.9 so that Theorems 2.3.9, 2.3.14, 2.3.15, and Theorem 2.3.17 are all equivalent.

The proof of "Theorem 2.3.17 $\Rightarrow$ Theorem 2.3.9": Define $F : Y \to 2^X \cup \{\emptyset\}$ by $F(y) = \{x \in X : \phi(x,y) > 0\}$ for each $y \in Y$. Then conditions of Theorem 2.3.17 are satisfied. Hence by Theorem 2.3.17, there exists $y^* \in Y$ such that $F(y^*) = \emptyset$, i.e., $\phi(x,y^*) \leq 0$ for all $x \in X$. □

As an immediate consequence of Theorem 2.3.17, we have the following result.

Corollary 2.3.18. Let X be a non-empty subset of a convex subset Y of a topological vector space E such that $X = \cup_{n=1}^{\infty} X_n$ and $Y = \cup_{n=1}^{\infty} Y_n$, where $\{X_n\}_{n=1}^{\infty}$ and $\{Y_n\}_{n=1}^{\infty}$ are increasing sequences of non-empty sets for which $X_i \subset Y_i$ and Y_i is compact convex for each $i = 1, 2, \cdots$. Suppose the mapping $F : Y \to 2^X \cup \{\emptyset\}$ is such that:

(a) for each $n \in \mathbb{N}$, $F : Y_n \to 2^{X_n} \cup \{\emptyset\}$ is transfer open inverse valued;

(b) for each $y \in Y$, $y \notin coF(y)$; and

(c) for each sequence $(y_n)_{n=1}^{\infty}$ in Y with $y_n \in Y_n$ for each $n \in \mathbb{N}$ which is escaping from Y relative to $\{Y_n\}_{n=1}^{\infty}$, there exist $n_0 \in \mathbb{N}$ and $x_{n_0} \in X_{n_0}$ such that $x_{n_0} \in F(y_{n_0})$.

Then there exists $y^* \in Y$ such that $F(y^*) = \emptyset$.

Proof. Suppose that there exists $n \in \mathbb{N}$, $A \in \mathcal{F}(X_n)$ and $y \in co(A)$ such that $y \notin \cup_{x \in A} cl_{Y_n}\{y' \in Y_n : x \notin F(y')\}$. Then $x \in F(y)$ for all $x \in A$ so that $y \in co(A) \subset coF(y)$ which contradicts (b). Therefore all conditions of Theorem 2.3.17 are satisfied and hence there exists $y^* \in Y$ such that $F(y^*) = \emptyset$. □

Corollary 2.3.18 generalizes Theorem 3.2 of Tan and Yu [505] and Theorem 7.10 of Border [48, p.35].

In what follows, we shall give two fixed point theorems and an existence theorem for maximal elements.

Theorem 2.3.19. Let X be a non-empty subset of an H-space $(Y, \{\Gamma_A\})$ such that $X = \cup_{n=1}^{\infty} X_n$ and $Y = \cup_{n=1}^{\infty} Y_n$, where $\{X_n\}_{n=1}^{\infty}$ and $\{Y_n\}_{n=1}^{\infty}$ are increasing sequences of non-empty sets for which $X_i \subset Y_i$ and Y_i is a non-empty compact and weakly H-convex subset of Y for each $i = 1, 2, \cdots$. Suppose the mapping $F : Y \to 2^X$ is such that:

(a) for each $n \in \mathbb{N}$, the mapping $F : Y_n \to 2^{X_n} \cup \{\emptyset\}$ is transfer open inverse valued;

(b) for each $y \in Y$, $F(y)$ is H-convex; and

(c) for each sequence $(y_n)_{n=1}^{\infty}$ in Y with $y_n \in Y_n$ for each $n \in \mathbb{N}$ which is escaping from Y relative to $\{Y_n\}_{n=1}^{\infty}$, there exists $n_0 \in \mathbb{N}$ and $x_{n_0} \in X_{n_0}$ such that $x_{n_0} \in F(y_{n_0})$.

Then there exists $y^* \in Y$ such that $y^* \in F(y^*)$.

Proof. For each $n \in \mathbb{N}$, since $(Y, \{\Gamma_A\})$ is an H-space and Y_n is weakly H-

convex, so that $(Y_n, \{\Gamma_A \cap Y_n\})$ is an H-space. If the condition (b) of Theorem 2.3.17 holds, then there is $\hat{y} \in Y$ such that $F(\hat{y}) = \emptyset$ which is a contradiction. Therefore condition (b) of Theorem 2.3.17 does not hold, i.e., there exist $n \in \mathbb{N}$, $A \in \mathcal{F}(Y_n)$ and $y^* \in \Gamma_A$ such that $y^* \notin cl_{Y_n}\{y \in Y_n : x \notin F(y)\}$ for all $x \in A$, hence $x \in F(y^*)$ for all $x \in A$. By (b), $F(y^*)$ is H-convex so that $y^* \in \Gamma_A \subset F(y^*)$. □

Theorem 2.3.20. Let X be a non-empty subset of an H-space $(Y, \{\Gamma_A\})$ such that $X = \cup_{n=1}^{\infty} X_n$ and $Y = \cup_{n=1}^{\infty} Y_n$, where $\{X_n\}_{n=1}^{\infty}$ and $\{Y_n\}_{n=1}^{\infty}$ are increasing sequences of non-empty sets for which $X_i \subset Y_i$ and Y_i is a non-empty compact and weakly H-convex subset of Y for each $i = 1, 2, \cdots$. Suppose the mapping $F : Y \to 2^X$ is such that:

(a) for each $n \in \mathbb{N}$, the mapping $F : Y_n \to 2^{X_n} \cup \{\emptyset\}$ is transfer open inverse valued; and

(b) for each sequence $(y_n)_{n=1}^{\infty}$ in Y with $y_n \in Y_n$ for each $n \in \mathbb{N}$ which is escaping from Y relative to $\{Y_n\}_{n=1}^{\infty}$, there exist $n_0 \in N$ and $x_{n_0} \in X_{n_0}$ such that $x_{n_0} \in F(y_{n_0})$.

Then there exists $y^* \in Y$ such that $y^* \in HcoF(y^*)$.

Proof. As in the preceding proof, there exist $n \in \mathbb{N}$, $A \in \mathcal{F}(X_n)$ and $y^* \in \Gamma_A$ such that $y^* \notin cl_{Y_n}\{y \in Y_n : x \notin F(y)\}$ for all $x \in A$, hence $x \in F(y^*)$ for all $x \in A$ and $y^* \in \Gamma_A \subset HcoF(y^*)$. □

Theorem 2.3.21. Let X be a non-empty subset of an H-space space $(Y, \{\Gamma_A\})$ such that $X = \cup_{n=1}^{\infty} X_n$ and $Y = \cup_{n=1}^{\infty} Y_n$, where $\{X_n\}_{n=1}^{\infty}$ and $\{Y_n\}_{n=1}^{\infty}$ are increasing sequences of non-empty sets for which $X_i \subset Y_i$ and Y_i is a non-empty compact and weakly H-convex subset of Y for each $i = 1, 2, \cdots$. Suppose the mapping $F : Y \to 2^X$ is such that:

(a) for each $n \in \mathbb{N}$, the mapping $F : Y_n \to 2^{X_n} \cup \{\emptyset\}$ is transfer open inverse valued;

(b) for each $y \in Y$, $F(y)$ is H-convex and $y \notin F(y)$; and

(c) for each sequence $(y_n)_{n=1}^{\infty}$ in Y with $y_n \in Y_n$ for each $n \in \mathbb{N}$ which is escaping from Y relative to $\{Y_n\}_{n=1}^{\infty}$, there exist $n_0 \in \mathbb{N}$ and $x_{n_0} \in X_{n_0}$ such that $x_{n_0} \in F(y_{n_0})$.

Then there exists $y^* \in Y$ such that $F(y^*) = \emptyset$.

Proof. Suppose $F(y) \neq \emptyset$ for all $y \in Y$, by Theorem 2.3.19, there exists $\hat{y} \in Y$ such that $\hat{y} \in F(\hat{y})$ which is a contradiction of condition (b). □

Finally, as an immediate consequence of Theorem 2.3.19, we have the following generalization of the Browder-Fan fixed point theorem (e.g, see Fan [157] or Browder [59]) which will play the most important role in the study of ecomomics and game theory throughout this book. Also its several variant versions will be given in Chapters 4, 7, 8 and 9.

Theorem 2.3.22 (Browder-Fan Fixed Point Theorem). Let X be a non-empty compact convex subset of a topological vector space E and $F : X \to 2^X$ is such that:

(a) for each $x \in X$, $F(x)$ is non-empty and convex; and

(b) F is transfer open inverse valued.

Then F has a fixed point.

As a special case of Theorem 2.3.22, we have the following original version of a fixed point theorem which was first given by Browder in [59] for set-valued mappings in topological vector spaces.

Theorem 2.3.23. Let X be a non-empty compact convex subset of a topological vector space E and $F : X \to 2^X$ is such that:

(a) for each $x \in X$, $F(x)$ is non-empty and convex; and

(b) $F^{-1}(y) := \{x \in X : y \in F(x)\}$ is open for each $y \in X$.

Then F has a fixed point.

But really, seven years before the paper [59] of Professor Browder, it was Professor Fan, who proved in [157] an equivalent form of the fixed point theorem above by using his extension to infinite dimensional topological vector spaces of the classical KKM lemma. Since then, this result is often called the Fann-Browder or Browder-Fan fixed point theorem in the literature.

The following example shows that Theorem 2.3.22 is really a generalization of the original version of the Browder-Fan fixed point theorem, i.e., Theorem 2.3.23.

Example 2.3.24. Let $X = [0,1]$ and define a mapping $F : X \to 2^{[0,1]}$ by

$$F(x) = \begin{cases} [x,1], & \text{if } x \text{ is rational} \\ [0,1], & \text{if } x \text{ is irrational} \end{cases}$$

Then it is clear that F is not open inverse valued but F is transfer open inverse valued.

As the Browder-Fan fixed point theorem is a very important tool in the study of nonlinear analysis, its various generalizations and equivalent forms haven been studied by many authors. In what follows, we only give some of its equivalent forms in compact cases for the convenience of our presentation. More general study could be found from Yuan [609] and related references therein.

The following theorem was first given by Tarafdar in [531].

Theorem 2.3.25. Let X be a non-empty compact convex subset of a topological vector space E, and $F : X \to 2^X$ such that:

(a) for each $x \in X$, $F(x)$ is non-empty and convex; and

(b) for each $y \in X$, the set $F^{-1}(y)$ contains an open subset O_y of X such that $\cup_{y \in X} O_y = X$.

Then F has a fixed point.

The following theorem has been proved by Ben-El-Mechaiekh et al. [35]:

Theorem 2.3.26. Let X be a non-empty compact convex subset of a topological vector space E and $F : X \to 2^X$ such that:

(a) for each $x \in X$, $F^{-1}(y)$ is convex for each $y \in X$; and

(b) there exists a mapping $G : X \to 2^X$ such that $G(x) \subset F(x)$ for each $x \in X$, we have:

(b_1) $G^{-1}(y)$ is non-empty for each $y \in X$; and

(b_2) $G(x)$ is open for each $x \in X$.

Then F has a fixed point in X.

The following theorem has been proved by Border [48]:

Theorem 2.3.27. Let X be a non-empty compact convex subset of a topological vector space E and $F : X \to 2^X$ is such that:

(a) for each $x \in X$, $x \notin coF(x)$; and

(b) for each $x \in X$, if $F(x) \neq \emptyset$, there exists $y \in X$ such that $x \in F^{-1}(y)$, where $intF^{-1}(y)$ is the topological interior of $F^{-1}(y)$ in X.

Then there exists $x_0 \in X$ such that $F(x_0) = \emptyset$.

We now prove that Theorems 2.3.25 to 2.3.27 stated above are all equivalent.

Theorem 2.3.28. Theorems 2.3.25 to 2.3.27 are equivalent.

Proof. For simplicity, we only carry out the proof that Theorem 2.3.26 implies Theorem 2.3.25. Suppose that conditions of Theorem 2.3.25 hold. We define a set-valued mapping $A : X \to 2^X$ by $A(x) = F^{-1}(x)$ for each $x \in X$. Then it is clear that $A^{-1}(x) = T(x)$ for each $x \in X$. Now define another mapping $G : X \to 2^X$ by $G(x) = O_x$ for each $x \in X$. It is easy to see that $G(x) \subset A(x)$ for each $x \in X$ and furthermore $B(x)$ is open in X. Moreover $B^{-1}(y)$ is non-empty for each $y \in X$. Thus Theorem 2.3.25 implies by Theorem 2.3.26. It is similar to show that Theorem 2.3.27 implies Theorem 2.3.26 and Theorem 2.3.25 implies Theorem 2.3.27. □

Indeed, we have the following simple fact about set-valued mappings with open inverse values.

Remark 2.3.29. If X and Y are topological spaces and $F : X \to 2^Y$ is a set-valued mapping with non-empty values, then the following two conditions are equivalent:

(1) F is transfer open inverse valued, i.e., for each $(x, y) \in X \times Y$ with $x \in F^{-1}(y)$, there exists some $y' \in Y$ such that $x \in intF^{-1}(y)$.

(2) for each $y \in Y$, $F^{-1}(y)$ contains an open subset O_y of X such that $\cup_{y\in Y} O_y = X$.

From Theorem 2.3.28, it is known that each type of the Browder-Fan fixed point theorem has an equivalent version of a maximal element (we recall that a point $x \in X$ is said to be a maximal element of a set-valued mapping F from a topological space X to another topological space Y if $F(x) = \emptyset$). The existence of maximal elements for mappings in topological (vector) spaces and its important applications to mathematical economics have been explained by many authors in both mathematics and economics. We will see this fact later in Chapter 4. Moreover, the non-compact version of the Browder-Fan fixed point theorem has also been given by many authors, e.g., see Ding and Tan [136], Tarafdar [533], Yuan [609] and references therein. More of non-compact version of the Browder-Fan fixed point theorem will be established in Section 3 of Chapter 4 for the purpose of abstract economic models.

2.4 Browder-Fan Fixed Point Theorems in Product Spaces

In this section, we shall study the generalization of the Browder-Fan fixed point theorem in product space for both couple-majorized and L_s-majorized mappings.

2.4.1 Fixed Point Theorems of Couple-Majorized Mappings

Let X and Y be two sets, $A : X \to 2^Y$ be a set-valued mapping and $S : Y \to X$ be a (single-valued) map (throughout this section, *map* always means a single-valued mapping). The composition of the mapping $A \circ S$ from Y to X is defined by $A \circ S(y) := A(S(y))$ for each $y \in Y$. Let U be a non-empty of a topological space X. Then $int_X U$ denotes the interior of U in X and we shall denote simply by $intU$ for convenience.

Definition 2.4.1. Let X be a topological space and Y be a non-empty set in a topological vector space. Then a set-valued mapping $A : X \to 2^Y$ is said

(1): to have *convex values* if $A(x)$ is convex in Y for all $x \in X$; and

(2): to have *open fibers* (or, *open inverse*, or *open lower sections*) if the set $A^{-1}(y) := \{x \in X : y \in A(x)\}$ is open in X for each $y \in Y$.

We also recall the following definition, which was first given by Deguire and Yuan in [126].

Definition 2.4.2. Let $X := \Pi_{j\in J} X_j$ and $Y = \Pi_{i\in I} Y_i$ be two product spaces of non-empty convex subsets $\{X_j\}_{j\in J}$ and $\{Y_i\}_{i\in I}$ in topological vector spaces E and F, respectively. The families $\{A_i\}_{i\in I}$ and $\{B_j\}_{j\in J}$ are said to be *couple-majorized* if that for any given $(i,j) \in I \times J$, and for each $(x,y) \in X \times Y$ with $A_i(x) \neq \emptyset$ and $B_j(y) \neq \emptyset$, there exist mappings $A_{i(x)} : X \to 2^{Y_i}$, $B_{j(y)} : Y \to 2^{X_j}$ and a non-empty open neighborhood $N(x)$ of x in X and a non-empty open neighborhood $N(y)$ of y in Y such that:

(a): both $A_{i(x)}$ and $B_{j(y)}$ have (maybe empty) convex values and open fibers;

(b): $A_i(z) \subset A_{i(x)}(z)$ and $B_j(w) \subset B_{j(y)}(w)$ for all $(z,w) \in N(x) \times N(y)$; and

(c): $(u_j, v_i) \notin (B_{j(y)}(v), A_{i(x)}(u))$ for all $(u,v) \in X \times Y$.

We also call the mapping pair $(A_{i(x)}, B_{j(y)})$ which satisfies conditions (a)-(c) above *a couple-majorant* of the mapping pair (A_i, B_j) at the point $(x,y) \in X \times Y$.

Let E be a topological vector space. Then a polytope of E is the convex hull of a non-empty finite subset of E. By the partition of unity for paracompact, the following result can be easily proved, for example, see Dugundji and Granas in [149], Michael [371], Yannelis and Prabhakar in [587] and Deguire and Lassonde in [124] for more recent results in this direction.

Lemma 2.4.3. Let X be a paracompact topological space and Y be a non-empty convex subset of a topological vector space E. Suppose $F : X \to 2^Y$ is a set-valued mapping with non-empty convex values and open fibers. Then there exists a continuous mapping $f : X \to Y$ such that $f(x) \in F(x)$ for each $x \in X$. In particular, if X is compact, there exists a polytope D in Y such that $f(X) := \cup_{x\in X} f(x) \subset D$.

Proof. For each $y \in Y$, the set $F^{-1}(y)$ is open in X and note that $F(x) \neq \emptyset$ for all $x \in X$. It follows that $\{F^{-1}(y)\}_{y\in Y}$ is an open covering of X. Since X is paracompact, there is a locally finite and open refinement $\{O_i\}_{i\in I}$ of $\{F^{-1}(y)\}_{y\in Y}$. Hence for each $\in I$, there is $y(i) \in Y$ such that $O_i \subset F^{-1}(y(i))$. Let $\{P_i\}_{i\in I}$ be a partition of unity subordinated to $\{O_i\}_{i\in I}$. The function $f : X \to Y$ defined by

$$f(x) := \Sigma_{i\in I} P_i(x) y(i)$$

for each $x \in X$ is continuous as it is locally a finite sum of continuous functions $\{P_i\}_{i\in I}$. For each $x \in X$, $P_i(x) \neq \emptyset$ implies that $x \in O_i \subset F^{-1}(y(i))$, so that $y(i) \in F(x)$. Note that $F(x)$ is non-empty and convex. It follows that $f(x) \in F(x)$ for all $x \in X$. If X is compact, by the proof above, there exists a finite number of $y_1, \cdots, y_n \in Y$ such that $X \subset \cup_{i=1}^n F^{-1}(y_i)$. Let D be the convex hull of $\{y_1, \cdots, y_n\}$. Then it is clear that the continuous selection f of F has the property that $f(x) \in D$ for all $x \in X$. □

We now have the following existence theorem of maximal elements for a pair of mappings which may not have convex values nor open fibers.

Theorem 2.4.4. Let X be a non-empty compact and convex subset of a topological vector space E and Y another non-empty convex subset of a topological vector space F. Suppose $A : X \to 2^Y$ and $B : Y \to 2^X$ are couple-majorized. Then there exists either $x_0 \in X$ such that $A(x_0) = \emptyset$ or $y_0 \in Y$ such that $B(y_0) = \emptyset$.

Proof. Suppose the contrary that for each $x \in X$ and $y \in Y$, $A(x) \neq \emptyset$ and $B(y) \neq \emptyset$. Then for each $(x, y) \in X \times Y$, there exists a non-empty open neighborhood $N(x)$ of x in X and an open neighborhood $N(y)$ of y in Y and two mappings $A_x : X \to 2^Y$ and $B_y : Y \to 2^Y$ such that:

(a) $A(z) \subset A_x(z)$ for each $z \in N(x)$ and $B(z) \subset B_y(z)$ for each $z \in N(y)$;

(b) $(u, v) \notin (B_y(v), A_x(u))$ for each $(u, v) \in X \times Y$;

(c) $A_x^{-1}(y)$ and $B_y^{-1}(x)$ are open in X and Y, respectively, for each $y \in Y$ and $x \in X$.

Since X is regular, for each $x \in X$ (resp., $y \in Y$), let $O(x)$ (resp., $O(y)$) be an open neighborhood of x in X (resp., $y \in Y$) such that $\overline{O(x)} \subset N(x)$ (resp., $\overline{O(y)} \subset N(y)$). Note that X is compact and $X = \cup_{x \in X} O(x)$, then there exist $x_1, \cdots, x_n \in X$ such that $X = \cup_{i=1}^n O(x_i)$. Define $A_i : X \to 2^Y$ for $i = 1, 2, \cdots, n$ by

$$A_i(x) = \begin{cases} A_{x_i}(x), & \text{if } x \in \overline{O(x_i)}; \\ Y, & \text{if } x \in X \setminus \overline{O(x_i)}. \end{cases}$$

Then we have

(a′) A_i is convex-valued;

(b′) $A(x) \subset A_i(x)$ for each $x \in X$; and

(c′) for each $y \in Y$, $A_i^{-1}(y)$ is open in X as

$$\begin{aligned} A_i^{-1}(y) &= \{x \in \overline{O(x_i)} : y \in A_{x_i}(x)\} \cup (X \setminus \overline{O(x_i)}) \\ &= (A_{x_i}^{-1}(y) \cap \overline{O(x_i)}) \cup (X \setminus \overline{O(x_i)}) \\ &= [A_{x_i}^{-1}(y) \cap \overline{O(x_i)})] \cap [\overline{O(x_i)} \cup (X \setminus \overline{O(x_i)}] \\ &= A_{x_i}^{-1}(y) \cup (X \setminus \overline{O(x_i)}), \end{aligned}$$

which is open by (c). Now define $A' : X \to 2^Y$ by $A'(x) = \cap_{i=1}^n A_i(x)$ for each $x \in X$. Clearly, we have:

(a″) A' is (maybe empty) convex-valued;

(b″) $A(x) \subset A'(x)$ for each $x \in X$; and

(c″) for each $y \in Y$, the set $(A')^{-1}(y)$ is open in X since

$$(A')^{-1}(y) = \{x \in X : y \in \cap_{i=1}^n A_i(x)\} = \cap_{i=1}^n A_i^{-1}(y).$$

Note that Y is also compact and $Y = \cup_{y\in Y} O(y)$, there exist $y_1, \cdots, y_m \in Y$ such that $Y = \cup_{i=1}^m O(y_i)$. Define $B_i : Y \to 2^X$ for $i = 1, 2, \cdots, m$ by

$$B_i(y) = \begin{cases} B_{y_i}(y), & \text{if } y \in \overline{O(y_i)}; \\ X, & \text{if } y \in Y \setminus \overline{O(y_i)}. \end{cases}$$

Now, similarly to the arguments used for the mapping A', we also have:

(i′) B_i is (maybe empty) convex-valued;

(ii′) $B(y) \subset B_i(y)$ for each $y \in Y$; and

(iii′) for each $x \in X$, $B_i^{-1}(x)$ is open in Y.

Now we define $B' : Y \to 2^X$ by $B'(x) = \cap_{i=1}^m B_i(y)$ for each $y \in Y$. Then we have:

(i″) B' is convex-valued;

(ii″) $B(y) \subset B'(y)$ for each $y \in Y$; and

(iii″) for each $x \in X$, the set $(B')^{-1}(x)$ is open in Y since

$$(B')^{-1}(x) = \{y \in Y : x \in \cap_{i=1}^n B_i(y)\} = \cap_{i=1}^n B_i^{-1}(x).$$

Moreover, we also have

(d) for each $x \in X$ and $y \in Y$, $x \notin B'(y)$ and $y \notin A'(x)$ by the definition of A' and B'; and

(e) both mappings A' and B' have non-empty values as $\emptyset \neq A(x) \subset A'(x)$ and $\emptyset \neq B(y) \subset B'(y)$ for each $x \in X$ and $y \in Y$.

As X is compact and A' has non-empty convex values and open fibers, Lemma 2.4.3 implies that there exists a polytope D in Y and a continuous (single-valued) function $f : X \to D$ such that $f(x) \in A'(x)$ for each $x \in X$. Now the mapping $F : X \to 2^X$ defined by $F(x) := B'(f(x))$ for each $x \in X$ has the following properties:

(1) $F(x)$ is non-empty convex for each $x \in X$; and

(2) $F^{-1}(y) := \{x \in X : y \in (B')^{-1}(f(x))\} = f^{-1}((B')^{-1}(y))$ is open in X for each $y \in X$ by (iii″).

Thus Lemma 2.4.3 implies that there exists $x_0 \in X$ such that $x_0 \in F(x_0) = B'(f(x_0))$. Let $y_0 = f(x_0)$. Then we have that $x_0 \in B'(y_0)$ and $y_0 \in A'(x_0)$ as $f(x_0) \in A'(x_0)$, but this contradicts with (d). Therefore the conclusion follows and we complete the proof. □

Let $X = Y$ in Theorem 2.4.4. As a direct application of the Browder-Fan fixed point theorem, we have the following exsitence of maximal elements for a set-valued mapping in topological vector spaces.

Theorem 2.4.5. Let X be a non-empty compact and convex subset of a topological vector space E and $A : X \to 2^X$ a set-valued mapping. Suppose that for each $x \in X$ with $A(x) \neq \emptyset$, there exists a non-empty open neighborhood $N(x)$ of x in X and a set-valued mapping $A_x : X \to 2^X$ which has (maybe empty) convex values such that:

(1) $A(z) \subset A_x(z)$ for all $z \in N(x)$;
(2) $A_x^{-1}(y) := \{z \in X : y \in A_x(z)\}$ is open for each $y \in X$; and
(3) $z \notin A_x(z)$ for all $z \in X$.

Then there exists $x_0 \in X$ such that $A(x_0) = \emptyset$.

Proof. By following the same idea used in the proof of Theorem 2.4.4, we will prove the conclusion. Suppose contrary that for each $x \in X$, $A(x) \neq \emptyset$. Then for each $x \in X$, there exists a non-empty open neighborhood $N(x)$ of x in X and a mapping $A_x : X \to 2^X$, which satisfies conditions (1)-(3) of Theorem 2.4.5. Note that X is compact. For each $x \in X$, let $O(x)$ be an open neighborhood of x in X such that $\overline{O(x)} \subset N(x)$. Then $X = \cup_{x \in X} O(x)$. By the compactness of X, there exist $x_1 \cdots x_n \in X$ such that $X = \cup_{i=1}^n O(x_i)$. Define $A_i : X \to 2^X$ for $i = 1, 2 \cdots n$ by

$$A_i(x) = \begin{cases} A_{x_i}(x), & \text{if } x \in \overline{O(x_i)} \\ X, & \text{if } x \in X \setminus \overline{O(x_i)}. \end{cases}$$

Then by the similar arguments used in the proof of Theorem 2.4.4, we have that:

(a′) A_i is non-empty convex-valued;
(b′) $A(x) \subset A_i(x)$ for each $x \in X$; and
(c′) for each $y \in X$, $A_i^{-1}(y)$ is open in X.

Now define $A' : X \to 2^X$ by $A'(x) = \cap_{i=1}^n A_i(x)$ for each $x \in X$. Clearly we also have:

(a″) A' is non-empty convex-valued;
(b″) $A(x) \subset A'(x)$ for each $x \in X$; and
(c″) for each $y \in X$, the set $(A')^{-1}(y)$ is open in X since

$$(A')^{-1}(y) = \{x \in X : y \in \cap_{i=1}^n A_i(x)\} = \cap_{i=1}^n A_i^{-1}(y).$$

As X is compact and A' has non-empty convex values and open fibers, Lemma 2.4.3 implies that A' has a fixed point. On the other hand, it is clear that $x \notin A'(x)$ for all $x \in X$ by its definition. This contradiction shows that there must exist $x_0 \in X$ such that $A(x_0) = \emptyset$. □

As another application of Theorem 2.4.4, we have the following existence theorem of maximal elements for two families which are coupled-majorized mappings in product spaces.

Theorem 2.4.6. Let $X = \Pi_{j \in J} X_j$ and $Y = \Pi_{i \in I} Y_i$, where $\{X_j\}_{j \in J}$ and $\{Y_i\}_{i \in I}$ are two families of non-empty compact convex subsets of topological vector spaces E_j and F_i, respectively for each $(i, j) \in I \times J$, where I and J are two arbitrary index sets. Suppose $A_i : X \to 2^{Y_i}$ and $B_j : Y \to 2^{X_j}$ are couple-majorized families such that

$$\cup_{j \in J} \{y \in Y : B_j(y) \neq \emptyset\} = \cup_{j \in J} int\{y \in Y : B_j(y) \neq \emptyset\}$$

and

$$\cup_{i\in I}\{x \in X : A_i(x) \neq \emptyset\} = \cup_{i\in I} int\{x \in X : A_i(x) \neq \emptyset\}.$$

Then there exists at least one $x \in X$ such that $A_i(x) = \emptyset$ for all $i \in I$ or there exists $y \in Y$ such that $B_j(y) = \emptyset$ for all $j \in J$.

Proof. For each $i \in I$, define $A'_i : X \to 2^Y$ by

$$A'_i(x) = \Pi_{j\in I, j\neq i} Y_j \times A_i(x)$$

for each $x \in X$. For each $x \in X$, define $I(x) = \{i \in I : A_i(x) \neq \emptyset\}$. Now we define $A : X \to 2^Y$ by

$$A(x) = \begin{cases} \cap_{i\in I(x)} A'_i(x), & \text{if } I(x) \neq \emptyset \\ \emptyset, & \text{if } I(x) = \emptyset. \end{cases}$$

Similarly, for each $j \in J$, define $B'_j : Y \to 2^{X_j}$ by

$$B'_i(y) = \Pi_{i\in J, i\neq j} X_j \times B_i(y)$$

for each $y \in Y$. For each $y \in Y$, define $J(y) = \{j \in J : B_j(y) \neq \emptyset\}$. Now we define $B : Y \to 2^X$ by

$$B(y) = \begin{cases} \cap_{j\in J(y)} B'_i(y), & \text{if } J(y) \neq \emptyset \\ \emptyset, & \text{if } J(y) = \emptyset. \end{cases}$$

Then we shall show that the pair (A, B) is couple-majorized. Let $(x, y) \in X \times Y$. We may assume that $I(x) \neq \emptyset$ and $J(y) \neq \emptyset$, otherwise the conclusion follows. Then $x \in \cup_{i\in I}\{x \in X : A_i(x) \neq \emptyset\}$ and $y \in \cup_{j\in J}\{y \in Y : B_j(y) \neq \emptyset\}$. Since (A_i, B_j) is couple-majorized, there exists an open neighborhood $N(x)$ of x and an open neighborhood $N(y)$ of y in Y and a pair of a couple-majorant $C_i : X \to 2^{Y_i}$ and $D_j : Y \to 2^{X_j}$ at (x, y) such that:

(a) $A_i(z) \subset C_i(z)$ for each $z \in N(x)$ and $B_j(z) \subset D_j(z)$ for each $z \in N(y)$;

(b) $(u_j, v_i) \notin (D_j(v), C_i(u))$ for each $(u, v) \in X \times Y$; and

(c) the set $C_i^{-1}(v_i)$ and $D_j^{-1}(u_j)$ are open in X and Y, respectively, for each $v_i \in Y_i$ and $u_j \in X_j$.

By our assumptions, without loss of generality, we may assume that $N(x) \subset int\{x \in X : A_i(x) \neq \emptyset\}$ and $N(y) \subset int\{y \in Y : B_j(y) \neq \emptyset\}$. Thus $A_i(z) \neq \emptyset$ for each $z \in N(x)$ and $B_j(z) \neq \emptyset$ for each $z \in N(y)$. Now we define $C_x : X \to 2^Y$ and $D_y : Y \to 2^X$ by

$$C_x(z) = \Pi_{j\in I, j\neq i} Y_j \times C_i(z)$$

for each $z \in X$ and

$$D_y(z) = \Pi_{i\in J, i\neq j} X_j \times D_j(z)$$

for each $z \in Y$. Then we claim that (C_x, D_y) is a couple-majorant of the pair (A, B) at $(x, y) \in X \times Y$. Indeed, we have

(i) $A(z) = \cap_{k\in I(z)} A'_k(z) \subset A'_i(z) \subset C_x(z)$ for each $z \in N(x)$;

(ii) $B(z) = \cap_{k\in J(z)} B'_k(z) \subset B'_i(z) \subset D_y(z)$ for each $z \in N(y)$;

(iii) $C_x^{-1}(v) = (C'_i)^{-1}(v)$ and $D_y^{-1}(u) = (D'_j)^{-1}(u)$ are open in X and Y, respectively, for each $u \in X$ and $v \in Y$; and

(iv) $(u, v) \notin D_y(v) \times C_x(u)$ for each $u \in X$ and $v \in Y$.

Therefore (C_x, D_y) is a couple-majorant of (A, B) at (x, y). Thus (A, B) satisfies all hypotheses of Theorem 2.4.4. By Theorem 2.4.4, there exists at least one $x \in X$ such that $A(x) = \emptyset$ or $y \in Y$ such that $B(y) = \emptyset$, i.e., $A_i(x) = \emptyset$ for all $i \in I$ or $B_j(y) = \emptyset$ for all $j \in J$ and the proof is completed. □

As an application of Theorem 2.4.5, we have the following existence of maximal elements for a family of set-valued mappings in topological vector spaaces.

Theorem 2.4.7. Let $X = \Pi_{i\in I} X_i$ be a family of non-empty compact convex subsets $\{X_i\}_{i\in I}$ of topological vector spaces E_i and I an arbitrary index set. Let $\{A_i\}_{i\in I}$ be a family of set-valued mappings such that $\cup_{i\in I}\{x \in X : A_i(x) \neq \emptyset\} = \cup_{i\in I} int\{x \in X : A_i(x) \neq \emptyset\}$, where for each $i \in I$, $A_i : X \to 2^{X_i}$ is a set-valued mapping. Suppose that for each $i \in I$ and $x \in X$ with $A_i(x) \neq \emptyset$, there exists a non-empty open neighborhood $N(x)$ of x in X and a set-valued mapping $A_{i,x} : X \to 2^{X_i}$ which takes (maybe empty) convex-values such that:

(1) $A_i(z) \subset A_{i,x}(z)$ for all $z \in N(x)$;

(2) $z_i \notin A_{i,x}(z)$ for all $z \in X$; and

(3) $A_{i,x}^{-1}(y) := \{z \in X : y \in A_{i,x}(z)\}$ is open in X for each $y \in X_i$.

Then there exists at least $\hat{x} \in X$ such that $A_i(\hat{x}) = \emptyset$ for all $i \in I$.

Proof. By following the similar used in the proof of Theorem 2.4.6, for each $i \in I$ we define $A'_i : X \to 2^X$ by

$$A'_i(x) = \Pi_{j\in I, j\neq i} X_j \times A_i(x)$$

for each $x \in X$. For each $x \in X$, define $I(x) = \{i \in I : A_i(x) \neq \emptyset\}$. Now define $A : X \to 2^X$ by

$$A(x) = \begin{cases} \cap_{i\in I(x)} A'_i(x), & \text{if } I(x) \neq \emptyset \\ \emptyset, & \text{if } I(x) = \emptyset. \end{cases}$$

Then we shall show that A satisfies all conditions of Theorem 2.4.5. Without loss of the generality, we may assume that $I(x) \neq \emptyset$ (otherwise the conclusion follows). Then $x \in \cup_{i\in I}\{x \in X : A_i(x) \neq \emptyset\}$ and there exists $i \in I$ such that $A_i(x) \neq \emptyset$. Thus there exists an open neighborhood $N(x)$ of x in X and a set-valued mapping $A_{i,x} : X \to 2^{X_i}$ which satisfies all conditions (1)-(3) of Theorem 2.4.7. By our assumptions, we can also assume that $N(x) \subset int\{x \in$

$X : A_i(x) \neq \emptyset\}$. Thus $A_i(z) \neq \emptyset$ for each $z \in N(x)$. Now define $A_x : X \to 2^X$ by

$$A_x(z) = \Pi_{j\in I, j\neq i} X_j \times A_{i,x}(z)$$

for each $z \in X$. Then it is clear that the mapping A_x has the following properties:

(i) $A(z) = \cap_{k\in I(z)} A'_k(z) \subset A'_i(z) \subset A_x(z)$ for each $z \in N(x)$, and A_x takes (maybe empty) convex values;

(ii) $A_x^{-1}(v) = (A'_{i,x})^{-1}(v)$ which is open in X for each $v \in X$; and

(iii) $z \notin A_x(z)$ for all $z \in X$.

Therefore we have shown that all hypotheses of Theorem 2.4.5 are satisfied. By Theorem 2.4.5, there exists $\hat{x} \in X$ such that

We recall that a coincidence theorem asserts that if $S : X \to 2^Y$ and $T : Y \to 2^X$ have nonempty values and satisfy certain other conditions then there exist $x_0 \in X$ and $y_0 \in Y$ such that $y_0 \in Sx_0$ and $x_0 \in Ty_0$.

Remark 2.4.8. When X is a compact topological space instead of having a linear structure and the mapping B_j is replaced by a continuous single-valued mapping S in Theorems 2.4.4 and 2.4.6, we will also establish existence theorems of maximal elements for L_S-majorized mappings (see Definition 2.4.13) in (not necessarily compact) product spaces. But first, as applications of existence theorems for maximal elements above, we shall derive some coincidence and fixed point theorems in compact product spaces. For the convenience, we first recall the following definition:

Definition 2.4.9. Let X be a Hausdorff topological space, and I any index set. For each $i \in I$, let Y_i be a non-empty convex subset of a Hausdorff topological vector space and $A_i : X \to 2^{Y_i}$. Then $\{A_i\}_{i\in I}$ is said to be a *KF family* if the following assumptions are satisfied:

(i) for each $i \in I$, $A_i(x)$ is convex for each $x \in X$; and

(ii) for each $x \in X$, there exists $i \in I$ such that $A_i(x) \neq \emptyset$; and

(iii) the set $A_i^{-1}(y_i)$ is open in X for all $y_i \in Y_i$ and $i \in I$.

Now we arrive in the place to have the following coincidence theorem in product spaces.

Theorem 2.4.10. Let I and J be any two index sets, $X = \Pi_{j\in I} X_j$ and $Y = \Pi_{i\in I} Y_i$ be product spaces of non-empty compact convex subset X_j and non-empty convex subset Y_i, each in a Hausdorff topological vector space. Suppose that both $\{A_i\}_{i\in I}$ and $\{B_j\}_{j\in J}$ are KF families. Then there exist $(i_0, j_0) \in I \times J$ and $(\hat{x}, \hat{y}) \in X \times Y$ such that $\hat{y}_{i_0} \in A_{i_0}(\hat{x})$ and $\hat{x}_{j_0} \in B_{j_0}(\hat{y})$.

Proof. Suppose the conclusion is not true. Then $\{A_i\}_{i\in I}$ and $\{B_j\}_{J\in J}$ are couple-majorized. By Theorem 2.4.6, there exist either $\hat{x} \in X$ such that $A_i(\hat{x}) = \emptyset$ for all $i \in I$ or there exists $\hat{y} \in Y$ such that $B_j(\hat{y}) = \emptyset$ for all $j \in J$, which contradicts that both $\{A_i\}_{i\in I}$ and $\{B_j\}_{j\in I}$ are KF families. Thus the conclusion follows. □

As an application of Theorem 2.4.7, we have the following component fixed point theorem which could be regarded as an extension of the Browder-Fan fixed point theorem in product spaces.

Theorem 2.4.11. Let I be any index set and $X = \Pi_{i\in I}X_i$ the product space of non-empty compact convex sets X_i in a topological vector space E_i for each $i \in I$. Suppose that $\{A_i : A_i : X \to 2^{X_i}\}_{i\in I}$ is a KF family. Then there exist $i_0 \in I$ and $\hat{x} \in X$ such that $\hat{x}_{i_0} \in A_{i_0}(\hat{x})$.

Proof. Suppose the conclusion were not true. By Theorem 2.4.7, there exists $\hat{x} \in X$ such that $A_i(\hat{x}) = \emptyset$ for all $i \in I$. Of course, this contradicts with that $\{A_i\}_{i\in I}$ is a KF family. Thus there must exist $\hat{x} \in X$ and $i_0 \in I$ such that $\hat{x}_{i_0} \in A_{i_0}(\hat{x})$ and the proof is completed. □

An an immediate consequence of Theorem 2.4.11, we shall derive Sion's minimax inequality for the version of two functions of Sion in [490].

Corollary 2.4.12 (Sion's Minimax Inequality). Let X be a non-empty compact convex subset of a topological vector space E and Y a non-empty compact convex subset of a topological vector space F. Suppose $f, g : X \times Y \to \mathbb{R} \cup \{-\infty, +\infty\}$ satisf that:

(i) $f(x,y) \le g(x,y)$ for each $(x,y) \in X \times Y$;

(ii) for each fixed $x \in X$, $y \mapsto f(x,y)$ is quasi-concave, and for each fixed $y \in Y$, $x \mapsto f(x,y)$ is lower semicontinuous; and

(iii) for each fixed $x \in X$, $y \mapsto g(x,y)$ is upper semicontinuous, and for each fixed $y \in Y$, $x \mapsto g(x,y)$ is quasi-convex.

Then we have

$$\min_{x\in X}\sup_{y\in Y} f(x,y) \le \sup_{y\in Y}\min_{x\in X} g(x,y).$$

Proof. Let $\lambda_0 := \min_{x\in X}\sup_{y\in Y} f(x,y)$. For each given $\epsilon > 0$, let $\lambda_\epsilon = \lambda_0 - \epsilon$ and define $A : X \to 2^Y$ and $B : Y \to 2^X$ by

$$A(x) = \{y \in Y : f(x,y) > \lambda_\epsilon\}$$

for each $x \in X$ and

$$B(y) = \{x \in X : g(x,y) < \lambda_\epsilon\}$$

for each $y \in Y$. Note that both A and B have (maybe empty) convex values and open fibers, and moreover they do not have any coincident point. Otherwise there exists $(x_0, y_0) \in X \times Y$ such that $\lambda_\epsilon < f(x_0, y_0) \le g(x_0, y_0) < \lambda_\epsilon$, which is impossible. By Theorem 2.4.10, we must have that either A or B is not a KF mapping. As $A(x) \ne \emptyset$ for each $x \in X$, there must exist $y_0 \in Y$ such that

$B(y_0) = \emptyset$. Therefore $\inf_{x\in X} g(x, y_0) \geq \lambda_0 - \epsilon$, so that $\sup_{y\in Y} \inf_{x\in X} g(x, y) \geq \lambda_0 - \epsilon$. As ϵ is arbitrary, we have $\sup_{y\in Y} \inf_{x\in X} g(x, y) \geq \lambda_0$. Then

$$\sup_{y\in Y} \inf_{x\in X} g(x, y) \geq \min_{x\in X} \sup_{y\in Y} f(x, y),$$

and the proof is complete. □

So far we have established Browder-Fan fixed point theorems in compact product spaces. In next section, we will prove the general non-compact versions of the Browder-Fan fixed point theorems in product spaces.

2.4.2 Browder-Fan Fixed Point Theorems of L_s-Majorized Mappings

The aim of this section is to discuss the Browder-Fan fixed point theorems in product spaces for L_s-majorized mappings in the setting of non-compact production spaces. As applications, the general minimax inequality theory for a family of functions also is developed. We first recall some definitions and notions.

Definition 2.4.13. Let $A_i : X \to 2^{Y_i}$ be a set-valued mapping for each $i \in I$. Then A_i is said to be

(i): *of class* L_S if (a) A_i is convex valued; (b) $y_i \notin A_i(S(y))$ for each for each $y \in Y$; and (c) $A_i^{-1}(y_i) := \{x \in X : y_i \in A_i(x)\}$ is open in X for each $y_i \in Y_i$; and

(ii): L_S*-majorized* if for each $x \in X$, there exists an open neighborhood $N(x)$ of x in X and a convex-valued mapping $B_x : X \to 2^{Y_i}$ such that (a): $A_i(z) \subset B_x(z)$ for each $z \in N(x)$; (b): $y_i \notin B_x(S(y))$ for each $y \in Y$ and (c): $B_x^{-1}(y_i)$ is open in X for each $y_i \in Y_i$. Here the mapping B_x is called an L_S-majorant of A_i at x.

Let X be a topological space and $Y = \Pi_{i\in I} Y_i$ be a product of non-empty convex sets Y_i, each in a Hausdorff topological vector space, where I is any index set. Suppose $S : Y \to X$ is a (single-valued) mapping and $A_i : X \to 2^{Y_i}$ is a set-valued mapping for each $i \in I$. Then we shall denote by $M_S(X, Y_i)_{i\in I}$ (resp., $L_S(X, Y_i)_{i\in I}$) the set of all families of $\{A_i\}_{i\in I}$ such that for each $i \in I$, $A_i : X \to Y_i$ is L_S-majorized (resp., is of class L_S). In the case that $X = \Pi_{i\in I} X_i$ and $Y_i = X_i$ and the map S is the identity mapping on Y, i.e., $S = I_X$, we shall write $M(X, X_i)_{i\in I}$ (resp., $L(X, X_i)_{i\in I}$)) in place of $M_{I_X}(X, X_i)_{i\in I}$ (resp., $L_{I_X}(X, X_i)_{i\in I}$) throughout this section. Moreover if the index set I is a singleton, the mapping A in $M_S(X, Y)$ (resp., A in $L_S(X, Y)$) is said to be L_S*-majorized* (resp., of class L_S) from X to Y.

It is clear that each class of L_S mapping is L_S-majorized, but the converse is not true in general. Let A_i be a non-empty subset of Y_i for any given but fixed

$i \in I$, we define $\Pi_{j\neq i, j\in I} Y_j \otimes A_i := \{y = (y_j)_{j\in I} \in \Pi_{j\in I} Y_i$ with the property that $y_i \in A_i\}$.

We now establish an existence theorem of maximal elements for L_S-majorized families. In this part, we first give an existence theorem for an L_S-majorized mapping defined in a compact topological space. Then as applications, the existence theorem for maximal elements for any arbitrary families which are L_S-majorized is derived.

Theorem 2.4.14. Let X be a compact Hausdorff topological space and Y be a non-empty convex subset of a Hausdorff topological vector space E, $S : Y \to X$ be continuous and $A : X \to 2^Y$ be L_S-majorized. Then there exists $x \in X$ such that $A(x) = \emptyset$.

Proof. Suppose the contrary that $A(x) \neq \emptyset$ for all $x \in X$. Then for each $x \in X$, then there exists a non-empty open neighborhood $N(x)$ of x in X and a convex-valued mapping $B_x : X \to 2^Y$ such that:

(a) $A(z) \subset B_x(z)$ for each $z \in N(x)$;

(b) $y \notin B_x(S(y))$ for each $y \in Y$; and

(c) $B_x^{-1}(y)$ is open in X for each $y \in Y$.

Since X is regular, for each $x \in X$, let $O(x)$ be an open neighborhood of x in X such that $\overline{O(x)} \subset N(x)$. Note that X is compact and $X = \cup_{x\in X} O(x)$, there exist $x_1, \cdots, x_n \in X$ such that $X = \cup_{i=1}^n O(x_i)$. Fix an arbitrary $i \in \{1, \cdots, n\}$. Define $B_i : X \to 2^Y$ by

$$B_i(x) = \begin{cases} B_{x_i}(x), & \text{if } x \in \overline{O(x_i)}; \\ Y, & \text{if } x \in X \setminus \overline{O(x_i)}, \end{cases}$$

then we have

(a′) B_i is convex-valued;

(b′) $A(x) \subset B_i(x)$ for each $x \in X$; and

(c′) for each $y \in Y$, $B_i^{-1}(y)$ is open in X since

$$\begin{aligned} B_i^{-1}(y) &= \{x \in \overline{O(x_i)} : y \in B_{x_i}(x)\} \cup (X \setminus \overline{O(x_i)}) \\ &= (B_{x_i}^{-1}(y) \cap \overline{O(x_i)}) \cup (X \setminus \overline{O(x_i)}) \\ &= [B_{x_i}^{-1}(y) \cup (X \setminus \overline{O(x_i)})] \cap [\overline{O(x_i)} \cup (X \setminus \overline{O(x_i)})] \\ &= B_{x_i}^{-1}(y) \cup (X \setminus \overline{O(x_i)}) \end{aligned}$$

which is open by (c). Now we define $B : X \to 2^Y$ by

$B(x) = \cap_{i=1}^n B_i(x)$

for each $x \in X$. Clearly,

(a″) B is convex-valued;

(b″) $A(x) \subset B(x)$ for each $x \in X$;

(c″) for each $y \in Y$, the set $B^{-1}(y)$ is open in X since

$$B^{-1}(y) = \{x \in X : y \in \cap_{i=1}^{n} B_i(x)\} = \cap_{i=1}^{n} B_i^{-1}(y);$$

and

(d″) for each $y \in Y$, $y \notin B(S(y))$: indeed if $y \in Y$, then $S(y) \in X = \cup_{i=1}^{n} O(x_i)$, so there exists $i \in \{1, \cdots, n\}$ such that $S(y) \in O(x_i)$. Thus $B_i(S(y)) = B_{x_i}(S(y))$ by the definition of B_i and hence $y \notin B(S(y))$ since $B(S(y)) \subset B_{x_i}(S(y))$.

By (c″), $\{B^{-1}(y) : y \in Y\}$ is an open cover of X; thus there exist $y_1, \cdots, y_n \in Y$ such that $X = \cup_{j=1}^{m} B^{-1}(y_j)$. Let $\{\beta_j : j = 1, \cdots, m\}$ be a continuous partition of X subordinated to $\{B^{-1}(y_j) : j = 1, \cdots, m\}$. Define $\phi : X \to Y$ by

$$\phi(x) = \Sigma_{j=1}^{m} \beta_j(x) y_j$$

for each $x \in X$. Then ϕ is continuous. If $x \in X$ and $\beta_j(x) \neq \emptyset$, then $x \in B^{-1}(y_j)$ which implies that $y_j \in B(x)$ so that $\phi(x) = \Sigma_{i=1}^{m} \beta_j(x) y_j \in B(x)$ for each $x \in X$. Therefore ϕ is a continuous selection of B. Let $Y_0 = co\{y_1, \cdots, y_m\}$. Then Y_0 is a non-empty compact subset of Y. Note that $\phi \circ S : Y \to Y_0$ is continuous, so that $\phi \circ S|_{Y_0} : Y_0 \to Y_0$ is a continuous self-map. By the Brouwer fixed point theorem, there exists $\hat{y} \in Y_0$ such such that $\hat{y} = \phi \circ S(\hat{y}) \in B(S(\hat{y}))$, which contradicts (d″) that $y \notin B(S(y))$ for each $y \in Y$. Hence there must exist $\hat{x} \in X$ such that $A(\hat{x}) = \emptyset$. □

As a consequence of Theorem 2.4.14, we have the following existence result for maximal elements of L_S-majorized mappings.

Theorem 2.4.15. Let X be a Hausdorff topological space and Y be a non-empty convex subset of a Hausdorff topological vector space E. Let $S : Y \to X$ be a compact and continuous map and $A : X \to 2^Y$ be L_S-majorized. Then there exists $\hat{x} \in X$ such that $A(\hat{x}) = \emptyset$.

Proof. Let $X_0 \subset X$ be non-empty and compact such that $S(Y) \subset X_0$. Consider the restriction of the mapping A to X_0, i.e., the mapping $A|_{X_0} : X_0 \to 2^Y$ which is L_S-majorized. By Theorem 2.4.14, there exists $\hat{x} \in X_0$ such that $A(\hat{x}) = \emptyset$. □

Theorem 2.4.16. Let X be a Hausdorff topological space and I any index set. For each $i \in I$, let Y_i be a non-empty convex subset of a topological vector space. Let $Y = \Pi_{i \in I} Y_i$ and $S : Y \to X$ be a compact continuous (single-valued) map and $\{A_i\}_{i \in I} \in M_S(X, Y_i)_{i \in I}$. Suppose that $\cup_{i \in I}\{x \in X : A_i(x) \neq \emptyset\} = \cup_{i \in I} int\{x \in X : A_i(x) \neq \emptyset\}$. Then there exists $\hat{x} \in X$ such that $A_i(\hat{x}) = \emptyset$ for each $i \in I$.

Proof. For each $i \in I$, define $A_i' : X \to 2^Y$ by $A_i'(x) := \Pi_{j \neq i, j \in I} Y_j \otimes A_i(x)$ for each $x \in X$. Since A_i is L_S-majorized from X to Y_i, so is A_i' from X to Y.

For each $x \in X$, define $I(x) = \{i \in I : A_i(x) \neq \emptyset\}$. Now we define $A : X \to 2^Y$ by

$$A(x) = \begin{cases} \cap_{i \in I(x)} A'_i(x), & \text{if } I(x) \neq \emptyset; \\ \emptyset, & \text{if } I(x) = \emptyset. \end{cases}$$

To complete the proof, it suffices to show that $I(x) = \emptyset$ for some $x \in X$. Suppose the contrary that $I(x) \neq \emptyset$ for each $x \in X$.

Fix an arbitrary $x \in X$. Since $I(x) \neq \emptyset$ by the assumption, $x \in \cup_{i \in I}\{x \in X : A_i(x) \neq \emptyset\}$. Thus there exists $i \in I$ such that $A_i(x) \neq \emptyset$. As A_i is L_S-majorized, there exists an open neighborhood $N(x)$ of x in X and an L_S majorant B_i of A_i at x such that:

(a) $A_i(z) \subset B_i(z)$ for each $z \in N(x)$;

(b) $y_i \notin B_i(S(y))$ for each $y \in Y$; and

(c) the set $B_i^{-1}(y_i)$ is open in X for each $y_i \in Y_i$.

Without loss of generality, we may assume that $N(x) \subset int\{x \in X : A_i(x) \neq \emptyset\}$. Thus $A_i(z) \neq \emptyset$ for each $z \in N(x)$. Now define $B_x : X \to 2^Y$ by $B_x(z) = \pi_i^{-1}(B_i(z))$ for each $z \in X$. We claim that B_x is an L_S-majorant of A at x. Indeed, we have

(i) $A(z) = \cap_{j \in I(x)} A'_j(z) \subset A'_i(z) \subset B_x(z)$ for each $z \in N(x)$;

(ii) $y_i \notin B(S(y))$ since $y_i \notin B_i(S(y))$ for each $y \in Y$ by (b); and

(iii) $B^{-1}(y) = (B'_i)^{-1}(y_i)$ which is open in X.

Therefore B_x is an L_S-majorant of A at x.

Thus A satisfies all hypotheses of Theorem 2.4.15. By Theorem 2.4.15, there exists $\hat{x} \in X$ such that $A(\hat{x}) = \emptyset$, i.e., $A_i(\hat{x}) = \emptyset$ for each $i \in I$. It follows that $I(\hat{x}) = \emptyset$ which contradicts our assumption. The conclusion follows. □

We remark that for a given (single-valued) map $S : Y \to X$ and $\{A_i\}_{i \in I} \in KF(X, Y_i)_{i \in I}$, for each $i \in I$, if $y_i \notin A_i(S(y))$ for all $y \in Y$, then the family $\{A_i\}_{i \in I} \in M_S(X, Y_i)_{i \in I}$. In the case I is a singleton set and $X = Y$, the mapping $A \in KF(X, X)$ is also called a KF mapping.

Before we give the existence of maximal elements for L_S-majorized family $\{A_i\}_{i \in I} \in M_S(X, Y_i)_{i \in I}$ when both X and S are not compact, we first need the following simple fact from [124].

Lemma 2.4.17. Let X be a compact Hausdorff topological space and $\{A_i\}_{i \in I} \in KF(X, Y_i)_{i \in I}$. Then there is a subset $C = \Pi_{i \in I} C_i$ of Y such that for each $x \in X$, there exists $i \in I$ such that $A_i(x) \cap C_i \neq \emptyset$. Moreover for each $i \in I$, C_i is a polytope and all of those polytope, except a finite number, consist of a single point.

Proof. Note that for each $x \in X$, there exists $i \in I$ such that $A_i(x) \neq \emptyset$, so that $X = \cup_{i \in I} \cup_{y_i \in Y_i} A_i^{-1}(y_i)$. Since X is compact, there exists a finite subset J of I such that for each j of J, there exists a finite subset $J_j = \{y_j^1, \cdots, y_j^{m_j}\}$ of

Y_j with $X = \cup_{j\in J} \cup_{i=1}^{m_j} A_j^{-1}(y_j^i)$. Take any fixed $y^0 = \{y_i^0\}_{i\in I} \in Y$ and define

$$C_j = \begin{cases} co(J_j), & \text{if } j \in J; \\ \{y_j^0\}, & \text{if } j \notin J. \end{cases}$$

Then $C = \Pi_{i\in I} C_i$ is the required set. □

The following selection theorem which shows that each L_S-majorized mapping in a paracompact space is majorized by a mapping of class L_S.

Lemma 2.4.18. Let X be a Hausdorff regular and paracompact topological space, I any index set, and $Y = \Pi_{i\in I} Y_i$, where each Y_i is a non-empty convex subset of a Hausdorff topological vector space for each $i \in I$. Let $S : Y \to X$ be a map and $\{A_i\} \in M_S(X, Y_i)_{i\in I}$. Then there exists a family $\{B_i\}_{i\in I} \in L_S(X, Y_i)_{i\in I}$ such that for each $i \in I$, $A_i(x) \subset B_i(x)$ for all $x \in X$.

Proof. Fix any $i \in I$. For each $x \in X$ and note that A_i is L_S-majorized, there exists a non-empty open neighborhood $N(x)$ of x in X and a mapping $B_x : X \to 2^{Y_i}$ such that:

(a) $A_i(z) \subset B_x(z)$ for each $z \in N(x)$;

(b) $y_i \notin B_x(S(y))$ for each $y \in Y$; and

(c) $B_x^{-1}(y_i)$ is open in X for each $y_i \in Y_i$.

Then $\cup_{x\in X} N(x)$ is an open cover of X. Note that X is paracompact and regular, let $\{O(x)\}_{x\in X}$ be a locally finite open cover of X such that $cl_X O(x) \subset N(x)$ for each $x \in X$ and $X = \cup_{x\in X} O(x)$. For each $x \in X$, define $B'_x : X \to 2^{Y_i}$ by

$$B'_x(z) = \begin{cases} B_x(z), & \text{if } z \in cl_X O(x); \\ Y_i, & \text{if } z \in X \setminus cl_X O(x). \end{cases}$$

Then it is easy to verify that

$$(B'_x)^{-1}(y_i) = (cl_X O(x) \cap B_x^{-1}(y_i)) \cup (X \setminus cl_X O(x)) = B_x^{-1}(y_i) \cup (X \setminus cl_X O(x))$$

which is open in X for each $y_i \in Y_i$. Now define $B_i : X \to 2^{Y_i}$ by

$$B_i(z) = \cap_{x\in X} B'_x(z)$$

for each $z \in X$. Then we can verify that B has the following properties:

(i) $B_i(x)$ is convex valued for each $x \in X$;

(ii) $y_i \notin B_i(S(y))$ for each $y \in Y$;

(iii) $B_i^{-1}(y_i)$ is, indeed, open in X by the locally finite property of the family $\{O(x)\}_{x\in X}$ (e.g., see the proof of Lemma 2 of Ding and Tan [138] or Yuan [609]); and

(iv) $A_i(x) \subset B_i(x)$ for each $x \in X$.

Thus B_i is of class L_S and the family $\{B_i\} \in L_S(X, Y_i)_{i\in I}$ and by the definition of B, we have that for each $i \in I$, $A_i(x) \subset B_i(x)$ for all $x \in X$. Thus the conclusion follows. □

In order to give the existence theorem of maximal elements for a family of class L_S-mappings which the map S is not compact, we need the following result which is an fixed point theorem as an application of Theorem 2.4.16.

Theorem 2.4.19. Let X be a Hausdorff topological space and I be any index set. For each $i \in I$, let Y_i be a non-empty convex subset of a Hausdorff topological vector space. Let $Y = \Pi_{i\in I} Y_i$, $S : Y \to X$ be compact continuous and $\{A_i\}_{i\in I} \in KF(X, Y_i)_{i\in I}$. Then there exists $\hat{y} \in Y$ and $i_0 \in I$ such that $\hat{y}_{i_0} \in A_{i_0}(S(\hat{y}))$.

Proof. Suppose the conclusion were false. Then $\{A_i\}_{i\in I} \in L_S(X, Y_i)_{i\in I}$ and the set $\{x \in X : A_i(x) \neq \emptyset\}$ is also open in X, so that the family $\{A_i\}_{i\in I}$ satisfies all hypotheses of Theorem 2.4.16. By Theorem 2.4.16, there exists $\hat{x} \in X$ such that $A_i(\hat{x}) = \emptyset$ for all $i \in I$, which contradicts our assumption that $\{A_i\}_{i\in I} \in KF(X, Y_i)_{i\in I}$. Therefore there must exist $\hat{y} \in Y$ and $i_0 \in I$ such that $\hat{y}_{i_0} \in A_{i_0}(S(\hat{y}))$. □

Now we have the following existence theorem of maximal elements for a family in which domains are not compact and the map S also need not be compact from Deguire et al. [125].

Theorem 2.4.20. Let X be a Hausdorff topological space and I be any index set. Let $Y = \Pi_{\in I} Y_i$ and $S : Y \to X$ continuous, where Y_i is a non-empty convex subset of a topological vector space for each $i \in I$ and $\{A_i\}_{i\in I} \in L_S(X, Y_i)_{i\in I}$. Suppose there exists a non-empty compact subset K of X and a non-empty compact convex subset C_i of Y_i for each $i \in I$ such that for each $x \in X \setminus K$, there exists $i \in I$ such that $A_i(x) \cap C_i \neq \emptyset$. Then there exists $x \in K$ such that $A_i(x) = \emptyset$ for all $i \in I$.

Proof. Suppose it is not true. Then for each $x \in K$, there exists $i \in I$ such that $A_i(x) \neq \emptyset$. By Lemma 2.4.17, there exists a subset $D = \Pi_{i\in I} D_i$ such that for each $x \in K$, there exists $i \in I$ such that $A_i(x) \cap D_i \neq \emptyset$, where D_i is a polytope for each $i \in I$. Note that S is continuous, so that the set $X_0 = S(D)$ is a non-empty compact subset of X. For each $i \in I$, we set $H_i = co(C_i \cup D_i)$, then H_i is a non-empty compact convex subset of Y_i. If $x \in X \setminus K$, there exists $i \in I$ such that $\emptyset \neq A_i(x) \cap C_i \subset A_i(x) \cap H_i$ by (I). If $x \in K$, by our assumption there exists $i \in I$ such that $\emptyset \neq A_i(x) \cap D_i \subset A_i(x) \cap H_i$. Thus in any case, for each $x \in X$, there exists $i \in I$ such that $A_i(x) \cap H_i \neq \emptyset$. For each $i \in I$, define $A_i' : X_0 \to 2^{H_i}$ by

$$A_i'(x) = A_i(x) \cap H_i$$

for each $x \in X_0$ and let $S_1 = S|_H$, which is the restriction of the continuous map S to $H := \Pi_{i\in I}H_i$. Let $X := X_0$ and $Y_i := H_i$ for each $i \in I$ in Theorem 2.4.19. Then the family $\{A_i'\}_{i\in I}$ and S_1 satisfy all hypotheses of Theorem 2.4.19. By Theorem 2.4.19, there exists $y \in H$ and $i_0 \in I$ such that $y_{i_0} \in A_{i_0}'(S_1(y))$, i.e., $y_{i_0} \in A_{i_0}(S(y)) \cap H_{i_0}$. Therefore $y_{i_0} \in A_{i_0}(S(y))$ which contradicts that $y_{i_0} \notin A_{i_0}(S(y))$ for all $y \in H$ since A_{i_0} is of class L_S. This contradiction shows that there must exist $x \in K$ such that $A_i(x) = \emptyset$ for all $i \in I$. □

As applications of Lemma 2.4.18 and Theorem 2.4.20, we have the following existence theorem of maximal elements for L_S-majorized family by assuming paracompactness instead of compactness for the domains.

Theorem 2.4.21. Let X be a Hausdorff regular and paracompact topological space and I any index set. For each $i \in I$, let Y_i be a non-empty convex subset of a Hausdorff topological vector space. Let $Y = \Pi_{i\in I}Y_i$, $S : Y \to X$ be a continuous mapping and $\{A_i\}_{i\in I} \in M_S(X, Y_i)_{i\in I}$. Suppose there exists a non-empty compact subset K of X and a non-empty compact convex subset C_i of Y_i for each $i \in I$ such that for each $x \in X \setminus K$, there exists $i \in I$ such that $A_i(x) \cap C_i \neq \emptyset$. Then there exists $x \in K$ such that $A_i(x) = \emptyset$ for all $i \in I$.

Proof. By Lemma 2.4.18, for each $i \in I$, there exists $B_i : X \to 2^{Y_i}$ which is of class L_S such that $A_i(x) \subset B_i(x)$ for each $x \in X$. Hence the family $\{B_i\}_{i\in I}$ satisfies all hypotheses of Theorem 2.4.20. By Theorem 2.4.20, there exists $x \in X$ such that $B_i(x) = \emptyset$ for each $i \in I$, so that $A_i(x) = \emptyset$ for each $i \in I$. □

Finally we are in turn to give component versions of Browder-Fan fixed point theorems and coincidence theorems in non-compact product spaces. In this part, as applications of existence theorems of maximal element for class L_S mappings in topological spaces, we shall give some component versions of Browder-Fan type fixed point theorems in product spaces for the family of KF mappings.

As an equivalent fixed point version of Theorem 2.4.20, we have the following fixed point theorem in product spaces.

Theorem 2.4.22. Let X be a Hausdorff topological space and I be any index set. For each $i \in I$, let Y_i be a non-empty convex subset of a Hausdorff topological vector space and $Y = \Pi_{i\in I}Y_i$. Suppose $S : Y \to X$ is continuous and $A_i : X \to 2^{Y_i}$ is convex-valued for each $i \in I$. Moreover, suppose there exists a non-empty compact subset K of X and a non-empty compact convex subset C_i of Y_i for each $i \in I$ such that:

(a) for each $x \in X \setminus K$, there exists $i \in I$ such that $A_i(x) \cap C_i \neq \emptyset$;

(b) the set $A_i^{-1}(y)$ is open in X for each $y \in Y_i$ and $i \in I$; and

(c) for each $x \in K$, there exists $i \in I$ such that $A_i(x) \neq \emptyset$.

Then there exists $\hat{y} \in Y$ and $i_0 \in I$ such that $\hat{y}_{i_0} \in A_{i_0}(S(\hat{y}))$.

Proof. Theorem 2.4.20 ⇒ Theorem 2.4.22: Suppose the conclusion of Theorem 2.4.22 were false. Then all hypotheses of Theorem 2.4.20 are satisfied. By Theorem 2.4.20, there exists $x \in K$ such that $A_i(x) = \emptyset$ for all $i \in I$, which contradicts condition (c). Hence the conclusion must be true. □

Theorem 2.4.22 ⇒ Theorem 2.4.20: Suppose the conclusion of Theorem 2.4.20 were false. Then all hypotheses of Theorem 2.4.22 are satisfied. By Theorem 2.4.22, there exists $\hat{y} \in Y$ and $i_0 \in I$ such that $\hat{y}_{i_0} \in A_{i_0}(S(\hat{y}))$ which contradicts the assumption that A_{i_0} is of class L_S. □

We wish to point out that the conclusion of Theorem 2.4.22 could not guarantee in general that $\hat{y}$ is a fixed point of the product mapping $\Pi_{i \in I} A_i(S)$ as we can only assure that there exists $i_0 \in I$ (not necessarily for all $i \in I$) such that $\hat{y}_{i_0} \in A_{i_0}(S(\hat{y}))$. As an application of Theorem 2.4.22, we have the following coincidence theorem for two families of set-valued mappings in product spaces.

Theorem 2.4.23. Let I be any index set, $X = \Pi_{i \in I} X_i$ and $Y = \Pi_{i \in I} Y_i$ be products of non-empty convex subsets X_i and Y_i, each in a Hausdorff topological vector space. For each $i \in I$, let $A_i : X \to 2^{Y_i}$ and $B_i : Y \to 2^{X_i}$. Suppose there exists a non-empty compact subset K of X and a non-empty compact subset L of Y and non-empty compact convex subsets C_i of Y_i and D_i of X_i for each $i \in I$ such that:

(i) for each $i \in I$, $A_i^{-1}(y)$ and $B_i^{-1}(x)$ are open in X and Y, respectively for each $(x, y) \in Y_i \times X_i$ for each $i \in I$;

(ii) for each $(x, y) \in (X \times Y) \setminus (K \times L)$, there exists $i \in I$ such that $A_i(x) \cap C_i \neq \emptyset$ and $B_i(y) \cap D_i \neq \emptyset$; and

(iii) for each $(x, y) \in K \times L$, there exists $i \in I$ such that $A_i(x) \neq \emptyset$ and $B_i(y) \neq \emptyset$.

Then there exist $(\hat{x}, \hat{y}) \in X \times Y$ and $i_0 \in I$ such that $\hat{y}_{i_0} \in A_{i_0}(\hat{x})$ and $\hat{x}_{i_0} \in B_{i_0}(\hat{y})$.

Proof. Let $M := K \times L$. For each $i \in I$, let $E_i = C_i \times D_i$ and define $W_i : X \times Y \to 2^{Y_i \times X_i}$ by

$$W_i(x, y) = A_i(x) \times B_i(y)$$

for each $(x, y) \in X \times Y$. Also define $S : X \times Y \to Y \times X$ by $S(x, y) = (y, x)$ for each $(x, y) \in X \times Y$. Then all hypotheses of Theorem 2.4.22 are satisfied. By Theorem 2.4.22, there exists $(\hat{x}, \hat{y}) \in X \times Y$ and $i_0 \in I$ such that $(\hat{x}_{i_0}, \hat{y}_{i_0}) \in W_{i_0}(\hat{x}, \hat{y})$. Thus the conclusion follows. □

Before we close this section, we will give some minimax inequalities for families of functions. In this part, as applications of existence theorems of maximal elements for a family of set-valued mappings, we shall first give an existence theorem of solutions for a system version of Ky Fan minimax inequalities in which functions are lower semicontinuous it its first variable and quasi-concave in its second variable. By introducing a new concept called L-majorized, we establish the system of Ky Fan minimax inequalities for a family of L-majorized functions which are not lower semicontinuous nor quasi-concave. In particular, our results improve well-known Ky Fan minimax inequality in several aspects.

As an application of Theorem 2.4.16, we have the following result Ky Fan minimax inequality for a family of functions.

Theorem 2.4.24. Let X be a Hausdorff topological space, I any index set and $Y = \Pi_{i\in I} Y_i$ be the product of non-empty convex set Y_i, each in a Hausdorff topological vector space. Suppose $S : Y \to X$ is compact continuous and for each $i \in I$, the function $f_i : X \times Y_i \to \mathbb{R}$ satisfies the following conditions:
(i) for each fixed $x \in X$, $y_i \mapsto f_i(x, y_i)$ is quasi-concave; and
(ii) for each fixed $y_i \in Y_i$, $x \mapsto f_i(x, y_i)$ is lower semicontinuous.
Then we have

(A): For each $\lambda \in \mathbb{R}$, at least one of the following statement is satisfied:
(1) there exists $x \in X$ such that $\sup_{i\in I} \sup_{y_i \in Y_i} f_i(x, y_i) \leq \lambda$, or
(2) there exists $y \in Y$ and $i \in I$ such that $f_i(S(y), y_i) > \lambda$.

and
(B): The following minimax inequalities holds:

$$\inf_{x\in X} \sup_{i\in I} \sup_{y_i\in Y_i} f_i(x, y_i) \leq \sup_{i\in I} \sup_{y\in Y} f_i(S(y), y_i).$$

Proof. (A): Fix arbitrary $i \in I$ and $\lambda \in \mathbb{R}$. Define $A_i : X \to 2^{Y_i}$ by

$$A_i(x) = \{y_i \in Y_i : f_i(x, y_i) > \lambda\}$$

for each $x \in X$. Then for each $y_i \in Y_i$, the set $A_i^{-1}(y_i)$ is open in X by (ii) and for each $x \in X$, $A_i(x)$ is convex valued by (i).

If the case (2) of (A) does not hold, then for each $y \in Y$ and $i \in I$, $y_i \notin A_i(S(y))$. Thus $\{A_i\}_{i\in I} \in L_S(X, Y_i)_{i\in I}$. By Theorem 2.4.16, there exists $x \in X$ such $A_i(x) = \emptyset$ for each $i \in I$, i.e., (1) of (A) holds.

(B): Let $\lambda_0 = \sup_{i\in I} \sup_{y\in Y} f_i(S(y), y_i)$, then (2) of (A) fails to hold, so that the (1) of (A) must hold, i.e., there exists $x_0 \in X$ such that $\sup_{i\in I} \sup_{y_i\in Y_i} f_i(x_0, y_i) \leq \lambda_0$. Therefore we have

$$\inf_{x\in X} \sup_{i\in I} \sup_{y_i\in Y_i} f_i(x, y_i) \leq \sup_{i\in I} \sup_{y\in Y} f_i(S(y), y_i). \qquad \square$$

As an analytical formulation of Theorem 2.4.20, we have the following result.

Theorem 2.4.25. Let X be a Hausdorff topological space, I any index set and $Y = \Pi_{i\in I} Y_i$ be the product of non-empty convex sets Y_i, each in a Hausdorff topological vector space, and $S : Y \to X$ be continuous. Suppose for each $i \in I$, the function $f_i : X \times Y_i \to \mathbb{R}$ satisfies the following conditions:
(i) for each fixed $x \in X$, $y_i \mapsto f_i(x, y_i)$ is quasi-concave; and

(ii) for each fixed $y_i \in Y_i$, $x \mapsto f_i(x, y_i)$ is lower semicontinuous.
Then for each $\lambda \in \mathbb{R}$, at least one of the following statement is satisfied:

(1) there exists $x \in X$ such that $\sup_{i\in I} \sup_{y_i\in Y_i} f_i(x, y_i) \leq \lambda$, or

(2) there exist $y \in Y$ and $i \in I$ such that $f_i(S(y), y_i)) > \lambda$, or

(3) for each given non-empty compact subset K of X and for each given non-empty compact convex subset C_i of Y_i, where $i \in I$, there exists $x \in X \setminus K$ such that

$$\sup_{i\in I} \sup_{y_i\in C_i} f_i(x, y_i) \leq \lambda.$$

Proof. Theorem 2.4.20 $\Rightarrow$ Theorem 2.4.25: For each fixed $i \in I$ and each given $\lambda \in \mathbb{R}$, define $A_i : X \to 2^{Y_i}$ by

$$A_i(x) = \{y_i \in Y_i : f_i(x, y_i) > \lambda\}$$

for each $x \in X$. Then conclusion follows from Theorem 2.4.20.

Theorem 2.4.25 $\Rightarrow$ Theorem 2.4.20: For each $i \in I$, define $f_i : X \times Y_i \to \mathbb{R}$ by

$$f_i(x, y) = \begin{cases} 1, & \text{if } y_i \in A_i(x); \\ 0, & \text{otherwise} \end{cases}$$

for each $(x, y_i) \in X \times Y_i$. By Theorem 2.4.25, with $\lambda = 0$, as (2) and (3) do not hold, thus (1) must hold, i.e., there exists $x \in X$ such that $\sup_{i\in I} \sup_{y_i\in Y_i} f_i(x, y_i) \leq 0$. It follows that $x \in K$ and by (1) and $A_i(x) = \emptyset$ for all $i \in I$. $\square$

We note that when I is a singleton set, Theorems 2.4.24 and 2.4.25 are non-linear alternative improved version of Ky Fan minimax inequality.

Before we study the system of Ky Fan minimax inequalities for a family of real-valued functions which may not be lower semicontinuous nor quasi-concave, we need the following definition.

Let X be a Hausdorff topological space and I be any index set. Let $Y = \Pi_{i\in I} Y_i$ be a product of non-empty convex sets Y_i, each in a Hausdforff topological vector space. Suppose for each $i \in I$, $f_i : X \times Y_i \to \mathbb{R} \cup \{-\infty, +\infty\}$ is a function. Then the family $\{f_i\}_{i\in I}$ is said to be *L-majorized* if the following condition (LC) is satisfied for each $i \in I$.

The Condition (LC): For each given $\lambda \in \mathbb{R}$, if there exists $(x, y_i) \in X \times Y_i$ such that $f_i(x, y_i) > \lambda$, then there exists a non-empty open neighborhood $N(x)$ of x in X and a real-valued function $f_i^x : X \times Y_i \to \mathbb{R} \cup \{-\infty, +\infty\}$ such that:

(a_1): $f_i(z, w_i) \leq f_i^x(z, w_i)$ for each $(z, w_i) \in N(x) \times Y_i$;

(a_2): for each fixed $z \in X$, $w_i \mapsto f_i^x(z, w_i)$ is quasi-concave on Y_i;

(a_3): for each fixed $w_i \in Y_i$, $z \mapsto f_i^x(z, w_i)$ is lower semicontinuous on X; and

(a_4): for each $w \in Y$, $f_i(S(w), w_i) \leq \lambda$ implies that $f_i^x(S(w), w_i) \leq \lambda$.

Now as an application of Theorem 2.4.16, we have the following existence theorem of solutions for Ky Fan minimax inequalities for a family of real-valued functions which do not have lower semicontinuity nor quasi-concavity.

Theorem 2.4.26. Let X be a Hausdorff topological space, I any index set and $Y = \Pi_{i\in I}Y_i$ the product of non-empty convex sets Y_i, each in a Hausdorff topological vector space. Suppose $S : Y \to X$ is compact continuous and for each $i \in I$, the function $f_i : X \times Y_i \to \mathbb{R}$ is L-majorized such that for each given $\lambda \in \mathbb{R}$, $\cup_{i\in I} \cup_{y_i\in Y_i} \{x \in X : f_i(x, y_i) > \lambda\} = \cup_{i\in I} int[\cup_{y_i\in Y_i}\{x \in X : f_i(x, y_i) > \lambda\}]$. Then we have:

(A): For each $\lambda \in \mathbb{R}$, at least one of the following statement is satisfied:
(1) there exists $x \in X$ such that $\sup_{i\in I} \sup_{y_i\in Y_i} f_i(x, y_i) \leq \lambda$, or
(2) there exists $y \in Y$ and $i \in I$ such that $f_i(S(y), y_i) > \lambda$.

(B): The following minimax inequality holds:

$$\inf_{x\in X} \sup_{i\in I} \sup_{y_i\in Y_i} f_i(x, y_i) \leq \sup_{i\in I} \sup_{y\in Y} f_i(S(y), y_i).$$

Proof. (A): For each given $\lambda \in \mathbb{R}$ and fixed $i \in I$, define $A_i : X \to 2^{Y_i}$ by

$$A_i(x) = \{y_i \in Y_i : f_i(x, y_i) > \lambda\}$$

for each $x \in X$. For each given $x \in X$, if $A_i(x) \neq \emptyset$, then there exists an non-empty open neighborhood $N(x)$ of x in X and a function $f_i^x : X \times Y_i \to \mathbb{R} \cup \{-\infty, +\infty\}$ having the properties (a_1)-(a_4). Now define another mapping $A_i^x : X \to 2^{Y_i}$ by

$$A_x^i(z) = \{y_i \in Y_i : f_i^x(z, y_i) > \lambda\}$$

for each $z \in X$. Then we have
(i): $A_i(z) \subset A_x^i(z)$ for each $z \in N_x$ by (a_1);
(ii): $A_x^i(z)$ is convex for each $z \in X$ by (a_2); and
(iii): $(A_x^i)^{-1}(y_i)$ is open in X for each $y_i \in Y_i$ by (a_3).

If the case (2) of (A) does not hold, then for each $y \in Y$ and $i \in I$, $y_i \notin A_i(S(y))$. By (a_4), $y_i \notin A_x^i(S(y))$ for each $y \in Y$. Thus A_x^i is a majorant of A_i at x. Therefore A_i is L_S-majorized and the family $\{A_i\}_{i\in I} \in M_S(X, Y_i)_{i\in I}$. Moreover, the set $\cup_{i\in I}\{x \in X : A_i(x) \neq \emptyset\} = \cup_{i\in I} int \cup_{y_i\in Y_i} \{x \in X : f_i(x, y_i) > \lambda\}$ is open in X by our assumption. Thus $\{A_i\}_{i\in I}$ satisfies all hypotheses of Theorem 2.4.16. By Theorem 2.4.16, there exists $x \in X$ such $A_i(x) = \emptyset$ for each $i \in I$, i.e., (1) of (A) holds.

(B): Let $\lambda_0 = \sup_{i\in I} \sup_{y\in Y} f_i(S(y), y_i)$, then (2) of (A) fails to hold, so that the (1) of (A) must hold, i.e., there exists $x_0 \in X$ such that $\sup_{i\in I} \sup_{y_i\in Y_i} f_i(x_0, y_i) \leq$

λ_0. Therefore we have

$$\inf_{x\in X}\sup_{i\in I}\sup_{y_i\in Y_i} f_i(x,y_i) \le \sup_{i\in I}\sup_{y\in Y} f_i(S(y),y_i). \qquad \square$$

When S is not compact, by a similar argument used in proving Theorems 2.4.25 and 2.4.26 and as an application of Theorem 2.4.21 instead of Theorem 2.4.20, we have the following existence theorem of Ky Fan minimax inequalities for a family of L-majorized functions, which is an analytical formulation of Theorem 2.4.21.

Theorem 2.4.27. Let X be a Hausdorff regular and paracompact topological space, I any index set, and $Y = \Pi_{i\in I}Y_i$ the product of non-empty convex sets Y_i, each in a Hausdorff topological vector space. Suppose $S : Y \to X$ is continuous and for each $i \in I$ and the function $f_i : X \times Y_i \to \mathbb{R}$ is L-majorized.

Then for each $\lambda \in \mathbb{R}$, at least one of the following statements is satisfied:
(1) there exists $x \in X$ such that $\sup_{i\in I}\sup_{y_i\in Y_i} f_i(x,y_i) \le \lambda$, or
(2) there exists $y \in Y$ and $i \in I$ such that $f_i(S(y),y_i)) > \lambda$, or
(3) for each given non-empty compact subset K of X and for each given non-empty compact convex subset C_i of Y_i, where $i \in I$, there exists $x \in X \setminus K$ such that

$$\sup_{i\in I}\sup_{y_i\in C_i} f_i(x,y_i) \le \lambda.$$

Proof. By combining arguments used in both Theorem 2.4.25 and Theorem 2.4.26, and by applying Theorem 2.4.21 instead of Theorem 2.4.20, the conclusions follows. $\square$

Let I be a singleton set, as a consequence of Theorem 2.4.24 above, we have the following:

Corollary 2.4.28. Let X be a Hausdorff topological space and Y be a non-empty convex set in a Hausdorff topological vector space. Suppose $S : Y \to X$ is compact continuous and $f : X \times Y \to \mathbb{R}$ satisfies the following conditions:
(i) for each fixed $x \in X$, $y \mapsto f(x,y)$ is quasi-concave;
(ii) for each fixed $y \in Y$, $x \mapsto f(x,y)$ is lower semicontinuous.
Then we have:

(A): For each $\lambda \in \mathbb{R}$, at least one of the following statement is satisfied:
(1) there exists $x \in X$ such that $\sup_{y\in Y} f(x,y) \le \lambda$, or
(2) there exists $y \in Y$ such that $f(S(y),y) > \lambda$.

(B): The following minimax inequality holds:

$$\inf_{x\in X}\sup_{y\in Y} f(x,y) \le \sup_{y\in Y} f(S(y),y).$$

As a special case of Theorem 2.4.25, we have the following minimax inequality.

Corollary 2.4.29. Let X be a Hausdorff topological space and Y be a non-empty convex set in a Hausdorff topological vector space, and $S : Y \to X$ be continuous. Suppose the function $f : X \times Y \to \mathbb{R}$ satisfies the following conditions:

(i) for each fixed $x \in X$, $y \mapsto f(x, y)$ is quasi-concave; and

(ii) for each fixed $y \in Y$, $x \mapsto f(x, y)$ is lower semicontinuous.

Then for each $\lambda \in \mathbb{R}$, at least one of the following statements is satisfied:

(1) there exists $x \in X$ such that $\sup_{y \in Y} f(x, y) \leq \lambda$, or

(2) there exist $y \in Y$ such that $f(S(y), y)) > \lambda$, or

(3) for each given non-empty compact subset K of X and for each given non-empty compact convex subset C of Y, there exists $x \in X \setminus K$ such that

$$\sup_{y \in C} f(x, y) \leq \lambda.$$

As special cases of Theorems 2.4.26 and 2.4.27, we also have the following result.

Corollary 2.4.30. Let X be a Hausdorff topological space and Y be a non-empty convex set in a Hausdorff topological vector space. Suppose $S : Y \to X$ is compact continuous and $f : X \times Y \to \mathbb{R}$ is L-majorized such that for each given $\lambda \in \mathbb{R}$, the set $\cup_{y \in Y} \{x \in X : f(x, y) > \lambda\}$ is open in X. Then we have that

(A): For each $\lambda \in \mathbb{R}$, at least one of the following statement is satisfied:

(1) there exists $x \in X$ such that $\sup_{y \in Y} f(x, y) \leq \lambda$, or

(2) there exists $y \in Y$ such that $f(S(y), y) > \lambda$.

(B): The following minimax inequality holds:

$$\inf_{x \in X} \sup_{y \in Y} f(x, y) \leq \sup_{y \in Y} f(S(y), y).$$

Corollary 2.4.31. Let X be a Hausdorff regular and paracompact topological space and Y a non-empty convex set in a Hausdorff topological vector space. Suppose $S : Y \to X$ is continuous and $f : X \times Y \to \mathbb{R}$ is L-majorized.

Then for each $\lambda \in \mathbb{R}$, at least one of the following statement is satisfied:

(1) there exists $x \in X$ such that $\sup_{y \in Y} f(x, y) \leq \lambda$, or

(2) there exist $y \in Y$ such that $f(S(y), y)) > \lambda$, or

(3) for each given non-empty compact subset K of X and for each given non-empty compact convex subset C of Y, there exists $x \in X \setminus K$ such that $\sup_{y \in C} f(x, y) \leq \lambda$.

All above Corollaries are the generalization of Fan's minimax inequalities in several ways.

As applications of Fan's lemma in [157] which is an infinite dimensional extension of the classical KKM lemma in [319], a number of minimax inequalities have been given by many authors and also its various applications to pure, applied mathematics, social sciences have been found. This subject is so rich and broad that this area of study definitely qualifies to be named KKM theory [417]. In the next several chapters, we shall see how important and useful the Ky Fan minimax inequality and its equivalent form of Browder-Fan fixed point theorem in the study of abstract economics, variational inequalities, game theorem, optimization problems, and mathematical economics is.

2.5 Generic Stability of Ky Fan Points

Let (X, d) be a compact metric space with the fixed point property for continuous mappings. In [178], Fort introduced the concept of essential fixed points of a continuous mapping f on X. He proved that (1) every continuous mapping on X can be arbitrarily approximated by a continuous mapping on X whose fixed points are all essential; and (2) if each fixed point of a continuous mapping f on X is essential, then the fixed point set $S(f) = \{x \in X : f(x) = x\}$ of f is stable in the sense that: for each $\epsilon > 0$, there is $\delta > 0$ such that for each continuous mapping g on X, if $\rho(f,g) = \sup\{d(f(x), g(x)) : x \in X\} < \delta$, then $h(S(f), S(g)) < \epsilon$ where h is the Hausdorff metric defined on all non-empty bounded closed subsets of X induced by the metric d; i.e., the fixed point set $S(g)$ of g is *close* to the fixed point set $S(f)$ of f whenever g is *close* to f which means the continuity of fixed point mapping S defined above.

In this section, the concepts of the *Ky Fan (KF) point* and *essential Ky Fan (KF) point* are introduced. We then study the stability of KF points which are the solutions of Ky Fan minimax inequalities in both compact and non-compact settings.

We shall recall some definitions. If X is a topological space, we shall denote by $K(X)$ and $\mathcal{P}_0(X)(= 2^X)$ the space of all non-empty compact subsets of X and the space of all non-empty subsets of X respectively, both endowed with the Vietoris topology (see, Klein and Thompson [317]).

Definition 2.5.1. Let X and Z be two topological space. Then we recall that a mapping $T : X \to 2^Z$ is said to be (i): upper (resp., lower) semicontinuous at $x \in X$, if for each open set G in Z with $G \supset T(x)$ (resp., $G \cap T(x) \neq \emptyset$), there exists an open neighborhood $O(x)$ of x in X such that $G \supset T(x')$ (resp., $G \cap T(x') \neq \emptyset$) for each $x' \in O(x)$; (ii): T is said to be *almost lower semicontinuous* at $x \in X$, if there exists $z \in T(x)$ such that for each open neighborhood

$N(z)$ of z in Z, there exists an open neighborhood $O(x)$ of x in X with the property that $N(z) \cap T(x') \neq \emptyset$ for each $x' \in O(x)$; (iii): T is an *usco* if T is upper semicontinuos with non-empty compact values; and (iv): T is *continuous* if T is both upper and lower semicontinuous.

Definition 2.5.2. A space X is said to be *Čech-complete* if it can be embedded as a G_δ subset of some compact Hausdorff space (e.g., see Engelking [154]). It is known that (i): A Čech-complete space is a Baire space; (ii): Locally compact spaces are also Čech-complete, because a non-compact locally compact space has a compactification with one-point remainder; (iii): The space of all irrational numbers with the topology of a subspace of the real line is an example of a Čech-complete space that is not locally compact and moreover (iv): Each completely metrizable space is Čech-complete. Thus, the Čech-complete spaces contain within them two important types of Baire spaces.

A space X is said to belong to the *class* $\mathcal{L}$ (see Kenderov [288]) if for each Čech-complete space Z, every usco mapping $S : Z \to K(X)$ is almost lower semicontinuous on some dense G_δ subset of Z.

Note that there are a number of spaces under which each usco set-valued mapping from X to Z is almost lower semicontinuous at the points of some dense G_δ subset of X. For example,

(a) X is a Baire space and Z is metrizable (e.g., see Fort [178, Theorem 2]);

(b) X is Čech-complete and Z is a Banach space with weak topology (which is non-metrizable if it is infinite dimensional) by Theorem 2 of Christensen [93]; and

(c) X is a Baire space and Z is the dual space of a Banach space with the weak*-topology provided Z has the Randon-Nikodym property (e.g., see Christensen and Kenderov [94]).

It follows from (a) and (b) above that the class $\mathcal{L}$ contains all metrizable spaces and all Banach spaces equipped with the weak topology. The class $\mathcal{L}$ has nice stability properties: it is closed under taking subspaces, countable products, countable sums of closed sets, and perfect images.

In order to study the stability of Ky Fan minimax inequalities in topological vector spaces, we first give the following general minimax inequality which is a non-compact version of Fan's minimax inequality and which is due to Ding and Tan [136].

Theorem 2.5.3. Let X be a non-empty convex subset of a topological vector space E and $f : X \times X \to \mathbb{R} \cup \{-\infty, +\infty\}$ be such that:

(1) for each fixed $x \in X$, $y \mapsto f(x,y)$ is compactly lower semicontinuous on X;

(2) for each $A \in \mathcal{F}(X)$ and for each $y \in co(A)$, $\min_{x\in A} f(x,y) \leq 0$; and

(3) there exists a non-empty compact and convex subset X_0 of X and a non-empty compact subset K of X such that for each $y \in X \setminus K$, there is an

$x \in co(X_0 \cup \{y\})$ with $f(x, y) > 0$.

Then there exists $\hat{y} \in X$ such that $\inf_{x \in X} f(x, \hat{y}) \leq 0$.

Proof. For each $x \in X$, let $K(x) = \{y \in K : f(x, y) \leq 0\}$. By (1), $K(x)$ is closed in K. We claim that the family $\{K(x) : x \in X\}$ has the finite intersection property. Indeed, let $\{x_1, \cdots, x_n\}$ be any non-empty finite subset of X and let $D = co(X_0 \cup \{x_1, \cdots, x_n\})$. Then D is a compact and convex subset of X. We first note that $f(x, x) \leq 0$ by (2) for each $x \in X$. Now define a set-valued mapping $F : D \to 2^D$ by

$$F(x) = \{y \in D : f(x, y) \leq 0\}$$

for each $x \in D$. Then it is clear that $F(x)$ is compact in D for each $x \in D$ and $co(A) \subset \cup_{x \in A} F(x)$ for each $A \in \mathcal{F}(D)$; otherwise, there would exist $A \in \mathcal{F}(D)$ and $y \in co(A)$ such that $y \notin \cup_{x \in A} F(x)$. It then follows that $f(x, y) > 0$ for all $x \in A$, which contradicts (2). By the classical KKM Lemma (e.g., see Theorem 2.2.17 or Corollary 2.2.18), it follows that $\cap_{x \in D} F(x) \neq \emptyset$. That is, there exists $\hat{y} \in D$ such that $f(x, \hat{y}) \leq 0$ for all $x \in D$. By (3), we must have $\hat{y} \in K$. Hence $\hat{y} \in \cap_{i=1}^{n} K(x_i)$. By the compact of K, it follows that $\cap_{x \in X} K(x) \neq \emptyset$. Take any $\hat{y} \in \cap_{x \in X} K(x)$, it then follows that $\hat{y} \in K$ and we have $\inf_{x \in X} f(x, \hat{y}) \leq 0$. The proof is complete. □

As a special case of Theorem 2.5.4 above, we have the following minimax inequality.

Theorem 2.5.4. Let X be a non-empty compact convex subset of a topological vector space and $f : X \times X \to \mathbb{R}$ be such that:

(i) $f(x, x) \leq 0$ for all $x \in X$;

(ii) for each fixed $x \in X$, $y \mapsto f(x, y)$ is lower semicontinuous; and

(iii) for each fixed $y \in X$, $x \mapsto f(x, y)$ is quasi-concave (i.e., for each $\lambda \in R$, the set $\{x \in X : f(x, y) > \lambda\}$ is convex).

Then there exists $\hat{y} \in X$ such that $f(x, \hat{y}) \leq 0$ for all $x \in X$.

It is clear that Theorem 2.5.4 is equivalent to the celebrated Ky Fan minimax inequality [164] stated as follows.

Theorem 2.5.5. Let X be a non-empty compact convex subset of a topological vector space and $f : X \times X \to \mathbb{R}$ be such that:

(a) for each fixed $x \in X$, $y \mapsto f(x, y)$ is lower semicontinuous; and

(b) for each fixed $y \in X$, $x \mapsto f(x, y)$ is quasi-concave.

Then $\inf_{y \in X} \sup_{x \in X} f(x, y) \leq \sup_{x \in X} f(x, x)$.

Following Tan et al. in [515], we have the following definition.

Definition 2.5.6. We shall call such a point $\hat{y}$ in Theorem 2.5.4 as *Ky Fan* *point* (in short, KF point) of f in X and denote by $S(f)$ the set of all KF points of f in X.

Thus $S(f)$ is non-empty by Theorem 2.5.4. Also, $S(f) = \cap_{x\in X}\{y \in X : f(x,y) \le 0\}$ by condition (ii) of Theorem 2.5.4 is closed in X and is thus also compact. Therefore, for each function $f : X \times X \to \mathbb{R}$ satisfying conditions (i), (ii) and (iii) of Theorem 2.5.4, one can associate a non-empty compact subset $S(f)$, the set of all solutions $y \in X$ of the inequality $\sup_{x\in X} f(x,y) \le 0$.

In this section, we shall first discuss the stability of $S(f)$ with f varying where f is a bounded real-valued function on $X \times X$ satisfying the conditions (i), (ii) and (iii) in Theorem 2.5.4 and X is a non-empty compact convex subset of a topological space. Next, if X is a Čech-complete space which belongs to the class $\mathcal{L}$ (see the definition below), we shall study the stability of the set $S(A,f) = \{y \in A : \sup_{x\in A} f(x,y) \le 0\}$ with both f and A varying, where $f : X \times X \to \mathbb{R}$ is bounded and lower semicontinuous and A is a non-empty compact subset of X. When X is a closed convex subset of a Frechet space, as an application, the stability of the set $S(A,f)$ is investigated, where $f : X\times X \to \mathbb{R}$ satisfies conditions (i) and (iii) of Theorem 2.5.4 and the subset A is, in addition, convex.

2.5.1 Stability Results in Compact Setting

Throughout this section, X denotes a non-empty compact convex subsets of a topological vector space. Let $L(X)$ be the family of all bounded real-valued functions on $X \times X$. For $f, g \in L(X)$, define

$$\rho(f,g) = \sup_{x,y\in X} |f(x,y) - g(x,y)|.$$

Clearly, $(L(X), \rho)$ is a complete metric space. Let

$$M = \{f \in L(X) : f \text{ satisfies conditions (i), (ii) and (iii) of Theorem 2.5.4}\}.$$

It is easy to show that M is closed in $L(X)$. Thus we have:

Lemma 2.5.7. (M,ρ) is a complete metric space.

Now for each $f \in M$, the set $S(f) = \{y \in X\colon \sup_{x\in X} f(x,y) \le 0\}$ is non-empty and compact by Theorem 2.5.4. Furthermore, we have:

Lemma 2.5.8. $F : M \to K(X)$ is upper semicontinuous.

Proof. Let $\{(f_\alpha, y_\alpha)\}_{\alpha\in\Gamma}$ be a net in graph S with $(f_\alpha, y_\alpha) \to (f_0, y_0) \in M \times X$, then $f_\alpha \to f_0$, $y_\alpha \to y_0$ and $f_\alpha(x, y_\alpha) \le 0$ for all $\alpha \in \Gamma$ and for all $x \in X$. Fix $x \in X$. Since $y \mapsto f_0(x,y)$ is lower semicontinuous at y_0, for any $\epsilon > 0$, there exists an open neighborhood $O(y_0)$ of y_0 in X such that for each $y' \in O(y_0)$, $f_0(x, y_0) < f_0(x, y') + \epsilon/4$. As $f_\alpha \to f_0$, there exists $\alpha_0 \in \Gamma$ such that for any $\alpha \geq \alpha_0$, $\rho(f_0, f_\alpha) < \epsilon/4$ so that for each $y' \in O(y_0)$, $f_\alpha(x, y_0) < f_0(x, y_0) + \epsilon/4 < f_0(x, y') + \epsilon/2 < f_\alpha(x, y') + 3\epsilon/4$. Since $y_\alpha \to y_0$, there exists $\alpha_1 \geq \alpha_0$ such that $y_{\alpha_1} \in O(y_0)$; it follows that $f_0(x, y_0) = f_0(x, y_0) - f_{\alpha_1}(x, y_0) + f_{\alpha_1}(x, y_0) - f_{\alpha_1}(x, y_{\alpha_1}) + f_{\alpha_1}(x, y_{\alpha_1}) < \rho(f_0, f_{\alpha_1}) + 3\epsilon/4 < \epsilon$. Since $\epsilon > 0$ is arbitrary, $f_0(x, y_0) \leq 0$ for all $x \in X$. This implies that $(f_0, y_0) \in \text{graph}S$ and hence graph S is closed in $M \times X$. Therefore S is upper semicontinuous since X is compact. □

Definition 2.5.9. For each $f \in M$, (i) a point $y \in S(f)$ is KF-essential relative to M if for each open neighborhood $N(y)$ of y in X, there exists an open neighborhood $O(f)$ of f in M such that $S(f') \cap N(y) \neq \emptyset$ for each $f' \in O(f)$; (ii) f is weakly essential relative to M if there exists $y \in S(f)$ which is KF-essential relative to M and (iii) f is essential relative to M if every $y \in S(f)$ is KF-essential relative to M.

The following result is due to Fort [178].

Lemma 2.5.10. If X is metrizable, Z is a Baire space and $S : Z \to K(X)$ is an usco mapping, then the set of points where S is lower semicontinuous is a dense G_δ set in Z.

Theorem 2.5.11. (i) S is almost lower semicontinuous at $f \in M$ if and only if f is weakly essential relative to M.

(ii) S is lower semicontinuous at $f \in M$ if and only if f is essential relative to M.

(iii) S is continuous at $f \in M$ if and only if f is essential relative to M.

Proof. (i) S is almost lower semicontinuous at $f \in M$ if and only if there exists $y \in S(f)$ such that y is KF-essential relative to M if and only if f is weakly essential relative to M.

(ii) S is lower semicontinuous at $f \in M$ if and only if each $y \in S(f)$ is KF-essential relative to M if and only if f is essential relative to M.

(iii) This follows from (ii) and Lemma 2.5.8. □

If X is metrizable by a metric d, then the Vietoris topology on $K(X)$ coincides with the topology generated by the Hausdorff metric h induced by d (e.g., see Corollary 4.2.3 of [317]). Then S is continuous at $f \in M$ if and only if for each $\epsilon > 0$, there is $\delta > 0$ such that for each $g \in M$, $h(S(f), S(g)) < \epsilon$ whenever $\rho(f, g) < \delta$; i.e., $S(f)$ is stable: $S(g)$ is "close" to $S(f)$ whenever g is "close" to f. Theorem 2.5.4 (iii) shows that $S(f)$ is stable if and only if f is essential relative to M.

We shall give a sufficient condition that $f \in M$ is essential relative to M:

Theorem 2.5.12. If $f \in M$ is such that $S(f)$ is a singleton set, then f is essential relative to M.

Proof. Suppose $S(f) = \{x\}$. Let G be any open set in X such that $S(f) \cap G \neq \emptyset$, then $x \in G$ so that $S(f) \subset G$. Since S is upper semicontinuous at f by Lemma 2.5.8, there is an open neighborhood $O(f)$ of f in M such that $S(f') \subset G$ for each $f' \in O(f)$; in particular, $G \cap S(f') \neq \emptyset$ for each $f' \in O(f)$. Thus S is lower semicontinuous at f. By Theorem 2.5.11 (ii), f is essential relative to M. □

Theorem 2.5.13. (i) Suppose that X belongs to class $\mathcal{L}$. Then there exists a dense G_δ subset Q of M such that f is weakly essential relative to M for each $f \in Q$.

(ii) Suppose that X is metrizable. Then there exists a dense G_δ subset Q of M such that f is essential relative to M for each $f \in Q$.

Proof. (i) Since M is a complete metric space, M is Čech-complete. By Lemma 2.5.8, the mapping $S : M \to K(X)$ is upper semicontinuous. Since X belongs to class $\mathcal{L}$, S is almost lower semicontinuous on some dense G_δ subset Q of M. By Theorem 2.5.11 (i), f is weakly essential relative to M for each $f \in Q$.

(ii) By Lemma 2.5.8 and Lemma 2.5.10, S is lower semicontinuous on some dense G_δ subset Q of M. By Theorem 2.5.11 (ii), f is essential relative to M for each $f \in Q$. □

We remark that if we define $M = \{f \in L(X) : f \text{ satisfies conditions (a) and (b) of Theorem 2.5.5 }\}$ and $S(f) = \{y \in X : \sup_{x\in X} f(x,y) \leq \sup_{x\in X} f(x,x)\}$ for each $f \in M$, then all the results in this section remain valid.

2.5.2 Stability Results in Non-Compact Setting

In Section 2.5.1, we studied the stability of the solution set $S(f)$ of f in X with f varying but X fixed. In this section we shall study the stability of the solution set $S(f) \cap A$ of f in A with both f and A varying.

Throughout this section, X denotes a topological space and $L(X)$ denotes the space of all bounded real-valued lower semicontinuous functions on $X \times X$. For each $f_1, f_2 \in L(X)$, let $\rho(f_1, f_2) = \sup_{(x,y)\in X\times X} |f_1(x,y) - f_2(x,y)|$, then clearly ρ is a metric on $L(X)$.

Remark 2.5.14. Let $Y = K(X) \times L(X)$. Now for each $u = (A, f) \in Y$, by Definition 2.5.6, a point y in A is called a KF point of f in A if $\sup_{x\in A} f(x,y) \leq 0$. Before we study the stability of the set $S(u)$ of KF points of f in A for $u = (A, f) \in Y$, we shall give several lemmas which will be used to prove our main results later. The proof of the following result is routine and is hence omitted.

Lemma 2.5.15. $(L(X), \rho)$ is a complete metric space.

Lemma 2.5.16. Suppose X is a non-empty subset of a topological vector space. If $\{A_\alpha\}_{\alpha\in\Gamma}$ is a net of compact and convex sets in $K(X)$ which converges to $A \in K(X)$ in the Vietoris topology, then A is also convex.

Proof. Suppose that A were not convex. Then there exist $x_1, x_2 \in A$ and $\lambda_1 \in (0,1)$ such that $\lambda_1 x_1 + (1-\lambda_1)x_2 \notin A$. Since A is compact, there exists an open set G in X containing A and an open neighborhood $O(\lambda_1 x_1 + (1-\lambda_1)x_2)$ of $\lambda_1 x_1 + (1-\lambda_1)x_2$ in X such that $O(\lambda_1 x_1 + (1-\lambda_1)x_2) \cap G = \emptyset$. Note that there exists an open neighborhood $O(x_1)$ of x_1 in X and an open neighborhood $O(x_2)$ of x_2 in X such that $\lambda_1 O(x_1) + (1-\lambda_1)O(x_2) \subset O(\lambda_1 x_1 + (1-\lambda_1)x_2)$. Since x_1, $x_2 \in A$ and $A_\alpha \to A$, there exists $\alpha_0 \in \Gamma$ such that for each $\alpha \geq \alpha_0$, $O(x_1) \cap A_\alpha \neq \emptyset$ and $O(x_2) \cap A_\alpha \neq \emptyset$. Since $G \supset A$, there exists $\alpha_1 \in \Gamma$ such that for each $\alpha \geq \alpha_1$, $G \supset A_\alpha$. Now let $\alpha_2 \in \Gamma$ be such that $\alpha_2 \geq \alpha_0$ and $\alpha_2 \geq \alpha_1$. Then for each $\alpha \geq \alpha_2$, $O(x_1) \cap A_\alpha \neq \emptyset$, $O(x_2) \cap A_\alpha \neq \emptyset$ and $A_\alpha \subset G$. Choose any $z_1 \in O(x_1) \cap A_{\alpha_2}$ and $z_2 \in O(x_2) \cap A_{\alpha_2}$. Since A_{α_2} is convex, $\lambda z_1 + (1-\lambda)z_2 \in A_{\alpha_2} \subset G$. But $\lambda z_1 + (1-\lambda)z_2 \in \lambda O(x_1) + (1-\lambda)O(x_2) \subset O(\lambda x_1 + (1-\lambda)x_2)$ which contradicts $O(\lambda_1 x_1 + (1-\lambda)x_2) \cap G = \emptyset$. Hence A must be convex. □

The following result is Lemma 3.3 of Beer [27]; as it was stated without a proof, we shall include its proof for completeness as follows.

Lemma 2.5.17. Let $\{A_\alpha\}_{\alpha\in\Gamma}$ be a net in $K(X)$ which converges to $A \in K(X)$ in the Vietoris topology. Then every net $\{x_\alpha\}_{\alpha\in\Gamma}$ with $x_\alpha \in A_\alpha$ for each $\alpha \in \Gamma$ has a cluster point in A.

Proof. Suppose that the net $\{x_\alpha\}_{\alpha\in\Gamma}$ has no cluster point in A. Then for each $x \in A$, there exists an open neighborhood $O(x)$ of x in X and an $\alpha(x) \in \Gamma$ such that $x_\alpha \notin O(x)$ for all $\alpha \geq \alpha(x)$. Since $A \subset \cup_{x\in A} O(x)$ and A is compact, there exist $x_1, x_2, \cdots, x_n \in A$ such that $A \subset \cup_{i=1}^n O(x_i)$. Now let α' be such that $\alpha' \geq \alpha(x_i)$ for $i = 1, 2, \cdots, n$. Then for each $\alpha \geq \alpha'$, $x_\alpha \notin O(x_i)$ for $i = 1, 2, \cdots, n$. Since $\cup_{i=1}^n O(x_i)$ is an open set which contains A and $A_\alpha \to A$ in the Vietoris topology, there exists $\alpha'' \in \Gamma$ such that for any $\alpha \geq \alpha''$, $x_\alpha \in A_\alpha \subset \cup_{i=1}^n O(x_i)$. Now let $\alpha''' \in \Gamma$ be such that $\alpha''' \geq \alpha'$ and $\alpha''' \geq \alpha''$; then $x_{\alpha'''} \notin O(x_i)$ for $i = 1, 2, \cdots, n$ which contradicts the fact that $x_{\alpha'''} \in A_\alpha \subset \cup_{i=1}^n O(x_i)$. Hence $\{x_\alpha\}_{\alpha\in\Gamma}$ has a cluster point in A. □

Now define the subspace M of Y by $M = \{(A, f) \in Y$: there exists $y \in A$ such that $\sup_{x\in A} f(x,y) \leq 0\}$. Then we define a mapping $S : M \to \mathcal{P}_0(X)$ by $S(u) = \{y \in A : \sup_{x\in A} f(x,y) \leq 0\}$ for each $u = (A, f) \in M$.

Lemma 2.5.18. M is closed in Y.

Proof. Suppose that $\{(A_\alpha, f_\alpha)\}_{\alpha\in\Gamma}$ is a net in M such that $(A_\alpha, f_\alpha) \to (A, f) \in Y$. For each $\alpha \in \Gamma$, let $y_\alpha \in A_\alpha$ be such that $\sup_{x\in A_\alpha} f_\alpha(x, y_\alpha) \leq 0$.

Since $A_\alpha \to A$ in the Vietoris topology, the net $\{y_\alpha\}_{\alpha\in\Gamma}$ has a cluster point $y_0 \in A$ by Lemma 2.5.17. Now we shall show that $\sup_{x\in A} f(x, y_0) \leq 0$. Suppose that this were not true, then there exists $\epsilon_0 > 0$ and $x_0 \in A$ such that $f(x_0, y_0) > \varepsilon_0$. Since f is lower semicontinuous at (x_0, y_0), there exists an open neighborhood $O(x_0)$ of x_0 in X and an open neighborhood $O(y_0)$ of point y_0 in X such that $f(x, y) > \epsilon_0$ for any $(x, y) \in O(x_0) \times O(y_0)$. Since $f_\alpha \to f$, there exists $\alpha_0 \in \Gamma$ such that for any $\alpha \geq \alpha_0$, $|f_\alpha(x, y) - f(x, y)| < \epsilon_0/2$ for all $(x, y) \in X \times X$, so that $f_\alpha(x, y) > f(x, y) - \epsilon_0/2$ for all $(x, y) \in X \times X$. Therefore $f_\alpha(x, y) > f(x, y) - \epsilon_0/2 > \epsilon_0 - \epsilon_0/2 = \epsilon_0/2$ for each $(x, y) \in O(x_0) \times O(y_0)$. As $A_\alpha \to A$, there exists $\alpha_1 \geq \alpha_0$ such that $O(x_0) \cap A_\alpha \neq \emptyset$ for all $\alpha \geq \alpha_1$. Note that $y_0 \in A$ is a cluster point of $\{y_\alpha\}_{\alpha\in\Gamma}$, there exists $\alpha_2 \geq \alpha_1$ such that $y_{\alpha_2} \in O(y_0)$. Choose any $x_{\alpha_2} \in O(x_0) \cap A_{\alpha_2}$, we have $f_{\alpha_2}(x_{\alpha_2}, y_{\alpha_2}) > \varepsilon_0/2$ which contradicts the choice of $y_{\alpha_2} \in A_{\alpha_2}$ that $\sup_{x\in A_{\alpha_2}} f_\alpha(x, y_\alpha) \leq 0$. Therefore we must have $\sup_{x\in A} f(x, y_0) \leq 0$. Hence $(A, f) \in M$ and M is closed in Y. □

Lemma 2.5.19. If X is Čech-complete, then M is Čech-complete.

Proof. The space $L(X)$ is Čech-complete since $L(X)$ is a complete metric space by Lemma 2.5.15. Since X is Čech-complete, $K(X)$ is also Čech-complete by Lemma 2.2 of Beer [27]. Therefore the product space $K(X) \times L(X)$ is Čech-complete by Theorem 3.9.8 of Engelking [154]. By Lemma 2.5.18 and Theorem 3.9.6 of Engelking [154], M is also Čech-complete. □

Lemma 2.5.20. $S(u) \in K(X)$ for each $u \in M$.

Proof. For each $u = (A, f) \in M$, since $S(u) \subset A$, it is sufficient to prove that $S(u)$ is closed in A. Let $(y_\alpha)_{\alpha\in\Gamma}$ be a net in $S(u)$ which converges to a point $y_0 \in A$. By the definition of S, we have $\sup_{x\in A} f(x, y_\alpha) \leq 0$ for each $\alpha \in \Gamma$. By the lower semicontinuity of $y \mapsto \sup_{x\in A} f(x, y)$, we have $\sup_{x\in A} f(x, y_0) \leq 0$. Hence $y_0 \in S(u)$ so that $S(u)$ is a closed subset of A. □.

Lemma 2.5.21. The correspondence $S : M \to K(X)$ is upper semicontinuous.

Proof. Suppose that S were not upper semicontinuous at some point $u = (A, f) \in M$, then there exists an open subset G of X with $G \supset S(u)$ and a net $\{u_\alpha\}_{\alpha\in\Gamma}$ in M with $u_\alpha \to u \in M$ such that for each $\alpha \in \Gamma$, there exists $y_\alpha \in S(u_\alpha)$ with $y_\alpha \notin G$. Denote $u_\alpha = (A_\alpha, f_\alpha)$ and $u = (A, f)$, then $f_\alpha \to f$ and $A_\alpha \to A$. Since $y_\alpha \in A_\alpha$ for each $\alpha \in \Gamma$, by Lemma 2.5.17, the net $\{y_\alpha\}_{\alpha\in\Gamma}$ has a cluster point $y_0 \in A$. Since $y_\alpha \notin G$ for each $\alpha \in \Gamma$, we have $y_0 \notin G$. Therefore $\sup_{x\in A} f(x, y_0) > 0$, so that there exists $\epsilon_0 > 0$ and $x_0 \in A$ such that $f(x_0, y_0) > \epsilon_0$. Since $(x, y) \mapsto f(x, y)$ is lower semicontinuous at (x_0, y_0), there exists an open neighborhood $N(x_0)$ of x_0 in X and an open neighborhood $N(y_0)$ of y_0 in X such that for each $(x, y) \in N(x_0) \times N(y_0)$, $f(x, y) > \epsilon_0$. Since $f_\alpha \to f$, there exists $\alpha_1 \in \Gamma$ such that for each $\alpha \geq \alpha_1$, $|f_\alpha(x, y) - f(x, y)| < \epsilon_0/2$ for all

$(x, y) \in X \times X$. Therefore $f_\alpha(x, y) > f(x, y) - \epsilon_0/2$ for all $(x, y) \in X \times X$. Since $N(x_0) \cap A \neq \emptyset$ and $A_\alpha \to A$, there exists $\alpha_2 \geq \alpha_1$ such that for each $\alpha \geq \alpha_2$, $N(x_0) \cap A_\alpha \neq \emptyset$. Note that because y_0 is a cluster point of the net $\{y_\alpha\}_{\alpha \in \Gamma}$ there exists $\alpha_3 \geq \alpha_2$ with $y_{\alpha_3} \in N(y_0)$. Now choose any $x_{\alpha_3} \in N(x_0) \cap A_{\alpha_3}$, we have $f_{\alpha_3}(x_{\alpha_3}, y_{\alpha_3}) > f(x_{\alpha_3}, y_{\alpha_3}) - \epsilon_0/2$. Therefore $f_{\alpha_3}(x_{\alpha_3}, y_{\alpha_3}) > f(x_{\alpha_3}, y_{\alpha_3}) - \epsilon_0/2 > \epsilon_0 - \epsilon_0/2 = \epsilon_0/2 > 0$ which contradicts the fact that $y_{\alpha_3} \in S(u_{\alpha_3})$. Therefore s must be upper semicontinuous. □

Now let M_1 be a non-empty closed subset of M.

Definition 2.5.22. For each $u \in M_1$, (i) a point $y \in S(u)$ is *KF-essential* relative to M_1 if for each open neighborhood $N(y)$ of y in X, there exists an open neighborhood $O(u)$ of u in M_1 such that $S(u') \cap N(y) \neq \emptyset$ for each $u' \in O(u)$; (ii) u is *weakly essential* relative to M_1 if there exists $y \in S(u)$ which is KF-essential relative to M_1 and (iii) u is *essential* relative to M_1 if every $y \in S(u)$ is KF-essential relative to M_1.

Theorem 2.5.23. (i) S is almost lower semicontinuous at $u \in M_1$ if and only if u is weakly essential relative to M_1.

(ii) S is lower semicontinuous at $u \in M_1$ if and only if u is essential relative to M_1.

(iii) S is continuous at $u \in M_1$ if and only if u is essential relative to M_1.

Proof. (i) S is almost lower semicontinuous at $u \in M_1$ if and only if there exists $y \in S(u)$ such that y is KF-essential relative to M_1 if and only if u is weakly essential relative to M_1.

(ii) S is lower semicontinuous at $u \in M_1$ if and only if each $y \in S(u)$ is KF-essential relative to M_1 if and only if u is essential relative to M_1.

(iii) This follows from (ii) and Lemma 2.5.21. □

A proof analogous to that of Theorem 2.5.12 and therefore omitted gives us the following result:

Theorem 2.5.24. If $u \in M_1$ is such that $S(u)$ is a singleton set, then u is essential relative to M_1.

Theorem 2.5.25. (i) Let X be Čech-complete and belong to the class $\mathcal{L}$. Then there exists a dense G_δ subset Q of M_1 such that u is weakly essential relative to M_1 for each $u \in Q$.

(ii) Let X be completely metrizable. Then there exists a dense G_δ subset Q of M_1 such that u is essential relative to M_1 for each $u \in Q$.

Proof. Note that S is an usco by Lemma 2.5.20 and Lemma 2.5.21.

(i) Since X is Čech-complete, Lemma 2.5.19 implies that M is also also Čech-complete. Since M_1 is closed in M, M_1 is also Čech-complete by Theorem 3.9.6 of Engelking [154]. Since X is of class $\mathcal{L}$, there is a dense G_δ subset Q of

M_1 such that S is almost lower semicontinuous at each $u \in Q$. By Theorem 2.5.23 (i), u is weakly essential relative to M_1 for each $u \in Q$.

(ii) By Lemma 2.5.10, there exists a dense G_δ subset Q of M_1 such that S is lower semicontinuous at each $u \in Q$. By Theorem 2.5.23 (ii), u is essential relative to M_1 for each $u \in Q$. □

If X is a complete metric space with metric d, then $K(X)$ is a complete metric space when equipped with the Hausdorff metric h induced by d. By Corollary 4.2.3 of Klein and Thompson [317, p.41], the Vietoris topology on $K(X)$ coincides with the topology induced by the Hausdorff metric h. By Lemma 2.5.15, it follows that $Y = K(X) \times L(X)$ and hence M and M_1 are also complete metric spaces when equipped with the metric D defined by

$$D(u, u') = \rho(f, f') + h(A, A')$$

for $u = (A, f)$ and $u' = (A', f')$. We note then, the mapping $S : M_1 \to K(X)$ is continuous at $u = (A, f) \in M_1$ if and only if for each $\epsilon > 0$, there is $\delta > 0$ such that $h(S(u), S(u')) < \epsilon$ whenever $u' \in M_1$ and $D(u, u') < \delta$; i.e., the solution set $S(u)$ of u is stable: $S(u')$ is close to $S(u)$ whenever u' is close to u for all $u' \in M_1$. Theorem 2.5.23 (iii) implies that if $u \in M_1$, then u is essential relative to M_1 if and only if the solution set $S(u)$ is stable.

Now let X be a non-empty closed and convex subset of a Frechet space E equipped with a translation invariant metric d. Denote

$CK(X) = \{A \in K(X) : A \text{ is convex }\}$,

$CL(X) = \{f \in L(X) : f \text{ satisfies (i) and (iii) of Theorem 2.5.4}\}$,

$M' = CK(X) \times CL(X)$.

The following is an application of the results obtained in this section.

Theorem 2.5.26. (i) M' is a non-empty closed subset of M.

(ii) There exists a dense G_δ subset Q of M' such that u is essential relative to M' for each $u \in Q$.

Proof. (i) Clearly M' is non-empty. If $u = (A, f) \in M'$, then by Ky Fan's minimax inequality Theorem 2.5.4, there exists $\hat{y} \in A$ such that $\sup_{x \in A} f(x, \hat{y}) \leq 0$; thus $u \in M$ so that $M' \subset M$. Now if $\{(A_n, f_n)\}_{n=1}^{\infty}$ is a sequence in M' such that $(A_n, f_n) \to (A, f) \in M$, then $f_n \to f$ and $A_n \to A$. Since for each $y \in X$, $x \mapsto f_n(x, y)$ is quasi-concave, it is also easy to see that $x \mapsto f(x, y)$ is also quasi-concave. By Lemma 2.5.16, A is also convex. Thus $(A, f) \in M'$ so that M' is closed in M.

Now (ii) follows from (i) and Theorem 2.5.25 (ii). □

Finally, we remark that if we define $M = \{(A, f) \in Y$: there exists $y \in A$ such that $\sup_{x \in A} f(x, y) \leq \sup_{x \in X} f(x, x)\}$, $S(u) = \{y \in A : \sup_{x \in A} f(x, y) \leq \sup_{x \in X} f(x, x)\}$ for each $u = (A, f) \in M$ and $CL(X) = \{f \in L(X) : f$ satisfies (a) and (b) of Theorem 2.5.5'$\}$, then all the results in this section remain valid.

2.5.3 Existence of Essential Components of KF Points

In this section, by following method given by Yu and Xiang [598], we discuss the existence of essential components of KF points. Throughout this section, let X be a non-empty compact and convex subset of a normed linear space E. Let M_2 be the collection of all functions $\phi : X \times X \to \mathbb{R}$ such that:

(1) for each fixed $x \in X$, $y \mapsto \phi(x, y)$ is lower semicontinuous;

(2) for each fixed $y \in X$, $x \mapsto \phi(x, y)$ is concave;

(3) $\phi(x, x) \leq 0$ for all $x \in X$; and

(4) $\sup_{(x,y)\in X\times X} |\phi(x, y)| < \infty$.

As above, we define a metric ρ on M_2 by $\rho(\phi, \psi) := \sup_{(x,y)\in X\times X} |\phi(x, y) - \psi(x, y)|$ for each $\phi, \psi \in M_2$. Clearly $M_2 \subset M$. By Theorem 2.5.4, for each $\phi \in M_2$, the set $\Phi(\phi)$ of all KY points for ϕ is non-empty and compact. We now have the following definition.

Definition 2.5.27. For each $\phi \in M_2$, let $S(\phi)$ be a non-empty closed subset of KY points set $F(\phi)$. The set $S(\phi)$ is said to be an *essential set* of $F(\phi)$ if for each open set $U \subset S(\phi)$, there exists $\delta > 0$ such that $F(\psi) \cap U \neq \emptyset$ for any $\psi \in M_2$ with $\rho(\phi, \psi) < \delta$. The subset $S(\phi)$ of $F(\phi)$ is said to be a minimal essential set of $F(\phi)$ if $S(\phi)$ is a minimal element of the family of all essential sets in $F(\phi)$ by the inclusion order relation.

We recall that the component of a point $y \in F(\phi)$ is the union of all connected subsets of $F(\phi)$ which contain the point y. Note that each component is a connected and closed subset of $F(\phi)$, thus it is also compact as $F(\phi)$ is compact. Clearly, two components of $F(\phi)$ either coincide or disjoint. Thus all components of $F(\phi)$ constitute a decomposition of $F(\phi)$ into connected pairwise disjoint compact subsets, i.e.,

$$F(\phi) := \cup_{\alpha\in\Lambda} C_\alpha(\phi), \tag{2.5.1}$$

where Λ is an index set; C_α is a non-empty connected compact and $C_\alpha(\phi) \cap C_\beta(\phi) = \emptyset$ for each $\alpha, \beta \in \Lambda$ with $\alpha \neq \beta$.

We also recall the following definition which was first given by Yu and Xiang [598].

Definition 2.5.28. For each $\phi \in M_2$, the subset $C_\alpha(\phi)$ in (2.5.1) is said to be an *essential component* of $F(\phi)$ if for each open set U containing $C_\alpha(\phi)$, there exists $\delta > 0$ such that $F(\psi) \cap U \neq \emptyset$ for any $\psi \in M_2$ with $\rho(\phi, \psi) < \delta$.

We now first prove that for each $\phi \in M_2$, there exists at least one essential connected component set for KY point set $F(\phi)$.

Theorem 2.5.29. For each $\phi \in M_2$, there exists at least one minimal essential set of $F(\phi)$.

Proof. By Lemma 2.5.21, the mapping $F : M_2 \to 2^X \setminus \{\emptyset\}$ is upper semicontinuous, i.e., for each open set $U \supset F(\phi)$, there exists $\delta > 0$ such that $U \supset F(\psi)$ for any $\psi \in M_2$ with $\rho(\phi, \psi) < \delta$. Hence $F(\phi)$ itself is an essential set. Let Φ denote the family of all essential sets of $F(\phi)$ ordered by the set inclusion relation. Then Φ is non-empty and every decreasing chain of elements in Φ has a lower bound as the intersection is still in Φ due to the compactness. Thus Zorn's Lemma implies that Φ has a minimal element and it is a minimal and essential subset of $F(\phi)$. □

Now we have the existence result which is essentially due to Yu and Xiang [598].

Theorem 2.5.30. For each $\phi \in M_2$, each minimal and essential set in $F(\phi)$ is connected.

Proof. We denote by $m(\phi)$ the minimal and essential set in $F(\phi)$. Suppose that $m(\phi)$ were not connected. Then there exists two non-empty closed sets $C_1(\phi)$ and $C_2(\phi)$ and two open sets V_1 and V_2 in X such that $V_1 \supset C_1(\phi)$ and $V_2 \supset C_2(\phi)$, and with the property that $m(\phi) = C_1(\phi) \cup C_2(\phi)$ and $V_1 \cap V_2 = \emptyset$. Since $m(\phi)$ is a minimal and essential set in $F(\phi)$, it implies that neither $C_1(\phi)$ nor $C_2(\phi)$ is essential. Thus there exist two open sets $O_1 \supset C_1(\phi)$ and $O_2 \supset C_2(\phi)$ such that for any $\delta > 0$, there exists $g_1, g_2 \in M_2$ with $\rho(\phi, g_1) < \delta$, $\rho(\phi, g_2) < \delta$, but $F(g_1) \cap O_1 = \emptyset$ and $F(g_2) \cap O_2 = \emptyset$. Let $W_1 := V_1 \cup O_1$ and $W_2 := V_2 \cup O_2$. It is clear that both W_1 and W_2 are open and $W_1 \supset C_1(\phi)$ and $W_2 \supset C_2(\phi)$. Without loss of generality, we may assume that $V_1 \supset \overline{W_1}$ and $V_2 \supset \overline{W_2}$. Since $W_1 \cup W_2 \supset m(\phi)$, there exists $\delta^* > 0$ such that $F(\psi) \cap (W_1 \cup W_2) \neq \emptyset$ for each $\psi \in M_2$ with $\rho(\phi, \psi) < \delta^*$. Thus, there exists $\psi_1, \psi_2 \in M_2$ with $\rho(\phi, \psi_1) < \delta^*$, $\rho(\phi, \psi_2) < \delta^*$, but $F(\psi_1) \cap W_1 = \emptyset$ and $F(\psi_2) \cap W_2 = \emptyset$. In order to finish the proof, we define a function $f : X \times X \to \mathbb{R}$ as follows:

$$f(x,y) = \begin{cases} \psi_1(x,y), & \text{if } x \in X \text{ and } y \in \overline{W_1}, \\ \psi_2(x,y), & \text{if } x \in X \text{ and } y \in \overline{W_2}, \\ \lambda(y)\psi_1(x,y) + \mu(y)\psi_2(x,y), & \text{if } x \in X \text{ and } y \in X \setminus (\overline{W_1} \cup \overline{W_2}). \end{cases}$$

where $\lambda(y) = \frac{d(y,\overline{W_2})}{d(y,\overline{W_1})+d(y,\overline{W_2})}$ and $\mu(y) = \frac{d(y,\overline{W_1})}{d(y,\overline{W_1})+d(y,\overline{W_2})}$. Note that $\lambda(y)$ and $\mu(y)$ are continuous, $\lambda(y) \geq 0$, $\mu(y) \geq 0$ and $\lambda(y) + \mu(y) = 1$ for all $y \in X$.

It can be routinely checked that (1): for each fixed $x \in X$, $y \mapsto f(x,y)$ is lower semicontinuous; (2): for each fixed $y \in X$, $x \mapsto f(x,y)$ is concave; (3): $f(x,x) \leq 0$; and (4): $\sup_{(x,y)\in X\times X} |f(x,y)| < \infty$. Thus $f \in M_2$. Note that $\rho(\phi, f) < \delta^*$ and $F(f) \cap (W_1 \cup W_2) \neq \emptyset$. Without loss of generality, we may assume that $F(f) \cap W_1 \neq \emptyset$. Take any $y^* \in F(f) \cap W_1$. Since $y^* \in W_1$, it follows that $\psi_1(x, y^*) = f(x, y^*) \leq 0$ for all $x \in X$. Thus $y^* \in F(\psi_1)$ and $F(\psi_1) \cap W_1 \neq \emptyset$, which contradicts the fact that $F(\psi_1) \cap W_1 = \emptyset$. This contradiction shows that $m(\phi)$ must be connected and the proof is complete. □

Theorem 2.5.31. For each $\phi \in M_2$, there exists at least one essential connected component of $F(\phi)$.

Proof. By Theorems 2.5.29 and 2.5.30, there exists at least one minimal and essential connected set $S(\phi)$. Choose any $y \in S(\phi)$ and let $C_\alpha(\phi)$ be component which contains the point y, we shall prove that $S(\phi) \subset C_\alpha(\phi)$; otherwise, there exists $C_\beta(\phi)$ such that $S(\phi) \cap C_\beta(\phi) \neq \emptyset$, where $\beta \neq \alpha$. By Corollary 6.1.10 of Engelking [154], the set $S(\phi) \cup C_\alpha(\phi) \cup C_\beta(\phi)$ is connected which contradicts the fact that $C_\alpha(\phi)$ is the union of all connected subsets of $F(\phi)$ which contain the point y. Now for each open set $O \supset C_\alpha(\phi)$ and $O \supset S(\phi)$, since $S(\phi)$ is essential, it follows that there exists $\delta > 0$ such that $F(\psi) \cap O \neq \emptyset$ for any $\psi \in M_2$ with $\rho(\phi, \psi) < \delta$. Hence $C_\alpha(\phi)$ must be essential and the proof is complete. □

We now give a sufficient condition that $\phi \in M_2$ is weakly essential or essential.

Theorem 2.5.32. If $\phi \in M_2$ is such that $F(\phi)$ is a totally disconnected set, then ϕ is weakly essential. In particular, if the set $F(\phi)$ is a singleton, then ϕ is essential.

Proof. Since $F(\phi)$ is a totally disconnected set, then $F(\phi) = \cup_{\alpha \in \Lambda} C_\alpha(\phi)$, where $C_\alpha(\phi)$ is a singleton set for each $\alpha \in \Lambda$. By Theorem 2.5.31, there exists $C_{\alpha_0}(\phi) = \{y_0\}$, which is an essential component of $F(\phi)$. It is clear that y_0 is essential and thus it is weakly essential. In the case, $F(\phi) = \{y\}$ is a singleton set, then y is essential and hence ϕ is essential and the proof is complete. □

Results established in this part, in particular Theorem 2.5.31, will be used in Chapter 8 to prove the existence of essential and connected components of Nash equilibra for noncooperative games by following the way of Yu and Xiang in [598].

2.6 Coincident Points in Topological Vector Spaces

The natural extension of fixed point theory is the study of coincidence points. Let X and Y be topological space and $S, T : X \to 2^Y$. The *coincidence problem* for (S, T) is to find $(x_0, y_0) \in X \times Y$ such that $y_0 \in S(x_0) \cap T(x_0)$. Geometrical problems of this type in an approximate context turn out to be intimately related to some basic problems arising in convex analysis. This important fact was discovered by J. von Neumann in 1937, who established a coincidence theorem in $\mathbb{R}^n$ which was then applied to prove his well-known minimax principle. Since then, geometrical problems of a similar kind (as well as their analytic counterparts) have attracted broad attention. Also, new applications in various mathematical areas have been found. In particular, since Eilenberg and Montgomery [152] studied coincidence theory in topological settings in 1946, this topic has been comprehensively developed by the contributions of Kakutani [280], Nash [380], Fan[156], Kneser [318], Gale [188], Debreu [115], Nikaido [388], Sion [490], Gorniewicz, Granas [202] and Kryszewski [203], Granas and Liu [208], Chang and Song [71], Ben-El-Medchaiek and Deguire [35], Ko and Tan [320], Powers [432] and other contributors. This topic has many applications in mathematics and other subjects, for example, see Aubin [10], Aubin and Cellina [14] and Zeidler [613].

In this section, we first consider the relations between Halpern's inward (resp., outward) mappings in [216] and Fan's inward (resp., outward) definitions [162]. Several facts involved in the study of fixed point theorems for non-self mappings are also exhibited. Next a general set-valued version of Fan's best approximation theorem [162] is given in topological vector space. As applications, a number of approximation theorems, fixed point theorems and coincidence theorems are given in topological vector spaces. These results improve or unify most of the well known results in Browder [61], Fan [157], [162], [164], [167], Komiya [326], Park [421], Reich [433], Halpern and Bergman [218], and Ha [214].

2.6.1 Some Facts

In this section, some facts concerning boundary conditions which are involved in the study of non-self mappings are exhibited. We first recal that if X and Y are topological spaces and $A : X \to 2^Y$, then A is continuous if A is both upper semicontinuous and lower semicontionous. If Y is a non-empty subset of a topological vector space, then A is convex valued (resp., closed convex valued) if $A(x)$ is convex (resp., closed and convex) for each $x \in X$.

The following definitions are due to Halpern (e.g., see [216]).

Definition 2.6.1. Let X be a non-empty subset of a vector space E and $y \in E$. Then the inward set $I_X(y)$ and the outward set $O_X(y)$ of X at y are defined by

$I_X(y) = \{x \in E :$ there exist $u \in X$ and $r > 0$ such that $x = y + r(u - y)\}$,

and

$O_X(y) = \{x \in E :$ there exists $u \in X$ and $r > 0$ such that $x = y - r(u - y)\}$.

If E is a topological vector space, then the closure of $I_X(y)$ and $O_X(y)$ in E, denoted by $\overline{I_X}(y)$ and $\overline{O_X}(y)$ respectively, are called the weakly inward set and weakly outward set of X at y respectively.

Definition 2.6.2. Let X be a non-empty subset of a topological vector space E. Then a mapping $T : X \to 2^E$ is called (i) *inward* (resp., *outward*) if for each $x \in X$, $T(x) \cap I_X(x) \neq \emptyset$ (resp., $T(x) \cap O_X(x) \neq \emptyset$); and (ii) *weakly inward* (resp., *weakly outward*) if for each $x \in X$, $T(x) \cap \overline{I_X}(x) \neq \emptyset$ (resp., $T(x) \cap \overline{O_X}(x) \neq \emptyset$).

We note that if T is a (weakly) inward mapping, then the mapping $G : X \to 2^E$ defined by $G(x) = 2x - T(x)$ for each $x \in X$ is (weakly) outward and vice versa. Also, x is a fixed point of T if and only if it is a fixed point of G. Hence fixed point results for (weakly) inward mappings are equivalent to such results for (weakly) outward mappings. Thus we shall mainly give details of proofs for (weakly) inward mappings.

Definition 2.6.3. Let X be a non-empty convex set in a (real or complex) vector space E. Following Fan [162], the algebraic boundary $\delta_E(X)$ of X in E is the set of all $x \in X$ for which there exists $y \in E$ such that $x + ry \notin X$ for all $r > 0$. If X is a subset of a topological vector space, the *topological boundary* $\partial_E(X)$ is the complement of $\text{int}_E X$ in E. It is easy to see that $\delta_E(X) \subset \partial_E(X)$ and in general $\delta_E(X) \neq \partial_E(X)$ as there exists a convex subset X of a topological vector space E such that $\partial_E(X) = X$ while $\delta_E(X) \neq X$, e.g., see [557, Example 4 in Chapter 3].

Definition 2.6.4. Let E and W be two topological vector spaces and X a non-empty subset of E. Let $G, F : X \to 2^W$. A point $x \in X$ is said to be a *coincidence point* of G and F if $G(x) \cap F(x) \neq \emptyset$.

Definition 2.6.5. Let E be a topological vector space and E^* be its continuous dual. The space E is said to have *sufficiently many continuous linear functionals* if for each $x \in E$ with $x \neq 0$, there exists $\phi \in E^*$ such that $Re\phi(x) \neq 0$, i.e., E^* separates points in E.

By the Hahn-Banach theorem, if E is a locally convex topological vector space, then E has sufficiently many continuous linear functionals. There are topological vector spaces with sufficiently many continuous linear functionals which are not locally convex, e.g., the Hardy space H^p, $0 < p < 1$.

Since most fixed point theorems of inward (outward) mappings depend on the boundary conditions of the domains, for example see Halpern and Bergman [218], Browder [57] and Fan [162] and Park [419], it is our purpose in this section to discuss the relations between various boundary conditions appearing in the literature. Following the idea of Fan [162], we first have the following result.

Proposition 2.6.6. Let X be a non-empty convex subset of a vector space E and $F : X \to 2^E$. Then the following two conditions (a) and (b) are equivalent:

(a) for each $x \in \delta_E(X)$, there exists $y \in X$, $u \in F(x)$ and $r > 0$ such that $u - x = r(y - x)$.

(b) for each $x \in X$, there exists $u \in F(x)$ and $r \in (0, 1)$ such that $rx + (1 -$

$r)u \in X$.

Proof. (a) $\Longrightarrow$ (b). Fix an arbitrary $x \in X$. If there exists $u \in F(x)$ such that $u \in X$, then because X is convex, $rx + (1-r)u \in X$ holds for every $r \in (0,1)$. Now assume $F(x) \subset E \setminus X$. (1) If $x \in \delta_E(X)$, then by (a), there exist $y \in X$ and $u \in F(x)$ and $r > 0$ such that $u - x = r(y - x)$. It follows that $ry + (1-r)x = u \notin X$. Since $y, x \in X$ and X is convex, we must have $r > 1$. Let $\lambda = \frac{r-1}{r}$, then $\lambda \in (0,1)$ and $\lambda x + (1-\lambda)u = y \in X$. (2) If $x \notin \delta_E(X)$, then by the definition of $\delta_E(X)$, for the point $u - x$, where $u \in F(x)$, there exist $r > 0$ such that $y = x + r(u-x) \in X$. Since $x, y \in X$ and $u = \frac{1}{r}y + \frac{r-1}{r}x \notin X$, we must have $r < 1$. Let $\lambda = 1 - r$, then we have $\lambda x + (1-\lambda)u = y \in X$. Thus condition (b) is verified.

(b) $\Longrightarrow$ (a). Suppose $x \in \delta_E(X)$. By (b), there exists $u \in F(x)$ and $r \in (0,1)$ such that $y = rx + (1-r)u$. If we take $\lambda = \frac{1}{1-r}$, then $\lambda > 0$ and $u - x = \lambda(y - x)$. □

Proposition 2.6.6 improves the result given by Fan [162] to set-valued mappings. Fan also gave the following result in [162].

Proposition 2.6.7. Let X be a non-empty convex subset of a vector space E. Suppose $F, G : X \to 2^E$. Then the following two conditions are equivalent:

(a) for each point $x \in \delta_E(X)$, there exist three points $y \in X$, $u \in F(x)$, $v \in G(x)$ and a real number $r > 0$ such that $y - x = r(u - v)$.

(b) for each $x \in X$, there exist three points $y \in X$, $u \in F(x)$, $v \in G(x)$ and a real number $r > 0$ such that $y - x = r(u - v)$.

By Proposition 2.6.6 and Proposition 2.6.7, we have the following:

Proposition 2.6.8. Let X be a non-empty convex subset of a vector space E. Suppose $F, G : X \to 2^E$. Then the following are equivalent:

(a) for each $x \in \delta_E(X)$, there exists $y \in X$, $u \in F(x)$, $v \in G(x)$ and $r > 0$ such that $y - x = r(u - v)$.

(b) for each $x \in X$, there exists $y \in X$, $u \in F(x)$, $v \in G(x)$ and $r > 0$ such that $y - x = r(u - v)$.

(c) for each $x \in X$, there exists $u \in F(x)$, $v \in G(x)$ and $r \in (0,1)$ such that $x + (1-r)(u - v) \in X$.

Proof. By Proposition 2.6.7, condition (a) is equivalent to condition (b).

Now define $W : X \to 2^E$ by $W(x) = \{x\} + F(x) - G(x)$ for each $x \in X$. Then it is clear that the condition (a) is equivalent to the following:

$(a)'$ for each $x \in \delta_E(X)$, there exists $y \in X$, $u \in W(x)$ and $r > 0$ such that $y - x = r(u - x)$.

Now by Proposition 2.6.6, the condition $(a)'$ is equivalent to the following condition:

$(c)'$ for each $y \in X$, there exists $u \in W(y)$ and $r \in (0,1)$ such that $ry + (1-$

$r)u \in X$.

By the definition of W, it is also obvious that the condition (c)' is equivalent to the condition (c). Therefore conditions (a), (b) and (c) are equivalent. □

We also have the following result.

Proposition 2.6.9. Let X be a non-empty subset of a vector space E and $F, G : X \to 2^E$. Then the following two conditions are equivalent:

(i) for each $x \in \delta_E(X)$ and $\phi \in E^*$ such that $Re\phi(x) \leq Re\phi(y)$ for all $y \in X$, there exists $u \in F(x)$ and $v \in G(x)$ with $Re\phi(u) \geq Re\phi(v)$.

(ii) for each $x \in X$ and $\phi \in E^*$ such that $Re\phi(x) \leq Re\phi(y)$ for all $y \in X$, there exists $u \in F(x)$ and $v \in G(x)$ with $Re\phi(u) \geq Re\phi(v)$.

Proof. We only need to show that (i) $\Longrightarrow$ (ii). Suppose $x \in X$. Let $\phi \in E^*$ be such that $Re\phi(x) \leq Re\phi(y)$ for all $y \in X$. If $x \notin \delta_E(X)$, then by the definition of $\delta_E(X)$, for each $y \in E$, there exist $r > 0$ such that $x + ry \in X$; it follows that $Re\phi(x) \leq Re\phi(x + ry)$, so that $Re\phi(y) \geq 0$ for all $y \in E$, and therefore ϕ is necessarily the zero linear functional. Since $Re\phi \equiv 0$, $Re\phi(u) \geq Re\phi(v)$ is satisfied for any $u \in F(x)$ and $v \in G(x)$. Next, if $x \in \delta_E(X)$, then for any $\phi \in E^*$ satisfying $Re\phi(x) \leq Re\phi(y)$ for all $y \in X$, by (i), there exist $u \in F(x)$ and $v \in G(x)$ such that $Re\phi(u) \geq Re\phi(v)$. Thus the condition (ii) is verified. □

Let X be a non-empty closed convex subset of a topological vector space E. We remark that:

(1) Corollary 6.3.1 of Rockafellar [444, p.47] implies that Fan's definition of *algebraic boundary* $\delta_E(X)$ of X in E is equivalent to the definition given by Browder [57, p.285]:

$\delta_E(X) = \{x \in X :$ there exists a finite dimensional flat M such that $x \in \partial_E(X \cap M)$;

and

(2) if $\text{int}_E(X) \neq \emptyset$, then $\partial_E(X) = \delta_E(X)$ (see, e.g., [557, Theorem 2.27(a)].

Therefore our Proposition 2.1.3 really generalizes Fan's result [167]. Since Halpern gave the definitions of inward (outward) mappings in his Ph.D. thesis [216] (e.g., see Halpern and Bergman [218]), many fixed point theorems were obtained. In [164], Fan also gave another definition for the inward (outward) mappings as follows.

Definition 2.6.10. Let X be a non-empty convex subset of a topological vector space E and $F : X \to 2^E$. The mapping F is an inward mapping (resp., outward mapping) if for each $x \in X$ and any continuous linear functional ϕ on E such that $Re\phi(x) \leq \inf_{y \in X} Re\phi(y)$, there exists a point $u \in F(x)$ such that $Re\phi(u) \geq Re\phi(x)$ (resp., $Re\phi(u) \leq Re\phi(x)$).

It is easy to see that the condition (a) (resp., (b)) of Proposition 2.6.2 implies condition (i) (resp., (ii)) of Proposition 2.6.9. Therefore Fan's definition of inward mappings (resp., outward mappings) in a topological vector space includes the definition of inward mappings (resp., outward mappings) for a single (or set-valued) mapping given by Halpern (also see Fan [162], and Halpern and Bergman [218]).

Let X be a non-empty convex subset of a topological vector space E and $W, T, F : X \to 2^E$. Consider the following conditions:

(a) for each $x \in X$, there is $y \in W(x)$ and $r \in [0, 1)$ such that $rx + (1 - r)y \in X$.

(b) for each $x \in X$, there is $y \in X$, $u \in T(x)$, $v \in F(x)$ and $r \in (0, 1]$ such that $y - x = r(u - v)$.

Then it is not difficult to see that T and F satisfy the condition (b) if and only if the mapping $W = I + T - F$ satisfies the condition (a); and moreover, $x \in W(x)$ if and only if $T(x) \cap F(x) \neq \emptyset$.

2.6.2 Fan's Best Approximation Theorems

In this section, we will give a general extension of Fan's best approximation theorem [162] in topological vector spaces. As applications, several coincidence theorems are derived which in turn imply some fixed point theorems. We first need the following result which is contained in the proof of Theorem 2 and the remark immediately following its proof in Ha [214].

Theorem 2.6.11. Let X be a non-empty compact subset of a topological vector space E which has sufficiently many continuous linear functionals and $F : X \to 2^E$ be upper semicontinuous with compact and convex values. If F has no fixed point, then there exists $\delta > 0$ and a continuous seminorm P on E such that $\inf_{x \in X} \inf_{u \in F(x)} P(x - u) \geq \delta$.

We shall need also the following Lemma 2 of Ha [213] (see also its topological version which is Lemma 2.9.3 in Section 9).

Theorem 2.6.12. Let Z be an n-simplex and let K be a non-empty compact convex subset of a topological vector space. If $A : Z \to 2^K$ is upper semicontinuous with closed and convex values and $p : K \to Z$ is continuous, then there exists $x_0 \in Z$ such that $x_0 \in p(A(x_0))$.

We shall now prove the following coincidence theorem in topological vector spaces.

Theorem 2.6.13. Let X be a contractible space and Y be a compact convex subset of a topological vector space E. Let $A : X \to 2^Y$ be upper semicontinuous with closed and convex values. Suppose that $B : Y \to 2^X$ is such that:

(a) $B^{-1}(x)$ is open for each $x \in X$; and
(b) for each open set O in Y, the set $\cap_{y\in O}B(y)$ is empty or contractible.

Then there exists $w_0 \in X$ and $z_0 \in Y$ such that $w_0 \in B(z_0)$ and $z_0 \in A(w_0)$.

Proof. We first show that there exists an n-simplex Δ_N and two functions $f : \Delta_N \to X$ and $\psi : Y \to \Delta_N$ such that $f(\psi(y)) \in B(y)$ for all $y \in Y$. Since Y is compact, by (a), there exists a finite subset $\{x_0, \cdots, x_n\}$ of X such that $Y = \cup_{i=0}^{n} B^{-1}(x_i)$. Now for each non-empty subset J of $N := \{0, \cdots, n\}$, we define

$$F_J = \begin{cases} \cap\{B(y) : y \in \cap_{j\in J}B^{-1}(x_j)\}, & \text{if } \cap_{j\in J}B^{-1}(x_j) \neq \emptyset, \\ X, & \text{otherwise.} \end{cases}$$

Note that if $y \in \cap_{j\in J}B^{-1}(x_j)$, then $\{x_j : j \in J\} \subset B(y)$. Therefore by (b), if $\cap_{j\in J}B^{-1}(x_j) \neq \emptyset$, then $F_J = \cap\{B(y) : y \in \cap_{j\in J}B^{-1}(x_j)\}$ is non-empty and contractible. It is clear that $F_J \subset F_{J'}$ whenever $\emptyset \neq J \subset J' \subset \{0, \cdots, n\}$. Thus F satisfies all hypotheses of Lemma 2.2.5. By Lemma 2.2.5, there is a continuous function $f : \Delta_N \to X$ such that $f(\Delta_J) \subset F_J$ for all $J \in \mathcal{F}(N)$. Let $\{\psi_i : i \in N\}$ be a continuous partition of unity subordinated to the covering $\{B^{-1}(x_i) : i \in N\}$, i.e., for each $i \in N$, $\psi_i : Y \to [0,1]$ is continuous, $\{y \in Y : \psi_i(y) \neq 0\} \subset B^{-1}(x_i)$ such that $\Sigma_{i=0}^{n}\psi_i(y) = 1$ for all $y \in Y$. Define $\psi : Y \to \Delta_N$ by $\psi(y) = (\psi_0(y), \psi_1(y), \cdots, \psi_n(y))$ for each $y \in Y$. Then $\psi(y) \in \Delta_{J(y)}$ for all $y \in Y$, where $J(y) = \{i \in \{0, \cdots, n\} : \psi_i(y) \neq 0\}$. Therefore $f(\psi(y)) \in f(\Delta_{J(y)}) \subset F_{J(y)} \subset B(y)$. On the other hand, since A is upper semicontinuous with closed and convex values and f is continuous, the composition $A \circ f : \Delta_N \to 2^Y$ is also upper semicontinuous with closed and convex values, and $\psi : Y \to \Delta_N$ is continuous. By Theorem 2.6.12, there exists $x_0 \in \Delta_N$ such that $x_0 \in \psi \circ (A \circ f(x_0))$. Let $w_0 := f(x_0)$, then $w_0 = f(x_0) \in f \circ (\psi \circ (A \circ f(x_0)) = f \circ (\psi \circ (A(w_0))$ so that there exists $z_0 \in A(w_0)$ is such that $w_0 = f \circ \psi(z_0) \in B(z_0)$. □

As a special case of Theorem 2.6.13, we have the following result which is Theorem 1 of Komiya [326].

Corollary 2.6.14. Let X be a non-empty convex subset of a topological vector space E and Y be a non-empty compact convex subset of a topological vector space W. Suppose $A : X \to 2^Y$ is upper semicontinuous with closed and convex values and $B : Y \to 2^X$ has convex values such that $B^{-1}(x)$ is open in Y for each $x \in X$. Then there exists $(x_0, y_0) \in X \times Y$ such that $x_0 \in B(y_0)$ and $y_0 \in A(x_0)$.

We now prove a set-valued generalization of Fan's best approximation theorem [162, Theorem 2] in topological vector spaces.

Theorem 2.6.15. Let E and W be two topological vector spaces. Let X be a non-empty compact convex subset of E and $G, F : X \to 2^W$. Suppose further there exists continuous $H : X \times W \to \mathbb{R}$ such that:

(i) for each fixed $x \in X$, the set $\{y \in W : H(x,y) < r\}$ is convex for each $r \in \mathbb{R}$;

(ii) G is continuous with compact values and the set $G^{-1}(y)$ is convex for each $y \in G(X)$, where $G(X)$ is a convex set; and

(iii) F is continuous with compact values.

Then there exists a point $x_0 \in X$ such that

$$\inf_{v\in G(X)}[\inf_{u\in F(x_0)} H(x_0, v-u)] = \inf_{v\in G(x_0)}[\inf_{u\in F(x_0)} H(x_0, v-u)].$$

Proof. Suppose that the conclusion were false. Define two mappings $A : G(X) \to 2^X$ and $B : X \to 2^{G(X)}$ by $A(y) = G^{-1}(y)$ for each $y \in G(X)$ and $B(x) = \{y \in G(X) : \inf_{u\in F(x)} H(x, y-u) < \inf_{v\in G(x)} \inf_{u\in F(x)} H(x, v-u)\}$ for each $x \in X$. Then by assumption, $B(x) \neq \emptyset$ for each $x \in X$. Since G is upper semicontinuous with closed values, the graph of G is closed in $E \times W$, graph A is also closed in $W \times E$. Since X is compact, the mapping $A : G(X) \to 2^X$ is upper semicontinuous and has closed convex values. Since F is continuous with compact values and $H : X \times W \to \mathbb{R}$ is continuous, Theorem 1 of Aubin [10, p.67] and Theorem 2 of Aubin and Ekeland [15, p.69] imply that $(x,y) \mapsto \inf_{u\in F(x)} H(x, y-u) - \inf_{v\in G(x)} \inf_{u\in F(x)} H(x, v-u)$ is continuous. It follows that for each $y \in G(X)$, the set $B^{-1}(y) = \{x \in X : \inf_{u\in F(x)} H(x, y-u) < \inf_{v\in G(x)} \inf_{u\in F(x)} H(x, v-u)\}$ is open in X. Since $G(X)$ is convex, by (i) $B(x)$ is convex for each $x \in X$. Therefore A and B satisfy all hypotheses of Corollary 2.6.14. By Corollary 2.6.14, there exist $x_0 \in X$ and $y_0 \in G(X)$ such that $x_0 \in A(y_0)$ and $y_0 \in B(x_0)$, i.e. $y_0 \in G(x_0)$ and

$$\inf_{u\in F(x_0)} H(x_0, y_0 - u) < \inf_{v\in G(x_0)}[\inf_{u\in F(x_0)} H(x_0, v-u)]$$

which is impossible. Therefore the conclusion must hold. □

If the mapping G in Theorem 2.6.15 is single-valued, then we have the following result.

Theorem 2.6.16. Let E and W be two topological vector spaces. Let X be a non-empty compact convex subset of E, $F : X \to 2^W$ and $G : X \to W$. Suppose further there exists a continuous function $H : X \times W \to \mathbb{R}$ such that:

(i) for each fixed $x \in X$, $y \to H(x,y)$ is convex;

(ii) G is continuous and the set $G^{-1}(y)$ is convex for each $y \in G(X)$, where $G(X)$ is convex; and

(iii) F is continuous with compact and convex values.

Then there exists a point $x_0 \in X$ such that

$$\inf_{w\in \overline{I_{G(X)}}(G(x_0))} \inf_{u\in F(x_0)} H(x_0, w-u) = \inf_{u\in F(x_0)} H(x_0, G(x_0) - u).$$

In particular, if $\inf_{u\in F(x_0)} H(x_0, G(x_0) - u) > 0$ and $H(x, ry) \le rH(x, y)$ for each $r \in (0,1)$ and each $(x, y) \in X \times W$, then $G(x_0) \in \delta_E(G(X))$.

Proof. By Theorem 2.6.15, there exists $x_0 \in X$ such that

$$\inf_{v\in G(X)} \inf_{u\in F(x_0)} H(x_0, v - u) = \inf_{u\in F(x_0)} H(x_0, G(x_0) - u).$$

we shall prove that

$$\inf_{w\in \overline{I_{G(X)}}(G(x_0))} \inf_{u\in F(x_0)} H(x_0, w - u) \ge \inf_{u\in F(x_0)} H(x_0, G(x_0) - u).$$

Fix an arbitrary $w \in I_{G(X)}(G(x_0)) \setminus G(X)$. As $G(X)$ is convex, there exist $z \in G(X)$ and $r > 1$ such that $w = G(x_0) + r(z - G(x_0))$. Suppose that

$$\inf_{u\in F(x_0)} H(x_0, w - u) < \inf_{u\in F(x_0)} H(x_0, G(x_0) - u).$$

Since $F(x_0)$ is compact and H is continuous, there exist $z_1, z_2 \in F(x_0)$ such that $H(x_0, w - z_1) = \inf_{u\in F(x_0)} H(x_0, w - u) < \inf_{u\in F(x_0)} H(x_0, G(x_0) - u) = H(x_0, G(x_0) - z_2)$. Let $\overline{z} = (1 - \frac{1}{r})z_2 + \frac{1}{r}z_1$, then $\overline{z} \in F(x_0)$ since $F(x_0)$ is also convex. Therefore we have

$$\begin{aligned} H(x_0, G(x_0) - z_2) &= \inf_{u\in F(x_0)} H(x_0, G(x_0) - u) = \inf_{v\in G(X)} \inf_{u\in F(x_0)} H(x_0, v - u) \\ &\le H(x_0, z - \overline{z}) = H(x_0, \frac{1}{r}w + (\frac{r-1}{r})G(x_0) - \frac{1}{r}z_1 - (\frac{r-1}{r})z_2) \\ &\le \frac{1}{r}H(x_0, w - z_1) + \frac{r-1}{r}H(x_0, G(x_0) - z_2) \\ &< H(x_0, G(x_0) - z_2) \end{aligned}$$

which is a contradiction. Thus we must have

$$\inf_{w\in I_{G(X)}(G(x_0))} \inf_{u\in F(x_0)} H(x_0, w - u) \ge \inf_{u\in F(x_0)} H(x_0, G(x_0) - u).$$

By the continuity of $w \to \inf_{u\in F(x_0)} H(x_0, w - u)$, we have

$$\inf_{w\in \overline{I_{G(X)}}(G(x_0))} \inf_{u\in F(x_0)} H(x_0, w - u) \ge \inf_{u\in F(x_0)} H(x_0, G(x_0) - u).$$

Hence

$$\inf_{w\in \overline{I_{G(X)}}(G(x_0))} \inf_{u\in F(x_0)} H(x_0, w - u) = \inf_{u\in F(x_0)} H(x_0, G(x_0) - u).$$

If $\inf_{u\in F(x_0)} H(x_0, G(x_0) - u) > 0$ and $H(x, ry) \le rH(x, y)$ for each $r \in (0,1)$ and for each, $(x, y) \in X \times W$, we shall show that $G(x_0) \in \delta_E(G(X))$. Note that $F(x_0)$ is compact, thus there exists a point $u_0 \in F(x_0)$ such that

$H(x_0, G(x_0)-u_0) = \inf_{u\in F(x_0)} H(x_0, G(x_0)-u)$. We first show that $u_0 \notin G(X)$. Suppose that $u_0 \in G(X)$. Then for any $r \in (0,1)$, $rG(x_0)+(1-r)u_0 \in G(X)$. It follows that

$$\begin{aligned} H(x_0, G(x_0)-u_0) &= \inf_{u\in F(x_0)} H(x_0, G(x_0)-u) = \inf_{v\in G(X)} \inf_{u\in F(x_0)} H(x_0, v-u) \\ &\le \inf_{u\in F(x_0)} H(x_0, rG(x_0)+(1-r)u_0-u) \\ &\le H(x_0, rG(x_0)+(1-r)u_0-u_0) \le H(x_0, r(G(x_0)-u_0)) \\ &\le rH(x_0, G(x_0)-u_0) < H(x_0, G(x_0)-u_0) \end{aligned}$$

which is a contradiction. Therefore $u_0 \notin G(X)$.

Now suppose that $G(x_0) \notin \delta_E(G(X))$, then by the definition of $\delta_E(G(X))$, for points $u_0 \in F(x_0)$ and $G(x_0) \in G(X)$, there exists $r > 0$ such that $x_1 = G(x_0) + r(u_0 - G(x_0)) \in G(X)$. Since $u_0 = \frac{1}{r}x_1 + \frac{r-1}{r}G(x_0) \notin G(X)$ and both x_1, $G(x_0) \in G(X)$ and $G(X)$ is convex, we must have $r < 1$. Therefore

$$\begin{aligned} H(x_0, G(x_0)-u_0) &= \inf_{u\in F(x_0)} H(x_0, G(x_0)-u) = \inf_{v\in G(X)} \inf_{u\in F(x_0)} H(x_0, v-u) \\ &\le \inf_{u\in F(x_0)} H(x_0, (1-r)G(x_0)+ru_0-u) \\ &\le H(x_0, (1-r)G(x_0)+ru_0-u_0) \\ &\le H(x_0, (1-r)G(x_0)-(1-r)u_0) \\ &\le (1-r)H(x_0, G(x_0)-u_0) \\ &< H(x_0, G(x_0)-u_0) \end{aligned}$$

which is a contradiction. Hence $G(x_0) \in \delta(G(X))$. □

Theorem 2.6.17. Let E and W be two topological vector spaces. Let X be a non-empty compact convex subset of E, $F : X \to 2^W$ and $G : X \to W$. Suppose further there exists a continuous function $H : X \times W \to \mathbb{R}$ such that:

(i) for each fixed $x \in X$, $y \to H(x,y)$ is convex;

(ii) G is continuous and the set $G^{-1}(y)$ is convex for each $y \in G(X)$, where $G(X)$ is convex; and

(iii) F is continuous with compact convex values.

Then there exists a point $x_0 \in X$ such that

$$\inf_{w\in \overline{O_{G(X)}}(G(x_0))} \inf_{u\in F(x_0)} H(x_0, w-u) = \inf_{u\in F(x_0)} H(x_0, G(x_0)-u).$$

In particular, if $\inf_{u\in F(x_0)} H(x_0, G(x_0)-u) > 0$, and $H(x, ry) \le rH(x,y)$ for each $r \in (0,1)$ and each $(x,y) \in X \times W$, then $G(x_0) \in \delta_E(G(X))$.

Proof. If we define the mappings $F_1 : X \to 2^W$ by $F_1(x) = 2G(x) - F(x)$ for each $x \in X$ and $H_1 : X \times W \to \mathbb{R}$ by $H_1(x,y) = H(x,-y)$ for each $(x,y) \in X \times W$. Then the mappings G, F_1 and H_1 satisfy all hypotheses of Theorem 2.6.16 and by the same argument used in Theorem 2.6.16, the conclusion follows. □

By Theorem 2.6.16 and Theorem 2.6.17, we have the following result.

Corollary 2.6.18. Let E be a topological vector space which has sufficiently many continuous linear functionals. Let X be a non-empty compact convex subset of E and $F : X \to 2^E$ be continuous with compact and convex values. If F satisfies the following condition (i) or (i)$'$, then F has a fixed point in X.

(i): For each $x \in \delta_E(X) \setminus F(x)$, there exists a real number $r \in (0,1)$ and $u \in F(x)$ such that $rx + (1-r)u \in \overline{I_X}(x)$ (resp., $rx + (1-r)u \in \overline{O_X}(x)$).

$(i)'$: For each $x \in \delta_E(X) \setminus F(x)$ and $u \in F(x)$, there exists a number r (real or complex, depending on whether the vector space E is real or complex) with $|r| < 1$ such that $rx + (1-r)u \in \overline{I_X}(x)$ (resp., $rx + (1-r)u \in \overline{O_X}(x)$).

Proof. Suppose that F has no fixed point, then by Theorem 2.6.11, there exist $\delta > 0$ and a continuous seminorm P on E such that $\inf_{u\in F(x)} P(x-u) \geq \delta$ for all $x \in X$. Define continuous mappings $H : X \times E \to \mathbb{R}$ and $G : X \to E$ by $H(x,y) = P(y)$ for each $(x,y) \in X \times E$ and $G(x) = x$ for each $x \in X$. By Theorem 2.6.16 (resp., Theorem 2.6.17), there exists $x_0 \in \delta_E(X)$ such that

$$\inf_{v\in\overline{I_X}(x_0)} \inf_{u\in F(x_0)} P(v-u) = \inf_{u\in F(x_0)} P(x_0-u) \geq \delta > 0$$

(resp.,

$$\inf_{v\in\overline{O_X}(x_0)} \inf_{u\in F(x_0)} P(v-u) = \inf_{u\in F(x_0)} P(x_0-u) \geq \delta > 0).$$

If F satisfies the condition (i), then $x_0 \in \delta_E(X) \setminus F(x_0)$. By (i), there exist $r \in (0,1)$ and $u_0 \in F(x_0)$ such that $rx_0 + (1-r)u_0 \in \overline{I_X}(x_0)$ (resp., $rx_0 + (1-r)u_0 \in \overline{O_X}(x_0)$) so that

$$\begin{aligned}
0 &< \inf_{u\in F(x_0)} P(x_0-u) \\
&= \inf_{v\in\overline{I_X}(x_0)} \inf_{u\in F(x_0)} P(v-u) (\text{ resp.}, = \inf_{v\in\overline{O_X}(x_0)} \inf_{u\in F(x_0)} P(v-u)) \\
&\leq \inf_{u\in F(x_0)} P(rx_0 + (1-r)u_0 - u)) \\
&\leq \inf_{u\in F(x_0)} P(rx_0 + (1-r)u_0 - (ru + (1-r)u_0)) \text{ (since } F(x_0) \text{ is convex)} \\
&= r \inf_{u\in F(x_0)} P(x_0-u) \text{ (since } P \text{ is a seminorm)} \\
&< \inf_{u\in F(x_0)} p(x_0-u)
\end{aligned}$$

which is a contradiction. Therefore F must have a fixed point in X.

Now suppose that F satisfies the condition (i)$'$, then $x_0 \in \delta_E(X) \setminus F(x_0)$. Since $F(x_0)$ is compact, there exists a point $u_1 \in F(x_0)$ such that $\inf_{u\in F(x_0)} P(x_0-$

$u) = P(x - u_1)$. By (i)$'$, there exists a number r with $|r| < 1$ such that $rx_0 + (1-r)u_1 \in \overline{I_X}(x_0)$ (resp., $\overline{O_X}(x_0)$). Thus

$$\begin{aligned}
0 &< \inf_{u\in F(x_0)} P(x_0 - u) \\
&= \inf_{v\in \overline{I_X}(x_0)} \inf_{u\in F(x_0)} P(v-u) \text{ (resp., } = \inf_{v\in \overline{O_X}(x_0)} \inf_{u\in F(x_0)} p(v-u)) \\
&\leq \inf_{u\in F(x_0)} P(rx_0 + (1-r)u_1 - u) \\
&\leq \inf_{u\in F(x_0)} P(rx_0 + (1-r)u_1 - u_1) \\
&\leq |r| P(x_0 - u_1) \\
&< P(x_0 - u_1) = \inf_{u\in F(x_0)} P(x_0 - u)
\end{aligned}$$

which is a contradiction. Therefore F must have a fixed point. □

We note that the condition of (i)$'$ is different from (i) since the number r in the condition (ii)$'$ may be real or complex.

Since for each non-empty subset X in a topological vector space E, its algebraic boundary $\delta_E(X)$ in E is usually smaller than the topological boundary $\partial_E(X)$, by Proposition 2.6.6, Corollary 2.6.18 generalizes Theorem 3 of Fan [162] which in turn improves Theorem 1 of Browder [61], Lemma 1.6 of Reich [433] and Theorem 4.1 of Halpern and Bergman [218] in the following ways: (1) the underlying space is a topological vector space instead of a locally convex space and (2) the mapping F is set-valued instead of single-valued and (3) the boundary condition of Corollary 2.6.18 is weaker than theirs. Moreover, Corollary 2.6.18 also generalizes Theorem 4 of Park [421] which in turn generalizes many fixed point theorems for single-valued or set-valued inward (outward) mappings in the literature.

2.6.3 Coincidence Theorems

In this section, as applications of Theorem 2.6.15, several coincidence theorems for set-valued inward and outward mappings are derived which in turn imply fixed point theorems of inward and outward set-valued mappings in topological vector spaces. Finally, a coincidence theorem in locally convex spaces is also given.

As an application of Theorem 2.6.15, we first have the following coincidence theorem in topological vector spaces.

Theorem 2.6.19. Let E and W be two topological vector spaces. Let X be a non-empty compact convex subset of E and $G, F : X \to 2^W$. Suppose further there exists a continuous function $H : X \times W \to \mathbb{R}$ such that:

(i) for each fixed $x \in X$, the set $\{y \in W : H(x,y) < r\}$ is convex for each $r \in \mathbb{R}$;

(ii) G is continuous with non-empty closed values and the set $G^{-1}(y)$ is convex for each $y \in G(X)$, where $G(X)$ is convex;

(iii) F is continuous with compact values.

(iv) for each $x \in X$, if $G(x) \cap F(x) = \emptyset$, there exists a point $y \in G(X)$ such that

$$\inf_{u \in F(x)} H(x, y-u) < \inf_{v \in G(x)} [\inf_{u \in F(x)} H(x, v-u)].$$

Then there exists a point $x_0 \in X$ such that $G(x_0) \cap F(x_0) \neq \emptyset$.

Proof. Suppose the conclusion is not true, i.e., for each $x \in X$, $G(x) \cap F(x) = \emptyset$. Note that G, F and H satisfy all hypotheses of Theorem 2.6.15. By Theorem 2.6.15, there exists $x_0 \in X$ such that

$$\inf_{v \in G(X)} [\inf_{u \in F(x_0)} H(x_0, v-u)] = \inf_{v \in G(x_0)} [\inf_{u \in F(x_0)} H(x_0, v-u)].$$

But by (iv), there exists $y \in G(X)$ such that

$$\begin{aligned} \inf_{u \in F(x_0)} H(x_0, y-u) &< \inf_{v \in G(x_0)} [\inf_{u \in F(x_0)} (x_0, v-u)] = \inf_{v \in G(X)} [\inf_{u \in F(x_0)} H(x_0, v-u)] \\ &\leq \inf_{u \in F(x_0)} H(x_0, y-u), \end{aligned}$$

which is a contradiction. □

Theorem 2.6.19 generalizes Theorem 4 of Sessa and Mehta [456] and Proposition 2.2 of Browder [61, p.4750] to topological vector spaces and set-valued mappings. We also note that Theorem 2.6.19 generalizes Theorem 2 of Fan [162, p.235] to topological vector spaces from normed linear spaces.

We now give some coincidence theorems for inward (resp., outward) and weakly inward (resp., weakly outward) mappings in topological vector spaces.

Theorem 2.6.20. Let E and W be two topological vector spaces. Let X be a non-empty compact convex subset of E, $F : X \to 2^W$ and $G : X \to W$. Suppose further there exists another continuous function $H : X \times W \to \mathbb{R}$ such that:

(i) for each fixed $x \in X$, $y \to H(x,y)$ is convex;

(ii) G is continuous and the set $G^{-1}(y)$ is convex for each $y \in G(X)$, where $G(X)$ is convex;

(iii) F is continuous with compact and convex values; and

(iv) for each $x \in X$ with $G(x) \notin F(x)$, there exists a point $y \in \overline{I_{G(X)}}(G(x))$ such that

$$\inf_{u \in F(x)} H(x, y-u) < \inf_{u \in F(x)} H(x, G(x)-u).$$

Then there exists a point $x_0 \in X$ such that $G(x_0) \in F(x_0)$.

Proof. Let $x \in X$ be such that $G(x) \notin F(x)$. By (iv) and continuity of $y \to \inf_{u \in F(x)} H(x, y-u)$, there exists y in $\overline{I_{G(X)}}(G(x))$ such that

$$\inf_{u \in F(x)} H(x, y-u) < \inf_{u \in F(x)} H(x, G(x)-u).$$

If y lies in $G(X)$, then the hypothesis (iv) of Theorem 2.6.19 is valid. If $y \notin G(X)$, since $y \in \overline{I_{G(X)}}(G(x))$ and $G(X)$ is convex, there exist $u_0 \in G(X)$ and $r \in (0,1)$, $u_0 = (1-r)G(x) + ry$. Now for any $u \in F(x)$, by (i) we have

$$H(x, u_0 - u) = H(x, (1-r)G(x) + ry - u) \leq (1-r)H(x, G(x)-u) + rH(x, y-u).$$

Since $F(x)$ is compact and convex, $H(x, \cdot)$ is continuous, there exist $u_1 \in F(x)$ and $u_2 \in F(x)$ such that $\inf_{u \in F(x)} H(x, G(x)-u) = H(x, G(x)-u_1)$ and

$$\inf_{u \in F(x)} H(x, y-u) = H(x, y-u_2).$$

Note that $F(x)$ is convex,

$$\begin{aligned}
\inf_{u \in F(x)} H(x, u_0 - u) \quad & \leq H(x, u_0 - ((1-r)u_1 + ru_2)) \\
& H(x, (1-r)(G(x)-u_1) + r(y-u_2)) \\
& \leq (1-r)H(x, G(x)-u_1) + rH(x, y-u_2) \\
& = \inf_{u \in F(x)} (1-r)H(x, G(x)-u) + \inf_{u \in F(x)} rH(x, y-u) \\
& < \inf_{u \in F(x)} H(x, G(x)-u),
\end{aligned}$$

Hence all the hypotheses of Theorem 2.6.19 hold so that there exists a point $x_0 \in X$ such that $G(x_0) \in F(x_0)$. □

The following result shows that in Theorem 2.6.20, $\overline{I_{G(X)}}(G(x))$ can be replaced by $\overline{O_{G(X)}}(G(x)$.

Theorem 2.6.21. Let E and W be two topological vector spaces. Let X be a non-empty compact convex subset of E, $F : X \to 2^W$ and $G : X \to W$. Suppose further there exists another continuous function $H : X \times W \to \mathbb{R}$ such that:

(i) for each fixed $x \in X$, $y \to H(x, y)$ is a convex function on W;

(ii) G is continuous and the set $G^{-1}(y)$ is convex for each $y \in G(X)$, where $G(X)$ is convex; and

(iii) F is continuous with compact and convex values.

(iv) for each $x \in X$ with $G(x) \notin F(x)$, there exists a point $y \in \overline{O_{G(X)}}(G(x))$ such that $\inf_{u \in F(x)} H(x, y-u) < \inf_{u \in F(x)} H(x, G(x)-u)$.

Then there exists a point $x_0 \in X$ such that $G(x_0) \in F(x_0)$.

Proof. Define $F_1 : X \to 2^W$ and $H_1 : X \times W \to \mathbb{R}$ by $F_1(x) = 2G(x) - F(x)$ and $H_1(x, y) = H(x, -y)$ for each $(x, y) \in X \times W$. Then G, F_1 and H_1 satisfy all conditions (i), (ii) and (iii) of Theorem 2.6.20. Let $x \in X$ be such that $G(x) \notin F_1(x)$, then $G(x) \notin F(x)$ so that by (iv), there exists $y \in \overline{O_{G(X)}}(G(x))$ such that $\inf_{u\in F(x)} H(x, y-u) < \inf_{u\in f(x)} H(x, G(x)-u)$. Let $z = 2G(x) - y$, then $z \in \overline{I_{G(X)}}(G(x))$. But then

$$\begin{aligned} \inf_{u\in F_1(x)} H_1(x, z-u) &= \inf_{u\in F(x)} H(x, y-u) \\ &< \inf_{u\in F(x)} H(x, G(x)-u) \\ &= \inf_{u\in F_1(x)} H_1(x, G(x)-u). \end{aligned}$$

This shows that G, F_1 and H_1 also satisfy the condition (iv) of Theorem 2.6.20. Therefore by Theorem 2.6.20, there exists a point $x_0 \in X$ such that $G(x_0) \in F_1(x_0)$ which implies that $G(x_0) \in F(x_0)$. □

By letting $E = W$ and $G = I_X$, the identity mapping on X in Theorem 2.6.20 and Theorem 2.6.21, respectively, we have the following result which generalizes Theorem 1 and Theorem 2 of Browder [61] in the following ways: (a) the underlying spaces are topological vector spaces instead of locally convex topological vector spaces and (2) the mapping F is set-valued instead of being single-valued.

Corollary 2.6.22. Let E be a topological vector space. Let X be a non-empty compact convex subset of E and $F : X \to 2^E$. Suppose further there exists a continuous function $H : X \times E \to \mathbb{R}$ such that:

(i) for each fixed $x \in X$, $y \to H(x, y)$ is convex;

(ii) F is continuous with compact and convex values; and

(iii) for each $x \in X$ with $x \notin F(x)$, there exists a point $y \in \overline{I_X(x)}$ (resp., $y \in \overline{O_X}(x)$) such that

$$\inf_{u\in F(x)} H(x, y-u) < \inf_{u\in F(x)} H(x, x-u).$$

Then there exists a point $x_0 \in X$ such that $x_0 \in F(x_0)$.

By Corollary 2.6.22, we have the following result.

Corollary 2.6.23. Let E be a topological vector space which has sufficiently many continuous linear functionals, X be a non-empty compact convex subset of E and $F : X \to 2^E$ be continuous with compact and convex values. Suppose further that for each $x \in X$, $F(x) \cap \overline{I_X}(x) \neq \emptyset$ (resp., $F(x) \cap \overline{O_X}(x) \neq \emptyset$). Then F has a fixed point.

Proof. Suppose that F has no fixed point, then by Theorem 2.6.11, there exist $\delta > 0$ and a continuous seminorm P on E such that $\inf_{u\in F(x)} P(x-u) \geq \delta$ for all $x \in X$. Now define $H : X \times E \to \mathbb{R}$ by $H(x, y) = P(y)$ for each $(x, y) \in X \times E$. Since $F(x) \cap \overline{I_X}(x) \neq \emptyset$ (resp., $F(x) \cap \overline{O_X}(x) \neq \emptyset$), F and

G satisfy the condition (iii) of Corollary 2.6.22. Clearly, F and H also satisfy the conditions (i) and (ii) of Corollary 2.6.22. Hence by Corollary 2.6.22, there exists $x \in X$ such that $x \in F(x)$ which is a contradiction. Thus F must have a fixed point in X. □

As another application of Theorem 2.6.15, we present another coincidence theorem in locally convex spaces.

Theorem 2.6.24. Let E be a topological vector space and W be a locally convex topological vector space. Let X be a non-empty compact convex subset of E and $G, F : X \to 2^W$ be such that:

(i) G is continuous with closed convex values and the set $G^{-1}(y)$ is convex for each $y \in G(X)$, where $G(X)$ is convex; and

(ii) F is continuous with compact convex values.

Then we have either

(1): there exists a point $x_0 \in X$ such that $G(x_0) \cap F(x_0) \neq \emptyset$, or there exists a point $x_0 \in X$ and a continuous seminorm P on W such that for all $y \in G(X)$,

$$\inf_{u \in F(x_0)} P(y - u) \geq \inf_{v \in G(x_0)} [\inf_{u \in F(x_0)} P(v - u)] > 0;$$

or

(2) if $F(x) \cap G(X) \neq \emptyset$ for all $x \in X$, then there exists a point $x_0 \in X$ such that $G(x_0) \cap F(x_0) \neq \emptyset$.

Proof. Case (1): Suppose for each $x \in X$, $G(x) \cap F(x) = \emptyset$. Let $x \in X$ be arbitrarily fixed. Since $G(x)$ is closed and convex and $F(x)$ is compact and convex, by separate theorem, we have $\delta_x > 0$ and a continuous seminorm P_x on W such that $\inf_{v \in G(x)} \inf_{u \in F(x)} P_x(v - u) \geq \delta_x$ for all $x \in X$. By the continuity of G and F, the mapping $y \mapsto \inf_{v \in G(y)} \inf_{u \in F(y)} P_x(v - u)$ is continuous at x so that there exists an open neighborhood $N(x)$ of x in X such that for each $z \in N(x)$, we have $\inf_{v \in G(z)} \inf_{u \in F(z)} P_x(v - u) \geq \delta_x$. Since the family $\{N(x) : x \in X\}$ is an open covering of the compact set X, there exists a finite subset $\{x_1, \cdots, x_n\}$ of X such that $\{N_{x_i} : 1 \leq i \leq m\}$ covers X. Let $P = \max\{P_{x_i} : 1 \leq i \leq m\}$ and $\delta = \min\{\frac{\delta_{x_i}}{2} : 1 \leq i \leq m\} > 0$. Then P is a continuous seminorm on W and $\inf_{v \in G(x)} \inf_{u \in F(x)} p(v - u) \geq \delta$ for all $x \in X$. Now we define $H : X \times W \to \mathbb{R}$ by $H(x, y) = P(y)$ for each $(x, y) \in X \times W$, then G, F and H satisfy all hypotheses of Theorem 2.6.15. By Theorem 2.6.15, there exists $x_0 \in X$ such that

$$\inf_{v \in G(X)} [\inf_{u \in F(x_0)} P(v - u)] = \inf_{v \in G(x_0)} [\inf_{u \in F(x_0)} P(v - u)] \geq \delta > 0,$$

which implies that conclusion of (1) holds.

Case (2): Now assume that $F(x) \cap G(X) \neq \emptyset$ for all $x \in X$. If $G(x) \cap F(x) = \emptyset$ for all $x \in X$, then by (1), there exist $x_0 \in X$ and a continuous seminorm P on W

such that for all $y \in G(x)$, $\inf_{u\in F(x)} P(y-u) \geq \inf_{v\in G(x)} \inf_{u\in F(x_0)} P(v-u) > 0$. Take any $y_0 \in F(x_0) \cap G(X)$, we have

$$0 = [\inf_{u\in F(x_0)} P(y_0 - u)] \geq \inf_{v\in G(x_0)} [\inf_{u\in F(x_0)} P(v-u)] > 0$$

which is a contradiction. Therefore conclusion (2) must hold. □

Theorem 2.6.24 improves Theorems 2 and 3 of Ha [214, p.15] to set-valued mappings which in turn improves Theorem 2 of Fan [162].

By the Remark of Ha [214, p.14] and the proof of Theorem 2.6.24, it is easy to see that Theorem 2.6.24 is still true if we assume that the space W is a topological vector space (not necessarily locally convex space) which has sufficiently many continuous linear functionals.

Before we close this section, we remark about Fan's best approximation theorem for the requirement of continuity for set-valued mappings. We first note that as a special case of Theorem 2.6.15, it is clear that Fan's type best approximation is true.

Theorem 2.6.25. Let C be a non-empty compact convex subset of a Banach space E and $F : C \to 2^E$ is a continuous set-valued mapping with non-empty convex and compact values. Then there exists $x_0 \in C$ such that $d(x_0, F(x_0)) = d(F(x_0), C)$, where we recall that for both sets A and B in E, $d(A, B) = \inf_{a\in A; b\in B} \|a - b\|$.

Proof. For example, it is a special case of Theorem 2.6.15 by taking $G(x) := x$ for each $x \in C$ and the mapping H is defined by $H(x, y) := \|x - y\|$ for each $(x, y) \in E \times E$. □

However, the following example of Waters [572] (see also Singh et al. [489]) shows that the continuity of F in Theorem 2.6.25 can not be replaced by upper semicontinuity alone.

Example 2.6.26. Let $E = R^2$ with the Euclidean norm and let $C = [0,1] \times \{0\}$. Clearly C is convex and compact. Now define a set-valued mapping $F : C \to 2^E$ by

$$F(a, 0) = \begin{cases} (0,1), & \text{if } a \neq 0; \\ L = \text{the line segment}[(0,1),(0,1)], & \text{if } a = 0. \end{cases}$$

Then $F(x)$ is non-empty compact and convex for each $x \in C$ and F is upper semicontinuous, but not lower semicontinuous due to the fact that for each $A \subset E$, we have

$$F^{-1}(A) = \begin{cases} \emptyset, & \text{if } A \cap L = \emptyset; \\ C, & \text{if } (0,1) \in A; \\ (0,0), & \text{if } (0,1) \notin A \text{ and } A \cap L \neq \emptyset. \end{cases}$$

However, for any $(a, 0) \in C$, we do have that $d((a,0), F(a,0)) > 1 = d(F(a,0), C)$ if $a \neq 0$; and $d((a,0), F(a,0)) = \frac{\sqrt{2}}{2} \neq d(F(0,0), C) = 0$ when $a = 0$. thus the conclusion of Theorem 2.6.25 is not satisfied.

Though Example 2.6.26 shows that one cannot dispense with the lower semicontinuity for Fan type best approximation; however, by using seminorm, it is possible for us to establish Fan's best approximation and related applications to the study of fixed point theorems for set-valued mappings which are only upper semicontinuous in locally convex topological vector spaces. The details will be carried in the Section 9 of this chapter.

2.7 Matching Theorems and Applications

By using a generalization of Fan's existence theorem for maximizable quasi-concave functions on convex spaces due to Park and Bae [423], we first prove some coincidence theorems for upper hemicontinuous non-self mappings in topological vector spaces with sufficiently many continuous linear functionals or in locally convex topological vector spaces. These results improve and unify many results in the literature (e.g, see Fan [167], Park [419], Ko and Tan [320], and references therein). Next, as applications of coincidence theorems, several matching theorems for closed coverings of convex sets are derived in locally convex topological vector spaces or topological vector spaces with sufficiently many continuous linear functionals which in turn imply Shapley's theorem [460].

2.7.1 Fan-Glicksberg Fixed Point Theorems in Topological Vector Spaces

The basic idea in this section is to apply the existence theorem for maximizable quasi-concave functions on topological vector space with sufficiently many continuous linear functionals. Several fixed point theorems for non-self mappings are given under weaker continuity and boundary hypothesis. For example, our fixed point theorems show that the hypotheses *the domain is paracompact* is superfluous for the existence of non-self upper hemicontinuous set-valued mappings, in fact, this *superfluous condition* is posed in much of the literature (e.g, see Fan [167], Lassonde [334], [167], Ko and Tan [320], and Browder [62]). In particular, the well-known Fan-Glicksberg fixed point theorem has been generalized into the non-compact setting in which the underlying space is a topological vector space with sufficiently many continuous linear functionals under weaker continuity and boundary hypotheses.

We recall that a *convex space* (e.g., see Lassonde [334, p.153]) is a non-empty convex set in a vector space with any topology that induces the Euclidean topology on the convex hulls of its finite subsets. A non-empty subset L of a convex space X is said to be *c-compact* if for each $S \in \mathcal{F}(X)$, there exists a non-empty compact set L_S with $L \cup S \subset L_S$. It is clear that a convex subset (resp., compact and convex subset) of a topological vector space is a convex space (resp., c-compact subset). For more details about convex spaces, we refer to Lassonde [334, p.153] and Dugundji [147, p.416].

By using the Ky Fan minimax inequality, Park and Bae [423] gave a generalization of the existence theorem for maximizable quasi-concave functions on convex spaces which in turn answered the question raised by Bellenger [29]. In this section, we first recall the following result which is essentially a consequence of the existence theorem for maximizable quasi-concave functions on convex spaces (e.g., see Theorem 1 of Park and Bae [423]).

Theorem 2.7.1. Let X be a convex space and suppose that:

(a) for each $x \in X$, $T(x)$ is a non-empty convex set of upper semicontinuous quasi-concave real functions on X;

(b) for each upper semicontinuous and quasi-concave real function f on X, the set $T^{-1}(f)$ is compactly open in X; and

(c) there exists a c-compact subset L of X and a non-empty compact subset K of X such that for each $z \in X \setminus K$ and $f \in T(z)$, $f(z) < \sup\{f(x) : x \in co(L \cup \{z\})\}$.

Then there exists $\hat{z} \in K$ and $\hat{f} \in T(\hat{z})$ such that $\hat{f}(\hat{z}) = \sup\{\hat{f}(x) : x \in X\}$.

Proof. In order to prove our conclusion, we first have the following '*Fact A*':

Let K be a compact subset of a normal space $\hat{K}$ and U an open subset of $\hat{K}$. Suppose $\alpha : K \to [0,1]$ is a continuous function such that $supp\alpha \subset U \cap K$. Then there exists a continuous extension $\hat{\alpha} : \hat{K} \to [0,1]$ of α such that $supp\hat{\alpha} \subset U$.

Indeed, by Tietze extension theorem [147], we have a continuous extension $\alpha_1 : \hat{K} \to [0,1]$ of α. As $supp\alpha \subset U$, there exists an open subset V of U such that $supp\alpha \subset V \subset \overline{V} \subset U$. Note that $\hat{K}$ is normal, by using Uryshon's Lemma [147], there is a continuous function $\beta : \hat{K} \to [0,1]$ such that $\beta(x) = 1$ for all $x \in supp\alpha$ and $\beta(x) = 0$ for each $x \in \hat{K} \setminus V$. Then the mapping $\hat{\alpha} = \alpha_1 \cdot \beta$ is the desired extension of α.

Now it is our turn to prove the conclusion of Theorem 2.7.1. Let $\hat{X}$ denote the set of all upper semicontinuous (in short, usc) and quasi-concave real functions defined on X. We first show that the restriction mapping $T|K$ has a selection $\phi : K \to \hat{X}$. As $T(x)$ is non-empty for each $x \in X$, there exists an $f_x \in \hat{X}$ such that $x \in T^{-1}(f_x)$. Then the family $\{T^{-1}(f_x) : x \in X\}$ is a compactly open cover of X. As K compact, we have a finite number of fucntions $f_1, f_2, \cdots f_n$ in $\hat{X}$ such that $K \subset \cup_{i=1}^n T^{-1}(f_i)$. Let $\{\alpha_i\}_{i=1}^n$ be a continuous partition of unity subordinate to $\{T^{-1}(f_i) \cap K\}_{i=1}^n$. Then we have:

(1) $supp\alpha_i \subset T^{-1}(f_i) \cap K$ and $\Sigma_{i=1}^n \alpha_i(x) = 1$ for all $x \in K$.

For each $x \in K$, define $\phi_x = \phi(x) = \Sigma_{i=1}^n \alpha_i(x) f_i$. Then by the condition (a) and (1) above, we have that $\phi_x \in T(x)$ and ϕ_x is usc and quasi-concave for each $x \in K$. Therefore we have that:

(2) for each $y \in X$, the funtion $x \mapsto \phi_x(y)$ is continuous on K;

(3) the function $x \mapsto \phi(x)$ is usc on K;

For any finite number of points $y_1, \cdots, y_n$ in X, let $\hat{L} = \overline{co}\{y_1, \cdots, y_n, L\}$, i.e., the closed convex hull of $\{y_1, \cdots, y_n, L\}$. Then $T|\hat{L}$ has a selection $\hat{\phi} : \hat{L} \to \hat{X}$ such that $\hat{\phi} = \phi$ on $\hat{L} \cap K$. As $\hat{L}$ is a compact subset of X, by conditions (a) and (b), we have a finite number of functions $f_{n+1}, f_{n+2}, \cdots, f_{n+k}$ in $\hat{X}$ such that $\hat{L} \subset \cup_{j=1}^k T^{-1}(f_{n+j})$. By the Fact A above again, there exists a continuous extension $\hat{\alpha} : \hat{L} \cup K \to [0,1]$ of α_i such that $supp\alpha_i \subset T^{-1}(f_i)$ for each $1 \leq i \leq n$. Let $L_1 := \{x \in \hat{L} : \Sigma_{i=1}^n \hat{\alpha}_i(x) = 0\}$ and and $U_{n+j} := T^{-1}(f_{n+j}) \cap (\hat{L} \setminus K)$ for each $1 \leq j \leq k$. Then $L_1 \subset \cup_{j=1}^k U_{n+j}$ and henece we have a continuous partition of unity $\{\alpha_{n+j} : 1 \leq j \leq k\}$ defined on L_1 satisfying

(4) $supp\alpha_{n+j} \subset U_{n+j} \cap L_1$ for each $1 \leq j \leq k$ and $\Sigma_{j=1}^n \alpha_{n+j}(x) = 1$ for all $x \in L_1$.

Now by using Fact A above, there exists a continuous extension $\hat{\alpha}_{n+j} : \hat{L} \cup K \to [0,1]$ of α_{n+j} such that $supp\hat{\alpha}_{n+j} \subset U_{n+j}$ for all $1 \leq j \leq k$. By (4) above, it follows that $\Sigma_{i=1}^{n+k} \hat{\alpha}_i(x) \neq 0$ for all $x \in \hat{L} \cup K$. Now for each $1 \leq i \leq n+k$, define $\beta_i : \hat{L} \cup K \to [0,1]$ by $\beta_i(x) = \frac{\hat{\alpha}_i(x)}{\Sigma_{j=1}^{n+k} \hat{\alpha}_j(x)}$ for each $x \in \hat{L} \cup K$. Then the continuous fucntion $\beta_i : \hat{L} \cup K \to [0,1]$ for each $1 \leq i \leq n+k$ has the following properties:

(5) $supp\beta_i \subset T^{-1}(f_i)$;

(6) for each $x \in K$, $\beta_i(x) = \alpha_i(x)$ for each $1 \leq i \leq n$ and $\beta_{n+j}(x) = 0$ for all $1 \leq j \leq k$; and

(7) $\Sigma_{i=1}^{n+k} \beta_i(x) = 1$ for all $x \in \hat{L} \cup K$.

For each $x \in \hat{L}$, we define $\hat{\phi}_x = \Sigma_{i=1}^{n+k} \beta_i(x) \cdot f_i$. By (5) and (7), it follows that $\hat{\phi}_x \in T(x)$ for each $x \in \hat{L}$ and $\hat{\phi}_x$ is an usc and quasi-concave function on X. Therefore the properties (2) and (3) still hold for $\hat{\phi}$ and $\hat{L}$ instead of ϕ and K, respectively. Note that by (6), $\hat{\phi}_x = \phi_x$ for each $x \in \hat{L} \cap K$.

In order to apply the Browder-Fan fixed point theorem (e.g., see Theorem 2.3.22), we define a set $A(y) = \{x \in K : \phi_x(y) \leq \phi_x(x)\}$ for each $y \in X$. Then $\{A(y) : y \in X\}$ have the finite intersection property. Note that by (2) and (3), each $A(y)$ is closed in K. For any finite number of points $y_1, \cdots, y_n$ from X, consider the closed convex hull $\hat{L} = \overline{co}\{y_1, \cdots, y_n, L\}$ and $\hat{\phi} : \hat{L} \to \hat{X}$ as defined above, we define a set $W(x) = \{y \in \hat{L} : \hat{\phi}_x(y) > \hat{\phi}_x(x)\}$ for each $x \in \hat{L}$. Then it follows that $W^{-1}(y) = \{x \in \hat{L} : \hat{\phi}_x(y) > \hat{\phi}_x(x)\}$ for each $y \in \hat{L}$. As $\hat{\phi}_x$ is quasi-concave, it follows that $W(x)$ is convex. Note that $x \notin W(x)$ for all $x \in \hat{L}$. Moreover, by (2) and (3), it follows that $W^{-1}(y)$ is open in $\hat{L}$. Therefore by the Browder-Fan fixed point theorem, there must exists an $\hat{x} \in \hat{L}$ such that $W(\hat{x}) = \emptyset$, or equivalently to say, $\hat{\phi}_{\hat{x}}(y) \leq \hat{\phi}_{\hat{x}}(\hat{x})$ for all $y \in \hat{L}$. This implies that $\hat{\phi}_{\hat{x}}(\hat{x}) = \max_{y \in \overline{co}\{\hat{x}, L\}} \hat{\phi}_{\hat{x}}(y)$. Since $\hat{\phi}_{\hat{x}} \in T(\hat{x})$, by the condition (c), we must have $\hat{x} \in K$. As $\hat{\phi}_{\hat{x}} = \phi_{\hat{x}}$, it follows that $\phi_{\hat{x}}(y) \leq \phi_{\hat{x}}(\hat{x})$ for all $y \in \hat{L}$, and hence we have that

$$\hat{x} \in \cap_{y \in \hat{L}} A(y) \subset \cap_{i=1}^{m} A(y_i).$$

Note that K is compact, it follows that $\hat{x} \in \cap\{A(y) : y \in X\}$. Let $f := \phi_{\hat{x}} \in T(\hat{x})$. Then $f(\hat{x}) = \max_{y \in X} f(y)$. This completes the proof. □

We note that Theorem 2.7.1 above generalizes Theorem 8 of Fan [167] and Theorem 1 of Bellenger [29] which in turn improves Theorem 0.1 of Simons [482] for the existence theorem for an upper semicontinuous quasi-concave real function which attains a global maximum on a given compact subset of a convex space X. Recently, Ding [132] generalized Theorem 2.7.1 to H-spaces by following Park and Bae's idea in [423].

Let X be a convex subset of a topological vector space E, We now state an equivalent form of Theorem 2.7.1 as follows.

Theorem 2.7.2. Let X be a convex space and Φ a non-empty convex set of lower semicontinuous convex real functions on X. Let S be a subset of $X \times \Phi$ such that:

(a) for each $\phi \in \Phi$, the section $\{x \in X : (x, \phi) \in \Phi\}$ is compactly open in X; and

(b) for each $x \in X$, the section $\{\phi \in \Phi : (x, \phi) \in S\}$ is non-empty and convex.

Then either

(I): there exists $(y, \phi) \in S$ such that $\phi(y) = \inf_{x \in X} \phi(x)$; or

(II): for each compact convex set L in X and each non-empty compact set K in X, there exists $(y, \phi) \in S$ such that $y \in X \setminus K$ and $\phi(y) = \inf_{x \in co(L \cup \{y\})} \phi(x)$.

Before we give generalizations of the Fan-Glicksberg fixed point theorem in topological vector spaces with sufficiently many continuous linear functionals or in locally convex topological vector spaces under weaker continuity assumptions, we first recall some facts about various continuity for set-valued mappings.

Definition 2.7.3. If E is a topological vector space, E^* is the dual space of all continuous linear functionals on E, the pairing between E^* and E is denoted by $\langle w, x \rangle$ for each $w \in E^*$ and $x \in E$. Let X be a non-empty subset of a topological vector space E. We recall that a mapping $f : X \to 2^E$ is said to be *upper hemicontinuous* if for each $\phi \in E^*$ and for each $\lambda \in \mathbb{R}$, the set $\{x \in X : \sup_{u \in f(x)} Re\langle \phi, u \rangle < \lambda\}$ is open in X.

We note that each upper semicontinuous mapping is upper hemicontinuous and the sum of two upper hemicontinuous mappings is again upper hemicontinuous.

Definition 2.7.4. By following Fan [162], a set-valued mapping $f : X \to 2^E$ is said to be *upper demicontinuous* on X if for each $x \in X$ and any open half-space H containing $f(x)$, there exists an open neighborhood N_x of x in X such that $f(u) \subset H$ for all $u \in N_x$, where we note that an open half-space H in E is a set of the form $H := \{v \in E : Re\phi(v) < t\}$ for some non-zero $\phi \in E^*$ and some real number t.

It is obvious that every upper semicontinuous mapping is upper demicontinuous, each upper demicontinuous mapping is upper hemicontinuous and the following examples from Shih and Tan [470] show that the converses do not hold in general.

Example 2.7.5. Let $E = \mathbb{R}^2$ and $X = \{t \in \mathbb{R} : t \in [0,1]\}$. Define $f, g : X \to 2^{\mathbb{R}^2}$ by

$$f(t) = \{(u,v)\} : (u-1)(v-1) \geq 1 \text{ and } u > 1\},$$

$$g(t) = \{(-z,z) \in \mathbb{R}^2 : 0 \leq z \leq t\sqrt{2}\}$$

for each $t \in X$. Then it is not hard to verify that f and g are both upper semicontinuous but $f + g$ is not upper demicontinuous.

Incidently, Example 2.7.5 also shows that an upper hemicontinuous mapping need not be upper demicontinuous, since $f + g$ is necessarily upper hemicontinuous.

We also observe that an upper demicontinuous may need not be upper semicontinuous as the following example shows.

Example 2.7.6. Define $f : \mathbb{R}^2 \to \mathbb{R}^2$ by

$$f(x,y) = \{(u,v) \in \mathbb{R}^2 : u \geq x \text{ and } v \geq y\}$$

for each $(x,y) \in \mathbb{R}^2$. Then f is upper demicontinuous, but not uper semicontinuous.

As a special case of both Corollaries 1.3.7 and 1.3.8 in Chapter 1, We know that if a set-valued mapping has compact values, then the concepts of upper hemicontinuity and upper demicontinuity coincide, i.e., we have the following statement (see also Shih and Tan [470]) which shows that the concepts of upper semicontinuity, upper demicontinuity and upper hemicontinuity are the same.

Theorem 2.7.7. Let X be a topological space, Z a non-empty compact subset of a real locally convex topological space E, and let $F : X \to 2^Z$ be such that each $F(x)$ is convex. Then the following statements are equivalent:

(1) F is upper semicontinuous.
(2) F is upper demicontinuous.
(3) F is upper hemicontinuous.

By Theorem 2.7.2, we have the following:

Theorem 2.7.8. Let E be a topological vector space which has sufficiently many continuous linear functionals, let X be a non-empty convex subset of E, X_0 a non-empty compact convex subset of X and K a non-empty compact subset of X. Let $F, G : X \to 2^E$ be upper hemicontinuous and such that:

(a) for each $x \in X$, $F(x)$ and $G(x)$ are closed convex at least, one of which is compact;

(b) for each $x \in K \cap \delta_E(X)$ and $\phi \in E^*$ with $Re\phi(x) \leq Re\phi(y)$ for all $y \in X$, there exist $u \in F(x)$ and $v \in G(x)$ such that $Re\phi(u) \geq Re\phi(v)$; and either

(c_1) suppose $\overline{co}K$ is compact (which is automatically satisfied if E is a complete locally convex topological vector space) for each $x \in X \setminus K$ and $\phi \in E^*$ such that $Re\phi(x) \leq Re\phi(y)$ for all $y \in co(K \cup \{x\})$ (in particular, $Re\phi(x) \leq Re\phi(y)$ for all $y \in co(\overline{co}K \cup \{x\})$), there exist $u \in F(x)$ and $v \in G(x)$ with $Re\phi(u) \geq Re\phi(v)$; or

(c_2) for each $x \in X \setminus K$ and $\phi \in E^*$ such that $Re\phi(x) \leq Re\phi(y)$ for all $y \in co(X_0 \cup \{x\})$, there exist $u \in F(x)$ and $v \in G(x)$ with $Re\phi(u) \geq Re\phi(v)$.

Then there exists a point $\hat{x} \in X$ such that for each $\phi \in E^*$ and each $t \in \mathbb{R}$, the following does not hold:

$$Re\phi(u) < t < Re\phi(v) \text{ for all } u \in F(\hat{x}) \text{ and } v \in G(\hat{x}).$$

Moreover,
(I): If $F(\hat{x})$ and $G(\hat{x})$ are both compact, then $F(\hat{x}) \cap G(\hat{x}) \neq \emptyset$.
(II): If E is a locally convex topological vector space, then $F(\hat{x}) \cap G(\hat{x}) \neq \emptyset$.

Proof. Note that by Proposition 2.6.9, it is clear that condition (b) of Theorem 2.7.8 is equivalent to the following condition:

$(b)'$: for any $x \in K$ and $\phi \in E^*$ such that $Re\phi(x) \leq Re\phi(y)$ for all $y \in X$, there exist $u \in F(x)$ and $v \in G(x)$ with $Re\phi(u) \geq Re\phi(v)$.

Now we follow the idea of Fan [167] and Ko and Tan [320] to prove our assertion. In order to apply Theorem 2.7.2, we take $\Phi = E^*$ and define the subset S of $X \times E^*$ as follows: $(x, \phi) \in S$ if and only if there exists a real number $t \in \mathbb{R}$ such that $Re\phi(u) < t < Re\phi(v)$ for all $u \in F(x)$ and all $v \in G(x)$.

(1): We first show that for each $\phi \in E^*$, the section $S(\phi) = \{x \in X : (x, \phi) \in S\}$ is open in X.

Indeed, let $x \in S(\phi)$, then $(x, \phi) \in S$ and hence there exists a real number t such that $Re\phi(u) < t < Re\phi(v)$ for all $u \in F(x)$ and all $v \in G(x)$.

First we assume that $F(x)$ is compact. Then there exists an $u_0 \in F(x)$ such that $Re\phi(u_0) = \sup_{u \in F(x)} Re\langle\phi, u\rangle$. Let ϵ_1, $\epsilon_2 > 0$ be such that $Re\phi(u_0) + \epsilon_1 + \epsilon_2 < t$. By upper hemicontinuity of F, there exists an open neighborhood N_1 of x in X such that for each $y \in N_1$,

$$\sup_{v \in F(y)} Re\langle\phi, v\rangle < \sup_{u \in F(x)} Re\langle\phi, u\rangle + \epsilon_1$$

so that $Re\phi(u) < Re\phi(u_0)+\epsilon_1$ for all $u \in F(y)$. Also, since $\sup_{v\in G(x)} Re\langle -\phi, v\rangle = -\inf_{v\in G(x)} Re\langle \phi, v\rangle \leq -t < +\infty$, by upper hemicontinuity of G, there exists an open neighborhood N_2 of x in X such that for each $y \in N_2$,

$$\sup_{v\in G(y)} Re\langle -\phi, v\rangle < \sup_{v\in G(x)} Re\langle -\phi, v\rangle + \epsilon_2,$$

so that $Re\phi(v) > t - \epsilon_2$ for all $v \in G(y)$. Let $N = N_1 \cap N_2$. Then N is an open neighborhood of x in X such that for each $y \in N$, for each $u \in F(y)$ and for each $v \in G(y)$,

$$Re\phi(u) < Re\phi(u_0) + \epsilon_1 < t - \epsilon_2 < Re\phi(v).$$

Therefore $N \subset S(\phi)$. Similarly if $G(x)$ is compact, we see that (by replacing ϕ by $-\phi$ and by interchanging F and G in the above argument) there exists an open neighborhood N' of x in X such that $N' \subset S(\phi)$. Therefore $S(\phi)$ is open in X. Thus the condition (b) of Theorem 2.7.2 is satisfied.

(2): For each $x \in X$, it is clear that the set $S(x) = \{\phi \in \Phi : (x, \phi) \in S\}$ is convex.

(3): Next we show that for each $z \in X \setminus K$ and $(z, \phi) \in S$, $Re\phi(z) > \inf\{Re\phi(x) : x \in co(\overline{co}K \cup \{z\})\}$ (resp., $Re\phi(z) > \inf\{Re\phi(x) : x \in co(X_0 \cup \{z\})\}$).

Note that if there exist $x \in X \setminus K$ and $\phi \in E^*$ such that $Re\phi(x) \leq Re\phi(y)$ for all $y \in co(K \cup \{x\})$, then it is clear that $Re\phi(x) \leq Re\phi(y)$ for all $y \in co(\overline{co}K \cup \{x\})$ by the linearity of real part of the linear continuous functional ϕ. Suppose the contrary that for some $z \in X \setminus K$ and $(z, \phi) \in S$ such that $Re\phi(z) = \inf\{Re\phi(x) : x \in co(\overline{co}K \cup \{z\})\}$ (resp., $Re\phi(z) = \inf\{Re\phi(x) : x \in co(X_0 \cup \{z\})\}$). Then by (c_1) (resp., (c_2)), there exist $u \in F(z)$ and $v \in G(z)$ such that $Re\phi(u) \geq Re\phi(v)$. This contradicts the assumption that $(z, \phi) \in S$. Thus the alternative (II) of Theorem 2.7.2 is false.

(4): Similarly, by $(b)'$ (which is equivalent to (b)), for each $z \in K$ and each $(z, \phi) \in S$, $Re\phi(z) > \inf_{x\in X} Re\phi(x)$. It follows from (3) that for each $z \in X$ and $(z, \phi) \in S$, $Re\phi(z) > \inf_{x\in X} Re\phi(x)$. Hence the alternative (I) of Theorem 2.7.2 is also not satisfied.

By Theorem 2.7.2, there must exist $\hat{x} \in X$ such that $\{\phi \in E^* : (\hat{x}, \phi) \in S\} = \emptyset$; i.e., for each $t \in \mathbb{R}$ and each $\phi \in E^*$, the following does not hold:

$$Re\phi(u) < t < Re\phi(v) \text{ for all } u \in F(\hat{x}) \text{ and } v \in G(\hat{x}). \tag{2.7.1}$$

Case (i): Suppose conclusion (I) were false, then $F(\hat{x}) \cap G(\hat{x}) = \emptyset$, so that $0 \notin F(\hat{x}) - G(\hat{x})$. Since both $F(\hat{x})$ and $G(\hat{x})$ are compact and convex, the set $D := F(\hat{x}) - G(\hat{x})$ is compact convex. Then for each $a \in D$, $a \neq 0$, as E^* separates points of E, there exists $\phi_a \in E^*$ such that $Re\phi_a(a) < 0$. Let O_a and U_a be disjoint open convex sets containing $Re\phi_a(a)$ and 0 respectively. Then $Re\phi_a^{-1}(O_a)$ and $Re\phi_a^{-1}(U_a)$ are disjoint open convex sets in E containing

a and 0 respectively. Since D is compact, there exist $a_1, \cdots, a_n \in D$ such that $D \subset \cup_{i=1}^n Re\phi_{a_i}^{-1}(O_{a_i})$. Let $U = \cap_{i=1}^n Re\phi_{a_i}^{-1}(U_{a_i})$, then U is an open convex set containing 0 such that $U \cap D = \emptyset$. By Theorem 3.4 of Rudin [449, p.58], there exists $\phi \in E^*$ and $r \in \mathbb{R}$ such that $Re\phi(a) \leq r < 0$ for all $a \in D$, i.e., $Re\phi(u) \leq r + Re\phi(v)$ for all $u \in F(\hat{x})$ and $v \in G(\hat{x})$. Let $t := r/2 + \inf_{w \in G(\hat{x})} Re\phi(w)$. Since $r < 0$, it follows that

$$\begin{aligned} R\phi(u) &\leq r + \inf_{w \in G(\hat{x})} Re\phi(w) \\ &< r/2 + \inf_{w \in G(\hat{x})} Re\phi(w) = t < Re\phi(v) \end{aligned}$$

which contradicts (2.7.1).
Case (ii): If E is a locally convex topological vector space, since $F(\hat{x})$ and $G(\hat{x})$ can not be strictly separated by a closed hyperplane in E and at least one of $F(\hat{x})$ and $G(\hat{x})$ is compact, we must have $F(\hat{x}) \cap F(\hat{x}) \neq \emptyset$ by Theorem 3.4 of Rudin [449]. □

From the proof (due to part (3)) of Theorem 2.7.8, the assumption that $\overline{co}K$ *is compact* in the assumption (c_1) is not needed if the underlying space E is a locally convex topological vector space, i.e., we have the following fixed point result.

Theorem 2.7.9. Let E be a locally convex topological vector space, let X be a non-empty convex subset of E, X_0 a non-empty compact convex subset of X and K a non-empty compact subset of X. Let $F, G : X \to 2^E$ be upper hemicontinuous and such that:

(a) for each $x \in X$, $F(x)$ and $G(x)$ are closed convex at least one of which is compact;

(b) for each $x \in K \cap \delta_E(X)$ and $\phi \in E^*$ with $Re\phi(x) \leq Re\phi(y)$ for all $y \in X$, there exist $u \in F(x)$ and $v \in G(x)$ such that $Re\phi(u) \geq Re\phi(v)$; and either

(c_1) for each $x \in X \setminus K$ and $\phi \in E^*$ such that $Re\phi(x) \leq Re\phi(y)$ for all $y \in co(K \cup \{x\})$ (in particular, Re $\phi(x) \leq Re\phi(y)$ for all $y \in co(\overline{co}K \cup \{x\})$), there exist $u \in F(x)$ and $v \in G(x)$ with $Re\phi(u) \geq Re\phi(v)$; or,

(c_2) for each $x \in X \setminus K$ and $\phi \in E^*$ such that $Re\phi(x) \leq Re\phi(y)$ for all $y \in co(X_0 \cup \{x\})$, there exist $u \in F(x)$ and $v \in G(x)$ with $Re\phi(u) \geq Re\phi(v)$.

Then $F(\hat{x}) \cap G(\hat{x}) \neq \emptyset$.

Proof. We first note that the assumptions of X, X_0, K, F and G remain unchanged in the completion $\overline{E}$ of E. Without loss of generality, we may assume that E is a complete locally convex topological vector space. Since $\overline{E}^* = E^*$, it follows that the conclusions of (1) and (2) still hold. By the completeness of E, the set $\overline{co}K$ is non-empty compact and convex since K is non-empty compact. Now, following the proof of Theorem 2.7.8, there must exist $\hat{x} \in X$ such that $F(\hat{x}) \cap G(\hat{x}) \neq \emptyset$.

Theorem 2.7.10. Let X be a non-empty convex subset of a locally convex topological vector space E. Let X_0 be a non-empty compact convex subset of X and K be a non-empty compact subset of X. Let $F : X \to 2^E$ be an upper hemicontinuous mapping with closed and convex values such that:

(a) for each $x \in K \cap \delta_E(X)$, $F(x) \cap X \neq \emptyset$; and either

(b) for each $x \in X \setminus K$, $F(x) \cap co(\overline{co}K \cup \{x\}) \neq \emptyset$ or,

(b)$'$ for each $x \in X \setminus K$, $F(x) \cap co(X_0 \cup \{x\}) \neq \emptyset$.

Then there exists $\hat{x} \in X$ such that $\hat{x} \in F(\hat{x})$.

Proof. Let $G = I_X$ be the identity mapping on X. Since $F(x) \cap X \neq \emptyset$ for all $x \in K \cap \delta(X)$, the condition (b) of Theorem 2.7.9 is satisfied. Also condition (b) (resp., $(b)'$) implies that condition (c_1) (resp., (c_2)) of Theorem 2.7.9 is satisfied. Hence the conclusion follows from Theorem 2.7.9. □

By Theorem 2.7.8, we have the following result.

Theorem 2.7.11. Let E be a topological vector space which has sufficiently many continuous linear functionals, let X be a non-empty convex subset of E and X_0 be a non-empty compact convex subset of X and K a non-empty compact subset of X. Let $F : X \to 2^E$ be upper hemicontinuous with compact convex values such that:

(a) for each $x \in K \cap \delta_E(X)$, $F(x) \cap X \neq \emptyset$;

(b) for each $x \in X \setminus K$, $F(x) \cap co(X_0 \cup \{x\}) \neq \emptyset$.

Then there exists $\hat{x} \in X$ such that $\hat{x} \in F(\hat{x})$.

We note that Theorem 2.7.10 improves the well-known Fan-Glicksberg fixed point theorem in the following ways: (a) the domain X need not be compact; (b) the mapping F is upper hemicontinuous instead of upper semicontinuous; (c) the mapping F need not have compact values (resp., the space E need not be locally convex) and (d) the mapping F need not be a self-map.

By Theorem 2.7.10, we have the following result.

Corollary 2.7.12. Let X be a non-empty compact convex subset of a locally convex topological vector space E. Let $F : X \to 2^E$ be upper hemicontinuous with closed and convex values such that for each $x \in \delta_E(X)$, $F(x) \cap X \neq \emptyset$. Then there exists $\hat{x} \in X$ such that $\hat{x} \in F(\hat{x})$.

Proof. The conclusion follows from Theorem 2.7.10 by taking $X = K$. □

Corresponding to Theorem 2.7.11, we have the following result.

Corollary 2.7.13. Let E be a topological vector space which has sufficiently many continuous linear functionals and X be a non-empty compact convex subset of E. Let $F : X \to 2^E$ be upper hemicontinuous with compact and convex values such that for each $x \in \delta_E(X)$, $F(x) \cap X \neq \emptyset$. Then there exists

$\hat{x} \in X$ such that $\hat{x} \in F(\hat{x})$.

Theorem 2.7.14. Let X be a non-empty convex set in a locally convex topological vector space E. Let X_0 be a non-empty compact convex subset of X and K be a non-empty compact subset of X. Let $F, G : X \to 2^E$ be upper hemicontinuous and such that:

(a) for each $x \in X$, $F(x)$ and $G(x)$ are closed convex at least one of which is compact;

(b) for each $x \in K \cap \delta_E(X)$, $(F(x) - G(x)) \cap \overline{\cup_{\lambda \geq 0} \lambda(X - x)} \neq \emptyset$ (resp., $(F(x) - G(x)) \cap \overline{\cup_{\lambda \leq 0} \lambda(X - x)} \neq \emptyset$); and either

(c) for each $x \in X \setminus K$, $(F(x) - G(x)) \cap \overline{\cup_{\lambda \geq 0} \lambda(\overline{co}K - x)} \neq \emptyset$ (resp., $(F(x) - G(x)) \cap \overline{\cup_{\lambda \leq 0} \lambda(\overline{co}K - x)} \neq \emptyset$) or,

(c)' for each $x \in X \setminus K$, $(F(x) - G(x)) \cap \overline{\cup_{\lambda \geq 0} \lambda(X_0 - x)} \neq \emptyset$ (resp., $(F(x) - G(x)) \cap \overline{\cup_{\lambda \leq 0} \lambda(X_0 - x)} \neq \emptyset$).

Then there exists $\hat{x} \in X$ such that $F(\hat{x}) \cap G(\hat{x}) \neq \emptyset$.

Proof. Let $x \in K \cap \delta(X)$ and $\phi \in E^*$ be such that $Re\phi(x) \leq Re\phi(y)$ for all $y \in X$. Since $(F(x) - G(x)) \cap \overline{\cup_{\lambda \geq 0} \lambda(X - x)} \neq \emptyset$, let $u \in F(x)$, $v \in G(x)$, $\{\lambda_\alpha\}_{\alpha \in \Gamma}$ be a net in $[0, \infty)$ and $\{x_\alpha\}_{\alpha \in \Gamma}$ be a net in X such that $\lambda_\alpha(x_\alpha - x) \to u - v$. It follows that

$$\lambda_\alpha Re\phi(x_\alpha - x) = Re\phi(\lambda_\alpha(x_\alpha - x)) \to Re\phi(u - v) = Re\phi(u) - Re\phi(v).$$

Since $\lambda_\alpha Re\phi(x_\alpha - x) \geq 0$ for each $\alpha \in \Gamma$, $Re\phi(u) \geq Re\phi(v)$. Thus the condition (b) of Theorem 2.7.8 is satisfied.

Next let $x \in X \setminus K$ and $\phi \in E^*$ be such that $Re\phi(x) \leq Re\phi(y)$ for all $y \in co(\overline{co}K \cup \{x\})$ (resp., $y \in co(X_0 \cup \{x\})$). By (c), since $(F(x) - G(x)) \cap \overline{\cup_{\lambda \geq 0} \lambda(K - x)} \neq \emptyset$ (resp., $(F(x) - G(x)) \cap \overline{\cup_{\lambda \geq 0} \lambda(X_0 - x)} \neq \emptyset$), let $u \in F(x)$, $v \in G(x)$, $\{\lambda_\alpha\}_{\alpha \in \Gamma}$ be a net in $[0, \infty)$ and $\{x_\alpha\}_{\alpha \in \Gamma}$ be a net in $\overline{co}K$ (resp., in $co(X_0 \cup \{x\})$) such that $\lambda_\alpha(x_\alpha - x) \to u - v$. It follows from $Re\phi(x) \leq Re\phi(x_\alpha)$ for all $\alpha \in \Gamma$ that $Re\phi(u) \geq Re\phi(v)$. Thus the condition (c_1) (resp., (c_2)) in Theorem 2.7.9 is satisfied. Therefore by Theorem 2.7.9, there exists $\hat{x} \in X$ such that $F(\hat{x}) \cap G(\hat{x}) \neq \emptyset$.

By interchanging the roles of F and G and by replacing the union "$\cup_{\lambda \geq 0}$" in both conditions (b) and (c) by "$\cup_{\lambda \leq 0}$", the proof is complete. □

Corresponding to Theorem 2.7.14, we also have the following conincidence.

Theorem 2.7.15. Let X be a non-empty convex set in a topological vector space E which has sufficiently many continuous linear functionals. Let X_0 be a non-empty compact convex subset of X and K a non-empty compact subset of X. Let $F, G : X \to 2^E$ be upper hemicontinuous and such that:

(a) for each $x \in X$, $F(x)$ and $G(x)$ are both compact and convex;

(b) for each $x \in K \cap \delta_E(X)$, $(F(x) - G(x)) \cap \overline{\cup_{\lambda \geq 0} \lambda(X - x)} \neq \emptyset$ (resp., $(F(x) - G(x)) \cap \overline{\cup_{\lambda \leq 0} \lambda(X - x)} \neq \emptyset$); and

(c) for each $x \in X \setminus K$, $(F(x) - G(x)) \cap \overline{\cup_{\lambda \geq 0} \lambda(X_0 - x)} \neq \emptyset$ (resp., $(F(x) - G(x)) \cap \overline{\cup_{\lambda \leq 0} \lambda(X_0 - x)} \neq \emptyset$).

Then there exists a point $\hat{x} \in X$ such that $F(\hat{x}) \cap G(\hat{x}) \neq \emptyset$.

As an immediate corollary to Theorem 2.7.14, we have the following fixed point result.

Corollary 2.7.16. Let X be a non-empty convex set in a locally convex topological vector space E, X_0 a non-empty compact convex subset of X, and K a non-empty compact subset of X. Let $F : X \to 2^E$ be upper hemicontinuous and such that:

(a) for each $x \in X$, $F(x)$ is closed and convex;

(b) for each $x \in K \cap \delta_E(X)$, $F(x) \cap [x + \overline{\cup_{\lambda \geq 0} \lambda(X - x)}] \neq \emptyset$ (resp., $F(x) \cap [x + \overline{\cup_{\lambda \leq 0} \lambda(X - x)}] \neq \emptyset$); and either

(c) for each $x \in X \setminus K$, $F(x) \cap [x + \overline{\cup_{\lambda \geq 0} \lambda(\overline{co} K - x)}] \neq \emptyset$ (resp., $F(x) \cap [x + \overline{\cup_{\lambda \leq 0} \lambda(\overline{co} K - x)}] \neq \emptyset$) or

(c)′ for each $x \in X \setminus K$, $F(x) \cap [x + \overline{\cup_{\lambda \geq 0} \lambda(X_0 - x)}] \neq \emptyset$ (resp., $F(x) \cap [x + \overline{\cup_{\lambda \leq 0} \lambda(X_0 - x)}] \neq \emptyset$).

Then there exists a point $\hat{x} \in X$ such that $\hat{x} \in F(\hat{x})$.

Proof. Let $G(x) = \{x\}$ for each $x \in X$. Then the conclusion follows from Theorem 2.7.14. □

Corollary 2.7.17. Let X be a non-empty convex set in a topological vector space E which has sufficiently many continuous linear functionals, X_0 be a non-empty compact convex subset of X. Let $F : X \to 2^E$ be upper hemicontinuous such that:

(a) for each $x \in X$, $F(x)$ is compact and convex;

(b) for each $x \in K \cap \delta_E(X)$, $F(x) \cap [x + \overline{\cup_{\lambda \geq 0} \lambda(X - x)}] \neq \emptyset$ (resp., $F(x) \cap [x + \overline{\cup_{\lambda \leq 0} \lambda(X - x)}] \neq \emptyset$); and

(c) for each $x \in X \setminus K$, $F(x) \cap [x + \overline{\cup_{\lambda \geq 0} \lambda(X_0 - x)}] \neq \emptyset$ (resp., $F(x) \cap [x + \overline{\cup_{\lambda \leq 0} \lambda(X_0 - x)}] \neq \emptyset$).

Then there exists a point $\hat{x} \in X$ such that $\hat{x} \in F(\hat{x})$.

As $F(x)$ is compact and convex in Corollary 2.7.17 it follows that F indeed is upper demicontinuous by Corollary 1.3.7.

We note that Theorem 2.7.8 and Theorem 2.7.14 generalize the corresponding results of Ko and Tan [320] and Ko and Tan [322].

2.7.2 Matching Theorems for Closed Coverings of Convex sets

In this section, as an application of Theorem 2.7.14, we shall consider matching theorems for closed coverings of a convex set.

Theorem 2.7.18. Let X be non-empty convex set of a real locally convex topological vector space E. Let X_0 be a non-empty compact convex subset of E and K a non-empty compact subset of X. Let $\{A_i : i \in I\}$ and $\{B_j : j \in J\}$ be two locally finite families of closed subsets of X and such that $\cup_{i\in I} A_I = \cup_{j\in J} B_j = X$. Let $\{C_i : i \in I\}$ and $\{D_j : j \in J\}$ be two families of non-empty subsets of E such that any finite union of the C_i's is contained in a compact convex subset of E. Let $S : X \to 2^E$ be upper hemicontinuous such that each $S(x)$ is a non-empty compact convex set. Suppose that for each $x \in (K \cap \delta(X)) \cup (X \setminus K)$, there exist $i \in I$ and $j \in J$ such that:
(i) $x \in A_i \cap B_i$; and
(ii) for each $x \in X$, setting $M(x) = \overline{co}(C_i + S(x)) - \overline{co}(D_j)$,

$$M(x) \cap \begin{cases} \overline{\cup_{\lambda \geq 0} \lambda(X - x)} \neq \emptyset, & \text{if } x \in K \cap \delta(X); \\ \overline{\cup_{\lambda \geq 0} \lambda(X_0 - x)} \neq \emptyset, & \text{if } x \in X \setminus K. \end{cases}$$

(resp.,

$$M(x) \cap \begin{cases} \overline{\cup_{\lambda \leq 0} \lambda(X - x)} \neq \emptyset), & \text{if } x \in K \cap \delta(X); \\ \overline{\cup_{\lambda \leq 0} \lambda(X_0 - x)} \neq \emptyset, & \text{if } x \in X \setminus K). \end{cases}$$

Then there exist two non-empty finite subsets I_0 of I and J_0 of J and a point $\hat{x} \in X$ such that

(a) $\hat{x} \in (\cap_{i\in I_0} A_i) \cap (\cap_{j \in J_0} B_j)$;
(b) $\overline{co}(\cup\{C_i : i \in I_0\}) + S(\hat{x})]$ meets the set $\overline{co}(\cup\{D_j : j \in J_0\})$.

Proof. For each $x \in X$, let $I(x) = \{i \in I : x \in A_i\}$ and $J(x) = \{j \in J : x \in B_j\}$. Then $I(x)$ and $J(x)$ are non-empty and finite since $\cup_{i\in I} A_i = \cup_{j\in J} B_j = X$ and $\{A_i : i \in I\}$ and $\{B_j : j \in J\}$ are locally finite. Define $F, G, H : X \to 2^E$ by $F(x) = \overline{co}(\cup\{C_i + S(x) : i \in I(x)\})$; $G(x) = \overline{co}(\cup\{D_j : j \in J(x)\})$ and $H(x) = \overline{co}(\cup\{C_i : i \in I(x)\})$. By hypothesis, for each $x \in X$, $H(x)$ and $S(x)$ are compact convex so that $F(x) = H(x) + S(x)$ is also compact convex. Since $\{A_i : i \in I\}$ is a locally finite family of closed subsets of X, for each $x \in X$, the set $U(x) = X \setminus \cup_{i \notin I(x)} A_i$ is an open neighborhood of x in X. Note that whenever $y \in U(x)$, $y \notin A_i$ for each $i \notin I(x)$ so that $I(y) \subset I(x)$ and therefore $H(y) \subset H(x)$. This shows that H is upper semicontinuous and hence $F = H + S$ is also upper hemicontinuous. Similarly we can show that G is upper semicontinuous (and hence upper hemicontinuous) on X. Thus the condition (a) of Theorem 2.7.14 is satisfied. By (i) and (ii), the conditions (b) and (c) of Theorem 2.7.14 are also satisfied. By Theorem 2.7.14, there exists $\hat{x} \in X$ such that $F(\hat{x}) \cap G(\hat{x}) \neq \emptyset$. Let $I_0 = I(\hat{x})$ and $J_0 = J(\hat{x})$, then I_0 and J_0 are non-empty and finite and the conclusions of the theorem hold. □

We note the proof above was motivated by Ko and Tan [322] which is a modification of Theorem 11 of Fan [167] and of Theorem 1 of Shih and Tan [469]. Also, Theorem 2.7.18 shows that Theorem 4 of Ko and Tan [322] is still true without assuming that X is paracompact. The following is an easy consequence of Theorem 2.7.18.

By the same proof as in Theorem 2.7.18, but by applying Theorem 2.7.15 instead of Theorem 2.7.14, and by assuming that the family $\{D_j\}_{j\in J}$ also has the same property as that of the family $\{C_i\}_{i\in I}$ (i.e., "any finite union of the D_j's is also contained in a compact convex subset of E"), Theorem 2.7.18 holds if the hypothesis underlying space E is weakened to a topological vector space with sufficiently many continuous linear functionals.

Theorem 2.7.19. Let X be a convex subset of a real locally convex topological vector space E, let X_0 be a non-empty compact convex subset of X and K be a non-empty compact subset of X. Let $\{A_i : i \in I\}$ and $\{B_j : j \in J\}$ be two locally finite families of closed subsets of X such that $\cup_{i\in I} A_i = \cup_{j\in J} B_j = X$. Let $\{C_i : i \in I\}$ and $\{D_j : j \in J\}$ be two families of non-empty subsets of E such that any finite union of C_i's is contained in a compact convex subset of E. Suppose that for each point $x \in (K \cap \delta(X)) \cup (X \setminus K)$, there exist $i \in I$ and $j \in J$ such that:

(i) $x \in A_i \cap B_j$;

(ii) setting $M := \overline{co}(C_i) - \overline{co}(D_j)$, then

$$M \cap \begin{cases} x + \overline{\cup_{\lambda \geq 0} \lambda(X - x)} \neq \emptyset, & \text{if } x \in K \cap \partial_E X; \\ x + \overline{\cup_{\lambda \geq 0} \lambda(X_0 - x)} \neq \emptyset, & \text{if } x \in X \setminus K); \end{cases}$$

(resp.,

$$M \cap \begin{cases} x + \overline{\cup_{\lambda \leq 0} \lambda(X - x)} \neq \emptyset), & \text{if } x \in K \cap \partial_E X; \\ x + \overline{\cup_{\lambda \leq 0} \lambda(X_0 - x)}) \neq \emptyset), & \text{if } x \in X \setminus K. \end{cases}$$

Then there exist non-empty finite subsets I_0 of I and J_0 of J such that

$$(\cap_{i\in I_0} A_i) \cap (\cap_{j\in J_0} B_j) \cap (\overline{co}(\cup_{i\in I_0} C_i) - \overline{co}(\cup_{j\in J_0} D_j)) \neq \emptyset.$$

Proof. Let $S : X \to 2^E$ be defined by $S(x) = \{-x\}$ for each $x \in X$. Then all hypotheses of Theorem 2.7.18 are satisfied so that there exist non-empty finite subsets I_0 of I and J_0 of J and a point $\hat{x} \in X$ such that

(a) $\hat{x} \in (\cap_{i\in I_0} A_i) \cap (\cap_{j\in J_0} B_j)$; and

(b) $(\overline{co}(\cup_{i\in I_0} C_i) - \hat{x}) \cap (\overline{co}(\cup_{j\in J_0} D_j) \neq \emptyset$. It follows that

$$\hat{x} \in (\cap_{i\in I_0} A_i) \cap (\cap_{j\in J_0} B_j) \cap (\overline{co}(\cup_{i\in I_0} C_i) - \overline{co}(\cup_{j\in J_0} D_j)). \qquad \square$$

If the space E in Theorem 2.7.19 is a topological vector space which has sufficiently many continuous linear functionals, by the same proof in Theorem 2.7.19, we have:

Theorem 2.7.20. Let X be a convex subset of a topological vector space E which has sufficiently many linear functionals. Let X_0 be a non-empty compact convex subset of X and K be a non-empty compact subset of X. Let $\{A_i : i \in I\}$ and $\{B_j : j \in J\}$ be two locally finite families of closed subsets of X such that $\cup_{i\in I} A_i = \cup_{j\in J} B_j = X$. Let $\{C_i : i \in I\}$ and $\{D_j : j \in J\}$ be two families of non-empty subsets of E such that any finite union of C_i's and D_j's is contained in a compact convex subset of E. Suppose that for each point $x \in (K \cap \delta(X)) \cup (X \setminus K)$, there exist $i \in I$ and $j \in J$ such that:
(i) $x \in A_i \cap B_j$; and
(ii) setting $M := \overline{co}(C_i) - \overline{co}(D_j)$, then

$$M \cap \begin{cases} x + \overline{\cup_{\lambda\geq 0}\lambda(X - x)} \neq \emptyset & \text{if } x \in K \cap \partial_E X; \\ x + \overline{\cup_{\lambda\geq 0}\lambda(X_0 - x)} \neq \emptyset, & \text{if } x \in X \setminus K; \end{cases}$$

(resp.,

$$M \cap \begin{cases} \cap x + \overline{\cup_{\lambda\leq 0}\lambda(X - x)} \neq \emptyset), & \text{if } x \in K \cap \partial_E X; \\ (x + \overline{\cup_{\lambda\leq 0}\lambda(X_0 - x)}) \neq \emptyset), & \text{if } x \in X \setminus K). \end{cases}$$

Then there exist non-empty finite subsets I_0 of I and J_0 of J such that

$$(\cap_{i\in I_0} A_i) \cap (\cap_{j\in J_0} B_j) \cap (\overline{co}(\cup_{i\in I_0} C_i) - \overline{co}(\cup_{j\in J_0} D_j)) \neq \emptyset.$$

We note that Theorems 2.7.18 and 2.7.19 show that Theorems 4 and 5 of Ko and Tan [322] hold without assuming that the set X is paracompact.

As an application of Theorem 2.7.19, we have the following result which is Theorem 13 of Fan in [167] and is also a generalization of Shapley's theorem [460].

Corollary 2.7.21. Let Δ be an n-dimensional simplex in a Euclidean space. Let $\mathcal{F}$ denote the family of all faces of Δ (of all dimensional $0, 1, \cdots, n$). For each $\tau \in \mathcal{F}$, let $p(\tau)$ and $q(\tau)$ be two given points in Δ, and let $A(\tau)$, $B(\tau)$ be two closed subsets of Δ such that:
(a) $\cup_{\tau\in\mathcal{F}} A(\tau) = \cup_{\tau\in\mathcal{F}} B(\tau) = \Delta$; and
(b) for each $\tau \in \mathcal{F}$ of dimensional $< n$ and for any point $x \in \tau$, there is a $\rho \in \mathcal{F}$ such that $x \in B(\rho)$ and $q(\rho) \in \tau$.

Then there exist two non-empty subfamilies $\mathcal{G}$ and $\mathcal{H}$ of $\mathcal{F}$ such that

(c) $[\cap_{\tau\in\mathcal{G}}A(\tau)]\cap[\cap_{\rho\in\mathcal{H}}B(\rho)]\neq\emptyset$, and

(d) $co(\{p(\tau):\tau\in\mathcal{G}\})\cap co(\{q(\rho):\rho\in\mathcal{H}\})\neq\emptyset$.

For a generalization of the above result, we refer to Theorem 6 of Ko and Tan ([322]).

Finally we have the following Fan-Glicksberg fixed point theorem which is a special case of Corollary 2.7.16 above.

Theorem 2.7.22 (Fan-Glicksberg Fixed Point Theorem). Let X be a non-empty compact and convex subset of a locally convex Hausdorff topological vector space E and $F: X\to 2^X$ be an upper semicontinuous mapping with non-empty closed and convex values. Then F has at least one fixed point in X.

2.8 Generic Stability of Coincident Points of Set-Valued Mappings

In [267], Jiang introduced the concept of essential fixed points for set-valued mappings and proved a corresponding approximation theorem. The concept of essentiality for fixed points is a stability property. In Fort [179] and Jiang [267], the stability of fixed points with respect to perturbations of mappings were studied. In this section, the concept of essential coincident points of set-valued mappings is given first. We then study the stability of coincident points and fixed points of set-valued mappings with perturbations of mappings and of constrained sets. Some new approximation theorems are also established.

2.8.1 Generic Stability of Coincident Points

Let (X,d) be a metric space and $K(X)$ be the space of all non-empty compact subsets of X equipped with the Hausdorff metric h which is induced by the metric d. For any $\epsilon>0$, $x_0\in X$ and $A\in K(X)$, let $U(\epsilon,A)=\{x\in X: d(u,x)<\epsilon \text{ for some } u\in A\}$ and $O(x_0,\epsilon)=\{x\in X: d(x_0,y)<\epsilon\}$. Let Y be a topological space. Recall that a subset $Q\subset Y$ is called a *residual set* if it is a countable intersection of open dense subsets of Y.

The following Lemma 2.8.1 is due to Fort [179, Theorem 2].

Lemma 2.8.1. Let X be a metric space, Y a topological space and $F: Y\to K(X)$ an usco mapping. Then the set of points where F is lower semicontinuous is a residual set in Y.

Lemma 2.8.2. Let X be a metric space, Y be a complete metric space and $F: Y\to K(X)$ be an usco mapping. Then the set of points where F is lower semicontinuous is a dense residual set in Y.

Proof. Since Y is complete, a residual set in Y is dense; the result now follows from Lemma 2.8.1. □

Throughout the remainder of this section, X denotes a complete metric space and $C = \{f : X \to K(X) : f$ is upper semicontinuous on X and $f(X) = \cup_{x\in X} f(x)$ is bounded $\}$. For each f, $f' \in C$, let $\rho(f, f') = \sup_{x\in X} h(f(x), f'(x))$. Clearly, ρ is a metric on C.

Lemma 2.8.3. (C, ρ) is a complete space.

Proof. Let $\{f_n\}_{n=1}^{\infty}$ be each Cauchy sequence in C, then for any $\epsilon > 0$, there is a positive integer $N(\epsilon)$ such that

$$\rho(f_n, f_m) = \sup_{x\in X} h(f_n(x), f_m(x)) < \epsilon$$

for any $n, m \geq N(\epsilon)$. It follows that for each $x \in X$, $\{f_n(x)\}_{n=1}^{\infty}$ is a Cauchy sequence in $K(X)$. Since $K(X)$ is complete, by Theorem 4.3.9 in [317], there is $f : X \to K(X)$ such that $\lim_{m\to\infty} f_m(x) = f(x)$.

For each $x \in X$ and each $\epsilon > 0$ and each $n \geq N(\epsilon)$, the inequality above implies that $f_n(x) \subset U(2\epsilon, f(x))$ and $f(x) \subset U(2\epsilon, f_n(x))$. Fix $n \geq N(\epsilon)$; since $f_n \in C$, there is $\delta > 0$ such that $f_n(x') \subset U(\epsilon, f_n(x))$ whenever $d(x, x') < \delta$. Thus,

$$f(x') \subset U(2\epsilon, f_n(x')) \subset U(3\epsilon, f_n(x)) \subset U(5\epsilon, f(x))$$

whenever $d(x, x') < \delta$. Therefore f is upper semicontinuous on X. It is easy to show that $f(X) = \cup_{x\in X} f(x)$ is bounded so that $f \in C$ and that $f_n \to f$. Hence C is complete. □

Set $Y = C \times C \times K(X)$. For each $y = (f, g, A) \in Y$, $y' = (f', g', A') \in Y$, let

$$D(y, y') = \rho(f, f') + \rho(g, g') + h(A, A')$$

Clearly D is a metric on Y. By Lemma 2.8.3, C is a complete metric space. By Theorem 4.3.9 in [317], $K(X)$ is a complete metric space. Hence Y is also a complete metric space.

Define $M = \{y = (f, g, A) \in Y$: there is $x \in A$ such that $f(x) \cap g(x) \neq \emptyset\}$. Then we have:

Lemma 2.8.4. (M, D) is a complete metric space.

Proof. Since $M \subset Y$ and Y is complete, it is sufficient to prove that M is closed in Y. Let $\{y_n\}_{n=1}^{\infty}$ be a sequence in M and $y_n \to y \in Y$. Set $y_n = (f_n, g_n, A_n)$, $n = 1, 2, \cdots$ and $y = (f, g, A)$; then $f_n \to f$, $g_n \to g$ and $A_n \to A$. For each $n = 1, 2, \cdots$, since $y_n \in M$, there is $x_n \in A_n$ such that $f_n(x) \cap g_n(x) \neq \emptyset$. Since A_n and A are compact and $A_n \to A$, by A.5.1 (ii) of

Mas-Colell [363, p.10], $\cup_{n=1}^{\infty} A_n \cup A$ is compact. Since $x_n \in A_n \subset \cup_{n=1}^{\infty} A_n \cup A$, we may assume without loss of generality that $x_n \to x \in \cup_{n=1}^{\infty} A_n \cup A$. If $x \notin A$, since A is compact, there is $a > 0$ such that $U(a, A) \cap O(x, a) = \emptyset$. Since $A_n \to A$ and $x_n \to x$, there is N_1 such that $A_n \subset U(a, A)$ and $x_n \in O(x, a)$ for all $n \geq N_1$, which contradicts the assumption that $x_n \in A_n$. Hence we must have $x \in A$.

If $f(x) \cap g(x) = \emptyset$, since $f(x)$ and $g(x)$ are compact, there is $b > 0$ such that $U(b, f(x)) \cap U(b, g(x)) = \emptyset$. Since $f_n \to f$ and $g_n \to g$, there is N_2 such that $f_n(u) \subset U(b/2, f(u))$ and $g_n(u) \subset U(b/2, g(u))$ for all $n \geq N_2$ and for all $u \in X$. Since f and g are upper semicontinuous at x and $x_n \to x$, there is $N_3 \geq N_2$ such that $f(x_n) \subset U(b/2, f(x))$ and $g(x_n) \subset U(b/2, g(x))$ for all $n \geq N_3$. Thus for all $n \geq N_3$, $f_n(x_n) \subset U(b/2, f(x_n)) \subset U(b, f(x))$ and $g_n(x_n) \subset U(b/2, g(x_n)) \subset U(b, g(x))$ which contradicts the assumption that $f_n(x_n) \cap g_n(x_n) \neq \emptyset$. Hence we must also have $f(x) \cap g(x) \neq \emptyset$. Therefore $y = (f, g, A) \in M$ so that M is closed in Y. □

For each $y = (f, g, A) \in M$, let $S(y) = \{x \in A : f(x) \cap g(x) \neq \emptyset\}$; note that $S(y) \neq \emptyset$.

Lemma 2.8.5. $S(y) \in K(X)$ for each $y \in M$.

Proof. Let $y = (f, g, A) \in M$ be given. Since $S(y) \subset A$ and A is compact, it is sufficient to prove that $S(y)$ is closed in A. Indeed, let $\{x_n\}_{n=1}^{\infty}$ be any sequence in $S(y)$ such that $x_n \to x \in A$; then $f(x_n) \cap g(x_n) \neq \emptyset$ for each $n = 1, 2, \cdots$. If $x \notin S(y)$, then $f(x) \cap g(x) = \emptyset$. Since $f(x)$ and $g(x)$ are compact, there is $\epsilon > 0$ such that $U(\epsilon_0, f(x)) \cap U(\epsilon_0, g(x)) = \emptyset$. Since f, g are upper semicontinuous at x and $x_n \to x$, there exists N such that $f(x_n) \subset U(\epsilon_0, f(x))$ and $g(x_n) \subset U(\epsilon_0, g(x))$ for all $n \geq N$. It follows that $f(x_n) \cap g(x_n) = \emptyset$ for all $n \geq N$, which is a contradiction. Therefore $x \in S(y)$ and hence $S(y)$ is closed in A. □.

By Lemma 2.8.5 the mapping $y \mapsto S(y)$ defines a mapping $S : M \to K(X)$.

Lemma 2.8.6. S is upper semicontinuous on M.

Proof. Suppose S is not upper semicontinuous at $y \in M$, then there exist $\epsilon_0 > 0$ and a sequence $\{y_n\}_{n=1}^{\infty}$ in M with $y_n \to y$ such that for each $n = 1, 2, \cdots$, there exists $x_n \in S(y_n)$ with $x_n \notin U(\epsilon_0, S(y))$. Let $y_n = (f_n, g_n, A_n)$ and $y = (f, g, A)$, then $f_n \to f$, $g_n \to g$ and $A_n \to A$. Since $x_n \in A_n \subset \cup_{n=1}^{\infty} A_n \cup A$ and $\cup_{n=1}^{\infty} A_n \cup A$ is compact, we may assume without loss of generality that $x_n \to x \in \cup_{n=1}^{\infty} A_n \cup A$. Note that we must have $x \notin U(\epsilon_0, S(y))$. Now the same argument as in the proof of Lemma 2.8.4 shows that $x \in A$ and $f(x) \cap g(x) \neq \emptyset$ so that $x \in S(y)$; this contradicts that $x \notin U(\epsilon_0, S(y))$. Therefore S must be upper semicontinuous. □

Let M_1 be a non-empty closed subset of M. Since M is complete, M_1 is also complete.

Definition 2.8.7. If $y_1 \in M_1$, then a point x in $S(y)$ is called an *essential coincident point* of y_1 with respect to M_1 provided that for any $\epsilon > 0$, there is $\delta > 0$ such that for any $y' \in M_1$ with $D(y_1, y') < \delta$, there exists $x' \in S(y')$ with $d(x, x') < \epsilon$. y is called essential with respect to M_1 if every $x \in S(y)$ is an essential coincident point of y with respect to M_1.

Theorem 2.8.8. S is lower semicontinuous at $y \in M_1$ if and only if y is essential with respect to M_1.

Proof. Suppose S is lower semicontinuous at $y \in M_1$. Then for any $\epsilon > 0$, there is $\delta > 0$ such that for any $y' \in M_1$ with $d(y, y') < \delta$, we have $S(y) \subset U(\epsilon, S(y'))$ so that for any $x \in S(y)$, there is $x' \in S(y')$ with $d(x, x') < \epsilon$. Thus every $x \in S(y)$ is an essential coincident point of y with respect to M_1 and hence y is essential with respect to M_1.

Conversely, suppose that y is essential with respect to M_1. If S were not lower semicontinuous at $y \in M_1$, then there exist $\epsilon_0 > 0$ and a sequence $\{y_n\}_{n=1}^{\infty}$ in M with $y_n \to y$ such that for each $n = 1, 2, \cdots$, there is $x_n \in S(y)$ with $x_n \notin U(\epsilon_0, S(y_n))$. Since $S(y)$ is compact, we may assume that $x_n \to x \in S(y)$. Since x is an essential coincident point of y with respect to M_1, $y_n \to y$ and $x_n \to x$, there is N such that $d(x_n, x) < \epsilon_0/2$ and $x \in U(\epsilon_0/2, S(y_n))$ for all $n \geq N$. Hence $x_n \in O(x, \epsilon_0/2) \subset U(\epsilon_0, S(y_n))$ for all $n \geq N$ which contradicts the assumption that $x_n \notin U(\epsilon_0, S(y_n))$ for all $n = 1, 2, \cdots$. Hence S must be lower semicontinuous at y. □

We shall prove the following approximation theorem.

Theorem 2.8.9. The set of essential points with respect to M_1 is a dense residual set in M_1. In particular, every point in M_1 can be arbitrarily approximated by essential points in M_1.

Proof. By Lemma 2.8.5 and Lemma 2.8.6, $S : M \to K(X)$ is an usco mapping. Since M_1 is complete, by Lemma 2.8.2, the set of points where S is lower semicontinuous is a dense residual set in M_1. By Theorem 2.8.8, the set of essential points in M_1 is a dense residual set in M_1. □

By Lemma 2.8.6, Theorems 2.8.8 and 2.8.9, we have the following result.

Theorem 2.8.10. S is continuous at $y \in M_1$ if and only if y is essential with respect to M_1. Moreover, the set of points at which S is continuous is a dense residual set in M_1.

We remark that S is continuous at $y \in M_1$, if and only if for each $\epsilon > 0$, there is $\delta > 0$ such that $h(S(y), S(y')) < \epsilon$ for each $y' \in M$ with $D(y, y') < \delta$. Theorem 2.8.10 implies that if $y = (f, g, A) \in M_1$, then y is essential with respect to M_1 if and only if its set $S(y)$ of coincident points is stable: $S(y')$ is close to $S(y)$ whenever y' is close to y.

We shall now give a sufficient condition that $y \in M_1$ is essential with respect to M_1.

Theorem 2.8.11. If $y \in M_1$ is such that $S(y)$ is a singleton set, then y is essential with respect to M_1.

Proof. Suppose $S(y) = \{x\}$. By Lemma 2.8.6, S is upper semicontinuous at y. Thus for any $\epsilon > 0$, there is $\delta > 0$ such that for each $y' \in M_1$, $D(y, y') < \delta$ implies $S(y') \subset U(\epsilon, S(y)) = O(x, \epsilon)$ so that $S(y) = \{x\} \subset U(\epsilon, S(y'))$. This shows that S is also lower semicontinuous at y. By Theorem 2.8.8, y is essential with respect to M_1. □

We note that if the given metric d on X is bounded, then the identity mapping $I : X \to X$ belongs to C.

Remark 2.8.12. If (X, d) is compact and if we take $M_1 = (\{I\} \times C \times \{X\}) \cap M = \{(I, f, X) \in Y$: there exists $x \in X$ such that $x \in f(x)\}$, then Theorem 2.8.10 reduces to Theorem 1 of Jiang [267] which is a set-valued generalization of Theorem 1 in [179]. In this case, if (I, f, X) is essential with respect to M_1, then every fixed point x of f in X is an essential fixed point of f (see [267]); i.e., for each $\epsilon > 0$, there exists $\delta > 0$ such that whenever $g : X \to K(X)$ is upper semicontinuous which has a fixed point in X and $\rho(f, g) < \delta$, there is a fixed point x' of g in X with $d(x, x') < \epsilon$. Hence our Theorem 3 generalizes Theorem 1 in [179] and Theorem 1 in [267] in several respects.

Recall that a function $g : X \to X$ is a *contraction* if there is a constant $k \in (0, 1)$ such that $d(g(x), g(y)) \leq kd(x, y)$ for all $x, y \in X$. If g is a contraction on X, then the classical Banach contraction mapping principle implies that g has a unique fixed point. This fact together with Theorem 2.8.11, yield:

Corollary 2.8.13. Suppose (X, d) is compact and $M_1 = \{(I, f, X) \in Y :$ there exist $x \in X$ such that $x \in f(x)\}$. If $g : X \to X$ is a contraction, then (I, g, X) is essential with respect to M_1.

In what follows, let X be a non-empty closed convex subset of a Banach space E and let A be a non-empty compact subset of X. Let $M_2 = \{(I, f, A) \in Y : f(x)$ is a compact convex subset of A for each $x \in X\}$.

Theorem 2.8.14. The set of points $y \in M_2$ which is essential with respect to M_2 is a dense residual set in M_2.

Proof. For each $(I, f, A) \in M_2$, $f : X \to K(X)$ is upper semicontinuous and $f(x)$ is a compact convex subset of A for each $x \in X$; it follows from Fan-Glicksberg fixed point Theorem 2.7.22 (e.g, see Fan [156] or Glicksberg [196]; or Smart [491, Theorem 9.2.3]) that there exists $x \in A$ such that $x \in f(x)$. This shows that $(I, f, A) \in M$. Thus by Theorem 2.8.9, we only need to show that M_2 is closed in M. Indeed, let $\{(I, f_n, A)\}_{n=1}^{\infty}$ be a sequence in M_2 such that $(I, f_n, A) \to (I, f, A) \in M$, then $\rho(f_n, f) = \sup_{x \in X} h(f_n(x), f(x)) \to 0$ as $n \to \infty$. Since for each $x \in X$, $f_n(x) \subset A$ for each $n = 1, 2, \cdots$, it follows that $f(x) = \cap_{n=1}^{\infty}(\cup_{k \geq n} f_k(x)) \subset A$ (e.g., see [317, Theorem 4.3.5, p.43]). To complete the proof it remains to be shown that $f(x)$ is convex for all $x \in X$. If this were false, there exist $x \in X$, $u_1, u_2 \in f(x)$ and $\lambda \in (0,1)$ such that $\lambda u_1 + (1-\lambda)u_2 \notin f(x)$. Since $f(x)$ is compact, there is $\epsilon_0 > 0$ with $O(\lambda u_1 + (1-\lambda)u_2, \epsilon_0) \cap U(\epsilon_0, f(x)) = \emptyset$. Since $\rho(f_n, f) \to 0$ as $n \to \infty$, there is N with $f(x) \subset U(\epsilon_0, f_N(x))$ and $f_N(x) \subset U(\epsilon_0, f(x))$. It follows that there exists $z_1, z_2 \in f_N(x)$ with $d(u_i, z_i) < \epsilon_0$ for $i = 1, 2$. Thus $d(\lambda u_1 + (1-\lambda)u_2, \lambda z_1 + (1-\lambda)z_2) \leq \lambda d(u_1, z_1) + (1-\lambda)d(u_2, z_2) < \epsilon_0$ so that $\lambda z_1 + (1-\lambda)z_2 \in O(\lambda u_1 + (1-\lambda)u_2, \epsilon_0)$. On the other hand, as $f_N(x)$ is convex, $\lambda z_1 + (1-\lambda)z_2 \in f_N(x) \subset U(\epsilon_0, f(x))$ which contradicts $O(\lambda u_1 + (1-\lambda)u_2), \epsilon_0) \cap U(\epsilon_0, f(x)) \neq \emptyset$. Hence $f(x)$ is convex for each $x \in X$. Therefore M_2 is closed in M. □.

2.8.2 Existence of Essential Components of Coincident Points

The study of the stability for both coincident and fixed points is one of the most important issues in nonlinear analysis. Since Kinoshita [311] introduced the notion of essential components for fixed points of single-valued continuous mappings and he proved that for any continuous mapping of the Hilbert cube into itself, there exists at least one essential component of its fixed points. Some new progress has been given by Tan et al. [514], Yu and Luo [597], Luo et al. [355], Isac and Yuan [258], and others. Recently Yu and Xiang [598] also establish the existence of essential components of Ky Fan points and equilibrium points.

In this section, by the concept of essential components of coincident points for set-valued mappings and following the ideas from Section 5 (see, Luo et al. [355], and Isac and Yuan [258]), our aim is to establish the general existence results of essentially connected components of coincident points for non-self, and upper semicontinuous set-valued mappings in normed spaces. As applications, the existence of essentially connected components of fixed points for non-self upper semicontinuous set-valued mappings are also given. These results improve and unify the corresponding results in the literature as special cases. Our study shows that it is possible to establish the stability results once we know the connectedness of coincidence or fixed point sets, without additional assumptions such as the objective mappings are either 'Lipschitz' or 'differential', and so on.

If X and Y are two Hausdorff topological spaces and $T : Y \to 2^X$ is a set-valued mapping. We also denote by $\delta(X)$ the algebraic boundary of X in E as given in Definition 2.6.3. Let X be a non-empty subset of a topological vector space E. We also recall that a non-self set-valued mapping $F : X \to 2^E$ is said to be a weakly inward (resp., outward) mapping if for each $x \in \delta(X)$, $F(x) \cap \overline{I_X(x)} \neq \emptyset$ (resp., $F(x) \cap \overline{O_X(x)} \neq \emptyset$) for all $x \in X$, where $I_X(x)$ (resp., $O_X(x)$) is the so-called inward (resp., outward) set of X at x and defined as $I_X(x) := \cup_{\lambda \geq 0} \lambda(X - x)$ (resp., $O_X(x) := \cup_{\lambda \leq 0} \lambda(X - x)$) for each $x \in X$.

Throughout this section, let X be a nonempty convex subset of a normed linear space E and let S be the set of all non-self and upper semicontinuous set-valued mappings with non-empty convex and compact values from X to E. For each pair $f, g \in S$, we can define a metric ρ_1 by

$$\rho_1(f, g) := \sup_{x \in X} h(f(x), g(x))$$

for each $f, g \in S$, where h is the Hausdorff metric defined on $K(E)$ through the norm $\|\cdot\|$ of E. Clearly, (S, ρ_1) is a metric space. In what follows, we set

$$Y := \{(f, g) \in S \times S : \text{ where } f - g \text{ is a weakly inward mapping}\}$$

and

$$Y' := \{(f, g) \in S \times S : \text{ where } f - g \text{ is a weakly outward mapping}\}.$$

Then (Y, ρ) and (Y', ρ) are metric spaces, where the metric ρ is defined by

$$\rho((f, g), (f', g')) := \rho_1(f, f') + \rho_1(g, g')$$

for each $(f, g), (f', g') \in Y$ (resp., in Y').

For any $y = (f, g) \in Y$, we define a (coincidence) set-valued mapping $F : Y \to 2^X$ by

$$F(y) := \{x \in X : f(x) \cap g(x) \neq \emptyset\}$$

for each $y = (f, g) \in Y$.

As a special case of Theorem 2.7.15, the following result says that the coincidence mapping $F : Y \to 2^X$ takes non-empty values.

Lemma 2.8.15. Let X be a nonempty compact convex subset of a normed space E. Suppose two set-valued mappings f and $g : X \to 2^E$ be upper semicontinuous mappings with nonempty, compact and convex values. For any $x \in \delta(X)$, the set $f(x) - g(x) \cap \overline{I_X(x)} \neq \emptyset$. Then there exists $x^* \in X$ such that $f(x^*) \cap g(x^*) \neq \emptyset$.

Proof. It is a special case of Theorem 2.7.15. □.

For each $y \in Y$, the component of a point $x \in F(y)$ is the union of all connected subsets of $F(y)$ which contain the point x. From Engelking [154], we know that components are connected closed subsets of $F(y)$ and thus they are also compact as $F(y)$ is compact. It is also easy to see that the components of two distinct points of $F(y)$ either coincide or are disjoint, so that all components constitute a decomposition of $F(y)$ into connected pairwise disjoint compact subsets, i.e.,

$$F(y) = \bigcup_{\alpha \in \Lambda} F_\alpha(y)$$

where Λ is an index set, for any $\alpha \in \Lambda$, $F_\alpha(y)$ is a nonempty connected compact and for any $\alpha, \beta \in \Lambda(\alpha \neq \beta)$, $F_\alpha(y) \cap F_\beta(y) = \emptyset$.

In order to study the stability of components of coincident points for set-valued mappings, by Definition 2.8.7 we first recall that for a given $y = (f, g) \in Y$, a coincident point $y \in F(y)$ is said to be essential if for each open neighborhood O of x in X, there exists $\delta > 0$ such that the coincident point set $F(y_1) \cap O \neq \emptyset$ for each $y_1 \in Y$ with $\rho(y, y_1) < \delta$. The pair $y = (f, g) \in Y$ is said to be *weakly essential* if there exists $x \in F(y)$ which is essential; and y is said to be *essential* if all its coincident points $x \in F(y)$ is essential.

Remark 2.8.16. We note that if E is metric space with the metric d, then the Vietoris topology on $K(X)$ coincides with the topology generated by the Hausdorff metric h induced by d (e.g., see Corollary 4.2.3 of Klein and Thompson [317]). Then the coincidence mapping $F : Y \to 2^X$ is continuous at $y = (f, g) \in Y$ if and only if for each $\epsilon > 0$, there is $\delta > 0$ such that for each $Y_1 \in Y$, $h(F(y), F(y_1)) < \epsilon$ whenever $\rho(y, y_1) < \delta$; i.e., $F(y)$ is stable in the sense that $F(y_1)$ converges ("*close*") to $F(y)$ whenever y_1 converges ("*close*") to y, i.e., $\lim_{y_1 \to y} F(y_1) = F(y)$. On the other hand, we do know that, in general, not all coincident set for any given pair $y = (f, g)$ enjoys this continuity as discussed in last section. Therefore, our aim in this part is to establish the general existence results of essentially connected components, we need the following notion (see also Kinoshita [311], or Luo et al. [355]).

Definition 2.8.17. For $y \in Y$, suppose the set $F(y) = \bigcup_{\alpha \in \Lambda} F_\alpha(y)$. Then a component $F_\alpha(y)$ for some $\alpha \in \Lambda$, is said to be an *essential component* of y if for each open set O containing $F_\alpha(y)$, there exists $\delta > 0$ such that for any $y' \in Y$ with $\rho(y, y') < \delta$, $F(y') \cap O \neq \emptyset$.

Remark 2.8.18. Definition 2.8.17 abive means that even though we could not expect the continuity for all coincident points for a pair of set-valued mappings in general, however, there is a case that maybe some of coincident points enjoy the continuous stability. In the rest part of this section, we will show that the existence of such nice coincident points for usc set-valued mappings.

In order to establish our existence theorem of essential components for coincident points of usc set-valued mappings in Y, we first need the following result which is a generalization of Theorem 2 in Luo et al. [355] in normed spaces.

Lemma 2.8.19. Let X be a non-empty compact convex set in a normed space $(E, \|\cdot\|)$. Then the set-valued mapping $F : Y \to 2^X$ is upper semicontinuous with nonempty compact valued.

Proof. For each $y = (f,g) \in Y$, we first show that $F(y) \subset X$ compact. As X is compact, it suffices to show that $F(y)$ is closed. Suppose $\{x_\alpha\} \subset F(y)$ is a net such that $x_\alpha \to x_0 \in X$. Since $x_\alpha \in F(y)$, it follows that $f(x_\alpha) \cap g(x_\alpha) \neq \emptyset$. Suppose now $f(x_0) \cap g(x_0) = \emptyset$. As both $f(x_0)$ and $g(x_0)$ are compact, there exists $\delta > 0$ such that $O(f(x_0), \delta) \cap O(g(x_0), \delta) = \emptyset$. By the upper semicontinuity of f and g and $x_\alpha \to x_0$, there is α_0 such that for any $\alpha > \alpha_0$, $f(x_\alpha) \subset O(f(x_0), \delta)$, $g(x_\alpha) \subset O(g(x_0), \delta)$, which contradicts with our assumption that $f(x_\alpha) \cap g(x_\alpha) = \emptyset$. Thus $x_0 \in F(y)$ and $F(y)$ is compact. By a fact that X is compact, in order to show that F is usc, it suffices to show that the graph of F, i.e., the set graph $F := \{(y,x) \in Y \times X : x \in F(y), y \in Y\}$ is closed in $Y \times X$. Suppose $\{(y_\alpha, x_\alpha)\}$ is a net in graph F such that $(y_\alpha, x_\alpha) \to (y_0, x_0)$. We denote by $y_\alpha := (f_\alpha, g_\alpha)$. Then $x_\alpha \in F(y_\alpha)$ and $f_\alpha(x_\alpha) \cap g_\alpha(x_\alpha) \neq \emptyset$. If $f_0(x_0) \cap g_0(x_0) = \emptyset$, since $f_0(x_0)$ and $g_0(x_0)$ are both compact, there exists a positive number $b \in (0,1)$ such that $O(f_0(x_0), b) \cap O(g_0(x_0), b) = \emptyset$. Since $f_\alpha \to f_0$ and $g_\alpha \to g_0$, there exists α_1 such that for each $\alpha \geq \alpha_1$, we have that $f_\alpha(u) \subset O(f_0(u), b/2)$ and $g_\alpha(u) \subset O(g_0(u), b/2)$ for all $u \in X$. Note that f_0 and g_0 are upper semicontinuous at x_0 and $x_\alpha \to x_)$, there exists $\alpha_2 \geq \alpha_1$ such that for each $\alpha \geq \alpha_2$, $f_0(x_\alpha) \subset O(f_0(x_0), b/2)$ and $g_0(x_\alpha) \subset O(g_0(x_0), b/2)$. Thus for all $\alpha \geq \alpha_2$, we have that $f_\alpha(x_\alpha) \subset O(f_0(x_0), b)$ and $g_\alpha(x_\alpha) \subset O(g_0(x_0), b)$ for all $\alpha \geq \alpha_2$. Since $O(f_0(x_0), b) \cap O(g_0(x_0), b) = \emptyset$, it follows that $f_\alpha(x_\alpha) \cap g_\alpha(x_\alpha) = \emptyset$, which is a contradiction. Hence we must also have $f_0(x_0) \cap g_0(x_0) \neq \emptyset$. Therefore the graph of F is closed and the mapping F is upper semicontinuous with nonempty compact values. □

We recall that for given non-empty subsets A and B of a metric space E, the Hausdorff metric h between A and B is defined by $h(A,B) := \inf\{\epsilon : A \subset O(B,\epsilon) \text{ and } B \subset O(A,\epsilon)\}$. Then we have the following simple fact (see Yu and Luo [597] and Luo et al. [355]).

Lemma 2.8.20. Let A, B and C be nonempty convex and bounded subsets of a normed linear space E. Then $h(A, \lambda B + \mu C) \leq \lambda h(A,B) + \mu h(A,C)$ where h is the Hausdorff metric defined on E, $\lambda \geq 0$ and $\mu \geq 0$ with $\lambda + \mu = 1$.

Proof. By the definition of Hausdorff metric $h(A,B)$, it suffices to prove that for any given $\epsilon_1 > 0$ and $\epsilon_2 > 0$ with $B \subset O(A, \epsilon_1)$ and $A \subset O(B, \epsilon_1)$, and $C \subset O(A, \epsilon_2)$ and $A \subset O(C, \epsilon_2)$, we have that $A \subset O(\lambda B + \mu C, \lambda\epsilon_1 + \mu\epsilon_2)$ and $\lambda B + \mu C \subset O(A, \lambda\epsilon_1 + \mu\epsilon_2)$. For any $a \in A$, as $A \subset O(B, \epsilon_1)$ and $A \subset O(C, \epsilon_2)$, there exist $b \in B$ and $c \in C$ such that $d(a,b) < \epsilon_1$ and $d(a,c) < \epsilon_2$. Note that $\lambda + \mu = 1$, it follows that

$$d(a, \lambda b + \mu c) = \|a - \lambda b - \mu c\| \leq \lambda\|a - b\| + \mu\|a - c\| \leq \lambda\epsilon_1 + \mu\epsilon_2,$$

which implies that $A \subset O(\lambda B + \mu C, \lambda\epsilon_1 + \mu\epsilon_2)$. By the convexity of B and

C and the similar argument used above, we can also verify that $\lambda B + \mu C \subset O(A, \lambda\epsilon_1 + \mu\epsilon_2)$ and thus the proof is completed. □.

We now have the main result of this section as follows, which says that for each $y \in Y$ (resp., Y_1'), there exists at least one essentially connected components of its coincident points (resp., fixed points) (see also Luo et al. [355], Isac and Yuan [258] and related references therein).

Theorem 2.8.21. Let X be a non-empty compact subset of a normed space E. For any $y \in Y$, there exists at least one essential connected component of $F(y)$.

Proof. For any given suppose that $F(y)$ is decomposed as follows:

$$F(y) = \bigcup_{\alpha \in \Lambda} F_\alpha(y)$$

where Λ is an index set, for any $\alpha \in \Lambda$, $F_\alpha(y)$ is a connected compact and for any $\alpha, \beta \in \Lambda(\alpha \neq \beta)$, $F_\alpha(x) \cap F_\beta(x) = \emptyset$. We shall prove that there exists at least one essential component of $F(y)$. Let us suppose otherwise there is no any essential connected component. Then for any $\alpha \in \Lambda$, there exists an open set $O_\alpha \supset F_\alpha(y)$ such that for any $\varepsilon > 0$, there is $y_\alpha \in Y$ with $\rho(y, y_\alpha) < \varepsilon$ such that $F(y_\alpha) \cap O_\alpha = \emptyset$. As $F(y)$ is compact, there exist two open and finite coverings $\{V_i\}_{i=1}^n$ and $\{W_i\}_{i=1}^n$ which satisfy the following conditions (e.g., see Kinoshita [311]): (1) $\overline{W}_i \subset V_i$; (2) $V_i \cap V_j = \emptyset$ for each $i \neq j$; and (3) V_i contains at least one $F_{\alpha_i}(y)$ with $O_{\alpha_i} \supset V_i \supset F_{\alpha_i}(y)$. By Lemma 2.8.19, F is upper semicontinuous at y and $\bigcup_{i=1}^n W_i \supset F(y)$ and $\bigcup_{i=1}^n W_i$ is open, then there exists a $\delta > 0$ such that $\bigcup_{i=1}^n W_i \supset F(y')$ for any $y' \in Y$ with $\rho(y, y') < \delta$. Thus there is $y_{\alpha_i} \in Y$ with $\rho(y, y_{\alpha_i}) < \delta$ such that $F(y_{\alpha_i}) \cap O_{\alpha_i} = \emptyset$.

Let $y = (f, g)$ and $y_{\alpha_i} = (f_{\alpha_i}, g_{\alpha_i})$, where $i = 1, \cdots, n$. We define two set-valued mappings $f^*, g^* : X \to 2^E$ by

$$f^*(x) = \begin{cases} f(x), & \text{if } x \in X \setminus \bigcup_{i=1}^n V_i, \\ f_i^*(x), & \text{if } x \in V_i \text{ for some } i = 1, 2, \cdots, n \end{cases}$$

and

$$g^*(x) = \begin{cases} g(x), & \text{if } x \in X \setminus \bigcup_{i=1}^n V_i, \\ g_i^*(x), & \text{if } x \in V_i \text{ for some } i = 1, 2, \cdots, n \end{cases}$$

where $f_i^*, g_i^* : X \to 2^E$ are set-valued mappings defined by

$$f_i^*(x) := \lambda_i(x) f(x) + \mu_i(x) f_{\alpha_i}(x)$$

and

$$g_i^*(x) := \lambda_i(x) g(x) + \mu_i(x) g_{\alpha_i}(x)$$

for each $x \in X$; and λ_i and μ_i are two functions from X to $\mathbb{R}$ defined by

$$\lambda_i(x) := \frac{d(x, \overline{W}_i)}{d(x, \overline{W}_i) + d(x, X \setminus \bigcup_{i=1}^n V_i)}$$

and

$$\mu_i(x) := \frac{d(x, X \setminus \bigcup_{i=1}^n V_i)}{d(x, \overline{W}_i) + d(x, X \setminus \bigcup_{i=1}^n V_i)}$$

for each $x \in X$. It is clear that λ_i and μ_i are continuous. Note that f, g, f_{α_i} and g_{α_i} are upper semicontinuous with compact values, it follows by Theorems 7.3.11, 7.3.14 and 7.3.15 of Klein and Thompson [317] that f_i^* and g_i* are upper semicontinuous. Secondly, for each given $x \in X$, if $x \in X \setminus \cup_{i=1}^n V_i$, then $f(x) = f_i^*(x)$ and $g(x) = g_i^*(x)$ if $x \in X \setminus \cup_{i=1}^n V_i$ for all $i = 1, 2, \cdots, n$; and $f_i^*(x) := f_{\alpha_i}(x)$ and $g_i^*(x) := g_{\alpha_i}(x)$ if $x \in \overline{W}_i$. Next we show that both f^* and g^* are upper semicontinuous on X. Without loss of generality, let G be any given open subset of E^* and suppose there exists $x_0 \in X$ such that $f^*(x_0) \subset G$. If there exists $i \in \{1, 2, \cdots, n\}$ By the upper semicontinuity of f_i^*, there exists a non-empty open neighbourhood $N(x_0)$ of x_0 in V_i such that $f^*(u) = f_i^*(u) \subset G$. As V_i is open in X, we may assume that $N(x_0)$ is also an open neighbrhood of X and thus f^* is upper semicontinuous at $x_0 \in V_i$ by the definition of upper semicontinuity. In the case $x_0 \in X \setminus \cup_{i=1}^n V_i$, by the facts that $f^*(x_0) = f_i^*(x_0) = f(x)$ for all $i = 1, 2, \cdots, n$, and the upper semicontinuity of f and f_i, it follows that for each $i = 1, 2, \cdots, n$, there exists a non-empty open neighbourhood $V_i(x_0)$ of x_0 in X such that $f_i^*(u) \subset G$ for all $u \in V_i(x_0)$ (as $f_i^*(x_0) \subset G$). Since f is also upper semicontinuous, there exists a non-empty open neighbourhood $V_0(x_0)$ of x_0 in X such that $f(u) \subset G$ for each $u \in V_0(x_0)$. Let $N(x_0) := \cap_{i=0}^n V_i(x_0)$. Then $N(x_0)$ is a non-empty open neighbourhood of x_0 in X. Moreover, it is easy to see that $f^*(u) \subset G$ for each $u \in N(x_0)$. Indeed, if $u \in N(x_0) \cap (X \setminus \cup_{i=1}^n V_i)$, then $f^*(u) = f(u) \subset G$; in the case $u \in N(x_0) \cap V_i$ for some $i = 1, \cdots, n$, then $f^*(u) = f_i^*(u) \subset G$. Therefore we have shown that for each open neighbourhood G of E^* with $f^*(x_0) \subset G$ for some $x_0 \in X$, there exists a non-empty open neighbourhood $N(x_0)$ of x_0 such that $f^*(u) \subset G$ for each $u \in N(x_0)$. This means f^* is upper semicontinuous. By the same reason, the mapping $g^* : X \to 2^E$ is also upper semicontinuous. Moreover both f^* and g^* take nonempty closed and convex values. Next we want to show that for any $x \in \delta(X)$, $f^* - g^* \cap \overline{I_X(x)} \neq \emptyset$. First, if $x \in \delta(X)$ and $x \in X \setminus \bigcup_{i=1}^n V_i$, it follows that $f^*(x) - g^*(x) = f(x) - g(x)$, and then we have that

$$(f^*(x) - g^*(x)) \cap \overline{I_X(x)} \neq \emptyset = (f(x) - g(x)) \cap \overline{I_X(x)} \neq \emptyset.$$

Secondly, if $x \in \delta(X)$ and $x \in \overline{W}_i$, it follows that $f^*(x) - g^*(x) = f_{\alpha_i}(x) - g_{\alpha_i}(x)$ and then we have that

$$(f^*(x) - g^*(x)) \cap \overline{I_X(x)} \neq \emptyset = (f_{\alpha_i}(x) - g_{\alpha_i}(x)) \cap \overline{I_X(x)} \neq \emptyset.$$

Finally, if $x \in \delta(X)$ and $x \in V_i \setminus \overline{W}_i$, it follows that

$$\begin{aligned} f^*(x) - g^*(x) &= (\lambda_i(x)f(x) + \mu_i f_{\alpha_i}(x)) - (\lambda_i(x)g(x) + \mu_i g_{\alpha_i}(x)) \\ &= \lambda_i(x)(f(x) - g(x)) + \mu_i(x)(f_{\alpha_i}(x) - g_{\alpha_i}(x)). \end{aligned}$$

Since X is convex, we have that $\overline{I_X(x)} \neq \emptyset$ is convex. Note that $(f(x) - g(x)) \cap \overline{I_X(x)} \neq \emptyset$, $(f_{\alpha_i}(x) - g_{\alpha_i}(x)) \cap \overline{I_X(x)} \neq \emptyset$, and $\lambda_i(x) + \mu_i(x) = 1$, it follows that

$$\begin{aligned} (f^*(x) - g^*(x)) \quad &\cap \quad \overline{I_X(x)} \\ &= \overline{I_X(x)} \neq \emptyset. \end{aligned}$$

Hence for any $x \in \delta(X)$, we show that

$$(f^*(x) - g^*(x)) \cap \overline{I_X(x)} \neq \emptyset$$

which means that $y^* = (f^*, g^*) \in Y$, so that $F(y^*) \neq \emptyset$. Note that $\rho(y, y_{\alpha_i}) < \delta$ for $i = 1, 2, \cdots, n$, it follows by Lemma 2.8.20 that

$$h(f(x), \lambda_i(x)f(x) + \mu_i(x)f_{\alpha_i}(x)) \leq h(f(x), f_{\alpha_i}(x))$$

and

$$h(g(x), \lambda_i(x)g(x) + \mu_i(x)g_{\alpha_i}(x)) \leq h(g(x), g_{\alpha_i}(x)).$$

Therefore $\rho(y, y^*) < \delta$ and $F(y^*) \subset \bigcup_{i=1}^n W_i$. Note that for any $x_0 \in F(y^*)$, there is an index i_0 such that $x_0 \in W_{i_0}$, and hence $x_0 \in W_{i_0} \subset \overline{W}_{i_0} \subset O_{\alpha_{i_0}}$. Therefore $f^*(x_0) = f_{\alpha_{i_0}}(x_0)$, $g^*(x_0) = g_{\alpha_{i_0}}(x_0)$ and $x_0 \in F(y_{\alpha_{i_0}})$. Thus contradicts our assumption that that $F(y_{\alpha_{i_0}}) \cap O_{\alpha_{i_0}} = \emptyset$. Hence there exists at least one essential connected component of $F(y)$ and completing the proof.□

Remark 2.8.22. By the same proof as used in Theorem 2.8.21, we can also prove the following existence of essential components of coincident points for weakly outward usc set-valued mappings in normed spaces.

Theorem 2.8.23. Let X be a non-empty compact subset of a normed space E. For any $y \in Y'$, there exists at least one essential connected component of $F(y)$.

Proof. The conclusion follows by the same proof used in Theorem 2.8.21. □.

Remark 2.8.24. For any $f \in S$, if g is the identity mapping, i.e., $g(x) = I(x) := x$ for all $x \in X$. Then for any $x \in \delta(X)$, the boundary condition becomes

$$(f(x) - g(x)) \cap \overline{I_X(x)} = (f(x) - x)) \cap \overline{I_X(x)}$$

In order to consider the existence of essentially connected components of fixed points for non-self upper semicontinuous mappings, let

$$Y_1 := \{f \in S : \text{ such that } f(x) \cap x + \overline{I_X(x)} \neq \emptyset\}$$

(resp.,

$$Y_1' := \{f \in S: \text{ such that } f(x) \cap x + \overline{O_X(x)} \neq \emptyset\}).$$

Then Y_1 (resp., Y_1') is a subset of Y (resp., Y_1'). As a special case of Theorem 2.8.21 (resp., 2.8.23), we have the following existence of essential components for fixed points of non-self mappings which are upper semicontinuous in normed spaces.

Theorem 2.8.25. Let X be a non-empty compact and convex subset of a normed space E. Then the set of fixed points for $f \in Y_1$ has, at least one essential component.

Theorems 2.8.21, 2.8.23 and 2.8.25 show that for weakly inward or outward set-valued mappings the set-valued version of the corresponding Theorem 3 of Kinoshita [311] holds, and Theorems 2.8.21 and 2.8.25 also include the corresponding results of Luo et al. [355] and Isac and Yuan [258] as special cases.

As another application of Theorems 2.8.21 and 2.8.23, we have the following stability of coincidence and fixed points for non-self upper semicontinuous set-valued mappings.

Theorem 2.8.26. Let X be a non-empty compact and convex subset of a normed space E. For any $y = (f, g) \in Y$ or Y' (resp., $y = f \in Y_1$ or Y_1'), we have the following result:

(i) if the coincident point (resp., fixed point set) $F(y)$ is a totally disconnected, then y is weakly essential; and

(ii) if the coincident point (resp., fixed point) set $F(y)$ is connected, then y is essential, i.e., the coincident points (resp., fixed points) of $y = (f, g) \in Y$ (resp., f) is stable.

Proof. It is immediately consequences of Theorems 2.8.21 and 2.8.23. □

Remark 2.8.27. Theorem 2.8.26(ii) shows that if the coincidence (resp., fixed point) set of any given pair $y = (f, g) \in Y$ or Y' (resp., $y = f \in Y_1$ or Y_1') is connected, then the coincidence (resp., fixed point) mapping F is continuous at y. More precisely, we have the following result.

Corollary 2.8.28. Let X be a non-empty compact subset of a normed space E. Suppose $f, g : X \to 2^E$ are two upper semicontinuous set-valued mappings with non-empty compact and convex values such that $f - g$ is either weakly inward or weakly outward. If the coincidence set of $y = (f, g)$ is connected, then all coincident points of $y = (f, g)$ is stable, i.e., the coincidence set mapping $F : Y(\text{resp., } Y') \to 2^X$ is continuous at $y = (f, g) \in Y$ (resp., Y'). In particular, for any given upper semicontinuous set-valued mapping $f : X \to 2^E$ which is either weakly inward or weakly outward with non-empty compact and convex values, if the fixed point set of f is connected, then all fixed points of f is essential in the sense that the fixed point set mapping $F : Y_1(\text{resp.}, Y_1') \to 2^E$ is continuous at $f \in Y_1$ (resp., Y_1').

Corollary 2.8.28 shows that once we know the connectedness of coincidence or fixed point set, the stability results can be established without traditional assumptions such as the objective mappings are either 'Lipschitz' or 'differential', and so on.

2.9 Ky Fan Section Theorems

In this section, extensions of Fan's section theorems are obtained. As applications, several minimax inequalities and fixed point theorems are derived involving in topological spaces (not necessarily having linear structures). One of these minimax inequality will be used to study Fan's alternative theorem in next section for set-valued upper semicontinuous mappings in locally convex vector spaces.

In 1961, Fan [157] proved the generalization of the classical KKM theorem in infinite dimensional Hausdorff topological vector spaces and established an elementary but very basic 'geometric lemma' for set-valued mappings. It was stated as:

Theorem 2.9.1. Let E be a Hausdorff topological vector space, $X \subset E$ be a non-empty compact convex subset and A be a subset of $X \times X$ such that:
(a) for each $y \in X$, the set $\{x \in X : (x, y) \in A\}$ is closed in X;
(b) for each $x \in X$, the set $\{y \in X : (x, y) \notin A\}$ is convex or empty; and
(c) for each $x \in X$, $(x, x) \in A$.

Then there exists a point $x_0 \in X$ such that $\{x_0\} \times X \subset A$.

In 1968, Browder [59] gave a fixed point form (by different techniques) of Fan's geometric lemma and it is now called Fan -Browder fixed point theorem. Fan's geometric lemma above was stated by himself [164] into the following equivalent form for convex sets (later it is called the Fan's section theorem).

Theorem 2.9.2. Let X be a non-empty compact convex subset of a Hausdorff topological vector space and let $B \subset X \times X$. Assume that:
(a) for each fixed $x \in X$, the section $\{y \in X : (x, y) \in B\}$ is open in X; and
(b) for each fixed $y \in X$, the set $\{x \in X : (x, y) \in B\}$ is non-empty and convex.

Then there exists a point $x_0 \in X$ such that $(x_0, x_0) \in B$.

Fan's theorem above has numerous connections with other areas of mathematics and serves to unify many results in the literature, in particular in the study of minimax inequality theory, fixed point theory for set-valued mappings, mathematical economics, game theory, and so on (e.g, see Aubin [10], a number of its generalizations and applications have been given by Bardaro and Ceppitelli [24], Ben-El-Mechaiekh et al. [35], Chang and Yang [72], Degundji and Granas [148], Ding and Tan [136], Fan [162]-[167], Granas and Liu [209], Ha [213]-[215], Horvath [237] Park [420], Shih and Tan [472], Tan and Yuan [507], Tarafdar [533], and others. It is our purpose in this section to give some generalizations of Fan's section theorems. Then as applications, some new Ky Fan minimax inequalities and fixed point theorems are given.

Now we explain some definitions and notations. The set of all real numbers is denoted by $\mathbb{R}$ and the set of natural numbers is denoted by $\mathbb{N}$. If X is a set, we shall denote by 2^X the family of all subsets of X. Let A be a subset of a topological space X. We shall denote by $\text{int}_X(A)$ the interior of A in X and by $cl_X(A)$ the closure of A in X. If A is a subset of a vector space, we shall denote by coA the convex hull of A. Let X and Y be two non-empty sets and $T : X \to 2^Y$ be a mapping. Then the graph of T, denoted by graph T is the set $\{(x, y) \in X \times Y : y \in T(x)\}$. Suppose X is a non-empty convex subset of a topological vector space E and $f : X \to \mathbb{R} \cup \{-\infty, +\infty\}$ is an extended valued function. Then f is said to be *quasi-concave* (resp., *quasi-convex*) if the set $\{x \in X : f(x) > \lambda\}$ (resp., the set $\{x \in X : f(x) < \lambda\}$) is convex for each $\lambda \in \mathbb{R}$. A topological space X is said to be *contractible* if the identity mapping I_X of X is homotopic to a constant function.

We recall that a non-empty space is siad to be *acyclic* if all of its reduced Čech homology groups over rationals vanish. In particular, each contractible space is acyclic and thus any non-empty convex or star-shaped set is acyclic.

By employing the same arguments of Eilenberg and Montgomery [152, p.106-107], and using Theorem 6.3 of Gòrniewicz [201, p.111] in his paper published in 1975, instead of the coincidence theorem used in [152, p.106-107], the following lemma was obtained by Shioji [475, p.188]. As it involves a long proof parallel to the one used by Eilenberg and Montgomery in their paper published in 1946, we thus omit its details here (see also Lemma 2.1 of Park et al. [424]).

Lemma 2.9.3 (Eilenberg and Montgomery, Gòrniewicz and Shioji and Park). Let Δ_N be an n-dimensional simplex with the Euclidean topology and Y be a compact topological space. Let $\psi : Y \to \Delta_N$ be a single-valued continuous mapping and $T : \Delta_N \to 2^Y$ be a set-valued upper semicontinuous mapping with non-empty compact acyclic (e.g., contractible) values. Then there exists $x_0 \in \Delta_N$ such that $x_0 \in \psi \circ T(x_0)$, where $\psi \circ T$ denotes the composition of mapping T with ψ.

As an application of Lemma 2.9.3, we have the following section theorem which generalizes Theorem 3 of Ha [213] which in turn improves Fan's section Theorem in [157] and [164].

Theorem 2.9.4. Let X be a topological space and Y be a non-empty convex subset of a Hausdorff topological vector space F. Suppose that A is a subset of $X \times Y$ and there exists a subset B of A and a non-empty compact subset K of X such that B is closed in $X \times Y$ such that:

(a) for each $y \in Y$, the set $\{x \in X : (x, y) \in A\}$ is closed in X;

(b) for each $x \in X$, the set $\{y \in Y : (x, y) \notin A\}$ is convex or empty; and

(c) for each $y \in Y$, the set $\{x \in K : (x, y) \in B\}$ is non-empty and contractible.

Then there exists $x_0 \in K$ such that $\{x_0\} \times Y \subset A$.

Proof. Following Ha [213], for each $y \in Y$, let $A(y) = \{x \in X : (x, y) \notin A\}$. Suppose that the conclusion of Theorem 2.9.4 were false. Then for each $x \in K$, there exists $y \in Y$ such that $(x, y) \notin A$, i.e., $x \in A(y)$, so that $K \subset \cup_{y \in Y} A(y)$. By (a) and the compactness of K, there exists a finite subset $\{y_1, y_2, \cdots, y_n\}$ of Y such that $K \subset \cup_{j=1}^n A(y_j)$. Let $\lambda_1, \lambda_2, \cdots, \lambda_n$ be non-negative real valued continuous functions on K such that for each $1 \leq j \leq n$, λ_j vanishes on $K \backslash A(y_j)$ and $\Sigma_{j=1}^n \lambda_j = 1$ for all $x \in K$. Define a function $p : K \to Y$ by

$$p(x) = \Sigma_{j=1}^n \lambda_j(x) y_j$$

for each $x \in K$. Note that for each $x \in K$ and $j = 1, 2, \cdots, n$, if $\lambda_j(x) > 0$, then $x \in A(y_j)$. By (b), we have that $(x, p(x)) \notin A$. Now let $Z := co(y_1, \cdots, y_n)$ and define a mapping $q : Z \to 2^K$ by

$$q(z) = \{x \in K : (x, z) \in B\}$$

for each $z \in Z$. Then q has non-empty closed values by (c). Note that B is closed in $X \times Y$ so that the graph of q is also closed in $Z \times K$ and q has non-empty compact contractible values. Since p is continuous, by Lemma 2.9.3, there exists $z_0 \in Z$ such that $z_0 \in p(q(z_0))$. Let $x_0 \in q(z_0)$ is such that $p(x_0) = z_0$. Then $(x_0, p(x_0)) \in B \subset A$ which contradicts that $(x, p(x)) \notin A$ for all $x \in X$. Thus there must exist $x_0 \in K$ such that $\{x_0\} \times Y \subset A$. $\square$

Theorem 2.9.4 improves Theorem 3 of Ha [213] in the following ways (i) the space X may not have linear structure and (ii) the set $\{x \in K : (x, y) \in B\}$ may not be convex.

As an equivalent form of Theorem 2.9.4, we have the following section theorem which improves Fan's section Theorem 2.9.2.

Theorem 2.9.5. Let X be a topological space and Y be a non-empty convex subset of a Hausdorff topological vector space F. Suppose that A_1 is a non-empty subset of $X \times Y$ and there exists a non-empty open subset B_1 of $X \times Y$ and a non-empty compact subset K of X such that:

(a) for each $y \in Y$, the set $\{x \in X : (x, y) \in A_1\}$ is open in X;

(b) for each $x \in X$, the set $\{y \in Y : (x, y) \in A_1\}$ is convex and for each $x \in K$, the set $\{y \in Y : (x, y) \in A_1\} \neq \emptyset$; and

(c) for each $y \in Y$, the set $\{x \in K : (x, y) \notin B_1\}$ is non-empty and contractible.

Then there exists $(x_0, y_0) \in A_1$ such that $(x_0, y_0) \notin B_1$.

Proof. Let $A := X \times Y \setminus A_1$ and $B := X \times Y \setminus B_1$. By condition (b), the conclusion of Theorem 2.9.4 does not hold. Note that A and B satisfy all the conditions (a) and (b) of Theorem 2.9.4. Thus B must not be a subset of A, so that $B_1 \not\supset A_1$. Therefore there exists $(x_0, y_0) \in A_1$ such that $(x_0, y_0) \notin B_1$. □

It is clear that Theorem 2.9.4 and Theorem 2.9.5 include Theorem 2.9.1 and Theorem 2.9.2 as special cases. Let $X = Y = K$ and $B = \Delta$ and $X = Y = K$ in Theorem 2.9.4, where $\Delta = \{(x, x) \in X \times X : x \in X\}$, then Theorem 2.9.4 reduces to Theorem 2.9.1. Let $X = Y = K$ and $B_1 = X \times X \setminus \Delta$ in Theorem 2.9.5. By Theorem 2.9.5, there exists $(x_0, y_0) \in A_1$ with $(x_0, y_0) \notin B_1$, i.e., $(x_0, y_0) \in \Delta$ so that $(x_0, x_0) \in A_1$ and Theorem 2.9.2 follows.

As an analytical form of Theorem 2.9.4, we have the following minimax inequality which is a generalization of Theorem 1 of Ha [215] which in turn improves the well-known Ky Fan minimax inequality in [164].

Theorem 2.9.6. Let X be a non-empty convex subset of a Hausdorff topological vector space E and Y be a compact topological space. Suppose $f : X \times Y \to \mathbb{R} \cup \{-\infty, +\infty\}$ is such that:

(a) for each $y \in Y$, $x \mapsto f(x, y)$ is quasi-convex in X;

(b) for each $x \in X$, $y \mapsto f(x, y)$ is upper semicontinuous; and

(c) suppose $T : X \to 2^Y$ is a set-valued mapping with non-empty contractible compact values.

Then there exists $y_0 \in Y$ such that

$$\inf_{x \in X, y \in T(x)} f(x, y) \leq \inf_{x \in X} f(x, y_0) \leq \max_{y \in Y} \inf_{x \in X} f(x, y).$$

Proof. In order to apply Theorem 2.9.4, let $\lambda = \inf_{x \in X, y \in T(x)} f(x, y)$, $A := \{(x, y) \in X \times Y : f(x, y) \geq \lambda\}$ and $B := \{(x, y) \in X \times Y : y \in T(x)\}$. Then we have:

(a') for each $y \in Y$, the set $\{x \in X : (x, y) \notin A\} = \{x \in X : f(x, y) < \lambda\}$ is convex or empty by (a);

(b') for each $x \in X$, the set $\{y \in Y : (x, y) \in A\} = \{y \in y : f(x, y) \geq \lambda\}$ is closed in X by (b);

(c') note that T is upper semicontinuous with closed values so that B is closed in $X \times Y$; and

(d') for each $x \in X$, the set $\{y \in Y : (x, y) \in B\} = \{y \in Y : y \in T(x)\} = T(x)$ which is non-empty and contractible.

It is clear that B=graph $T \subset A$. Now let $K := Y$ exchange both X and Y in Theorem 2.9.4. By Theorem 2.9.4, there exists $y_0 \in Y$ such that $X \times \{y_0\} \subset A$. From the definition of A, we have

$$\inf_{x \in X, y \in T(x)} f(x,y) \leq \inf_{x \in X} f(x, y_0) \leq \max_{y \in Y} \inf_{x \in X} f(x,y). \qquad \square$$

As an application of Theorem 2.9.6, we have the following best approximation theorem which generalizes Theorem 2 of Ha in [215], which in fact can be used to improve Theorem 1 of Fan [162] by using his argument, hence we omit it here.

Theorem 2.9.7. Let X be a non-empty compact convex subset of a Hausdfoff topological vector space E and Y be a compact topological space. Suppose $T : X \to 2^Y$ is upper semicontinuous with non-empty contractible compact values and the extended real-valued function $g : X \times Y \to \mathbb{R} \cup \{-\infty, +\infty\}$ is such that:

(a) g is (jointly) lower semicontinuous and for each fixed $x \in X$, $y \mapsto g(x,y)$ is upper semicontonus (hence, $y \mapsto g(x,y)$ is continuous for each fixed $x \in X$); and

(b) $x \mapsto g(x,y)$ is quasi-convex for each fixed $y \in Y$.

Then there exists $x_0 \in X$ and $y_0 \in T(x_0)$ such that

$$g(x_0, y_0) \leq \inf_{x \in X} g(x, y_0).$$

Proof. Define a mapping $f : X \times Y \to \mathbb{R} \cup \{-\infty, +\infty\}$ by

$$f(x,y) = g(x,y) - \inf_{z \in X} g(z,y)$$

for each $(x,y) \in X \times Y$. Clearly f satisfies the condition (a) of Theorem 2.9.6. Note that $y \mapsto g(x,y)$ is continuous for each fixed $x \in X$, therefore f is lower semicontinuous by (a). Hence by Theorem 2.9.6, there exists $y_0 \in Y$ such that

$$\inf_{x \in X, y \in T(x)} f(x,y) \leq \inf_{x \in X} f(x, y_0) = \inf_{x \in X} [g(x, y_0) - \inf_{z \in X} g(z, y_0)] = 0.$$

Note that T has a closed graph which is compact in $X \times Y$, and $x \mapsto f(x,y)$ is lower semicontinuous for each fixed $y \in Y$, so that the infimum of left side of (1) above is attained. Thus there exists $y_0 \in T(x_0)$ such that $g(x_0, y_0) \leq \inf_{x \in X} g(x, y_0)$. $\square$

As another application of Theorem 2.9.4, we have the following minimax inequality.

Theorem 2.9.8. Let X be a topological space and Y be a non-empty convex subset of a Hausdorff topological vector space F. Suppose $f : X \times Y \to \mathbb{R} \cup \{-\infty, +\infty\}$ is lower semicontinuous such that:

(a) for each $x \in X$, $y \mapsto f(x,y)$ is quasi-concave; and

(b) for each non-empty compact subset K of X, the set $\{x \in K : f(x,y) \le t\}$ is contractible for each $t \in \mathbb{R}$.

Then we have,

$$\inf_{x\in X}\sup_{y\in Y} f(x,y) = \inf_{K\subset X}\sup_{y\in Y}\min_{x\in K} f(x,y) \tag{2.9.1}$$

where the infimum on the right side of (2.9.1) is taken over the non-empty compact subsets K of X. If in addition, X is compact, then

$$\min_{x\in X}\sup_{y\in Y} f(x,y) = \sup_{y\in Y}\min_{x\in X} f(x,y).$$

Proof. It is clear that $\inf_{x\in X}\sup_{y\in Y} f(x,y) \le \sup_{y\in Y}\inf_{x\in X} f(x,y)$. To prove (2.9.1), we can assume that $\inf_{K\subset X}\sup_{y\in Y}\inf_{x\in K} f(x,y) < +\infty$. Choose $t \in \mathbb{R}$ with $t > \inf_{K\subset X}\sup_{y\in Y}\inf_{x\in K} f(x,y)$ and let $A = B = \{(x,y) \in X \times Y : f(x,y) \le t\}$. It is clear that A satisfies the conditions (a) and (b) of Theorem 2.9.4. Let K be a non-empty compact subset of X such that $t > \sup_{y\in Y}\inf_{x\in K} f(x,y)$. Then for each $y \in Y$, the set $\{x \in K : f(x,y) \le t\}$ is contractible and non-empty by the choice of K. By Theorem 2.9.4, there exists $x_0 \in K$ such that $\{x_0\} \times Y \subset A$, that is, $f(x_0,y) \le t$ for all $y \in Y$. Thus $\inf_{x\in X}\sup_{y\in Y} f(x,y) \le t$ and hence the equality of (2) holds. In addition, if X is compact, then we have:

$$\min_{x\in X}\sup_{y\in Y} f(x,y) = \inf_{K\subset X}\sup_{y\in Y}\min_{x\in K} f(x,y) = \sup_{y\in Y}\min_{x\in X} f(x,y). \quad \square$$

As an immediate consequence of Theorem 2.9.8, we have the following corollary which is Theorem 4 of Ha in [213].

Corollary 2.9.9 (Ha [213]). Let X be a non-empty convex subset of a topological space E and Y be a non-empty convex subset of a Hausdorff topological vector space F. Suppose $f : X \times Y \to \mathbb{R} \cup \{-\infty, +\infty\}$ is lower semicontinuous such that:

(a) for each $x \in X$, $y \mapsto f(x,y)$ is quasi-concave; and

(b) for each $y \in Y$, $x \mapsto f(x,y)$ is quasi-convex;

then

$$\inf_{x\in X}\sup_{y\in Y} f(x,y) = \inf_{K\subset X}\sup_{y\in Y}\min_{x\in K} f(x,y) \tag{2.9.2}$$

where the infimum on the right side of (2.9.2) is taken over all non-empty compact convex subset K of X. If in addition, X is compact, then

$$\inf_{x\in X}\sup_{y\in Y} f(x,y) = \sup_{y\in Y}\min_{x\in X} f(x,y).$$

Proof. Let $\mathcal{F}_1$ and $\mathcal{F}_2$ be the family of all non-empty compact convex subsets of X and non-empty compact subsets of X respectively. It is clear $\mathcal{F}_1 \subset \mathcal{F}_2$.

Note that

$$\inf_{x\in X}\sup_{y\in Y} f(x,y) \geq \inf_{K\in\mathcal{F}_1}\sup_{y\in Y}\min_{x\in K} f(x,y)$$

and

$$\inf_{Z\in\mathcal{F}_2}\sup_{y\in Y}\min_{z\in Z} f(x,y) \leq \inf_{K\in\mathcal{F}_1}\sup_{y\in Y}\min_{x\in K} f(x,y).$$

Hence the conclusion follows by (2.9.1) of Theorem 2.9.8. □

Finally as applications of Lemma 2.9.3, we shall give some extensions of Ky Fan's fixed point theorems in [162]. We first have the following result.

Theorem 2.9.10. Let X be a non-empty compact and convex subset of a Hausdorff topological vector space E and F be a Hausdorff topological vector space with sufficient continuous linear functionals. Suppose $T: X \to 2^F$ is upper semicontinuous with non-empty closed convex values and $g: X \to F$ is a (single-valued) continuous mapping such that:

(a) $T(x) \cap g(X) \neq \emptyset$ for each $x \in X$; and

(b) for given continuous seminorm p on F and $x \in X$, the set $\{z \in Y : p(g(z) - g(x)) \leq \delta\}$ is convex, where $\delta \in \mathbb{R}$, and $g^{-1}(T(z))$ is contractible for each $z \in X$.

Then there exists $x_0 \in X$ such that $g(x_0) \in T(x_0)$.

Proof. Following the argument of Ha [214, p.13-14] Suppose the conclusion were false. Then for each $x \in X$, $0 \notin g(x) - T(x)$ and so there exists $\delta_x > 0$ and a continuous seminorm p_x on F such that $p_x(g(x) - u) > 2\delta_x$ for all $u \in T(x)$. By the upper semicontinuity of T and the continuity of g, there exists an open neighborhood $N(x)$ of x in X such that $p_x(g(z) - v) > \delta_x$ for all $z \in N(x)$ and $v \in T(x)$. Since the family $\{N(x) : x \in X\}$ is an open cover of the compact set X, there exists a finite subset $\{x_1, \cdots, x_m\}$ of X such that $\cup\{N(x_i) : i = 1, 2, \cdots, m\} \supset X$. Now define a seminorm $p : F \to \mathbb{R}$ by $p(z) = \max_{i=1,2,\cdots,m}\{p_{x_i}(z)\}$ for each for each $z \in F$. Let $\delta = \min_{i=1,2,\cdots,m}\{\delta_{x_i}\}$. Then p is a continuous seminorm on F and

$$p(g(x) - u) > \delta$$

for all $x \in X$ and $u \in T(x)$. Let $B(z) = \{x \in X : p(g(z) - g(x)) < \delta\}$ for each $z \in X$. Then the family $\{B(z) : z \in X\}$ is an open cover of the compact set X so that there exists a finite subset $\{z_1, \cdots, z_n\}$ of X such that $\cup_{j=1}^n B(z_j) \supset X$. Let $\{\lambda_i : i = 1, 2, \cdots, n\}$ be a continuous partition of unity on X subordinated to the cover $\{B(z_j) : j = 1, 2, \cdots, n\}$, that is $\lambda_1, \cdots, \lambda_n$ are non-negative real valued continuous functions on X such that $\lambda_j(x) = 0$ for each $x \in X \setminus B(z_j)$ and $\Sigma_{i=1}^n \lambda_j(x) = 1$ for all $x \in X$. Let $Z = co(z_j : j = 1, 2, \cdots, n)$. Then $Z \subset X$. Define a function $\Phi : X \to Z$ by

$$\Phi(x) = \Sigma_{j=1}^n \lambda_j(x) z_j$$

for each $x \in X$. Note that for each $x \in X$, and $j = 1, 2, \cdots, n$, if $\lambda_j(x) \neq 0$, then $x \in B(z_j)$ and so that $p(g(z_j) - g(x)) < \delta$. For each $x \in X$, let $C_x = \{z \in Y : p(z - g(x)) \leq \delta\}$. Then C is a non-empty closed subset of C and $g(z_j) \in C$ if $\lambda_j(x) \neq 0$. Since $g^{-1}(C)$ is convex by (c), so that $g(\Phi(x)) \in C$, i.e., $p(g(\Phi(x)) - g(x)) \leq \delta$. Now define $h : Z \to 2^X$ by

$$h(z) = \{x \in X : g(x) \in T(z)\}$$

for each $z \in Z$. Then h is a set-valued mapping with non-empty and contractible values by (a) and (b). Moreover it is easy to verify that the graph of h is closed in $Z \times X$ and so that h is upper semicontinuous on Z. By Lemma 2.9.3, there exists $z_0 \in Z \subset X$ and $x_0 \in X$ such that $z_0 = \Phi(x_0)$ and $x_0 \in h(z_0)$, this means that $g(x_0) \in T(z_0)$. Note that $p(g(\Phi(x)) - g(x))) \leq \delta$ for all $x \in X$. Thus $p(g(z_0) - g(x_0)) \leq \delta$, which contradicts (1). Therefore there must exist $x_0 \in X$ such that $g(x_0) \in T(x_0)$. □

Remark 2.9.11. The condition (b) of Theorem 2.9.10 is satisfies if $g^{-1}(C)$ is convex (maybe empty) for each closed convex C of F. Thus Theorem 2.9.10 improves Theorem 2 of Ha [214] in the sense that $g^{-1}(C)$ may not be convex for each closed convex subset C of F.

Theorem 2.9.12. Let X be a non-empty compact and convex subset of a Hausdorff topological vector space E and F be a Hausdorff topological vector space with sufficient continuous linear functionals. Suppose $f, g : X \to F$ are two (single-valued) continuous mapping such that $g(X)$ is convex and $g^{-1}(y)$ is contractible for each $y \in g(X)$. Then one of the following must hold:

(a) there exists a point $x_0 \in X$ such that $g(x_0) = f(x_0)$;

or

(b) there exists a point $\hat{x} \in X$ such that

$$0 < p(g(\hat{x}) - f(\hat{x})) \leq p(y - f(\hat{x})).$$

Proof. Assume that the conclusion is false. Then $g(x) \neq f(x)$ for all $x \in X$. By using the similar arguments as in the proof of Theorem 2.9.10 above, there exist $\delta > 0$ and a continuous seminorm p on F such that $p(g(x) - f(x)) > 0$ for all $x \in X$. Moreover for each $x \in X$, there is $y \in g(X)$ such that $p(g(x) - f(x)) > p(y - f(x))$ (since we assume that (b) does not hold). Now for each $y \in g(X)$, let $A(y) = \{x \in X : p(g(x) - f(x)) > p(y - f(x))\}$. Then $\{A(y) : y \in g(X)\}$ is an open cover of the compact set X. Thus there exists a finite subset $\{y_1, \cdots, y_m\}$ in $g(X)$ such that $\cup_{i=1}^m A(y_i) \supset X$. Let $\{\lambda_i, \cdots, \lambda_m\}$ be the continuous partition of unity on X subordinated to $\{A(y_i) : i = 1, 2, \cdots, m\}$ and $Z := co(y_i : i = 12, \cdots, m)$. Define a mapping $\Psi : X \to Z$ by

$$\Psi(x) = \Sigma_{i=1}^m \lambda_i(x) y_i$$

for each $x \in X$. Now for each $x \in X$, if $\lambda_i(x) \neq \emptyset$ for some $i = 1, 2, \cdots, m$, then $x \in A(y_i)$ so that $p(g(x) - f(x)) > p(y_i - f(x))$. Hence $p(g(x) - f(x)) > p(\Psi(x) - f(x))$ for all $x \in X$. For each $z \in Z$, let $h(z) = g^{-1}(z)$. Then clearly h is an upper semicontonuous set-valued mappings on Z with non-empty contractibe values. By Lemma 2.9.3, there exist $z_0 \in Z \subset g(X)$ (since $g(X)$ is convex) and $x_0 \in X$ such that $z_0 = \Psi(x_0)$ and $x_0 \in h(z_0)$, i.e., $z_0 = g(x_0)$. Note that $p(g(x) - f(x)) > p(\Psi(x) - f(x))$ for all $x \in X$, we must have that

$$p(g(x_0) - f(x_0) > p(z_0 - f(x_0)) = p(g(x_0) - f(x_0))$$

which is impossible. Therefore the conclusion must hold. □

Remark 2.9.13. Theorem 2.9.12 shows that Theorem 3 of Ha [214] still true in the case the set $g^{-1}(y)$ may not be convex. When $E = F$ and $g : X \to X$ is the identity map, Theorem 7 reduces to Theorem 2 of Fan in [162].

Finally as an immediate consequence of Theorem 2.9.12, we have the following result.

Theorem 2.9.14. Let X be a non-empty compact and convex subset of a Hausdorff topological vector space E and F be a Hausdorff topological vector space with sufficient continuous linear functionals. Suppose $f, g : X \to F$ are two (single-valued) continuous mapping such that:

(a) $g(X)$ is convex and $g^{-1}(y)$ is contractible for each $y \in g(X)$; and

(b) for each $x \in X$, there exists a number λ (real or complex, depending on whether the vector space F is real or complex) such that $|\lambda| < 1$ and $\lambda g(x) + (1 - \lambda) f(x) \in g(X)$.

Then there exists $x_0 \in X$ such that $f(x_0) = g(x_0)$.

Proof. Suppose that $g(x) \neq f(x)$ for all $x \in X$. By Theorem 2.9.12, there is a point $x_0 \in X$ and a continuous seminorm p on F satisfies (5). Now there exists a number λ with $|\lambda| < 1$ and $y = \lambda g(x_0) + (1 - \lambda) f(x_0) \in g(X)$. Substituting this y in (5), it follows that

$$0 < p(g(x_0) - f(x_0)) \leq |\lambda| p(g(x_0) - f(x_0)),$$

which contradicts $|\lambda| < 1$.

Remark 2.9.15. Let X and Y be non-empty sets and $f : X \times Y \to \mathbb{R}$ be a function. A minimax problem is to find certain conditions such that the following holds:

$$\inf_{y \in Y} \sup_{x \in X} f(x, y) = \sup_{x \in X} \inf_{y \in Y} f(x, y).$$

It is well-known that the above minimax problem could be solved by fixed point theorems under certain circumstance. But it was Wu in [579], who first recognized the importance of the *connectedness* in the study of minimax theory.

This idea then was picked by Terkerson in [550]. By a refined method, it was Tuy, who derived a generalized version of Sion's classical minimax theorem [490] (see also Geraghty and Lin [194]). Independently, inspired by Joó's paper [274], the *method of level sets* was developed by Joó and his Hungarian compatriot Stachó [276] and Komornik [328]. For example, by introducing the concept of the *interval space*, it was Stachó [499], who established an intersection theorem which was used by Komornik [328] to derive a generalization of Ha's minimax theorem in [213]. All these results were unified by Kindler and Trost [310]. Following this line, many minimax theorems which only involve the *connectedness* instead of convexity were obtained by Chang et al. [67], Horvath [237]-[238], Komiya [327], Lin and Quan [343], Kindler [305]-[308], König [329], and Simons [483]-[485] et al. We will present the study of topological minimax theorems in Chapter 3.

However we would like to note that our minimax inequalities and fixed point theorems in this section are proved by Fan's section theorems. Therefore minimax inequalities obtained in this section are independent of those results given by the authors mentioned above.

2.10 Fan's Best Approximation and Fan-Glicksberg Fixed Point Theorems in Locally Convex H-Spaces

As an application of minimax inequality in section 9, i.e., Theorem 2.9.6, we will obtain Fan's alternative theorem for set-valued upper semicontinuous mappings in locally convex topological vector spaces. This result generalizes the corresponding results given in section 6 where we show that set-valued mappings are continuous in locally convex topological spaces. We also remark that most results in this section hold if we assume the underlying spaces are Hausdorff topological vector spaces with sufficiently continuous linear functionals instead being locally convex topological vector spaces.

2.10.1 Fan's Best Approximation and Fixed Points in Locally Convex Spaces

Throughout this section, let (E,T) denote a topological vector space with the topology T and we denote by $W = W(E,E^*)$ the weak topology of E and by $\mathcal{P} = \mathcal{P}(E,T)$ the family of all continuous semi-norms on (E,T). If X is a subset of (E,T), we denote by (X,T) and (X,W) the set X equipped with the relative topology of T to X and the relative topology of the weak topology W of E to X, respectively.

In order to obtain Fan's best approximation for upper semicontinuous set-valued mappings, we first state the following form of Ky Fan minimax inequality which is a special case of Theorem 2.9.6 (see also Theorem 1 of Ha [215]).

Lemma 2.10.1. Let X and Y be non-empty convex convex subsets of Hausdorff topological vector spaces E and F, respectively. Let $F : X \to 2^Y$ be an upper semicontinuous mapping with non-empty closed and convex values and $g : X \times \mathbb{R}$ be such that:

(1) for each fixed $x \in X$, $g(x, y)$ is lower semicontinuous in Y; and

(2) for each fixed $y \in Y$, $g(x, y)$ is quasi-concave in X.

Then

$$\inf_{y\in Y} \sup_{x\in X} g(x,y) \leq \sup_{u\in F(x), x\in X} g(x,u).$$

For the convenience or our readers, we also need the following simple fact (e.g., see Lemma 4 of Ding and Tan [139]) and thus we do not include its proof here.

Lemma 2.10.2. Let X be a non-empty subset of a Hausdorff topological vector space (E,T) whose topological dual E^* separates points of E. For a given $p \in \mathcal{P}$ and $F : (X, W) \to 2^{(E,T)}$ is an usc mapping such that $F(x)$ is T-compact for each $x \in X$. Then the function $V : (X, W) \to \mathbb{R}$ defined by

$$V(x) = d_p(x, F(x)) = \inf_{u\in F(x)} p(x-u)$$

for each $x \in X$ is weakly lower semicontinuous.

As an application of Lemma 2.10.1, we have the following Fan's best approximation theorem for upper semicontinuous set-valued mappings due to Ding and Tan in [139].

Theorem 2.10.3. Let X be a non-empty W-compact convex subset of a locally convex Hausdorff topological vector space (E,T). Suppose that $F : (X, W) \to 2^{(E,T)}$ is upper semicontinuous with non-empty T-compact convex values. Then either F has a fixed point in X or there exists $x_0 \in X$ and $u_0 \in F(x_0)$ and a weakly continuous semi-norm p on E such that

$$0 < p(x_0 - u_0) \leq p(x - u_0)$$

for all $x \in \overline{I_X}(x_0)$ (resp.,

$$0 < p(x_0 - u_0) \leq p(x - u_0)$$

for all $x \in \overline{O_X}(x_0)$).

Proof. Suppose F has no fixed point in X. By following the proof of Ding and Tan [139], for each $x \in X$, $0 \notin x - F(x)$ and $x - F(x)$ is a T-compact convex set in E. By Theorem 3.4 of Rudin [449], there exists $\delta_x > 0$ and a linear function $p_x \in E^*$ such that

$$d_{p_x}(x, F(x)) = \inf_{u \in F(x)} |p_x(x-u)| > \delta_x.$$

Note that the mapping $z \mapsto d_{p_x}(z, F(z))$ is weakly lower semicontinuous by Lemma 2.10.2 above, there exists a W-neighborhood $N(x)$ of x in X such that

$$d_{p_x}(z, F(z)) > \delta_x$$

for all $z \in N(x)$. Note that $X = \cup_{x \in X} N(x)$ and X is W-compact, there exists a non-empty finite subset $\{x_1, \cdots, x_n\}$ of X such that $X = \cup_{i=1}^n N(x_i)$. Let $p = \max\{|p_{x_i}| : i = 1, 2, \cdots, n\}$ and $\delta = \min\{\delta_{x_i} : i = 1, 2, \cdots, n\}$. Then p is a weakly continuous semi-norm on E. For each $x \in X$, there exists x_{i_0} such that $x \in N(x_{i_0})$. It follows that for each $u \in F(x)$,

$$p(x-u) = \max_{1 \le i \le n} |p_{x_i}(x-u)| \ge |p_{x_0}(x-u)| \ge d_{p_{x_0}}(x, F(x)),$$

so that

$$d_p(x, F(x) = \inf\{p(x-u) : u \in F(x)\} \ge d_{p_{x_{i_0}}}(x, F(x)) > \delta_{x_{i_0}} \ge \delta.$$

Hence we have $d_p(x, F(x)) > \delta$ for all $x \in X$. Now define the function $g : (X, W) \times (E, T) \to \mathbb{R}$ by

$$g(x, y) = \min_{u \in X} p(u-y) - p(x-y)$$

for each $(x, y) \in X \times E$. It is clear that $g : (X, W) \times (E, T) \to \mathbb{R}$ is continuous. Thus the condition (1) of Lemma 2.10.1 is satisfied. As F is usc with T-compact values, it follows that $F(X)$ is also T-compact. Note that the assumption on X, F and the graph of F remain unchanged in the completion of E, without loss of generality, we may assume that E is complete. Let $Y = \overline{co}(F(X))$. Then Y is T-compact and convex in (E, T). By applying Lemma 2.10.1 above, we have

$$\inf_{y \in Y} \sup_{x \in X} g(x, y) \le \sup_{u \in F(x), x \in X} g(x, u).$$

Since for each $y \in Y$, we have

$$\sup_{x \in X} g(x, y) = \sup_{x \in X} [\min_{u \in X} p(u-y) - p(x-y)] = \min_{u \in X} p(u-y) - \min_{x \in X} p(x-y) = 0,$$

it follows that $\sup_{u \in F(x), x \in X} g(x, u) \ge 0$. Note that g is continuous and the graph of F is compact in $(X, W) \times (E, T)$, there exists $(x_0, u_0) \in Gr(F)$ such that

$$g(x_0, u_0) = \min_{x \in X} p(x-u_0) - p(x_0-u_0) \ge 0.$$

It follows that $u_0 \in F(x_0)$ and

$$\delta < d_p(x_0, F(x_0)) \leq p(x_0 - u_0) \leq p(x - u_0)$$

for all $x \in X$. For each $x \in I_X(x_0)$, there exists $v \in X$ and $r > 1$ such that $x = x_0 + r(v - x_0)$. Suppose that $p(x - u_0) < p(x_0 - u_0)$. Since $v = \frac{1}{r}x + (1 - \frac{1}{r})x_0 \in X$, it follows by inequality above, we have

$$\begin{aligned} p(x_0 - u_0) &\leq p(v - u_0) = p(\frac{1}{r}x + (1 - \frac{1}{r})x_0 - u_0) \\ &\leq \frac{1}{r}p(x - u_0) + (1 - \frac{1}{r})p(x_0 - u_0) < p(x_0 - u_0) \end{aligned}$$

which is a contradiction. Therefore we must have

$$p(x_0 - u_0) \leq p(x - u_0)$$

for all $x \in I_X(x_0)$. Since p is continuous, we have

$$0 < p(x_0 - u_0) \leq p(x - u_0)$$

for all $x \in \overline{I_X}(x_0)$.

For the outward case, we define a a mapping $F' : (X, W) \to 2^{(E,T)}$ by

$$F'(x) = 2x - F(x)$$

for each $x \in X$. The conclusion follows by the same arguments used above for the mapping F' instead of F and thus the proof is complete. □

As an equivalent version of Theorem 2.10.3, we have the following result (Ding and Tan [139, Theorem 5]),

Theorem 2.10.4. Let X be a non-empty W-compact convex subset of a locally convex Hausdorff topological vector space (E, T). Suppose $F : (X, W) \to 2^{(E,T)}$ is an upper semicontinuous set-valued mapping with non-empty T-compact convex values. If for each weakly continuous semi-norm p of E, each $x \in X$ with $d_p(x, F(x)) > 0$ and for each $u \in f(x)$, we have

$$d_p(u, \overline{I_X}(x)) < p(x - u) \text{ (resp., } d_p(u, \overline{O_X}(x)) < p(x - u)).$$

Then F has a fixed point in X.

As a consequence of Theorem 2.10.3, we also have the following fixed point theorem for non self-mapping set-valued correspondences, which is Theorem 5 in [139].

Theorem 2.10.5. Let X be a non-empty W-compact convex subset of a locally convex Hausdorff topological vector space (E, T). Suppose $F : (X, W) \to$

$2^{(E,T)}$ is an upper semicontinuous set-valued mapping with non-empty T-compact convex values. Suppose that for each $x \in \partial_{(E,W)}X \setminus F(x)$ and $u \in F(x)$, there exists a number λ (real or complex, depending on whether the space E is real or complex) with $\lambda| < 1$ such that $\lambda x + (1-\lambda)u \in \overline{I_X}(x)$ (resp., $\lambda x + (1-\lambda)u \in \overline{O_X}(x)$), where $\partial_{(E,W)}X$ denotes the boundary of X in (E, W). Then F has a fixed point in X.

Proof. Suppose F has no fixed point in X. By Theorem 2.10.3, there exist $x_0 \in X$, $u_0 \in F(x_0)$ and a weakly continuous semi-norm p on E such that

$$0 < p(x_0 - u_0) \leq p(x - u_0)$$

for all $x \in \overline{I_X}(x_0)$ (resp., for all $x \in \overline{O_X}(x_0)$).

If $x_0 \in int_{(E,W)}X$, then we have $I_X(x_0) = E$ (resp., $O_X(x_0) = E$) and hence $x = \frac{1}{2}x_0 + \frac{1}{2}u_0 \in E = I_X(x_0)$ (resp., $x \in O_X(x_0)$) so that

$$0 < p(x_0 - u_0) \leq p(x - u_0) = \frac{1}{2}p(x_0 - u_0)$$

which is a contradiction.

In the case $x_0 \in \partial_{(E,W)}X$, then $x_0 \in \partial_{(E,W)}X \setminus F(x_0)$ so that by the condition (1) of Theorem 2.10.5, there exists λ with $|\lambda| < 1$ such that $x = \lambda x_0 + (1-\lambda)u_0 \in \overline{I_X}(x_0)$ (resp.,$x \in \overline{O_X}(x_0)$). It then follows that

$$0 < p(x_0 - u_0) \leq p(x - u_0) \leq |\lambda| p(x_0 - u_0)$$

which is again a contradiction. Therefore F must have a fixed point in X. □

Theorem 2.10.5 improves and generalizes Theorem 4 of Ha [215], Theorem 3.1 of Reich [435], Theorem 4 of Park [419] and Theorem 3 of Fan [162].

2.10.2 Fan-Glicksberg Fixed Point Theorems in Locally Convex H-Spaces

The aim of this section is to establish fixed point theorems for upper semicontinuous set-valued mappings in locally convex H-spaces which include locally convex topological vector spaces as a special class.

In order to prove Fan-Glicksberg fixed point theorem for upper semicontinuous set-valued mappings in locally convex H-spaces, we first recall the following definition (see Wu [575], or Yuan [610]).

Definition 2.10.6. An H-space (X, Γ_A) is said to be a *locally convex H-space* if X is a uniform topological space with the uniform structure $\mathcal{U}$ and there exists a base family $\{V_i : i \in I\}$ for $\mathcal{U}$ such that for each $V_i \in \{V_i : i \in I\}$, the set $V_i(x) := \{y \in X : (y, x) \in V_i\}$ is H-convex for each $x \in X$.

We remark that Definition 2.10.6 above includes so-called *l.c.*-space given by Horvath [239] as a special case (see also Tarafdar [539] and [543]). We do know that each non-empty convex subset S in locally convex spaces has a uniform structure $\mathcal{U}$ induced by the family of seminorms; on the other hand, each convex S in topological vector spaces has a natural H-structure Γ_A given by $\Gamma_A := co(A)$ for each non-empty finite subset $A \in \mathcal{F}(S)$. Thus locally convex H-spaces include locally convex topological vector spaces as a special case.

In the setting of metric spaces, we have the following definition of locally convex H-spaces for metric spaces the corresponding to Definition 2.10.6.

Definition 2.10.7. Let the metric space (Y, Γ) be an H-space with the metric d. Then Y is said to be a *locally metrically convex H-space* if for each $r > 0$, the set $\{y \in Y : d(y, E) < r\}$ is H-convex for each H-convex set E in Y and also all open balls are H-convex.

As an application of Lemma 2.2.5, we first state the following existence of continuous selection (e.g., see Tarafdar [536]).

Lemma 2.10.8. Let X be a compact topological space and (Y, Γ) an H-space. Suppose $F : X \to 2^Y \setminus \{\emptyset\}$ is a set-valued mapping with non-empty H-convex values such that $X \subset \cup_{y \in Y} int(F^{-1}(y))$, where $int(F^{-1}(y))$ denotes the interior of $F^{-1}(y)$ in X for each $y \in Y$. Then there exist some $n \in N$, and two (single-valued) continuous mappings $g : \Delta_N \to Y$ and $\phi : X \to \Delta_N$ such that $f(x) := g(\phi(x)) \in F(x)$ for all $x \in X$, where $N = \{0, 1, \cdots, n\}$.

Proof. Since X is compact, there exists a finite subset $\{y_0, y_1, \cdots, y_n\}$ of Y such that $\cup_{i=0}^n int(F^{-1}(y_i)) = X$. For each non-empty subset $J \subset N = \{0, 1, \cdots, n\}$, let $\Gamma_J = \Gamma_{A_J}$, where $A_J = \{y_j : j \in J\}$. By Lemma 2.2.5, there exists a (single-valued) continuous mapping $g : \Delta_N \to Y$ such that $g(\Delta_J) \subset \Gamma_J$ for each subset J of N. Now let $\{p_i : i = 0, 1, \cdots, n\}$ be a partition of unity the corresponding to the finite covering $\{intF^{-1}(y_i) : i = 0, 1, \cdots, n\}$ of X, that is, p_i is a real valued continuous function defined on X such that $p_i(x)$ vanishes outside $int(F^{-1}(y_i))$ and $p_i(x) \in [0, 1]$ for all $x \in X$ such that $\Sigma_{i=0}^n p_i(x) = 1$ for each $x \in X$. We now define a mapping $\phi : X \to \Delta_N$ by $\phi(x) := \Sigma_{i=0}^n \phi_i(x) e_i$ for each $x \in X$. Then it is easy to verify that $\phi(x) \in \Delta_J(x)$ for each $x \in X$, where $J(x) := \{i \in N : \psi_i(x) \neq 0\}$. By the H-convexity of $F(x)$, we do have that $\Gamma_{J(x)} \subset F(x)$ and thus $f(x) := g(\phi(x)) \in F(x)$ for each $x \in X$. This completes the proof. □

By using both Lemma 2.10.8 and Lemma 2.9.3, we can prove the Fan-Glicksberg fixed point theorem for upper semicontinuous set-valued mappings with non-empty closed acyclic (e.g., contractible) values in locally convex H-spaces, which is a generalization of Fan-Glicksberg fixed point theorem in locally convex Hausdorff topological vector spaces (see also Wu [575], and Yuan [610]). The following version is first given by Wu [575].

Theorem 2.10.9. Let $(X, \Gamma_A\}$ be a compact Hausdorff locally convex H-space. Suppose $F : X \to 2^X$ is an upper semicontinuous set-valued mapping with non-empty closed acyclic values. Then F has at least one fixed point, i.e., there exists $x \in X$ such that $x \in F(x)$.

Proof. Since (X, Γ_A) is a locally convex Hausdorff H-space, X has a uniform structure $\mathcal{U}$ and without loss of generality, suppose $\{V_i : i \in I\}$ is a symmetric and open base family for the uniform structure $\mathcal{U}$. Then for each base V_i and $x \in X$, the set $V_i(x) := \{y \in X : (y, x) \in V_i\} = \{y \in X : (x, y) \in V_i\}$ is open and H-convex and $\cap_{i\in I} V_i = \Delta = \{(x, x) : x \in X\}$. For each V_i, we define a set-valued mapping $T_i : X \to 2^X$ by $T_i(x) := \{y \in X : (y, x) \in V_i) = V_i(x)$ for each $x \in X$. Then it is clear T_i is a set-valued mapping from X to X with non-empty H-convex values and $\cup_{y\in X} int(T_i^{-1}(y)) \subset X$ by the fact that for each $y \in X$, the set $T_i^{-1}(y) = \{x \in X : (x, y) \in V_i\}$ is open in X, so that $int(T_i^{-1}(y)) = T_i^{-1}(y)$. By Lemma 2.10.8 there exist some $n \in N$ and two (single-valued) continuous mappings $g_i : \Delta_N \to X$ and $\phi_i : X \to \Delta_N$ such that $f_i(x) := g_i(\phi_i(x)) \in T_i(x)$ for each $x \in X$, where Δ_N is a standard n-dimensional simplex and $N = \{0, 1, \cdots, n\}$. Now we define another set-valued mapping $P_i : \Delta_N \to 2^X$ by $P_i(x) := F(g_i(x))$ for each $x \in \Delta_N$. Then P_i is also an upper semicontinuous set-valued mapping with non-empty closed and acyclic values. By Lemma 2.9.3, the mapping $\phi_i \circ P_i : \Delta_N \to 2^{\Delta_N}$ has a fixed point, i.e., there exists $u_{v_i} \in \Delta_N$ such that $u_{v_i} \in \phi_i(P_i(u_{v_i}))$. Then there exists $x_{v_i} \in P_i(u_{v_i}) = F(g_i(u_{v_i})) \subset X$ and $u_{v_i} = \phi_i(x_{v_i})$. Let $y_{v_i} = g_i(u_{v_i}) = g_i(\phi_i(x_{v_i})) = f(x_{v_i}) \in T_i(x_{v_i})$. Thus $x_{v_i} \in F(y_{v_i})$, i.e., $(x_{v_i}, y_{v_i}) \in$ graph F and $(x_{v_i}, y_{v_i}) \in V_i$, where graph F denotes the graph of the mapping F. Since $X \times X$ is compact, we may assume that $\{x_{v_i}, y_{v_i}\}_{i\in I}$ converges to (x_0, y_0). Note that F is an upper semicontinuous mapping with closed values, it follows that graph F is closed and thus $(x_0, y_0) \in$ graph F, i.e., $y_0 \in F(x_0)$. In order to finish the proof, it suffices to show that $x_0 = y_0$. Note that $\{V_i : i \in I\}$ is an open symmetric base family for the uniform structure $\mathcal{U}$, for any given $U \in \mathcal{U}$, there exists $V_u \in \{V_i : i \in I\}$ such that $V_U \circ V_U \circ V_U \circ V_U \subset U$. On the other hand, $\lim_{i\in I} x_{v_i} = x_0$ and $\lim_{i\in I} y_{v_i} = y_0$, without loss of generality, there exists $V_0 \in \{V_i : i \in I\}$ and $V_0 \subset V_U$ such that $(x_{v_i}, x_0) \in V_U$ for $i \in I$ such that $V_i \subset V_0$. Now for any $A \in \{V_i : i \in I\}$ with $A \subset V_U$, note that $(x_A, y_A) \in A$, it follows that $(x_0, y_A) \in A \circ V_U \subset V_0 \circ V_U \subset V_U \circ V_U$. Then it is easy to verify that $(X_0, y_0) \subset V_U \circ V_U \circ V_U \circ V_U \subset U$. This implies that $(x_0, y_0) \in \cap_{U\in\mathcal{U}} U$. Note that $\cap_{U\in\mathcal{U}} U = \Delta$, it follows that $x_0 = y_0$ and thus x_0 is a fixed point of F. The proof is complete. □

Remark 2.10.10. We wish to note that Theorem 2.10.9 includes most existing Fan-Glicksberg fixed point theorems for upper semicontinuous set-valued mappings from Tarafdar [539], and Tarafdar and Watson [543] as special cases. Secondly, we wish to mention that in next Section 12, more general fixed point theory for set-valued mappings in the setting of a larger class for abstract convexity called MC-spaces and locally convex MC-spaces will be established. In particular, fixed point theorms of the Browder-Fan type for set-valued mappings with transfer open inversed values and fixed point theorems of the Fan-Glicksberg fixed point theorems for upper semicontinuous set-valued mappings with closed acyclic (e.g., convex) values will be given in MC-spaces , locally convex MC-spaces and locally G-convex spaces. . The results given in Section 12 could be regarded as a unified theory for fixed point theorems of set-valued analysis in the setting of MC-spaces and G-convex spaces since they include H-spaces, hyperconvex metric spaces (its definition will be given in the next section), locally convex topological vector spaces (resp., locally convex H-spaces, hyperconvex metric spaces) as a special class.

Remark 2.10.11. Though we have established several fixed point theorems for upper semicontinuous (resp., hemicontinuous) set-valued mappings in topological vector spaces which have *sufficient enough continuous linear functionals*, it is clear that a Hausdorff topological vector space may not have enough continuous linear functionals (e.g, the Hardy space H^p for $p \in (0,1)$), thus Schauder's famous conjecture which asks '*if a continuous (even single-valued) mapping defined on a non-empty compact and convex subset of a Hausdorff topologocal vector space has a fixed point*' remains open since 1934. For more historic details on Schauder's conjecture and its development, the interested readers are referred to Dugundji and Granas [149], Granas [204], Idzik [249], Park [412] and references therein.

Remark 2.10.12. The celebrated Ky Fan minimax inequality has numerous applications in various and diverse fileds of mathematics and applied science. Also, there have appeared a number of equivalent formulations and generalizations. For example, this chapter gives us a clear picture and relationships among Brouwer fixed point theorem [56], Sperner lemma [498], KKM lemma, Fan's geometric lemma [157], Browder-Fan fixed point theorem, Fan's matching theorem in Section 7, Fan-Glicksberg fixed point theorem, Fan's section theorem, Fan's best approximation theorem, and minimax theorem in section 9; their many equivalent forms to Ky Fan minimax inequality, which is regarded as one of the most important principles in nonlinear analsysis. In this line, more details have been illustrated by Aubin [13], Aubin and Ekeland [15], Border [48], Park [417], Yuan [609] and references therein. We all know that nonlinear analysis began early this century with the birth of Brouwer fixed point. However, Ky Fan minimax inequality is a more convenient form as it can be used to impliy *a whole series of existence theorems* for solutions of nonlinear equations

theory, best approximations and KKM-map principle.
or inclusions in Aubin [13, p.135] in a very general form. Such theorems are very useful in many applications, in particular in optimization theory, mathematical economics, and game theory as we will see through Chapters 7 to 9. The book of Singh et al. [489] also gives an extensively study among fixed point

2.11 KKM Principle in Hyperconvex Metric Spaces and Some Applications

In this chapter, we study the KKM theory and its various applications for a special class of H-spaces, so-called hyperconvex metric spaces. First we give Fan-Glicksberg fixed point theorem for upper semicontinuous set-valued mappings in hyperconvex metric spaces. Then, by using some special properties of hyperconvex metric spaces instead of applying results established in sections 2 and 3 of this chapter, we first directly derive the characterization of the KKM principle in hyperconvex metric spaces. The characterization for a family of subsets with the finite intersection property in hyperconvex metric spaces is then given. As applications, we give hyperconvex versions of Fan's celebrated minimax principle and Fan's best approximation theorem for set-valued mappings, which in turn are applied to derive the Browder-Fan fixed point theorem for set-valued mappings with open inversed values. Finally the existence theorems for saddle points, intersection theorems, and Nash equilibria are given. Our results unify and improve the corresponding results in the literature as special cases.

2.11.1 Hyperconvex Metric Spaces and Fan-Glicksberg Fixed Point Theorems

Sine [486] and Soardi [494] proved independently that the fixed point property for nonexpansive mappings holds in bounded hyperconvex spaces. Ever since hyperconvex metric spaces have been used widely, and many interesting results for nonexpansive mappings have been established in the framework of hyperconvex structures, e.g., see Baillon [21], Goebel and Kirk [198], Khamsi [292], Sine [486]-[487] and others. Recently, Khasmi [292] established the hyperconvex version of the famous KKM-Fan principle in hyperconvex metric spaces; Kirk [312] gave a constructive fixed point theorem which arises from interval analysis in compact hyperconvex metric spaces, and also Kirk and Shin [313] established a number of fixed point theorems for both condensing and nonexpansive mappings in hyperconvex spaces. In this section, we first show that each hyperconvex metric space is indeed a complete locally metrically H-space (thus, it is a locally convex H-space) and then using fixed point theorem established in the previous section, i.e., Theorem 2.10.9, we give a hyperconvex version of fixed point theorem for upper semicontinuous set-valued mappings with closed acyclic values in hyperconvex metric spaces, which, in turn, unifies the corresponding fixed point theory of set-valued mappings in existing literature.

We first recall the following definition of a hyperconvex metric space due to Aronszajn and Panitchpakdi [7].

Definition 2.11.1. A metric space (X, d) is said to be a *hyperconvex* space if for any connection of points $\{x_\alpha\}$ of X and any collection $\{r_\alpha\}$ of non-negative real numbers with $d(x_\alpha, x_\beta) \leq r_\alpha + r_\beta$, we have $\cap_\alpha B(x_\alpha, r_\alpha) \neq \emptyset$, where $B(x, r)$ denotes the closed ball centered at $x \in X$ with the radius r.

This definition can be seen as equivalently a binary intersection property plus metric convexity in the sense that for each given $x, y \in X$ and $\alpha \in [0, 1]$, there exists $z \in X$ such that $d(x, z) = \alpha d(x, y)$ and $d(y, z) = (1-\alpha)d(x, y)$. The corresponding linear theory for hyperconvex spaces can be found in Lacey [332]. Sine [487] points out the nonlinear theory is still developing. Hyperconvex spaces can have quite strange aspects. For example, a hyperconvex subset need not be convex in even $\mathbb{R}^2$ (with the l_∞ norm). Also convex sets in linear spaces may fail to be hyperconvex (but for this one must go to at least $\mathbb{R}^3$, e.g., see Sine [487]). Some nice spaces such as Hilbert space may fail to be hyperconvex. However hyperconvexity enjoys some properties similar to convexity and others much like compactness. More convincing analogies hold for hyperconvex sets which are ball intersections. The Nachbin-Kelley-Goodner and Hasumi Theorem in Lacey [332, p.92] (see also Isbell [263], Nachbin [377] or Kelley [285]) says that a Banach space is hyperconvex if and only if it is linearly isometric to $C(K)$, where $C(K)$ is the space of all continuous real functions defined on some Stonian space K (i.e., K is Hausdorff, compact and extremely disconnected). Thus the space $l_\infty(I)$ for any set I and the space $L_\infty(\mu)$ for a finite measure μ are examples of hyperconvex spaces. Secondly, order intervals in L_∞ are hyperconvex, but weak compact convex sets of L_∞ need not be hyperconvex. Moreover we note that hyperconvex sets of L_∞ may not be convex (e.g., see Lin and Sine [346, p.943]). We also remark that any hyperconvex space (indeed any metric space) embeds isometrically in some $l_\infty(I)$, e.g., see Lacey [332] or Khamsi [292, p.300-301].

The relationship between hyperconvex metric spaces and nonexpansive mappings is an important one as shown independently by the work of Sine [486] and Soardi [494]. The main trust of this section is in the same direction, i.e., our aim is to give a comprehensive study of KKM theory in hyperconvex spaces and its related applications to fixed point theorems, to characterize KKM principle in hyperconvex spaces, to obtain Fan's minimax principle in hyperconvex spaces, the existence of saddle points, the intersection of sets and the existence of Nash equilibria in game theory. In order to study the KKM theory in hyperconvex spaces, we first recall some notions and basic facts about hyperconvex spaces which will be used later in this chapter.

Definition 2.11.2. Let (M, d) be a metric space. Assume that a subset A of M is bounded. We will use the following notions throughout this chapter.

(1) $co(A) = \cap\{B \subset M : B$ is a closed ball such that $A \subset B\}$.

(2) $\mathcal{A}(M) = \{A \subset M : A = co(A)\}$, i.e., $A \in \mathcal{A}(M)$ if and only if A is an intersection of closed balls. In this case, we shall say that A is an *admis-*

sible subset of M. We also note that if M is a hyperconvex space, then each admissible set in M is also hyperconvex.

For the convenience of our study, we also have the some facts about hyperconvex metric spaces whose proofs can be found from Aronszajn and Panitchpakdi [7], and Lacey [332] (see also Isbell [263], Nachbin [377], Kelley [285] and Khamsi [292]).

Proposition 2.11.3. Let M be a metric space. Then we have the follows:
(1) there exists an index set I and a natural isometric embedding from M to $l^\infty(I)$.
(2) If M is hyperconvex, the it is complete.
(3) M is hyperconvex if and only if for each metric space N which contains isometrically M, there exists a nonexpansive retract $r: N \to M$; i.e., r is nonexpansive and $r(x) = x$ for each $x \in M$. In particular if N is a normed space, for any non-empty finite points $y_1, y_2, \cdots, y_n \in M$, we have that $r(conv\{y_1, \cdots, y_n\}) \subset co\{y_1, y_2, \cdots, y_n\}$, where $co\{y_1, y_2, \cdots, y_n\}$ is an admissible set defined by Definition 2.11.2 above.
(4) M is hyperconvex if and only if for each metric space N which is contained metrically in any space D, and any nonexpansive mapping $T: N \to M$, there exists an extension $T^*: D \to M$ which is nonexpansive, i.e., $T(x) = T^*(x)$ for each $x \in N$.

Let M be a metric space and consider the natural embedding into $l^\infty(I)$ given by statement (1) in Proposition 2.11.3. We denote by $M_\infty := co(M) \in \mathcal{A}(l^\infty(I))$. Clearly M_∞ is a hyperconvex subset of $l^\infty(I)$ and M_∞ is also a convex subset of the linear space $l^\infty(I)$.

Let X be a non-empty set. We shall denote by $\mathcal{F}(X)$ and 2^X the family of all non-empty finite subsets of X and the family of all subsets of X, respectively. If A is a subset of a linear space E, the notion '$conv(A)$' always means the *convex hull* of A.

It follows by Proposition 2.11.3 that a hyperconvex space is a nonexpansive retract of any metric space in which it is embedded. Since any metric space can be embedded in a Banach space, it follows that any hyperconvex metric space is contractible. Note that if $\{B(y_i, r_i)\}_{i\in I}$ is any collection of closed balls in a hyperconvex space, then $\cap_{i\in I} B(y_i, r_i)$ itself is a hyperconvex space. Therefore for a hyperconvex space (Y, d), the set $\cap_{i\in I} B(y_i, r_i)$ is contractible, or empty. For any non-empty finite subset $A \in \mathcal{F}(Y)$, the admissible set of A (also called ball hull of A by Sine [487]) defined by

$$F(A) := \cap\{B(y, r) : A \subset B(y, r)\}$$

defines an H-structure in Y and thus each hyperconvex metric space (Y, d) is an H-space with the H-structure given by ball hull above.

Each hyperconvex metric space, indeed, is a complete locally metrically convex H-space as shown by Lemma 2.11.4 below. This fact was first given by Horvath [240].

Lemma 2.11.4. Each hyperconvex metric space (Y, d) is a complete locally metrically convex H-space (of course, (Y, d) is a locally convex H-spaces).

Proof. By following Horvath [240], for each $y \in Y$, $F((y)) = \{y\}$ which, of course, is an F-set. Weshow that for any H- convex set E and any $r > 0$, the set $\{y \in Y : d(y, e) < r\}$ is an H-set. Let $y_0, \cdots, y_n \in Y$ such that $d(y_i, E) < r$ and $y_0 \in F((y_1, \cdots, y_n))$, i.e., y_0 belongs to any closed ball containing $\{y_1, \cdots, y_n\}$. We want to show that $d(y_0, E) < r$. Take any fixed points $\hat{y}_1, \cdots, \hat{y}_n \in E$ such that $d(y_i, \hat{y}_i) < r$ for $i = 1, \cdots, n$. Since $F(\{\hat{y}_1, \cdots, \hat{y}_n\}) \subset E$, it suffices to show that $d(y_0, F(\{\hat{y}_1, \cdots, \hat{y}_n\})) < r$. By the definition of $F(\{\hat{y}_1, \cdots, \hat{y}_n\})$, $F(\{\hat{y}_1, \cdots, \hat{y}_n\}) = \cap_{i \in I} B(u_i, r_i)$, where $(u_i, r_i) \in Y \times R_+$. Choose $r' \in (0, r)$ such that $d(y_i, \hat{y}_i) < r'$ for $i = 1, \cdots, n$. As $\{\hat{y}_1, \cdots, \hat{y}_n\} \subset B_i(u_i, r_i)$ for $i \in I$, it follows that $\{y_1, \cdots, y_n\} \subset B(u_i, r_i + r)$ for $i \in I$ and $y_0 \in \cap_{i \in I} B(u_i, r_i + r')$. By Sine's Lemma in [487, p.864], there exists a retraction $R : \cap_{i \in I} B(u_i, r_i + r') \to \cap_{i \in I} B(u_i, r_i)$ such that $d(y, R(y)) \leq r'$ for any $y \in \cap_{i \in I} B(u_i, r_i + r')$. Then $R(y_0) \in F(\{\hat{y}_1 \cdots, \hat{y}_n\})$ and $d(y_0, R(y_0)) \leq r' < r$. This completes the proof. □

By the fact that each hyperconvex metric space is a locally convex H-space (thus, an H-space), we have the following hyperconvex version of Fan-Glicksberg fixed point theorem for set-valued upper semicontinuous with non-empty closed acyclic values as an application of Theorem 2.10.9.

Theorem 2.11.5. Let $X \in \mathcal{A}(M)$ be a compact subset of a hyperconvex metric space M. Suppose $F : X \to 2^X$ is a upper semicontinuous set-valued mappings with non-empty closed admissible values. Then F has at least one fixed point, i.e., there exists $x \in X$ such that $x \in F(x)$.

Proof. Note that each hyperconvex metric space is complete. It follows by Theorem 8.3.5 of Engelking [154] M has a uniform structure. Thus X is a locally convex H-space by Lemma 2.11.4. Then the conclusion follows by Theorem 2.10.9 and this completes the proof. □

We would like to note that Theorem 2.11.5 improves the corresponding fixed point theorms given by Horvath [240], Kirk and Shin [313], and Khamsi [292] as special cases.

2.11.2 KKM Principle

In this section, our purpose is to first establish a characterization of the Knaster-Kuratowski and Mazurkiewicz principle in hyperconvex metric spaces and then a characterization for a family of subsets with the finite intersection property in hyperconvex metric spaces. As applications, we give hyperconvex versions of Fan's celebrated minimax principle and Fan's best approximation theorem for set-valued mappings. These in turn are applied to give the Browder-Fan fixed point theorem in hyperconvex metric spaces for set-valued mappings with open inversed values.

As Lemma 2.11.4 shows that each hyperconvec metric space is, indeed an H-spaces, then the KKM theory developed in previous few sections can apply to hyperconvex metric spaces directly. However, by using the special feature of the hyperconvex metric spaces as given in Proposition 2.11.3, Khamsi [292] first established a hyperconvex version of KKM principle as a direct application of Brouwer fixed point theorem (without any help of theory developed in H-spaces by Horvath [239], Bardaro and Ceppitelli [24], and others); and as applications, he also developed hyperconvex version of Fan's best approximation theorem for single-valued mappings and the Schauder-Tychonoff fixed point theorem. The main trust of this section is in the same direction. Our main goal is give a characterization of generalzied metric KKM mapping (see its meaning given in Definition 2.11.6 below) by fully using the structure of hyperconvex meric spaces and Brouwer fixed point theorem. As applications, we give a comprehensive study of KKM theory in hyperconvex spaces and its related applications to fixed point theorems, to characterize KKM principle in hyperconvex spaces, to obtain Fan's minimax principle in hyperconvex spaces, the existence of saddle points, the intersection of sets and the existence of Nash equilibria in game theory.

Let (M,d) be a metric space. Following Khamsi [292], a subset $S \subset M$ is said to be *finitely metrically closed* if for each $F \in \mathcal{F}(M)$, the set $co(F) \cap S$ is closed. Note that $co(F)$ is always defined and belongs to $\mathcal{A}(M)$. Thus if S is closed in M, it is obviously finitely metrically closed. We also recall that a family $\{A_\alpha\}_{\alpha \in D}$ of M has the *finite intersection property* if the intersection of each non-empty finite sub-family is not empty.

Definition 2.11.6. Let X be any non-empty set and let M be a metric space. A set-valued mapping $G : X \to 2^M \setminus \{\emptyset\}$ is said to be a *generalized metric KKM* mapping (in short, GMKKM) if for each non-empty finite set $\{x_1, \cdots, x_n\} \subset X$, there exists a sequence of points of M, $y_1, \cdots, y_n$, not necessarily all different, such that for each subsequenct $\{y_{i_1}, \cdots, y_{i_k}\}$, we have

$$co\{y_{i_j} : j = 1, \cdots, k\} \subset \cup_{j=1}^{k} G(x_{i_j}).$$

In partiuclar, the mapping $G : X \to 2^M$ is said to be a *metric KKM* (in short, MKKM) mapping if for each finite subset $F \in \mathcal{F}(X)$, we have $co(F) \subset \cup_{x \in F} G(x)$.

As shown by Lemma 2.11.4, each hyperconvex metric space is an H-space and thus each GMKKM mapping is a GHKKM mapping, the following characterization of the generalized in hyperconvex metric spaces could be derived as an application of Theorem 2.2.15; however, we give its a complete proof by using some special properties of hyperconvexity metric spaces as stated in Proposition 2.11.3 instead of applying Theorem 2.2.15, which was first given by Kirk et al. [314].

Theorem 2.11.7. Let X be a non-empty set and let M be a hyperconvex metric space. Suppose $G : X \to 2^M \setminus \{\emptyset\}$ is a set-valued mapping with finitely metrically closed values. Then the family $\{G(x) : x \in X\}$ has the finite intersection property if and only if the mapping G is a generalized metric KKM mapping.

Proof. Necessity: By following Kirk et al. [314] if the family $\{G(x) : x \in X\}$ has the finite intersection property, then for each finite subset $\{x_1, \cdots, x_n\} \subset X$, $\cap_{i=1}^n G(x_i) \neq \emptyset$. Take any point $x^* \in \cap_{i=1}^n G(x_i)$ and set $y_i \equiv x^*$ for $i = 1, \cdots, n$. Then for any $1 \leq k \leq n$ and any subsequenct $y_{i_1}, \cdots, y_{i_k}$, it follows that $co(\{y_{i_j} : j = 1, \cdots, k\}) = co(\{x^*\}) = \{x^*\} \subset \cup_{i=1}^k G(x_{i_j})$. This proves that G is a GMKKM mapping.

Sufficiency: Suppose that $G : X \to 2^M \setminus \{\emptyset\}$ is a GMKKM mapping and suppose the family $\{G(x) : x \in X\}$ does not have the finite intersection property. Then there exists a nonempty finite set $\{x_1, \cdots, x_n\}$ for which $\cap_{i=1}^n G(x_i) = \emptyset$. Since G is a GMKKM mapping there exist the corresponding points $y_1, \cdots, y_n$ of M such that for each subsequenct $y_{i_1}, \cdots, y_{i_k}$, we have $co(\{y_{i_1}, \cdots, y_{i_k}\}) \subset \cup_{j=1}^k G(x_{i_j})$. Since M is a hyperconvex metric space there exists a nonexpansive retraction $r : M_\infty \to M$. In particular if we identify M with its isometric copy in the Banach space M_∞ then r maps the linear span L of points $\{y_1, \cdots, y_n\}$ into M. Let $Y := co(\{y_1, \cdots, y_n\})$ and $S := conv(\{y_1, \cdots, y_n\})$. Then $r(S) \subset M$ (indeed, $r(S) \subset Y$) and $r(x) = x$ for each $x \in M$. The assumption that $G(x)$ is finitely metrically closed for each $x \in X$ implies that $Y \cap G(x_i)$ is closed for $i = 1, \cdots, n$. Note also that $Y \cap G(x_i) \neq \emptyset$ since, in particular, $y_i \in Y \cap G(x_i)$. However $\cap_{i=1}^n G(x_i) = \emptyset$, so for each $s \in S$ there exists $i_s \in \{1, \cdots, n\}$ such that $r(s) \notin Y \cap G(x_{i_s})$. Hence $dist(r(s), Y \cap G(x_{i_s})) > 0$. Therefore if the mapping $f : S \to [0, \infty)$ is defined by setting

$$f(s) := \sum_{i=1}^{n} dist(r(s), Y \cap G(x_i))$$

for each $s \in S$, it must be the case that $f(s) > 0$ for each $s \in S$ and also f is obviously continuous.

Now define a (single-valued) mapping $F : S \to S$ by setting

$$F(s) := \frac{1}{f(s)} \sum_{i=1}^{n} dist(r(s), Y \cap G(x_i)) y_i$$

for each $s \in S$. Then F is also continuous. Since S is a bounded closed and convex subset of the finite dimensional space L, by Brouwer's fixed point theorem there exists $s_0 \in S$ such that $F(s_0) = s_0$; i.e.,

$$s_0 = F(s_0) = \frac{1}{f(s_0)} \sum_{i=1}^{n} dist(r(s_0), Y \cap G(x_i))y_i. \tag{2.11.1}$$

If

$$I := \{i_1, \cdots, i_k\} = \{i \in \{1, \cdots, n\} : dist(r(s_0), Y \cap G(x_i)) > 0\}, \tag{2.11.2}$$

then $I \neq \emptyset$, and for each $i \in I$, $r(s_0) \notin Y \cap G(x_i)$. Note that $r(S) \subseteq Y$ by Proposition 2.11.3(3), $r(s_0) \in Y$. (Indeed, $s_0 \in conv\{y_1, \cdots, y_n\}$. To see this, note that B is a closed ball centered at a point of M which contains the set $\{y_{i_j} : j = 1, \cdots, n\}$ since r is nonexpansive and leaves points of M fixed, it follows that $r(y) \in B$. This in turn implies $r(y) \in co(\{y_{i_1}, \cdots, y_{i_k}\}))$. Thus it must be the case that $r(s_0) \notin G(x_i)$ for each $i \in I$, i.e.,

$$r(s_0) \notin \cup_{i \in I} G(x_i). \tag{2.11.3}$$

Then by definition of F, we have that

$$s_0 = F(s_0) = \frac{1}{f(s_0)} \sum_{j=1}^{k} dist(r(s_0), Y \cap G(x_{i_j}))y_{i_j} \in conv(\{y_{i_1}, \cdots, y_{i_k}\}).$$

This in turn implies $r(s_0) \in co(\{y_{i_1}, \cdots, y_{i_k}\})$. Thus we are able to conclude that $r(s_0) \in co(\{y_{i_1}, \cdots, y_{i_k}\}) \subseteq \cup_{j=1}^{k} G(x_{i_j})$ and this contradicts (2.11.3), completing proof. □

As an application of Theorem 2.11.7, we have the following characterization for the existence of a non-empty intersection for the values of a set-valued mapping in an hyperconvex metric space.

Theorem 2.11.8. Let X be a non-empty set and M be a hyperconvex metric space. Suppose $G : X \to 2^M \setminus \{\emptyset\}$ is a set-valued mapping with non-empty closed values and there exists $x_0 \in X$ such that $G(x_0)$ is compact. Then $\cap_{x \in X} G(x) \neq \emptyset$ if and only if the mapping G is a generalized metric KKM mapping.

Proof. Necessity: Since $\cap_{x \in X} G(x) \neq \emptyset$, it follows that the family $\{G(x) : x \in X\}$ has the finite intersection property. Since $G(x)$ is closed for each $x \in X$, it is finitely metrically closed. By Theorem 2.11.7, G is a generalized metric KKM mapping.

Sufficiency: Since G is a generalized metric KKM mapping, it follows by Theorem 2.11.7 that the family $\{G(x) : x \in X\}$ has the finite intersection property. Rewriting this to $\{G(x) \cap G(x_0) : x \in X\}$ and noting $G(x_0)$ is compact, we have

$$\emptyset \neq \cap_{x \in X} G(x) \cap G(x_0) = G(x_0) \cap_{x \in X} G(x) = \cap_{x \in X} G(x).$$

This completes the proof. □

As a special cases of Theorem 2.11.7, we also have the following result which includes Theorem 3 of Khamsi [292, p.303] as a special case.

Corollary 2.11.9. Let X be a non-empty subset of a hyperconvex metric space M. Suppose $G : X \to 2^M \setminus \{\emptyset\}$ is a metric KKM mapping with finitely metrically closed values. Then the family $\{G(x) : x \in X\}$ has the finite intersection property.

By Theorem 2.11.7 and Lemma 2.2.10, we have the following characterization of the finite intersection property of set-valued mappings.

Theorem 2.11.10. Let X be a non-empty set and let M be a hyperconvex metric space. Suppose $G : X \to 2^M \setminus \{\emptyset\}$ is a set-valued mapping such that G is transfer closed valued and there exists a finite subset X_0 of X such that the set $\cap_{x \in X_0} clG(x)$ is non-empty and compact. Then the intersection $\cap_{x \in X} G(x)$ is non-empty if and only if the mapping clG is a generalized metric KKM mapping.

Proof. By Theorem 2.11.7, we have that the family $\{clG(x) : x \in X\}$ has the finite intersection property if and only if clG is a GMKKM mapping. Since there exists a non-empty finite subset X_0 of X such that $\cap_{x \in X_0} clG(x)$ is non-empty compact, it follows that $\cap_{x \in X} clG(x)$ is non-empty if and only if clG is a GMKKM mapping. As an application of Lemma 2.2.10 (by noting that G is transfer closed valued), it follows that $\cap_{x \in X} G(x)$ is non-empty if and only if clG is a generalized metric KKM mapping. □

Theorems 2.11.7, 2.11.8 and 2.11.10 tell us that in hyperconvex metric spaces the finite intersection property for a family of subsets is equivalent to the existence of a non-empty intersection for a set-valued mapping which is a generalized metric KKM mapping. In order to study in which situations such generalized metric KKM mappings can be derived, we now introduce the following definitions.

Definition 2.11.11. Let $X \in \mathcal{F}(M)$ be a non-empty subset of a hyperconvex space M. Then:

(1) the function $f : X \to [-\infty, +\infty]$ is said to be *hyper quasi-convex* (resp., *concave*) if the set $\{x \in X : f(x) \leq \lambda\}$ (resp., the set $\{x \in X : f(x) \geq \lambda\}$) for each $\lambda \in \mathbb{R}$ is an admissible set in M.

(2) the function $\psi(x, y) : X \times X \to [-\infty, +\infty]$ is said to be *hyper diagonal quasi-convex* (resp., *concave*) in y if for each $A \in \mathcal{F}(X)$ and any $y_0 \in co(A)$, we have that $\psi(y_0, y_0) \leq \max_{y_i \in A} \psi(y_0, y_i)$ (resp., $\psi(y_0, y_0) \geq \min_{y_i \in A} \psi(y_0, y_i)$).

(3) $\psi(x, y)$ is said to be *hyper γ-diagonal quasi-convex* (resp., *concave*) in y for some $\gamma \in [-\infty, +\infty]$ if for any $A \in \mathcal{F}(X)$ and $y_0 \in co(A)$, we have that $\gamma \leq \max_{y_i \in A} \psi(y_0, y_i)$ (resp., $\gamma \geq \min_{y_i \in A} \psi(y_0, y_i)$).

Remark 2.11.12. We note that the inequality ' $\leq$' (resp., ' $\geq$') could be replaced equivalently by the strict inequality ' $<$' (reps., ' $>$') in the Definition 2.11.11(1). It is clear that if $\psi(x, y)$ is hyper diagonal quasi-convex (resp., concave) in y, then $\psi(x, y)$ must be hyper γ-diagonally quasi-convex (resp., concave) in y, where $\gamma = \inf_{x \in X} \psi(x, x)$ (resp, $\gamma = \sup_{x \in X} \psi(x, x)$).

Definition 2.11.13. Let $X \in \mathcal{F}(M)$ be a non-empty set of a hyperconvex metric space M and $\gamma \in (-\infty, +\infty]$. Suppose $\psi : M \times X \to (-\infty, +\infty]$ is a function. Then ψ is said to be *hyper γ-generalized quasi-convex* (resp., *concave*) in M if for each non-empty finite subset $\{x_1, \cdots, x_n\} \subset X$, there exists a sequence $y_1, \cdots, y_n$ in M such that for each subsequenct $y_{i_1}, \cdots, y_{i_k}$ and any $x_0 \in co\{x_{i_1}, \cdots, x_{i_k}\}$, we have $\gamma \leq \max_{1 \leq j \leq k} \psi(x_0, y_{i_j})$ (resp., $\gamma \geq \min_{1 \leq j \leq k} \psi(x_0, y_{i_j})$).

Lemma 2.11.14. Let X be a non-empty set, M a hyperconvex metric space and let $\gamma \in \mathbb{R}$ be a given real number. Suppose $\psi : M \times X \to (-\infty, +\infty]$ is a mapping. Then the following two statements are equivalent:

(1) the mapping $G : X \to 2^M \backslash \{\emptyset\}$ defined by $G(x) := \{y \in M : \psi(y, x) \leq \gamma\}$ (resp., $G(x) := \{y \in M : \psi(y, x) \geq \gamma\}$) for each $x \in X$ is a generalized metric KKM mapping.

(2) the function ψ is hyper γ-generalized quasi-concave (resp., convex) in M.

Proof. We only prove the conclusion for the case ψ is hyper γ-generalized quasi-concave. The convex case is proved similarly.

(1) $\Rightarrow$ (2): As $G : X \to 2^M \setminus \{\emptyset\}$ is a GMKKM mapping, for each finite set $\{x_1, \cdots, x_n\} \subset X$, there exists $y_1, \cdots, y_n$ in M such that for any subsequenct set $y_{i_1}, \cdots, y_{i_k}$ and any $y_0 \in co\{y_{i_1}, \cdots, y_{i_k}\}$, we have that $y_0 \in \cup_{j=1}^k G(x_{i_j})$. Hence there exists some $m \in \{1, \cdots, k\}$ such that $y_0 \in G(x_{i_m})$. It follows that $\psi(y_0, x_{i_m}) \leq \gamma$. Therefore we have that $\min_{1 \leq j \leq k} \psi(y_0, x_{i_j}) \leq \gamma$, which implies that ψ is hyper γ-generalized quasi-concave in M.

(2) $\Rightarrow$ (1): As ψ is hyper γ-generalized quasi-concave in M, for any non-empty finite set $\{x_1, \cdots, x_n\} \subset X$, there exists $y_1, \cdots, y_n$ in M such that for any subsequencet $y_{i_1}, \cdots, y_{i_k}$ and $y_0 \in co\{y_{i_1}, \cdots, y_{i_k}\}$ we have that $\min_{1 \leq j \leq k} \psi(y_0, x_{i_j}) \leq \gamma$. Hence there exists some $m \in \{1, \cdots, k\}$ such that $\psi(y_0, x_{i_m}) \leq \gamma$. This implies that $y_0 \in G(x_{i_m})$. By the arbitrariness of $y_0 \in co\{y_{i_1}, \cdots, y_{i_k}\}$, it follows that $co\{y_{i_1}, \cdots, y_{i_k}\} \subset \cup_{j=1}^k G(x_{i_j})$ which implies that $G : X \to 2^M \setminus \{\emptyset\}$ is a generalized metric KKM mapping. □

We now give the following minimax inequality which is a hyperconvex version of Fan's celebrated minimax inequality principle.

Theorem 2.11.15. Let $X \in \mathcal{A}(M)$ be a compact subset of a hyperconvex space M. Suppose $\psi : X \times X \to (-\infty, +\infty]$ is a mapping such that

(1) for each fixed $x \in X$, $y \mapsto \psi(x, y)$ is hyper 0-generalized quasi-concave; and

(2) for each fixed $y \in X$, $x \mapsto \psi(x, y)$ is lower semicontinous.

Then there exists $x_0 \in X$ such that $\sup_{y \in X} \psi(x_0, y) \leq 0$.

Proof. We define a mapping $G : X \to 2^X$ by

$$G(x) := \{y \in X : \psi(y, x) \leq 0\}$$

for each $x \in X$. Then it is clear that G is a GMKKM mapping by Lemma 2.11.14. Now by Theorem 2.11.8, it follows that $\cap_{x \in X} G(x) \neq \emptyset$. Take any point $x_0 \in \cap_{x \in X} G(x)$, then we have that $\sup_{y \in X} \psi(x_0, y) \leq 0$ and the proof is completed. □

As an application of the generalized metric MKKM principle established above, we give the following hyperconvex version of Fan's best approximation in hyperconvex spaces for set-valued mappings.

Theorem 2.11.16. Let M be a hyperconvex metric space and $X \in \mathcal{A}(M)$ be a non-empty compact subset of M. Suppose $F : X \to \mathcal{A}(M)$ is a set-valued continuous mapping. Then there exists $x_0 \in X$ such that

$$d(x_0, F(x_0)) = \inf_{x \in X} d(x, F(x_0)).$$

In particular if $F(x_0)$ is compact and $x_0 \notin F(x_0)$, x_0 must be a boundary point of X (i.e., $x_0 \in Bd(X)$).

Proof. Define a mapping $G : X \to 2^M \setminus \{\emptyset\}$ by

$$G(x) := \{y \in X : d(y, F(y)) \leq d(x, F(y))\}$$

for each $x \in X$. As F is continuous, $G(x)$ is closed and non-empty for each $x \in X$. Now we claim that G is a metric KKM mapping. Suppose it were not. Then there exists a non-empty and finite subset $\{x_1, \cdots, x_n\}$ and $y \in co(\{x_i : i = 1, \cdots, n\})$ such that

$$d(x_i, F(y)) < d(y, F(y))$$

for $i = 1, \cdots, n$. Let $\epsilon > 0$ such that $d(x_i, F(y)) \leq d(y, F(y)) - \epsilon$ for $i = 1, \cdots, n$. Let $r = dist(y, F(y)) - \epsilon$. Then for $i = 1, \cdots, n$, $x_i \in F(y) + r$, where $F(y) + r := \cup\{B(a; r) : a \in F(y)\}$ and $B(a, r)$ is a closed ball centered at a with the radius r. Note that $F(y) \in \mathcal{A}(M)$, it follows that $F(y) + r \in \mathcal{A}(M)$ (e.g., see Sine [487, p.864]). Thus $co\{x_1, x_2, \cdots, x_n\} \subset F(y) + r$. This in turn implies $y \in F(y) + r$, that is, $dist(y, F(y)) \leq r = dist(y, F(y)) - \epsilon$, which is impossible. Therefore G must be a metric KKM-mapping.

Note that X is compact, we have that $\cap_{x \in X} G(x) \neq \emptyset$. Take any point $x_0 \in \cap_{x \in X} G(x)$, then it is clear that $d(x_0, F(x_0)) \leq d(x, F(x_0))$ for all $x \in X$, which implies that $d(x_0, F(x_0)) = \inf_{x \in X} d(x, F(x_0))$.

If $F(x_0)$ is compact and $x_0 \notin F(x_0)$, then there exists $u_0 \in F(x_0)$ such that $d(x_0, u_0) = d(x_0, F(x_0)) > 0$. We now claim that x_0 is a boundary point of X. Suppose it were not, i.e., $x_0 \in intX$, where $intX$ is the set of all relatively interior points of X in M. Then there exists a postive number $r > 0$ small enough such that $B(x_0, r) \subset intX \subset X$ and

$$0 < r < d(x_0, F(x_0)) \leq d(z, F(x_0))$$

for all $z \in B(x_0, r)$. Note that $d(x_0, u_0) = d(x_0, F(x_0)) = r + d(x_0, F(x_0)) - r$ and $d(x_0, F(x_0)) - r > 0$, it follows by the hyperconvexity of M that

$$B(x_0, r) \cap B(u_0, d(x_0, F(x_0)) - r) \neq \emptyset.$$

Since $u_0 \in F(x_0)$ and thus $B(u_0, d(x_0, F(x_0)) - r) \subset B(F(x_0), d(x_0, F(x_0)) - r)$, there exists $y_0 \in B(x_0, r) \cap B(F(x_0), d(x_0, F(x_0)) - r)$ such that $d(y_0, F(x_0)) \leq d(x_0, F(x_0)) - r$. However, we do have that $d(y_0, F(x_0)) \geq d(x_0, F(x_0)) > 0$, which is a contraction. Therefore x_0 must be a boundary point of X, i.e., $x_0 \in Bd(X)$ and completes the proof. □

As a special case of our best approximation theorem, we have the following best approximation result for single-valued mappings, first given by Khamsi [18].

Corollary 2.11.17. Let M be a hyperconvex metric space and $X \in \mathcal{A}(M)$ be a non-empty compact subset of M. Suppose $F : X \to M$ is a single-valued continuous mapping. Then there exists $x_0 \in X$ such that

$$d(x_0, F(x_0)) = \inf_{x \in X} d(x, F(x_0)).$$

2.11.3 Intersection Theorems

In what follows, we shall give a dual form of above KKM principle in hyperconvex space in the sense that the family $\{G(x) : x \in X\}$ of Theorem 2.11.7 still has a finite intersection property if $G(x)$ takes finitely metrically open values instead of taking finitely metrically closed values for each $x \in X$.

Theorem 2.11.18 (The dual form of the Fan-KKM principle in hyperconvex space.) Let X be a non-empty subset of a hyperconvex metric space M. Suppose $G : X \to 2^M \setminus \{\emptyset\}$ is a MKKM mapping with finitely metrically open values. Then the family $\{G(x) : x \in X\}$ has the finite intersection property.

Proof. As each hyperconvex metric space is an H-space by Lemma 2.11.4, the conclusion follows by Proposition 2.2.6 and the proof is complete. □

As applications of both Theorems 2.11.7 and 2.11.18, we have the following Ky Fan matching theorem for both closed and open covers in hyperconvex spaces.

Theorem 2.11.19. Let $X \in \mathcal{A}(M)$ be a non-empty subset of a hyperconvx metric space M. Suppose $\{A_i\}_{i=1}^n$ is a family of closed subsets of X such that $\cup_{i=1}^n A_i = X$ and $x_1, \cdots, x_n$ are (not necessarily distinct) n points of X. Then there exist k indices $i_1 < i_2 < \cdots < i_k$ between 1 and n such that $co\{x_{i_1}, x_{i_2}, \cdots, x_{i_k}\} \cap (\cap_{j=1}^k A_{i_j}) \neq \emptyset$.

Proof. Let $X_0 = \{x_1, x_2, \cdots, x_n\}$ and we define a set-valued mapping $F : X_0 \to 2^X$ by $F(x_i) := X \setminus A_i$ for each $i = 1, 2, \cdots, n$. Suppose the conclusion is false. Then $F(x)$ is open for each $x \in X_0$ and for each non-empty finite subset $B \subset X_0$, we have $co\{x : x \in B\} \subset \cup\{F(x) : x \in B\}$. By Theorem 2.11.18, it follows that $\cap_{i=1}^n F(x_i) \neq \emptyset$. But this means that $\cup_{i=1}^n A_i \neq X$, which contradicts our hypothesis and this contradiction completes our proof. □

As an application of Theorem 2.11.7, we also have the following Ky Fan matching theorem for open covers in hyperconvex spaces.

Theorem 2.11.20. Let $X \in \mathcal{A}(M)$ be a non-empty subset of a hyperconvx metric space M. Suppose $\{A_i\}_{i=1}^n$ is a family of open subsets of X such that $\cup_{i=1}^n A_i = X$ and $x_1, \cdots, x_n$ are (not necessarily distinct) n points of X. Then there exist k indices $i_1 < i_2 < \cdots < i_k$ between 1 and n such that $co\{x_{i_1}, x_{i_2}, \cdots, x_{i_k}\} \cap (\cap_{j=1}^k A_{i_j}) \neq \emptyset$.

Proof. Let $X_0 = \{x_1, x_2, \cdots, x_n\}$ and we define a set-valued mapping $F : X_0 \to 2^X$ by $F(x_i) := X \setminus A_i$ for each $i = 1, 2, \cdots, n$. Suppose the conclusion is false. Then $F(x)$ is closed for each $x \in X_0$ and for each non-empty finite subset $B \subset X_0$, we have $co\{x : x \in B\} \subset \cup\{F(x) : x \in B\}$. By Theorem 2.11.7, it follows that $\cap_{i=1}^n F(x_i) \neq \emptyset$. But this means that $\cup_{i=1}^n A_i \neq X$, which contradicts our hypothesis and this contradiction completes our proof. □

We note that Theorem 2.11.20 need not to assume that X is compact as required by Theorem 2.11.19.

In this part, as applications of Ky Fan matching theorems for closed covers, we shall establish some fixed point theorems for set-valued mappings and some existence of intersection theorems which are hyperconvex versions of the corresponding results given by Alexandroff-Pasynkoff and others.

First we have the following fixed point theorem for set-valued mappings in hyperconvex spaces.

Theorem 2.11.21. Let $X \in \mathcal{A}(M)$ be a non-empty subset of a hyperconvex metric space M. Suppose $A : X \to 2^X$ is a set-valued mapping with closed values (maybe taking empty values for some point $x \in X$) and there exists n points $x_1, \cdots, x_n$ (not necessarily distinct) of X such that $\cup_{i=1}^n A(x_i) = X$ and the set $A^{-1}(y) := \{x \in X : y \in A(x)\}$ for each $y \in X$ is admissible. Then A has a fixed point, i.e., there exists $w \in X$ such that $w \in A(w)$.

Proof. By Theorem 2.11.19, there exist k indices $i_1 < i_2 < \cdots < i_k$ between 1 and n such that $co\{x_{i_1}, x_{i_2}, \cdots, x_{i_k}\} \cap (\cap_{j=1}^k A(x_{i_j})) \neq \emptyset$. Take any $w \in co\{x_{i_1}, x_{i_2}, \cdots, x_{i_k}\} \cap \cap_{j=1}^k A(x_{i_j})$. Then $x_{i_j} \in A^{-1}(w)$ for $j = 1, \cdots, k$. By the fact that $A^{-1}(w)$ is admissible, it follows that $w \in co\{x_{i_1}, \cdots, x_{i_k}\} \subset A(w)$. Therefore w is a fixed point of A and the proof is complete. □

As an application of Theorem 2.11.20, we also have the following Browder-Fan type fixed point theorm for set-valued mappings in hyperconvex spaces.

Theorem 2.11.22. Let $X \in \mathcal{A}(M)$ be a non-empty subset of a hyperconvex metric space M. Suppose $A : X \to 2^X$ is a set-valued mapping with open values (maybe taking empty values for some point $x \in X$) and there exist n points $x_1, \cdots, x_n$ (not necessarily distinct) of X such that $\cup_{i=1}^n A(x_i) = X$ and the set $A^{-1}(y) := \{x \in X : y \in A(x)\}$ for each $y \in X$ is admissible. Then A has a fixed point, i.e., there exists $w \in X$ such that $w \in A(w)$.

Proof. By Theorem 2.11.20, there exist k indices $i_1 < i_2 < \cdots < i_k$ between 1 and n such that $co\{x_{i_1}, x_{i_2}, \cdots, x_{i_k}\} \cap (\cap_{j=1}^k A(x_{i_j})) \neq \emptyset$. Take any $w \in co\{x_{i_1}, x_{i_2}, \cdots, x_{i_k}\} \cap (\cap_{j=1}^k A(x_{i_j}))$. Then $x_{i_j} \in A^{-1}(w)$ for $j = 1, \cdots, k$. By the fact that $A^{-1}(w)$ is admissible, it follows that $w \in co\{x_{i_1}, \cdots, x_{i_k}\} \subset A(w)$. Therefore w is a fixed point of A and the proof is complete. □

As equivalent forms of Theorems 2.11.21 and 2.11.22, we have the following fixed point theorems for set-valued mappings in hyperconvex spaces.

Theorem 2.11.23. Let $X \in \mathcal{A}(M)$ be a subset of a hyperconvex space M. Suppose $F : X \to 2^X$ is a set-valued mapping with non-empty admissible values. There exists a non-empty finite subset C of X such that $F(x) \cap C \neq \emptyset$ for each $x \in X$ and $F^{-1}(y)$ is closed for each $y \in X$. Then F has a fixed point, i.e., there exists $w \in X$ such that $w \in F(w)$.

Proof. Define $A : X \to 2^X$ by $A(x) := F^{-1}(x)$ for each $x \in X$. Then it is easy to verify that A satisfies all hypotheses of Theorem 2.11.21. Thus the conclusion follows by Theorem 2.11.21 and the proof is complete. □

Theorem 2.11.24. Let $X \in \mathcal{A}(M)$ be a subset of a hyperconvex space

M. Suppose $F : X \to 2^X$ is a set-valued mapping with non-empty admissible values. There exists a non-empty finite subset C of X such that $F(x) \cap C \neq \emptyset$ for each $x \in X$ and $F^{-1}(y)$ is open for each $y \in X$. Then F has a fixed point, i.e., there exists $w \in X$ such that $w \in F(w)$.

Proof. Define $A : X \to 2^X$ by $A(x) := F^{-1}(x)$ for each $x \in X$. Then it is easy to verify that A satisfies all hypotheses of Theorem 2.11.22. Thus the conclusion follows by Theorem 2.11.22 and the proof is complete. □

The following are consequences of Theorems 2.11.23 and 2.11.24.

Theorem 2.11.25. Let $X \in \mathcal{A}(M)$ be a non-empty subset of a hyperconvex space M. Suppose $G : X \to 2^X$ is a upper semicontinuous set-valued mapping with non-empty closed admissible valuesand there exists a non-empty finite subset C of X such that $G(x) \cap C \neq \emptyset$ for each $x \in X$. Then G has a fixed point, i.e., there exists $w \in X$ such that $w \in G(w)$.

Proof. As G is upper semicontinuous with non-empty closed values, it follows that the graph of G is closed and thus the set $G^{-1}(y)$ is closed for each $y \in X$. Therefore all hypotheses of Theorem 2.11.23 are satisfied. The conclusion follows from Theorem 2.11.23 and the proof is complete. □

Theorem 2.11.26. Let $X \in \mathcal{A}(M)$ be a compact subset of a hyperconvex space M. Suppose $F : X \to 2^X$ is a set-valued mapping with non-empty admissible values and $F^{-1}(y)$ is open for each $y \in X$. Then F has a fixed point, i.e., there exists $w \in X$ such that $w \in F(w)$.

Proof. Note that $F^{-1}(y)$ is open for each $y \in X$ and $F(x) \neq \emptyset$ for all $x \in X$, it follows that $\cup_{y \in X} F^{-1}(y) = X$. As X is compact, there exists a finite subset C of X such that $X = \cup_{y \in C} F^{-1}(y)$. Therefore for each $x \in X$, we have that $F(x) \cap C \neq \emptyset$. Thus F satisfes all hypotheses of Theorem 2.11.24. The conclusion follows by Theorem 2.11.24 and completing the proof. □

Remark 2.11.27. Theorem 2.11.26 is a hyperconvex version of the famous Browder-Fan fixed point theorm. In next section, we will establish a more general form for non-compact hyperconvex spaces. We also note that both Theorems 2.11.21 and 2.11.22 are hyperconvex versions of the corresponding fixed points given by Kim [297] and Lassonde [335].

In the remainder of this section, as applications of both Theorems 2.11.18 and 2.11.20, we will establish some intersection theorems which are hyperconvex versions of the corresponding results due to Alexandroff-Pasynkoff [1] and others.

Theorem 2.11.28. Let $X \in \mathcal{A}(M)$ be a non-empty subset of a hyperconvex metric space M. Suppose $\{A_i : i = 1, 2, \cdots, n\}$ is a closed cover of X (i.e., A_i is a closed subset of X for $i = 1, \cdots, n$ and $\cup_{i=1}^n A_i = X$) and there exist n points

$\{x_i, i \in I = (1, \cdots, n)\}$ (not necessarily distinct) of X such that

$$co\{x_j : i \in I \text{ and } j \neq i\} \subset A_i$$

for each $i \in I$. Then we have $\cap_{i=1}^n A_i \neq \emptyset$.

Proof. By Theorem 2.11.19, it follows that there exists a non-empty subset $J \subset I$ such that $co\{x_j : j \in J\} \cap (\cap_{j \in J} A_j) \neq \emptyset$. On the other hand, our assumption implies that $co\{x_j : j \in J\} \subset \cap\{A_i : i \in I \setminus J\}$. Therefore, it follows that there exists $x_0 \in co\{x_j : j \in J\}$ such that $x_0 \in \cap_{i \in I} A_i$, i.e., $\cap_{i=1}^n A_i \neq \emptyset$ and the proof is complete. □

Theorem 2.11.29. Let $X \in \mathcal{A}(M)$ be a non-empty subset of a hyperconvex metric space M. Suppose $\{A_i : i = 1, 2, \cdots, n\}$ is an open cover of X (i.e., A_i is an open subset of X for $i = 1, \cdots, n$ and $\cup_{i=1}^n A_i = X$) and there exist n points $\{x_i, i \in I = (1, \cdots, n)\}$ (not necessarily distinct) of X such that

$$co\{x_j : i \in I \text{ and } j \neq i\} \subset A_i$$

for each $i \in I$. Then we have $\cap_{i=1}^n A_i \neq \emptyset$.

Proof. By Theorem 2.11.20, it follows that there exists a non-empty subset $J \subset I$ such that $co\{x_j : j \in J\} \cap \cap_{j \in J} A_j \neq \emptyset$. On the other hand, our assumption implies that $co\{x_j : j \in J\} \subset \cap\{A_i : i \in I \setminus J\}$. Therefore, it follows that there exists $x_0 \in co\{x_j : j \in J\}$ such that $x_0 \in \cap_{i \in I} A_i$, i.e., $\cap_{i=1}^n A_i \neq \emptyset$ and the proof is complete. □

Theorem 2.11.30. Let $X \in \mathcal{A}(M)$ be a non-empty subset of a hyperconvex metric space M. Suppose $\{A_i : i = 1, 2, \cdots, n\}$ is a closed cover of X such that $\cap_{j \in J} A_j \neq \emptyset$ for each $J \subset I$ with $J \neq I$, where $I = \{1, 2, \cdots, n\}$. Then $\cap_{i=1}^n A_i \neq \emptyset$.

Proof. For each $i \in I = \{1, \cdots, n\}$. Take any $x_i \in \cap\{A_j : j \in I \text{ and } j \neq i\}$. Then by Theorem 2.11.19, it follows that there exists a non-empty subset $J \subset I$ such that $co\{x_j : j \in J\} \cap (\cap_{j \in J} A_j) \neq \emptyset$. On the other hand, our assumption implies that $\{x_j : j \in J\} \subset \cap\{A_i : i \in I \setminus J\}$. Therefore, it follows that there exists $x_0 \in \{x_j : j \in J\}$ such that $x_0 \in \cap_{i \in I} A_i$, i.e., $\cap_{i=1}^n A_i \neq \emptyset$ and the proof is complete. □

Theorem 2.11.31. Let $X \in \mathcal{A}(M)$ be a non-empty subset of a hyperconvex metric space M. Suppose $\{A_i : i = 1, 2, \cdots, n\}$ is an open cover of X such that $\cap_{j \in J} A_j \neq \emptyset$ for each $J \subset I$ with $J \neq I$, where $I = \{1, 2, \cdots, n\}$. Then $\cap_{i=1}^n A_i \neq \emptyset$.

Proof. For each $i \in I = \{1, \cdots, n\}$. Take any $x_i \in \cap\{A_j : j \in I \text{ and } j \neq i\}$. Then by Theorem 2.11.20, it follows that there exists a non-empty subset $J \subset I$ such that $co\{x_j : j \in J\} \cap \cap_{j \in J} A_j \neq \emptyset$. On the other hand, our assumption implies that $\{x_j : j \in J\} \subset \cap\{A_i : i \in I \setminus J\}$. Therefore, it follows that there exists $x_0 \in \{x_j : j \in J\}$ such that $x_0 \in \cap_{i\in I} A_i$, i.e., $\cap_{i=1}^n A_i \neq \emptyset$ and the proof is complete. □

2.11.4 Fixed Point Theorems

In this section, we shall use the generalized metric KKM principle and Fan's best approximation theorem in hyperconvex metric space established in section 2 above, to derive hyperconvex versions of the Browder-Fan fixed point theorem and Schauder-Tychonoff fixed point theorms for set-valued mappings.

Theorem 2.11.32. Let $X \in \mathcal{A}(M)$ be a compact subset of a hyperconvex metric space M. Suppose that $F : X \to 2^X$ is a set-valued mapping with non-empty values such that:
(1) F is transfer open inversed valued; and
(2) for each $x \in X$, $F(x)$ is admissible.
Then there exists $x_0 \in X$ such that $x_0 \in F(x_0)$.

Proof. Define a set-valued mapping $G : X \to 2^X$ by

$$G(x) = X \setminus F^{-1}(x)$$

for each $x \in X$. Since $F(x)$ is non-empty and transfer closed valued, $X = \cup_{y\in X} F^{-1}(y)$. Therefore we have $\cap_{x\in X} G(x) = \emptyset$. By Theorem 2.11.10, G must not be a metric KKM mapping, thus there exists a subset $\{x_1, \cdots, x_n\} \subset X$ and a point $x_0 \in co\{x_1, \cdots, x_n\}$ such that $x_0 \notin G(x_i)$ for $i = 1, 2, \cdots, n$. It follows that $x_i \in F(x_0)$ for all $i = 1, \cdots, n$. Note that $F(x_0)$ is admissible, so it follows that $x_0 \in co\{x_i, i = 1, \cdots, n\} \subset F(x_0)$. Therefore x_0 is a fixed point of F and thus the proof is completed. □

We note that Theorem 2.11.32 was also derived by Park [418]. As an application of Theorem 2.11.32, we have the following result which is Fan's geometric lemma [157] in hyperconvex metric space.

Theorem 2.11.33. Let $X \in \mathcal{A}(M)$ be a non-empty compact hyperconvex subset of a hyperconvex metric space M. Suppose C is a non-empty subset of $X \times X$ such that:
(1) for each fixed $x \in X$, the set $\{y \in X : (x, y) \notin C\}$ is empty or admissible;
(2) for each fixed $y \in X$, the set $\{x \in X : (x, y) \in C\}$ is closed; and
(3) for each $x \in X$, $(x, x) \in C$.
Then there exists $x_0 \in X$ such that $\{x_0\} \times X \subset C$.

Proof. In order to apply Theorem 2.11.32, we define a set-valued mapping $F : X \to 2^X$ by

$$F(x) := \{y \in X : (x, y) \notin C\}$$

for each $x \in X$. Then we can verify that F satisfies all the hypotheses of Theorem 2.11.32. Since F has no fixed point, by our condition (3), it follows that there must exist $x_0 \in X$ such that $F(x_0) = \emptyset$. Thus we must have $\{x_0\} \times X \subset C$ by the definition of F and the proof is completed. □

As another application of Theorem 2.11.33, we have the following result.

Theorem 2.11.34. Let $X \in \mathcal{A}(M)$ be a non-empty compact subset of a hyperconvex metric space M. Suppose $F : X \to 2^X$ is a set-valued mapping such that:

(1) for each $y \in X$, $X \setminus F^{-1}(y)$ is admissible;

(2) for each $x \in X$, the set $F(x)$ is closed; and

(3) for each $x \in X$, $x \in F(x)$.

Then there exists $x_0 \in X$ such that $x_0 \in \cap_{x \in X} F(x)$.

Proof. In order to apply Theorem 2.11.32, let $C := \{(x, y) \in X \times X : x \in F(y)\}$. Then we have that

(1) for each $x \in X$, the set $\{y \in X : (x, y) \notin C\} = \{y \in X : x \notin F(y)\} = X \setminus F^{-1}(x)$ which is admissible by condition (1).

(2) for each fixed $y \in X$, the set $\{x \in X : (x, y) \in C\} = F(y)$ is closed in X by the hypothesis (2).

(3) for each $x \in X$, it is clear that $(x, x) \in X$.

Thus all hypotheses of Theorem 2.11.33 are satisfied and it follows that there exists $x_0 \in X$ with $\{x_0\} \times X \subset C$. This implies that $x_0 \in \cap_{x \in X} F(x)$ and the proof is completed. □

Remark 2.11.35. Both Theorems 2.11.33 and 2.11.34 are equivalent to each other. To see this, it suffices to prove Theorem 2.11.33 from Theorem 2.11.34. Define $F : X \to 2^X$ by $F(y) = \{x \in X : (x, y) \in C\}$ for each $y \in X$. Then the hypotheses in Theorem 2.11.32.5.2 ensure that F satisfies all the hypotheses of Theorem 2.11.34. By Theorem 2.11.34, $\cap_{x \in X} F(x) \neq \emptyset$. Take any $x_0 \in \cap_{x \in X} F(x)$, and we have $\{x_0\} \times X \subset C$, as required.

As an equivalent statement of Theorem 2.11.32, we have the following existence theorem for maximal elements in hyperconvex spaces.

Theorem 2.11.36. Let $X \in \mathcal{A}(M)$ be a compact hyperconvex subset of a hyperconvex metric space M. Assume $F : X \to 2^X$ is a set-valued mapping such that:

(1) F is transfer open inversed valued; and

(2) for each $x \in X$, $x \notin F(x)$ and $F(x)$ is admissible.

Then there exists $x_0 \in X$ such that $F(x_0) = \emptyset$.

As an application of Theorem 2.11.16, we have the following Schauder-Tychonoff fixed point theorem for set-valued mappings.

Theorem 2.11.37. Let $X \in \mathcal{F}(M)$ be a compact subset of a hyperconvex metric space M. Suppose $F : X \to \mathcal{A}(M)$ is a set-valued continuous mapping with non-empty closed values such that for each $x \in X$ with $x \notin F(x)$, there exists $z \in X$ such that $d(z, F(x)) < d(x, F(x))$. Then F has a fixed point in X. In particular when F takes compact values, the conclusion of Theorem 2.11.37 still holds if for all $x \in Bd(X)$ with $x \notin F(x)$, there exists $z \in X$ such that $d(z, F(x)) < d(x, F(x))$.

Proof. By Theorem 2.11.16, there exists $x_0 \in X$ such that

$$d(x_0, F(x_0)) = \inf_{x \in X} d(x, F(x_0)).$$

We claim that x_0 is a fixed point of F. Indeed, assume it were not true, i.e., $x_0 \notin F(x_0)$, it follows that $d(x_0, F(x_0)) > 0$. Then by assumption, there exists $z_0 \in X$ such that $d(z_0, F(x_0)) < d(x_0, F(x_0))$. On the other hand, note that $d(z_0, F(x_0)) \geq d(x_0, F(x_0)) > 0$. This implies that $0 < d(z_0, F(x_0)) < d(z_0, F(x_0))$, which is impossible and thus x_0 must be a fixed point of F. When F takes compact values and $x_0 \notin F(x_0)$, it follows by the particular case of Theorem 2.11.16, $x_0 \in Bd(X)$ and thus the conclusion follows by the same argument above and we complete the proof. □

As an immediate consequence of Theorem 2.11.37, we have the following fixed point theorem which is Corollary 3.5 of Kirk and Shin [313, p.180].

Corollary 2.11.38. Let $X \in \mathcal{F}(M)$ be a compact subset of a hyperconvex metric space M. Suppose $F : X \to \mathcal{A}(M)$ is a set-valued continuous mapping with non-empty closed values for which $F(x) \cap X \neq \emptyset$ for all $x \in X$. Then F has a fixed point in X.

Proof. Note that $F(x) \cap X \neq \emptyset$ for each $x \in X$. It follows that for any $x \in X$ with $x \notin F(x)$, if we take $z \in F(x) \cap X$, we have $d(z, F(x)) = 0 < d(x, F(x))$. Thus all hypotheses of Theorem 2.11.37 are satisfied and the conclusion follows. □

Remark 2.11.39. Theorem 2.11.37 improves Theorem 6 of Khamsi [292] in the following ways: (1) the mapping F is a set-valued mapping; and (2) our boundary condition is weaker than the corresponding condition given in [292, p.305].

Before closing this part, we will prove one generalization of Browder-Fan fixed point theorem for set-valued mappings in hyperconvex. In order to do so, we first need the following Ky Fan matching theorem for open covers in hyperconvex spaces.

Theorem 2.11.40. Let $X \in \mathcal{A}(M)$ be a non-empty compact subset of a hyperconvex space M and $A : X \to 2^X$ be a set-valued mapping such that:

(1) $A(x)$ is open for each $x \in X$; and

(2) $A(X) := \cup_{x \in X} A(x) = X$.

Then there exists a non-empty finite subset $\{x_1, \cdots, x_n\} \subset X$ and an $x_0 \in co\{x_1, \cdots, x_n\}$ such that $x_0 \in \cap_{i=1}^n A(x_i)$.

Proof. Define a set-valued mapping $F : X \to 2^X$ by $F(x) := X \setminus A(x)$ for each $x \in X$. Suppose the conclusion is false. Then for each finite subset $\{x_1, \cdots, x_n\} \subset X$, it follows that $co\{x_1, \cdots, x_n\} \subset \cup_{i=1}^n F(x_i)$. Therefore F is a MKKM mapping with closed values. As X is compact, by Theorem 2.11.7, we have that $\cap_{x \in X} F(x) \neq \emptyset$. But this implies $\cup_{x\in X} A(x) = A(X) \neq X$, which is against our assumption and this completes our proof. $\square$

As an application of Theorem 2.11.40, we have the following corollary which is Theorem 1 of Park [418].

Corollary 2.11.41. Let $X \in \mathcal{A}(M)$ be a non-empty compact subset of a hyperconvex space M and Y be any given topological space. Suppose $A : X \to 2^Y$ is a set-valued mapping such that:

(1) $A(x)$ is open for each $x \in X$; and

(2) $A(X) := \cup_{x\in X} A(x) = Y$.

Then for each (single-valued) continuous mapping $f : X \to Y$, there exists a non-empty finite subset $\{x_1, \cdots, x_n\} \subset X$ and an $x_0 \in co\{x_1, \cdots, x_n\}$ such that $f(x_0) \in \cap_{i=1}^n A(x_i)$.

Proof. Define a set-valued mapping $A : X \to 2^X$ by $F(x) = f^{-1}(A(x))$ for each $x \in X$. This definition is well defined as $A(X) = Y$. Then the mapping F satisfies all hypotheses of Theorem 2.11.40. By Theorem 2.11.40, there exists a finite subset $\{x_1, \cdots, x_n\} \subset X$ and an $x_0 \in co\{x_1, \cdots, x_n\}$ such that $x_0 \in \cap_{i=1}^n f^{-1}(A(x_i))$, which implies that $f(x_0) \in \cap_{i=1}^n A(x_i)$ and thus completing the proof. $\square$

As an application of Theorem 2.11.40, we have the following Browder-Fan fixed point theorem for set-valued mappings in hyperconvex spaces.

Theorem 2.11.42. Let $X \in \mathcal{A}(M)$ be a compact subset of a hyperconvex metric space M. Suppose there exist two set-valued mappings $F_1, F_2 : X \to 2^X \setminus \{\emptyset\}$ such that:

(1) $F_1(x) \subset F_2(x)$ for each $x \in X$;

(2) $X = \cup\{intF_1^{-1}(y) : y \in X\}$ (i.e., F_1 is transfer open inversed valued; and

(3) $F_2(x)$ is admissible for each $x \in X$.

Then there exists $x_0 \in X$ such that $x_0 \in F_2(x_0)$.

Proof. Define a set-valued mapping $A : X \to 2^X$ by $A(x) := intF_1^{-1}(x)$ for each $x \in X$. Then we have that (1): $A(x)$ is open for each $x \in X$; (2): $\cup_{x\in X} A(x) = A(X) = X$ by condition (2); and (3): $A^{-1}(x) \subset F_2(x)$ for each $x \in X$ by condition (1). By Theorem 2.11.40, there exists a non-empty finite subset $\{x_1, \cdots, x_n\} \subset X$ and an $x_0 \in co\{x_1, \cdots, x_n\}$ such that $x_0 \in \cap_{i=1}^n A(x_i)$.

Therefore $x_i \in A^{-1}(x_0) \subset F_2(x_0)$ for $i = 1, \cdots, n$. By the condition $F(x_0)$ is admissible, it follows that $x_0 \in co\{x_1, \cdots, x_n\} \subset F_2(x_0)$, which is a fixed point of F_2 and we complete the proof. □

We would like to note that Theorem 2.11.42 unifies Theorem 3 of Park [418] and Theorem 3.1 of Kirk et al. [314] for the Browder-Fan fixed point theorems in hyperconvex spaces. Befor closing this part, we would like to metion that by using the notion of noncompactness measure, and non-compact version of the KKM principle in hyperconvex metric space can be established. For more details, see Kirk et al. [314] and thus we omit its details here.

2.11.5 Existence of Saddle Points and Nash Equilibria

In this section, we use our generalized metric KKM principle in hyperconvex spaces to give some applications to the existence for saddle points, intersection theorems, and the existence of Nash equilibria in game theory. We first have the following result.

Theorem 2.11.43. Let $X \in \mathcal{A}(M)$ be a non-empty compact subset of a hyperconvex metric space M. Suppose $\psi : X \times X \to (-\infty, +\infty)$ is a function such that:

(1) for each fixed $y \in X$, $\psi(x, y)$ is lower semicontinuous in x; and for each fixed $x \in X$, $\psi(x, y)$ is hyper 0-generalized quasi-concave in y; and

(2) for each fixed $x \in X$, $\psi(x, y)$ is upper semicontinuous in y; and for each fixed $x \in X$, $\psi(x, y)$ is hyper 0-generalized quasi-convex in x.

Then ψ has at least one saddle point in $X \times X$, i.e., there exists $(x_0, y_0) \in X \times X$ such that

$$\max_{y\in X} \inf_{x\in X} \psi(x,y) = \psi(x_0,y_0) = \min_{x\in X} \sup_{y\in X} \psi(x,y) = 0.$$

Proof. By our assumptions, ψ satisfies all the hypotheses of Theorem 2.11.15, it follows that there exists $x_0 \in X$ such that

$$\sup_{y\in X} \psi(x_0, y) \leq 0. \tag{2.11.4}$$

Let $\lambda(x, y) = -\psi(x, y)$ for each $(x, y) \in X \times X$, then $\lambda : X \times X \to (-\infty, +\infty)$ is a mapping which also satisfies all hypotheses of Theorem 2.11.15. By Theorem 2.11.15, it follows that there exists $y_0 \in X$ such that

$$\sup_{x\in X} \lambda(x, y_0) \leq 0. \tag{2.11.5}$$

By combining (2.11.4) and (2.11.5), it follows that

$$\max_{y\in X} \min_{x\in X} \psi(x,y) = \psi(x_0,y_0) = \min_{x\in X} \sup_{y\in X} \psi(x,y) = 0,$$

thus completing the proof. □

In what follows, we shall establish one intersection theorem and the existence of Nash equilibria in hyperconvex metric spaces. We first recall the following notion: Given a Cartesian product $X = \Pi_{i=1}^n X_i$ of topological spaces, let $X^j = \Pi_{i\neq j} X_i$ and let $p_i : X \to X_i$ and $p^i : X \to X^i$ denote the natural projections; write $p_i(x) = x_i$ and $p^i(x) = x^i$. Given $x, y \in X$, we let $(y_i, x^i) = (x_1, \cdots, x_{i-1}, y_i, x_{i+1}, \cdots, x_n)$.

The following intersection theorem is a hyperconvex version of Fan's intersection theorem.

Theorem 2.11.44. Let $X_1, \cdots, X_n$ be non-empty compact sets in hyperconvex metric spaces M_i such that $X_i \in \mathcal{A}(M_i)$ for $i = 1, \cdots, n$ and let $A_1, A_2, \cdots, A_n$ be n subsets of X (where, $X = \Pi_{i=1}^n X_i$) such that:

(1) for each fixed $x \in X$ and for each $i = 1, \cdots, n$, the set

$$A_i(x) := \{y \in X : (y_i, x^i) \in A_i\}$$

is a non-empty and admissible set; and

(2) for each fixed $y \in X$ and each $i = 1, \cdots, n$, the set

$$A^i(y) := \{x \in X : (y_i, x^i) \in A_i\}$$

is open.
Then the intersection $\cap_{i=1}^n A_i \neq \emptyset$.

Proof. Define a set-valued mapping $G : X \to 2^X$ by

$$G(x) := X \setminus \cap_{i=1}^n A^i(x)$$

for each $x \in X$. Then we can show that G is not a metric KKM mapping. Indeed, by condition (1), for each $x \in X$, there exists $y \in X$ such that $(y_i, x^i) \in A_i$ for all $i = 1, 2, \cdots, n$. This implies that $x \in \cap_{i=1}^n A^i(y)$, which means that $X \subset \cup_{y\in X}(\cap_{i=1}^n A^i(y))$. Therefore $\cap_{y\in X} G(y) = \emptyset$ and also by Corollary 2.11.9, G must not be a metric KKM mapping. Thus there exists a non-empty finite subset $\{x(1), x(2), \cdots, x(m)\} \subset X$ for which there exists $w \in co\{x(1), \cdots, x(m)\}$ with the property that $w \notin G(x(j))$, i.e., $w \in \cap_{i=1}^n A^i(x(j))$ for all $j = 1, \cdots, m$. Note that the set $A_i(w)$ is admissible and $x(j) \in A_i(w)$ for all $j = 1, 2, \cdots, m$. It follows that $w \in co\{x(1), x(2), \cdots, x(m)\} \subset A_i(w)$. Thus we have that $w \in A_i(w)$ for all $i = 1, 2, \cdots, n$, i.e., $w \in \cap_{i=1}^n A_i \neq \emptyset$ and the proof is completed. □.

As an immediate corollary of Theorem 2.11.44, we have the following existence of Nash equilibria in hyperconvex spaces.

Theorem 2.11.45. Let $X_1, \cdots, X_n \in \mathcal{A}(M)$ be non-empty compact sets in a hyperconvex metric space M. Let $f_1, \cdots, f_n$ be n real-valued continuous function defined on $X = \Pi_{i=1}^n X_i$ such that for each $y \in X$ and for each $i = 1, \cdots, n$, the function $x_i \mapsto f_i(x_i, y^i)$ is hyper quasi-concave on X_i. Then there exists a point $y_0 \in X$ such that $f_i(y_0) = \max_{x_i \in X_i} f_i(x_i, y_0^i)$ for $i = 1, \cdots, n$.

Proof. For any given $\epsilon > 0$, define for each $i = 1, 2, \cdots, n$,

$$A_i^\epsilon = \{z \in X : f_i(z) > \max_{x_i \in X_i} f_i(x_i, y^i) - \epsilon\}.$$

It follows that for each fixed $x \in X$, there exists $y \in X$ such that $(y_i, x^i) \in A_i^\epsilon$ for all $i = 1, \cdots, n$ and the set A_i^ϵ is non-empty and admissible, by the hyper quasi-concavity of the mapping $x_i \mapsto f_i(x_i, y^i)$ for each $y \in X$. Secondly, for each $y \in X$ and for each $i = 1, \cdots, n$, we note that $x \in A_i^\epsilon(y)$ if and only if $(y_i, x^i) \in A_i^\epsilon$. Further A_i^ϵ is open by the upper semicontinuity of the mapping $y^i \mapsto f_i(x_i, y^i)$ for each fixed $x \in X$. Thus all the hypotheses of Theorem 2.11.44 are satisfied and so there exists $x(\epsilon) \in X$ such that $x(\epsilon) \in \cap_{i=1}^n A_i^\epsilon$. As X is compact, without loss of the generality, we may assume that $x(\epsilon) \mapsto y_0 \in X$ as $\epsilon \mapsto 0^+$. The conclusion then follows by the continuity of f_i for $i = 1, 2, \cdots, n$, and the proof is completed. □

Remark 2.11.46. By following a similar methodology and arguments to those used by Fan [167] (see also Park [417] and references therein for more recent study), many variations and generalizations of the results given in this section can be established. Since the ideas are basically the same as those used in Fan [167], we will not include further discussion here. Finally we point out that by using the notion of noncompactness measure given in Section 6, we can also obtain non-compact versions of results in this section. As the arguments are not particularly difficult we will omit the details.

2.12 Fixed Point Theorems of Set-Valued Mappings in Abstract Convex Spaces

The aim of this section is to establish a general fixed point theory for set-valued analysis that unifies all previous results developed in this chapter.

In nonlinear functional analysis, there are basically two types of topological fixed point theorems. The first one is so-called the Browder-Fan fixed point theorem that claims a set-valued self-mapping defined in a compact convex subset of a Hausdorff topological vector spaces has at least one fixed point if the set-valued mapping has open inverse values. The second type is the Fan-Glicksberg fixed point theorem that says that an upper semicontinuous set-valued self-mapping defined in a compact convex subset of Hausdorff locally convex topological vector spaces has at least one fixed point. The celebrated Fan-Glicksberg fixed point theorem is so general that includes many fixed point theorems such as

Kakutani fixed point theorem for upper semicontinuous set-valued in Euclidean spaces, Tychonoff, Schauder, Bohnenlus-Karlin and Brouwder, and many other fixed point theorems for continuous (single-valued) mappings in locally convex topological vector spaces, normed spaces, Banach spaces and Euclidean spaces as special cases. The fixed point theory in nonlinear analysis has been extensively studied and developed this century and its importance in nonlinear analysis and the essential relationships with other problems in nonlinear analysis can be found Zeidler's series books [613, Vol.I and Chapter 77 of Vol.IV] (and see also Goebel and Kirk's book [198] from the point of view of metric spaces). As we said before, the Schauder conjecture raised in 1930s in fixed point theory is still unsolvable yet today (see Granas [204], and Dugundji and Granas [148] for more details), which asks if a (single-valued) continuous self-mapping defined in a compact and convex subset of a Hausdorff topological vector spaces has any fixed point.

On the other hand, as many problems in nonlinear analysis (in particular, in the study of economic model and game theory) can be formulated equivalently to the existence of fixed points for a (typically, upper semicontinuous) set-valued mapping in the setting of spaces which, no longer, have linear structures such as locally convex topological vector spaces enjoy, it is necessary build some new theory for Fan-Glicksberg type fixed point theorems which can cover spaces which are not linear. In this way, there are a few papers, e.g., see Horvath [236]-[240], Khamsi [292], Kirk and Shin [313], Park [418], Park and Kim [427], Sine [486]-[488], Tarafdar [534]-[539], Tarafdar and Watson [543], Wu [575] and others. They try to extend Fan-Glicksberg fixed point theorem in the setting of general abstract convexity notions. However, all results so far obtained in this direction require to assume either the both lower and upper continuity of set-valued mappings or underlying spaces need to be a hyperconvex metric space, or be a locally convex H-space in [575], or be an l.c.space (called also, locally H-convex spaces by Tarafdar [539]) given by Horvath [239].

The aim of this section is to establish fixed point theory of the Fan-Glicksberg type fixed point theorems for upper semicontinuous set-valued mappings with non-empty closed and acyclic (e.g., contractible) values in the setting of so-called locally G-convex spaces, which contains all locally convex H-spaces, l.c.spaces (locally H-convex spaces), hyperconvex metric space, and in particular, locally convex topological spaces as special cases. Thus our fixed point theorem shows that celebrated Fan-Glicksberg type fixed point theorem holds in locally G-convex spaces, specially for locally convex H-spaces and locally H-convex spaces, hyperconvex metric spaces, which shows that Fan-Glicksberg type fixed point theorem holds in hyperconvex metric spaces (e.g., see the corresponding fixed point theorems given by Kirk and Shin [313, p.180] under the assumption of continuity for set-valued mappings).

As we will see in next few chapters, many results have been devoted to prove the existence of equilibria in abstract economics by assuming convexity conditions on the strategy spaces as well as on the constraint correspondences. These convexity restrictions allow to apply well known fixed point such as the Browder-Fan fixed point theorem, the Fan-Glicksberg fixed point and others in locally convex spaces or its generalization such as H-spaces. In order to cover situations such as strategy spaces or the images of the preference mappings are not necessarily convex or even H-convex by using fixed point theory, the aim of this chapter is to introduce a new abstract convexity notion called the '*MC-Space*' due to Llinares [350], which generalizes usual convexity as well as other abstract convexity structures such as simplicial convexity of Bielawski [46], H-spaces of Horvath [239]), G-convex space due to Park and Kim [427] and others. Then we establish general topological fixed point theory in MC-spaces, in particular, the Browder-Fan fixed point theorem for set-valued mappings with open inverse values and the Fan-Glicksberg fixed point theorem for upper semicontinuous set-valued with closed acyclic values. Hopefully these new result will allow us to cover more general models for the study of existence of equilibria.

2.12.1 MC-Space: An Abstract Structure

In this section, the notion of abstract convexity called the 'MC-spaces' due to Llinares [348] will be introduced, which will play as a powerful tool in the study of nonlinear analysis, in particular, for the study of fixed point theory in nonlinear set-valued analysis. In general, an abstract convexity on a set X we mean a family $\mathcal{C} = \{A_i\}_{i \in I}$ of subsets of X which is stable under arbitrary intersections in the sense that $\cap_{i \in J} A_i \in \mathcal{C}$ for each $J \subset I$ and also the family $\mathcal{C}$ contains the empty set $\emptyset$ and X itself. The abstract convexity structure that we are going to introduce is based on the idea of replacing linear segments, which joins any pair of points up (resp., the convex hull of a finite set of points) in the usual convexity by multi-paths (resp., set) which will play their role (and thus denoted by 'MC' standing for the meaning of multi-connections). In particular, the idea is to assign to any finite family of points, a family of functions whose composition is continuous. The image of this composition generates a set associated to the finite family of points in the similar way in which the convex hull operator associates a set to each finite family of points. Formally we have the following definition of MC-spaces due to Llinares [348](see also [350]).

Definition 2.12.1. A topological space X is said to be an *MC-space* if for any non-empty finite subset $A = \{a_0, a_1, \cdots, a_n\}$ of X, there exists a family of elements $B = \{b_0, \cdots, b_n\}$ (not necessarily different) in X and a family of functions $P_i^A : X \times [0,1] \to X$ for $i = 0, 1, \cdots, n$ such that

(1) $P_i^A(x, 0) = x$ and $P_i^A(x, 1) = b_i$ for each $x \in X$; and

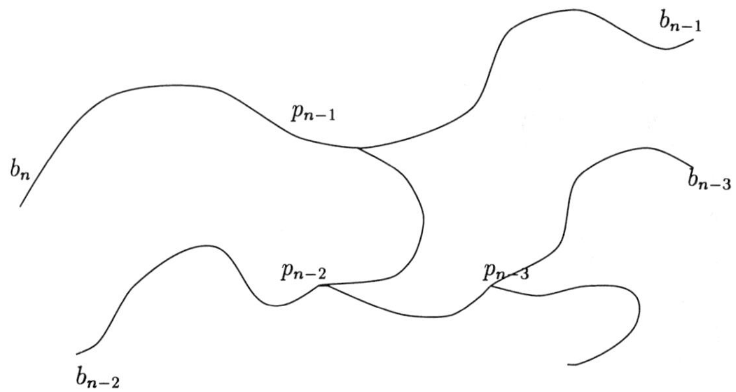

Figure 2.1: The multi-connection structuce and construction of the fucntion G_A for MC-spaces.

(2) the function $G_A : [0,1]^n \to X$ defined by

$$G_A(t_0, t_1, \cdots, t_{n-1}) := P_0^A(\cdots(P_{n-1}^A(P_n^A(a_n, 1), t_{n-1}), t_{n-2})\cdots), t_0)$$

for each $(t_0, t_1, \cdots, t_{n-1}) \in [0,1]^n$ is continuous.

Throughout this chapter we say that X has an MC-structure if and only if X has an MC-structure.

Remark 2.12.2. From Definition 2.12.1, if $P_i^A(x,t)$ is continuous in t, then $P_i^A(x,[0,1])$ represents a continuous path which joins x and b_i. These paths depend the points which are considered, as well as on the finite subsets A which contain them. The function G_A can be interpreted as follows: $P_{n-1}^A(b_n, t_{n-1}) = P_{n-1}$ represents a point of the path which joins b_n with b_{n-1}; $P_{n-2}^A(p_{n-1}, t_{n-2}) = p_{n-2}$ is a point in the path which joins p_{n-1} with b_{n-2}, and so on. Therefore, G_A can be regarded as a composition of these paths and it can be considered as an abstract convex combination of the finite set A, the Figure 2.1 from Llinares [350] illustrates the construction of the function G_A for MC-spaces. Secondly, we wish to emphasize that the MC-structure, in general, does not only depend on finite subsets, but on the way in which the finite subsets are ordered. The disadvance of the notion MC-spaces is that one has to explicitely put in the definition that for each finite subset a definit ordering has been fixed (a rather unnatural condition); otherwsie one may runs into troubles. For more deatils, see Llinares [348]-[350] and related references therein.

From an MC-structure it is possible to define an abstract convexity given by the family of those sets which are stable under function G_A. To define this convexity we need some notions from Llinares [348].

Definition 2.12.3. Let Z be a subset of an MC-space X and let $\mathcal{F}(X)$ denote the family of all non-empty finite subsets of X. Then for all $A \in \mathcal{F}(X)$, such that $A \cap Z \neq \emptyset$, if $A \cap Z = \{a_{i_0}, \cdots, a_{i_m}\}$, where $i_0 < i_1 < \cdots, < i_m$, the '*restriction of functions*' G_A to Z is a mapping $G_{A|Z} : [0,1]^m \to X$ defined by

$$G_{A|Z}(t) := P_{i_0}^A(\cdots P_{i_{m-1}}^A(P_{i_m}^A(a_{i_m}, 1), t_{m_1}) \cdots), t_0),$$

for each $t \in [0,1]^m$, where $P_{i_k}^A$ are functions associated with the elements $a_{i_k} \in A \cap Z$.

We now can define MC-sets which corresponds to the generalization of concepts for usual convexity and H-convex sets.

Definition 2.12.4. A subset Z of an MC-space X is said to be an MC-set (or say, MC-convex) if for each $A \in \mathcal{F}(X)$, $\emptyset \neq A \cap Z = \{a_0, a_1, \cdots, a_m\}$, then $G_{A|Z}([0,1]^m) \subset Z$, where $m + 1 = |A \cap Z|$ denotes the cardinality of the set $A \cap Z$.

Since the family of MC-sets is still an MC-set under the arbitrary intersections, it defines an abstract convexity on X. Moreover we can define the 'MC-hull' operator in the usual way as:

$$C_{mc}(Z) := \cap\{B : Z \subset B \text{ is an } MC\text{-set}\}.$$

Then it is obvious that for each $A \in \mathcal{F}(X)$, and $m = |A \cap Z| > 0$, then $G_{A|Z}([0,1]^m) \subset C_{mc}(Z)$.

Remark 2.12.5. Note that if X is a convex set of a topological vector space E, we can define functions $P_i^A(x,t) := (1-t)x + ta_i$, which leads to an MC-structure on X. In this case, $b_i = a_i$ represents the segments joining a_i and x when $t \in [0,1]$ and the image of the composition $G_A([0,1]^n)$ coincides with the convex hull of $A = \{a_0, a_1, \cdots, a_m\}$. Secondly, if C is a convex set and X a set such that there exists an homeomorphism $h : X \to C$. Then we can define an MC-structure on X as follows: For each $A \in \mathcal{F}(X)$, define function $P_i^A : X \times [0,1] \to X$ by

$$P_i^A(x,t) = h^{-1}((1-t)h(x) + t \cdot h(a_i))$$

for each $(x,t) \in X \times [0,1]$. Thus MC-structures are generalizations of convex sets. Other abstract convexity structures which are generalized by the notion of MC-structures include simplicial convexity of Bielawski [46], H-spaces of Horvath [239]) and so on, e.g., see Llinares [348] for more details on how the MC-structures include those convexity mentioned above as special cases. For example, we recall that a topological space X is contractible if there exists a point $x^* \in X$ and a continuous function $H : X \times [0,1] \to X$ such that $H(x,1) = x^*$ and $H(x,0) = x$ for all $x \in X$. Then it is easy to verify that each H-space is an MC-space.

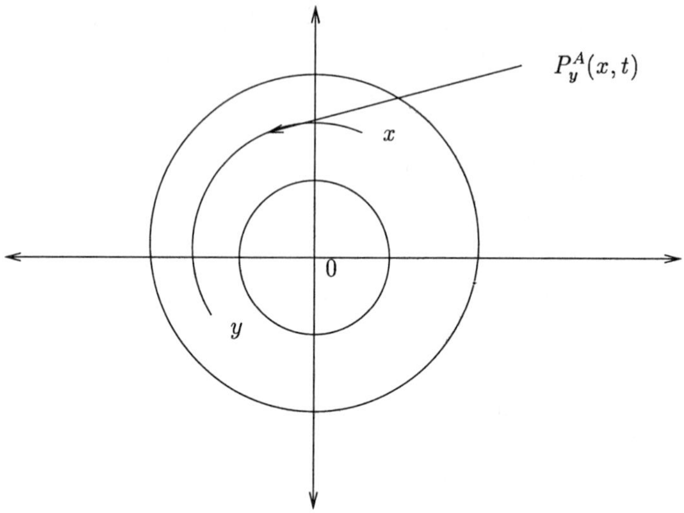

Figure 2.2: A non-contractible set with an MC-structure.

Example 2.12.6 (see Figure 2.2) from [350] show that a non-contractible set (and thus not homeomorphic to a convex hull) where an MC-structure can be defined.

Example 2.12.6. Let X be a subset of R^2 defined by $X := \{x \in R^2 : 0 < a \leq \|x\| \leq b, a, b, \in R\}$. Consider the complex representation: $x = \rho_x e^{i\alpha x}$ and $y = \rho_y e^{i\alpha y}$ for any $A \in \mathcal{F}(X)$. The functions P_i^A can be defined by $P_y^A : X \times [0,1] \to X$ by

$$p_y^A(x,t) = ((1-t)\rho_x + t\rho_y)e^{i(1-t)\alpha x + t\alpha y)}$$

for each $(x,t) \in X \times [0,1]$.

We wish to point out that the concept of the MC-space seems to be a very powerful notion in the study of nonlinear analysis and applications. For example, Llinares [349] gave a unified treatment of the existence for maximal elements in binary relations, which covers both ways (convexity and acyclicity) on the problems of existence of maximal elements for non-transitive binary relations, and this result has been recently generalized to the setting of non-compact spaces by Isac and Yuan [259]. Secondly Llinares established a characterization of the existence of maximal elements of a binary relation defined on a compact topological spaces in terms of MC-spaces. Precisely, he proved that a compact topological space with an acyclic binary relation can be given an MC-spaces, and also there exists a topological space with non-acyclic and non-convex preference. Furthermore as applications of the Browder-Fan fixed point theorem in

MC-spaces, some existence results for abstract economics and generalized games in the framework of MC-spaces are given in [350]. In this section, however, our goal is to develop the general fixed point theory for set-valued mappings in MC-spaces.

2.12.2 Browder-Fan Fixed Points in MC-Spaces

The aim of this section is to establish general Browder-Fan type fixed point theorem in MC-spaces.

The following result was first established by Llinares [350].

Theorem 2.12.7. Let $\{X_\alpha\}_{\alpha\in I}$ be a family of non-empty compact MC-spaces, where I is a countable or uncountable index set. If $X = \Pi_{\alpha\in I}X_\alpha$ and $T_\alpha : X \to 2^{X_\alpha} \setminus \{\emptyset\}$ is an open inverse valued mapping, then for each $\alpha \in I$, there exists a non-empty finite subset B_α of X_α and a continuous mapping $f_\alpha : X \to X_\alpha$ such that:
(1): there exists $x^* \in X$ such that $x^*_\alpha = f_\alpha(x^*)$ for each $\alpha \in I$; and
(2): $f_\alpha(x) \in G_{B_\alpha|T_\alpha(x)}([0,1]^m)$ for each $x \in X$.

Proof. Note that T_α is open inverse valued for each $\alpha \in I$ and X is compact, it follows by Remark 2.3.29 that X has a finite open cover $\{O^i_{x_\alpha} : i = 0,1,\cdots,n_\alpha\}$, i.e., $X = \cup_{i=0}^{n_\alpha} O^i_{x_\alpha}$. Let $\{\psi^\alpha_i\}_{i=0}^{n_\alpha}$ be the continuous finite partition of unity subordinated to this open cover. Then we have that $\psi^\alpha_i(x) \geq 0$ and $\Sigma_i\psi^\alpha_i(x) = 1$ and if $\psi^\alpha_i(x) > 0$, then $x \in O^i_{x_\alpha}$. Let $\psi^\alpha(x) := (\psi^\alpha_0(x),\cdots,\psi^\alpha_{n_\alpha}(x))$ and set $J_\alpha(x) := \{i \in I : \psi^\alpha_i(x) > 0\}$ for each $x \in X$. Then for each $i \in J_\alpha(x)$, it follows that $x \in O^i_{x_\alpha} \subset T^{-1}_\alpha(x_\alpha)^i)$, so that $x^i_\alpha \in T_\alpha(x)$. For each $\alpha \in I$, let $B_\alpha = \{x^i_\alpha : i = 0,\cdots,n\}$. Since X_α is an MC-space, it follows that there exist functions $P^{B_\alpha}_i : X_\alpha \times [0,1] \to X_\alpha$ such that their composition mapping $G_{B_\alpha} : [0,1]^{n_\alpha} \to X_\alpha$ is continuous. Then we define another mapping $f_\alpha : X \to X_\alpha$ by

$$f_\alpha(x) := G_{B_\alpha}(\Gamma^\alpha(\psi^\alpha(x)))$$

for each $x \in X$, where $\Gamma^\alpha := (\Gamma^\alpha_0,\cdots,\Gamma^\alpha_{n_\alpha-1}) : \Delta_{n_\alpha} \to [0,1]^{n_\alpha}$ is a (single-valued) mapping and the mapping $\Gamma^\alpha_i : \Delta_{n_\alpha} \to [0,1]$ is given by

$$\Gamma^\alpha_i(z) := \begin{cases} 0, & \text{if } z_i = 0 \\ \frac{z_i}{\Sigma^n_{j=i}z_j}, & \text{if } z_i \neq 0 \end{cases}$$

for $i = 0,1,\cdots,n_\alpha - 1$. By the continuity of $\Gamma^\alpha \circ \psi^\alpha$ (e.g., see Llinares [349]) and G_{B_α}, it follows that the composition function f_α is also continuous. Furthermore, we can verify that $f_\alpha(x) \in G_{B_\alpha|T_\alpha(x)}([0,1]^m)$ for each $x \in X$. Let $\Delta := \Pi_{\alpha\in I}\Delta_{n_\alpha}$.

Now define the mapping $f : X \to X$ by

$$f(x) := \{f_\alpha(x)\}_{\alpha\in I}$$

for each $x \in X$; define a mapping $\psi : X \to \Delta$ by

$$\psi(x) := \{\psi^\alpha(x)\}_{\alpha\in I}$$

for each $x \in X$; and define the mapping $\phi : \Delta \to X$ by

$$\phi(x) := \{G_{B_\alpha}(\Gamma^\alpha(z_\alpha))\}_{\alpha\in I}$$

for each $x \in \Delta$, where $\Delta = \Pi_{\alpha\in I}\Delta_{n_\alpha}$. It is clear that $f = \phi \circ \psi$, i.e., $f(x) = \phi(\psi(x))$ for each $x \in X$. Then Tychonoff's fixed point theorem implies that the continuous mapping $\psi \circ \phi : \Delta \to \Delta$ has at least one fixed point $z_0 \in \Delta$, i.e., $z_0 = \psi(\phi(z_0))$. Let $x^* = \phi(z_0)$, it follows that $x^*_\alpha = f_\alpha(x^*)$ for each $\alpha \in I$ and the proof is complete. □

As an application of Theorem 2.12.7, we have the following Browder-Fan fixed point theorem in MC-spaces which, in turn includes the corresponding famous Browder-Fan fixed point theorems in topological vector spaces for set-valued with open inversed values as a special case.

Theorem 2.12.8. Let $\{X_\alpha\}_{\alpha\in I}$ be a family of non-empty compact MC-spaces, where I is a countable or uncountable index set and $X = \Pi_{\alpha\in I}X_\alpha$. Suppose for each $\alpha \in I$, $T_\alpha : X \to 2^{X_\alpha} \setminus \{\emptyset\}$ is an open inverse valued mapping with MC-set values (i.e., the set $T_\alpha(x)$ is MC-convex for all $x \in X$). Then there exists $x^* \in X$ such that $x^*_\alpha \in T_\alpha(x^*)$ for each $\alpha \in I$,

Proof. By Theorem 2.12.7, for each $\alpha \in I$, there exists a single-valued continuous mapping $f_\alpha : X \to X_\alpha$ such that $f_\alpha(x) \in G_{B_\alpha|T_\alpha(x)}[(0,1]^m)$. Since $T_\alpha(x)$ is an MC-set, it follows that $f_\alpha(x) \in G_{B_\alpha|T_\alpha(x))}([0,1]^m) \subset T_\alpha(x)$. Therefore f_α is a continuous selection of T_α and there exists a point $x^* \in X$ such that $x^*_\alpha \in T_\alpha(x^*)$ for each $\alpha \in I$ and thus $x^* \in T(x^*)$, where $T(x) := \{T_\alpha(x)\}_{\alpha\in I}$, i.e., $x^*_\alpha \in T_\alpha(x^*)$ for all $\alpha \in I$. This completes the proof. □

As a special case of Theorem 2.12.8, we have the following Browder-Fan fixed point theorem in MC-spaces.

Theorem 2.12.9. Let X be a non-empty compact MC-space and $T : X \to 2^X \setminus \{\emptyset\}$ is an open inverse valued mapping with MC-set values. Then T has a fixed point $x^* \in X$.

Note that each hyperconver metric space is an H-space, which, in turn, is an MC-space, we then have the following fixed point theorems.

Theorem 2.12.10. Let X be a non-empty compact H-space and $T : X \to 2^X \setminus \{\emptyset\}$ is an open inverse valued mapping with H-convex values. Then T has a fixed point $x^* \in X$.

Theorem 2.12.11. Let $X \in \mathcal{A}(M)$ be a compact admissible subset of a hyperconvex metric space M and $T : X \to 2^X \setminus \{\emptyset\}$ is an open inverse valued mapping with H-convex values. Then T has a fixed point $x^* \in X$.

As a special case of Theorem 2.12.10, we have the following Browder-Fan fixed point theorem in topological vector spaces.

Theorem 2.12.12 (Browder-Fan Fixed Point Theorem). Let X be a non-empty compact convex subset of a Hausdorff topological vector space and $T : X \to 2^X \setminus \{\emptyset\}$ is an open inverse valued mapping with convex values. Then T has a fixed point $x^* \in X$.

We also note that as applications of Browder-Fan fixed point theorems in MC-spaces, some general existence theory for equilibra of generalzied games without linear structures in the setting of MC-spaces has been developed by Llinares in [350].

2.12.3 Fan-Glicksberg Fixed Points in MC-Spaces

The aim of this section is to develop a fixed point theory for upper semicontinuous set-valued mappings with closed acyclic values in locally convex MC-spaces which inlcude locally convex H-spaces and locally convex topological vector spaces as special cases. In order to do so, we first introduce the following definition.

Definition 2.12.13. A MC-space X is said to be a locally convex MC-space if X is a uniform topological space with the uniform structure $\mathcal{U}$ and there exists a base family $\{V_i : i \in I\}$ for $\mathcal{U}$ such that for each $V_i \in \{V_i : i \in I\}$, the set $V_i(x) := \{y \in X : (y, x) \in V_i\}$ is an MC-set for each $x \in X$.

From Definition 2.12.13, it is clear that our Definition 2.12.13 includes locally convex H-spaces due to Horvath [239], and Wu [575] as a special class since each H-space has an MC-structure which, in turn, includes locally convex topological vector spaces a special case.

We note that the proof of Theorem 2.12.7 indeed carries out the following selection result and its proof is thus omitted here.

Lemma 2.12.14. Let X be a non-empty compact MC-space and $T : X \to 2^X$ is an open inverse valued mapping with non-empty MC-values. Then there exist some positive integer $n \in N$ and two single-valued continuous mappings $\phi : \Delta_N \to X$ and $\psi : X \to \Delta$ such that the mapping $f : X \to X$ defined by $f(x) := \phi(\psi(x))$ for each $x \in X$ is a continuous selection of T, i.e., $f(x) \in T(x)$ for all $x \in X$, where $N = \{0, 1, \cdots, n\}$.

We now have the following Fan-Glicksberg fixed point theorem for upper semicontinuous set-valued mappings with non-empty closed acyclic values in MC-spaces, which was first given by Yuan [610].

Theorem 2.12.15. Let X be a compact locally convex MC-space and

$F : X \to 2^X$ be an upper semicontinious set-valued mappings with non-empty closed acyclic values. Then F has a fixed point $x^* \in X$.

Proof. Since (X, Γ_A) is a locally convex Hausdorff MC-space, X has a uniform structure $\mathcal{U}$ and without loss of generality, suppose $\{V_i : i \in I\}$ is a symmetric and open base family for the uniform structure $\mathcal{U}$. Then for each base V_i and $x \in X$, the set $V_i(x) := \{y \in X : (y, x) \in V_i\} = \{y \in X : (x, y) \in V_i\}$ is an open MS-set, and the $\cap_{i \in I} V_i = \Delta = \{(x, x) : x \in X\}$. For each V_i, we define a set-valued mapping $T_i : X \to 2^X$ by $T_i(x) := \{y \in X : (y, x) \in V_i) = V_i(x)$ for each $x \in X$. Then it is clear T_i is a set-valued mapping from X to X with non-empty MC-convex values and $\cup_{y \in X} int(T_i^{-1}(y)) \subset X$ by the fact that for each $y \in X$, the set $T_i^{-1}(y) = \{x \in X : (x, y) \in V_i\}$ is open in X, so that $int(T_i^{-1}(y)) = T_i^{-1}(y)$. By Lemma 2.12.14, there exists some $n \in N$ and two (single-valued) continuous mappings $g_i : \Delta_N \to X$ and $\phi_i : X \to \Delta_N$ such that $f_i(x) := g_i(\phi_i(x)) \in T_i(x)$ for each $x \in X$, where Δ_N is a standard n-dimensional simplex and $N = \{0, 1, \cdots, n\}$. Now we define another set-valued mapping $P_i : \Delta_N \to 2^X$ by $P_i(x) := F(g_i(x))$ for each $x \in \Delta_N$. Then P_i is also an upper semicontinuous set-valued mapping with non-empty closed and acyclic values. By Lemma 2.9.3, the mapping $\phi_i \circ P_i : \Delta_N \to 2^{\Delta_N}$ has a fixed point, i.e., there exists $u_{v_i} \in \Delta_N$ such that $u_{v_i} \in \phi_i(P_i(u_{v_i}))$. Then there exists $x_{v_i} \in P_i(u_{v_i}) = F(g_i(u_{v_i})) \subset X$ and $u_{v_i} = \phi_i(x_{v_i})$. Let $y_{v_i} = g_i(u_{v_i}) = g_i(\phi_i(x_{v_i})) = f(x_{v_i}) \in T_i(x_{v_i})$. Thus $x_{v_i} \in F(y_{v_i})$, i.e., $(x_{v_i}, y_{v_i}) \in$ graph F and $(x_{v_i}, y_{v_i}) \in V_i$, where graph F denotes the graph of the mapping F. Since $X \times X$ is compact, we may assume that $\{x_{v_i}, y_{v_i}\}_{i \in I}$ converges to (x_0, y_0). Note that F is an upper semicontinuous mapping with closed values, it follows that graph F is closed and thus $(x_0, y_0) \in$ graph F, i.e., $y_0 \in F(x_0)$. In order to finish the proof, it suffices to show that $x_0 = y_0$. Note that $\{V_i : i \in I\}$ is an open symmetric base family for the uniform structure $\mathcal{U}$, for any given $U \in \mathcal{U}$, there exists $V_u \in \{V_i : i \in I\}$ such that $V_U \circ V_U \circ V_U \circ V_U \subset U$. On the other hand, $\lim_{i \in I} x_{v_i} = x_0$ and $\lim_{i \in I} y_{v_i} = y_0$, without loss of generality, there exists $V_0 \in \{V_i : i \in I\}$ and $V_0 \subset V_U$ such that $(x_{v_i}, x_0) \in V_U$ for $i \in I$ such that $V_i \subset V_0$. Now for any $A \in \{V_i : i \in I\}$ with $A \subset V_U$, note that $(x_A, y_A) \in A$, it follows that $(x_0, y_A) \in A \circ V_U \subset V_0 \circ V_U \subset V_U \circ V_U$. Then it is easy to very that $(X_0, y_0) \subset V_U \circ V_U \circ V_U \circ V_U \subset U$. This implies that $(x_0, y_0) \in \cap_{U \in \mathcal{U}} U$. Note that $\cap_{U \in \mathcal{U}} U = \Delta$, it follows that $x_0 = y_0$ and thus x_0 is a fixed point of F. The proof is complete. □

Remark 2.12.16. Note that if X is a compact and convex subset of a locally convex Hausdorff topological vector space E and let $\Gamma_A = co(A)$ for each $A \in \mathcal{F}(X)$. Then (X, Γ_A) is a locally convex H-space. Thus Theorem 2.12.12 is a really generalization of celebrated Fan-Glicksberg fixed point theorem for upper semicontinuous set-valued mappings with non-empty closed and acyclic (e.g., convex) values in both locally convex H spaces and locally convex topological vector spaces.

Note that each hyperconvex metric space is a locally H-convex space by Lemma 2.11.4, it follows by Theorem 2.1.2.15 that we have the following Fan-Glicksberg type fixed point theorem in hyperconvex metric spaces.

Theorem 2.12.17. Let X be a compact locally convex H-space and $F : X \to 2^X$ be an upper semicontinuous set-valued mapping with non-empty closed acyclic values. Then F has a fixed point $x^* \in X$, i.e., $x^* \in F(x^*)$.

In their paper [313, p.180], Kirk and Shin prove a fixed point theorem for continuous set-valued mappings and they ask whether their result remains true under the assumption that the mapping is upper semicontinuous rather than continuous. Theorem 2.12.17 answers their question in the affirmatively.

Since each hyperconvex metric space is, indeed, a locally convex H-space by Lemma 2.11.4, we have the following fixed point theorem.

Theorem 2.12.18. Let $X \in \mathcal{A}(M)$ be a compact admissible set of a hyperconvex metric space M and $F : X \to 2^X$ be an upper semicontinuous set-valued mapping with non-empty closed acyclic values. Then F has a fixed point $x^* \in X$.

Finally we have the following famous fixed point theorem due to Fan and Glicksberg in locally convex topological vector spaces.

Theorem 2.12.19 (Fan-Glicksberg Fixed Point Theorem). Let X be a compact convex subset of a Hausdorff locally convex topological vector space and $F : X \to 2^X$ be an upper semicontinuous set-valued mapping with non-empty closed acyclic (e.g., closed) values. Then F has a fixed point $x^* \in X$.

2.12.4 Fan-Glicksberg Fixed Points in G-Convex Spaces

In this section, we will establish Fan-Glicksberg type fixed point theorem for upper semicontinuous set-valued mappings with closed acyclic values in the setting of another abstract convex structure - called a locally G-convex space which generalizes usual convexity such as locally convex H-spaces, l.c.spaces (locally H-convex spaces), hyperconvex metric spaces and locally convex topological spaces.

Throughout this section, we denote by $\mathcal{F}(X)$ and 2^X the family of all non-empty finite subsets of X and the family of all subsets of X. All topological spaces are assumed to be Hausdorff unless specified for the simplicity of our discussion.

We have seen in last few years, many papers have been devoted to prove the existence of fixed points for set-valued mappings in nonlinear analysis without traditional linear structures. As a result, several generalizations of convexity notions such as hyperconvex metric spaces, H-spaces, G-convex spaces have been introduced and studies by Aronszajn and Panitchpakdi [7], Bardaro and Ceppitelli [24], Bielawaski [46], Horvath [236]-[240], Kirk et al. [314], Park [414]-[418], Park and Kim [427], Tarafadr [539], Tarafdar and Watson [543], Wu [575], Yuan [610] and many others. Recently, in order to cover the general economic models without linear convex structures, Park and Kim [427] introduces another abstract convexity notion-called G-convex spaces which include many abstract convexity notions such as H-convex spaces as a special case. Formally we recall the following definition which is first given by Park and Kim in [427].

Definition 2.12.20. A *generalized convex space*, or say, a *G-convex* space $(X, D; \Gamma)$ consists of a topological space X, a non-empty subset D of X and a function $\Gamma : \mathcal{F}(X) \to X \setminus \{\emptyset\}$ such that (1): for each $A, B \in \mathcal{F}(X)$, $\Gamma(A) \subset \Gamma(B)$ if $A \subset B$; and (2): for each $A \in \mathcal{F}(X)$ with $|A| = n + 1$, there exists a continuous function $\phi_A : \Delta_n \to \Gamma(A)$ such that $\phi_A(\Delta_J) \subset \Gamma(J)$ for each $\emptyset \neq J \subset \{0, 1, \cdots, n\}$, where $A = \{x_0, x_1, \cdots, x_n\}$, and Δ_J denotes the face of Δ_n the corresponding to the sub-index of J in $\{0, 1, 2, \cdots, n\}$; and (3): A subset C of the G-convex space $(X, D; \Gamma)$ is said to be *G-convex* if for each $A \in \mathcal{F}(D)$, $\Gamma_A \subset C$ for all $A \subset C$.

For the convenience of our discussion, we also denote $\Gamma(A)$ by Γ_A or Γ_N if there is no confusion throughout this section for each $A = \{x_0, x_1, x_2, \cdots, x_n\} \in \mathcal{F}(X)$, where N is the set of all index for the set A, i.e., $N = \{0, 1, 2, \cdots, n\}$. A space X is said to have a G-convex structure if and only of X is a G-convex space. By Lemma 2.2.5, it is clear that the notion of G-convex spaces includes the corresponding H-spaces (see its definition below) as a special class, which, in turn, includes topological vector spaces as special cases. In order to establish our fixed point theorem, we now introduce another notion called locally G-convex which include locally H-convex spaces, hyperconvex metric spaces (their definitions will be given in next section shortly), and locally convex spaces as special cases.

Definition 2.12.21. A G-convex X is said to be a *locally G-convex space* if X is a uniform topological space with uniformity $\mathcal{U}$ which has an open base $\beta := \{V_i : i \in I\}$ of symmetric encourages such that for each $V \in \beta$, the set $V(x) := \{y \in X : (y, x) \in V\}$ is a G-convex set for each $x \in X$.

Recall that a non-empty space is said to be acyclic if all of its reduced Čech homology groups over rational vanish. In particular, any contractible space is acyclic, and thus any non-empty convex or star-shaped set is acyclic. Let X and Y be two topological spaces. We recall that a set-valued mapping $F : X \to 2^Y$ is said to have transfer open inversed values, if $X = \cup_{y \in Y} int(F^{-1}(y))$, where $int(F^{-1}(y))$ denotes the relative interior of the set $F^{-1}(y) := \{x \in X : y \in F(x)\}$ for each $y \in Y$ in X; and F is said to be upper semicontinuous if the set

$\{x \in X : F(x) \subset U\}$ is open for each open neighborhood U in Y. When Y is a G-convex space, the mapping $T : X \to 2^Y$ is said to have G-convex values if $T(x)$ is a G-convex set for each $x \in X$.

We first have the following selection result of for a set-valued mapping with transfer open inversed values, which includes the corresponding Lemma 2.2 of Tarafdar [536] as a special case.

Lemma 2.12.22. Let Y be a compact topological space and X be a G-convex space. Suppose $T : Y \to 2^X \setminus \{\emptyset\}$ is an open inverse valued mapping with G-convex values. Then there exist some positive integer $n \in N$ and two single-valued continuous mappings $\phi : \Delta_n \to X$ and $\psi : Y \to \Delta_n$ such that the mapping $f : Y \to X$ defined by $f(y) := \phi(\psi(y))$ for each $x \in Y$ is a continuous selection of T, i.e., $f(y) \in T(y)$ for all $y \in Y$, where Δ_n is the standard n-dimensional simplex with vertices $e_0, e_1, \cdots, e_n$.

Proof. Since Y is compact and T has open inversed values, it follows that Y has a finite open cover $\{O_i : i = 0, 1, \cdots, n\}$, where $O_i \subset int(F^{-1}(x_i))$ for some $x_i \in X$. Let $A = \{x_0, x_1, \cdots, x_n\}$. Suppose $\{\psi_i\}_{i=0}^n$ is a partition of unity the corresponding to the finite covering $\{O_i : i = 0, 1, \cdots, n\}$, that is, $\psi_i(y) \geq 0$ and $\Sigma_i \psi_i(y) = 1$ and if $\psi_i(y) > 0$, then $y \in O_i$. Set $J(y) := \{i \in I : \psi_i(y) > 0\}$ for each $y \in Y$. Then for each $i \in J(y)$, it follows that $y \in O_i \subset T^{-1}(x_i)$, so that $x_i \in T(y)$. This means $A \cap T(x) \neq \emptyset$ for all $x \in X$. Secondly, the mapping $\psi : Y \to \Delta_n$ defined by $\psi(y) := \Sigma_{i=0}^n \psi_i(y) e_i$ for each $y \in Y$ is continuous. Since X is a G-convex space, there exists a continuous mapping $\phi : \Delta_n \to \Gamma_A$ such that $\phi(\Delta_J) \subset \Gamma_J$ for each non-empty finite subset J in $\{0, 1, \cdots, n\}$. Then the mapping $f : Y \to X$ defined by $f(y) := \phi(\psi(y))$ for each $y \in Y$, is continuous. Since $T(y)$ is G-convex, it follows that $f(y) := \phi(\psi(y)) \subset \phi(\Delta_{J(y)}) \subset \Gamma_{J(y)} \subset T(y)$ for each $y \in Y$ (as $\Gamma_{J(y)} \subset T(y)$), i.e., $f(y) \in T(y)$ for each $y \in Y$ and thus this completes the proof. □

By the same argument used in the proof of Theorem 2.12.15, we have the following fixed point theorem in locally G-convex spaces given by Yuan [611], which is a generalization of the Fan-Glicksberg type fixed point theorems for upper semicontinuous set-valued mappings with non-empty closed acyclic values given in several existing literature (e.g., see Kirk and Shin [313], Park [414]-[418], Tarafdar [534], Wu [575] and others in locally convex spaces).

Theorem 2.12.23. Let X be a compact locally G-convex space and $F : X \to 2^X$ be an upper semicontinious set-valued mappings with non-empty closed acyclic values. Then F has a fixed point, i.e., there exists $x_0 \in X$ such that $x_0 \in F(x_0)$.

Proof. Since (X, Γ_A) is a locally Hausdorff G-convex space, X has a uniform structure $\mathcal{U}$ and without loss of generality, suppose $\mathcal{V} := \{V_i : i \in I\}$ is a symmetric and open base family for the uniform structure $\mathcal{U}$. Then for each base

V_i and $x \in X$, the set $V_i(x) := \{y \in X : (y,x) \in V_i\} = \{y \in X : (x,y) \in V_i\}$ is an open G-convex set, and the $\cap_{i\in I} V_i = \Delta = \{(x,x) : x \in X\}$. For each V_i, we define a set-valued mapping $T_i : X \to 2^X$ by $T_i(x) := \{y \in X : (y,x) \in V_i) = V_i(x)$ for each $x \in X$. Then it is clear T_i is a set-valued mapping from X to X with non-empty G-convex values and $\cup_{y\in X} int(T_i^{-1}(y)) \subset X$ by the fact that for each $y \in X$, the set $T_i^{-1}(y) = \{x \in X : (x,y) \in V_i\}$ is open in X, so that $int(T_i^{-1}(y)) = T_i^{-1}(y)$. By Lemma 2.12.22, there exist some positive integer $n \in N$ and two (single-valued) continuous mappings $g_i : \Delta_n \to X$ and $\phi_i : X \to \Delta_n$ such that $f_i(x) := g_i(\phi_i(x)) \in T_i(x)$ for each $x \in X$, where Δ_n is a standard n-dimensional simplex. Now we define another set-valued mapping $P_i : \Delta_n \to 2^X$ by $P_i(x) := F(g_i(x))$ for each $x \in \Delta_n$. Then P_i is an upper semicontinuous set-valued mapping with non-empty closed and acyclic values. By Lemma 2.9.3, the mapping $\phi_i \circ P_i : \Delta_n \to 2^{\Delta_n}$ has a fixed point, i.e., there exists $u_{v_i} \in \Delta_n$ such that $u_{v_i} \in \phi_i(P_i(u_{v_i}))$. Then there exists $x_{v_i} \in P_i(u_{v_i}) = F(g_i(u_{v_i})) \subset X$ and $u_{v_i} = \phi_i(x_{v_i})$. Let $y_{v_i} = g_i(u_{v_i}) = g_i(\phi_i(x_{v_i})) = f_i(x_{v_i}) \in T_i(x_{v_i})$. Thus $x_{v_i} \in F(y_{v_i})$, i.e., $(x_{v_i}, y_{v_i}) \in$ graph F and $(x_{v_i}, y_{v_i}) \in V_i$, where graph F denotes the graph of the mapping F. Since $X \times X$ is compact, we may assume that $\{x_{v_i}, y_{v_i}\}_{i\in I}$ converges to (x_0, y_0). Note that F is an upper semicontinuous mapping with non-empty closed values, it follows that the graph of F is closed and thus $(x_0, y_0) \in$ graph F, i.e., $y_0 \in F(x_0)$. In order to finish the proof, it suffices to show that $x_0 = y_0$. Since $\{V_i : i \in I\}$ is an open symmetric base family for the uniform structure $\mathcal{U}$ and $\lim_{i\in I}(x_{v_i}, y_{v_i}) = (x_0, y_0)$, without loss of generality, there exists $V_0 \in \{V_i : i \in I\}$ such that $(x_{v_j}, x_0) \in V_j$ and $(y_{v_j}, y_0) \in V_j$ for all $V_j \in \{V_i : i \in I\}$ with $V_j \subset V_0$. Now for any $U \in \{V_i : i \in I\}$ with $U \subset V_0$, let $V_u \in \{V_i : i \in I\}$ with $V_u \subset V_0$ be such that $V_u \circ V_u \circ V_u \subset U$. Since $(x_{v_u}, y_{v_u}) \in V_u$, it follows that $(x_0, y_{v_u}) \in V_u \circ V_u$ and thus $(x_0, y_0) \subset V_u \circ V_u \circ V_u \subset U$. Therefore we have show that $(x_0, y_0) \in U$ for all $U \in \{V_i : i \in I\}$ with $U \subset V_0$. Note that $\cap\{U : U \in \mathcal{V}$, and $U \subset V_0\} = \cap\{V_i : i \in I\} = \Delta$. This implies that $x_0 = y_0$ and thus x_0 is a fixed point of F. This completes the proof. □

It is clear each locally convex H-space is a locally G-convex space. Note that if X is a compact convex subset of a locally convex Hausdorff topological vector space E and let $\Gamma_A := co(A)$ for each $A \in \mathcal{F}(X)$. Then (X, Γ_A) is a locally convex H-space. Thus Theorem 2.12.23 is a really generalization of celebrated Fan-Glicksberg fixed point theorem for upper semicontinuous set-valued mappings with non-empty closed and acyclic (e.g., contractible) values in both locally convex H-spaces, l.c.spaces (locally H-convex spaces) and locally convex topological vector spaces. Theorem 2.12.23 shows that Theorem 2.1 of Tarafdar [539] is true when mapping F is only upper semicontinous. Theorem 2.12.23 also generalizes Theorem 3.2 of Tarafdar and Watson [543] and the corresponding Fan-Glicksberg type fixed point theorems given by Wu [575] to locally G-convex spaces which include locally H-convex space and locally convex H-spaces as special classes.

2.12.5 Generic Stability of Fixed Points in Hyperconvex Metric Spaces

As we know that the stability study of fixed point theory is also a very important topic in nonlinear analysis. The aim of this section is to estabilish generic stability of fixed points for upper semicontinuous set-valued mappings in hyperconvex metric spaces. Our results show that almost all of upper semicontinous set-valued mappings defined in compact hyperconvex metric spaces have stable fixed points from viewpoint of Baire category theory.

Let (X, d) be a metric space and $K(X)$ be the space of all non-empty compact subsets of X equipped with the Hausdorff metric h which is induced by the metric d. For any $\epsilon > 0$, $x_0 \in X$ and $A \in K(X)$, let $U(\epsilon, A) = \{x \in X : d(u, x) < \epsilon \text{ for some } u \in A\}$ and $O(x_0, \epsilon) = \{x \in X : d(x_0, y) < \epsilon\}$. Let Y be a topological space. Recall that a subset $Q \subset Y$ is called a residual set if it is a countable intersection of open dense subsets of Y.

In order to study generic stability of fixed points for upper semicontinuous set-valued mappings in hyperconvex metric spaces, throughout the rest part of this section, we denote by X a hyperconvex metric space (and thus it is complete) and $\mathcal{A}(X)$ denotes the collection of all ball intersections (i.e., admissible sets) in X. By Theorem 15 of Sine [488] (see also Lemma 2 of Kirk [312]), we have the following fact.

Lemma 2.12.24. Let X be a hyperconvex metric space. Then the space $(\mathcal{A}(X), h)$ is a hyperconvex metric space with the Hausdorff metric h (thus $(\mathcal{A}(X), h)$ is a complete metric space).

Let X be a hyperconvex metric space and $K\mathcal{A}(X)$ the collection of all compact admissible sets of X. By Lemma 2.12.24 and Theorem 4.3.9 of Klein and Thompson [317], it follows that $(K\mathcal{A}(X), h)$ is still a complete metric space.

Let $C := \{f : X \to K\mathcal{A}(X) : f$ is upper semicontinuous with non-empty compact and admissible values on $X\}$ and for each f, $f' \in C$, define $\rho(f, f') := \sup_{x \in X} h(f(x), f'(x))$. Clearly, ρ is a metric on C and we also have the following result.

Lemma 2.12.25. The space (C, ρ) is complete.

Proof. Let $\{f_n\}_{n=1}^{\infty}$ be each Cauchy sequence in C, then for any $\epsilon > 0$, there is a positive integer $N(\epsilon)$ such that $\rho(f_n, f_m) = \sup_{x \in X} h(f_n(x), f_m(x)) < \epsilon$ for any $n, m \geq N(\epsilon)$. It follows that for each $x \in X$, $\{f_n(x)\}_{n=1}^{\infty}$ is a Cauchy sequence in $\mathcal{A}(X)$. Note that $K\mathcal{A}(X)$ is complete by Lemma 2.12.24 and Theorem 4.3.9 of Klein and Thompson [317], it follows that there is $f : X \to \mathcal{A}(X)$ such that $\lim_{m \to \infty} f_m(x) = f(x)$. for all $x \in X$. Now for each given $x \in X$, each $\epsilon > 0$ and each $n \geq N(\epsilon)$, the inequality above implies that $f_n(x) \subset U(2\epsilon, f(x))$

and $f(x) \subset U(2\epsilon, f_n(x))$. Fix $n \geq N(\epsilon)$; since $f_n \in C$, there is $\delta > 0$ such that $f_n(x') \subset U(\epsilon, f_n(x))$ whenever $d(x,x') < \delta$. Thus, $f(x') \subset U(2\epsilon, f_n(x')) \subset U(3\epsilon, f_n(x)) \subset U(5\epsilon, f(x))$ whenever $d(x,x') < \delta$. Therefore f is upper semicontinuous on X, so that $f \in C$ and that $f_n \to f$. Hence C is complete. □

Throughout this section, we set $Y := C \times K\mathcal{A}(X)$ and for each $y = (f, A) \in Y$, $y' = (f', A') \in Y$, define a metric D on Y by

$$D(y, y') = \rho(f, f') + h(A, A')$$

Then it follows by Lemma 2.12.25, Lemma 2.12.24 and Theorem 4.3.9 of Klein and Thompson [317] again that Y is a complete metric space.

Define $M := \{y = (f, A) \in Y$: there is $x \in A$ such that $x \in f(x)\}$. Then M is, indeed, a subset of Y and we do have the following result.

Lemma 2.12.26. The metric space (M, D) is complete.

Proof. Since $M \subset Y$ and Y is complete, it is sufficient to prove that M is closed in Y. Let $\{y_n\}_{n=1}^{\infty}$ be a sequence in M and $y_n \to y \in Y$. Set $y_n = (f_n, A_n)$, $n = 1, 2, \cdots$ and $y = (f, A)$; then $f_n \to f$ and $A_n \to A$. For each $n = 1, 2, \cdots$, since $y_n \in M$, there is $x_n \in A_n$ such that $x_n \in f_n(x_n)$. Since A_n and A are compact and $A_n \to A$, by A.5.1 (ii) of Mas-Colell [363, p.10], $\cup_{n=1}^{\infty} A_n \cup A$ is compact. Since $x_n \in A_n \subset \cup_{n=1}^{\infty} A_n \cup A$, we may assume without loss of generality that $x_n \to x \in \cup_{n=1}^{\infty} A_n \cup A$. If $x \notin A$, since A is compact, there is $a > 0$ such that $U(a, A) \cap O(x, a) = \emptyset$. Since $A_n \to A$ and $x_n \to x$, there is N_1 such that $A_n \subset U(a, A)$ and $x_n \in O(x, a)$ for all $n \geq N_1$, which contradicts the assumption that $x_n \in A_n$. Hence we must have $x \in A$. If $x \notin f(x)$, since $f(x)$ is compact, there is $b > 0$ such that $U(b, f(x)) \cap U(b, x) = \emptyset$. Since $f_n \to f$, there is N_2 such that $f_n(u) \subset U(b/2, f(u))$ for all $n \geq N_2$ and for all $u \in X$. Since f is upper semicontinuous at x and $x_n \to x$, there is $N_3 \geq N_2$ such that $f(x_n) \subset U(b/2, f(x))$ and $x_n \subset U(b/2, x)$ for all $n \geq N_3$. Thus for all $n \geq N_3$, $f_n(x_n) \subset U(b/2, f(x_n)) \subset U(b, f(x))$ and $x_n \subset U(b, x)$ which contradicts the assumption that $f_n(x_n) \cap x_n \neq \emptyset$. Hence we must also have $x \in f(x)$. Therefore $y = (f, A) \in M$ so that M is closed in Y. □

Now for each $y = (f, A) \in M$, define a fixed point mapping $S : M \to K(X)$ by $S(y) := \{x \in A : x \in f(x)\}$. Then $S(y) \neq \emptyset$ and following fact is true.

Lemma 2.12.27. The fixed point set $S(y) \in K(X)$ for each $y \in M$.

Proof. Let $y = (f, A) \in M$ be given. Since that $x_n \to x \in \cup_{n=1}^{\infty} A_n \cup A$. If $x \notin A$, since A is compact, there is $a > 0$ such that $U(a, A) \cap O(x, a) = \emptyset$. Since $A_n \to A$ and $x_n \to x$, there is N_1 such that $A_n \subset U(a, A)$ and $x_n \in O(x, a)$ for all $n \geq N_1$, which contradicts the assumption that $x_n \in A_n$. Hence we must have $x \in A$. If $x \notin f(x)$, since $f(x)$ is compact, there is

$b > 0$ such that $U(b, f(x)) \cap U(b, x) = \emptyset$. Since $f_n \to f$ and $g_n \to g$, there is N_2 such that $f_n(u) \subset U(b/2, f(u))$ all $n \geq N_2$ and for all $u \in X$. Since f is upper semicontinuous at x and $x_n \to x$, there is $N_3 \geq N_2$ such that $f(x_n) \subset U(b/2, f(x))$ and $x_n \subset U(b/2, x)$ for all $n \geq N_3$. Thus for all $n \geq N_3$, $f_n(x_n) \subset U(b/2, f(x_n)) \subset U(b, f(x))$ and $x_n \subset U(b, gx)$ which contradicts the assumption that $x_n \in f_n(x_n)$. Hence we must also have $x \in f(x)$. Therefore $y = (f, A) \in M$ so that M is closed in Y. □

By Lemma 2.12.27, the fixed point mapping $y \mapsto S(y)$ defines a set-valued correspindence from M to $K(X)$; and moreove S is upper semicontinuous.

Lemma 2.12.28. The mapping S is upper semicontinuous on M.

Proof. Suppose S is not upper semicontinuous at $y \in M$, then there exist $\epsilon_0 > 0$ and a sequence $\{y_n\}_{n=1}^{\infty}$ in M with $y_n \to y$ such that for each $n = 1, 2, \cdots$, there exists $x_n \in S(y_n)$ with $x_n \notin U(\epsilon_0, S(y))$. Let $y_n = (f_n, A_n)$ and $y = (f, A)$, then $f_n \to f$ and $A_n \to A$. Since $x_n \in A_n \subset \cup_{n=1}^{\infty} A_n \cup A$ and $\cup_{n=1}^{\infty} A_n \cup A$ is compact, we may assume without loss of generality that $x_n \to x \in \cup_{n=1}^{\infty} A_n \cup A$. Note that we must have $x \notin U(\epsilon_0, S(y))$. Now the same argument as in the proof of Lemma 2.12.26 shows that $x \in A$ and $x \in f(x)$ so that $x \in S(y)$; this contradicts that $x \notin U(\epsilon_0, S(y))$. Therefore S must be upper semicontinuous. □

In order to study the stability of fixed points in hyperconvex spaces, and the corresponding Definition 2.8.7, we need the following notion.

Definition 2.12.29. Let M_1 be a non-empty closed subset of M (since M is complete, so is M_1). For any given point $y_1 \in M_1$, a point x in $S(y)$ is said to be an *essential fixed point* of y_1 with respect to M_1 provided that for any $\epsilon > 0$, there is $\delta > 0$ such that for any $y' \in M_1$ with $D(y_1, y') < \delta$, there exists $x' \in S(y')$ with $d(x, x') < \epsilon$. The point y is said to be *essential* with respect to M_1 if every $x \in S(y)$ is an essential fixed point of y with respect to M_1.

Remark 2.12.30. Definition 2.12.29 says that if x is an essential fixed point of the mapping $f \in C$. Then for any $\epsilon > 0$, there exists a $\delta > 0$ such that every mapping $f' \in C$ with $\sup_{x \in X}\{d(f(x), f'(x)) : x \in X\} < \delta$ has at least on fixed point in the ϵ-neighborhood of the point y; in another word, the fixed point x of f is continuous (or say, stable) respect to the mapping f.

Though, in general the fixed point set of any given mapping $f \in C$ is not stable, we do have the following characteristic of stability for fixed points of upper semicontinuous set-valued mappings in C.

Theorem 2.12.31. The mapping S is lower semicontinuous at $y \in M_1$ if and only if y is essential with respect to M_1.

Proof. Suppose S is lower semicontinuous at $y \in M_1$. Then for any $\epsilon > 0$, there is $\delta > 0$ such that for any $y' \in M_1$ with $d(y, y') < \delta$, we have $S(y) \subset U(\epsilon, S(y'))$ so that for any $x \in S(y)$, there is $x' \in S(y')$ with $d(x, x') < \epsilon$. Thus every $x \in S(y)$ is an essential fixed point of y with respect to M_1 and hence y is essential with respect to M_1.

Conversely, suppose that y is essential with respect to M_1. If S were not lower semicontinuous at $y \in M_1$, then there exist $\epsilon_0 > 0$ and a sequence $\{y_n\}_{n=1}^{\infty}$ in M with $y_n \to y$ such that for each $n = 1, 2, \cdots$, there is $x_n \in S(y)$ with $x_n \notin U(\epsilon_0, S(y_n))$. Since $S(y)$ is compact, we may assume that $x_n \to x \in S(y)$. Since x is an essential fixed point of y with respect to M_1, $y_n \to y$ and $x_n \to x$, there is N such that $d(x_n, x) < \epsilon_0/2$ and $x \in U(\epsilon_0/2, S(y_n))$ for all $n \geq N$. Hence $x_n \in O(x, \epsilon_0/2) \subset U(\epsilon_0, S(y_n))$ for all $n \geq N$ which contradicts the assumption that $x_n \notin U(\epsilon_0, S(y_n))$ for all $n = 1, 2, \cdots$. Hence S must be lower semicontinuous at y. □

The following result says that if an upper semicontinuous set-valued mapping with compact and admissible values has fixed points defined in hyperconvex metric spaces, then it can be approximated arbitrarily by an upper semicontinuous set-valued mapping with compact and admissible values whose fixed point set is stable.

Theorem 2.12.32. The set of essential points with respect to M_1 is a dense residual set in M_1. In particular, every point in M_1 can be arbitrarily approximated by an essential points in M_1.

Proof. By Lemma 2.12.27 and Lemma 2.12.28, the mapping $S : M \to K(X)$ is an usco mapping. Since M_1 is complete, by Lemma 2.8.2, the set of points where S is lower semicontinuous is a dense residual set in M_1. By Theorem 2.12.31, the set of essential points in M_1 is a dense residual set in M_1. □

By Lemma 2.12.28, Theorems 2.12.31 and 2.12.32, we have the following:

Theorem 2.12.33. The mapping S is continuous at $y \in M_1$ if and only if y is essential with respect to M_1. Moreover, the set of points at which S is continuous is a dense residual set in M_1.

Remark 2.12.34. We remark that S is continuous at $y \in M_1$, if and only if for each $\epsilon > 0$, there is $\delta > 0$ such that $h(S(y), S(y')) < \epsilon$ for each $y' \in M$ with $D(y, y') < \delta$. Theorem 2.12.33 implies that if $y = (f, A) \in M_1$, then y is essential with respect to M_1 if and only if its set $S(y)$ of fixed points is stable in the sense that: $S(y')$ is close to $S(y)$ whenever y' is close to y.

We shall now give a sufficient condition that $y \in M_1$ is essential with respect to M_1:

Theorem 2.12.35. If $y \in M_1$ is such that $S(y)$ is a singleton set, then y is essential with respect to M_1.

Proof. Suppose $S(y) = \{x\}$. It follows by Lemma 2.12.28 that S is upper semicontinuous at y. Thus for any $\epsilon > 0$, there is $\delta > 0$ such that for each $y' \in M_1$, $D(y, y') < \delta$ implies $S(y') \subset U(\epsilon, S(y)) = O(x, \epsilon)$ so that $S(y) = \{x\} \subset U(\epsilon, S(y'))$. This shows that S is also lower semicontinuous at y. By Theorem 2.12.31, y is essential with respect to M_1. □

Finally as applications of Theorem 2.12.33, we have the following stability result of fixed points in hyperconvex metric spaces for upper semicontinuous set-valued mappings.

Theorem 2.12.36. Let X be a compact hyperconvex metric space and C_1 be the space consisting of all upper semicontinuous set-valued mappings from X to itself with non-empty admissible values. Then there exists a dense residual subset C_1 of C such that:

(1) the fixed point set of each mapping in C_1 is essential; and

(2) any given $f \in C$ and $\epsilon > 0$, there exists $\delta > 0$ such that for each $f' \in C_1$ with
$\sup_{x \in X} h(f(x), f'(x)) < \delta$, we have that $h(S(f), S(f')) < \epsilon$.

Proof. By Theorem 2.12.18, it follows that $C_1 \times \{X\}$ is a non-empty closed subset of M. Thus the conclusion follows by Theorem 2.12.33. This completes the proof. □

Remark 2.12.37. Theorem 2.12.36 show that almost all of fixed points of upper semicontinuous set-valued mappings defined in compact hyperconvex metric spaces are stable in the sense that each upper semicontinuous set-valued mapping with non-empty compact and admissible values (and thus having fixed points) defined in hyperconvex metric spaces can be approximated arbitrarily by an essentially upper semicontinuous set-valued mapping with non-empty compact and admissible values whose fixed point set is stable.

Remark 2.12.38. By introducing so-called 'locally generalzied convex spaces'- which is another kind of general abstract convexity due to Ben-El-Mechaiekh et al. in [31], Himmelberg's fixed point theorem [233] has been generalized by replacing the usual convexity in topological vector spaces with an abstract topological notion of convexity that generalizes classic convexity as well as several metric convexity structures found of existing literature in [31]. Secondly, the Browder-Fan and Fan-Glicksberg fixed point theorems developed in MC-spaces and locally MC-spaces, respectively in this section also indicate that it is possible to deveop the general KKM theory in the setting of MC-spaces, or in the setting of some abstract convexity such as those introduced in the literature due to Ben-El-Mechaiekh [31], Horvath and Llinares [241], Llinares [348] and others. For the time being, not many results are available and we think it is also worthwhile to pay attention in this direction. Some preliminary results could be found from Llinares [349]-[350], Yuan [610] and references therein.

Chapter 3

Topological Intersection and Minimax Inequality Theory

In this chapter, we shall study topologcal intersection and minimax theorems from the point of view of so-called *connectedness* which is a pure topologically property. This subject was opened by Wu in 1958 (e.g., see Wu in [579]). The study of the minimax inequality used in this chapter is quite different from the method used in Chapter 2 as all arguments applying in this part seem to be *elementary*. As applications, topological versions of minimax inequalities developed in Chapter 2 will be established.

3.1 Introduction

It is clear that many existence problems in mathematics can be reduced to the following so called '*Intersection Problem*': Let Y be a non-empty set, X an index set and $\{F(x) : x \in X\}$ a family of non-empty subsets of Y. Now the question is:

When does the family $\{F(x) : x \in X\}$ *have a non-empty intersection, i.e.,* $\cap_{x\in X} F(x) \neq \emptyset$*?*

It is convenient to formulate the intersection problem above in terms of correspondences. More precisely, let X and Y be two non-empty sets and $F : X \to 2^Y$ a correspondence with non-empty values. A single-valued mapping $f : X \to Y \setminus \{\emptyset\}$ is said to be a selector for F if $f(x) \in F(x)$ for all $x \in X$. Thus the intersection question is when does the correspondence F possess a constant selector ?

One of those mathematical problems which has the very close relationship with intersection problms is the following so-caled minimax problem formulated as follows:

Let X and Y be non-empty sets and $f : X \times Y \to \mathbb{R}$ be a function. A minimax problem is to find certain conditions such that the following holds:

$$\inf_{y\in Y} \sup_{x\in X} f(x,y) = \sup_{x\in X} \inf_{y\in Y} f(x,y). \tag{3.1.1}$$

The original motivation for the study of minimax theorems was, of course, von Neumann's work on the game theory strategies in 1928. After a generalization of von Nemuann's original results for matrices appeared, the study of minimax theorems becames progressively more remote from game theory, and minimax theory started becoming objects of study in their own right.

The importance of *connectedness* in the study of minimax theory was first recognized by Wu [579]. Then this idea was picked by Terkerson [550]. By a refined method, it was Tuy, who derived a generalized version of Sion's classical minimax theorem in [490] (see also Geraghty and Lin [194]). Independently, inspired by Joó's paper [274], the *method of level sets* was developed by Joó and his Hungarian compatriots Stachó [276] and Komornik [328]. For example, by introducing the concept of the *interval space*, it was Stachó [499], who established an intersection theorem which was used by Komornik [328] to derive a generalization of Ha's minimax theorem [213]. All these results were unified by Kindler and Trost [310]. Following this line, many minimax theorems which only involve the *connectedness* instead of convexity were obtained by Komiya [327], Horvath [237], Lin and Quan [344], Kindler [305]-[308], König [329], and Simons [483]- [485], and many others. We only mention a few names here, for the historical trace of the development of minimax theory starting from von Neumann's work, we refer the reader to Simons' recent survey paper [485]. There have been various types of minimax theorems as mentioned by Professor Simons, for example, such as (i): *topological minimax theorems* in which various connectedness hypotheses are assumed for X, Y and the function f; (ii) *quantitative minimax theorems* in which no special properties are assumed for X and Y, but various quantitative properties are assumed for f; (iii) *mixed minimax theorems* in which the quantitative and the topological properties are mixed; and (iv) *unified metaminimax theorems* which include the minimax theorems of types (i)-(iii) above as special cases. Using the idea that the minimax theorems can be reduced to the equivalent existence problems of non-empty intersection of set-valued mappings (e.g., see Remark 5 of Kindler [305, p.1008]), it was Kindler, who was the first person to give a number of topological (resp., abstract set theory) characterization of the existence of non-empty intersection theorems [305]-[307], by unifying the ideas of Wu [579], Cheng and Lin [85]-[87], Terkerson [550], Tuy [563], Joó [274]-[275], Joó and Stachó [276], Komyia [327], Komornik [328], König [329], Simons [483]-[485], Horvath [238], Thompson and Yuan [551] and many others. As applications of his intersection theorems for set-valued mappings, many minimax theorems are derived.

In this chapter, by the connectedness of topological spaces, we offer several sufficient conditions to guarantee the existence of the non-empty intersection for a family of sets in topological spaces. As applications, topological characterizations for the existence of a non-empty intersection of a set-valued mapping in topological spaces are given, which in turn are applied to derive topological fixed point theorem for upper semicontinuous set-valued mappings. Our results include the corresponding results of Chang et al. [67], Kindler [305], Joó and Sövegjártó [274] and Joó [275] as special cases. The idea behind this section is quite simple, i.e., based on the the following property of connectedness (e.g., see Theorem 6.1.1 of Engelking [154].

The Connectedness Fact: *Let X be a non-empty connected topological space and both A and B are open (resp., closed) subsets X such that $X = A \cup B$ and $A \cap B = \emptyset$, then A or B must be empty.*

This simple idea was first used by Wu [579] in the study of the minimax theory. Since the sufficient and necessary existence of the equality (3.1.1) can be reduced to the equivalent sufficient and necessary existence problem of the non-empty intersection of corresponding set-valued mappings (e.g., see Kindler [305], it is very important to study the topological characterization for the non-empty intersection of set-valued mappings. In particular, our arguments in this chapter are motivated by the connectedness approach developed in the study of minimax theory and their applications in game theory and mathematical economics due to Wu [579], Tuy [563], Terkelsen [550], Joó [274], Kindler and Trost [310], Komornik [328], Stachó [499], König [329], Simons [485], Chang et al. [67] and many others.

There are many ways to study the intersection problems. However, the way we use through this chapter is totally elementary and different from the way of applying KKM theory as developed in Chapter 2. Before we conclude this section, we recall some notations and definitions that will be used throughout this chapter. If A is a subset of a topological space X, then $cl_X A$ denotes the closure of A and $A^C := \{x \in X : x \notin A\}$ denotes the complement of A in X. Let $F : X \to 2^Y$ be a correspondence and A a non-empty subset of Y. Then (1) $F^* : Y \to 2^X$ is a set-valued mapping defined by $F^*(y) = \{x \in X : y \notin F(x)\}$

for each $y \in Y$, which is called *the dual of* F; (2) F is said to have *open inverse values* if the set $F^{-1}(y) = \{x \in X : y \in F(x)\}$ is open in X for each $y \in Y$; and (3) $F_A : X \to 2^Y$ is a mapping defined by $F_A(x) = F(x) \cap A$ for each $x \in X$.

3.2 Characterization of Topological Intersection Theorems

In this section, using the idea developed by Kindler [305], we shall study the charactertization of topological intersection theorems for set-valued mappings in topological spaces. Our results include corresponding results as special cases. As applications, and topological fixed points are derived which are different from the Fan-Glicksberg fixed point theorem in locally convex topological spaces.

Recently, Kindler [305] gave a topological characterization of the existence of a constant selector for an upper semicontinuous correspondence Φ, i.e., a pure topological condition is given that is both necessary and sufficient for $\cap_{x\in X}\Phi(x)$ to be non-empty. But he assumed that the mapping Φ has compact values, and we shall see that this condition in fact is not necessary. In order to generalize Kindler's result in [305] to the case that the Φ mapping may not have compact values, we first recall the following fact, which is the Remark 1 of Kindler [305].

Lemma 3.2.1. Let X be a topological space, Y a non-empty set and $\Phi : X \to 2^Y$ be a correspondence. Then the following are equivalent:
(a) the set $\cap_{y\in B}\Phi^*(y)$ is connected or empty for each $B \in 2^Y$.
(b) for each $x_1, x_2 \in X$, there exists a connected set $C \supset \{x_1, x_2\}$ such that

$$\Phi(x) \subset \Phi(x_1) \cup \Phi(x_2) \text{ for all } x \in C.$$

(c) for each pair $(x_1, x_2) \in X \times X$, the set $\{x \in X : \Phi(x) \subset \Phi(x_1) \cup \Phi(x_2)\}$ is connected.

We also need the following simple result; for completeness, we include its proof here.

Lemma 3.2.2. Let X and Y be two topological spaces and A is a non-empty closed (resp., open) subset of Y. If the mapping $\Phi : X \to 2^Y$ is upper (resp., lower) semicontinuous, then the mapping $\Phi_A : X \to 2^Y$ defined by $\Phi_A(x) = \Phi(x) \cap A$ is also upper (resp., lower) semicontinuous.

Proof. Let G be an open (resp., closed) subset of Y and $x \in X$ such that $\Phi(x) \cap A \subset G$. Then $\Phi(x) \subset G \cup A^C$. Note that Φ is upper (resp., lower) semicontinuous at x and A is closed (resp., open), so that there exists an open (resp., closed) subset $N(x)$ of x such that for each $z \in N(x)$, $\Phi(z) \subset G \cup A^C$, i.e., $\Phi(z) \cap A \subset [G \cup A^C] \cap A = G \cap A \subset G$ for each $z \in N(x)$. Therefore the mapping Φ_A is upper (resp., lower) semicontinuous at $x \in X$. $\square$

Now we have the following sufficient conditions which guarantee the existence of 2-points intersection for upper semicontinuous set-valued mappings which may have closed or open values in topological spaces.

Theorem 3.2.3. Let X and Y be two topological spaces and $\Phi : X \to 2^Y$ be a set-valued mapping with non-empty values. Suppose Φ satisfies the following conditions:

(i) Φ is upper semicontinuous with non-empty closed (resp., open) connected values; and

(ii) the set $\cap_{y \in B} \Phi^*(y)$ is connected or empty for each $B \in 2^Y$.

Then the family $\{\Phi(x) : x \in X\}$ has the 2-points intersection property, i.e., $\Phi(x_1) \cap \Phi(x_2) \neq \emptyset$ for each $x_1, x_2 \in X$.

Proof. Suppose there exist $x_1, x_2 \in X$ with $x_1 \neq x_2$ and $\Phi(x_1) \cap \Phi(x_2) \neq \emptyset$, then we shall get a contradiction. Let $C = \{x \in X : \Phi(x) \subset \Phi(x_1) \cup \Phi(x_2)\}$. By (ii) and Lemma 3.2.1, the set C is connected in X. For $i = 1, 2$, let $M_i = \{x \in C : \Phi(x) \subset \Phi(x_i)\}$. Note that for each $x \in X$, $\Phi(x)$ is non-empty connected, both $\Phi(x_1)$ and $\Phi(x_2)$ are closed (resp., open) and $\Phi(x_1) \cap \Phi(x_2) = \emptyset$. Therefore for each $x \in C$, if $\Phi(x) \subset \Phi(x_1) \cup \Phi(x_2)$, then it follows that $\Phi(x) \subset \Phi(x_1)$ or $\Phi(x) \subset \Phi(x_2)$ by the connectedness fact above. Thus we have that $M_1 \cup M_2 = C$. Note that $\Phi(x_1) \cap \Phi(x_2) = \emptyset$, so that $M_1 \cap M_2 = \emptyset$. Since Φ is upper semicontinuous and the set $Y \setminus \Phi(x_i)$ is open (resp., closed) in Y, the set $M_i' = \{x \in X : \Phi(x) \subset \Phi(x_i)\} = \{x \in X : \Phi(x) \subset X \setminus \Phi(x_{3-i})\}$ is open in X. Note that $M_i' \cap C = M_i$, so that M_i is open in C. Since C is connected, both M_1 and M_2 are non-empty open in C, $M_1 \cap M_2 = \emptyset$ and $C = M_1 \cup M_2$, which is impossible by the connectedness fact. Therefore $\Phi(x_1) \cap \Phi(x_2) \neq \emptyset$ for each $x_1, x_2 \in X$. $\square$

If the mapping Φ is lower semicontinuous instead of being upper semicontinuous in Theorem 3.2.3, we also have the following intersection result.

Theorem 3.2.4. Let X and Y be two topological spaces and $\Phi : X \to 2^Y$ be a set-valued mapping with non-empty values. Suppose Φ satisfies the following conditions:

(i) Φ is lower semicontinuous with non-empty closed (resp., open) connected values; and

(ii) the set $\cap_{y \in B} \Phi^*(y)$ is connected or empty for each $B \in 2^Y$.

Then the family $\{\Phi(x) : x \in X\}$ has the 2-points intersection property, i.e., $\Phi(x_1) \cap \Phi(x_2) \neq \emptyset$ for each $x_1, x_2 \in X$.

Proof. Following the same idea in the proof of Theorem 3.2.3: if the conclusion were false, then there exist $x_1, x_2 \in X$ with $x_1 \neq x_2$, and $\Phi(x_1) \cap \Phi(x_2) \neq \emptyset$, thus we get a contradiction. Let $C = \{x \in X : \Phi(x) \subset \Phi(x_1) \cup \Phi(x_2)\}$. By (ii) and Lemma 3.2.1, the set C is connected in X. For $i = 1, 2$, let $M_i = \{x \in C : \Phi(x) \subset \Phi(x_i)\}$. Then $M_1 \cup M_2 = C$. Note that $\Phi(x_1) \cap \Phi(x_2) = \emptyset$, so that $M_1 \cap M_2 = \emptyset$. Since Φ is lower semicontinuous and the set $Y \setminus \Phi(x_i)$ is open (resp., closed) in Y, the set $M_i' = \{x \in X : \Phi(x) \subset \Phi(x_i)\} = \{x \in X : \Phi(x) \subset X \setminus \Phi(x_{3-i})\}$ is closed in X. Note that $M_i' \cap C = M_i$, so that M_i is closed in C. Since C is connected, both M_1 and M_2 are non-empty closed in C, $M_1 \cap M_2 = \emptyset$ and $C = M_1 \cup M_2$, which is impossible by the connectedness fact. Therefore $\Phi(x_1) \cap \Phi(x_2) \neq \emptyset$ for each $x_1, x_2 \in X$. □

As applications of Theorems 3.2.3 and 3.2.4, we have the following sufficient conditions to guarantee the existence of the finite intersection property for upper (resp., lower) semicontinuous set-valued mappings in topological spaces.

Theorem 3.2.5. Let X and Y be two topological spaces and $\Phi : X \to 2^Y$ be a set-valued mapping with non-empty values. Suppose Φ satisfies the following conditions:

(i) Φ is upper semicontinuous with non-empty closed values;

(ii) the set $\cap_{x \in A} \Phi(x)$ is connected or empty for each $A \in \mathcal{F}(X)$; and

(iii) the set $\cap_{y \in B} \Phi^*(y)$ is connected or empty for each $B \in 2^Y$.

Then the family $\{\Phi(x) : x \in X\}$ has the finite intersection property, i.e., $\cap_{i=1}^n \Phi(x_i) \neq \emptyset$ for each $x_1, \cdots, x_n \in X$, where $n \in \mathbb{N}$.

Proof. We shall prove the conclusion by the induction. Without loss of generality, suppose $\cap_{i=1}^n \Phi(x_i) \neq \emptyset$ for each $x_1, x_2, \cdots, x_n \in X$, where $n \in \mathbb{N}$ with $n \geq 2$. We shall show that

$$\cap_{i=1}^{n+1} \Phi(x_i) \neq \emptyset$$

for each $x_1, \cdots, x_{n+1} \in X$ with $n \in \mathbb{N}$ and $n \geq 2$. Define a mapping $\Phi' : X \to 2^Y$ by $\Phi'(x) = \cap_{i=1}^{n-1} \Phi(x_i) \cap \Phi(x)$ for each $x \in X$. Then Φ' is upper semicontinuous with non-empty closed values by Lemma 3.2.2. By (ii), it is clear that the mapping Φ' satisfies conditions (i) and (ii) of Theorem 3.2.3. By Theorem 3.2.3, $\Phi'(x') \cap \Phi'(x'') \neq \emptyset$ for each $x', x'' \in X$. Now take $x' = x_n$ and $x'' = x_{n+1}$, we have $\cap_{i=1}^{n+1} \Phi(x_i) \neq \emptyset$. Therefore $\cap_{i=1}^{n+1} \Phi(x_i) \neq \emptyset$ for each $x_1, x_2, \cdots, x_n \in X$, where $n \in \mathbb{N}$. Thus the family $\{\Phi(x) : x \in X\}$ has the finite intersection property. □.

By Theorem 3.2.4 and the same argument of Theorem 3.2.5, we have the following result.

Theorem 3.2.6. Let X and Y be two topological spaces and $\Phi : X \to 2^Y$ be a set-valued mapping with non-empty values. Suppose Φ satisfies the following conditions:

(i) Φ is lower semicontinuous with non-empty open values;

(ii) the set $\cap_{x\in A}\Phi(x)$ is connected or empty for each $A \in \mathcal{F}(X)$; and
(iii) the set $\cap_{y\in B}\Phi^*(y)$ is connected or empty for each $B \in 2^Y$.

Then the family $\{\Phi(x) : x \in X\}$ has the finite intersection property, i.e., $\cap_{i=1}^{n}\Phi(x_i) \neq \emptyset$ for each $x_1, \cdots, x_n \in X$, where $n \in \mathbb{N}$.

Proof. By Theorem 3.2.4 and following the same proof of Theorem 3.2.5, we may suppose $\cap_{i=1}^{n}\Phi(x_i) \neq \emptyset$ for $x_1, x_2, \cdots, x_n \in X$, where $n \in \mathbb{N}$ with $n \geq 2$. We shall show that $\cap_{i=1}^{n+1}\Phi(x_i) \neq \emptyset$ for each $x_1, \cdots, x_{n+1} \in X$. Define a mapping $\Phi' : X \to 2^Y$ by

$$\Phi'(x) = \cap_{i=1}^{n-1}\Phi(x_i) \cap \Phi(x)$$

for each $x \in X$. Then Φ' is lower semicontinuous with non-empty open values by Lemma 3.2.2 and the mapping Φ' satisfies conditions (i) and (ii) of Theorem 3.2.4. By Theorem 3.2.4, $\Phi'(x') \cap \Phi'(x'') \neq \emptyset$ for each $x', x'' \in X$. Now take $x' = x_n$ and $x'' = x_{n+1}$, we have $\cap_{i=1}^{n+1}\Phi(x_i) \neq \emptyset$. Therefore $\cap_{i=1}^{n+1}\Phi(x_i) \neq \emptyset$ for each $x_1, x_2, \cdots, x_n \in X$, where $n \in \mathbb{N}$. Thus the family $\{\Phi(x) : x \in X\}$ has finite intersection property. □.

If the lower semicontinuous mapping Φ in Theorem 3.2.6 has the open inverse property, i.e., the set $\Phi^{-1}(y) := \{x \in X : y \in \Phi(x)\}$ is open in X for each $y \in Y$, the following theorem shows that the conclusion of Theorem 3.2.6 still holds if the mapping Φ has non-empty closed values instead of having open values.

Theorem 3.2.7. Let X and Y be two topological spaces and $\Phi : X \to 2^Y$ be a set-valued mapping with non-empty values. Suppose Φ satisfies the following conditions:
(i) Φ has non-empty closed values and the set $\Phi^{-1}(y) = \{x \in X : y \in \Phi(x)\}$ is open in X for each $y \in Y$ (so that Φ is lower semicontinuous);
(ii) the set $\cap_{x\in A}\Phi(x)$ is connected or empty for each $A \in \mathcal{F}(X)$; and
(iii) the set $\cap_{y\in B}\Phi^*(y)$ is connected or empty for each $B \in 2^Y$.

Then the family $\{\Phi(x) : x \in X\}$ has the finite intersection property, i.e., $\cap_{i=1}^{n}\Phi(x_i) \neq \emptyset$ for each $x_1, \cdots, x_n \in X$, where $n \in \mathbb{N}$.

Proof. By (i) Φ is lower semicontinuous, so that $\Phi(x_1) \cap \Phi(x_2) \neq \emptyset$ for each $x_1, x_2 \in X$ by Theorem 3.2.4. Now we shall show that the family $\{\Phi(x) : x \in X\}$ has the finite intersection property by the induction. Without loss of generality, we may assume that $\cap_{i=1}^{n}\Phi(x_i) \neq \emptyset$ for each $x_1, x_2, \cdots .x_n \in X$, where $n \in \mathbb{N}$ with $n \geq 2$. We shall show that $\cap_{i=1}^{n+1}\Phi(x_i) \neq \emptyset$ for each $x_1, x_2, \cdots, x_{n+1} \in X$. If it were false, then there exist $x_1, x_2, \cdots, x_n \in X$ such that $\cap_{i=1}^{n+1}\Phi(x_i) = \emptyset$. Now define the mapping $\Phi_1 : X \to 2^Y$ by

$$\Phi_1(x) = \cap_{i=3}^{n+1}\Phi(x_i) \cap \Phi(x)$$

for each $x \in X$. Then $\Phi_1(x) \neq \emptyset$ for each $x \in X$ and $\Phi_1(x_1) \cap \Phi_1(x_2) = \emptyset$ by our assumption. By (iii) and Lemma 3.2.1, the set $C := \{x \in X : \Phi_1(x) \subset \Phi_1(x_1) \cup$

$\Phi_1(x_2)\}$ is non-empty and connected in X. Let $M_i = \{x \in C : \Phi_1(x) \subset \Phi_1(x_i)\}$ for each $i = 1, 2$.

First, we claim that $C = M_1 \cup M_2$. It is clear that $M_1 \cup M_2 \subset C$. Now suppose $x \in C$, then $\Phi_1(x) \subset \Phi_1(x_1) \cup \Phi_1(x_2)$. Note that $\Phi_1(x)$ is non-empty and connected, both $\Phi_1(x_1)$ and $\Phi_1(x_2)$ are non-empty, disjoint and closed in X, we must have $\Phi_1(x) \subset \Phi_1(x_1)$ or $\Phi_1(x) \subset \Phi_1(x_2)$ by the connectedness fact. Therefore for each $x \in X$, $x \in M_1$ or $x \in M_2$, so that $C = M_1 \cup M_2$.

Secondly, we claim that both M_1 and M_2 are closed in C. Suppose it were false, we may assume that M_2 is not closed in C. Taking $x_0 \in [cl_C M_2 \setminus M_2] \cap M_1$, then there exists a net $\{x_\alpha\}_{\alpha \in I}$ in M_2 such that $x_\alpha \to x_0$. Since $x_0 \in M_1$, we have $\Phi_1(x_0) \subset \Phi_1(x_1)$. Note that $x_\alpha \in M_2$, so that $\Phi_1(x_\alpha) \subset \Phi_1(x_2)$ for all $\alpha \in I$. Taking $y_0 \in \Phi_1(x_0)$, then $y_0 \in \Phi_1(x_1)$. Note that $\Phi_1(x_1) \cap \Phi_1(x_2) = \emptyset$ and $\Phi_1(x_1) = \Phi(x_1) \cap \cap_{i=3}^{n+1} \Phi(x_i)$ and $\Phi_1(x_2) = \Phi(x_2) \cap \cap_{i=3}^{n+1} \Phi(x_i)$. We must have $y_0 \notin \Phi_1(x_\alpha)$. Therefore $y_0 \notin \Phi(x_\alpha)$. Thus $\{x_\alpha\}_{\alpha \in I} \subset X \setminus \Phi^{-1}(y_0)$. Note that $x_\alpha \to x_0$ and the set $X \setminus \Phi^{-1}(y_0)$ is closed by (i), so that $x_0 \in X \setminus \Phi^{-1}(y_0)$, i.e., $y_0 \notin \Phi(x_0)$, which contradicts that $y_0 \in \Phi_1(x_0) = \Phi(x_0) \cap \cap_{i=3}^{n-1} \Phi(x_i)$. Therefore both M_1 and M_2 are closed in C. Now we have $C = M_1 \cup M_2$, C is non-empty and connected, both M_1 and M_2 are non-empty, disjoint and closed in C, which is impossible by the connectedness fact. This shows that $\Phi_1(x_1) \cap \Phi_1(x_2)$ must be non-empty. Thus the family $\{\Phi(x) : x \in X\}$ has the finite intersection property. □.

We would like to point out that our Theorems 3.2.5 and 3.2.6 improve Theorem 1 of Chang et al. [67] in the sense that the mapping Φ may be upper or lower semicontinuous instead of having open inverse values. Theorem 3.2.7 shows that Theorem 1 of Chang et al. [67] still holds in which the space X is not a W-space. We recall that a topological space is said to be a *W-space* if there exists a family $\{C_A : A \in \mathcal{F}(X)\}$ of non-empty connected subsets of X, indexed by $A \in \mathcal{F}(X)$ such that $A \subset C_A$). Moreover Theorems A and B of Joó [274] are immediate consequences of Theorems 3.2.6 and 3.2.7, respectively.

As an application of Theorem 3.2.4, we have the following topological characterization for the existence of constant selector of upper semicontinuous set-valued mappings which may not have compact values in topological spaces.

Theorem 3.2.8. Let X and Y be two non-empty sets and $\Phi : X \to 2^Y$ be a set-valued mapping with non-empty values. Then Φ has a constant selector (i.e., $\cap_{x\in X} \Phi(x) \neq \emptyset$) if and only if there exist topologies $\mathcal{F}_X$ and $\mathcal{F}_Y$ on X and Y respectively such that:

(i) Φ is upper semicontinuous with non-empty closed values;
(ii) there exists $A_0 \in \mathcal{F}(X)$ such that $\cap_{x\in A_0} \Phi(x)$ is compact;
(iii) the set $\cap_{x\in A} \Phi(x)$ is connected or empty for each $A \in \mathcal{F}(X)$; and
(iv) the set $\cap_{y\in B} \Phi^*(y)$ is connected or empty for each $B \in 2^Y$.

Proof. Necessity: Following Kindler [305], choose $\hat{y} \in \cap_{x\in X} \Phi(x)$. Let

$(X, \mathcal{F}_X)$ and $(Y, \mathcal{F}_Y)$ be topological spaces with the topology $\mathcal{F}_X = \{\emptyset, X\}$ and $\mathcal{F}_Y = \{G \subset Y : \hat{y} \notin G\} \cup \{Y\}$. Then we shall show that Φ satisfies all conditions (i)-(iv). Note that $(X, \mathcal{F}_X)$ is a trivial space, and conditions (i) and (iv) are satisfied automatically. Now for each $G \in \mathcal{F}_Y$, let $U = \{x \in X : \Phi(x) \subset G\}$. If $U \neq \emptyset$, then $U = X$ so that Φ is upper semicontinuous. For each $G \in \mathcal{F}_Y$, if $\hat{y} \in G$, then G is closed and connected,so that $\Phi(x)$ is closed for each $x \in X$. By the definition of $\mathcal{F}_Y$, it is clear that $(Y, \mathcal{F}_Y)$ is compact. Therefore condition (iii) is also satisfied.

Sufficiency: By Theorem 3.2.5, the family $\{\Phi(x) : x \in X\}$ has the finite intersection property. Let $\Phi_1(x) = \cap_{x \in A_0} \Phi(x) \cap \Phi(x)$ for each $x \in X$, then the family $\{\Phi_1(x) : x \in X\}$ is a compact family which has the finite intersection property. Therefore $\cap_{x \in X} \Phi(x) \neq \emptyset$. □

Theorem 3.2.8 improves Theorem 1* of Kindler [305] in the sense that the space Y may not compact and the mapping Φ need not have compact values (compare Remark 1 of Kindler [305]).

As an immediate consequence of Theorem 3.2.6, we have the following Theorem (see also Theorem 2* of Kindler [305]):

Theorem 3.2.9. Let X and Y be two non-empty sets and $\Phi : X \to 2^Y$ be a set-valued mapping with non-empty values. Then $\cap_{x \in A} \Phi(x) \neq \emptyset$ for each $A \in \mathcal{F}(X)$ if and only if there exist topologies $\mathcal{F}_X$ and $\mathcal{F}_Y$ on X and Y respectively such that:

(i) Φ is lower semicontinuous with non-empty open values;
(ii) the set $\cap_{x \in A} \Phi(x)$ is connected or empty for each $A \in \mathcal{F}(X)$; and
(iii) the set $\cap_{y \in B} \Phi^*(y)$ is connected or empty for each $B \in 2^Y$.

Proof. Necessity: Following Kindler [305], take the topology $\mathcal{F}_X = \{\emptyset, X\}$ for X and the topology

$$\mathcal{F}_Y = \{G \subset Y : \exists A \in \mathcal{F}(X) \text{ and } G \supset \cap_{x \in A} \Phi(x)\} \cup \{\emptyset\}$$

for Y. Then condition (i) and (iii) are obviously satisfied. It remains to be shown hat the set $C := \cap_{x \in A} \Phi(x)$ is $\mathcal{F}_Y$-connected for each $A \in \mathcal{F}_Y$. To see it, let $C = G_1 \cup G_2$ with $G_1, G_2 \in \mathcal{F}_Y \setminus \{\emptyset\}$. Then there exist $A_i \in \mathcal{F}(X)$ with $G_i \supset \cap_{x \in A_i} \Phi(x)$ for $i = 1, 2$. Then $C \cap G_1 \cap G_2 \supset \cap_{x \in A \cup A_1 \cup A_2} \Phi(x) \neq \emptyset$. Thus $\cap_{x \in A} \Phi(x)$ is $\mathcal{F}_Y$-connected for each $A \in \mathcal{F}(X)$.

Sufficiency: It follows from Theorem 3.2.6. □

As an application of Theorem 3.2.8, we have the following fixed point theorem which improves the fixed point theorem of Joó and Sövergjártó [277] (see also Theorem A of Joó [274] and Proposition 2 of Stachó [499]).

Theorem 3.2.10. Let X be a topological space. Suppose that the mapping $\Phi : X \to 2^X$ is upper semicontinuous with non-empty compact and closed values such that:

(1) the set $\cap_{x\in A}\Phi(x)$ is connected or empty for each $A \in \mathcal{F}(X)$ and

(2) for each $x, y \in X$, there exists a non-empty connect set $C_{\{x,y\}}$ containing $\{x, y\}$ in X such that $\Phi(z) \subset \Phi(x) \cup \Phi(y)$ for each $z \in C_{\{x,y\}}$. Then there exists $x_0 \in X$ such that $x_0 \in \Phi(x_0)$.

Proof. By Lemma 3.2.1, condition (2) implies that the set $\cap_{y\in B}\Phi^*(y)$ is connected or empty for each $B \in 2^Y$. Therefore all hypotheses of Theorem 3.2.8 are satisfied. By Theorem 3.2.8, $\cap_{x\in X}\Phi(x) \neq \emptyset$. Taking $x_0 \in \cap_{x\in X}\Phi(x)$, then $x_0 \in X$ so that $x_0 \in \Phi(x_0)$. □

3.3 Topological Intersection of Set-Valued Mappings without Closed Graphs

As we mention in the introduction, the original motivation for the study of minimax theorems was, of course, von Neumann's work on the game theory of strategies in 1928. As time went on, these generalizations became progressively more remote from game theory, and minimax theory object of study in its own right. The importance of *connectedness* in the study of minimax theory was first recognized by Wu [579], then this idea was picked up by many people.

In this section, topological intersection theorems for two types of set-valued mappings without closed graphs have been established. These theorems answer the question of Kindler [305] in the affirmative when the topological space has the 'connection structure' in the sense of Tuy [563]. As applications, topological fixed point theorems, minimax inequalities and vector valued variational inequalities are given.

In his paper [305], Professor Kindler proved the following topological existence theorem of constant selector for set-valued mappings (e.g., see Corollary 1 of Kindler [305], and see also Proposition 2 of Stachó [499]). We state it as the following form.

Theorem 3.3.1. Let X and Y be topological spaces and $F : X \to 2^Y$ be a set-valued mapping with non-empty values such that:

(i) the graph of F is closed in $X \times Y$;

(ii) Y is compact;

(iii) for each $A \in \mathcal{F}(X)$, if the set $\cap_{x\in A}F(x)$ is non-empty, then $\cap_{x\in A}F(x)$ is connected; and

(iv) for each $x_1, x_2 \in X$, the set $\{x \in X : F(x) \subset F(x_1) \cup F(x_2)\}$ is connected.

Then F has a constant selector, i.e., $\cap_{x\in X}F(x) \neq \emptyset$.

Definition 3.3.2. We recall that a topological space X is called an *interval space* (e.g., see Stachó [499]) if there exists a mapping $[\cdot,\cdot] : X \times X \to$ {the family of all non-empty connected subsets of $X\}$ such that $x_1, x_2 \in [x_1, x_2] = [x_2, x_1]$ for all $x_1 x_2 \in X$.

As condition (iv) of Theorem 3.3.1 above implies that X is an interval space (the inverval structure mapping $[\cdot, \cdot]$ can be defined by $[x_1, x_2] := \{x \in X : F(x) \subset F(x_1) \cup F(x_2)\}$ for each $x_1, x_2 \in X$), thus Proposition 2 of Stachó in [499] indeed includes Theorem 3.3.1 above as a special case.

If the graph of F in Theorem 3.3.1 above is closed, it then implies the following property (i)$'$:

(i)$'$ $F(x)$ and $F^{-1}(y) := \{x \in X : y \in F(x)\}$ are closed for each $x \in X$ and $y \in Y$.

It is also clear that the converse does not hold in general, i.e., the condition (i)$'$ above could not guarantee that the graph of F is closed. On the other hand, motivated by the existence of minimax theorems for separately upper (or lower) semicontinuous functions (e.g, see Remark 4.4 and question 4.5 of Kindler and Trost [310, p.45]; Remark of Joó [275] and König [329, p.57]), Kindler asked the following question in [305, p.1007]:

Does the conclusion of Theorem 3.3.1 above remain true if condition (i) is replaced by condition (i)$'$?

Being motivated by the above question, we first prove some topological intersection theorems for set-valued mappings without closed graphs. Then we show that one of our intersection theorems answers Kindler's question in the affirmative when underlying space X has the so-called *α-connectedness structure* in the sense of Tuy [563, p.145]. Indeed, our result shows that the positive answer for Kindler's question remains true even if the space Y may be not compact provided the space X is a non-empty convex subset of a vector space E without any topology. As applications, topological fixed points and minimax theorems are derived. Finally, some topological variational inequalities are also given. Our results improve and unify many corresponding results in the literature (e.g, see Chang et al. [67], Joó [274], Geraghty and Lin [194], Komornik [328], Kindler [305] and so on).

In orer to study Kindler's question, we first have the following topological intersection theorem.

Theorem 3.3.3. Let X and Y be both topological spaces. Suppose that $F : X \to 2^Y$ is a set-valued mapping with non-empty compact values such that:

(1) for each $x, y \in X$, there exists a continuous mapping $u_{x,y} : [0,1] \to X$ with $u_{x,y}(0) = x$, $u_{x,y}(1) = y$ and $F((u_{x,y}(t)) \subset F(u_{x,y}(t_1)) \cup F(u_{x,y}(t_2))$ for each $t \in [t_1, t_2] \subset [0,1]$ (resp., the set $\{x \in X : F(z) \subset F(x) \cup F(y)\}$ is connected);

(2) for each $A \in \mathcal{F}(X)$, if the set $\cap_{x \in A} F(x)$ is non-empty, then $\cap_{x \in A} F(x)$ is connected; and

(3) for each $y \in Y$, the set $F^{-1}(y) = \{x \in X : y \in F(x)\}$ is closed (resp., $F^{-1}(y)$ is open (and thus F is lower semicontinuous)) in X.

Then $\cap_{x\in X} F(x) \neq \emptyset$.

Proof. We shall first prove that $F(x) \cap F(y) \neq \emptyset$ for each $x, y \in X$. Suppose $F(x) \cap F(y) = \emptyset$. By (1), there exists a continuous function $u_{x,y} : [0,1] \to X$ such that $F(u_{x,y}(t)) \subset F(x) \cup F(y)$ for each $t \in [0,1]$. Let $M_1 = \{t \in [0,1] : F(u_{x,y}(t)) \subset F(x)\}$ and $M_2 = \{t \in [0,1] : F(u_{x,y}(t)) \subset F(y)\}$. Then M_1 and M_2 are non-empty and disjoint. Note that for each $t \in [0,1]$, $F(u_{x,y}(t)) \subset F(x) \cup F(y)$, $F(u_{x,y}(t))$ is connected, and $F(x)$ and $F(y)$ are both closed and disjoint, so that $F(u_{x,y}(t)) \subset F(x)$ or $F(u_{x,y}(t)) \subset F(y)$. Thus t must be in M_1 or M_2. Therefore $M_1 \cup M_2 = [0,1]$.

Next we shall prove that both M_1 and M_2 are open in $[0,1]$. Let $t_0 \in M_1$ and $x_0 = u_{x,y}(t_0)$. Then $F(x_0) \subset F(x)$. For each $z \in F(y)$, we have $z \notin F(x_0)$, i.e., $x_0 \in X \setminus F^{-1}(z)$. By (3), $X \setminus F^{-1}(z)$ is open in X, and thus there exists a non-empty open neighborhood V_z of x_0 such that $z \notin F(x')$ for each $x' \in V_z$. Let $U_z = u_{x,y}^{-1}(V_z)$. Then U_z is an open neighborhood of t_0 in $[0,1]$. Thus there exist $\lambda'_z, \lambda''_z \in [0,1]$ with $\lambda'_z < t_0 < \lambda''_z$ such that $t_0 \in [\lambda'_z, \lambda''_z] \subset U_z$. Therefore $z \notin F(u_{x,y}(\lambda'_z)) \cup F(u_{x,y}(\lambda''_z))$. Since $F(u_{x,y}(\lambda'_z))$ and $F(u_{x,y}(\lambda''_z))$ are compact, there exist non-empty open neighborhoods V'_z and V''_z of z such that $z' \notin F(u_{x,y}(\lambda'_z))$ and $z'' \notin F(u_{x,y}(\lambda''_z))$ for all $z' \in V'_z$ and $z'' \in V''_z$. Let $V_z = V'_z \cap V''_z$. Then V_z is a non-empty open neighborhood of z such that $z' \notin F(u_{x,y}(\lambda'_z)) \cup F(u_{x,y}(\lambda''_z))$ for each $z' \in V_z$. Taking over $z \in F(y)$, then the family $\{V_z : z \in F(y)\}$ is an open cover of $F(y)$. By the compactness of $F(y)$, there exists a finite number of subsets $\{V_{z_1}, \cdots, V_{z_n}\}$ of the family $\{V_z : z \in F(y))$ such that $\cup_{i=1}^n V_{z_i} \supset F(y)$. Now let $\lambda' := max\{\lambda'_{z_i} : i = 1, \cdots, n\}$ and $\lambda'' := min\{\lambda''_{z_i} : i = 1, \cdots, n\}$. Then $\lambda' < t_0 < \lambda''$ and (λ', λ'') is a non-empty open neighborhood of t_0 in $[0,1]$. Moreover, for each $t \in (\lambda', \lambda'')$, $z \notin F(u_{x,y}(t))$ for all $z \in F(y)$. Thus $(\lambda', \lambda'') \subset M_1$. It follows that M_1 is open in $[0,1]$. Similarly, the set M_2 is also open in $[0,1]$. Thus the segment $[0,1]$ is the union of two non-empty, disjoint open subsets M_1 and M_2, which is impossible. Therefore we must have $F(x) \cap F(y) \neq \emptyset$ for each $x, y \in X$. (Resp., if the set $F^{-1}(y) = \{x \in X : y \in F(x)\}$ is open in X for each $y \in Y$, we shall be able to show that $F(x) \cap F(x) \neq \emptyset$ for each $x, y \in X$. If not, there exist $x_1, x_2 \in X$ such that $F(x_1) \cap F(x_2) = \emptyset$. Let $C = \{x \in X : F(x) \subset F(x_1) \cup F(x_2)\}$ and $M_i = \{x \in C : F(x) \subset F(x_i)\}$ for $i = 1, 2$. Then C is connected and $M_1 \cup M_2 = C$ by condition (1). Note that $F(x_1) \cap F(x_2) = \emptyset$, so that $M_1 \cap M_2 = \emptyset$. By (3), F is lower semicontinuous and $F(x_i)$ is closed, hence the set $M'_i = \{x \in X : F(x) \subset F(x_i)\} = \{x \in X : F(x) \subset X \setminus F(x_{3-i})\}$ is closed in X. Note that $M'_i \cap C = M_i$, so that M_i is closed in C. As C is connected, both M_1 and M_2 are non-empty closed in C, $M_1 \cap M_2 = \emptyset$ and $C = M_1 \cup M_2$, which is a contraction. This contradiction shows that $F(x_1) \cap F(x_2) \neq \emptyset$ for each $x_1, x_2 \in X$).

Finally, we shall prove that $\cap_{x\in A}F(x) \neq \emptyset$ for each $A \in \mathcal{F}(X)$ by the induction. Without loss of generality, we may assume that $\cap_{i=1}^{n} F(x_i) \neq \emptyset$ for each $x_1, \cdots, x_n \in X$, where $n \in \mathbb{N}$ and $n \geq 2$. Define a mapping $F_1 : X \to 2^Y$ by $F_1(x) = \cap_{i=1}^{n-1} F(x_i) \cap F(x)$ for each $x \in X$. Then $F_1(x) \neq \emptyset$ for each $x \in X$. Moreover it is easy to verify that F_1 satisfies all hypotheses (1)-(3). Thus $F_1(x') \cap F_1(x'') \neq \emptyset$ for each $x', x'' \in X$ by the earlier proof above. Let $x' = x_n$ and $x'' = x_{n+1}$, then $\cap_{i=1}^{n+1} F(x_i) \neq \emptyset$. Therefore the compact family $\{F(x) : x \in X\}$ has finite intersection property, so that $\cap_{x\in X} F(x) \neq \emptyset$. □

Remark 3.3.4. We would like to point out that condition (1) of Theorem 3.3.3 implies that X is an interval space with a practically interval mapping $[\cdot, \cdot]$ defined by $[x, y] := \{u_{x,y}(t) : t \in [0, 1]\}$ for each $x, y \in X$. However we do not know whether Theorem 3.3.3 is still true if X is a general interval space such that condition (1) of Theorem 3.3.1 is replaced by the following:

(1)′ $F(z) \subset F(x) \cup F(y)$ for all $z \in [x, y]$ for each $x, y \in X$.

We also remark that Theorem 3.3.3 not only improves corresponding results of Theorem 1 of Chang et al. [67], but it also shows their condition that X *is a W-space* is superfluous.

Let X be a non-empty convex subset of a vector space E. In what follows, we shall denote by $[x_1, x_2]$ the line segment $\{tx_1 + (1-t)x_2 : t \in [0, 1]\}$, equipped with the Euclidean topology (Of course, each topological vector space E with this structure $[x, y]$ is an interval space). A function $f : X \to \mathbb{R} \cup \{-\infty, +\infty\}$ is said to *be segment upper (resp., lower) semicontinuous* if the function $t \to f(tx_1 + (1-t)x_2)$ is upper (resp., lower) semicontinuous on $[0, 1]$ for each given $x_1, x_2 \in X$.

When X is a non-empty subset of a vector space, the following theorem can be shown to follow from Theorem 3.3.3. However we provide a direct proof here.

Theorem 3.3.5. Let X be a convex of a vector space E and Y be a topological space. Suppose that $F : X \to 2^Y$ is a set-valued mapping with non-empty compact values such that:

(1) for each $x, y \in X$, $F(z) \subset F(x) \cup F(y)$ for each $z \in [x, y]$;

(2) for each $A \in \mathcal{F}(X)$, if the set $\cap_{x\in A}F(x)$ is non-empty, then $\cap_{x\in A}F(x)$ is connected; and

(3) for each $y \in Y$, the set $F^{-1}(y)$ is closed in the line segment $[x_1, x_2]$ for each $x_1, x_2 \in X$.

Then $\cap_{x\in X}F(x) \neq \emptyset$.

Proof. Step 1: we shall first prove that $F(x_1) \cap F(x_2) \neq \emptyset$ for each $x_1, x_2 \in X$. If it were false, then there exist $x_1, x_2 \in X$ with $x_1 \neq x_2$ and $F(x_1) \cap F(x_2) = \emptyset$. Now let $M_i := \{x \in [x_1, x_2] : F(x) \subset F(x_i)\}$ for $i = 1, 2$. Note that $F(x)$ is non-empty connected for each $x \in X$, we must have $[x_1, x_2] = M_1 \cup M_2$ by condition (1). For the simplicity, we shall denote the set $[x_1, x_2]\setminus\{x_2\}$ by the set $[x_1, x_2) = (x_2, x_1]$. Now we claim that both M_1 and M_2 are open in $[x_1, x_2]$. If not, without loss of generality, we assume M_1 is not open in $[x_1, x_2]$. Then there exists a point $x_0 \in M_1$ and a convergent net $\{x_\alpha\}_{\alpha\in I}$ in M_2 such that $x_\alpha \to x_0$. In order to prove $\cap_{x\in M_2} F(x) \neq \emptyset$. It suffices to prove the compact family $\{F(z) : z \in M_2\}$ has the finite intersection property. For each $z_1, z_2 \in M_2$, there exist $\lambda_1, \lambda_2 \in [0, 1]$ such that $z_i = (1 - \lambda_i)x_1 + \lambda_i x_2$ for $i = 1, 2$. Without loss of generality, we may assume that $\lambda_1 \leq \lambda_2$, so that $[x_1, z_1] \subset [x_1, z_2]$. Note that $z_1 \in [x_1, z_2]$, so that $F(z_1) \subset F(x_1) \cup F(z_2)$ by condition (1). Since $F(z_1) \subset F(x_2)$ and $F(x_1) \cap F(x_2) = \emptyset$, so that $F(z_1) \subset F(z_2)$. Therefore $F(z_1) \cap F(z_2) \neq \emptyset$ for each $z_1, z_2 \in M_2$. Next we shall prove that $\cap_{z\in A} F(z) \neq \emptyset$ for each $A \in \mathcal{F}((x_0, x_2])$. Assume that $\cap_{i=1}^n F(x_i) \neq \emptyset$ for each $x_1, x_2, \cdots, x_n \in X$ and $n \geq 2$. Now define a mapping $F_1 : X \to 2^Y$ by $F_1(x) = \cap_{i=1}^{n-1} F(x_i) \cap F(x)$ for each $x \in X$. Then it is clear that F_1 satisfies all hypotheses (1)-(3). By repeating the same arguments above, we have that $F_1(x') \cap F_1(x'') \neq \emptyset$ for each $x', x'' \in (x_0, x_2] = M_2$. Let $x' = x_n$ and $x'' = x_{n+1}$. Then $\cap_{i=1}^{n+1} F(x_i) \neq \emptyset$. Thus the compact family $\{F(x) : x \in (x_0, x_2]\}$ has the finite intersection property. Therefore $\cap_{z\in(x_0,x_2]} F(z) \neq \emptyset$. Taking $y_0 \in \cap_{z\in(x_0,x_2]} F(z)$, then $y_0 \in F(x)$ for each $x \in (x_0, x_2]$. Therefore $(x_0, x_2] \subset F^{-1}(y_0)$. Since $\{x_\alpha\}_{\alpha\in I}$ is a net in M_2 such that $x_\alpha \to x_0 \in M_1$, so that $x_\alpha \in F^{-1}(y_0)$. Note that $F^{-1}(y_0)$ is closed in $[x_1, x_2]$ by (3), we have that $x_0 \in F^{-1}(y_0)$, i.e., $y_0 \in F(x_0)$. Therefore $y_0 \in F(x_0) \cap F(x_2)$, i.e., $F(x_0) \cap F(x_2) \neq \emptyset$ which contradicts that $F(x_0) \cap F(x_2) \subset F(x_1) \cap F(x_2) = \emptyset$. This contradiction shows that $F(x_1) \cap F(x_2) \neq \emptyset$ for each $x_1, x_2 \in X$.

Step 2: by the induction, we shall prove that $\cap_{x\in A} F(x) \neq \emptyset$ for each $A \in \mathcal{F}(X)$. Without loss of generality, we may assume that $\cap_{i=1}^n F(x_i) \neq \emptyset$ for each $x_1, \cdots, x_n \in X$, where $n \in \mathbb{N}$ and $n \geq 2$. Define a mapping $F_2 : X \to 2^Y$ by $F_2(x) = \cap_{i=1}^{n-1} F(x_i) \cap F(x)$ for each $x \in X$. Then $F_2(x) \neq \emptyset$ for each $x \in X$. Moreover it is clear that the mapping F_2 satisfies all hypotheses (1)-(3). Therefore $F_2(x') \cap F_1(x'') \neq \emptyset$ for each $x', x'' \in X$ by the proof of step (1). Let $x' = x_n$ and $x'' = x_{n+1}$. Then $\cap_{i=1}^{n+1} F(x_i) \neq \emptyset$. Therefore the compact family $\{F(x) : x \in X\}$ has the finite intersection property, so that $\cap_{x\in X} F(x) \neq \emptyset$. □

Let C and D be non-empty subsets of two Hausdorff topological spaces X and Y, respectively. Assume that $F : C \times D \to \mathbb{R} \cup \{-\infty, +\infty\}$ is an extended real-valued function. For each $\alpha \in \mathbb{R}$ and $x \in C$, we set $D_\alpha = \{y \in D : F(x, y) \geq \alpha\}$. By following Tuy in [563, p.145-46], we have the following definition.

Definition 3.3.6. The set C is said to have *α-connected structure* (associated with F and D) if

(1) the set $\cap_{a\in A} D_\alpha(a)$ is connected for each $A \in \mathcal{F}(C)$; and

(2) for each $a, b \in C$, there exists a continuous mapping $u : [0, 1] \to C$ such that $u(0) = a$ and $u(1) = b$ and for any λ, μ' and μ in $[0, 1]$, we have

$D_\alpha(u(\lambda)) \subset D_\alpha(u(\mu)) \cup D_\alpha(u(\mu'))$ when $\mu' \leq \lambda \leq u$.

Remark 3.3.7. Theorem 3.3.3 answers Kindler's problem above in the affirmative when the underlying topological space X has the α-connected structure above. When X is not a topology space, Theorem 3.3.5 also shows that Kindler's question above still holds positively when the underlying space X is a non-empty convex subset of a vector space and the topological space Y may not be compact. But we still do not know if the above result remains true in case X is a topological space instead of having α-connected structure in the sense of Definition 3.3.6 above or of being a convex subset of a vector space.

Let Y be a regular space and $F : X \to Y$ be upper semicontinuous with closed values. Then F has a closed graph, which in turn implies that $F^{-1}(y) = \{x \in X : y \in F(x)\}$ is closed in X for each $y \in Y$. As consequences of Theorem 3.3.3, we have the following:

Corollary 3.3.8. Let X and Y be both topological spaces. Suppose that $F : X \to 2^Y$ is a set-valued mapping with non-empty compact values such that:

(1) for each $x, y \in X$, there exists a continuous mapping $u_{x,y} : [0,1] \to X$ with $u_{x,y}(0) = x$, $u_{x,y}(1) = y$ and $F((u_{x,y}(t)) \subset F(u_{x,y}(t_1)) \cup F(u_{x,y}(t_2))$ for each $t \in [t_1, t_2] \subset [0,1]$;

(2) for each $A \in \mathcal{F}(X)$, if the set $\cap_{x\in A} F(x)$ is non-empty, then $\cap_{x\in A} F(x)$ is connected; and

(3) the graph of F is closed in $X \times Y$.

Then $\cap_{x\in X} F(x) \neq \emptyset$.

As condition (3) of Theorem 3.3.3 could not, in general, guarantee that the graph of F is closed, thus Theorem 3.3.3 is a topological intersection theorem for set-valued mappings which may not have closed graphs.

As applications of Theorems 3.3.1 and 3.3.5, we have the following topological fixed point theorems:

Theorem 3.3.9. Let X be a topological space. Suppose that $F : X \to 2^X$ is a set-valued mapping with non-empty compact values such that:

(1) for each $x, y \in X$, there exists a continuous mapping $u : [0,1] \to X$ with $u(0) = x$, $u(1) = y$ and $F((u(t)) \subset F(u(t_1)) \cup F(u(t_2))$ for each $t \in [t_1, t_2] \subset [0,1]$ (resp., the set $\{x \in X : F(x) \subset F(x) \cup F(y)\}$ is connected);

(2) for each $A \in \mathcal{F}(X)$, if the set $\cap_{x\in A} F(x)$ is non-empty, then $\cap_{x\in A} F(x)$ is connected; and

(3) for each $y \in X$, the set $F^{-1}(y)$ is closed (resp., $F^{-1}(y)$ is open) in X.

Then there exists $x_0 \in X$ such that $x_0 \in F(x_0)$.

Proof. By Theorem 3.3.3, $\cap_{x\in X} F(x) \neq \emptyset$. Take any fixed $x_0 \in \cap_{x\in X} F(x)$. Then $x_0 \in F(x_0)$. □

Similarly, as an application of Theorem 3.3.5, we have the following fixed point result.

Theorem 3.3.10. Let X be a convex of a vector space E (which may not be a topological vector space). Suppose that $F : X \to 2^X$ is a set-valued mapping with non-empty compact values such that:

(1) for each $x, y \in X$, $F(z) \subset F(x) \cup F(y)$ for each $z \in [x, y]$, where $[x, y]$ is a line segment;

(2) for each $A \in \mathcal{F}(X)$, if the set $\cap_{x\in A} F(x)$ is non-empty, then $\cap_{x\in A} F(x)$ is connected; and

(3) for each $y \in X$, the set $F^{-1}(y)$ is closed in the line segment $[x_1, x_2]$ for each $x_1, x_2 \in X$.

Then there exists $x_0 \in X$ such that $x_0 \in F(x_0)$.

Let $P := \cap_{x\in X} F(x)$ in Theorem 3.3.5 and Corollary 3.3.8. Then P is, in fact a non-empty fixed point set of the mapping F. We also note that Theorem 3.3.5 and Corollary 3.3.8 improve and unify the corresponding fixed point theorems of Joó [274]-[275] and Proposition 2 of Stachó [499] in several aspects.

As another application of the topological intersection Theorem 3.3.5 above, we have the following topological minimax theorem.

Theorem 3.3.11. Let X be a compact topological space and Y be a non-empty convex subset of a vector space E. Suppose $f : X \times Y \to \mathbb{R} \cup \{-\infty, +\infty\}$ is such that:

(1) the set $\cap_{y\in A}\{x \in X : f(x, y) \geq a\}$ is connected or empty for each $A \in \mathcal{F}(Y)$, where $a := \inf_{y\in Y} \max_{x\in X} f(x, y)$;

(2) for each fixed $y \in Y$, the mapping $x \mapsto f(x, y)$ is upper semicontinuous;

(3) for each fixed $x \in X$, the set $\{y \in Y : f(x, y) < a\}$ is convex in Y; and

(4) for each fixed $x \in Y$, the mapping $y \mapsto f(x, y)$ is segment upper semicontinuous.

Then

$$\max_{x\in X} \inf_{y\in Y} f(x, y) = \inf_{y\in Y} \max_{x\in X} f(x, y).$$

Proof. We define a set-valued mapping $F : Y \to 2^X$ by

$$F(y) = \{x \in X : f(x, y) \geq a\}$$

for each $y \in Y$. Since X is compact, $F(x)$ is non-empty and closed for each $y \in Y$ by (2). Moreover we have:

(i) the set $\cap_{y\in A} F(y)$ is connected or empty for each $A \in \mathcal{F}(Y)$ by (1);

(ii) for each $y_1, y_2 \in Y$, $F(z) \subset F(y_1) \cup F(y_2)$ for each $z \in [y_1, y_2]$ by (3); and

(iii) for each $x \in X$, the set $F^{-1}(x) = \{y \in Y : x \in F(y)\} = \{y \in Y : f(x, y) \geq a\}$ which is closed in the segment $[y_1, y_2]$ for each given $y_1, y_2 \in Y$ by (4).

Therefore F satisfies all hypotheses of Theorem 3.3.5. By Theorem 3.3.5, there exists $x_0 \in X$ such that $x_0 \in \cap_{y \in Y} F(y)$. Thus $f(x_0, y) \geq \inf_{y \in Y} \max_{x \in X} f(x, y)$ for all $y \in Y$, so that

$$\max_{x \in X} \inf_{y \in Y} f(x, y) \geq \inf_{y \in Y} f(x_0, y) \geq \inf_{y \in Y} \max_{x \in X} f(x, y).$$

Note that $\max_{x \in X} \inf_{y \in Y} f(x, y) \leq \inf_{y \in Y} \max_{x \in X} f(x, y)$ holds in general, we must have

$$\max_{x \in X} \inf_{y \in Y} f(x, y) = \inf_{y \in Y} \max_{x \in X} f(x, y).$$

Theorem 3.3.11 improves Theorem 3 of Komornik [328] in the sense that X may not be an interval space. Theorem 3.3.11 also includes the corresponding results of Geraghty and Lin [194] and Example 5.1 of Kindler [305] as special cases. For other extensive study of topological minimax theorems by employing connectedness, we refer to Kindler [305], König [329], Simons [485] and his survey paper [485].

Before we conclude this section, we would like to give the following topological variational inequalities.

Theorem 3.3.12. Let X be a non-empty convex subset of a vector space and Y be a compact topological space. Let (E, C) be a topological Riesz space, where C is a closed cone with $C^o \neq \emptyset$, where C^o denotes the *interior* of C. Suppose $r \in E$ and $f : X \times Y \to E$ satisfies the following conditions:

(1) the set $\{y \in Y : f(x, y) \notin r + C^o\}$ is non-empty and closed in Y for each $x \in X$;

(2) for each $y \in Y$, the set $\{x \in X : f(x, y) \notin r + C^o\}$ is closed in each line segment and is convex in X; and

(3) the set $\cap_{x \in A}\{y \in Y : f(x, y) \notin r + C^o\}$ is connected or empty for each $A \in \mathcal{F}(X)$.

Then there exists $y_0 \in Y$ such that

$$f(x, y_0) \notin r + C^o \text{ for all } x \in X.$$

Proof. We define a mapping $F : X \to 2^Y$ by

$$F(x) = \{y \in Y : f(x, y) \notin r + C^o\}$$

for each $x \in X$. Then it is easy to verify that the mapping F satisfies all hypotheses of Theorem 3.3.5. By Theorem 3.3.5, $\cap_{x \in X} F(x) \neq \emptyset$. Taking $y_0 \in \cap_{x \in X} F(x)$, then we have $f(x, y_0) \notin r + C^o$ for all $x \in X$. □

As an immediate consequence of Theorem 3.3.12, we have the following.

Corollary 3.3.13. Let X be a non-empty convex subset of a vector space and Y be a compact topological space. Suppose $f : X \times Y \to \mathbb{R}$ is a real-valued function such that:

(1) for each $x \in X$, there exists $y \in Y$ such that $f(x, y) \leq 0$;

(2) for each fixed $y \in Y$, $x \mapsto f(x, y)$ is segment lower semicontinuous in X;

(3) for each fixed $x \in X$, $y \mapsto f(x, y)$ is lower semicontinuous in Y;

(4) for each $A \in \mathcal{F}(X)$, the set $\{y \in Y : f(x, y) \leq 0\}$ is connected or empty; and

(5) for each $y \in Y$, the set $\{x \in X : f(x, y) \leq 0\}$ is convex.

Then there exists $y_0 \in Y$ such that

$$f(x, y_0) \leq 0 \text{ for all } x \in X.$$

Before we conclude this section, we would like to mention that how the concept of *connectedness* and related *topological intersection theorems* play important roles in the study of some problems in functional analysis could be found from a recent paper of Ricceri in [437], Simons's paper [485], and references therein.

3.4 Topological Intersection of Two Set-Valued Mappings

It is our purpose in this section to study some sufficient conditions for the non-empty intersection of two set-valued mappings acting between topological spaces. As applications, we first establish some topological minimax inequalities for two functions in which one of them is separately lower (or upper) semicontinuous. As applications, these topological intersection theorems for two set-valued mappings will be used to derive minimax inequalities without separately lower (or upper) semicontinuity in next section.

The idea behind this section is quite simple and the same as last section, i.e., it is, based on the property of connectedness above. This simple idea was first used by Wu [579] in the study of minimax problem. For the convenience of our discussion, we restate the following lemma again which is the Remark 1 of Kindler [305] (see also Lemma 3.2.1).

Lemma 3.4.1. Let X be a topological space, Y a non-empty set and $F : X \to 2^Y$ a correspondence. Then the following are equivalent:

(a) The set $\cap_{y \in B} F^*(y)$ is connected or empty for each $B \in 2^Y$.

(b) for each $x_1, x_2 \in X$, there exists a connected set $C \supset \{x_1, x_2\}$ such that

$$F(x) \subset F(x_1) \cup F(x_2) \text{ for all } x \in C.$$

(c) for each pair $(x_1, x_2) \in X \times X$, the set $\{x \in X : F(x) \subset F(x_1) \cup F(x_2)\}$ is connected.

We also need the following simple result; for completeness, we include its proof as follows.

Lemma 3.4.2. Let X and Y be two topological spaces and A a non-empty subset of Y. Suppose the mapping $F : X \to 2^Y$ has open inverse values. Then $F_A^{-1}(y)$ is open in X for $y \in Y$.

Proof. For each $y \in Y$, let $x \in F_A^{-1}(y)$. Then $y \in F_A(x) = F(x) \cap A$. Let $O_x = F^{-1}(y)$. Note that for each $z \in O_x$, we have $y \in F(z)$ and $y \in A$, so that $y \in F_A(z)$. Hence $x \in O_z \subset F_A^{-1}(y)$. $\square$.

Now we have the following sufficient conditions which guarantee the existence of 2-points intersection for two set-valued mappings in topological spaces.

Theorem 3.4.3. Let X and Y be two topological spaces and $F, G : X \to 2^Y$ be both set-valued mappings with non-empty values. Suppose the following conditions are satisfied:

(i) for each $x \in X$, $F(x) \subset G(x)$ and $G(x)$ is closed (respectively, open) and $F(x)$ is connected;

(ii) the set $\{x \in X : G(x) \subset G(x_1) \cup G(x_2)\}$ is connected for each given $x_1, x_2 \in X$; and

(iii) the set $F^{-1}(y)$ is open for each $y \in Y$.

Then the family $\{G(x) : x \in X\}$ has the 2-points intersection property, i.e., $G(x_1) \cap G(x_2) \neq \emptyset$ for each $x_1, x_2 \in X$.

Proof. If not, suppose there exists $x_1, x_2 \in X$ such that $G(x_1) \cap G(x_2) = \emptyset$. Let $C = \{x \in X : G(x) \subset G(x_1) \cup G(x_2)\}$. By (ii), C is connected in X. For $i = 1, 2$, let $M_i^F = \{x \in C : F(x) \subset G(x_i)\}$ and $M_i^G = \{x \in C : G(x) \subset G(x_i)\}$. Then $\emptyset \neq M_i^G \subset M_i^F$ for each $i = 1, 2$. We now show that $C = M_1^F \cup M_2^F$. Note that for each $x \in C$, $F(x) \subset G(x) \subset G(x_1) \cup G(x_2)$, $F(x)$ is non-empty connected and $G(x_1)$ and $G(x_2)$ are both disjoint closed (respectively, open) in Y, so that $F(x) \subset G(x_1)$ or $F(x) \subset G(x_2)$ by the connectedness fact above. Therefore for each $x \in C$, we have $F(x) \subset G(x_1)$ or $F(x) \subset G(x_2)$. Thus $C = M_1^F \cup M_2^F = M_1^G \cup M_2^G$.

Next we claim that both M_i^F for $i = 1, 2$ are closed in C. If not, without loss of generality, we may assume that M_2^F is not closed in C. Then there exist $x_0 \in M_1^F \setminus M_2^F$ and a convergent net $\{x_\alpha\}_{\alpha \in \Gamma}$ in M_2^F such that $x_\alpha \to x_0$. Note that $x_0 \in M_1^F$ and $x_\alpha \in M_2^F$. Take any fixed $y_0 \in F(x_0)$, then $y_0 \notin G(x_\alpha)$ for each $\alpha \in \Gamma$. Thus $x_\alpha \in X \setminus F^{-1}(y_0)$. Since the set $X \setminus F^{-1}(y_0)$ is closed in X by (iii) and $x_\alpha \to x_0$, we have that $x_0 \in X \setminus F^{-1}(y_0)$, i.e., $y_0 \notin F(x_0)$ which is impossible by our choice of that $y_0 \in F(x_0)$. Therefore we have proved that $M_1^F \cup M_2^F = C$, where M_1^F and M_2^F are both non-empty and disjoint closed in

X, and C is non-empty connected, which is impossible by the connectedness fact above. This contradiction shows that $G(x_1) \cap G(x_2) \neq \emptyset$. □

As an application of Theorem 3.4.3, we have the following topological finite intersection theorem for two set-valued mappings in topological spaces.

Theorem 3.4.4. Let X and Y be two topological spaces and $F, G : X \to 2^Y$ be both set-valued mappings with non-empty values. Suppose the following conditions are satisfied:

(1) for each $x \in X$, $F(x) \subset G(x)$ and $\cap_{x \in A} G(x)$ is empty or closed (respectively, open) and the set $\cap_{x \in A} F(x)$ is empty or connected for each $A \in \mathcal{F}(X)$;

(2) $\cap_{y \in B} G^*(y)$ is connected or empty for each $B \in 2^Y$ (or equivalently to say, the set $\{x \in X : G(x) \subset G(x_1) \cup G(x_2)\}$ is empty or connected for each $x_1, x_2 \in X$ by Lemma 3.4.1);

(3) the set $F^{-1}(y)$ is open for each $y \in Y$; and

(4) for each $A \in \mathcal{F}(X)$, if $\cap_{x \in A} G(x) \neq \emptyset$ implies that $\cap_{x \in A} F(x) \neq \emptyset$.

Then the family $\{G(x) : x \in X\}$ has the finite intersection property, i.e., $\cap_{x \in A} G(x) \neq \emptyset$ for each $A \in \mathcal{F}(X)$.

Proof. We shall prove the conclusion by induction. By Theorem 3.4.3, without loss of generality, we may assume that $\cap_{i=1}^{n} G(x_i) \neq \emptyset$ for each $x_1, x_2, \cdots, x_n \in X$, where $n \geq 2$. Now we shall prove that $\cap_{i=1}^{n+1} G(x_i) \neq \emptyset$ for any $n+1$ elements $x_1, x_2, \cdots, x_{n+1}$ of X. We define $F', G' : X \to 2^Y$ by

$$F'(x) := \cap_{i=1}^{n-1} F(x_i) \cap F(x)$$

and

$$G'(x) := \cap_{i=1}^{n-1} G(x_i) \cap G(x)$$

for each $x \in X$. Then for each $x \in X$, we have

(i) $F'(x)$ is non-empty by (4) and our induction hypothesis above and $F'(x) \subset G'(x)$;

(ii) $G'(x)$ is closed (respectively, open) by (1);

(iii) the set $\{x \in X : G'(x) \subset G'(x_1) \cup G'(x_2)\}$ is connected by (2) and Lemma 3.4.1; and

(iv) $(F')^{-1}(y)$ is open in X for each $y \in Y$ by (3) and Lemma 3.4.2.

Thus F' and G' satisfy all hypotheses of Theorem 3.4.3. By Theorem 3.4.3, $G'(x') \cap G'(x'') \neq \emptyset$ for each $x', x'' \in X$. Let $x' = x_n$ and $x'' = x_{n+1}$. Then $G'(x_n) \cap G'(x_{n+1}) \neq \emptyset$, i.e., $\cap_{i=1}^{n+1} G(x_i) \neq \emptyset$. Thus the family $\{G(x) : x \in X\}$ has the finite intersection property. □

We also have the following result.

Theorem 3.4.5. Let X and Y be two topological spaces and $F, G : X \to 2^Y$ be both set-valued mappings with non-empty values. Suppose the following conditions are satisfied:

(i) for each $x \in X$, $F(x) \subset G(x)$ and $\cap_{x\in A}G(x)$ is empty or closed (respectively, open) for each $A \in \mathcal{F}(X)$;

(ii) the set $\cap_{x\in A}F(x)$ is empty or connected for each $A \in \mathcal{F}(X)$;

(iii) the set $\{x \in X : F(x) \subset F(x_1) \cup F(x_2)\}$ is connected for each $x_1, x_2 \in X$ (equivalently to say, set $\cap_{y\in B}F^*(y)$ is connected or empty for each $B \in 2^Y$ by Lemma 3.4.1); and

(iv) the set $F^{-1}(y)$ is open for each $y \in Y$;

(v) $\cap_{x\in A}F(x) \neq \emptyset$ when $\cap_{x\in A}G(x) \neq \emptyset$ for each $A \in \mathcal{F}(X)$.

Then the family $\{G(x) : x \in X\}$ has the finite intersection property, i.e., $\cap_{x\in A}G(x) \neq \emptyset$ for each $A \in \mathcal{F}(X)$.

Proof. We shall first prove that $G(x_1) \cap G(x_2) \neq \emptyset$ for each $x_1, x_2 \in X$. If not, suppose there exist $x_1, x_2 \in X$ such that $G(x_1) \cap G(x_2) = \emptyset$. Then we shall derive a contradiction. Let $C := \{x \in X : F(x) \subset F(x_1) \cup F(x_2)\}$. By (iii), C is connected in X. or $i = 1, 2$, let $M_i = \{x \in C : F(x) \subset F(x_i)\}$. Then both M_1 and M_2 are non-empty. Note that for each $x \in C$, $F(x)$ is connected and $F(x) \subset F(x_1) \cup F(x_2) \subset G(x_1) \cup G(x_2)$, and $G(x_1)$ and $G(x_2)$ are non-empty closed (respectively, open) and therefore $F(x) \subset G(x_1)$ or $F(x) \subset G(x_2)$. Without loss of generality, we may assume that $F(x) \subset G(x_1)$. Now we claim that $F(x) \subset F(x_1)$. If not, note that $F(x) \subset F(x_1) \cup F(x_2)$. Then there exists $y_0 \in F(x)$ such that $y_0 \in F(x_2) \subset G(x_2)$. Therefore $y_0 \in F(x) \cap G(x_2) \subset G(x_1) \cap G(x_2)$, which contradicts our assumption that $G(x_1) \cap G(x_2) = \emptyset$. Thus $F(x) \subset F(x_1)$. Therefore we have proved that $F(x) \subset F(x_1)$ or $F(x) \subset F(x_2)$, so that $C = M_1 \cup M_2$. Next we claim that both M_1 and M_2 are closed in C. If not, we may assume that M_2 is not closed in C. Then there exist $x_0 \in M_1 \setminus M_2$ and a net $\{x_\alpha\}_{\alpha\in\Gamma}$ in M_2 such that $x_\alpha \to x_0$. Note that since $x_0 \in M_1$ and $x_\alpha \in M_2$. Take any fixed $y_0 \in F(x_0)$, then $y_0 \notin F(x_\alpha)$ for each $\alpha \in \Gamma$. Thus $x_\alpha \in X \setminus F^{-1}(y_0)$. Note that the set $X \setminus F^{-1}(y_0)$ is closed in X by (iv) and $x_\alpha \to x_0$, we have that $x_0 \in X \setminus f^{-1}(y_0)$, i.e., $y_0 \notin F(x_0)$ which is impossible. Thus both M_1 and M_2 must be closed in C. Therefore we have that $M_1 \cup M_2 = C$, where M_1 and M_2 are both non-empty and closed in C and C is non-empty connected, which is a contradiction, by the connectedness fact. This contradiction shows $G(x_1) \cap G(x_2) \neq \emptyset$.

Secondly, we shall prove by induction that the family $\{G(x) : x \in X\}$ has the finite intersection property. We may assume that $\cap_{i=1}^{n}G(x_i) \neq \emptyset$ for each $x_1, x_2, \cdots, x_n \in X$, where $n \geq 2$. Now we prove that $\cap_{i=1}^{n+1}G(x_i) \neq \emptyset$ for any $n+1$ elements $x_1, x_2, \cdots, x_{n+1}$ of X. We define $F', G' : X \to 2^Y$ by

$$F'(x) := \cap_{i=1}^{n-1}F(x_i) \cap F(x)$$

and

$$G'(x) := \cap_{i=1}^{n-1}G(x_i) \cap G(x)$$

for each $x \in X$. Then for each $x \in X$, we have

(1) $F'(x) \subset G'(x)$ and $F'(x)$ is non-empty by (v) and our assumption;

(2) $G'(x)$ is closed (respectively, open) by (i);

(3) the set $\{x \in X : G'(x) \subset G'(x_1) \cup G'(x_2)\}$ is connected by (iii) and Lemma 3.4.1; and

(4) $(F')^{-1}(y)$ is open in X for each $y \in Y$ by (iv) and Lemma 3.4.2.

Thus F' and G' satisfy all hypotheses of Theorem 3.4.5. By the proof above again, $G'(x') \cap G'(x'') \neq \emptyset$ for each $x', x'' \in X$. Let $x' = x_n$ and $x'' = x_{n+1}$. Then $G'(x_n) \cap G'(x_{n+1}) \neq \emptyset$, i.e., $\cap_{i=1}^{n+1} G(x_i) \neq \emptyset$. Thus the family $\{G(x) : x \in X\}$ has the finite intersection property. □

Remark 3.4.6. Recently, Kindler in [305]-[308] has established a number of topological intersection theorems which involve only one set-valued mapping (see also König [329] and references therein), hence their results are not comparable with our Theorems 3.4.4 and 3.4.5. Let X and Y be two topological spaces and $F : X \to 2^Y$ a set-valued mapping. Chang et al. introduced the following Condition (P) for the mapping F in [66].

The Condition (P): For any $x_0 \in Y$ and any net $\{x_\alpha\}_{\alpha \in \Gamma}$ of X with $x_\alpha \to x_0$, if $y_0 \notin F(x_\alpha)$ for all $\alpha \in \Gamma$, then $y_0 \notin F(x_0)$.

We wish to point out that condition (P) of F is, in fact, equivalent to saying that the mapping F has open inverse values (i.e., the set $F^{-1}(y)$ is open in X for each $y \in Y$). Hence our Theorems 3.4.4 and 3.4.5 include Theorem 1 of Chang et al. [66]-[67] as a special case.

The following example from Thompson and Yuan [551] shows that the conclusion of Theorems 3.4.4 (resp., Theorem 3.4.5) is not true if we withdraw the condition (4) (respectively, (v)):

Example 3.4.7. Let $X = [0, 2\pi)$ and $Y = \{e^{i\theta} : 0 \leq \theta < 2\pi\}$. Define $F, G : X \to 2^Y$ by

$$F(\theta) = \{e^{i\psi} : \theta - 1 < \psi < \theta + 1\}$$

for each $\theta \in X$ and

$$G(\theta) = \begin{cases} \{e^{i\psi} : \quad -1 \leq \psi \leq \frac{2\pi}{3} + 1\} & \text{if } \theta \in [0, \frac{2\pi}{3}]; \\ \{e^{i\psi} : \frac{2\pi}{3} - 1 \leq \psi \leq \frac{4\pi}{3} + 1\} & \text{if } \theta \in (\frac{2\pi}{3}, \frac{4\pi}{3}]; \\ \{e^{i\psi} : \quad \frac{4\pi}{3} - 1 \leq \psi \leq 1\} & \text{if } \theta \in (\frac{4\pi}{3}, 2\pi) \end{cases}$$

for each $\theta \in X$. Then it is easy to verify that

(1): $F(\theta) \subset G(\theta)$ and $\cap_{\theta \in A} G(\theta)$ is closed and connected or empty for each $A \in \mathcal{F}(X)$;

(2): $F^{-1}(e^{i\psi}) = \{\theta : \psi - 1 < \theta < \psi + 1\}$, which is open in X for each $e^{i\psi} \in Y$; and

(3): the set $\{\theta \in X : G(\theta) \subset G(\theta_1) \cup G(\theta_2)\}$ is connected.

Note that for each $\theta_1, \theta_2 \in X$, $G(\theta_1) \cap G(\theta_2) \neq \emptyset$, but $F(\theta_i) \cap F(\theta_j) = \emptyset$ for each $\theta_i \in [0, \frac{2\pi}{3}]$ and $\theta_j \in (\frac{2\pi}{3}, \frac{4\pi}{3}]$ or $\theta_j \in (\frac{3\pi}{3}, 2\pi)$, i.e., condition (4) (respectively, (v)) of Theorem 3.4.4 (resp., Theorem 3.4.5) does not hold. Thus the family $\{G(\theta) : \theta \in X\}$ does not have the finite intersection property, e.g., $G(\frac{\pi}{6}) \cap G(\frac{2\pi}{3} + \frac{\pi}{6})) \cap G(\frac{4\pi}{3} + \frac{\pi}{6}) = \emptyset$.

Letting $F = G$ in Theorem 3.4.4, we have the following corollary which includes Theorem 1 of Chang et al. [66]-[67] as special cases.

Corollary 3.4.8. Let X and Y be two topological spaces and $G : X \to 2^Y$ be a set-valued mapping with non-empty values. Suppose the following conditions are satisfied:

(i) the set $\cap_{x \in A} G(x)$ is closed (respectively, open) connected or empty for each $A \in \mathcal{F}(X)$;

(ii) the set $\{x \in X : G(x) \subset G(x_1) \cup G(x_2)\}$ is empty or connected for each $x_1, x_2 \in X$ (or equivalently to say, $\cap_{y \in B} G^*(y)$ is connected or empty for each $B \in 2^Y$); and

(iii) the set $G^{-1}(y)$ is open in X for each $y \in Y$.

Then the family $\{G(x) : x \in X\}$ has the finite intersection property.

3.5 Topological Minimax Inequalities

In this section, as applications of topological intersections proved in the last three sections, several topological minimax inequalities will be established.

Let X and Y be topological spaces and $a : X \times Y \to \mathbb{R} \cup \{-\infty, +\infty\}$ be a function. We set

$$a^* = \inf_{y \in Y} \sup_{x \in X} a(x, y), \qquad a_* = \sup_{x \in X} \inf_{y \in Y} a(x, y)$$

and

$$\hat{a}^* = \sup_{A \in \mathcal{F}(x)} \inf_{y \in Y} \sup_{x \in A} a(x, y).$$

Then it is clear that $a_* \leq \hat{a}^* \leq a^*$. We say that a fulfills

(1): a minimax relation (in short, MM) if $a_* = a^*$;

(2): the preminimax relation (in short, PMM) if $\hat{a}^* = a_*$; and

(3): the minimum minimax relation (in short, MMM) if $\sup_{x \in X} a(x, \hat{y}) = a_*$ for some $\hat{y} \in Y$.

In this section, as applications of the topological intersection theorems above, we shall study topological minimax inequalities for "two-function versions" of the minimax theorems which are generalizations of Sion's minimax theorem [490] (see also Fan [158]).

Theorem 3.5.1. Let X and Y be topological spaces. Let $a, b : X \times Y \to \mathbb{R} \cup \{-\infty, +\infty\}$ be two functions such that for each $\lambda > b_*$:

(1) $a(x, y) \le b(x, y)$ for each $(x, y) \in X \times Y$;

(2) for each fixed $y \in Y$, $x \mapsto b(x, y)$ is upper semicontinuous and for each fixed $x \in X$, $y \mapsto b(x, y)$ is upper semicontinuous;

(3) for each fixed $x \in X$, $y \mapsto a(x, y)$ is lower semicontinuous;

(4) the set $\cap_{x \in A}\{y \in Y : b(x, y) < \lambda\}$ is empty or connected for each $A \in \mathcal{F}(X)$;

(5) the set $\cap_{y \in B}\{x \in X : b(x, y) \ge \lambda\}$ is empty or connected for each $B \in 2^Y$; and

(6) there exist $\lambda_0 > b_*$ and $A_0 \in \mathcal{F}(X)$ such that the set $\cap_{x \in A_0}\{y \in Y : a(x, y) \le \lambda_0\}$ is empty or compact.

Then we have

$$\inf_{y \in Y} \sup_{x \in X} a(x, y) \le \sup_{x \in X} \inf_{y \in Y} b(x, y).$$

Proof. For each $\lambda > b_*$, we shall prove that the family $\{y \in Y : a(x, y) \le \lambda\}_{x \in X}$ has the finite intersection property. We define $F_\lambda, G_\lambda : X \to 2^Y$ by

$$F_\lambda(x) := \{y \in Y : b(x, y) < \lambda\} \text{ and } G_\lambda(x) := \{y \in Y : a(x, y) \le \lambda\}$$

for each $x \in X$. Then we have that

(i) $F_\lambda(x) \neq \emptyset$ and $F_\lambda(x) \subset G_\lambda(x)$ by (1);

(ii) $F_\lambda(x)$ is open for each $x \in X$ and the set $F_\lambda^{-1}(y)$ is open in X for each $y \in Y$ by (2);

(iii) the set $\cap_{x \in A} F_\lambda(x)$ is empty or connected for each $A \in \mathcal{F}(X)$ by (4); and

(iv) the set $\{x \in X : F_\lambda(x) \subset F_\lambda(x_1) \cup F_\lambda(x_2)\}$ is empty or connected for each $x_1, x_2 \in X$ by (5) and Lemma 3.4.1.

Thus F_λ satisfies all hypotheses of Corollary 3.4.7. By Corollary 3.4.7, the family $\{F_\lambda(x) : x \in X\}$ has the finite intersection property, hence so does the family $\{G_\lambda(x) : x \in X\}$. Therefore for each $\lambda > b_*$, $\{y \in Y : a(x, y) \le \lambda\}_{x \in X}$ has the finite intersection property. Note that for each $\lambda \in (b_*, \lambda_0)$, $\{y \in Y : a(x, y) \le \lambda\} \subset \{y \in Y : a(x, y) \le \lambda_0\}$ for each $x \in X$ and the set $\cap_{x \in A_0}\{y \in Y : a(x, y) \le \lambda_0\}$ is non-empty and compact by (6), so that $\cap_{x \in X}\{y \in Y : a(x, y) \le \lambda\} \neq \emptyset$. For each $\lambda \in (b_*, \lambda_0)$, taking any $y_0 \in \cap_{x \in X}\{y \in Y : a(x, y) \le \lambda\}$, then $\inf_{y \in Y} \sup_{x \in X} a(x, y) \le \sup_{x \in X} a(x, y_0) \le \lambda$. Hence for each $\lambda \in (b_*, \lambda_0)$, $\inf_{y \in Y} \sup_{x \in X} a(x, y) \le \lambda$. Therefore we must have the following inequality,

$$\inf_{y \in Y} \sup_{x \in X} a(x, y) \le b_* = \sup_{x \in X} \inf_{Y \in Y} b(x, y). \qquad \square$$

As an immediate consequence of Theorem 3.5.1, we have the following corollary which includes and improves the corresponding results of Bardaro and Ceppitelli [24], Chang et al. [66]-[67], Fan [158] and Geraghty and Lin [194].

Corollary 3.5.2. Let X be a topological space and Y a compact topological space. Let $a, b : X \times Y \to \mathbb{R} \cup \{-\infty, +\infty\}$ be two functions such that for each $\lambda > b_*$:

(1) $a(x,y) \leq b(x,y)$ for each $(x,y) \in X \times Y$;

(2) for each fixed $y \in Y$, $x \mapsto b(x,y)$ is upper semicontinuous and for each fixed $x \in X$, $y \mapsto b(x,y)$ is upper semicontinuous;

(3) for each fixed $y \in Y$, $x \mapsto a(x,y)$ is lower semicontinuous; and

(4) the set $\cap_{x\in A}\{y \in Y : b(x,y) < \lambda\}$ is empty or connected for each $A \in \mathcal{F}(X)$;

(5) the set $\cap_{y\in B}\{x \in X : b(x,y) \geq \lambda\}$ is empty or connected for each $B \in 2^Y$.

Then

$$\inf_{y\in Y} \sup_{x\in X} a(x,y) \leq \sup_{x\in X} \inf_{y\in Y} b(x,y)$$

Proof. Since Y is compact, condition (5) of Theorem 3.5.1 is automatically satisfied. Hence the conclusion follows from Theorem 3.5.1. □.

Let $a(x,y) = b(x,y)$ for each $(x,y) \in X \times Y$ in Corollary 3.5.2. Then Corollary 3.5.2 implies the MMM result, i.e., there exists $y_0 \in Y$ such that $\sup_{x\in X} a(x,y_0) = \sup_{x\in X} \inf_{y\in Y} a(x,y)$.

By a similar proof to that of Theorem 3.5.1, we also have the following minimax inequality.

Theorem 3.5.3. Let X and Y be topological spaces. Let $a, b : X \times Y \to \mathbb{R} \cup \{-\infty, +\infty\}$ be two functions such that for each $\lambda < a^*$:

(1) $a(x,y) \leq b(x,y)$ for each $(x,y) \in X \times Y$;

(2) for each fixed $y \in Y$, $x \mapsto a(x,y)$ is lower semicontinuous and for each fixed $x \in X$, $y \mapsto a(x,y)$ is lower semicontinuous;

(3) for each fixed $y \in Y$, $x \mapsto b(x,y)$ is upper semicontinuous;

(4) the set $\cap_{y\in A}\{x \in X : a(x,y) > \lambda\}$ is empty or connected for each $A \in \mathcal{F}(Y)$;

(5) the set $\cap_{x\in B}\{y \in Y : a(x,y) \leq \lambda\}$ is empty or connected for each $B \in 2^X$; and

(6) there exist $\lambda_0 < a^*$ and $A_0 \in \mathcal{F}(Y)$ such that the set $\cap_{y\in A_0}\{x \in X : b(x,y) \geq \lambda_0\}$ is empty or compact.

Then we have

$$\inf_{y\in Y} \sup_{x\in X} a(x,y) \leq \sup_{x\in X} \inf_{y\in Y} b(x,y).$$

Proof. The idea is similar to that in Theorem 3.5.1. For each $\lambda < a^*$, we wish to show that the family $\{x \in X : b(x,y) \geq \lambda\}_{y \in Y}$ has the finite intersection property. Now define $F_\lambda, G_\lambda : Y \to 2^X$ by

$$F_\lambda(y) := \{x \in X : a(x,y) > \lambda\} \text{ and } G_\lambda(y) := \{x \in X : b(x,y) \geq \lambda\}$$

for each $y \in Y$. Then it is easy to verify that

(i) $F_\lambda(y) \neq \emptyset$ and $F_\lambda(y) \subset G_\lambda(y)$ by (1);

(ii) $F_\lambda(y)$ is open for each $y \in Y$ and the set $F_\lambda^{-1}(x)$ is open in Y for each $x \in X$ by (2);

(iii) the set $\cap_{y \in A} F_\lambda(y)$ is empty or connected for each $A \in \mathcal{F}(Y)$ by (4); and

(iv) the set $\{y \in Y : F_\lambda(y) \subset F_\lambda(y_1) \cup F_\lambda(y_2)\}$ is empty or connected by (5) and Lemma 3.4.1 for each $y_1, y_2 \in Y$.

Thus F_λ satisfies all hypotheses of Corollary 3.4.7 (by exchanging X and Y). By Corollary 3.4.7, the family $\{F_\lambda(y) : y \in Y\}$ has the finite intersection property, hence so does the family $\{G_\lambda(y) : y \in Y\}$. Hence for each $\lambda < a^*$, $\{x \in X : a(x,y) > \lambda\}_{y \in Y}$ has the finite intersection property. Note that for each $\lambda \in (\lambda_0, a^*)$, $\{x \in X : a(x,y) > \lambda\} \subset \{x \in X : b(x,y) \geq \lambda_0\}$ for each $y \in Y$ and the set $\cap_{y \in A_0}\{x \in X : b(x,y) \geq \lambda_0\}$ is non-empty and compact by (6), so that $\cap_{y \in Y}\{x \in X : b(x,y) \geq \lambda\} \neq \emptyset$. For each $\lambda \in (\lambda_0, a^*)$, taking any $x_0 \in \cap_{y \in Y}\{x \in X : b(x,y) \geq \lambda\}$. Then $\sup_{x \in X} \inf_{y \in Y} b(x,y) \geq \inf_{y \in Y} b(x_0, y) \geq \lambda$. Hence for each $\lambda \in (\lambda_0, a^*)$, $\sup_{x \in X} \inf_{y \in Y} b(x,y) \geq \lambda$. Therefore we must have the following inequality,

$$\sup_{x \in X} \inf_{y \in Y} b(x,y) \geq a^* = \inf_{y \in Y} \sup_{x \in X} a(x,y). \qquad \square$$

By Lemma 3.4.1 and Theorem 3.5.3, it is not difficult to derive the following corollary which includes Theorem 3 of Chang [67] as a special case.

Corollary 3.5.4. Let X and Y be topological spaces. Let $a, b : X \times Y \to \mathbb{R} \cup \{-\infty, +\infty\}$ be two functions such that:

(1) $a(x,y) \leq b(x,y)$ for each $(x,y) \in X \times Y$;

(2) for each fixed $y \in Y$, $x \mapsto a(x,y)$ is lower semicontinuous and for each fixed $x \in X$, $y \mapsto a(x,y)$ is lower semicontinuous;

(3) for each fixed $y \in Y$, $x \mapsto b(x,y)$ is upper semicontinuous;

(4) for each $\lambda < a^*$, the set $\cap_{y \in A}\{x \in X : a(x,y) > \lambda\}$ is empty or connected for each $A \in \mathcal{F}(Y)$;

(5) for each $y_1, y_2 \in Y$, there exists a connected subset C_{y_1,y_2} of Y with $\{y_1, y_2\} \subset C_{\{y_1,y_2\}}$ such that

$$a(x,y) \leq a(x,y_1) \vee a(x,y_2) \text{ for all } (x,y) \in X \times C_{\{y_1,y_2\}},$$

where "$\vee$" stands for "maximum"; and

(6) there exist $\lambda_0 < a^*$ and $A_0 \in \mathcal{F}(Y)$ such that the set $\cap_{y \in A_0}\{x \in X : b(x,y) \geq \lambda_0\}$ is empty or compact.

Then

$$\inf_{y\in Y}\sup_{x\in X} a(x,y) \leq \sup_{x\in X}\inf_{y\in Y} b(x,y).$$

By Lemma 5 of Geraghty and Lin [194], it is clear that Corollary 3.5.4 also includes Corollary 2 of Chang et al. [67] as a special case. We would like to point out that the proof of Theorems 3.5.1 and 3.5.3, actually produce the following PMM theorems:

Theorem 3.5.5. Let X and Y be topological spaces. Let $a, b : X \times Y \to \mathbb{R} \cup \{-\infty, +\infty\}$ be two functions such that for each $\lambda > b_*$:

(1) $a(x,y) \leq b(x,y)$ for each $(x,y) \in X \times Y$;

(2) for each fixed $y \in Y$, $x \mapsto b(x,y)$ is upper semicontinuous and for each fixed $x \in X$, $y \mapsto b(x,y)$ is upper semicontinuous;

(3) the set $\cap_{y\in B}\{x \in X : b(x,y) \geq \lambda\}$ is empty or connected for each $B \in 2^Y$; and

(4) the set $\cap_{x\in A}\{y \in Y : b(x,y) < \lambda\}$ is empty or connected for each $A \in \mathcal{F}(X)$.

Then we have

$$\sup_{A\in\mathcal{F}(X)}\inf_{y\in Y}\sup_{x\in A} a(x,y) \leq \sup_{x\in X}\inf_{y\in Y} b(x,y) \leq \sup_{A\in\mathcal{F}(X)}\inf_{y\in Y}\sup_{x\in A} b(x,y).$$

Proof. By the proof of Theorem 3.5.1, for each $\lambda > b_*$, the family $\{y \in Y : b(x,y) < \lambda\}_{x\in X}$ has the finite intersection property, and hence so does the family $\{y \in Y : a(x,y) < \lambda\}_{x\in X}$. Thus for each $A \in \mathcal{F}(X)$, taking any fixed $y_0 \in \cap_{x\in A}\{y \in Y : a(x,y) < \lambda\}$, we have $\inf_{y\in Y}\sup_{x\in A} a(x,y) \leq \sup_{x\in A} a(x,y_0) < \lambda$. Therefore $\sup_{A\in\mathcal{F}(X)}\inf_{y\in Y}\sup_{x\in A} a(x,y) \leq \lambda$ for each $\lambda > b_*$, so that

$$\sup_{A\in\mathcal{F}(X)}\inf_{y\in Y}\sup_{x\in A} a(x,y) \leq b_* = \sup_{x\in X}\inf_{y\in Y} b(x,y).$$

Note that $\sup_{x\in X}\inf_{y\in Y} b(x,y) \leq \sup_{A\in\mathcal{F}(X)}\inf_{y\in Y}\sup_{x\in A} b(x,y)$. Then the conclusion follows. □

Theorem 3.5.6. Let X and Y be topological spaces. Let $a, b : X \times Y \to \mathbb{R} \cup \{-\infty, +\infty\}$ be two functions such that for each $\lambda < a^*$:

(1) $a(x,y) \leq b(x,y)$ for each $(x,y) \in X \times Y$;

(2) for each fixed $y \in Y$, $x \mapsto a(x,y)$ is lower semicontinuous and for each fixed $x \in X$, $y \mapsto a(x,y)$ is lower semicontinuous;

(3) the set $\cap_{x\in B}\{y \in Y : a(x,y) \leq \lambda\}$ is empty or connected for each $B \in 2^X$; and

(4) the set $\cap_{y\in A}\{x \in X : a(x,y) > \lambda\}$ is empty or connected for each $A \in \mathcal{F}(Y)$.

Then we have

$$\sup_{A\in\mathcal{F}(X)}\inf_{y\in Y}\sup_{x\in A}a(x,y)\leq\inf_{y\in Y}\sup_{x\in X}a(x,y)\leq\inf_{A\in\mathcal{F}(Y)}\sup_{x\in X}\inf_{y\in A}b(x,y).$$

Proof. From the proof of Theorem 3.5.3, for each $\lambda < a^*$, the family $\{x \in X : b(x,y) \geq \lambda\}_{y\in Y}$ has the finite intersection property. For each $A \in \mathcal{F}(Y)$, there exists $x_0 \in \cap_{y\in A}\{x \in X : b(x,y) \geq \lambda\}$ such that $\sup_{x\in X}\inf_{y\in A} b(x,y) \geq \inf_{y\in Y} b(x_0,y) \geq \lambda$. Therefore $\inf_{A\in\mathcal{F}(Y)}\sup_{x\in X}\inf_{y\in A} b(x,y) \geq \lambda$ for each $\lambda > a^*$. Hence

$$\inf_{A\in\mathcal{F}(Y)}\sup_{x\in X}\inf_{y\in A}b(x,y)\geq a^*.$$

Note that

$$\sup_{A\in\mathcal{F}(X)}\inf_{y\in Y}\sup_{x\in A}a(x,y)\leq\inf_{y\in Y}\sup_{x\in X}a(x,y).$$

The conclusion follows. □

Before we conclude this section, we recall the following result which is Consequence 2.4 of König [329] (compare also Lemma 5 of Geraghty and Lin [194]).

Lemma 3.5.7. Let X be a non-empty set, Y a topological space, $f : X \times Y \to \mathbb{R}$ a function and I a non-empty open interval of $\mathbb{R}$. Consider the following properties:

(a) for each $\lambda \in I$, the set $\cap_{x\in A}\{y \in Y : f(x,y) \leq \lambda\}$ is empty or connected for each $A \in \mathcal{F}(X)$; and

(b) for each $\lambda \in I$, the set $\cap_{x\in A}\{y \in Y : f(x,y) < \lambda\}$ is empty or connected for each $A \in \mathcal{F}(X)$.

Then we have that

(1): the property (a) implies property (b) is always true; and

(2): if Y is a compact Hausdorff topological space and for each fixed $x \in X$, $y \mapsto f(x,y)$ is lower semicontinuous, then property (b) implies property (a) (i.e., the property (a) is equivalent to (b) in this case).

Remark 3.5.8. By Lemma 3.5.7, it is clear all conclusions from Theorems and Corollaries 3.4.4 to 3.5.6 still hold if the " $<$" (respectively, " $>$") is replaced by " $\leq$" (or " $\geq$") in each condition 4.

Before we conclude this section, we shall study minimax inequalities for functions which are not separately semicontinuous. So far, as applications of topological intersection theorems for one set-valued mapping, we have given some minimax inequalities of two functions in which one of them is separately upper or lower semicontinuous. In this section, as applications of our topological intersection theorems for two set-valued mappings, we shall also give some minimax inequalities for two functions without separately semicontinuous conditions but we need some other assumption (i.e., condition (5) in both Theorems 3.5.9 and 3.5.10).

As an application of Theorem 3.4.4, we have the following result.

Theorem 3.5.9. Let X and Y be topological spaces. Let $a, b : X \times Y \to \mathbb{R} \cup \{-\infty, +\infty\}$ be two functions such that for each $\lambda > b_*$:

(1) $a(x,y) \leq b(x,y)$ for each $(x,y) \in X \times Y$;

(2) for each fixed $y \in Y$, $x \mapsto b(x,y)$ is upper semicontinuous and for each fixed $x \in X$, $y \mapsto a(x,y)$ is lower semicontinuous;

(3) the set $\cap_{x\in A}\{y \in Y : b(x,y) < \lambda\}$ is empty or connected for each $A \in \mathcal{F}(X)$;

(4) the set $\cap_{y\in B}\{x \in X : a(x,y) > \lambda\}$ is empty or connected for each $B \in 2^Y$;

(5) for each $A \in \mathcal{F}(X)$, if $\cap_{x\in A}\{y \in Y : a(x,y) \leq \lambda\} \neq \emptyset$, then $\cap_{x\in A}\{y \in Y : b(x,y) < \lambda\} \neq \emptyset$; and

(6) there exist $\lambda_0 > b_*$ and $A_0 \in \mathcal{F}(X)$ such that the set $\cap_{x\in A_0}\{y \in Y : a(x,y) \leq \lambda_0\}$ is empty or compact.

Then we have

$$\inf_{y\in Y} \sup_{x\in X} a(x,y) \leq \sup_{x\in X} \inf_{y\in Y} b(x,y).$$

Proof. Using the same idea as in the proof of Theorem 3.5.1, for each $\lambda > b_*$, we shall first prove that the family $\{y \in Y : a(x,y) \leq \lambda\}_{x\in X}$ has the finite intersection property. We define $F_\lambda, G_\lambda : X \to 2^Y$ by

$$F_\lambda(x) := \{y \in Y : b(x,y) < \lambda\} \text{ and } G_\lambda(x) := \{y \in Y : a(x,y) \leq \lambda\}$$

for each $x \in X$. Then it is easy to check that both mapping F_λ and G_λ satisfy all hypotheses of Theorem 3.4.4. By Theorem 3.4.4, the family $\{G_\lambda(x) : x \in X\}$ has the finite intersection property. Hence each $\lambda > b_*$, $\{y \in Y : a(x,y) \leq \lambda\}_{x\in X}$ has the finite intersection property. Note that for each $\lambda \in (b_*, \lambda_0)$, $\{y \in Y : a(x,y) \leq \lambda\} \subset \{y \in Y : a(x,y) \leq \lambda_0\}$ for each $x \in X$ and the set $\cap_{x\in A_0}\{y \in Y : a(x,y) \leq \lambda_0\}$ is non-empty and compact by (6), so that $\cap_{x\in X}\{y \in Y : a(x,y) \leq \lambda\} \neq \emptyset$. For each $\lambda \in (b_*, \lambda_0)$, taking any $y_0 \in \cap_{x\in X}\{y \in Y : a(x,y) \leq \lambda\}$, then $\inf_{y\in Y} \sup_{x\in X} a(x,y) \leq \sup_{x\in X} a(x,y_0) \leq \lambda$. Hence for each $\lambda \in (b_*, \lambda_0)$, $\inf_{y\in Y} \sup_{x\in X} a(x,y) \leq \lambda$. Therefore we must have the following inequality,

$$\inf_{y\in Y} \sup_{x\in X} a(x,y) \leq b_* = \sup_{x\in X} \inf_{Y\in Y} b(x,y) \qquad \square$$

As an application of Theorem 3.4.5 instead of Theorem 3.4.4, and by the same proof as in Theorem 3.5.9, we have the following.

Theorem 3.5.10. Let X and Y be topological spaces. Let $a, b : X \times Y \to \mathbb{R} \cup \{-\infty, +\infty\}$ be two functions such that for each $\lambda > b_*$:

(1) $a(x,y) \leq b(x,y)$ for each $(x,y) \in X \times Y$;

(2) for each fixed $y \in Y$, $x \mapsto b(x,y)$ is upper semicontinuous and for each fixed $x \in X$, $y \mapsto a(x,y)$ is lower semicontinuous;

(3) the set $\cap_{x\in A}\{y \in Y : b(x,y) < \lambda\}$ is empty or connected for each $A \in \mathcal{F}(X)$;

(4) the set $\cap_{y\in B}\{x \in X : b(x,y) \geq \lambda\}$ is empty or connected for each $B \in 2^Y$;

(5) for each $A \in \mathcal{F}(X)$, if $\cap_{x\in A}\{y \in Y : a(x,y) \leq \lambda\} \neq \emptyset$, then $\cap_{x\in A}\{y \in Y : b(x,y) < \lambda\} \neq \emptyset$; and

(6) there exist $\lambda_0 > b_*$ and $A_0 \in \mathcal{F}(X)$ such that the set $\cap_{x\in A_0}\{y \in Y : a(x,y) \leq \lambda_0\}$ is empty or compact.

Then we have

$$\inf_{y\in Y}\sup_{x\in X} a(x,y) \leq \sup_{x\in X}\inf_{y\in Y} b(x,y).$$

Proof. For each $\lambda > b_*$, we define $F_\lambda, G_\lambda : X \to 2^Y$ by

$$F_\lambda(x) := \{y \in Y : b(x,y) < \lambda\} \text{ and } G_\lambda(x) := \{y \in Y : a(x,y) \leq \lambda\}$$

for each $x \in X$. By employing Theorem 3.4.5 instead of Theorem 3.4.4 and the same proof as in Theorem 3.5.9, the conclusion follows. □

Remark 3.5.11. We do not know if the conclusions of Theorems 3.5.9 and 3.5.10 are still valid without condition (5). Moreover we note that Theorem 3.5.9 is a topological version of corresponding minimax inequalities in topological vector spaces given by Fan [158] and Sion [490] except condition (5). For some other generalizations of minimax inequalities for more than two functions in topological vector spaces, we refer the reader to Ben-El-Mechaiekh et al. [35]-[36], Simons [485], and references therein. For other work on topological intersection theorems and their applications to minimax inequalities, we refer the reader to Horvath [238], Joó and Stachó [276], Kindler [305]-[308], Komornik [328], Komiya [327], König [329], Simons [483]-[485], Stachó [499], and references therein.

In this chapter, we only study some existence of topological intersection theorems and its equivalent topological minimax theorems. No applications in other subjects have been carried out. However, we still believe that topological minimax theorems will play another very important role in the study of nonlinear problems, for some recent development in this direction, please see Ricceri [436]-[437], and related references therein.

Chapter 4

Abstract Economic Models and Generalized Games

4.1 Introduction

An abstract economy (resp., a generalized game) is a family of quadruples

$$\Gamma = (X_i; A_i, B_i; P_i)_{i\in I}$$

where I is a (possible uncountable) set of agents (resp., players) such that for each $i \in I$, X_i is a non-empty commodity (resp., strategic) set of a topological vector space and $A_i, B_i : X = \Pi_{j\in I}X_j \to 2^{X_i}$ are constraint correspondences and $P_i : X \to 2^{X_i}$ is a preference correspondence . When $I = \{1, \cdots, N\}$ where N is a positive integer, $\Gamma = (X_i; A_i, B_i; P_i)_{i\in I}$ is also called an N-person game. An equilibrium of Γ is a point $\hat{x} \in X$ such that for each $i \in I$,

$$\hat{x}_i = \pi_i(\hat{x}) \in \overline{B_i}(\hat{x}) \text{ and } A_i(\hat{x}) \cap P_i(\hat{x}) = \emptyset.$$

The aim of this chapter is to study generalized games (resp., abstract economics) model above. By developing the so-called approximation technique, we establish general existence results of equilibria for the above model in which preference correspondences may not have lower semicontinuity, and constraint correspondences are only lower or upper semicontinuous. In order to expose clearly the idea and method used in this chapter, we will introduce a new class of set-valued correspondences or so-called $\mathcal{L}$-majorized mappings whose roots go back to Gale and Mas-Colell's pioneering work [191] and [193]. Their idea has been carried forward by Shafer and Sonnenschein [459] and further developed by Borglin and Keiding in their important paper [51].

Since the existence of equilibria in an abstract economy with compact strategy sets in $\mathbb{R}^n$ was proved in a seminal paper of Debreu [115] (see also Debreu [118] for more details) which is classical Arrow-Debreu-McKenzie model of exchange economics under the perfect information competition, there have been many generalizations of Debreu's theorem. For example, Aumann [18] has extended Debreu's model to the case that the set of agents is a measure space. Secondly, Bewley [45] has generalized the commodity space of exchange economy model to an infinite dimensional space setting. In particular, by using the idea of maximal elements for set-valued mappings which is an equivalent version of powerful topological fixed point theorems in nonlinear analysis, it was Gale and Mas-Colell [191] and [193], who established a remarkable existence theorem of equilibria for general exchange economy model in which preference correspondences do not require any transitive nor complete assumptions. In this direction, Shafer and Sonnenschein [459] have furthermore generalized Gale and Mas-Colell's general existence result to the general theoretic setting the so-called 'abstract economy', or 'generalized game' due to a simple fact that the existence problem of Nash-equilibria in game theory (e.g., see Nash [380]-[383]) could be formulated as a special case of the existence of equilibria for abstract economy model. Since then there have been many generalizations of Debreu's theorem by Arrow and Debreu [8], Borglin and Keiding [51] and others. These works generalize Debreu's theorem by considering preference correspondences that are not necessarily transitive or total, by allowing externalities in consumption and by assuming that the commodity space is not necessarily finite-dimensional. In these papers, the domain (and or codomain) of the preference and constraint correspondences is assumed to be compact or paracompact.

Following the works of Sonnenschein [495], Gale and Mas-Colell [191] and Borglin and Keiding [51] on non-ordered preference relations, many theorems on the existence of maximal elements of preference relations which may not be transitive or complete, have been proved by Aliprantis and Brown [2], Bergstrom [42], Kim [299], Mehta and Tarafdar [370], Shafer and Sonnenschein [459], Sonnenschein [495], Tan and Yuan [507], Tarafdar [534], Toussaint [560], Tulcea [562], Yannelis [586] and Yannelis and Prabhakar [587] and others. However, most of these existence theorems for maximal elements and equilibrium points deal with preference correspondences which have open lower sections or are majorized by correspondences with open lower sections. Note that every correspondence with open lower sections must be lower semicontinuous but the converse is not true in general. Moreover, in most cases, preference and constraint correspondences may be upper semicontinuous (or majorized by upper semicontinuous correspondences) instead of being lower semicontinuous (or being majorized by lower semicontinuous), or the preference and constraint mappings are condensing. Furthermore, in the study of equilibrium theory in most economic models, the feasible sets or the budget constraints are generally not (weakly) compact in infinite dimensional commodity spaces and are not convex in the case of the indivisibility of commodities and the underlying spaces do not have a linear structure. Thus, relaxation of the convexity of choice sets and generalizations of spaces enable us to deal with the existence of maximal elements and equilibrium points even though commodities are indivisible.

Therefore it is necessary and important to study the existence of equilibria for generalized games in which the preference and constraint correspondences need not have open lower sections nor open upper sections and also the underlying spaces need not have any linear structure and so on.

The objective of this chapter is to systematically study the existence of maximal elements and equilibria for generalized games under various hypotheses. In particular, the question raised by Yannelis and Prabhakar [587] is answered in the affirmative with weaker assumptions. The essential idea behind these existence theorems for equilibria of generalized games is to reduce them first to qualitative games and then to the existence problem of maximal elements for preference correspondences. Since the existence of maximal elements of the correspondences have equivalent formulations in fixed point theorems, which can be derived from Ky Fan's minimax inequalities, the results in Chapter 2 are applicable.

More precisely, this Chapter is organized as follows. In Section 2, we first give the existence theorems of maximal elements and equilibria for non-compact abstract economics (resp., generalized games) in topological spaces which have the so-called H-structure.

It is our purpose in Section 3 to present an outline to show that how KKM theory can be used to study the general exitence of equilibria for abstact economics. In order to do so, we introduce a more general kind of KF mapping, and then, another non-compact version of Fan-browder fied point is given which will be used to obtain the existence of maximal elements which, in turn is used to establish the existence of equilibria for abstrcat economics in topological vector spaces and locally convex topological vector spaces.

In Sections 4 and 5, as applications of approximative equilibria for generalized games, several existence theorems for equilibria of generalized games in which preference correspondences are lower semicontinuous and the domain spaces are non-compact in locally convex topological vector spaces are given. By developing the so-called *approximative method* which was first motivated by Tulcea [561], we establish existence theorems for equilibria of generalized games in which the constraint correspondences are upper semicontinuous. In particular, the results in this section answer the question raised by Yannelis and Prabhakar [587] in the affirmative with weaker assumptions.

In Section 6, by introducing another kind of mapping in topological mappings, it enables us to establish existence of equilibria for abstract economics whutout the assumeptions of compactness and paracompactness which are often used in the literature. as a result, the results in this section also answer Kim's question in [299] and Yannelis and Prabhakar's question in [587] positively and ven in weaker hypotheses.

In Section 7, the concept of $\mathcal{U}$-majorized mapping is first introduced. Then the existence theorems for equilibria of generalized games in which the constraint correspondences are $\mathcal{U}$-majorized are given.

In Section 8, several existence theorems for equilibria of generalized games in which the constraint correspondences are Φ-condensing are given.

In Section 9, by Michael's selection theorem, we consider the existence theorems for equilibria of generalized games in which the underlying spaces are Frechet spaces.

Finally in Section 10, we first discuss some properties of set-valued mappings in finite dimensional spaces. As applications, fixed point theorems and the existence theorems of equilibria for generalized games are given in finite dimensional spaces.

Moreover, we remark that the existence theorems for equilibria of non-compact generalized games in this chapter will be applied to give the existence theorems of non-compact quasi-variational and generalized quasi-variational inequalities in Chapter 5.

4.2 Abstract Economics in H-Spaces

In this section, we first introduce the notions of Ky Fan mapping and Ky Fan-majorized mapping (in short, KF mapping and KF-majorized mapping) in H-spaces. Then a selection theorem is derived which is applied to give an existence theorem for maximal elements of KF-majorized mappings in H-spaces (which need not have a linear structure). As an application of a maximal element existence theorem, we prove the existence of equilibria for one-person games and qualitative games. The existence theorem for qualitative games is then applied to give existence theorems of equilibria for N-person games.

We now give some notions and definitions. If A is a non-empty subset of a topological vector space E and $S, T : A \to 2^E \cup \{\emptyset\}$ are correspondences, then coT, $T \cap S : A \to 2^E \cup \{\emptyset\}$ are correspondences defined by $(coT)(x) = coT(x)$ and $(T \cap S)(x) = T(x) \cap S(x)$ for each $x \in A$, respectively. If X and Y are topological spaces and $T : X \to 2^Y \cup \{\emptyset\}$ is a correspondence, then (1) T is said to be upper semicontinuous at $x \in X$ if for any open subset U of Y containing

$T(x)$, the set $\{z \in X : T(z) \subset U\}$ is an open neighborhood of x in X; (2) T is upper semicontinuous (on X) if T is upper semicontinuous at x for each $x \in X$; (3) the correspondence $\overline{T} : X \to 2^Y$ is defined by $\overline{T}(x) = \{y \in Y : (x, y) \in cl_{X\times Y} graph T\}$ (which is also called the adherence mapping of T); and (4) the mapping $clT : X \to 2^Y$ is defined by $clT(x) = cl_Y(T(x))$ for each $x \in X$. It is east to see that $clT(x) \subset \overline{T}(x)$ for each $x \in X$.

We remark here that throughout Chapter 4, an upper (or lower) semicontinuous correspondence is not require to be non-empty valued.

Definition 4.2.1. Let X and Y be two topological spaces and $F : X \to 2^Y \cup \{\emptyset\}$. Then F is said to be *compact* if for each $x \in X$, there exists a neighborhood V_x of x in X such that $F(V_x) = \cup_{z \in V_x} F(z)$ is relatively compact in Y. If X is a subset of a topological vector space E, X is said to have property K if for each compact subset B of X, the convex hull of B is relatively compact in X.

Definition 4.2.2. Let X be a topological space and Y a subset of an H-space E, let $\theta : X \to E$ be a mapping and $\psi : X \to 2^Y \cup \{\emptyset\}$ be a correspondence. Then (1) ψ is said to be of *class* KF_θ (resp., $KF_{\theta,C}$) or ψ is a KF_θ (respectively, $KF_{\theta,C}$) correspondence if for each $x \in X$, $Hco\psi(x) \subset Y$ and $\theta(x) \notin Hco\psi(x)$ and for each $y \in Y$, $\psi^{-1}(y) = \{x \in X : y \in \psi(x)\}$ is open (respectively, compactly open) in X; (2) a mapping $\psi_x : X \to 2^Y$ is said to be a KF_θ-*majorant* (respectively, $KF_{\theta,C}$-majorant) of ψ at $x \in X$ if there exists an open neighborhood N_x of x in X such that (a): for each $z \in N_x$, $\psi(z) \subset \psi_x(z)$ and $\theta(z) \notin Hco\psi_x(z)$, (b): for each $z \in X$, $Hco\psi_x(z) \subset Y$ (this condition is redundant if Y is an H-convex subset of E) and (c): for each $y \in Y$, $\psi_x^{-1}(y)$ is open (respectively, compactly open) in X; (3) ψ is KF_θ-*majorized* (respectively, $KF_{\theta,C}$-majorized) if for each $x \in X$ with $\psi(x) \neq \emptyset$, there exists a KF_θ-majoriant (resp., $KF_{\theta,C}$-majorant) of ψ at x.

If the underlying space E is a topological vector space, it is clear that the notions of the correspondence ψ being of class KF_θ or KF_θ-majorized and the correspondence ψ_x being a KF_θ-majorant of ψ at x generalize the corresponding notions of L_θ^*-correspondence or L_θ^*-majorized and correspondence ψ_x being an L_θ^*-majorant of ψ at x respectively introduced by Ding and Tan [138] which in turn generalize corresponding notions given by Borglin and Keiding [51], Yannelis and Prabhakar [587] and Tulcea [561]. For other kinds of mapping, we refer to Tan and Yuan [507] (which is a generalization of Ding and Tan [137]), Deguire et al.[125], Ben-El-Mechaiekh and Deguire [34], and Deguire and Lassonde [124]. In this section, we shall deal with either (I): $X = Y$ and $\theta = I_X$, the identity mapping on X; or (II): $X = \Pi_{i \in I} X_i$ and $\theta = \pi_j : X \to X_j$ is the projection of X onto X_j. In these cases, we shall write KF (respectively, KF_C) in place of KF_θ (respectively, $KF_{\theta,C}$).

The following example from Kim [299] shows that a KF-majorized mapping that is not of class KF.

Example 4.2.3. Let $X = [0,1]$ and $\phi : X \to 2^X \cup \{\emptyset\}$ be defined by

$$\phi(x) = \begin{cases} \{y \in X : y \in [0, x^3]\}, & \text{if } x \in (0,1), \\ \emptyset, & \text{otherwise.} \end{cases}$$

Then ϕ is not of class KF since $\phi^{-1}(y)$ is not open in X for each $y \in (0,1)$. For each $x \in (0,1)$, let $N_x = X$ which is an open neighborhood of x in X and define $\phi_x : X \to 2^X \cup \{\emptyset\}$ by

$$\phi_x(z) = \begin{cases} \{y \in X : y \in [0, x)\}, & \text{if } z \in (0,1), \\ \emptyset, & \text{otherwise.} \end{cases}$$

Then it is clear that ϕ_x is a KF-majorant of ϕ at x and ϕ is KF-majorized correspondence (see also Kim [299, p.799]).

Now we have the following result.

Lemma 4.2.4. Let X be a topological space and Y be an H-space. Let $\psi : X \to 2^Y \cup \{\emptyset\}$ be a correspondence with compactly open lower sections (i.e., $\psi^{-1}(y)$ is compactly open in X for each $y \in Y$). Define the correspondence $\phi : X \to 2^Y \cup \{\emptyset\}$ by $\phi(x) = Hco\psi(x)$ for each $x \in X$. Then ϕ also has compactly open lower sections.

Proof. For each $y_0 \in Y$ and each non-empty compact subset C of X, we need to prove that $\phi^{-1}(y_0) \cap C$ is open in C. Let $x_0 \in \phi^{-1}(y_0) \cap C$. Since $y_0 \in \phi(x_0) = Hco\psi(x_0)$, there exists a finite subset $Y_0 = \{y_1, \cdots, y_n\}$ of Y such that $y_0 \in HcoY_0$ by Lemma 1 of Tarafdar [534], where $y_i \in \psi(x_0)$ for each $i = 1, \cdots, n$. For each $i = 1, \cdots, n$, the set $\psi^{-1}(y_i)$ is compactly open in X and $x_0 \in \psi^{-1}(y_i)$. Let $U = \cap_{i=1}^n \psi^{-1}(y_i)$. Then $x_0 \in U \cap C$ and $U \cap C$ is open in C. Now for any $x \in U \cap C$, then $y_i \in \psi(x)$ for all $i = 1, \cdots, n$. Hence $Y_0 \subset \psi(x)$ which implies that $y_0 \in HcoY_0 \subset Hco\psi(x) = \phi(x)$. Therefore $x \in \phi^{-1}(y_0) \cap C$ for all $x \in U \cap C$. Consequently, $\phi^{-1}(y_0) \cap C$ is open in C. □.

By Lemma 4.2.4, we have the following selection theorem for KF-majorized mappings.

Lemma 4.2.5. Let X be a regular topological space and Y a non-empty subset of an H-space E. Let $\theta : X \to E$ and $P : X \to 2^Y \cup \{\emptyset\}$ be KF_θ-majorized (respectively, $KF_{\theta,C}$-majorized). If each open subset of X containing the set $B = \{x \in X : P(x) \neq \emptyset\}$ is paracompact, then there exists a correspondence $\phi : X \to 2^Y \cup \{\emptyset\}$ of class KF_θ (respectively, $KF_{\theta,C}$) such that $P(x) \subset \phi(x)$ and $\phi(x)$ is H-convex for each $x \in X$.

Proof. Since P is KF_θ-majorized (respectively, $KF_{\theta,C}$-majorized), for each $x \in B$, let N_x be an open neighborhood of x in X and $\phi_x : X \to 2^Y \cup \{\emptyset\}$ be such that (1) for each $z \in N_x$, $P(z) \subset \phi_x(z)$ and $\theta(z) \notin Hco\phi_x(z)$; (2) for each $z \in X$, $Hco\phi_x(z) \subset Y$ and (3) for each $y \in Y$, $\phi_x^{-1}(y)$ is compactly open in X. Since X is regular, for each $x \in B$ there exists an open neighborhood G_x of x in X such that $cl_X G_x \subset N_x$. Let $G = \cup_{x \in B} G_x$, then G is an open subset of X which contains $B = \{x \in X : P(x) \neq \emptyset\}$ so that G is paracompact by assumption. By Theorem VIII.1.4 of Dugundji [147] the open covering $\{G_x\}$ of G has an open precise neighborhood-finite refinement $\{G'_x\}$. Fix an arbitrary $x \in B$ and define $\phi'_x : G \to 2^Y \cup \{\emptyset\}$ by

$$\phi'_x(z) = \begin{cases} Hco\phi_x(z), & \text{if } z \in G \cap cl_X G'_x, \\ Y, & \text{if } z \in G \setminus cl_X G'_x, \end{cases}$$

then we have: For each $y \in Y$,

$$\begin{aligned} (\phi'_x)^{-1}(y) &= \{z \in G : y \in \phi'_x(z)\} \\ &= \{z \in G \cap cl_X G'_x : y \in \phi'_x(z)\} \cup \{z \in G \setminus cl_X G'_x : y \in \phi'_x(z)\} \\ &= \{z \in G \cap cl_X G'_x : y \in Hco\phi_x(z)\} \cup (G \setminus cl_X G'_x) \\ &= [(G \cap cl_X G'_x) \cap ((Hco\phi_x)^{-1}(y)] \cup (G \setminus cl_X G'_x) \\ &= (G \cap (Hco\phi_x)^{-1}(y)) \cup (G \setminus cl_X G'_x). \end{aligned}$$

It follows that for each non-empty compact subset C of X, $(\psi'_x)^{-1}(y) \cap C = (G \cap ((Hco\phi_x)^{-1}(y)) \cap C) \cup ((G \setminus cl_X G'_x) \cap C)$ is open in C by (3) and Lemma 4.2.4 above.

Now define $\phi : X \to 2^Y \cup \{\emptyset\}$ by

$$\phi(z) = \begin{cases} \cap_{x \in B} \phi'_x(z), & \text{if } z \in G, \\ \emptyset, & \text{if } z \in X \setminus G. \end{cases}$$

Let $z \in X$ be given, clearly (2) implies that $Hco\phi(z) \subset Y$. If $z \in X \setminus G$, then $\phi(z) = \emptyset$ so that $\theta(z) \notin Hco\phi(z)$. If $z \in G$, then $z \in G \cap cl_X G'_x$ for some $x \in B$ so that $\phi'_x(z) = Hco\phi_x(z)$ and hence $\phi(z) \subset Hco\phi_x(z)$. As $\theta(z) \notin Hco\phi'_x(z)$ by (1) we must have $\theta(z) \notin Hco\phi(z)$. Therefore $\theta(z) \notin Hco\phi(z)$ for all $z \in X$. Now we show that for each $y \in Y$, $\phi^{-1}(y)$ is compactly open in X. Indeed, let $y \in Y$, C be a non-empty compact subset of X and $u \in \phi^{-1}(y) \cap C = \{z \in X : y \in \phi(z)\} \cap C = \{z \in G : y \in \phi(z)\} \cap C$ be arbitrarily fixed. Since $\{G'_x\}$ is a neighborhood-finite refinement, there exists an open neighborhood M_u of u in G such that $\{x \in B : M_u \cap G'_x \neq \emptyset\} = \{x_1, \cdots, x_n\}$. Note that for each $x \in B$ with $x \notin \{x_1, \cdots, x_n\}$, $\emptyset = M_u \cap G'_x = M_u \cap cl_X G'_x$ so that $\phi'(z) = Y$ for all $z \in M_u$. Thus we have $\phi(z) = \cap_{x \in B} \phi'_x(z) = \cap_{i=1}^n \phi'_{x_i}(z)$ for all $z \in M_u$. It follows that

$$\begin{aligned} \phi^{-1}(y) &= \{z \in X : y \in \phi(z)\} = \{z \in G : y \in \cap_{x \in B} \phi'_x(z)\} \\ &\supset \{z \in M_u : y \in \cap_{x \in B} \phi'_x(z)\} = \{z \in M_u : y \in \cap_{i=1}^n \phi'_{x_i}(z)\} \\ &= M_u \cap \{z \in G : y \in \cap_{i=1}^n \phi'_{x_i}(z)\} = M_u \cap [\cap_{i=1}^n (\phi'_{x_i})^{-1}(y)]. \end{aligned}$$

But $M'_u = M_u \cap [\cap_{i=1}^n (\phi'_{x_i})^{-1}(y)] \cap C$ is an open neighborhood of u in C such that $M'_u \subset \phi^{-1}(y) \cap C$ since $(\phi'_{x_i})^{-1}(y)$ is compactly open in X. This shows that for each $y \in Y$, $\phi^{-1}(y)$ is compactly open in X. Therefore ϕ is of class KF_θ.

Now we shall show that $P(z) \subset \phi(z)$ for each $z \in X$. Indeed, let $z \in X$ with $P(z) \neq \emptyset$. Note that $z \in G$. For each $x \in B$, if $z \in G \setminus cl_X G'_x$, then $\phi'_x(z) = Y \supset P(z)$ and if $z \in G \cap cl_X G'_x$, we have $z \in cl_X G'_x \subset cl_X G_x \subset N_x$ so that by (1), $P(z) \subset \phi_x(z) \subset \phi'_x(z)$. It follows that $P(z) \subset \phi'_x(z)$ for each $x \in B$ so that $P(z) \subset \cap_{x \in B} \phi'_x(z) = \phi(z)$. Finally we replace ϕ by $Hco\phi$ and, by Lemma 4.2.4, the result follows. □

We note that Lemma 4.2.5 generalize Lemma 2 of Ding and Tan [138], which in turn improves Lemma 1 of Ding et al. [141].

Now by Lemma 4.2.5, we have the following existence theorem for KF_θ correspondences in topological spaces which generalizes Theorem 4.1 of Tan and Yu [505], and Corollary 1 of Borglin and Keiding [51].

Theorem 4.2.6. Let X be an H-space such that $X = \cup_{n=1}^\infty C_n$, where $\{C_n\}_{n=1}^\infty$ is an increasing sequence of non-empty compact and weakly H-convex subsets. Suppose $P : X \to 2^X \cup \{\emptyset\}$ is KF_C-majorized. If for each sequence $(y_n)_{n=1}^\infty$ in X with $y_n \in C_n$ for each $n = 1, 2, \cdots$ which is escaping from X relative to $\{C_n\}_{n=1}^\infty$, there exist $n_0 \in \mathbb{N}$ and $x_{n_0} \in C_{n_0}$ such that $x_{n_0} \in P(y_{n_0})$, then there exists $\hat{x} \in X$ such that $P(\hat{x}) = \emptyset$.

Proof. Suppose the conclusion were not true, then $X = \{x \in X : P(x) \neq \emptyset\}$. First we note that since X is regular and σ-compact, X is paracompact by Corollary 33.15 of Cullen [112, p.341]. Hence by Lemma 4.2.5, there exists a correspondence $\psi : X \to 2^X$ of class KF_C such that $P(x) \subset \psi(x)$ and $\psi(x)$ is H-convex for all $x \in X$. Note that conditions (i) and (ii) of Theorem 4.3.17 are satisfied by ψ. By assumption, for each sequence $(y_n)_{n=1}^\infty$ in X with $y_n \in C_n$ for each $n = 1, 2, \cdots$ which is escaping from X relative to $\{C_n\}_{n=1}^\infty$, there exist $n_0 \in \mathbb{N}$ and $x_{n_0} \in C_{n_0}$ such that $x_{n_0} \in P(y_{n_0}) \subset \psi(y_{n_0})$. This shows that condition (iii) of Theorem 4.3.17 is also satisfied by ψ. Hence by Theorem 4.3.17, there exists $\hat{y} \in X$ such that $\psi(\hat{y}) = \emptyset$. It follows that $P(\hat{y}) = \emptyset$ which is a contradiction. Hence the conclusion must hold. □

Definition 4.2.7. A *one-person game* is a quadruple $(X; A, B; P)$ where X is a topological space, $A, B : X \to 2^X \cup \{\emptyset\}$ are constraint correspondences and $P : X \to 2^X \cup \{\emptyset\}$ is a preference correspondence. An equilibrium point for $(X; A, B; P)$ is a point $x^* \in X$ such that $x^* \in \overline{B}(x^*)$ and $A(x^*) \cap P(x^*) = \emptyset$.

Definition 4.2.8. According to Gale and Mas-Colell [192], a *qualitative game* is a family $\Gamma = (X_i, P_i)_{i \in I}$ of ordered pairs (X_i, P_i) where for each $i \in I$, X_i is a topological space, $P_i : X = \Pi_{j \in I} X_j \to 2^{X_i} \cup \{\emptyset\}$ is a preference correspondence. A point $x^* \in X$ is said to be an equilibrium point of the qualitative game Γ if $P_i(x^*) = \emptyset$ for all $i \in I$.

Definition 4.2.9. Let I be a (finite or infinite) set of players (agents). A *generalized game* (an *abstract economy*) is a family $\Gamma = (X_i; A_i, B_i; P_i)_{i\in I}$ of quadruples $(X_i; A_i, A_i; P_i)$ where for each $i \in I$, X_i is a topological space, $A_i, B_i : X := \Pi_{j\in I}X_j \to 2^{X_i} \cup \{\emptyset\}$ are constraint correspondences and $P_i : X \to 2^{X_i} \cup \{\emptyset\}$ is a preference correspondence. An *equilibrium point* for G is a point $x^* \in X$ such that for each $i \in I$, $x_i^* = \pi_i(x^*) \in \overline{B_i}(x^*)$ and $A_i(x^*) \cap P_i(x^*) = \emptyset$ where $\pi_i : X \to X_i$ is the projection.

We remark that when $\overline{B_i}(x^*) = cl_{X_i}B_i(x^*)$ (which is the case when B_i has a closed graph in $X \times X_i$; in particular, when $cl_{X_i}B_i$ is upper semicontinuous with closed values) for each $i \in I$, our definition of equilibrium point for a generalized game (an abstract economy coincides with that of Ding and Tan [138].

As an application of Theorem 4.2.6, we have the following existence theorem of equilibria for one person-games as follows.

Theorem 4.2.10. Let $(X; A, B; P)$ be a one-person game such that X is an H-space and $X = \cup_{n=1}^{\infty} C_n$ where $\{C_n\}_{n=1}^{\infty}$ is an increasing sequence of compact and weakly H-convex subsets and the following conditions are satisfied:

(1) for each $x \in X$, $A(x)$ is non-empty and $HcoA_i(x) \subset B_i(x)$;

(2) for each $y \in X$, $A^{-1}(y)$ is compactly open in X;

(3) $A \cap P$ is KF_C-majorized; and

(4) for each sequence $(y_n)_{n=1}^{\infty}$ in X with $y_n \in C_n$ for each $n = 1, 2, \cdots$ which is escaping from X relative to $\{C_n\}_{n=1}^{\infty}$, there exist $n \in \mathbb{N}$ and $x_n \in C_n$ such that $x_n \in (A(y_n) \cap P(y_n)) \cap C_n$.

Then (X, A, B, P) has an equilibrium point, i.e., there exists $\hat{x} \in X$ such that $\hat{x} \in \overline{B_i}(\hat{x})$ and $A(\hat{x}) \cap P(\hat{x}) = \emptyset$.

Proof. Let $F = \{x \in X : x \in \overline{B}(x)\}$, then F is closed in X. Define $\Psi : X \to 2^X \cup \{\emptyset\}$ by

$$\Psi(x) = \begin{cases} A(x) \cap P(x), & \text{if } x \in F, \\ A(x), & \text{if } x \notin F. \end{cases}$$

Let $x \in X$ be such that $\Psi(x) \neq \emptyset$. If $x \notin F$, then $X \setminus F$ is an open neighborhood of x such that for each $z \in X \setminus F$, $z \notin \overline{B}(z)$. Now define $\Phi_x : X \to 2^X \cup \{\emptyset\}$ by $\Phi_x(z) = A(z)$ for each $z \in X$ and $N_x = X \setminus F$, then N_x is an open neighborhood of x in X such that

(i) $\Psi(z) \subset \Phi_x(z)$ and $z \notin Hco\Phi_x(z)$ for each $z \in N_x$, and

(ii) $\Phi_x^{-1}(y) = A^{-1}(y)$ is compactly open in X.

Therefore Φ_x is an L_C-majorant of Ψ at x. If $x \in F$, then $\Psi(x) = A(x) \cap P(x) \neq \emptyset$. Since $A \cap P$ is KF_C-majorized, there exist an open neighborhood N_x of x in

X and a correspondence $\Phi_x : X \to 2^X$ such that $\Psi(z) = A(z) \cap P(z) \subset \Phi_x(z)$ and $z \notin Hco\Phi_x(z)$ for each $z \in N_x$, and $\Phi_x^{-1}(y)$ is compactly open in X for each $y \in X$. Define the mapping $\Phi'_x : X \to 2^X \cup \{\emptyset\}$ by

$$\Phi'_x(z) = \begin{cases} A(z) \cap \Phi_x(z), & \text{if } z \in F, \\ A(z), & \text{if } z \notin F. \end{cases}$$

Note that as $(A \cap P)(z) \subset \Phi_x(z)$ for each $z \in N_x$, we have $\Psi(z) \subset \Phi'_x(z)$. It is easy to see that $z \notin Hco\Phi'_x(z)$ for all $z \in X$.

Moreover, for any $y \in X$, the set $(\Phi'_x)^{-1}(y) = [\Phi_x^{-1}(y) \cup (X \setminus F)] \cap A^{-1}(y)$ is compactly open in X. Therefore Φ'_x is a KF_C-majorant of Ψ at x. Hence, Ψ is a KF_C-majorized correspondence.

By (4), for each sequence $(y_n)_{n=1}^{\infty}$ in X with $y_n \in C_n$ for each $n = 1, 2, \cdots$ which is escaping from X relative to $\{C_n\}_{n=1}^{\infty}$ there exist $n \in \mathbb{N}$ and $x_n \in C_n$ with $x_n \in Hco(A(y_n) \cap P(y_n)) \cap C_n \subset \Psi(y_n) \cap C_n$. Hence by Theorem 4.2.6, there exists $\hat{x} \in X$ such that $\Psi(\hat{x}) = \emptyset$; since $A(\hat{x}) \neq \emptyset$ by (1), we must have $\hat{x} \in \overline{B}(\hat{x})$ and $A(\hat{x}) \cap P(\hat{x}) = \emptyset$. □

As another application of Theorem 4.2.6, we have the following existence theorem for equilibria for qualitative games.

Theorem 4.2.11. Let $(X_i, P_i)_{i=1}^N$ be a qualitative game such that for each $i = 1, \cdots, N$, $X_i = \cup_{j=1}^{\infty} C_{i,j}$ is an H-space (so that $X = \Pi_{i \in I} X_i$ is also an H-spaces), where $\{C_{i,j}\}_{j=1}^{\infty}$ is an increasing sequence of non-empty compact and weakly H-convex subsets. Suppose the following conditions are satisfied:

(1) for each $i = 1, \cdots, N$, $P_i : X \to 2^{X_i} \cup \{\emptyset\}$ is KF_C-majorized;

(2) for each $i = 1, \cdots, N$, the set $F_i = \{x \in X : P_i(x) \neq \emptyset\}$ is open in X; and

(3) for each sequence $(y_n)_{n=1}^{\infty}$ in X which is escaping from X relative to $\{C_n\}_{n=1}^{\infty}$ where $C_n = \Pi_{i=1}^N C_{i,n}$ for each $n = 1, 2, \cdots$, there exist $n_0 \in N$ and $x_{n_0} \in C_{n_0}$ such that $\pi_i(x_{n_0}) \in P_i(y_{n_0})$ for each $i \in I(y_{n_0})$, where $I(x) = \{j \in \{1, \cdots, N\} : P_j(x) \neq \emptyset\}$.

Then $(X_i, P_i)_{i=1}^N$ has an equilibrium point $\hat{x} \in X$; i.e. $P_i(\hat{x}) = \emptyset$ for all $i = 1, \cdots, N$.

Proof. Suppose the conclusion were false; then for each $x \in X$, $I(x) \neq \emptyset$. For each $i = 1, \cdots N$, define $P'_i : X \to 2^X \cup \{\emptyset\}$ by $P'_i(x) = \pi_i^{-1}(P_i(x))$ for each $x \in X$. Define $P : X \to 2^X \cup \{\emptyset\}$ by $P(x) = \cap_{i \in I(x)} P'_i(x)$ for each $x \in X$. Suppose $x \in X$ and fix one $i \in I(x)$. By (1), there exist a correspondence $\psi_i : X \to 2^{X_i} \cup \{\emptyset\}$ and an open neighborhood N_x of x in X such that

(a) for each $z \in N_x$, $P_i(z) \subset \psi_i(z)$ and $\pi_i(z) \notin Hco\psi_i(z)$; and

(b) for each $y \in X_i$, $\psi_i^{-1}(y)$ is compactly open in X.

By (2), we may assume that $N_x \subset F_i$ so that $P_i(z) \neq \emptyset$ for all $z \in N_x$ and hence $i \in I(z)$ for all $z \in N_x$. Define $\psi_x : X \to 2^X$ by $\psi_x(z) = \pi_i^{-1}(\psi_i(z))$ for each

$z \in X$. Now if $z \in N_x$, then by (a),

$$P(z) = \cap_{j \in I(z)} P'_j(z) \subset P'_i(z) = \pi_i^{-1}(P_i(z)) = \psi_x(z)$$

and $z \notin Hco\psi_x(z)$. Moreover, if $y \in X$, then

$$\begin{aligned} \psi_x^{-1}(y) \quad &= \{z \in X : y \in \psi_x(z)\} \\ &= \{z \in X : \pi_i(y) \in \psi_i(z)\} \\ &= \psi_i^{-1}(\pi_i(y)) \end{aligned}$$

is compactly open in X by (b). This shows that ψ_x is a KF_{I_X}-majorant of P at x. Hence P is KF_{I_X}-majorized.

Finally by (3), for each sequence $(y_n)_{n=1}^{\infty}$ in X with $y_n \in C_n$ for each $n = 1, 2, \cdots$ which is escaping from X relative to $\{C_n\}_{n=1}^{\infty}$, there exist $n_0 \in \mathbb{N}$ and $x_{n_0} \in C_{n_0}$ such that $\pi_i(x_{n_0}) \in P_i(y_{n_0})$ for each $i \in I(y_{n_0})$, it follows that

$$x_{n_0} \in \cap_{i \in I(y_{n_0})} \pi_i^{-1}((P_i(y_{n_0})) = \cap_{i \in I(y_{n_0})} P'_i(y_{n_0}) = P(y_{n_0}).$$

Note that $X = \cup_{n=1}^{\infty} C_n$ and C_n is also a compact H-space for $n = 1, 2, \cdots$. Hence by Theorem 4.2.6, there exists $\hat{x} \in X$ such that $P(\hat{x}) = \emptyset$ which contradicts our assumption. Therefore the conclusion must hold. □

Theorem 4.2.12. Let $(X_i, A_i, B_i; P_i)_{i=1}^N$ be an N-person game such that for each $i = 1, \cdots, N$, $X_i = \cup_{j=1}^{\infty} C_{i,j}$ is an H-space (so that $X = \Pi_{i=1}^N X_i$ is also an H-space), where $\{C_{i,j}\}_{j=1}^{\infty}$ is an increasing sequence of non-empty compact and weakly H-convex subsets of X_i. Suppose the following conditions are satisfied:

(1) for each $i = 1, \cdots, N$ and for each $x \in X$, $A_i(x)$ is non-empty and $HcoA_i(x) \subset B_i(x)$;

(2) for each $i = 1, \cdots, N$ and for each $y \in X_i$, $A_i^{-1}(y)$ is compactly open in X;

(3) for each $i = 1, \cdots N$, the correspondence $A_i \cap P_i : X \to 2^{X_i} \cup \{\emptyset\}$ is KF_C-majorized;

(4) for each $i = 1, \cdots, N$, the set $F_i = \{x \in X : (A_i \cap P_i)(x) \neq \emptyset\}$ is open in X; and

(5) for each sequence $(y_n)_{n=1}^{\infty}$ in X with $y_n \in C_n$ for each $n = 1, 2, \cdots$ which is escaping from X relative to $\{C_n\}_{n=1}^{\infty}$ where $C_n = \Pi_{i=1}^N C_{i,n}$ for each $n = 1, 2, \cdots$, there exist $n_0 \in N$ and $x_{n_0} \in C_n$ such that for each $i = 1, \cdots, N$, $\pi_i(x_{n_0}) \in A_i(y_{n_0}) \cap P_i(y_{n_0})$ if $A_i(y_{n_0}) \cap P_i(y_{n_0}) \neq \emptyset$ and $\pi_i(y_{n_0}) \in cl_{X_i} B_i(y_{n_0})$ and $\pi_i(x_{n_0}) \in A_i(y_{n_0})$ if $\pi_i(y_{n_0}) \notin \overline{B_i}(y_{n_0})$.

Then $(X_i; A_i, B_i; P_i)_{i=1}^N$ has an equilibrium point $\hat{x}$ in X; i.e. for each $= 1, \cdots, N$, $\pi_i(\hat{x}) \in \overline{B_i}(\hat{x})$ and $A_i(\hat{x}) \cap P_i(\hat{x}) = \emptyset$.

Proof. For each $i = 1, \cdots, N$, let $G_i = \{x \in X : \pi_i(x) \in \overline{B_i}(x)\}$, then G_i is closed in X; define define $Q_i : X \to 2^{X_i}$ by

$$Q_i(x) = \begin{cases} (A_i \cap P_i)(x), & \text{if } x \in G_i; \\ A_i(x), & \text{if } x \notin G_i. \end{cases}$$

We shall show that $(X_i, Q_i)_{i=1}^N$ is a qualitative game satisfying the hypotheses of Theorem 4.2.11. Fix any $i \in \{1, \cdots, N\}$. The set

$$\begin{aligned}\{x \in X : Q_i(x) \neq \emptyset\} &= \{x \in G_i : (A_i \cap P_i)(x) \neq \emptyset\} \cup \{x \in X \setminus G_i : A_i(x) \neq \emptyset\} \\ &= (G_i \cap F_i) \cup (X \setminus G_i) \\ &= F_i \cup (X \setminus G_i)\end{aligned}$$

is compactly open in X by (4). Thus condition (2) of Theorem 4.2.11 is satisfied. Now let $x \in X$ be such that $Q_i(x) \neq \emptyset$.

Case 1: Suppose $x \notin G_i$.

Let $\psi_x = A_i$ and $N_x = X \setminus G_i$, then N_x is an open neighborhood of x in X such that

(i) for each $z \in N_x$, $Q_i(z) = A_i(z) = \psi_x(z)$ and $\pi_i(z) \notin HcoA_i(z) = Hco\psi_x(z)$ by (1); and

(ii) for each $y \in X_i$, $\psi_x^{-1}(y) = A_i^{-1}(y)$ is compactly open in X by (2).

Thus ψ_x is a KF_C-majorant of Q_i at x.

Case 2: Suppose $x \in G_i$.

Since $Q_i(x) = (A_i \cap P_i)(x) \neq \emptyset$, by (3), there exist $\psi_x : X \to 2^{X_i} \cup \{\emptyset\}$ and an open neighborhood N_x of x in X such that

(i) for each $z \in N_x$, $(A_i \cap P_i)(z) \subset \psi_x(z)$ and $\pi_i(z) \notin Hco\psi_x(z)$; and

(ii) for each $y \in X_i$, $\psi_x^{-1}(y)$ is compactly open in X.

Define $\psi'_x : X \to 2^{X_i} \cup \{\emptyset\}$ by

$$\psi'_x(z) = \begin{cases} A_i(z) \cap \psi_x(z), & \text{if } z \in G_i, \\ A_i(z), & \text{if } z \notin G_i. \end{cases}$$

Then for each $z \in N_x$, we have $Q_i(z) \subset \psi'_x(z)$ and $\pi_i(z) \notin Hco\psi'_x(z)$ by (i) and (1). Moreover for each $y \in X_i$, the set $(\psi'_x)^{-1}(y) = [\psi_x^{-1}(y) \cup (X \setminus G_i)] \cap A_i^{-1}(y)$ is compactly open in X by (ii) and (2). Thus ψ'_x is a KF_C-majorant of Q_i at x.

This shows that Q_i is KF_C-majorized so that condition (1) of Theorem 4.2.11 is also satisfied.

Finally, let $(y_n)_{n=1}^\infty$ be a sequence in X with $y_n \in C_n$ for each $n = 1, 2, \cdots$ which is escaping from X relative to $\{C_n\}_{n=1}^\infty$. By (5), there exist $n_0 \in \mathbb{N}$ and $x_{n_0} \in C_{n_0}$ such that $\pi_i(x_{n_0}) \in Q_i(y_{n_0})$ for each $i = 1, \cdots, N$ with $Q_i(y_{n_0}) \neq \emptyset$. Hence condition (3) of Theorem 4.2.11 is also satisfied.

By Theorem 4.2.11, $(X_i, Q_i)_{i=1}^N$ has an equilibrium point $\hat{x} \in X$, i.e. $Q_i(\hat{x}) = \emptyset$ for all $i = 1, \cdots, N$. Since $A_i(\hat{x}) \neq \emptyset$ for each $i = 1, \cdots, N$, we must have $\pi_i(\hat{x}) \in cl_{X_i} B_i(\hat{x})$ and $A_i(\hat{x}) \cap P_i(\hat{x}) = \emptyset$ for all $i = 1, \cdots, N$. □

4.3 The Study of Abstract Economics by KKM Theory

In this section, from the view point of the Browder-Fan fixed point theorem, we give a unified study for the existence of maximal elements and equilibria for abstract economics. Precisely, by our generalization of the Browder-Fan type fixed point theorem, the existence of maximal elements and equilibria of qualitative games and abstract economics have been established in topological vector spaces. Secondly, by introducing the notion of *approximate equilibria*, existence theorems of equilibria for abstract economics are obtained when constraint mappings are lower semicontinuous instead of having open lower sections in locally convex topological vector spaces. Our results unify and improve most of the corresponding results in the study of general equilibrium theory in the last three decades.

4.3.1 Introduction

Since the existence of equilibria in an abstract economy with compact strategy sets in $\mathbb{R}^n$ was proved in a seminal paper by Debreu in 1952, there have been many generalizations of Debreu's theorems. During last three decades, the classical Arrow and Debreu's existence theorem of Walrasian equilibria in [8] has been generalized in many directions. In finite dimensional spaces, Gale and Mas-Colell [191] proved the existence of a competitive equilibrium without assuming the preference correspondences being total or transitive. Shafer and Sonnenschein [459] obtained results in the same direction and they proved the so-called 'Arrow and Debreu Lemma' for abstract economics for the case where preference correspondences may not be total or transitive. For infinitely dimensional strategy spaces and with finite or infinite many agents, the existence results of equilibria of generalized games were given by Aliprants and Brown [2], Allen [4], Aubin [10], Bewley [45], Border [48], Borglin and Keiding [51], Chang [77], Deguire and Lassonde [124], Ding and Tan [136], Dugundji and Granas [149], Flam [176], Florenzano [177], Hildenbrand and Sonnenschein [230], Ichiishi [245]-[247], Keiding [284], Khan and Papageorgiou [293], Khan and Yannelis [295], Kim [299], Kim and Richter [301], Kim et al. [302], Mas-Colell [362]-[363], Mas-Colell and Richard [364], Mas-Colell and Zame [365] and many others. Recently, several different kinds of mappings have been introduced such as the class of L_F (respectively, L^*, $\mathcal{L}$, $\mathcal{L}(X,Y,\theta)$), L_F-majorized (respectively, L^*, $\mathcal{L}$, $\mathcal{L}(X,Y,\theta)$-majorized), e.g., see Ding and Tan [137], Kim [299], Tan and Yuan [507], Tian [552], Tulcea [561]-[562] and others. Note that all these definitions are motivated by the Browder-Fan fixed point theorem (e.g., see Fan [157] and Browder [59]), that is, in order to give the existence theorems of maximal elements, we can reduce the existence problem of maximal elements into the equivalent existence problem of fixed points of corresponding mappings. This idea motivates us to give a general definition of Fan (in short, F) set-valued mappings which enable us to use fixed point theorems of the Browder-Fan type to prove the existence of maximal elements. Moreover, our F-majorized mappings unify all other kinds of mappings mentioned above. It is our objective in this section to give a unified approach to study the existence of maximal elements and equilibria for qualitative games, generalized games in topological vector spaces, and locally convex topological vector spaces.

Precisely, we shall first introduce the notions of the correspondence of class F, F-majorant of ϕ at x, and F-majorized correspondences which include corresponding definitions of Ding and Tan [136], Ding and Tarafdar [143], Tan and Yuan [511], and Tulcea [561] as special cases. We then discuss some fixed point theorems of the Browder-Fan type. These results improve corresponding results of Ben-El-Mechaiekh et al. [36], Border [48], Ding and Tan [136], Dugundji and Granas [149], Tarafdar [533] and so on. By employing fixed point theorems, an existence theorem of maximal elements for F-majorized mappings is obtained which generalizes the corresponding results of Borglin and Keiding [51], Ding and Tan [138], Tarafdar [534], Toussaint [560], Tulcea [561]-[562] and others. As applications of our existence theorem for maximal elements, we prove equilibrium existence theorems for a non-compact one-person game and for a non-compact qualitative game with an infinite number of players and with F-majorized correspondences. The latter result is applied to obtain an equilibrium existence theorem for a non-compact generalized game with an infinite number of players and with KF correspondences. Finally, by introducing the concept *approximate equilibria*, the existence theorems of equilibria for one-person games and generalized games are obtained when constraint mappings are lower semicontinuous instead of having open inverse values in locally convex spaces. Our results unify and improve corresponding results in the literature.

Let A be a subset of a topological space X. We recal that A is said to be *compactly open* in X if for each non-empty compact subset C of X, $A \cap C$ is open in C. If $S : A \to 2^E$ is a set-valued mapping (or say, a correspondence, or a set-valued mapping or a multifunction), then $domS$ denotes the domain of S, i.e., $domS := \{x \in X : S(x) \neq \emptyset\}$.

Definition 4.3.1. Let X be a topological space and Y be non-empty subset of a topological vector space E and $\theta : X \to Y$ be a (single-valued) function. Then two mappings $\psi, \phi : X \to 2^Y$ are called a *Fan pair* with respect to θ (in short, F_θ-pair) if (a) for each $x \in X$, $\theta(x) \notin co\phi(x) \subset Y$ and $\psi(x) \subset \phi(x)$; and (b) for each $y \in Y$, $\psi^{-1}(y)$ is compactly open in X.

Definition 4.3.2. Let $P : X \to 2^Y$ be a correspondence and $\theta : X \to Y$ be a (single-valued) function. Then (1) P is said to be a *Ky-Fan mapping* with respect to θ (in short, F_θ) if there exists a F_θ-pair $\psi, \phi : X \to 2^Y$ such that

(a) $domP \subset dom\psi$ and (b) $P(x) \subset \phi(x)$ for each $x \in X$; (2) $(\phi_x, \psi_x; N_x)$ is a *F_θ-majorant* of P at x if $\phi_x, \psi_x : X \to 2^Y$ are two mappings and N_x is an open neighborhood of x in X such that (a) for each $z \in N_x$, $\theta(z) \notin co\phi_x(z) \subset Y$ and $\{z \in N_x : P(z)\} \subset \{z \in N_x : \psi_x(z) \neq \emptyset\}$, (b) for each $z \in X$, $\psi_x(z) \subset \phi_x(z)$ and and (c) for each $y \in Y$, $\psi_x^{-1}(y)$ is compactly open in X; (3) P is said to be *F_θ-majorized* if (a) for each $x \in X$ with $P(x) \neq \emptyset$, there exists an F_θ-majorant (ϕ_x, ψ_x, N_x) of ϕ at x and (b) for each $A \in \mathcal{F}(domP)$, $domP \cap (\cap_{x \in A} N_x) \subset \{z \in \cap_{x \in A} N_x : \cap_{x \in A} \psi_x(z) \neq \emptyset\}$.

We note that our notions of the mapping P being of class KF_θ and KF_θ-majorized correspondence include all notions of the correspondence of class KF_θ and KF_θ-majorized mappings, respectively introduced in Section 2 when the H-space E reduces to a topological vector space, and by Ding and Tan [136], Tan and Yuan [511], which, in turn, generalize the notions of $\phi \in C(X, Y, \theta)$ and C-majorized mappings introduced by Tulcea [561].

In this section, we shall deal mainly with either the case (I): Y is a non-empty subset of the topological space X and $\theta = I_Y$, the identity mapping on Y, or the case (II): $X = \Pi_{i \in I} X_i$ and $\theta = \pi_j : X \to X_j$ is the projection of X onto X_j and $Y = X_j$ is a topological space. In both cases (I) and (II), we shall write F in place of F_θ.

4.3.2 Browder-Fan Fixed Point Theorems

As an application of Fan's lemma, we will establish a general fixed point theorem to transfer compactly open inverse valued mappings in topological vector spaces. These results will be used to give the existence of maximal elements in the next section. For the convenience of the reader, we state Theorem 2.2.8 (see Fan [157]) again, which is a generalization of the classical KKM principle [319] into infinite dimensional vector spaces as follows.

Lemma 4.3.3 (Fan's Lemma). Let X and Y be non-empty subsets in a topological vector space E. Suppose a mapping $F : X \to 2^Y$ has closed values such that:

(1) for each $A \in \mathcal{F}(X)$, $co(A) \subset \cup_{x \in A} F(x)$; and

(2) there exists $x_0 \in X$ such that $F(x_0)$ is compact.

Then $\cap_{x \in X} F(x) \neq \emptyset$.

Fan's lemma is so powerful because of the fact that not only the Browder-Fan fixed point theorem is an immediate consequence of Fan's lemma, but also it can be used to prove the Fan-Glicksberg fixed point theorem (e.g., see Fan [167]). In order to give a general non-compact version of the Browder-Fan fixed point theorem for set-valued mappings (which has been developed in Section 4 of Chapter 2), we introduce the following coercive condition (C).

The Condition (C)**:** Let X be a non-empty convex subset of a topological vector space E. A mapping $P : X \to 2^X$ is said to satisfy '*the condition* C' if there exists a non-empty compact and convex subset X_0 and a non-empty compact (not necessarily convex) subset K of X such that for each $y \in X \setminus K$, there exists $x \in co(X_0 \cup \{y\})$ with $y \notin cl_X[X \setminus (coP)^{-1}(x)]$.

It is clear that if the set X is compact and convex, then each mapping $P : X \to 2^X$ also satisfies condition (C) above.

The following fixed point essentially due to Ding and Tarafdar [143] is a generalization of Theorem 2.3.19 in several aspects. Here we include its proof by employing Fan's Lemma directly.

Theorem 4.3.4 (Browder-Fan Fixed Point Theorem). Let X be a non-empty convex subset of a topological vector space and $F, G : X \to 2^X$ be such that:

(1) for each $x \in X$, $F(x) \subset G(x)$ and F satisfies the condition (C), i.e., there exists a non-empty compact and convex subset X_0 and a non-empty compact (not necessarily convex) subset K of X such that for each $y \in X \setminus K$, there exists $x \in co(X_0 \cup \{y\})$ with $y \notin cl_X[X \setminus (coG)^{-1}(x)]$; and

(2) for each $y \in X$, $F^{-1}(y)$ is compactly open in X and $F(x) \neq \emptyset$ for each $x \in K$.

Then there exists $x \in X$ such that $x \in coG(x)$

Proof. Suppose the conclusion were false. Then for each $x \in X$, $x \notin coG(x)$. By (2), it is easy to verify that $(coF)^{-1}(y)$ is compactly open in X for each $y \in X$. Now define $T, S : X \to 2^X$ by,

$$T(x) = cl_X[X \setminus (coG)^{-1}(x)] \cap K$$

and,

$$S(x) = [X \setminus (coF)^{-1}(x)] \cap K$$

for each $x \in X$. We shall show that the family $\{T(x) : x \in X\}$ has the finite intersection property. Let $\{x_1, \cdots, x_n\}$ be any non-empty finite subset of X and set $D := co(X_0 \cup \{x_1, \cdots, x_n\})$. Then D is non-empty compact and convex since X_0 is non-empty compact. Define another two mappings $T_0, S_0 : D \to 2^D$ by,

$$T_0(x) = cl_X[(X \setminus (coG)^{-1}(x)] \cap D$$

and

$$S_0(x) = [X \setminus (coF)^{-1}(x)] \cap D$$

for each $x \in D$. We have that:

(a) for each $x \in D$, $S_0(x)$ is closed in D since $(coF)^{-1}(x)$ is compactly open in X;

(b) for each $x \in X$, $T_0(x)$ is compact in D and $T_0(x) \subset S_0(x)$ by (1); and

(c) the mapping T_0 is a KKM mapping, i.e., for each non-empty finite subset A of D, $co(A) \subset \cup_{x\in A} T_0(x)$. Indeed, if it were false, then there exists $\{u_1, \cdots, u_m\} \subset D$, $\lambda_1, \cdots, \lambda_m \in (0,1]$ with $\Sigma_{i=1}^m \lambda_i = 1$ such that $\Sigma_{i=1}^m \lambda_i u_i \notin \cup_{i=1}^m (X \setminus (coG)^{-1}(u_i)) \cap D$. It follows that $\Sigma_{i=1}^m \lambda_i u_i \in \cap_{i=1}^m (coG)^{-1}(u_i) \cap D$. Therefore $\Sigma_{i=1}^m \lambda_i u_i \in co(G(\Sigma_{i=1}^m \lambda_i u_i)$ which contradicts that the mapping coG has no fixed point.

By Fan's lemma above, $\cap_{x\in D} T_0(x) \neq \emptyset$. Take any $\hat{y} \in \cap_{x\in D} T_0(x) = \cap_{x\in D} cl_X[(X \setminus (coG)^{-1})] \cap D$. Note that $co(X_0 \cup \{y\}) \subset D$ so that $\hat{y} \in K$ by condition (4). Therefore $\hat{y} \in \cap_{i=1}^n cl_X[(X \setminus (coG)^{-1}(x_i)] \cap K = \cap_{i=1}^n T(x_i)$. Hence the family $\{T(x) : x \in X\}$ has the finite intersection property. By the compactness of K, $\cap_{x\in X} T(x) \neq \emptyset$. Since $T(x) \subset S(x)$for each $x \in X$, so that

$$\cap_{x\in X} S(x) = \cap_{x\in X}[X \setminus (coF)^{-1}(x)] \cap K \neq \emptyset.$$

Taking any $\hat{y} \in \cap_{x\in X} S(x)$, then $\hat{y} \in K \cap (X \setminus [\cup_{x\in X} (coF)^{-1}(x)]$, so that $y \notin \cup_{x\in X}(coF)^{-1}(x)$. But the hypothesis (2) implies that $K \subset \cup_{x\in X}(coF)^{-1}(x)$ which is a contradiction. Therefore there must exist $x_0 \in X$ such that $x_0 \in coG(x_0)$. □

Remark 4.3.5. If the mapping G has compactly open inverse values in Theorem 2.2, then condition (C) is equivalent to the following statement:

(C′): *there exist a non-empty compact convex subset X_0 of X and a non-empty compact (not necessarily convex) subset K of X such that for each $y \in X \setminus K$, there is $x \in co(X_0 \cup \{y\})$ with $x \in coG(y)$.*

It is trivial that $(C) \Rightarrow (C')$. We prove that $(C') \Rightarrow (C)$. In fact, for each $y \in X \setminus K$, let $D = co(X_0 \cup \{y\})$. Then D is non-empty compact and convex. Since $G^{-1}(x)$ is compactly open, so is $(coG)^{-1}(x)$. Thus the set $[X \setminus (coG)^{-1}(x)]$ is compactly closed in X and $cl_X[X \setminus (coG)^{-1}(x)] \cap D = [X \setminus (coG)^{-1}(x)] \cap D$. Note that $y \in D$ and $y \notin [X \setminus (coG)^{-1}(x)]$. Thus $y \notin cl_X[X \setminus (coG)^{-1}(x)]$.

Remark 4.3.6. Theorem 4.3.4 was first proved by Ding and Tarafdar [143] under the same hypotheses except that condition (C) was replaced by the following:

(C″): *there exists a non-empty compact convex subset X_0 of X and a non-empty compact (not necessarily convex) compact subset K of X such that for each $y \in X \setminus K$, there is $x \in co(X_0 \cup \{y\})$ with $y \notin cl_C[(X \setminus (coG)^{-1}(x)) \cap C]$ for any non-empty compact subset C of X.*

We note that condition (C″) implies condition (C). Indeed, suppose (C″) holds, for each $y \in X \setminus K$, let $C = co(X_0 \cup \{y\})$. Then C is compact and convex in X. Note that $cl_C[(X \setminus (coG)^{-1}) \cap C] = [cl_X(X \setminus (coG)^{-1})] \cap C$ and $y \in C$. Note that $y \notin cl_C[(X \setminus (coG)^{-1}) \cap C]$, we must have $y \notin cl_X[X \setminus (coG)^{-1}(x)]$. Thus Theorem 2.2 also includes Theorem 3.1 of Ding and Tarafdar [16]. In fact we have shown that both conditions (C) and (C″) are equivalent to each other.

By Remark 4.3.5 and Theorem 4.3.4, we have the following corollary which is Lemma 1 of Ding and Tan [136].

Corollary 4.3.7. Let X be a non-empty subset of a topological vector space E and $G : X \to 2^X$ be such that:

(1) for each $y \in X$, $G^{-1}(y)$ is compactly open in X and $G(x) \neq \emptyset$ for each $x \in X$; and

(2) G satisfies condition (C).

Then there exists $x_0 \in X$ such that $x_0 \in coG(x_0)$.

As an application of Theorem 4.3.4, we have the following result.

Theorem 4.3.8. Let X be a non-empty convex subset of a topological vector space and $G : X \to 2^X$ be a mapping. Suppose there exists a non-empty compact convex subset X_0 of X and a non-empty compact (not necessarily convex) compact subset K of X such that:

(1) $K \subset \cup_{y \in X} int_K\{(coG)^{-1}(y)\}$, where $int_K\{(coG)^{-1}(y)\}$ denotes the relative interior of $(coG)^{-1}(y)$ in K; and

(2) for each $y \in X \setminus K$, there is $x \in co(X_0 \cup \{y\})$ with $y \notin cl_X[X \setminus (coG)^{-1}(x)]$.

Then there exists $x \in X$ such that $x \in coG(x)$.

Proof. For each $y \in X$, let $O_y := int_K(coG)^{-1}(y)$. Then O_y is a compactly open (maybe empty) subset of X and $K \subset \cup_{y \in X} O_y$ by (1). Now define a mapping $F : X \to 2^X$ by

$$F(x) = \{y \in X : x \in O_y\}$$

for each $x \in X$. Then we have that (a) $F(x) \subset coG(x)$ for each $x \in X$; (b) for each $x \in X$, $x \in K$ and $F(x) \neq \emptyset$ by (1); and (c) the set $F^{-1}(y)$ is compactly open in X for each $y \in X$. Therefore F and G satisfy all hypotheses of Theorem 4.3.4 by (2). By Theorem 4.3.4, the conclusion follows. □

Now an immediate consequence of Theorem 4.3.8, we have the following fixed point theorem for transfer open inverse valued mappings.

Theorem 4.3.9. Let X be a non-empty convex subset of a topological vector space and $F, G : X \to 2^X$ be mappings such that:

(1) $F(x) \subset G(x)$ for each $x \in X$;

(2) there exist a non-empty compact convex subset X_0 of X and a non-empty compact (not necessarily convex) compact subset K of X such that for each $y \in X \setminus K$, there is $x \in co(X_0 \cup \{y\})$ with $y \notin cl_X[X \setminus (coG)^{-1}(x)]$; and

(3) F is transfer compactly open inverse valued on K and $F(x) \neq \emptyset$ for each $x \in K$.

Then there exists $x \in X$ such that $x \in coG(x)$.

Proof. Since $coF(x) \neq \emptyset$ for each $x \in K$, $K \subset \cup_{y \in K}\{F^{-1}(y)\}$. Let $O_y := K \cap int_X F^{-1}(y)(= int_K F^{-1}(y))$ for each $y \in X$, then O_y is compactly (maybe empty) open in X. Note that K is compact, so that O_y is open in K for each $y \in X$. Since F is transfer compactly open inverse valued on K, for each $x \in K$, there must exist some $y' \in X$ and a non-empty open subset N_x in K such that $N_x \subset (coF)^{-1}(y)$. Since $N_x \subset K$ we have $N_x \subset int_K(coF)^{-1}(y) = O_y$. Hence $K \subset \cup_{y \in X} O_x$. Now combining (3), all hypotheses of Theorem 4.3.8 are satisfied. By Theorem 4.3.8, there exists $x_0 \in X$ such that $x_0 \in coG(x_0)$. □

Example 2.3.2 also shows that Theorem 4.3.9 is really a generalization of the Browder-Fan fixed point Theorem 2.3.19.

A subset of a topological space X is called a *k-test* set if its intersection with each non-empty compact set C in X is closed in C. A topological space X is called a *k-space* if each k-test set is closed (or equivalently, a subset B of X is open in X if and only if B is compactly open in X, e.g., see Wilansky [573, p.142] or Dugundji [147, p.248]). In Theorem 4.3.4, for each $y \in X$, the set $P^{-1}(y)$ is required to be compactly open in X, while in Border [48], Browder [59], Tarafdar [533], the set $P^{-1}(y)$ is required to be open in X. This generalization would be vacuous if every topological vector space is a k-space. However, this is not the case: the topological vector space $\mathbb{R}^{\mathbb{R}}$ is not a k-space (e.g., see Kelley [286, p.240] or Wilansky [573, p.143]).

4.3.3 Existence of Maximal Elements for F-majorized Mappings

In order to give the existence of maximal elements for F-majorized mappings in topological vector spaces, we first have the selection theorem for F-majorized mappings.

Theorem 4.3.10. Let X be a regular topological vector space and Y be a non-empty subset of a vector space E. Let $\theta : X \to E$ and $P : X \to 2^Y$ be F_θ-majorized. If each open subset of X containing the set $B = \{x \in X : P(x) \neq \emptyset\}$ is paracompact, then there exists a F_θ-pair $\psi, \phi : X \to 2^Y$ such that $P(x) \subset \phi(x)$ for each $x \in X$ and $domP \subset dom\psi$.

Proof. Since P is $F_{\theta,C}$-majorized, for each $x \in B$, let N_x be an open neighborhood of x in X and $\psi_x, \phi_x : X \to 2^Y$ be such that (1): for each $z \in N_x$, $\theta(z) \notin co(\phi_x(z))$ and $\{z \in N_x : P(z) \neq \emptyset\} \subset \{z \in N_x : \psi_x(z) \neq \emptyset\}$; (2): for each $z \in X$, $P(z) \subset \phi_x(z)$, $\psi_x(z) \subset \phi_x(z)$ and $co\phi_x(z) \subset Y$; (3): for each $y \in Y$, $\psi_x^{-1}(y)$ is compactly open in X; and (4): for each $A \in \mathcal{F}(domP)$, $domP \cap (\cap_{x \in A} N_x) \subset \{z \in \cap_{x \in A} N_x : \cap_{x \in A}\psi_x(z) \neq \emptyset\}$. Note that X is regular, for each $x \in B$ there exists an open neighborhood G_x of x in X such that $cl_X G_x \subset N_x$. Let $G = \cup_{x \in B} G_x$, then G is an open subset of X which contains

$B = \{x \in X : P(x) \neq \emptyset\}$ so that G is paracompact by assumption. By Theorem VIII.1.4 of Dugundji [147, p.162], the open covering $\{G_x\}$ of G has an open precise neighborhood-finite refinement $\{G'_x\}$. Given any $x \in B$, we define $\psi'_x, \phi'_x : G \to 2^Y$ by

$$\psi'_x(z) = \begin{cases} \psi_x(z), & \text{if } z \in G \cap cl_X G'_x, \\ Y, & \text{if } z \in G \setminus cl_X G'_x, \end{cases}$$

and

$$\phi'_x(z) = \begin{cases} \phi_x(z), & \text{if } z \in G \cap cl_X G'_x, \\ Y, & \text{if } z \in G \setminus cl_X G'_x, \end{cases}$$

then we have
(i) by (2), for each $z \in G$, $\psi'_x(z) \subset \phi'_x(z)$;

(ii) $DomP \subset Dom\psi'_x$ and $P(z) \subset \phi'_x(z)$; and

(iii) for each $y \in Y$, the set

$$\begin{aligned} (\psi'_x)^{-1}(y) &= \{z \in G : y \in \psi'_x(z)\} \\ &= \{z \in G \cap cl_X G'_x : y \in \psi'_x(z)\} \cup \{z \in G \setminus cl_X G'_x : y \in \psi'_x(z)\} \\ &= \{z \in G \cap cl_X G'_x : y \in \psi_x(z)\} \cup \{z \in G \setminus cl_X G'_x : y \in Y\} \\ &= [(G \cap cl_X G'_x) \cap \psi_x^{-1}(y)] \cup (G \setminus cl_X G'_x) \\ &= (G \cap \psi_x^{-1}(y)) \cup (G \setminus cl_X G'_x). \end{aligned}$$

It follows that for each non-empty compact subset C of X, $(\psi'_x)^{-1}(y) \cap C = (G \cap \psi_x^{-1}(y) \cap C) \cup ((G \setminus cl_X G'_x) \cap C)$ is open in C by (3). Now define $\psi, \phi : X \to 2^Y$ by

$$\psi(z) = \begin{cases} \cap_{x \in B} \psi'_x(z), & \text{if } z \in G, \\ \emptyset, & \text{if } z \in X \setminus G; \end{cases}$$

and

$$\phi(z) = \begin{cases} \cap_{x \in B} \phi'_x(z), & \text{if } z \in G, \\ \emptyset, & \text{if } z \in X \setminus G. \end{cases}$$

Let $z \in X$ be given. Clearly (2) implies $\psi(z) \subset \phi(z)$ and $co\phi(z) \subset Y$. If $z \in X \setminus G$, then $\phi(z) = \emptyset$ so that $\theta(z) \notin co\phi(z)$; if $z \in G$, then $z \in G \cap cl_X G'_x$ for some $x \in B$ so that $\phi'_x(z) = \phi_x(z)$ and hence $\phi(z) \subset \phi_x(z)$. As $\theta(z) \notin co\phi'_x(z)$ by (1) we must have $\theta(z) \notin co\phi(z)$. Therefore $\theta(z) \notin co\phi(z)$ for all $z \in X$. Now we show that for each $y \in Y$, $\psi^{-1}(y)$ is compactly open in X. Indeed, let $y \in Y$ be such that $\psi^{-1}(y) \neq \emptyset$ and C be a compact subset of X; fix an arbitrary $u \in \psi^{-1}(y) \cap C = \{z \in X : y \in \psi(z)\} \cap C = \{z \in G : y \in \psi(z)\} \cap C$. Since $\{G'_x\}$ is a neighborhood-finite refinement, there exists an open neighborhood M_u of u in G such that $\{x \in B : M_u \cap G'_x \neq \emptyset\} = \{x_1, \cdots, x_n\}$. Note that for each $x \in B$ with $x \notin \{x_1, \cdots, x_n\}$, $\emptyset = M_u \cap G'_x = M_u \cap cl_X G'_x$ so that $\psi'_x(z) = Y$ for all $z \in M_u$. Thus we have $\psi(z) = \cap_{x \in B}\psi'_x(z) = \cap_{i=1}^n \psi'_{x_i}(z)$ for all $z \in M_u$. It follows that

$$\begin{aligned}\psi^{-1}(y) &= \{z \in X : y \in \psi(z)\} = \{z \in G : y \in \cap_{x\in B}\psi'_x(z)\} \\ &\supset \{z \in M_u : y \in \cap_{x\in B}\psi'_x(z)\} = \{z \in M_u : y \in \cap_{i=1}^n \psi'_{x_i}(z)\} \\ &= M_u \cap \{z \in G : y \in \cap_{i=1}^n \psi'_{x_i}(z)\} \\ &= M_u \cap [\cap_{i=1}^n (\psi'_{x_i})^{-1}(y)].\end{aligned}$$

But $M'_u = M_u \cap [\cap_{i=1}^n (\psi'_{x_i})^{-1}(y)] \cap C$ is an open neighborhood of u in C such that $M'_u \subset \psi^{-1}(y) \cap C$ since $(\psi_{x_i}{}')^{-1}(y)$ is compactly open in X. This shows that for each $y \in Y$, $\psi^{-1}(y)$ is compactly open in X. Next we claim that

$$\{z \in X : P(z) \neq \emptyset\} \subset \{z \in X : \psi(z) \neq \emptyset\}.$$

Indeed, for each $w \in X$ with $P(w) \neq \emptyset$, we must have $w \in G$. Since $\{G'_x\}$ is neighborhood-finite, the set $\{x \in B : w \in cl_X G'_x\}$ is finite, say, $= \{x'_1, \cdots, x'_m\}$ so that if $x \notin \{x'_1, \cdots, x'_m\}$, then $w \notin cl_X G'_x$ and $\phi'_x(w) = \psi'_x(w) = Y$. Thus we have

$$\psi(w) = \cap_{x\in B}\psi'_x(w) = \cap_{i=1}^m \psi_{x'_i}(w).$$

Since $w \in \cap_{i=1}^m cl_X G_{x'_i} \subset \cap_{i=1}^m N_{x'_i}$, it follows from (4) that $\psi(w) \neq \emptyset$. Hence $\{z \in X : P(z) \neq \emptyset\} \subset \{z \in X : \psi(z) \neq \emptyset\}$. Thus (ψ, ϕ) is a F_θ-pair. Finally, it is clear that $P(x) \subset \phi(x)$ for each $x \in X$ by the definition of ϕ. □

As an application of Theorem 4.3.10 and Theorem 4.3.4, we have the following existence theorem of maximal elements.

Theorem 4.3.11. Let X be a non-empty paracompact convex subset of a topological vector space. Suppose the mapping $P : X \to 2^X$ is F_{I_X}-majorized and satisfies condition (C). Then there exists an $\hat{x} \in K$ such that $P(\hat{x}) = \emptyset$, i.e., $\hat{x}$ is a maximal element of P.

Proof. Suppose that the conclusion does not hold, then $P(x) \neq \emptyset$ for all $x \in X$ and $domP = X$ is paracompact. By Theorem 4.3.10, there exists a F-pair $\psi, \phi : X \to 2^X$ such that (a): $domP \subset dom\psi = X$; (b): $P(x) \subset \phi(x)$ for each $x \in X$; and (c): $x \notin co\phi(x)$ for each $x \in X$. Note that P satisfies condition (C), so does ϕ. Therefore the mappings (ψ, ϕ) satisfy all hypotheses of Theorem 4.3.8. By Theorem 4.3.8, there exists a point $x \in X$ such that $x \in co\phi(x)$ which is a contradiction. Thus there must exist $x_0 \in X$ such that $P(x_0) = \emptyset$. By condition (C), $\hat{x}$ must be in K. □

Theorem 4.3.11 generalizes most existence theorems of maximal elements in the literature, e.g., see Theorem 4.1 of Ding and Tarafdar [143], Theorem 3.3 of Tan and Yuan [511], Corollary 1 of Borglin and Keiding [51], Theorem 2.2 of Toussaint [560], Theorem 2 of Tulcea [562] and so on.

4.3.4 Equilibria in Topological Vector Spaces

Let I be a (possibly infinite) set of agents. For each $i \in I$, let its choice or strategy set X_i be a topological space. Let $X = \Pi_{i\in I} X_i$. For each $i \in I$, let $P_i : X \to 2^{X_i}$ be a correspondence. The collection $\Gamma = (X_i, P_i)_{i\in I}$ will be called a qualitative game. A point $\hat{x} \in X$ is said to be an equilibrium of the game Γ if $P_i(\hat{x}) = \emptyset$ for all $i \in I$.

Theorem 4.3.12 (One-Person Game). Let X be a non-empty convex subset of a topological vector space and mappings $A, B, P : X \to 2^X$ be such that:

(i) for each $x \in X$, $A(x)$ is non-empty and $co(A(x)) \subset B(x)$;
(ii) for each $y \in X$, $A^{-1}(y)$ is compactly open in X; and
(iii) $A \cap P$ is of class F and satisfies condition (C).

Then there exists a point $\hat{x} \in K$ such that $\hat{x} \in \overline{B}(\hat{x})$ and $A(\hat{x}) \cap P(\hat{x}) = \emptyset$.

Proof. Let $M = \{x \in X : x \notin \overline{B}(x)\}$, then M is open in X. Define $Q : X \to 2^X$ by

$$Q(x) = \begin{cases} A(x) \cap P(x), & \text{if } x \notin M, \\ A(x), & \text{if } x \in M. \end{cases}$$

Since $A \cap P$ is of class F, there exist two mappings $\psi, \phi : X \to 2^X$ such that (a) for each $x \in X$, $x \notin co\phi(x)$, $\psi(x) \subset \phi(x)$ and $A(x) \cap P(x) \subset \phi(x)$; (b) for each $y \in X$, $\psi^{-1}(y)$ is compactly open in X; and (c) $\{x \in X : A \cap P(x) \neq \emptyset\} \subset \{x \in X : \psi(x) \neq \emptyset\}$. Now define $\Psi, \Phi : X \to 2^X \cup \{\emptyset\}$ by

$$\Psi(x) = \begin{cases} \psi(x), & \text{if } x \notin M, \\ A(x), & \text{if } x \in M. \end{cases}$$

and

$$\Phi(x) = \begin{cases} \phi(x), & \text{if } x \notin M, \\ A(x), & \text{if } x \in M. \end{cases}$$

Then clearly for each $x \in X$, $\Psi(x) \subset \Phi(x)$, $Q(x) \subset \Phi(x)$ and $\{x \in X : Q(x) \neq \emptyset\} \subset \{x \in X : \Psi(x) \neq \emptyset\}$. If $y \in X$, then it is easy to verify that $\Phi^{-1}(y) = [M \cup \phi^{-1}(y)] \cap A^{-1}(y)$ is compactly open in X by (ii) and (b). Finally, if $x \in M$, then $x \notin \overline{B}(x)$, it follows from (i) that $x \notin coA(x) = co\Phi(x)$, and if $x \notin M$, then $x \notin co\phi(x) = co\Phi(x)$ by (a). This shows that Ψ, Φ are of F-pair. By (iii), Φ also satisfies condition (C). Hence Φ satisfies all hypotheses of Theorem 4.3.11. Thus there exists a point $\hat{x} \in X$ such that $\Phi(\hat{x}) = \emptyset$, so that $Q(\hat{x}) = \emptyset$. Since for each $x \in X$, $A(x) \neq \emptyset$, we must have $\hat{x} \in \overline{B}(\hat{x})$ and $A(\hat{x}) \cap P(\hat{x}) = \emptyset$. □

As another application of Theorem 4.3.11, we have the following existence theorem of equilibria for qualitative games.

Theorem 4.3.13 (Qualitative Game). Let $\Gamma = (X_i, P_i)_{i\in I}$ be a qualitative game. Suppose the following conditions are satisfied for each $i \in I$:

(a) X_i is a non-empty convex subset of a topological vector space E_i;

(b) $P_i : X \to 2^{X_i}$ is F-majorized;

(c) $\cup_{i\in I}\{x \in X : P_i(x) \neq \emptyset\} = \cup_{i\in I} int_X\{x \in X : P_i(x) \neq \emptyset\}$; and

(d) the mapping $\hat{P} : X \to 2^X$ defined by $\hat{P}(x) = \Pi_{i\in I} P_i(x)$ for each $x \in X$ satisfies the condition (C) (i.e., there exist a nonempty compact convex subset X_0 of X and a nonempty compact subset K of X such that for each $y \in X \setminus K$, there is an $x = (x_i) \in \text{co}(X_0 \cup \{y\})$ with $y \notin cl_X[(X \setminus (\text{co}\, P_i)^{-1}(x_i))]$ for each $i \in I$).

Then Γ has an equilibrium point in X (in fact $x \in K$).

Proof. For each $x \in X$, let $I(x) = \{i \in I : P_i(x) \neq \emptyset\}$. Define a correspondence $P : X \to 2^Y$ by

$$P(x) = \begin{cases} \cap_{i\in I(x)} P_i'(x), & \text{if } I(x) \neq \emptyset, \\ \emptyset, & \text{if } I(x) = \emptyset, \end{cases}$$

where $P_i'(x) = \Pi_{j\neq i, j\in I} X_j \otimes P_i(x)$ for each $x \in X$. Then for each $x \in X$, $P(x) \neq \emptyset$ if and only if $I(x) \neq \emptyset$. We shall show that P is F-majorized. For each $x \in X$ with $P(x) \neq \emptyset$, by (c) let $i(x) \in I$ be such that $x \in int_X\{z \in X : P_{i(x)}(z) \neq \emptyset\}$ and by (b) let $N(x)$ be an open neighborhood of x in X and $\phi_{i(x)}, \psi_{i(x)} : X \to 2^{X_i}$ be correspondences such that

(i) $\{z \in N(x) : P_{i(x)}(z) \neq \emptyset\} \subset \{z \in N(x) : \psi_{i(x)}(z)\}$ and $z_{i(x)} \notin co\phi_{i(x)}(z)$;

(ii) for each $z \in X$, $\psi_{i(x)}(z) \subset \phi_{i(x)}(z)$; and

(iii) for each $y \in X_{i(x)}$, $\psi^{-1}_{i(x)}(y)$ is compactly open in X; and

(iv) for each finite subset $\{x_1, \dots, x_n\}$ of $\{x \in X : P(x) \neq \emptyset\}$ with $i(x_1) = \cdots = i(x_n)$, we have

$$\{z \in \cap_{j=1}^n N(x_j) : \cap_{j=1}^n \psi_{i(x_j)}(z) \neq \emptyset\} \supset \{z \in \cap_{j=1}^n N(x_j) : P_i(z) \neq \emptyset\}.$$

Without loss of generality we may assume that

$$\text{N(x)} \subset \text{int}_X\{z \in X : P_{i(x)}(z) \neq \emptyset\}$$

so that $P_{i(x)}(z) \neq \emptyset$ and hence $i(x) \in I(z)$ for all $z \in N(x)$. Let $x \in X$ be such that $P(x) \neq \emptyset$. Now define two mappings $\phi'_{i(x)}, \psi'_{i(x)} : X \to 2^X$ by

$$\psi'_{i(x)}(z) = \prod_{j\in I, j\neq i(x)} X_j \otimes \psi_{i(x)}(z)$$

and

$$\phi'_{i(x)}(z) = \prod\nolimits_{j\in I, j\neq i(x)} X_j \otimes \phi_{i(x)}(z)$$

for each $z \in X$. Then we have:

(a') for each $z \in N(x)$, by (i),

$$\begin{aligned} P(z) = \cap_{i\in I(z)} P_i'(z) \subset P_{i(x)}'(z) &= \Pi_{j\in I, j\neq i(x)} X_j \otimes P_{i(x)}(z) \\ &\subset \Pi_{j\in I, j\neq i(x)} X_j \otimes \phi_{i(x)}(z) = \phi_{i(x)}'(z), \end{aligned}$$

and $z_{i(x)} \notin co\phi_{i(x)}'(z)$;
(b') for each $z \in X$, by (ii), $\psi_{i(x)}'(z) \subset co\phi_{i(x)}'(z)$;
(c') for each $y \in X$, $(\psi_{i(x)}')^{-1}(y) = \psi_{i(x)}^{-1}(y_{i(x)})$ is compactly open in X by (iii); and
(d') for any finite subset A of $\{x \in X : P(x) \neq \emptyset\}$, let $\cup\{I(x) : x \in A\} = \{i_1, \ldots, i_k\}$ where $i_1, \cdots, i_k$ are all distinct and for each $t = 1, \cdots, k$. Let $A_t = \{x \in A : i(x) = i_t\}$. Note that for each $z \in X$,

$$\begin{aligned} \cap_{x\in A}\psi_{i(x)}'(z) &= \cap_{x\in A}\Pi_{j\in I, j\neq i(x)} X_j \otimes \psi_{i(x)}(z) \\ &= \cap_{t=1}^{k}\Pi_{j\in I, j\neq i_t} X_j \otimes (\cap_{x\in A_t}\psi_{i(x)}(z)), \end{aligned}$$

so that for each $z \in \cap_{x\in A} N(x)$, if $\cap_{x\in A}\psi_{i(x)}'(z) = \emptyset$, then there exists $m \in \{1, \ldots, k\}$ such that $\cap_{x\in A_m}\psi_{i(x)}(z) = \emptyset$; it follows from (iv) that $P_{i(x)}(z) = \emptyset$. Thus

$$\{z \in \cap_{x\in A} N(x) : \cap_{x\in A}\psi_{i(x)}(z) \neq \emptyset\} \supset \{z \in \cap_{x\in A} N(x) : P_{i(x)}(z) \neq \emptyset\}.$$

This shows that P is F-majorized. Moreover, by the definition of P, it is clear that $\hat{P}(x) \subset P(x)$ for each $x \in X$, so that (d) implies that P also satisfies condition (C). By Theorem 4.3.11, there exists an $\hat{x} \in X$ such that $P(\hat{x}) = \emptyset$. This implies that $I(\hat{x}) = \emptyset$ and therefore $P_i(\hat{x}) = \emptyset$ for all $i \in I$. □

Theorem 4.3.12 improves Theorem 5.2 of Ding and Tarafdar [143] which in turn improves Theorem 4.2 of Tan and Yuan [511]. In Theorem 4.3.12, if X_i is compact for each $i \in I$, then $X = \Pi_{i\in I} X_i$ is also compact. By letting $K = X$, condition (4) of Theorem 4.3.13 is satisfied trivially. Hence Theorem 4.3.12 generalizes Theorem 2.4 of Toussaint [560] and Proposition 3 of Tulcea [562] and others.

As an application of Theorem 4.3.13, we shall deduce the following equilibrium existence theorem for a non-compact generalized game with an infinite number of players.

Theorem 4.3.14 (Generalized Game). Let $\Gamma = (X_i; A_i, B_i; P_i)_{i\in I}$ be a generalized game. Suppose that the following conditions are satisfied for each $i \in I$:
(i) X_i is a non-empty convex subset of a topological vector space;
(ii) for each $x \in X$, $A_i(x)$ is non-empty, $A_i(x) \subset coB_i(x)$; and
(iii) for each $y \in X_i$, $A_i^{-1}(y)$ is compactly open in X;
(iv) for each $i \in I$, $A_i \cap P_i$ is F-majorized;

(v) $E_i = \{x \in X : A_i(x) \cap P_i(x) \neq \emptyset\}$ is open in X; and

(vi) the mapping $A \cap P : X \to 2^X$ defined by $A \cap P(x) = \Pi_{i \in I} A_i \cap P_i(x)$ for each $x \in X$ satisfies condition (C) (i.e., there exist a nonempty compact convex subset X_0 of X and a nonempty compact subset K of X such that for each $y \in X \setminus K$, there is an $x \in \mathrm{co}(X_0 \cup \{y\})$ with $y \notin cl_X[(X \setminus (\mathrm{co}(A_i \cap P_i))^{-1}(x_i))]$ for each $i \in I$).

Then Γ has an equilibrium $\hat{x}$ in X (in fact, $\hat{x} \in X$).

Proof. For each $i \in I$, let $F_i = \{x \in X : x_i \notin \overline{B}_i(x)\}$, then F_i is open in X. For each $i \in I$, define $Q_i : X \to 2^{X_i}$ by

$$Q_i(x) = \begin{cases} (A_i \cap P_i)(x), & \text{if } x \notin F_i, \\ A_i(x), & \text{if } x \in F_i. \end{cases}$$

We shall prove that the qualitative game $\Gamma = (X_i, Q_i)_{i \in I}$ satisfies all hypotheses of Theorem 4.3.13. For each $i \in I$, we have that the set

$$\begin{aligned} \{x \in X : Q_i(x) \neq \emptyset\} &= \{x \in F_i : Q_i(x) \neq \emptyset\} \cup \{x \in X \setminus F_i : Q_i(x) \neq \emptyset\} \\ &= F_i \cup \{x \in X \setminus F_i : (A_i \cap P_i)(x) \neq \emptyset\} \\ &= F_i \cup [(X \setminus F_i) \cap E_i] \\ &= F_i \cup E_i \end{aligned}$$

is open in X and hence condition (3) of Theorem 4.3.13 is satisfied. By (4), for each $x \in E_i$, there exist an open neighborhood N_x of x in X and correspondences $\psi_x, \varphi_x : X \to 2^{X_i}$ such that

(a) for each $z \in N_x$, $z_i \notin co\varphi_x(z)$;

(b) for each $z \in X$, $\psi_x(z) \subset \varphi_x(z)$;

(c) for each $y \in X$, $\psi_x^{-1}(y)$ is compactly open in X; and

(d) for each nonempty finite set $A \subset E_i$,

$$\{z \in \cap_{x \in A} N_x : \cap_{x \in A} \psi_x(z) \neq \emptyset\} \supset \{z \in \cap_{x \in A} N_x : A_i(x) \cap P_i(x) \neq \emptyset\}.$$

Now for each $x \in X$ with $Q_i(x) \neq \emptyset$, let

$$M(x) = \begin{cases} F_i, & \text{if } x \in F_i, \\ N_x, & \text{if } x \notin F_i, \end{cases}$$

and define correspondences Φ_x, $\Psi_x : X \to 2^{X_i}$ by

$$\Phi_x(z) = \begin{cases} \varphi_x(z), & \text{if } z \notin F_i, \\ A_i(z), & \text{if } z \in F_i, \end{cases}$$

$$\Psi_x(z) = \begin{cases} \psi_x(z), & \text{if } z \notin F_i, \\ A_i(z), & \text{if } z \in F_i. \end{cases}$$

Then for each $x \in X$ with $Q_i(x) \neq \emptyset$, $M(x)$ is an open neighborhood of x such that

(i) for each $z \in M(x)$ and $z_i \notin \Phi_x(z)$ by (2) and (a);

(ii) for each $z \in X$, $\Psi_x(z) \subset \Phi_x(z)$ by (b); and

(iii) for each $y \in X_i$, we have

$$\begin{aligned}
\Psi_x^{-1}(y) &= \{z \in X \setminus F_i : y \in \Psi_x(z)\} \cup \{z \in F_i : y \in \Psi_x(z)\} \\
&= \{z \in X \setminus F_i : y \in \psi_x(z)\} \cup \{z \in F_i : y \in A_i(z)\} \\
&= [(X \setminus F_i) \cap \psi_x^{-1}(y)] \cup (F_i \cap A_i^{-1}(y)) \\
&= [F_i \cup \psi_x^{-1}(y)] \cap A_i^{-1}(y)
\end{aligned}$$

is compactly open in X by (3), (c) and F_i being open in X. Therefore $(\Psi_x, \Phi_x, M(x))$ is an F-majorant of Q_i at $x \in X$. Now let A be a finite subset of $\{x \in X : Q_i(x) \neq \emptyset\}$, then $A = A_1 \cup A_2$ where $A_1 = \{x \in A : x \in F_i\}$ and $A_2 = \{x \in A : x \notin F_i\}$. We first assume that $A_1 = \emptyset$, then by (d),

$$\begin{aligned}
\{z \in \cap_{x \in A} M(x) : \quad & \cap_{x \in A} \Psi_x(z) \neq \emptyset\} \\
= \ & \{z \in \cap_{x \in A_2} M(x) \cap F_i : \cap_{x \in A_2} \Psi_x(z) \neq \emptyset\} \\
\cup \ & \{z \in \cap_{x \in A_2} M(x) \setminus F_i : \cap_{x \in A_2} \Psi_x(z) \neq \emptyset\} \\
= \ & \{z \in \cap_{x \in A_2} M(x) \cap F_i : A_i(z) \neq \emptyset\} \\
\cup \ & \{z \in \cap_{x \in A_2} M(x) \setminus F_i : \cap_{x \in A_2} \psi_x(z) \neq \emptyset\} \\
\supset \ & \{z \in \cap_{x \in A_2} M(x) \cap F_i : A_i(z) \neq \emptyset\} \\
\cup \ & \{z \in \cap_{x \in A_2} M(x) \setminus F_i : A_i(z) \cap P_i(z) \neq \emptyset\} \text{ by (d)} \\
= \ & \{z \in \cap_{x \in A_2} M(x) \cap F_i : Q_i(z) \neq \emptyset\} \\
\cup \ & \{z \in \cap_{x \in A_2} M(x) \setminus F_i : Q_i(z) \neq \emptyset\} \\
= \ & \{z \in \cap_{x \in A} M(x) : Q_i(z) \neq \emptyset\}.
\end{aligned}$$

Secondly, suppose that $A_1 \neq \emptyset$, then we have

$$\begin{aligned}
\{z \in \cap_{x \in A} M(x) : \quad & \cap_{x \in A} \Psi_x(z) \neq \emptyset\} \\
= \ & \{z \in \cap_{x \in A_1} M(x) \cap \cap_{x \in A_2} M(x) : \cap_{x \in A} \Psi_x(z) \neq \emptyset\} \\
= \ & \{z \in F_i \cap \cap_{x \in A_2} M(x) : \cap_{x \in A} \Psi_x(z) \neq \emptyset\} \\
= \ & \{z \in F_i \cap \cap_{x \in A_2} N_x : \cap_{x \in A} \psi_x(z) \neq \emptyset\} \\
= \ & \{z \in F_i \cap \cap_{x \in A_2} M(x) : A_i(z) \neq \emptyset\} \\
= \ & \{z \in \cap_{x \in A} M(x) : Q_i(z) \neq \emptyset\},
\end{aligned}$$

since $\Psi_x(z) = \Phi_x(z) = A_i(z)$ for each $z \in F_i$. This shows that for each $i \in I$, Q_i is F-majorized.

By (vi), $A_i \cap P_i$ satisfies condition (C), so is the mapping Q_i for each $i \in I$. Thus $(X_i, Q_i)_{i \in I}$ satisfies all hypotheses of Theorem 4.3.13. By Theorem 4.3.13, there exists a point $\hat{x} \in X$ such that $Q_i(\hat{x}) = \emptyset$ for all $i \in I$. By (2) and the definition of Q_i, this implies that for each $i \in I$, $\hat{x}_i \in \overline{B}_i(\hat{x})$ and $A_i(\hat{x}) \cap P_i(\hat{x}) = \emptyset$.

As an immediate consequence of Theorem 4.3.14, we have the following result.

Corollary 4.3.15. Let $\Gamma = (X_i, A_i, B_i, P_i)_{i \in I}$ be a generalized game such that $X = \Pi_{i \in I} X_i$ is paracompact. Suppose that the following conditions are satisfied for each $i \in I$:

(i) X_i is a non-empty convex subset of a topological vector space;
(ii) for each $x \in X$ $A_i(x)$ is non-empty and $coA_i(x) \subset coB_i(x)$;
(iii) for each $y \in X_i$, $A_i^{-1}(y)$ and $P_i^{-1}(y)$ are open in X;
(iv) for each $x \in X$, $x_i \notin coP_i(x)$; and
(v) the mapping $A \cap P$ satisfies condition C.

Then Γ has an equilibrium in X.

Proof. Since $\{x \in X : (A_i \cap P_i)(x) \neq \emptyset\} = \cup_{y \in X_i}(A_i^{-1}(y) \cap P_i^{-1}(y))$, by (iii), conditions (iii) and (v), all hypotheses of Theorem 4.3.14 are satisfied. By Theorem 4.3.14, the conclusion follows. □

Corollary 4.3.15 generalizes Theorem 2.5 of Toussaint [560], Corollary 2 of Tulcea [561] (also Corollary 2 in [562]).

Remark 4.3.16. Note that the notion of L_C-majorized correspondences in Ding and Tarafdar [143] is a special case of the notion of F-majorized correspondences, Theorem 4.3 improves Theorem 5.3 of Ding and Tarafdar [143] which in turn generalizes Theorem Theorem 3 of Tulcea [562] (which is also Theorem 4 of Tulcea [561]).

Remark 4.3.17. It is clear that the following two coercive conditions implies condition (C):

$(*)'$: There exists a nonempty closed compact subset K of X and a point $x^0 = (x_i^0)_{i \in I} \in X$ such that $x_i^0 \in \text{co}(A_i(y) \cap P_i(y))$ for all $i \in I$ and for all $y \in X \setminus K$;

and

$(*)$: There exist a nonempty compact convex subset X_0 of X and a nonempty compact subset K of X such that for each $y \in X \setminus K$, there is an $x \in \text{co}(X_0 \cup \{y\})$ with $y \notin \text{cl}_C((X \setminus (\text{co}(A_i \cap P_i))^{-1}(x_i)) \cap C)$ for each $i \in I$ and for any nonempty compact subset C of X.

Therefore Theorem 4.3.14 generalizes Theorem 4.3 of Tan and Yuan [507] to $A_i \cap P_i$ being F-majorized for each $i \in I$. Theorem 4.3.14 also improves and generalizes Theorem 3 of Tulcea [561] and Theorem 4 of Tulcea [562]. Since the coercive condition (v) of Corollary 4.4 of Tan and Yuan [507] implies the coercive condition (C), therefore Corollary 4.3.15 also generalizes Corollary 4.4

of Tan and Yuan [511] and corresponding results of Ding and Tan [143] and others.

4.3.5 The Concept of Approximate Equilibria

In this section, we shall employ the so-called approximation technique preliminarily used by Tuclea [562] to study the existence of approximate equilibria. Then as applications, some existence theorems of equilibria in locally convex topological vector spaces are obtained. We first have the following definition.

Definition 4.3.18. Let $\Gamma = (X_i, A_i, B_i, P_i)_{i\in I}$ be a generalized game. Γ is said to have *approximate equilibria* if for each for non-empty open convex neighborhood V_i of the topological vector space E_i for each $i \in I$, the generalized game $\Gamma' = (X_i, A_i, B_{V_i}, P_i)_{i\in I}$ has an equilibrium point, i.e., there exists a point $x = (x_i)_{i\in I} \in X$ such that $A_i(x) \cap P_i(x) = \emptyset$ and $x_i \in \overline{B_{V_i}}(x)$, where the mapping $B_{V_i} : X \to 2^{X_i}$ is defined by

$$B_{V_i}(x) = (B_i(x) + V_i) \cap X_i$$

for each $x \in X$ and for each $i \in I$.

Now we shall prove the existence theorem of approximate equilibria for generalized games in topological vector spaces when constraint mapping A_i is lower semicontinuous instead of having open inverse values for each $i \in I$.

Theorem 4.3.19. Let X be a non-empty convex subset of a topological vector space E. Let $A, B, P : X \to 2^X$ be such that:

(i) A is lower semicontinuous such that for each $x \in X$, $A(x)$ is non-empty and $coA(x) \subset B(x)$; and

(ii) $A \cap P$ is of class F and satisfies condition C.

Then the one person game $(X; A, B; P)$ has approximate equilibria i.e., for each non-empty open convex neighborhood V of E, there exists a point $x_V \in K$ such that $x_V \in \overline{B_V}(x_V)$ and $A(x_V) \cap P(x_V) = \emptyset$, where $B_V(x) = (B(x) + V) \cap X$ for each $x \in X$.

Proof. Let V be an open convex neighborhood of zero in E. Define the correspondence $A_V, B_V : X \to 2^X$ by $A_V(x) = (A(x) + V) \cap X$, $B_V(x) = (B(x) + V) \cap X$ for each $x \in X$. Then A_V has an open graph in $X \times X$ by (i) and Lemma 4.1 of Chang [77] (or see Tuclea [561, p.7]) such that for each $x \in X$, $A_V(x) \subset B_V(x)$. Let $F_V = \{x \in X : x \notin \overline{B_V}(x)\}$, then F is open in X. Define $Q_V : X \to 2^X \cup \{\emptyset\}$ by

$$Q_V(x) = \begin{cases} A(x) \cap P(x), & \text{if } x \notin F_V, \\ A_V(x), & \text{if } x \in F_V. \end{cases}$$

By (ii), since $A \cap P$ is of class F, there exists a F-pair $\psi, \phi : X \to 2^X$ such that (a) each $x \in X$, $x \notin co(\phi(x)$, $A \cap P(x) \subset \phi(x)$ and $\psi(x) \subset \phi(x)$, (b) $domA \cap P \subset dom\psi$ and (c) $\phi^{-1}(y)$ is compactly open in X for each $y \in X$. Define $\Psi_V, \Phi_V : X \to 2^X$ by

$$\Psi_V(x) = \begin{cases} \psi(x), & \text{if } x \notin F_V, \\ A_V, & \text{if } x \in F_V, \end{cases}$$

and

$$\Phi_V(x) = \begin{cases} \phi(x), & \text{if } x \notin F_V, \\ A_V, & \text{if } x \in F_V, \end{cases}$$

Then clearly for each $x \in X$, $\Psi_V(x) \subset \Phi_V(x)$, $Q(x) \subset \Phi_V(x)$, and $domQ \subset dom\Psi_V$. If $y \in X$, then it is clear that $\Phi_V^{-1}(y) = \{x \in F_V : y \in A_V(x)\} \cup \{x \in X \setminus F_V : y \in \phi(x)\} = [F_V \cup \phi^{-1}(y)] \cap A_V^{-1}(y)$ is compactly open in X by (c) and the fact that A_V has an open graph. Therefore for each $V \in \mathcal{B}$, Q_V is of class F. Finally note that $A \cap P$ satisfies condition (C), so is Q. Thus Φ_V satisfies condition (C). By Theorem 4.3.11, there exists $\hat{x} \in X$ such that $\Psi_V(\hat{x}) = \emptyset$, so that $Q(x) = \emptyset$. Since for each $x \in X$, $A(x) \neq \emptyset$, we must have that $\hat{x} \in \overline{B_V}(\hat{x})$ and $A(\hat{x}) \cap P(\hat{x}) = \emptyset$; that is, the one person game $(X; A, \overline{B_V}; P)$ has an equilibrium point in X. □

Theorem 4.3.20. Let $\Gamma = (X_i, A_i, B_i, P_i)_{i \in I}$ be an abstract economy such that $X = \prod_{i \in I} X_i$ is paracompact. Suppose that the following conditions are satisfied for each $i \in I$:

(1) X_i is a nonempty convex subset of a topological vector space E_i;

(2) A_i is lower semicontinuous with non-empty values and for each $x \in X$, $\text{co}(A_i(x)) \subset \overline{B}_i(x)$;

(3) $A_i \cap P_i$ is F-majorized;

(4) $E^i = \{x \in X : (A_i \cap P_i)(x) \neq \emptyset\}$ is open in X; and

(5) the mapping $A \cap P$ satisfies condition (C).

Then Γ has approximate equilibria, i.e., for give any $V = \prod_{i \in I} V_i$ where for each $i \in I$, V_i is a convex open neighborhood of zero in E_i, there exists a point $\hat{x}_V = (\hat{x}_{V_i})_{i \in I} \in K$ such that for each $i \in I$, $\hat{x}_{V_i} \in \overline{B}_{V_i}(\hat{x}_V)$ and $A_i(\hat{x}_V) \cap P_i(\hat{x}_V) = \emptyset$, where $B_{V_i}(x) = (B_i(x) + V_i) \cap X$ for each $i \in I$ and for each $x \in X$.

Proof. Let $V = \prod_{i \in I} V_i$ be given. For each $i \in I$, define the correspondences $A_{V_i}, B_{V_i} : X \to 2^{X_i}$ by

$$A_{V_i}(x) = (\text{co}(A_i(x)) + V_i) \cap X_i,$$

and

$$B_{V_i}(x) = (B_i(x) + V_i) \cap X_i$$

for each $x \in X$. Then, by (2) and Lemma 4.1 of Chang [77] (or see Tulcea [561, p.7]), A_{V_i} has an open graph in $X \times X_i$ which implies that A_{V_i} has open lower sections (e.g., see Bergstrom et al. [43, p.265-266]), i.e. $(A_{V_i})^{-1}(y)$ is open in X for each $y \in X_i$. For each $i \in I$, define the correspondence $Q_{V_i} : X \to 2^{X_i}$ by

$$Q_{V_i} = \begin{cases} (A_i \cap P_i)(x), & \text{if } x \notin F_{V_i}, \\ A_{V_i}(x), & \text{if } x \in F_{V_i}, \end{cases}$$

where $F_{V_i} = \{x \in X : x_i \notin \overline{B}_{V_i}(x)\}$. Clearly, F_{V_i} is open in X. By using same argument as in the proof of Theorem 4.3, we can prove that the qualitative game $\Gamma = (X_i, Q_{V_i})_{i \in I}$ satisfies that for each $i \in I$, Q_{V_i} is F-majorized. Note that by the definition of Q_{V_i}, we have that for each $i \in I$ and for each $y \in X$, $A_i(y) \cap P_i(y) \subset A_{V_i}(y) \subset Q_{V_i}(y)$ and so Q_{V_i} also satisfies condition (C) by (5). Hence all hypotheses of Theorem 4.3.11 are satisfied. By Theorem 4.3.11, there exists a point $\hat{x}_V = (\hat{x}_{V_i})_{i \in I} \in X$ such that $Q_{V_i}(\hat{x}_V) = \emptyset$ for all $i \in I$. Since for each $i \in I$ and for each $x \in X$, $A_{V_i}(x) \neq \emptyset$ by (2), we must have $\hat{x}_{V_i} \in \overline{B}_{V_i}$ and $A_i(\hat{x}_V) \cap P_i(\hat{x}_V) = \emptyset$ for each $i \in I$. □

Remark 4.3.21. Theorem 4.3.20 generalizes Theorem 5.5 of Ding and Tarafdar [143] which in turn improves Theorem 5.2 of Tan and Yuan [511] in the following two aspects: (a) for each $i \in I$, $A_i \cap P_i$ is F-majorized; (b) the coercive condition (C) is weaker that the coercive condition (e) of Theorem 5.2 of Tan and Yuan [507].

In order to give the existence theorem of equilibria for generalized games in locally convex topological vector spaces, we recall the following result.

Lemma 4.3.22. Let X be a topological space, Y be a nonempty subset of a topological vector space E, $\mathcal{B}$ be a neighborhood base at zero in E and $B : X \to 2^Y$. For each $V \in \mathcal{B}$, let $B_V : X \to 2^Y$ be defined by $B_V(x) = (B(x)+V) \cap Y$ for each $x \in X$. If $\hat{x} \in X$ and $\hat{y} \in Y$ are such that $\hat{y} \in \bigcap_{V \in \mathcal{B}} B_V(x)$, then $\hat{y} \in \overline{B}(\hat{x})$.

Proof. Suppose $\hat{y} \notin \overline{B}(\hat{x})$, then $(\hat{x}, \hat{y}) \notin cl_{X \times Y} graph B$. Let U be an open neighborhood of $\hat{x}$ in X and $V \in \mathcal{B}$ be such that

$$(U \times (\hat{y} + V)) \cap graph B = \emptyset. \tag{4.3.1}$$

Choose $W \in \mathcal{B}$ such that $W - W \subset V$. Since $\hat{y} \in \overline{B_W}(\hat{x})$, by assumption, $(\hat{x}, \hat{y}) \in cl_{X \times Y} graph B_W$ so that $U \times (\hat{y} + W)) \cap graph B_W \neq \emptyset$. Take any $x \in U$ and $w_1 \in W$ with $(x, \hat{y} + w_1) \in graph B_W$ so that $\hat{y} + w_1 \in B_W(x) = (B(x) + W) \cap Y$. Let $z \in B(x)$ and $w_2 \in W$ be such that $y + w_1 = z + w_2 \in Y$, it follows that $z = \hat{y} + w_1 - w_2 \in \hat{y} + W - W \subset \hat{y} + V$ so that $(\hat{y} + V) \cap B(x) \neq \emptyset$ where $x \in U$. This contradicts (4.3.1). Thus we must have $\hat{y} \in \overline{B}(\hat{x})$. □

Now we have the following result.

Theorem 4.3.23. Let $\Gamma = (X_i, A_i, B_i, P_i)_{i\in I}$ be an abstract economy such that $X = \prod_{i\in I} X_i$ is paracompact. Suppose the following conditions are satisfied for each $i \in I$:

(1) X_i is a nonempty convex subset of a locally convex topological vector space E_i;

(2) A_i is lower semicontinuous such that for each $x \in X$, $A_i(x)$ is nonempty and $\mathrm{co}(A_i(x)) \subset \overline{B}_i(x)$;

(3) $A_i \cap P_i$ is F-majorized;

(4) the set $E^i = \{x \in X : (A_i \cap P_i)(x) \neq \emptyset\}$ is open in X;

(5) the mapping $A_i \cap P_i$ satisfies condition (C), i.e., there exist a non-empty compact convex subset X_0 and a non-empty compact subset K (not necessarily convex) of X such that for each $y \in X \setminus K$, $y \notin cl_X[X \setminus (coA_i \cap P_i)^{-1}(x)]$.

Then Γ has an equilibrium point in X.

Proof. For each $i \in I$, let $\mathcal{B}_i$ be the collection of all open convex neighbourhoods of zero in E_i and let $\mathcal{B} = \prod_{i\in I} \mathcal{B}_i$. Given $V = \prod_{i\in I} V_i \in \mathcal{B}$ where $V_i \in \mathcal{B}_i$ for each $i \in I$. By Theorem 4.3.20, there exists an $\hat{x}_V \in K$ such that $\hat{x}_{V_i} \in \overline{B}_{V_i}(\hat{x}_V)$ and $A_i(\hat{x}_V) \cap P_i(\hat{x}_V) = \emptyset$ for each $i \in I$, where $B_{V_i}(x) = (B_i(x) + V_i) \cap X_i$ for each $x \in X$. It follows that the set

$$D_V = \{x \in K : x_i \in \overline{B}_{V_i}(x) \text{ and } A_i(x) \cap P_i(x) = \emptyset \text{ for all } i \in I\}$$

is a nonempty closed subset of K by (4) and hence D_V is compact. Now we claim that the family $(D_V)_{V\in\mathcal{B}}$ has the finite intersection property. Let $\{V^1, \cdots, V^n\}$ be any finite subset of $\mathcal{B}$. For each $k = 1, \cdots, n$, let $V^k = \prod_{i\in I} V_i^k$ where $V_i^k \in \mathcal{B}_i$ for each $i \in I$. Let $V = \prod_{i\in I}(\bigcap_{k=1}^{n} V_i^k)$. Then $D_V \neq \emptyset$. Obviously, $D_V \subset \bigcap_{k=1}^{n} D_{V^k}$ so that $\bigcap_{k=1}^{n} D_{V^k} \neq \emptyset$. Therefore the family $\{D_V : V \in \mathcal{B}\}$ has the finite intersection property. Since K is compact, $\cap\{D_V : V \in \mathcal{B}\} \neq \emptyset$. Now take any $\hat{x} \in \cap\{D_V : V \in \mathcal{B}\}$, then for each $i \in I$, $\hat{x}_i \in \mathcal{B}_{V_i}(\hat{x})$ for each $V_i \in \mathcal{B}_i$ and $A_i(\hat{x}) \cap P_i(\hat{x}) = \emptyset$. By Lemma 4.3.22, it follows that for each $i \in I$, $\hat{x}_i \in \overline{B}_i(\hat{x})$. □

Remark 4.3.24. Theorem 4.3.23 improves and generalizes Theorem 5.6 of Ding and Tarafdar [143] which in turn improves 5.4 of Tan and Yuan [507] to the case that for each $i \in I$, $A_i \cap P_i$ is F-majorized and hence Theorem 4.3.23 also answer the question presented by Tan and Yuan [507] in the affirmative.

Corollary 4.3.25. Let $\Gamma = (X_i, A_i, B_i, P_i)_{i\in I}$ be an abstract economy such that $X = \prod_{i\in I} X_i$ is paracompact. Suppose the following conditions are satisfied for each $i \in I$:

(1) X_i is a nonempty convex subset of a locally convex topological vector space;

(2) A_i has an open graph (resp., is lower semicontinuous) such that for each $x \in X$ $A_i(x)$ is nonempty and $\text{co}(A_i(x)) \subset \overline{B}_i(x)$;

(3) P_i is lower semicontinuous (resp., has an open graph);

(4) $A_i \cap P_i$ is F-majorized; and

(5) the mapping $A \cap P$ satisfies condition (C).

Then Γ has an equilibrium point.

Proof. Since A_i has an open graph (resp., is lower semicontinuous) and P_i is lower semicontinuous (resp., has an open graph), the correspondence $A_i \cap P_i : X \to 2^{X_i}$ is also lower semicontinuous (e.g., see Bergstrom et al. [43]), so that the set $E^i = \{x \in X : (A_i \cap P_i)(x) \neq \emptyset\}$ is open in X. By Theorem 4.3.23, Γ has an equilibrium point in X. □

Corollary 4.3.26. Let $\Gamma = (X_i, A_i, B_i, P_i)_{i \in I}$ be an abstract economy such that such that for each $i \in I$:

(1) X_i is a nonempty compact convex subset of a locally convex topological vector space;

(2) A_i has an open graph (resp., is lower semicontinuous) such that for each $x \in X$, $A_i(x)$ is nonempty and $\text{co}(A_i(x)) \subset \overline{B}_i(x)$;

(3) P_i is lower semicontinuous (resp., has an open graph); and

(4) P satisfies condition (C).

Then Γ has an equilibrium point in X.

Proof. By Corollary 4.3.25 with $X_0 = K = X = \prod_{i \in I} X_i$, the conclusion holds. □

Note that Corollaries 4.3.25 and 4.3.26 generalize Corollaries 5.5 and 5.6 of Tan and Yuan [507] to F-majorized correspondences and generalize Corollary 3 of Borglin and Keiding [51], Theorem 4.1 of Chang [77] and the Theorem of Shafer and Sonnensche [459] in several aspects. Finally we would like to point out that most variational inequalities could be obtained as applications of existence theorems of equilibria for generalized games in topological vector spaces and locally convex topological vector spaces. This idea was first used by Tulcea [562]. For recent developments in this direction, we refer to Tarafdar and Yuan [546] and references therein. Furthermore for the study of other kinds of existence theorems of equilibria for generalized games, we refer to Deguire and Lassonde [13], Tan and Yuan [49], Tarafdar [51] and references therein. Before we conclude this section, we would like to point out that the study of general existence of equilibria for abstact economics and related topics as applications of KKM theory can be illustrated by the following diagram and related notions which do not appear in this section but can be found from Yuan [608] and references therein.

The KKM Principle

$\Uparrow$

Fan's Geometry Lemma

$\Uparrow$

Browder-Fan Fixed Point Theorem

$\Uparrow$

Maximal Element Theorems of KF-Mappings

$\Uparrow$

Selection Theorems KF-Majorized Mappings

$\Uparrow$

Maximal Element Theorems of KF-Majorized Mappings

$\Uparrow$

One-Person Game

$\Uparrow$

Qualitative Game

$\Uparrow$

Abstract Economics (resp., Generalized Game) $\Longleftarrow$ Nash Equilibria

$\Uparrow$

Approximate Equilibria

$\nearrow\nearrow$ $\nwarrow\nwarrow$

Economics Models with LSC Mappings or with USC Mappings

4.4 Approximate Equilibria in Topological Vector Spaces

In this section, by developing an *approximation* method which idea was first motivated by Tulcea [561] and which was also preliminarily used in section 2 above, we obtain an equilibrium existence theorem for a generalized game (abstract economy) in which the constraint correspondences are not assumed to have open graphs nor open lower sections (which are generally assumed in the literature, e.g., see Ding, Kim and Tan [141], Ding and Tan [138] and Yannelis and Prabhakar [587] and the references therein). Our result generalizes the corresponding results of Shafer and Sonnenschein [459], Borglin and Keiding [51], Yannelis and Prabhakar [587], Tulcea [561] and Chang [77] in several ways.

Note that each KF-majorized mapping is a special case of F-majorized mapping. As special cases of Theorem 4.3.11 and Theorem 4.3.14 above, we have the following:

Lemma 4.4.1. Let X be a non-empty paracompact convex subset of a topological vector space and $P : X \to 2^X \cup \{\emptyset\}$ be KF_C-majorized. Suppose that there exist a non-empty compact convex subset X_0 of X and a non-empty compact subset K of X such that for each $y \in X \backslash K$, there is an $x \in co(X_0 \cup \{y\})$ with $x \in coP(y)$. Then there exists an $\hat{x} \in K$ such that $P(\hat{x}) = \emptyset$.

Lemma 4.4.2. Let $\Gamma = (X_i, P_i)_{i \in I}$ be a qualitative game such that $X = \Pi_{i \in I} X_i$ is paracompact. Suppose the following conditions are satisfied:

(i) for each $i \in I$, X_i is a non-empty convex subset of a topological vector space;

(ii) for each $i \in I$, $P_i : X \to 2^{X_i} \cup \{\emptyset\}$ is KF_C-majorized;

(iii) $\cup_{i \in I} \{x \in X : P_i(x) \neq \emptyset\} = \cup_{i \in I} int_X \{x \in X : P_i(x) \neq \emptyset\}$; and

(iv) there exist a non-empty compact convex subset X_0 of X and a non-empty compact subset K of X such that for each $y \in X \setminus K$, there is an $x \in co(X_0 \cup \{y\})$ with $x_i \in coP_i(y)$ for all $i \in I$.

Then Γ has an equilibrium point in K.

The following is an existence of *approximate* equilibrium points for the one-person game.

Theorem 4.4.3. Let X be a non-empty paracompact convex subset of a topological vector space E and $A, B, P : X \to 2^X \cup \{\emptyset\}$ be such that:

(i) A is lower semicontinuous on X and for each $x \in X$, $A(x)$ is non-empty and $coA(x) \subset B(x)$;

(ii) $A \cap P$ is KF_C-majorized; and

(iii) there exists a non-empty compact convex subset X_0 of X and a non-empty compact subset K of X such that for each $y \in X \setminus K$, $co(X_0 \cup \{y\}) \cap co(A(y) \cap P(y)) \neq \emptyset$.

Then for each open convex neighborhood V of 0 in E, the one-person game $(X; A, \overline{B_V}; P)$ has an equilibrium point in K, i.e., there exists a point $x_V \in K$ such that $x_V \in \overline{B_V}(x_V)$ and $A(x_V) \cap P(x_V) = \emptyset$ where $B_V(x) = (B(x) + V) \cap X$ for each $x \in X$.

Proof. Let V be an open convex neighborhood of 0 in E and define $A_V, B_V : X \to 2^X$ by $A_V(x) = (A(x) + V) \cap X$ and $B_V(x) = (B(x) + V) \cap X$ for each $x \in X$. Since A is lower semicontinuous, A_V has an open graph in $X \times X$ by Lemma 4.1 of Chang [77] or Tulcea [561]. By (i), $A_V(x) \subset B_V(x)$ for each $x \in X$. Let $F_V = \{x \in X : x \notin \overline{B_V}(x)\}$, then F_V is open in X. Define $\Psi_V : X \to 2^X \{\emptyset\}$ by

$$\Psi_V = \begin{cases} A(x) \cap P(x), & \text{if } x \notin F_V, \\ A_V(x), & \text{if } x \in F_V. \end{cases}$$

Suppose $x \in X$ is such that $\Psi_V(x) \neq \emptyset$.

Case 1: Suppose $x \in F_V$. Let $\Phi_x = A_V$ and $N_x = F_V$, then N_x is an open neighborhood of x in X such that (a) for each $z \in N_x$, $\Psi_V(z) = A_V(z) = \Phi_x(z)$, $z \notin \overline{B_V}(z)$ so that $z \notin A_V(z) = \Phi_x(z)$; (b) for each $y \in X$, $\Phi_x^{-1}(y) = A_V^{-1}(y)$ is

open in X since A_V has an open graph in $X \times X$. Thus Φ_x is a KF_C-majorant of Ψ_V at x.
Case 2: Suppose $x \notin F_V$. Then $\Psi_V(x) = A(x) \cap P(x) \neq \emptyset$. Since $A \cap P$ is KF_C-majorized by (ii), there exists an open neighborhood N_x of x in X and a correspondence $\Phi'_x : X \to 2^X$ such that (a) $\Psi_V(z) = A(z) \cap P(z) \subset \Phi'_x(z)$ and $z \notin co\Phi'_x(z)$ for each $z \in N_x$ and (b) for each $y \in X$, $(\Phi'_x)^{-1}(y)$ is compactly open in X. Define $\Phi_x : X \to 2^X$ by

$$\Phi_x = \begin{cases} A_V(z) \cap \Phi'_x(z), & \text{if } z \notin F_V, \\ A_V(z), & \text{if } z \in F_V. \end{cases}$$

Note that (i) for each $z \in N_x$, clearly $\Psi_x^{-1}(y) \subset \Phi_x(z)$ and it is easy to see that $z \notin co\Phi_x(z)$ and (ii) for each $y \in x$, $\Phi_x^{-1}(y) = [F_V \cup (\Phi'_x)^{-1}y)] \cap A_V^{-1}(y)$ is compactly open in X. Hence Φ_x is a KF_C-majorant of Ψ_V at x.

Therefore Ψ_V is a KF_C-majorized correspondence. Moreover by (*iii*), for each $y \in X \setminus K$, there exists $x \in co(X_0 \cup \{y\}) \cap co(A(y) \cap P(y))$ so that $x \in co(A(y) \cap P(y)) \subset co\Psi_V(y)$. Thus by Lemma 4.4.1, there exists $\hat{x} \in K$ such that $\Psi_V(\hat{x}) = \emptyset$. Since $A(\hat{x}) \neq \emptyset$ by (i), we must have $\hat{x} \in \overline{B_V}(\hat{x})$ and $A(\hat{x}) \cap P(\hat{x}) = \emptyset$. □

A proof similar to that of Theorem 4.4.3 and therefore its proof omitted gives the following.

Theorem 4.4.4. Let $(X; A, B; P)$ be a one-person game such that $X = \cup_{n=1}^{\infty} C_n$, where $\{C_n\}_{n=1}^{\infty}$ is an increasing sequence of non-empty compact convex subsets in a topological vector space E and the following conditions are satisfied:

(1) A is lower semicontinuous on X and for each $x \in X$, $A(x)$ is non-empty and $coA_i(x) \subset B_i(x)$;

(2) $A \cap P$ is KF_C-majorized; and

(3) for each sequence $(y_n)_{n=1}^{\infty}$ in X with $y_n \in C_n$ for each $n = 1, 2, \cdots$ which is escaping from X relative to $\{C_n\}_{n=1}^{\infty}$, there exist $n \in N$ and $x_n \in C_n$ such that $x_n \in co(A(y_n) \cap P(y_n)) \cap C_n$.

Then for each open convex neighborhood V of zero in E, the one-person game $(X; A, \overline{B_V}; P)$ has an equilibrium point in X, i.e., there exists a point $x_V \in X$ such that $x_V \in \overline{B_V}(x_V)$ and $A(x_V) \cap P(x_V) = \emptyset$ where $B_V(x) = (B(x) + V) \cap X$ for each $x \in X$.

The following is an existence theorem for *approximate* equilibrium points for a generalized game.

Theorem 4.4.5. Let I be any (countable or uncountable) set. For each $i \in I$, let X_i be a non-empty convex subset of a topological vector space E_i and $A_i, B_i, P_i : X = \Pi_{j \in I} X_j \to 2^{X_i} \cup \{\emptyset\}$ be such that:

(a) A_i is lower semicontinuous and for each $x \in X$, $A_i(x)$ is non-empty and $coA_i(x) \subset B_i(x)$;

(b) $A_i \cap P_i$ is KF_C-majorized; and
(c) the set $E^i = \{x \in X : (A_i \cap P_i)(x) \neq \emptyset\}$ is open in X.

Suppose that X is paracompact and that there exist a non-empty compact convex subset X_0 of X and a non-empty compact subset K of X such that for each $y \in X \setminus K$, there is an $x \in co(X_0 \cup \{y\})$ with $x_i \in co(A_i(y) \cap P_i(y))$ for all $i \in I$. Then given any $V = \Pi_{i \in I} V_i$ where for each $i \in I$, V_i is an open convex neighborhood of zero in E_i, the generalized game $\Gamma_V = (X_i; A_i, B_{V_i}; P_i)_{i \in I}$ has an equilibrium point in K; i.e., there exists a point $x_V = (x_{V_i})_{i \in I} \in K$ such that $x_{V_i} \in \overline{B_{V_i}}(x_V)$ and $A_i(x_V) \cap P_i(x_V) = \emptyset$ for each $i \in I$, where $B_{V_i} = (B_i(x) + V_i) \cap X_i$ for each $x \in X$ and each $i \in I$.

Proof. Let $V = \Pi_{i \in I} V_i$ be given where, for each $i \in I$, V_i is an open convex neighborhood of zero in E_i. Fix any $i \in I$ and define $A_{V_i}, B_{V_i} : X \to 2^{X_i}$ by $A_{V_i}(x) = (coA_i(x) + V_i) \cap X_i$ and $B_{V_i}(x) = (B_i(x) + V_i) \cap X_i$ for each $x \in X$. By (a), A_i is lower semicontinuous so that coA_i is also lower semicontinuous by Proposition 2.6 of Michael [371, p.366], it follows from Lemma 4.1 of Chang [77] or from Yannelis [586] that A_{V_i} has an open graph in $X \times X$. Now let $F_{V_i} = \{x \in X : x_i \notin \overline{B_{V_i}}(x)\}$, then F_{V_i} is open in X. Define the mapping $Q_{V_i} : X \to 2^{X_i}$ by

$$Q_{V_i}(x) = \begin{cases} (A_i \cap P_i)(x), & \text{if } x \notin F_{V_i}\,, \\ A_i(x), & \text{if } x \in F_{V_i}. \end{cases}$$

We shall prove that the qualitative game $\mathcal{T} = (X_i, Q_{V_i})_{i \in I}$ satisfies all conditions of Lemma 4.4.2. First we note that for each $i \in I$, the set

$$\begin{aligned} \{x \in X : Q_{V_i}(x) \neq \emptyset\} &= F_{V_i} \cup \{x \in X \setminus F_{V_i} : A_i(x) \cap P_i(x) \neq \emptyset\} \\ &= F_{V_i} \cup ((X \setminus F_{V_i}) \cap E^i) = F_{V_i} \cup E^i \end{aligned}$$

is open in X by (c). Let $x \in X$ be such that $Q_{V_i}(x) \neq \emptyset$. We consider the following two cases:
Case 1: $x \in F_{V_i}$. Let $\Psi_x = A_{V_i}$ and $N_x = F_{V_i}$, then N_x is an open neighborhood of x in X such that (i) $Q_{V_i}(z) \subset \Psi_x(z)$ and by (b), $z_i \notin co\Psi_x(z)$ for each $z \in N_x$; (ii) $co\Psi_x(z) \subset X_i$ for each $z \in X$ by (b) and (iii) $\Psi_x^{-1}(y) = A_{V_i}^{-1}(y)$ is open in X for all $y \in X_i$ since A_{V_i} has an open graph. Therefore, Ψ_x is a KF_C-majorant of Q_{V_i} at x.
Case 2: $x \notin F_{V_i}$. Since $Q_{V_i}(x) = (A_i \cap P_i)(x) \neq \emptyset$ and $A_i \cap P_i$ is KF_C-majorized, there exist an open neighborhood N_x of x in X and a correspondence $\phi_x : X \to 2^{X_i} \cup \{\emptyset\}$ such that (i) $(A_i \cap P_i)(z) \subset \phi_x(z)$ and $z_i \notin co\phi_x(z)$ for each $z \in N_x$ and (iii) $\phi_x^{-1}(y)$ is compactly open in X for each $y \in X_i$. Define $\Psi_x : X \to 2^{X_i} \cup \{\emptyset\}$ by

$$\Psi_x(z) = \begin{cases} A_{V_i}(z) \cap \phi_x(z), & \text{if } z \notin F_{V_i}, \\ A_{V_i}(z), & \text{if } z \in F_{V_i}. \end{cases}$$

Note that as $(A_i \cap P_i)(z) \subset \phi_x(z)$ and $A_i(z) \subset A_{V_i}(z)$ for each $z \in N_x$, we have $Q_{V_i}(z) \subset \Psi_x(z)$ and $co\Psi_x(z) \subset X_i$. It is easy to see that $z_i \notin co\Psi_x(z)$ for all $z \in X$. Moreover, for any $y \in X_i$, the set

$$\begin{aligned}
\Psi_x^{-1}(y) &= \{z \in X : y \in \Psi_x(z)\} \\
&= \{z \in X \setminus F_{V_i} : y \in \Psi_x(z)\} \cup \{z \in F_{V_i} : y \in \Psi_x(z)\} \\
&= \{z \in X \setminus F_{V_i} : y \in A_{V_i}(z) \cap \phi_x(z)\} \cup \{z \in F_{V_i} : y \in A_{V_i}(z)\} \\
&= [(X \setminus F_{V_i}) \cap A_{V_i}^{-1}(y) \cap \phi_x^{-1}(y)] \cup [F_{V_i} \cap A_{V_i}^{-1}(y)] \\
&= [\phi_x^{-1}(y) \cup F_{V_i}] \cap A_{V_i}^{-1}(y)
\end{aligned}$$

is compactly open in X. Therefore, Ψ_x is a KF_C-majorant of Q_{V_i} at x.

Hence Q_{V_i} is a KF_C-majorized correspondence. Now by assumption, there exists a non-empty compact convex subset X_0 of X and a non-empty compact subset K of X such that for each $y \in X \setminus K$, there is an $x \in co(X_0 \cup \{y\})$ with $x_i \in co(A_i(y) \cap P_i(y))$ for all $i \in I$. Note that if $y \notin F_{V_i}$, then $x_i \in co(A_i(y) \cap P_i(y)) \subset coQ_{V_i}(y)$ and if $y \in F_{V_i}$, then $x_i \in co(A_i(y) \cap P_i(y)) \subset co(A_i(x)) = coQ_{V_i}(y)$. Thus for each $i \in I$, $x_i \in coQ_{V_i}(y)$. Moreover the set $\{x \in X : Q_{V_i}(x) \neq \emptyset\} = F_{V_i} \cup \{x \in X \setminus F_{V_i} : (A_i \cap P_i)(x) \neq \emptyset\} = F_{V_i} \cup E^i$ is open in X by condition (c). Therefore all hypotheses of Lemma 4.4.2 are satisfied, so that by Lemma 4.4.2, there exists a point $x_V = (x_{V_i})_{i \in I} \in K$ such that $Q_{V_i}(x_V) = \emptyset$ for all $i \in I$. Since for each $i \in I$, $A_i(x)$ is non-empty, we must have $x_{V_i} \in \overline{B_{V_i}}(x_V)$ and $A_i(x_V) \cap P_i(x_V) = \emptyset$. □

A proof similar to that of Theorem 4.4.5 gives the following result and is thus omitted.

Theorem 4.4.6. Let $(X_i, A_i, B_i, P_i)_{i=1}^N$ be an N-person generalized game such that for each $i = 1, \cdots, N$, $X_i = \cup_{j=1}^{\infty} C_{i,j}$ where $\{C_{i,j}\}_{j=1}^{\infty}$ is an increasing sequence of non-empty compact convex subsets of a topological vector space E_i. Suppose the following conditions are satisfied:

(1) for each $i = 1, 2, \cdots, N$, A_i is lower semicontinuous and for each $x \in X = \Pi_{j \in I} X_j$, $A_i(x)$ is non-empty and $coA_i(x) \subset B_i(x)$;

(2) for each $i = 1, 2, \cdots, N$, $A_i \cap P_i$ is is KF_C-majorized;

(3) for each $i = 1, 2, \cdots, N$, the set $F_i = \{x \in X : A_i(x) \cap P_i(x) \neq \emptyset\}$ is open in X; and

(4) for each sequence $(y_n)_{n=1}^{\infty}$ in X which is escaping from X relative to $\{C_n\}_{n=1}^{\infty}$ where $C_n = \Pi_{i \in I} C_{i,n}$ for each $n = 1, 2, \cdots$, there exist $n \in \mathbb{N}$ and $x_n \in C_n$ such that $\pi_i(x_n) \in A_i(y_n) \cap P_i(y_n)$ for all $i = 1, 2, \cdots, N$.

Then given any $V = \Pi_{i \in I} V_i$ where for each $i = 1, 2, \cdots, N$, V_i is an open convex neighborhood of zero in E_i, the generalized game $\Gamma_V = (X_i; A_i, B_{V_i}; P_i)_{i \in I}$ has an equilibrium point in X; i.e., there exists a point $x_V = (x_{V_i})_{i \in I} \in X$ such that $x_{V_i} \in \overline{B_{V_i}}(x_V)$ and $A_i(x_V) \cap P_i(x_V) = \emptyset$ for each $i = 1, 2, \cdots, N$, where $B_{V_i} = (B_i(x) + V_i) \cap X_i$ for each $x \in X$ and each $i = 1, 2, \cdots, N$.

4.5 Abstract Economics in Locally Convex Spaces

In this section, by constructing an approximate generalized game which is associated with a given generalized game (this procedure is called *approximation method* in this book), existence theorems for equilibria of generalized games are obtained for both KF- and L-majorized mappings in locally convex spaces. In these theorems, the strategy spaces may be non-compact sets in infinite-dimensional locally convex topological vector spaces. The number of agents may be uncountably infinite and the preference correspondences may be non-total or non-transitive and may not have open lower (or upper) sections. Our results generalize many existence theorems for equilibria of generalized games by relaxing the compactness of strategy spaces and by weakening the continuity of constraint and preference correspondences. In particular, the question raised by Yannelis and Prabhakar [587] in 1983 is answered in the affirmative with weaker assumptions.

4.5.1 Equilibria of Abstract Economics for KF-Majorized Mappings in Locally Convex Spaces

For the convenience of our study, we first recall the following fact, which is Lemma 4.3.22 in section 3 and re-write it as:

Lemma 4.5.1. Let X be a topological space, Y a non-empty subset of a topological vector space E, $\mathcal{B}$ a base for the zero neighborhoods in E and $B : X \to 2^Y$. For each $V \in \mathcal{B}$, Let $B_V : X \to 2^Y$ be defined by $B_V(x) = (B(x)+V)\cap Y$ for each $x \in X$. If $\hat{x} \in X$ and $\hat{y} \in Y$ are such that $\hat{y} \in \cap_{V\in\mathcal{B}}\overline{B_V}(\hat{x})$, then $\hat{y} \in \overline{B}(\hat{x})$.

Using Theorem 4.4.5 and Lemma 4.5.1, we shall present one of our main results in this section as follows.

Theorem 4.5.2. Let $\mathcal{G} = (X_i; A_i, B_i; P_i)_{i\in I}$ be a generalized game such that $X = \Pi_{i\in I}X_i$ is paracompact. Suppose the following conditions are satisfied:

(a) for each $i \in I$, X_i is a non-empty convex subset of a locally convex topological vector space E_i;

(b) for each $i \in I$, $A_i : X \to 2^{X_i}$ is lower semicontinuous such that for each $x \in X$, $A_i(x)$ is non-empty and $coA_i(x) \subset B_i(x)$;

(c) for each $i \in I$, $A_i \cap P_i$ is KF_C-majorized;

(d) for each $i \in I$, the set $E^i = \{x \in X : (A_i \cap P_i)(x) \neq \emptyset\}$ is open in X; and

(e) there exist a non-empty compact convex subset X_0 of X and a non-empty compact subset K of X such that for each $y \in X\backslash K$, there is an $x \in co(X_0\cup\{y\})$

with $x_i \in co(A_i(y) \cap P_i(y))$ for all $i \in I$.

Then $\mathcal{G}$ has an equilibrium point in K, i.e. there exists a point $\hat{x} = (\hat{x}_i)_{i \in I} \in K$ such that for each $i \in I$, $\hat{x}_i \in \overline{B_i}(\hat{x})$ and $A_i(\hat{x}) \cap P_i(\hat{x}) = \emptyset$.

Proof. For each $i \in I$, let $\mathcal{B}_i$ be the collection of all open convex neighborhoods of zero in E_i and $\mathcal{B} = \Pi_{i \in I}\mathcal{B}_i$. Given any $V \in \mathcal{B}$, let $V = \Pi_{j \in I}V_j$, where $V_j \in \mathcal{B}_j$ for each $j \in I$. By Theorem 4.4.5, there exists $\hat{x}_V \in K$ such that $\hat{x}_{V_i} \in \overline{B_{V_i}}(\hat{x}_V)$ and $A_i(\hat{x}_V) \cap P_i(\hat{x}_V) = \emptyset$ for each $i \in I$, where $B_{V_i}(x) = (B_i(x) + V_i) \cap X_i$ for each $x \in X$. It follows that the set $Q_V := \{x \in K : x_i \in \overline{B_{V_i}}(x) \text{ and } A_i(x) \cap P_i(x) = \emptyset\}$ is a non-empty closed subset of K by (d).

Now we want to prove $\{Q_V\}_{V \in \mathcal{B}}$ has the finite intersection property. Let $\{V_1, \cdots, V_n\}$ be any finite subset of $\mathcal{B}$. For each $i = 1, \cdots, n$, let $V_i = \Pi_{j \in I}V_{ij}$ where $V_{ij} \in \mathcal{B}_j$ for each $j \in I$; let $V = \Pi_{j \in I}(\cap_{i=1}^n V_{ij})$, then $Q_V \neq \emptyset$. Clearly $Q_V \subset \cap_{i=1}^n Q_{V_i}$ so that $\cap_{i=1}^n Q_{V_i} \neq \emptyset$. Therefore the family $\{Q_V : V \in \mathcal{B}\}$ has the finite intersection property. Since K is compact, $\cap_{V \in \mathcal{B}} Q_V \neq \emptyset$. Now take any $\hat{x} \in \cap_{V \in \mathcal{B}} Q_V$, then for each $i \in I$, $\hat{x}_i \in \overline{B_{V_i}}(\hat{x})$ for each $V_i \in \mathcal{B}_i$ and $A_i(\hat{x}) \cap P_i(\hat{x}) = \emptyset$. By Lemma 4.5.1, we also have for each $i \in I$, $\hat{x}_i \in \overline{B_i}(\hat{x})$. □

By Theorem 4.4.6 and Lemma 4.5.1, we also have the following result which corresponds to Theorem 4.5.2.

Theorem 4.5.3. Let $(X_i, A_i, B_i, P_i)_{i=1}^N$ be an N-person generalized game such that for each $i = 1, 2, \cdots, N$, $X_i = \cup_{j=1}^\infty C_{i,j}$ where $\{C_{i,j}\}_{j=1}^\infty$ is an increasing sequence of non-empty compact convex subsets of a locally convex topological vector space E_i. Suppose the following conditions are satisfied:

(1) for each $i = 1, 2, \cdots, N$, and for each $x \in X = \Pi_{j \in I}X_j$, $A_i(x)$ is non-empty and $coA_i(x) \subset B_i(x)$;

(2) for each $i = 1, 2, \cdots, N$, A_i is lower semicontinuous;

(3) for each $i = 1, 2, \cdots, N$, the correspondence $A_i \cap P_i : X \to 2^{X_i} \cup \{\emptyset\}$ is KF_C-majorized;

(4) for each $i = 1, 2, \cdots, N$, the set $F_i = \{x \in X : A_i(x) \cap P_i(x) \neq \emptyset\}$ is open in X; and

(5) for each sequence $(y_n)_{n=1}^\infty$ in X which is escaping from X relative to $\{C_n\}_{n=1}^\infty$ where $C_n = \Pi_{i \in I}C_{i,n}$ for each $n = 1, 2, \cdots$, there exists $n \in \mathbb{N}$ and $x_n \in C_n$ such that $\pi_i(x_n) \in A_i(y_n) \cap P_i(y_n)$ for all $i = 1, 2, \cdots, N$.

Then $(X_i, A_i, B_i, P_i)_{i=1}^N$ has an equilibrium point $\hat{x} \in X$, i.e., for each $i = 1, 2, \cdots, N$, $\pi_i(\hat{x}) \in \overline{B_i}(\hat{x})$ and $A_i(\hat{x}) \cap P_i(\hat{x}) = \emptyset$.

Corollary 4.5.4. Let $\mathcal{G} = (X_i; A_i, B_i; P_i)_{i \in I}$ be a generalized game such that $X = \Pi_{i \in I}X_i$ is paracompact. Suppose the following conditions are satisfied:

(a) for each $i \in I$, X_i is a non-empty convex subset of a locally convex topological vector space E_i;

(b) for each $i \in I$ and for each $x \in X$, $A_i(x)$ is non-empty and $coA_i(x) \subset B_i(x)$;

(c) for each $i \in I$, A_i and P_i have open lower sections;

(d) for each $i \in I$, $A_i \cap P_i$ is KF_C-majorized; and

(e) there exist a non-empty compact convex subset X_0 of X and a non-empty compact subset K of X such that for each $y \in X \backslash K$, there is an $x \in co(X_0 \cup \{y\})$ with $x_i \in co(A_i(y) \cap P_i(y))$ for all $i \in I$.

Then $\mathcal{G}$ has an equilibrium point in K, i.e., there exists point $\hat{x} \in K$ such that for each $i \in I$, $\hat{x}_i \in \overline{B_i}(\hat{x})$ and $A_i(\hat{x}) \cap P_i(\hat{x}) = \emptyset$.

Proof. By (c), the mapping $A_i : X \to 2^{X_i}$ is lower semicontinuous and the set $E^i = \{x \in X : A_i(x) \cap P_i(x) \neq \emptyset\}$ is open in X. Therefore all the hypotheses in Theorem 4.5.2 are satisfied, so that the conclusion follows. □

Corollary 4.5.5. Let $(X_i, A_i, B_i, P_i)_{i=}^N$ be an N-person generalized game such that for each $i = 1, 2, \cdots, N$, $X_i = \cup_{j=1}^{\infty} C_{i,j}$ where $\{C_{i,j}\}_{j=1}^{\infty}$ is an increasing sequence of non-empty compact convex subsets of a locally convex topological vector space E_i. Suppose the following conditions are satisfied:

(1) for each $i = 1, 2, \cdots, N$, and for each $x \in X = \Pi_{j \in I} X_j$, $A_i(x)$ is non-empty and $coA_i(x) \subset B_i(x)$;

(2) for each $i = 1, 2, \cdots, N$, both A_i and P_i have open lower sections;

(3) for each $i = 1, 2, \cdots, N$, $A_i \cap P_i$ is KF_C-majorized; and

(4) for each sequence $(y_n)_{n=1}^{\infty}$ in X which is escaping from X relative to $\{C_n\}_{n=1}^{\infty}$ where $C_n = \Pi_{i \in I} C_{i,n}$ for each $n = 1, 2, \cdots$, there exist $n \in \mathbb{N}$ and $x_n \in C_n$ such that $\pi_i(x_n) \in A_i(y_n) \cap P_i(y_n)$ for all $i = 1, 2, \cdots, N$.

Then $(X_i, A_i, B_i, P_i)_{i=1}^N$ has an equilibrium point $\hat{x} \in X$, i.e., for each $i = 1, 2, \cdots, N$, $\pi_i(\hat{x}) \in \overline{B_i}(\hat{x})$ and $A_i(\hat{x}) \cap P_i(\hat{x}) = \emptyset$.

Corollary 4.5.4 (resp., Corollary 4.5.5) improves Theorem 6.1 of Yannelis and Prabhakar [587] in the following ways: (i) the index I need not be countable, (ii) for each $i \in I$, the set X_i need not be metrizable and (iii) for each $i \in I$, $A_i \cap P_i$ need not be of class KF.

Corollary 4.5.6. Let $\mathcal{G} = (X_i; A_i, B_i; P_i)_{i \in I}$ be a generalized game such that $X = \Pi_{i \in I} X_i$ is paracompact. Suppose the following conditions are satisfied:

(a) for each $i \in I$, X_i is a non-empty convex subset of a locally convex topological vector space;

(b) for each $i \in I$ and for each $x \in X$, $A_i(x)$ is non-empty and $coA_i(x) \subset B_i(x)$;

(c) for each $i \in I$, A_i has an open graph in $X \times X_i$ (respectively, is lower semicontinuous) and P_i is lower semicontinuous (respectively, has an open graph in $X \times X_i$);

(d) for each $i \in I$, $A_i \cap P_i$ is KF_C-majorized; and

(e) there exist a non-empty compact convex subset X_0 of X and a non-empty compact subset K of X such that for each $y \in X \backslash K$, there is an $x \in co(X_0 \cup \{y\})$ with $x_i \in co(A_i(y) \cap P_i(y))$ for all $i \in I$.
Then $\mathcal{G}$ has an equilibrium point in K, i.e., there exists point $\hat{x} \in X$ such that for each $i \in I$, $\hat{x}_i \in \overline{B_i}(\hat{x})$ and $A_i(\hat{x}) \cap P_i(\hat{x}) = \emptyset$.

Proof. For each $i \in I$, since A_i has an open graph in $X \times X_i$ (respectively, is lower semicontinuous) and P_i is lower semicontinuous (respectively, has an open graph in $X \times X_i$), the mapping $A_i \cap P_i : X \to 2^{X_i} \cup \{\emptyset\}$ is also lower semicontinuous by Lemma 4.2 of Yannelis [586], so that the set $E^i = \{x \in X : A_i(x) \cap P_i(x) \neq \emptyset\}$ is open in X. Therefore all conditions of Theorem 4.4.3 are satisfied and the conclusion follows. □

Corollary 4.5.7. Let $(X_i; A_i, B_i; P_i)_{i=1}^N$ be an N-person game such that for each $i = 1, 2, \cdots, N$, $X_i = \cup_{j=1}^{\infty} C_{i,j}$ where $\{C_{i,j}\}_{j=1}^{\infty}$ is an increasing sequence of non-empty compact convex subsets of a locally convex topological vector space E_i. Suppose the following conditions are satisfied:

(1) for each $i = 1, 2, \cdots, N$ and for each $x \in X = \Pi_{j \in I} X_j$, $A_i(x)$ is non-empty and $coA_i(x) \subset B_i(x)$;

(2) for each $i = 1, 2, \cdots, N$, A_i has an open graph in $X \times X_i$ (respectively, is lower semicontinuou) and P_i is lower semicontinuous (respectively, has an open graph in $X \times X_i$);

(3) for each $i = 1, 2, \cdots, N$, the correspondence $A_i \cap P_i : X \to 2^{X_i} \cup \{\emptyset\}$ is KF_C-majorized; and

(4) for each sequence $(y_n)_{n=1}^{\infty}$ in X which is escaping from X relative to $\{C_n\}_{n=1}^{\infty}$ where $C_n = \Pi_{i \in I} C_{i,n}$ for each $n = 1, 2, \cdots$, there exists $n \in \mathbb{N}$ and $x_n \in C_n$ such that $\pi_i(x_n) \in A_i(y_n) \cap P_i(y_n)$ for all $i = 1, 2, \cdots, N$.

Then $(X_i, A_i, B_i, P_i)_{i=1}^N$ has an equilibrium point $\hat{x} \in X$, i.e., for each $i = 1, 2, \cdots, N$, $\pi_i(\hat{x}) \in \overline{B_i}(\hat{x})$ and $A_i(\hat{x}) \cap P_i(\hat{x}) = \emptyset$.

Corollary 4.5.6 (respectively, 4.5.7) generalizes Corollary 3 of Borglin and Keiding [51], Theorem 4.1 of Chang [77], Theorem of Shafer and Sonnenschein [459] and Theorem 5 of Tulcea [561].

Remark 4.5.8. As shown above, we have proved the existence theorem for equilibria of generalized games with non-compact and infinite dimensional strategy spaces, an infinite numbers of agents, and non-total and non-transitive constraint and preference correspondences which need not have open graphs or open lower or upper sections. Since it is well known that if a correspondence has an open graph, then it has open upper and lower sections (see Bergstrom et al. [43, p.266], and thus the correspondences having open graphs are lower semicontinuous. However, a continuous correspondence need not have open lower or upper sections in general (see Yannleis and Prabhakar [587, p.237]). Also, in infinite settings, the set of feasible allocations generally is not compact in the commodity spaces. Our result generalizes many results in the literature by relaxing the compactness of strategy spaces and the openness of graphs or lower (upper) sections of constraint correspondences.

In 1983, Yannelis and Prabhakar [587] gave the following existence theorem for equilibria of generalized games.

Theorem 4.5.9 (Yannelis and Prabhakar(1983)). Let $\Gamma = (X_i; A_i; P_i)_{i\in I}$ be a generalized game satisfying for each $i \in I$ (where I is countable):

(i) X_i is a non-empty compact, convex and metrizable subset of a locally convex topological vector space;

(ii): the mapping $A_i : X(= \Pi_{i\in I}X_i) \to 2^{X_i}$ satisfies that $clA_i(x) = \overline{A_i}(x)$ for each $x \in X$ (so that the mapping clA_i is upper semicontinuous);

(iii): A_i and P_i have open lower sections; and

(iv): $x_i \notin coP_i(x)$ for all $x \in X$.

Then Γ has an equilibrium, i.e., there exists $\hat{x} \in X$ is such that $\hat{x}_i \in clA_i(\hat{x})$ and $A_i(\hat{x}) \cap P_i(\hat{x}) = \emptyset$.

In addition, Yannelis and Prabhakar asked that '*if Theorem 4.5.9 can be extended to non-metrizable subsets without introducing additional assumptions?*'

Theorem 4.5.2 improves Theorem 4.5.9 of Yannelis and Prabhakar in the following ways: (i) the index set I need not be countable, (ii) for each $i \in I$, the set X_i need not be metrizable and need not compact (iii) for each $i \in I$, both A_i and P_i need not have open lower sections. Therefore the question raised by Yannelis and Prabhakar [587, p.242] is answered in the affirmative with weaker assumptions.

Definition 4.5.10. Let X and Y be topological spaces. A mapping $T : X \to 2^Y \cup \{\emptyset\}$ is said to be *quasi-regular* if

(i) it has open lower sections (i.e., for each $y \in Y$, $T^{-1}(y)$ is open in X);

(ii) $T(x)$ is non-empty and convex for each $x \in X$; and

(iii) $\overline{T}(x) = cl_Y T(x)$ for all $x \in X$.

The correspondence T is said to be *regular* if it is quasi-regular and has an open graph.

Definition 4.5.11. Let X be a non-empty set, Y a non-empty subset of a topological vector space E and $F : X \to 2^Y$. A family $(f_j)_{j\in J}$ of the correspondences between X and Y, indexed by a non-empty filtering set J (we denote by $\leq$ the order relation in J), is an *upper approximating* family for the mapping F (e.g., see [562, p.269]) if

(A_I): $F(x) \subset f_j(x)$ for all $x \in X$ and all $j \in J$;

(A_{II}): for each $j \in J$ there is $j^* \in J$ such that for each $h \in J$ with $h \geq j^*$, $f_h(x) \subset f_j(x)$ for each $x \in X$; and

(A_{III}): for each $x \in X$ and $V \in \mathcal{B}$, where $\mathcal{B}$ is a base for the zero neighborhoods in E, there is $j_{x,V} \in J$ such that $f_h(x) \subset F(x) + V$ if $h \in J$ and $j_{x,V} \leq h$.

From (A_I)-(A_{III}), it is easy to deduce that:

(A_{IV}): for each $x \in X$ and $k \in J$, $F(x) \subset \cap_{j\in J} f_j(x) = \cap_{k\leq j, k\in J} f_j(x) \subset clF(x) \subset \overline{F}(x)$.

If X is a subset of a topological vector space E, we recall that X is said to *have the property* K proprty! if for every compact subset B of X, the convex hull of B is relatively compact in X.

By Theorem 3 and its Remark in Tulcea [562, p.280 and p.281-282], we have the following result.

Lemma 4.5.12. Let $(X_i)_{i\in I}$ be a family of paracompact spaces and $(Y_i)_{i\in I}$ be a family of non-empty closed convex subsets, each in a locally convex topological vector space and each having property K. For each $i \in I$, let $F_i : X_i \to 2^{Y_i}$ be such that F_i is compact and upper semicontinuous with compact convex values. Then there is a common filtering set J (independent of $i \in I$) such that for each $i \in I$, there is a family $(f_{ij})_{j\in J}$ of the correspondences between X_i and Y_i with the following properties:

(a) for each $j \in J$, f_{ij} is regular;

(b) $(f_{ij})_{j\in J}$ and $(\overline{f_{ij}})_{j\in J}$ are upper approximating families for F_i and

(c) for each $j \in J$, $\overline{f_{ij}}$ is continuous if Y_i is compact.

By the above approximation theorem for upper semicontinuous correspondences (Tulcea [562, Theorem 3, p.280]), we can also prove the following existence theorem for equilibria of generalized games in which the constraint correspondences are upper semicontinuous.

Theorem 4.5.13. Let $\mathcal{G} = (X_i; A_i, B_i; P_i)_{i\in I}$ be a generalized game such that $X = \Pi_{i\in I} X_i$ is paracompact. Suppose the following conditions are satisfied:

(a) for each $i \in I$, X_i is a non-empty closed convex subset of a locally convex topological vector space E_i and X_i has the property K;

(b) for each $i \in I$, B_i is compact and upper semicontinuous with non-empty compact convex values and $A_i(x) \subset B_i(x)$ for each $x \in X$;

(c) for each $i \in I$, P_i is lower semicontinuous and KF_C-majorized;

(d) for each $i \in I$, $E^i = \{x \in X : (A_i \cap P_i)(x) \neq \emptyset\}$ is open in X; and

(e) there exist a non-empty compact convex subset X_0 of X and a non-empty compact subset K of X such that for each $y \in X \backslash K$, there is an $x \in co(X_0 \cup \{y\})$ with $x_i \in co(A_i(y) \cap P_i(y))$ for all $i \in I$.

Then there exists $\bar{x} \in K$ such that for each $i \in I$, $\bar{x}_i \in \overline{B_i}(\bar{x})$ and $A_i(\bar{x}) \cap P_i(\bar{x}) = \emptyset$.

Proof. By Lemma 4.5.12, there is a common filtering set J such that for every $i \in I$, there exists a family $(B_{ij})_{j\in J}$ of regular correspondences between X and X_i such that both $(B_{ij})_{j\in J}$ and $(\overline{B_{ij}})_{j\in J}$ are upper approximating of families for B_i. Let $j \in J$ be arbitrarily fixed. The game $\mathcal{G}_j = (X_i; B_{ij}, \overline{B_{ij}}; P_i)_{i\in I}$ satisfies all hypotheses of Theorem 4.5.2. Hence $\mathcal{G}_j$ has an equilibrium $\bar{x}^j \in K$ such that $B_{ij}(\bar{x}^j) \cap P_i(\bar{x}^j) = \emptyset$, and $\pi_i(\bar{x}^j) \in \overline{B_{ij}}(\bar{x}^j)$ for all $i \in I$.

Since $(\bar{x}^j)_{j\in J}$ is a net in the compact set K, without loss of generality we may assume that $(\bar{x}^j)_{j\in J}$ converges to $x^* \in K$. Then for each $i \in I$, $\pi_i(x^*) = \lim_{j\in J}\pi_i(\bar{x}^j)$. Note that for every $j \in J$ and $x \in X$, $A_i(x) \subset B_i(x) \subset B_{ij}(x)$, we have $A_i(\bar{x}^j) \cap P_i(\bar{x}^j) = \emptyset$ for all $i \in I$. By condition (d), for every $i \in I$, $A_i(x^*) \cap P_i(x^*) = \emptyset$. Since $\bar{x}^j$ is an equilibrium point of Γ_j and B_{ij} is regular, for each $x \in X$, $clB_{ij}(x) = \overline{B_{ij}}(x)$, therefore $(\bar{x}^j)_i \in cl(B_{ij}(\bar{x}^j)) = \overline{B_{ij}}(\bar{x}^j)$. As $\overline{B_{ij}}$ has a closed graph, $(x^*, x_i^*) \in graph\overline{B_{ij}}$ for every $i \in I$. For each $i \in I$, since $(\overline{B_{i\,j}})_{j\in J}$ is also an upper approximation family for B_i, $\cap_{j\in J}\overline{B_{i\,j}}(x) \subset \overline{B_i}(x)$ for each $x \in X$ so that $(x^*, x_i^*) \in graph\overline{B_i}$. Therefore for each $i \in I$, $A_i(x^*) \cap P_i(x^*) = \emptyset$ and $\pi_i(x^*) \in \overline{B_i}(x^*)$. □

Corresponding to Theorem 4.5.13, we have the following result.

Theorem 4.5.14. Let $(X_i; A_i; P_i)_{i=1}^N$ be an N-person generalized game such that for each $i = 1, 2, \cdots, N$, $X_i = \cup_{j=1}^\infty C_{i,j}$ is closed and has property K, where $\{C_{i,j}\}_{j=1}^\infty$ is an increasing sequence of non-empty compact convex subsets of a locally convex topological vector space E_i. Suppose the following conditions are satisfied:

(1) for each $i = 1, 2, \cdots, N$, $A_i : X \to 2^{X_i}$ is compact and upper semicontinuous with non-empty compact and convex values;

(2) for each $i = 1, 2, \cdots, N$, the correspondence $P_i : X \to 2^{X_i} \cup \{\emptyset\}$ is lower semicontinuous and KF_C-majorized;

(3) for each $i = 1, 2, \cdots, N$, the set $F_i = \{x \in X : A_i(x) \cap P_i(x) \neq \emptyset\}$ is open in X; and

(4) for each sequence $(y_n)_{n=1}^\infty$ in X which is escaping from X relative to $\{C_n\}_{n=1}^\infty$ where $C_n = \Pi_{i\in I}C_{i,n}$ for each $n = 1, 2, \cdots$, there exist $n \in \mathbb{N}$ and $x_n \in C_n$ such that $\pi_i(x_n) \in A_i(y_n) \cap P_i(y_n)$ for all $i \in I$.

Then $(X_i; A_i; P_i)_{i=1}^N$ has an equilibrium point $\hat{x} \in X$, i.e., for each $i = 1, 2, \cdots, N$, $\pi_i(\hat{x}) \in A_i(\hat{x})$ and $A_i(\hat{x}) \cap P_i(\hat{x}) = \emptyset$.

If X_i is non-empty compact convex in Theorem 4.5.13, we have the following existence result.

Corollary 4.5.15. Let $\mathcal{G} = (X_i; A_i; P_i)_{i\in I}$ be a generalized game and let $X = \Pi_{i\in I}X_i$. Suppose the following conditions are satisfied for each $i \in I$:

(a) X_i is a non-empty compact convex subset of the locally convex topological vector space E_i;

(b) $A_i : X \to 2^{X_i}$ is upper semicontinuous with non-empty compact and convex values for each $x \in X$;

(c) $P_i : X \to 2^{X_i} \cup \{\emptyset\}$ is lower semicontinuous and KF_C-majorized; and
(d) for each $i \in I$, $E^i = \{x \in X; (A_i \cap P_i)(x) \neq \emptyset\}$ is open in X.

Then there exists an $\hat{x} \in X$ such that for each $i \in I$, $\hat{x}_i \in A_i(\hat{x})$ and $A_i(\hat{x}) \cap P_i(\hat{x}) = \emptyset$.

Corollary 4.5.15 also generalizes Theorem 5 of Tulcea [562] and Theorem of Shafer and Sonnenschein [459, p.374].

By Corollary 4.5.15, we obtain the well-known Fan-Glicksberg fixed point theorem (see Fan [156] or Glicksberg [196]) for upper semicontinuous correspondence in locally convex topological vector space.

Corollary 4.5.16. Let X be a compact and convex subset of a locally convex topological vector space E and let $A : X \to 2^X$ be upper semicontinuous with non-empty closed and convex values. Then A has a fixed point.

Proof. Let $I = \{1\}$ and define $P : X \to 2^X \cup \{\emptyset\}$ by $P_i(x) = \emptyset$ for each $x \in X$ in Corollary 4.4.7, then the conclusion is true. □

The following example shows that condition (d) *for each* $i \in I$, $E^i = \{x \in X : (A_i \cap P_i)(x) \neq \emptyset\}$ *is open in* X of Theorem 4.5.13 is essential.

Example 4.5.17. Let $I = \{1\}$ and $X = [0,1]$. Define $A, P : X \to 2^X \cup \{\emptyset\}$ by

$$A(x) = \begin{cases} [1/2, 1], & \text{if } x \in [0, 1/2), \\ [0, 1], & \text{if } x = 1/2, \\ [0, 1/2], & \text{if } x \in (1/2, 1], \end{cases}$$

and

$$P(x) = \begin{cases} \emptyset, & \text{if } x = 0, \\ [0, x), & \text{if } x \in (0, 1]. \end{cases}$$

Then A is upper semicontinuous with non-empty closed convex values and the fixed point set of A is the singleton set $\{1/2\}$. The correspondence P has convex values with open lower sections since for each $y \in [0,1]$, $P^{-1}(y) = (y, 1]$ which is open in X. Therefore A, P and X satisfy all conditions of Corollary 4.5.15 except that $E = \{x \in [0,1] : A(x) \cap P(x) \neq \emptyset\} = [1/2, 1]$ is closed but not open in $[0,1]$. But $A(1/2) \cap P(1/2) \neq \emptyset$, i.e., the generalized game $\Gamma = ([0,1]; A; P)$ has no equilibrium point.

4.5.2 Equilibria of Abstract Economics for L-Majorized Mappings in Locally Convex Spaces

As indicated by the title, the aim of this section is to study the existence of equilibria of abstract economics involving L-majorized mappings inder general setting of non-compact conditions and in which strategic (resp., commodity) spaces are non-compact convex sets in locally topological vector spaces. By using the approximate technique described in Section 4 and approximation for set-valued mappings, we will establish general existence results of equilibria for abstract economics in which preference correspondences may not have lower semicontinuity, and constraint correspondences are only lower or upper semiontinuous. In particular, the underlying setting of non-compact generalized games is general enough to cover corresponding non-compact generalized game models in the existing literature as special cases. In order to exhibit clearly the idea and method used in this section, we only deal with so-called L-majorized mappings whose roots go back to Gale and Mas-Colell's pioneering work [191] and [193] which has been carried forward by Shafer and Sonnenschein in [459] and further developed by Borglin and Keiding in their important paper [51].

Now we recall the definition of so-called $L_{\theta,C}$-majorized mappings as follows.

Definition 4.5.18. Let X be a topological space, Y be a non-empty subset of a vector space E, $\theta : X \to E$ be a (single-valued) mapping and $\phi : X \to 2^Y$ be a correspondence. Then

(1) ϕ is said to be of *class* $L_{\theta,C}$ if (a) for each $x \in X$, $co\phi(x) \subset Y$ and $\theta(x) \notin co\phi(x)$ and (b) for each $y \in Y$, $\phi^{-1}(y)$ is compactly open in X;

(2) (ϕ_x, N_x) is an $L_{\theta,C}$-*majorant* of of ϕ at x if $\phi_x : X \to 2^Y$ and N_x is an open neighborhood of x in X such that (a) for each $z \in N_x$, $\phi(z) \subset \phi_x(z)$; (b) for each $z \in N_x$, $\theta(z) \notin co\phi_x(z)$ and (c) for each $y \in Y$, $\phi_x^{-1}(y)$ is compactly open in X; and

(3) ϕ is said to be $L_{\theta,C}$-*majorized* if for each $x \in X$ with $\phi(x) \neq \emptyset$, there exists an L_θ-majorant (ϕ_x, N_x) of ϕ at x.

Remark 4.5.19. We remark that our definition of $L_{\theta,C}$-majorized mapping above is stronger than KF-majorized mappings introduced in Section 3, however it enables us to establish general theory for the existence of equilibria for generalized games under some very general hypotheses and we believe that this new type of class is worthy to be studied as indeed, L-majorized mappings include most set-valued mappings in the literature as special cases (e.g., see Borgin and Keiding [51], Ding and Tan [136], Yannelis and Prabhakar [587], and so on). For example, our notion of the mapping P being of class $L_{\theta,C}$ and $L_{\theta,C}$-majorized has been used by Ding and Tan [138], Ding and Tarafdar [143], Tan and Yuan [507], and Zhou [621] which, in turn generalize the notions of $\phi \in C(X,Y,\theta)$ and C-majorized mappings introduced by Borglin and Keiding [51], Toussiant [560], Tulcea [561]-[562], and related to Yannelis and Prabhakar [587] as special cases. In particular, our notion of $L_{I_X,C}$-majorized mappings includes self set-valued mappings which have open inverse values and with fixed point free property defined in convex subsets of topological vector spaces as special cases.

Let $X = \Pi_{i \in I} X_i$ be a product of non-empty convex sets X_i in a Hausdorff topological vector space V_i for each $i \in I$, where I be any (possibly countable or uncountable) index set. and $A_i : X \to 2^{X_i}$ is a set-valued mapping for each $i \in I$. We shall denote by $M_S(X, X_i)_{i \in I}$ (resp., $L_S(X, X_i)_{i \in I}$) the set of all families of $\{A_i\}_{i \in I}$ such that for each $i \in I$, $A_i : X \to X_i$ is L-majorized (resp., is of class L). Moreover if the index set I is a singleton, the mapping A in $M_S(X, X)$ (resp., A in $L_S(X, X)$) is said to be L-majorized (resp., of class L) from X to Y. It is clear that each class of L mapping is L-majorized, but the converse is not true in general. In this section, we shall deal mainly with either the case (I): $X = Y$, which is a non-empty convex subset of the topological vector space E and $\theta = I_X$, the identity map on X, or the case (II): $X = \Pi_{i \in I} X_i$ and $\theta = \pi_j : X \to X_j$ is the projection of X onto X_j and $Y = X_j$ is a non-empty convex subset of a topological vector space. In both cases (I) and (II), we shall write L in place of $L_{\theta,C}$.

We first have the following existence of L-class mappings for L-majorized mappings in topological vector spaces.

Theorem 4.5.20. Let X be a regular topological space and Y be a non-empty subset of a vector space E. Let $\theta : X \to E$ and $P : X \to 2^Y$ be $L_{\theta,C}$-majorized. If the domain set B of the mapping P is open and paracompact, then there exists an $L_{\theta,C}$-class mapping $\phi : X \to 2^Y$ such that $P(x) \subset \phi(x)$ for each $x \in X$.

Proof. Since P is $L_{\theta,C}$-majorized, for each $x \in B$, let N_x be an open neighborhood of x in X and $\phi_x : X \to 2^Y$ be a set-valued mapping such that (1) for each $z \in N_x$, $\theta(z) \notin co(\phi_x(z))$; (2) for each $z \in N(x)$, $P(z) \subset \phi_x(z)$, and $co\phi_x(z) \subset Y$; (3) for each $y \in Y$, $\phi_x^{-1}(y)$ is compactly open in X. Note that X is regular, for each $x \in B$, there exists an open neighborhood G_x of x in X such that $cl_X G_x \subset N_x$. Let $G = \cup_{x \in B} G_x$. Then G is an open subset of X which contains $B = \{x \in X : P(x) \neq \emptyset\}$. Note that B is open in X, without loss of the generality, we may assume that $G = B$ (otherwise, taking the set $G \cap B$, which is indeed the same set as B). By the paracompactness assumption on B and Theorem VIII.1.4 of Dugundji [147], the open covering $\{G_x\}$ of G has an open precise neighborhood-finite refinement $\{G'_x\}$. Given any $x \in B$, we define a set-valued mapping $\phi'_x : G \to 2^Y$ by

$$\phi'_x(z) = \begin{cases} \phi_x(z), & \text{if } z \in G \cap cl_X G'_x, \\ Y, & \text{if } z \in G \setminus cl_X G'_x, \end{cases}$$

then we have that (i) for each $y \in Y$, the set

$$\begin{aligned}(\phi'_x)^{-1}(y) &= \{z \in G : y \in \phi'_x(z)\} \\ &= \{z \in G \cap cl_X G'_x : y \in \phi'_x(z)\} \cup \{z \in G \setminus cl_X G'_x : y \in \phi'_x(z)\} \\ &= \{z \in G \cap cl_X G'_x : y \in \phi_x(z)\} \cup \{z \in G \setminus cl_X G'_x : y \in Y\} \\ &= [(G \cap cl_X G'_x) \cap \phi_x^{-1}(y)] \cup (G \setminus cl_X G'_x) \\ &= (G \cap \phi_x^{-1}(y)) \cup (G \setminus cl_X G'_x).\end{aligned}$$

It follows that for each non-empty compact subset C of X, $(\phi'_x)^{-1}(y) \cap C = (G \cap \phi_x^{-1}(y) \cap C) \cup ((G \setminus cl_X G'_x) \cap C)$ is open in C by (3). Now define $\phi : X \to 2^Y$ by

$$\phi(z) = \begin{cases} \cap_{x \in B} \phi'_x(z), & \text{if } z \in G, \\ \emptyset, & \text{if } z \in X \setminus G. \end{cases}$$

Let $z \in X$ be given. Clearly (2) implies $P(z) \subset \phi(z)$ and $co\phi(z) \subset Y$ for all $z \in X$. As $\theta(z) \notin co\phi'_x(z)$ by (1) we must have $\theta(z) \notin co\phi(z)$. Therefore $\theta(z) \notin co\phi(z)$ for all $z \in X$. Now we show that for each $y \in Y$, $\phi^{-1}(y)$ is compactly open in X. Indeed, let $y \in Y$ be such that $\phi^{-1}(y) \neq \emptyset$ and C be a compact subset of X; fix an arbitrary $u \in \phi^{-1}(y) \cap C = \{z \in X : y \in \phi(z)\} \cap C = \{z \in G : y \in \phi(z)\} \cap C$. Since $\{G'_x\}$ is a neighborhood-finite refinement, there exists an open neighborhood M_u of u in G such that $\{x \in B : M_u \cap G'_x \neq \emptyset\} = \{x_1, \cdots, x_n\}$. Note that for each $x \in B$ with $x \notin \{x_1, \cdots, x_n\}$, $\emptyset = M_u \cap G'_x = M_u \cap cl_X G'_x$ so that $\phi'_x(z) = Y$ for all $z \in M_u$. Thus we have $\phi(z) = \cap_{x \in B} \phi'_x(z) = \cap_{i=1}^n \phi'_{x_i}(z)$ for all $z \in M_u$. It follows that

$$\begin{aligned}\phi^{-1}(y) &= \{z \in X : y \in \phi(z)\} = \{z \in G : y \in \cap_{x \in B} \phi'_x(z)\} \\ &\supset \{z \in M_u : y \in \cap_{x \in B} \phi'_x(z)\} = \{z \in M_u : y \in \cap_{i=1}^n \phi'_{x_i}(z)\} \\ &= M_u \cap \{z \in G : y \in \cap_{i=1}^n \phi'_{x_i}(z)\} \\ &= M_u \cap [\cap_{i=1}^n (\phi'_{x_i})^{-1}(y)].\end{aligned}$$

But $M'_u = M_u \cap [\cap_{i=1}^n (\psi'_{x_i})^{-1}(y)] \cap C$ is an open neighborhood of u in C such that $M'_u \subset \phi^{-1}(y) \cap C$ since $(\psi_{x_i}{}')^{-1}(y)$ is compactly open in X. This shows that for each $y \in Y$, $\phi^{-1}(y)$ is compactly open in X. Thus ϕ is of class L_θ and we complete the proof. □

We now establish an existence theorem of equilibria for qualitative games in which preference mappings are L-majorized mappings under the general non-compact setting as applications of the Browder-Fan fixed point theorem above. In order to do so, we first have the following existence of maximal elements for L-majorized mappings in topological vector spaces.

Theorem 4.5.21. Let X be a non-empty compact convex subset of a Hausdorff topological vector space E and $A : X \to 2^X$ be an L-majorized set-valued mapping. Then there exists $x \in X$ such that $A(x) = \emptyset$.

Proof. Suppose the contrary that $A(x) \neq \emptyset$ for all $x \in X$. By Theorem 4.5.20, there exists an L class mapping $\phi : X \to 2^X \setminus \{\emptyset\}$ such that for each $x \in X$, $A(x) \subset \phi(x)$, where ϕ has the following properties: (1) $\phi(x)$ is non-empty and convex for each $x \in X$; (2) $\phi^{-1}(y)$ is open in X for each $y \in X$; and (3) $x \notin \phi(x)$ for all $x \in X$. On the other hand, by the Browder-Fan fixed point theorem, i.e., Theorem 2.3.22, it follows that there exists at least one point $\hat{x} \in X$ such that $\hat{x} \in \phi(\hat{x})$, which is impossible. Thus there must exist some point $x \in X$ such that $A(x) = \emptyset$. □.

Let $X = \Pi_{i \in I} X_i$ be a product space, where X_i is a non-empty convex subset of a Hausdorff topological vector space for each $i \in I$ and $A_i : X \to 2^{X_i}$ is a set-valued mapping. We recall that $\{A_i\}_{i \in I}$ is said to be a *KF family* if the following assumptions are satisfied:

(i) for each $i \in I$, $A_i(x)$ is convex (may be empty) for each $x \in X$; and
(ii) for each $x \in X$, there exists $i \in I$ such that $A_i(x) \neq \emptyset$; and
(iii) for each $i \in I$, the set $A_i^{-1}(y_i)$ is compactly open in X for each $y_i \in X_i$.

As we did in Section 4 of Chapter 2, we denote by $KF(X, X_i)_{i \in I}$ the set of all KF families. For each $i \in I$, if $x_i \notin A_i(x)$ for all $x \in X$, then the family $\{A_i\}_{i \in I} \in M_S(X, X_i)_{i \in I}$. In the case I is a singleton, the mapping $A \in KF(X, X)$ is also called a KF mapping.

Theorem 4.5.22. Let $X = \Pi_{i \in I} X_i$ be a product space, where each $i \in I$, X_i is non-empty compact convex subset of a topological vector space. Suppose that the family $\{A_i\}_{i \in I} \in M_S(X, X_i)_{i \in I}$ is such that $\cup_{i \in I}\{x \in X : A_i(x) \neq \emptyset\} = \cup_{i \in I} int\{x \in X : A_i(x) \neq \emptyset\}$. Then there exists $\hat{x} \in X$ such that $A_i(\hat{x}) = \emptyset$ for each $i \in I$.

Proof. For each $i \in I$, define $A_i' : X \to 2^X$ by $A_i'(x) = \pi_i^{-1}(A_i(x))$ for each $x \in X$. Since A_i is L-majorized from X to X_i, so is A_i' from X to X. For each $x \in X$, define $I(x) = \{i \in I : A_i(x) \neq \emptyset\}$. Now we define a set-valued mapping $A : X \to 2^X$ by

$$A(x) = \begin{cases} \cap_{i \in I(x)} A_i'(x), & \text{if } I(x) \neq \emptyset; \\ \emptyset, & \text{if } I(x) = \emptyset. \end{cases}$$

To complete the proof, it suffices to show that $I(x) = \emptyset$ for some $x \in X$. Suppose the contrary that $I(x) \neq \emptyset$ for each $x \in X$. Fix an arbitrary $x \in X$. Since $I(x) \neq \emptyset$ by the assumption, $x \in \cup_{i \in I}\{x \in X : A_i(x) \neq \emptyset\}$. Thus there exists $i \in I$ such that $A_i(x) \neq \emptyset$. As A_i is L-majorized, there exist an open neighborhood $N(x)$ of x in X and an L majorant B_i of A_i at x such that (a) $A_i(z) \subset B_i(z)$ for each $z \in N(x)$; (b) $y_i \notin B_i(y)$ for each $y = (y_i)_{i \in I} \in X$; and (c) the set $B_i^{-1}(y_i)$ is open in X for each $y_i \in X_i$.

Without loss of generality, we may assume that $N(x) \subset int\{x \in X : A_i(x) \neq \emptyset\}$. Thus $A_i(z) \neq \emptyset$ for each $z \in N(x)$. Now define $B_x : X \to 2^X$ by $B_x(z) = \pi_i^{-1}(B_i(z))$ for each $z \in X$. We claim that B_x is an L-majorant of A at x. Indeed, we have that (i) $A(z) = \cap_{j\in I(x)} A'_j(z) \subset A'_i(z) \subset B_x(z)$ for each $z \in N(x)$; (ii) $y_i \notin B(y)$ since $y_i \notin B_i(y)$ for each $y \in X$ by (b); and (iii) $B^{-1}(y) = (B'_i)^{-1}(y_i)$, which is open in X for each $y \in X$. Therefore B_x is an L-majorant of A at x. Thus A satisfies all hypotheses of Theorem 4.5.21. By Theorem 4.5.21, there exists $\hat{x} \in X$ such that $A(\hat{x}) = \emptyset$, i.e., $A_i(\hat{x}) = \emptyset$ for each $i \in I$. It follows that $I(\hat{x}) = \emptyset$ which contradicts our assumption. Thus the conclusion follows and we complete the proof. □

Now we have the following existence theorem of maximal elements for a family in which domains are not compact.

Theorem 4.5.23. Let $X = \Pi_{i\in I} X_i$ be a product space and I any index set, where X_i is a non-empty convex subset of a topological vector space for each $i \in I$ and $\{A_i\}_{i\in I} \in L_S(X, X_i)_{i\in I}$. Suppose there exist a non-empty compact subset K (not necessarily convex) of X and a non-empty compact convex subset C_i of X_i for each $i \in I$ with the property that for each $x \in X \setminus K$, there exists $i \in I$ such that $A_i(x) \cap C_i \neq \emptyset$. Then there exists $x \in K$ such that $A_i(x) = \emptyset$ for all $i \in I$.

Proof. Suppose it is not true. Then for each $x \in K$, there exists $i \in I$ such that $A_i(x) \neq \emptyset$. By Lemma 2.4.17, there exists a subset $D = \Pi_{i\in I} D_i$ such that for each $x \in K$, there exists $i \in I$ such that $A_i(x) \cap D_i \neq \emptyset$, where D_i is a non-empty polytope (thus it is non-empty compact and convex) for each $i \in I$. It is clear that D is a non-empty compact subset of X. For each $i \in I$, we set $H_i = co(C_i \cup D_i)$, then H_i is a non-empty compact and convex subset of X_i. If $x \in X \setminus K$, then there exists $i \in I$ such that $\emptyset \neq A_i(x) \cap C_i \subset A_i(x) \cap H_i$ by our hypothesis above. If $x \in K$, by our argument above, there exists $i \in I$ such that $\emptyset \neq A_i(x) \cap D_i \subset A_i(x) \cap H_i$. Now for each $i \in I$, define a set-valued mapping $A'_i : X \to 2^{H_i}$ by $A'_i(x) = A_i(x) \cap H_i$ for each $x \in X$. Thus in any case, for each $x \in X$, there exists $i \in I$ such that

$$A_i(x) \cap H_i \neq \emptyset. \tag{4.5.1}$$

Now set $H := \Pi_{i\in I} H_i$ and $X_i := H_i$ for each $i \in I$ in Theorem 4.5.22. Then the family $\{A'_i\}_{i\in I}$ satisfies all hypotheses of Theorem 4.5.22. By Theorem 4.5.22, there exists $x_0 \in H$ such that $A'_i(x_0) = \emptyset$ for all $i \in I$, which contradicts with (4.5.1). Therefore there must exist $x \in K$ such that $A_i(x) = \emptyset$ for all $i \in I$ and we complete the proof. □

Now as applications of Theorem 4.5.20 Theorem 4.5.23, we have the following existence theorem of maximal elements for L-majorized family in topological vector spaces under non-compact setting.

Theorem 4.5.24. Let $X = \Pi_{i\in I} X_i$ be a product space and I any index set, where X_i is a non-empty convex subset of a topological vector space for each $i \in I$. Suppose that the family $\{A_i\}_{i\in I} \in M_S(X, X_i)_{i\in I}$ is such that the set $\{x \in X : A_i(x) \neq \emptyset\}$ is open and paracompact for each $i \in I$. Moreover, there exists a non-empty compact subset K (not necessarily convex) of X and a non-empty compact convex subset C_i of X_i for each $i \in I$ with the property that for each $x \in X \setminus K$, there exists $j \in I$ (not necessarily for all $j \in I$) such that $A_j(x) \cap C_j \neq \emptyset$. Then there exists $x \in K$ such that $A_i(x) = \emptyset$ for all $i \in I$.

Proof. For each $i \in I$, by the assumption of Theorem 4.5.24, it follows that by Theorem 4.5.20, there exists a set-valued mapping $B_i : X \to 2^{Y_i}$, which is of class L such that $A_i(x) \subset B_i(x)$ for each $x \in X$. Hence the family $\{B_i\}_{i \in I}$ satisfies all hypotheses of Theorem 4.5.23. By Theorem 4.5.23, there exists $x \in X$ such that $B_i(x) = \emptyset$ for each $i \in I$, so that $A_i(x) = \emptyset$ for each $i \in I$ as $A_i(x) \subset B_i(x)$. □

In order to study the general existence of equilibria for non-compact generalized games with non-compact hypothesis, which is weaker than those non-compact condition in the existing literature, we first state the following notion called 'non-compact condition' (in short, WC).

The Non-Compact Condition (WC)**:** Let $X = \Pi_{i \in I} X_i$ be a product space and I any possibly countable or uncountable set, where X_i is a non-empty convex subset of a topological vector space for each $i \in I$. Suppose that $\{A_i\}_{i \in I}$ is a family of mappings such that for each $i \in I$, A_i is a set-valued correspondence from X to X_i. Then the family $\{A_i\}_{i \in I}$ is said to satisfy the condition (WC) if there there exists a non-empty compact subset K (not necessarily convex) of X and a non-empty compact convex subset C_i of X_i for each $i \in I$ with the property that for each $x \in X \setminus K$, there exists $j \in I$ (not necessarily for all index $j \in I$) such that a non-empty compact convex subset C_i of X_i for each $i \in I$ with the property that for each $x \in X \setminus K$, there exists $j \in I$ (not necessarily for all index $j \in I$) such that $A_j(x) \cap C_j \neq \emptyset$.

It is clear that above non-compact (WC) condition is much weaker than corresponding non-compact conditions for generalized games used by Ding and Tan [19], Ding and Tarafdar [20], Tan and Yuan [36-39], Tarafdar [30] and related references therein. The significant difference between our non-compact condition (WC) above and the corresponding one used in the litearture, e.g., see condition (vii) of Theorem 4 due to Ding and Tan [19, p.235] is that "for each point $x \in X \setminus K$, our (WC) condition only requires to find one index $j \in I$ (not necessarily for all index $j \in I$) such that $A_j(x) \cap C_j \neq \emptyset$"; but Ding and Tan's non-compact condition needs to have the property that '$A_i(x) \cap co(C_i \cup \{x\})$ is non-empty for all $i \in I$'!. Therefore, our existence theorems of equilibria for non-compact generalized games (below) include those corresponding existing results in the literature mentioned above as special cases.

We also recall that a topological space X is said to be a *perfectly normal space* if X is a normal space and verey closed subset of X is a G_δ-set (recall that a G_δ-set is countable intersections of open sets) (e.g., see Engelking [154, p.45]). Clearly, a normal space X is perfectly normal if and only if every open subset of X is an F_σ-set.

No we have the following existence theorem for 'approximate' equilibria of a non-compact generalized game.

Theorem 4.5.25. Let $G = (X_i, A_i, B_i, P_i)_{i\in I}$ be a generalized game such that $X = \Pi_{i\in I}X_i$ is perfectly normal and paracompact, where I is any (possibly, countable or uncountable) set of players. Suppose that following conditions are satisfied for each each $i \in I$:

(a) the strategic set X_i is a non-empty convex subset of a topological vector space E_i;

(b) the constraint mapping A_i is lower semicontinuous and for each $x \in X$, $A_i(x)$ is non-empty and $coA_i(x) \subset B_i(x)$;

(c) the intersection mapping $A_i \cap P_i$ is L-majorized;

(d) the domain set $E^i = \{x \in X : A_i(x) \cap P_i(x) \neq \emptyset\}$ is open in X; and

(e) the mapping family $\{A_i \cap P_i\}_{i\in I}$ satisfies the weaker non-compact condition (WC), i.e., there exists a non-empty compact subset K (not necessarily convex) of X and a non-empty compact convex subset C_i of X_i for each $i \in I$ with the property that for each $x \in X \setminus K$, there exists $j \in I$ (not necessarily for all index $j \in I$) such that $A_j \cap P_j(x) \cap C_j \neq \emptyset$.

Then given any $V = \Pi_{i\in I}V_i$ where for each $i \in I$, V_i is an open convex neighborhood of zero in E_i, the generalized game $\Gamma_V = (X_i; A_i, B_{V_i}; P_i)_{i\in I}$ has an equilibrium point in K; i.e. there exists a point $x_V = (x_{V_i})_{i\in I} \in K$ such that $x_{V_i} \in \overline{B_{V_i}}(x_V)$ and $A_i(x_V) \cap P_i(x_V) = \emptyset$ for each $i \in I$, where $B_{V_i} = (B_i(x) + V_i) \cap X_i$ for each $x \in X$ and each $i \in I$.

Proof. Let $V = \Pi_{i\in I}V_i$ be given where for each $i \in I$, V_i is an open convex neighborhood of zero in E_i. Fix any $i \in I$ and define $A_{V_i}, B_{V_i} : X \to 2^{X_i}$ by $A_{V_i}(x) = (coA_i(x) + V_i) \cap X_i$ and $B_{V_i}(x) = (B_i(x) + V_i) \cap X_i$ for each $x \in X$. By (b), A_i is lower semicontinuous so that coA_i is also lower semicontinuous by Proposition 2.6 of Michael [371, p.366], it follows from Lemma 4.1 of Chang [65, p.244] or from Tulcea [561, p.7] that A_{V_i} has an open graph in $X \times X$. Now let $F_{V_i} = \{x \in X : x_i \notin \overline{B_{V_i}}(x)\}$, then F_{V_i} is open in X. Define a set-valued mapping $Q_{V_i} : X \to 2^{X_i}$ by

$$Q_{V_i}(x) = \begin{cases} (A_i \cap P_i)(x), & \text{if } x \notin F_{V_i}, \\ A_i(x), & \text{if } x \in F_{V_i}. \end{cases}$$

We shall prove that the qualitative game $\mathcal{T} = (X_i, Q_{V_i})_{i\in I}$ satisfies all conditions of Theorem 4.5.24. First note that for each $i \in I$, the set $\{x \in X : Q_{V_i}(x) \neq \emptyset\} = F_{V_i} \cup \{x \in X \setminus F_{V_i} : A_i(x) \cap P_i(x) \neq \emptyset\} = F_{V_i} \cup E^i$ is open in X. As X is perfectly normal paracompact, it follows that the set $\{X \in X : Q_{V_i}(x) \neq \emptyset\}$ is

also paracompact for each $i \in I$. Secondly let x be any given point in X. We consider the following two cases.
Case 1: Suppose that $x \in F_{V_i}$. Let $\Psi_x = A_{V_i}$ and $N_x = F_{V_i}$, then N_x is an open neighborhood of x in X such that (i) $Q_{V_i}(z) \subset \Psi_x(z)$ for each $z \in X$; and by (c), $\pi_i(z) \notin co\Psi_x(z)$ for each $z \in N_x$; (ii) $co\Psi_x(z) \subset X_i$ for each $z \in X$ by (c) and (iii) $\Psi_x^{-1}(y) = A_{V_i}^{-1}(y)$ is open in X for all $y \in X_i$ since A_{V_i} has an open graph. Therefore, Ψ_x is an L-majorant of Q_{V_i} at x.
Case 2: Suppose that $x \notin F_{V_i}$. Since $Q_{V_i}(x) = (A_i \cap P_i)(x) \neq \emptyset$ and $A_i \cap P_i$ is $\mathcal{L}$-majorized, there exist an open neighborhood N_x of x in X and a correspondence $\phi_x : X \to 2^{X_i}$ such that (i) for each $z \in X$, $(A_i \cap P_i)(z) \subset \phi_x(z)$; (ii) $\pi_i(z) \notin co\phi_x(z)$ for each $z \in N_x$; (iii) $co\phi_x(z) \subset X_i$ for each $z \in X$ and (iv) $\phi_x^{-1}(y)$ is open in X for each $y \in X_i$. Define $\Psi_x : X \to 2^{X_i}$ by

$$\Psi_x(z) = \begin{cases} A_{V_i}(z) \cap \phi_x(z), & \text{if } z \notin F_{V_i}, \\ A_{V_i}(z), & \text{if } z \in F_{V_i}. \end{cases}$$

Note that as $(A_i \cap P_i)(z) \subset \phi_x(z)$ and $A_i(z) \subset A_{V_i}(z)$ for each $z \in N_x$, we have $Q_{V_i}(z) \subset \Psi_x(z)$ and $co\Psi_x(z) \subset X_i$. It is easy to see that $\pi_i(z) \notin co\Psi_x(z)$ for all $z \in X$. Moreover, for any $y \in X_i$, the set

$$\begin{aligned} \Psi_x^{-1}(y) &= \{z \in X : y \in \Psi_x(z)\} \\ &= \{z \in X \setminus F_{V_i} : y \in \Psi_x(z)\} \cup \{z \in F_{V_i} : y \in \Psi_x(z)\} \\ &= \{z \in X \setminus F_{V_i} : y \in A_{V_i}(z) \cap \phi_x(z)\} \cup \{z \in F_{V_i} : y \in A_{V_i}(z)\} \\ &= [(X \setminus F_{V_i}) \cap A_{V_i}^{-1}(y) \cap \phi_x^{-1}(y)] \cup [F_{V_i} \cap A_{V_i}^{-1}(y)] \\ &= [\phi_x^{-1}(y) \cup F_{V_i}] \cap A_{V_i}^{-1}(y) \end{aligned}$$

is compactly open in X. Therefore, Ψ_x is an L-majorant of Q_{V_i} at point x.

Hence Q_{V_i} is an $\mathcal{L}$-majorized correspondence. Now by the assumption (e), it follows that all hypotheses of Theorem 4.5.24 are satisfied. By Theorem 4.5.24, there exists a point $x_V = (x_{V_i})_{i \in I} \in K$ such that $Q_{V_i}(x_V) = \emptyset$ for all $i \in I$. Since for each $i \in I$, $A_i(x)$ is non-empty, we must have $x_{V_i} \in \overline{B_{V_i}}(x_V)$ and $A_i(x_V) \cap P_i(x_V) = \emptyset$. □

Remark 4.5.26. Under the hypothesis of perfectly normal and paracompactness, Theorem 4.5.25 shows that approximate equilibria still exist for non-compact generalized games which is Theorem 3.2 of Tan and Yuan [511].

By Theorem 4.5.25 and Lemma 4.3.22, we have the following mainly existence theorem of equilibria for generalized games in which constraint correspondences are only lower semicontinuous instead of having open lower sections in locally convex topological vector spaces.

Theorem 4.5.27. Let $G = (X_i; A_i, B_i; P_i)_{i \in I}$ be a generalized game such that $X = \Pi_{i \in I} X_i$ is perfectly normal and paracompact. Suppose that the following conditions are satisfied for each $i \in I$:

(a) X_i is a non-empty convex subset of a locally convex topological vector space E_i;

(b) $A_i : X \to 2^{X_i}$ is lower semicontinuous such that for each $x \in X$, $A_i(x)$ is non-empty and $coA_i(x) \subset B_i(x)$;

(c) the mapping $A_i \cap P_i$ is L-majorized;

(d) the set $E^i = \{x \in X : (A_i \cap P_i)(x) \neq \emptyset\}$ is open in X; and

(e) the mapping family $\{A_i \cap P_i\}_{i \in I}$ satisfies the weaker non-compact condition (WC), i.e., there there exist a non-empty compact subset K (not necessarily convex) of X and a non-empty compact convex subset C_i of X_i for each $i \in I$ with the property that for each $x \in X \setminus K$, there exists $j \in I$ (not necessarily for all index $j \in I$) such that $A_j(x) \cap C_j \neq \emptyset$.

Then $\mathcal{G}$ has an equilibrium point in K, i.e. there exists a point $\hat{x} = (\hat{x}_i)_{i \in I} \in K$ such that for each $i \in I$, $\hat{x}_i \in \overline{B_i}(\hat{x})$ and $A_i(\hat{x}) \cap P_i(\hat{x}) = \emptyset$.

Proof. For each $i \in I$, let $\mathcal{B}_i$ be the collection of all open convex neighborhoods of zero in E_i and $\mathcal{B} = \Pi_{i \in I}\mathcal{B}_i$. Given any $V \in \mathcal{B}$, let $V = \Pi_{j \in I}V_j$, where $V_j \in \mathcal{B}_j$ for each $j \in I$. By Theorem 4.5.25, there exists a $\hat{x}_V \in K$ such that $\hat{x}_{V_i} \in \overline{B_{V_i}}(\hat{x}_V)$ and $A_i(\hat{x}_V) \cap P_i(\hat{x}_V) = \emptyset$ for each $i \in I$, where $B_{V_i}(x) = \overline{(B_i(x) + V_i)} \cap X_i$ for each $x \in X$. It follows that the set $Q_V := \{x \in K : x_i \in \overline{B_{V_i}}(x) \text{ and } A_i(x) \cap P_i(x) = \emptyset\}$ is a non-empty closed subset of K by (d). Now we want to prove that the family $\{Q_V\}_{V \in \mathcal{B}}$ has the finite intersection property. Let $\{V_1, \cdots, V_n\}$ be any finite subset of $\mathcal{B}$. For each $i = 1, \cdots, n$, let $V_i = \Pi_{j \in I}V_{ij}$ where $V_{ij} \in \mathcal{B}_i$ for each $j \in I$; let $V = \Pi_{j \in I}(\cap_{i=1}^n V_{ij})$, then $Q_V \neq \emptyset$. Clearly $Q_V \subset \cap_{i=1}^n Q_{V_i}$ so that $\cap_{i=1}^n Q_{V_i} \neq \emptyset$. Therefore the family $\{Q_V : V \in \mathcal{B}\}$ has the finite intersection property. Since K is compact, $\cap_{V \in \mathcal{B}} Q_V \neq \emptyset$. Now take any $\hat{x} \in \cap_{V \in \mathcal{B}} Q_V$, then for each $i \in I$, $\hat{x}_i \in \overline{B_{V_i}}(\hat{x})$ for each $V_i \in \mathcal{B}_i$ and $A_i(\hat{x}) \cap P_i(\hat{x}) = \emptyset$. By Lemma 4.3.22, we also have for each $i \in I$, $\hat{x}_i \in \overline{B_i}(\hat{x})$.
□

Remark 4.5.28. Under the perfectly normal hypothesis, Theorem 3.3 of Tan and Yuan [511] still holds under our much weaker non-compact condition (WC). For example, Theorem 4.5.25 improves Theorem 4 of Ding and Tan [138] in the following: (a) the mapping A_i may not have any open inverse values; (b) the non-compact condition is weaker than their corresponding non-compact condition (vii). Theorem 4.5.25 also improves Corollaries 2 and 3 of Zhou [621] in the following ways: (a) the mapping A_i may not have open lower sections; (b) the mapping clA_i is not necessarily upper semicontinuous; (c) the preference mapping P_i may not have an open graph; (d) the strategic set X_i may not be metrizable; and (e) the index set I may be uncountable.

Corollary 4.5.29. Let $\mathcal{G} = (X_i; A_i, B_i; P_i)_{i \in I}$ be a generalized game such that $X = \Pi_{i \in I}X_i$ is perfectly normal and paracompact. Suppose that the following conditions are satisfied for each $i \in I$:

(a) X_i is a non-empty convex subset of a locally convex topological vector space E_i;

(b) for each $x \in X$, $A_i(x)$ is non-empty and $coA_i(x) \subset B_i(x)$;

(c) A_i and P_i have open lower sections;

(d) $A_i \cap P_i$ is L-majorized; and

(e) the mapping family $\{A_i \cap P_i\}_{i \in I}$ satisfies the weaker non-compact condition (WC), i.e., there there exist a non-empty compact subset K (not necessarily convex) of X and a non-empty compact convex subset C_i of X_i for each $i \in I$ with the property that for each $x \in X \setminus K$, there exists $j \in I$ (not necessarily for all index $j \in I$) such that $A_j(x) \cap C_j \neq \emptyset$.

Then $\mathcal{G}$ has an equilibrium point in K, i.e., there exists point $\hat{x} \in K$ such that for each $i \in I$, $\hat{x}_i \in \overline{B_i}(\hat{x})$ and $A_i(\hat{x}) \cap P_i(\hat{x}) = \emptyset$.

Proof. By (c), the mapping $A_i : X \to 2^{X_i}$ is lower semicontinuous and the set $E^i = \{x \in X : A_i(x) \cap P_i(x) \neq \emptyset\}$ is open in X. Therefore all the hypotheses in Theorem 4.5.27 are satisfied, so that the conclusion follows by Theorem 4.5.27. □

Corollary 4.5.30. Let $G = (X_i, A_i, B_i, P_i)_{i \in I}$ be a generalized game such that $X = \Pi_{i \in I} X_i$ is perfectly normal and paracompact. Suppose that the following conditions are satisfied for each $i \in I$:

(a) X_i is a non-empty convex subset of a locally convex topological vector space;

(b) for each $x \in X$, $A_i(x)$ is non-empty and $coA_i(x) \subset B_i(x)$;

(c) the mapping A_i has an open graph in $X \times X_i$ (resp., is lower semicontinuous) and P_i is lower semicontinuous (resp., has an open graph in $X \times X_i$);

(d) the mapping $A_i \cap P_i$ is L-majorized; and

(e) the mapping family $\{A_i \cap P_i\}_{i \in I}$ satisfies the weaker non-compact condition (WC), i.e., there exist a non-empty compact subset K (not necessarily convex) of X and a non-empty compact and convex subset C_i of X_i for each $i \in I$ with the poverty that for each $x \in X \setminus K$, there exists $j \in I$ such that $co(A_j(x) \cap P_j(x)) \cap C_j \neq \emptyset$.

Then $\mathcal{G}$ has an equilibrium point in K, i.e., there exists point $\hat{x} \in K$ such that for each $i \in I$, $\hat{x}_i \in \overline{B_i}(\hat{x})$ and $A_i(\hat{x}) \cap P_i(\hat{x}) = \emptyset$

Proof. For each $i \in I$, since A_i has an open graph in $X \times X_i$ (resp., is lower semicontinuous) and P_i is lower semicontinuous (respectively, has open graph in $X \times X_i$), the mapping $A_i \cap P_i : X \to 2^{X_i}$ is also lower semicontinuous by Lemma 4.2 of Yannelis [586, p.103], so that the set $E^i = \{x \in X : A_i(x) \cap P_i(x) \neq \emptyset\}$ is open in X. Therefore all conditions of Theorem 4.5.27 are satisfied and the conclusion follows. □

We also remark that Corollaries 4.5.29 and 4.5.30 improve or generalize corresponding existence theorems of equilibria for generalized games given by

Borglin and Keiding [51], Ding et al. [141], Ding and Tan [138], Gale and Mas-Colell [191], Shafer and Sonnenschein [459], and Yannelis and Prabhakar [587] in several aspects.

As a compact and a metrizable versions of Theorem 4.4, we have the following two results.

Corollary 4.5.31. Let $G = (X_i; A_i, B_i; P_i)_{i \in I}$ be a generalized game such that $X = \Pi_{i \in I} X_i$ is perfecetly normal, where I may not be countable. Suppose that the following conditions are satisfied for each $i \in I$:

(a) strategic set X_i is a non-empty compact and convex subset of a locally convex topological vector space E_i;

(b) the constraint mapping $A_i : X \to 2^{X_i}$ is lower semicontinuous such that for each $x \in X$, $A_i(x)$ is non-empty and $coA_i(x) \subset B_i(x)$;

(c) the mapping $A_i \cap P_i$ is L-majorized; and

(d) the set $E^i = \{x \in X : (A_i \cap P_i)(x) \neq \emptyset\}$ is open in X.

Then $\mathcal{G}$ has an equilibrium point in K, i.e., there exists point $\hat{x} \in K$ such that for each $i \in I$, $\hat{x}_i \in \overline{B_i}(\hat{x})$ and $A_i(\hat{x}) \cap P_i(\hat{x}) = \emptyset$

Proof. As X_i is compact, it follows that $X = \Pi_{i \in I} X_i$ is compact and paracompact and thus condition (e) of Theorem 4.5.27 is automatically satisfied. Then the conclusion follows by Theorem 4.5.27 □

Corollary 4.5.32. Let $G = (X_i; A_i, B_i; P_i)_{i \in I}$ be a generalized game and I is countable. Suppose that the following conditions are satisfied for each $i \in I$:

(a) X_i is a non-empty compact and metrizable convex subset of a locally convex topological vector space E_i;

(b) $A_i : X \to 2^{X_i}$ is lower semicontinuous such that for each $x \in X$, $A_i(x)$ is non-empty and $coA_i(x) \subset B_i(x)$;

(c) the mapping $A_i \cap P_i$ is L-majorized; and

(d) the set $E^i = \{x \in X : (A_i \cap P_i)(x) \neq \emptyset\}$ is open in X.

Then $\mathcal{G}$ has an equilibrium point in K, i.e., there exists point $\hat{x} \in K$ such that for each $i \in I$, $\hat{x}_i \in \overline{B_i}(\hat{x})$ and $A_i(\hat{x}) \cap P_i(\hat{x}) = \emptyset$

Proof. As X_i is metrizable for each $i \in I$ and I is countable, it follows that $X = \Pi_{i \in I} X_i$ is still metrizable and thus it is perfectly normal and paracompact by Theorem 4.1.13 of Engelking [154, p.254]. Thus all hypotheses of Theorem 4.5.27 are satisfied and the conclusion follows. □

As each metrizable space is perfectly normal and paracompact, our Theorem 4.5.27 is general enough to include many corresponding existence theorems of equilibria for (even non-compact) generalized games in the literature as special cases. For example, if the set I of players is countable and the strategic set X_i is metrizable, as a special case of Theorem 4.5.27, we have the following.

Corollary 4.5.33. Let $G = (X_i, A_i, B_i, P_i)_{i \in I}$ be a generalized game, where the set I of players is countable. Suppose that the following conditions are satisfied for each $i \in I$:

(a) the strategic set X_i is a non-empty metrizable convex subset of a locally convex topological vector space;

(b) the mapping A_i is lower semicontinuous and for each $x \in X$, $A_i(x)$ is non-empty and $coA_i(x) \subset B_i(x)$;

(c) the mapping $A_i \cap P_i$ is L-majorized;

(d) the set $E^i = \{x \in X : A_i \cap P_i(x) \neq \emptyset\}$ is open in X; and

(e) the mapping family $\{A_i \cap P_i\}_{i \in I}$ satisfies the weaker non-compact condition (WC), i.e., there exist a non-empty compact subset K (not necessarily convex) of X and a non-empty compact and convex subset C_i of X_i for each $i \in I$ with the property that for each $x \in X \setminus K$, there exists $j \in I$ (not necessarily for all $j \in I$) such that $co(A_j(x) \cap P_j(x)) \cap C_j \neq \emptyset$.

Then $\mathcal{G}$ has an equilibrium point in K, i.e., there exists point $\hat{x} \in K$ such that for each $i \in I$, $\hat{x}_i \in \overline{B_i}(\hat{x})$ and $A_i(\hat{x}) \cap P_i(\hat{x}) = \emptyset$

Finally we are ready to prove the existence of equilibria for non-compact generalized games in which preference mappings are not necessarily lower semicontinuous and constraint mappings are only either lower or upper semicontinuous under the general non-compact condition (WC).

Theorem 4.5.34. Let $\Gamma = (X_i; F_i; P_i)_{i \in I}$ be a generalized game such that X is perfectly normal and paracompact and each X_i has the property K, where I may be an uncountable set of players. Suppose that the following conditions are satisfied for each $i \in I$:

(i) the strategic set X_i is a non-empty closed convex set in a Hausdorff locally topological vector space;

(ii) the constraint mapping F_i is compact and upper semicontinuous with non-empty closed and convex values;

(iii) the preference mapping $P_i : X \to 2^{X_i}$ is such that the mapping $F_i \cap P_i$ is L-majorized;

(iv) the set $E_i = \{x \in X : F_i(x) \cap P_i(x) \neq \emptyset\}$ is open in X.

(v) the mapping family $\{F_i \cap P_i\}_{i \in I}$ satisfies the weaker non-compact condition (WC), i.e., there exist a non-empty compact subset K (not necessarily convex) of X and a non-empty compact and convex subset C_i of X_i for each $i \in I$ with the property that for each $x \in X \setminus K$, there exists $j \in I$ such that $co(F_j(x) \cap P_j(x)) \cap C_j \neq \emptyset$.

Then $\mathcal{G}$ has an equilibrium point in K, i.e., there exists point $\hat{x} \in K$ such that for each $i \in I$, $\hat{x}_i \in \overline{B_i}(\hat{x})$ and $A_i(\hat{x}) \cap P_i(\hat{x}) = \emptyset$

Proof. By Lemma 4.5.12, there is a common filtering set J such that for every $i \in I$, there exists a family $(F_{ij})_{j \in J}$ of regular correspondences between X and X_i, such that both $(F_{ij})_{j \in J}$ and $(\overline{F_{ij}})_{j \in J}$ are upper approximating families for F_i. Note that E_i is paracompact and $F_i \cap P_i$ is L-majorized on E_i from E_i to X_i. As X is perfectly normal and paracompact, it follows that the set E_i is also paracompact. By Theorem 4.5.20, it follows that there exists an L-class mapping $\phi_i : E_i \to 2^{X_i}$ such that (a): $(F_i \cap P_i)(x) \subset \phi_i(x)$ and $x_i \notin co\phi_i(x)$ for all $x \in X$; and (b): $\phi_i^{-1}(y)$ is compactly open in X for each $y \in X_i$. For each $j \in J$, define a set-valued mapping $\phi_{ij} : X \to 2^{X_i}$ by $\phi_{ij}(x) = F_{ij}(x) \cap \phi_i(x)$ for each $x \in X$. Note that F_{ij} is regular, it follows that F_{ij} has an open graph and thus the mapping ϕ_{ij} is of class L. Secondly, for each $i \in I$, the set $\{x \in X : F_{ij}(x) \cap \phi_{ij}(x) \neq \emptyset\} = \cup_{y \in X_i} F_{ij}^{-1}(y) \cap \phi_i^{-1}(y)$, which is open in X. Therefore the game $\Gamma_j = (X_i; F_{ij}, \overline{F_{ij}}, \phi_{ij})_{i \in I}$ satisfies all hypotheses of Theorem 4.5.27 above. By Theorem 4.5.27, it follows that the Γ_j has an equilibrium $\hat{x}^j \in K$ such that $F_{ij}(\hat{x}^j) \cap \phi_i(\hat{x}^j) = \emptyset$, and $\pi_i(\hat{x}^j) \in \overline{F_{ij}}(\hat{x}^j)$ for all $i \in I$. Since $F_i(\hat{x}^j) \subset F_{ij}(\hat{x}^j)$, it follows that $F_i(\hat{x}^j) \cap \phi_i(\hat{x}^j) = \emptyset$. Therefore $\{x^j\}_{j \in J} \subset E_i^c$, where the set $E_i^c = \{x \in X : x \notin E_i\}$ is closed in X by condition (iv).

On the other hand, note that $(\hat{x}^j)_{j \in J}$ is a net in the compact set K, without loss of generality, we may assume that $(\hat{x}^j)_{j \in J}$ converges to $x^* \in K$. Then for each $i \in I$, $x_i^* = \lim_{j \in J} \hat{x}_i^j$. As $x^* \in E_i^c$ for all $i \in I$, it follows that $F_i(x^*) \cap P_i(x^*) = \emptyset$. Since $\hat{x}^j$ is an equilibrium point of Γ_j and $\overline{F_{ij}}$ is regular, for each $x \in X$, $clF_{ij}(x) = \overline{F_{ij}}(x)$, therefore $(\hat{x}_i^j) \in cl(B_{ij}(\hat{x}^j)) = \overline{F_{ij}}(\hat{x}^j)$. As $\overline{F_{ij}}$ has a closed graph, it follows that $(x^*, x_i^*) \in graph\overline{F_{ij}}$ for every $i \in I$. For each $i \in I$, since $cl(F_{ij})_{j \in J}$ is an upper approximation family for F_i, it follows that $\cap_{j \in J} \overline{F_{ij}}(x) \subset \overline{F_i}(x)$ for each $x \in X$ so that $(x^*, x_i^*) \in graph\overline{F_i}$. Therefore for each $i \in I$, $F_i(x^*) \cap P_i(x^*) = \emptyset$ and $\pi_i(x^*) \in \overline{F_i}(x^*)$ and this completes the proof. □

Remark 4.5.35. Theorem 4.5.34 generalizes and improves the corresponding Corollary 1 of Zhou [621] in the following aspects: (a) X_i may not be compact nor metrizable; and (b) the set of players may not be countable. Theorem 4.5.34 also improves and generalizes corresponding results in the existing literature.

Note that each metrizable space is perfectly normal and paracompact by Theorem 4.1.13 of [154, p.254], as an application of Theorem 4.5.34, we have the following.

Theorem 4.5.36. Let $\Gamma = (X_i; F_i; P_i)_{i \in I}$ be a generalized game such that X is metrizable, and each X_i has the property K, where I is countable. Suppose that the following conditions are satisfied for each $i \in I$:

(i) the strategic set X_i is a non-empty closed convex set (indeed, it is metrizable) in a Hausdorff locally topological vector space;

(ii) the constraint mapping F_i is compact and upper semicontinuous with non-empty closed and convex values;

(iii) the mapping $F_i \cap P_i$ is L-majorized;

(iv) the set $E_i = \{x \in X : F_i(x) \cap P_i(x) \neq \emptyset\}$ is open in X; and

(v) the mapping family $\{F_i \cap P_i\}_{i \in I}$ satisfies the weaker non-compact condition (WC), i.e., there exists a non-empty compact subset K (not necessarily convex) of X and a non-empty compact and convex subset C_i of X_i for each $i \in I$ with the property that for each $x \in X \setminus K$, there exists $j \in I$ (not necessarily for all $j \in I$) such that $co(F_j(x) \cap P_j(x)) \cap C_j \neq \emptyset$.

Then Γ has an equilibrium $\hat{x}$ in K such that $\hat{x}_i \in F_i(\hat{x})$ and $F_i(\hat{x}) \cap P_i(\hat{x}) = \emptyset$ for all $i \in I$.

Remark 4.5.37. We first note that the assumption X is metrizable in Theorem 4.5.36 is equivalent to saying that 'the index set I is countable and X_i is metrizable for each $i \in I$'. Recently, Zhou [621] proved some existence theorems of equilibria for generalized games by assuming that constraint mappings have upper semicontinuity plus they also satisfy so-called ϵ-CS and (CS)-properties , which look somehow like variant versions of fixed-point free property for set-valued mappings; and in this way, the further generalization of the so-called LSC-property was introduced by Zheng [619]. Secondly, our Theorems 4.5.34 and 4.5.36 generalize the corresponding Corollary 1 of Zhou [621] in the following ways that the strategic sets may not be compact and the set of players (resp., agents) may not be countable.

As compact versions of both Theorem 4.5.34 and Theorem 4.5.36, we have the following corollaries.

Corollary 4.5.38. Let $\Gamma = (X_i; F_i; P_i)_{i\in I}$ be a generalized game such that $X = \Pi_{i\in I} x_i$ is perfectly normal and the index set I may be uncountable. Suppose that the following conditions are satisfied for each $i \in I$;

(i) the strategic set X_i is a non-empty compact and convex set in a Hausdorff locally topological vector space;

(ii) the constraint mapping F_i is upper semicontinuous with non-empty closed and convex values;

(iii) the mapping $F_i \cap P_i$ is L-majorized; and

(iv) the set $E_i = \{x \in X : F_i(x) \cap P_i(x) \neq \emptyset\}$ is open in X.

Then Γ has an equilibrium $\hat{x}$ in K such that $\hat{x}_i \in F_i(\hat{x})$ and $F_i(\hat{x}) \cap P_i(\hat{x}) = \emptyset$ for all $i \in I$.

Corollary 4.5.38 generalizes Corollary 1 of Zhou [621] to a case that the set of players (resp., agents) may not be countable.

Corollary 4.5.39. Let $\Gamma = (X_i; F_i; P_i)_{i\in I}$ be a generalized game, where the I is a countable set of players. Suppose that the following conditions are satisfied for each $i \in I$:

(i) the strategic set X_i is a metrizable compact and convex set in a Hausdorff locally topological vector space;

(i) the constraint mapping F_i is upper semicontinuous with non-empty closed and convex values;

(ii) the mapping $F_i \cap P_i$ is L-majorized; and
(iii) the set $E_i = \{x \in X : F_i(x) \cap P_i(x) \neq \emptyset\}$ is open in X.

Then Γ has an equilibrium $\hat{x}$ in X such that $\hat{x}_i \in F_i(\hat{x})$ and $F_i(\hat{x}) \cap P_i(\hat{x}) = \emptyset$ for all $i \in I$.

Remark 4.5.40. We also note that Example 4.5.17 shows that Theorems 4.5.34 and 4.5.36, Corollaries 4.5.38, and 4.5.39 do not hold if we withdraw the hypothesis ' (iv) for each $i \in I$, the set $E^i = \{x \in X : (A_i \cap P_i)(x) \neq \emptyset\}$ is open in X'.

Finally we note that Theorem 4.5.36 shows that the corresponding existence results of equilibria for generalized games due to Zhou [621] (e.g., see Corollary 1) still hold when the set of players (resp., agents) is uncountable instead of countable and the strategic (resp., commodity) spaces may not be compact; however, we need one extra hypothesis of the so-called 'paracompactness' (of course, this hypothesis is automatically satisfied when the space X is metrizable), but we do not know whether the conclusions of both Theorem 4.5.27 and Theorem 4.5.34 (resp., Theorem 4.5.36) still hold without the hypothesis of perfectly normal paracompactness on the set X.

We also mention that by using Michael type continuous selection theorems and their generalized versions plus Himmelberg fixed point theorem in [233], which is a different approach compared with the approximate method used in this section, some existence of equilibria for generalized games (resp., abstract economics) have been established by Wu [576], Wu and Shen [577], Zheng [619] and Zhou [621]. As a result, both results are independent of each other.

4.6 Abstract Economics without Compactness and Paracompactness

The purpose of this section has two parts. We first prove the general existence theorems for qualitative games and generalized games with infinite (countable or uncountable) sets of players without compact and paracompact hypotheses in the topological vector spaces. These results answer questions raised by Kim [299] and Yannelis and Prabhakar [587] again in the affirmative with weaker hypotheses. Second, we obtain an equilibrium existence theorem for generalized games in which the constraint correspondences are lower semicontinuous instead of having an open graph or open lower sections. Our results generalize the corresponding results of Arrow and Debreu [8], Shafer and Sonnenschein [459], Yannelis and Prabhakar [587], Chang [77], Ding et al. [141] and Kim [299] in several ways. We also note that all results in this section are given in non-compact and non-paracompact settings.

As most results in the existing literature on the existence of equilibria for generalized games (or, abstract economice), the underlying spaces (resp., choice sets or commodity spaces) are always compact and convex. However, several authors in Mehta [367], Tan and Yuan [511], Tulcea [561] generalize the underlying spaces to be non-compact cases, but they still need conditions that all underlying spaces are paracompact. It is well known that if a correspondence has an open graph, then it has open upper and lower sections and thus it must be lower semicontinuous. However, a continuous correspondence need not have open lower or upper sections in general. Also, in the infinite settings, the set of feasible allocations generally is not compact in the commodity spaces. The motivations for economists continually to be interested in setting forth conditions for the existence of equilibria come from the importance of generalized games (also called abstract economics) in the study of markets and other general games and from the restrictions of the existing theorems. Since we will encounter many kinds of preferences in various economic situations, it is important that we consider several types of preferences and obtain some existence results for such correspondences in non-compact and non-pararcompact settings.

We first recall the following definition was first introduced by Kim [299].

Definition 4.6.1. Let X be a topological space, Y be a non-empty subset of a vector space E, $\theta : X \to E$ be a mapping and $\phi : X \to 2^Y$ be a correspondence. Then (1) ϕ is said to be of *class* $\mathcal{L}_\theta$ if (a) for each $x \in X$, $co\phi(x) \subset Y$ and $\theta(x) \notin co\phi(x)$ and (b) for each $y \in Y$, $\phi^{-1}(y)$ is open in X; (2) (ϕ_x, N_x) is an $\mathcal{L}_\theta$*-majorant* of of ϕ at x if $\phi_x : X \to 2^Y$ and N_x is an open neighborhood of x in X such that (a) for each $z \in X$, $\phi(z) \subset \phi_x(z)$; (b) for each $z \in N_x$, $\theta(z) \notin co\phi_x(z)$ and (c) for each $y \in Y$, $\phi_x^{-1}(y)$ is open in X; (3) ϕ is said to be $\mathcal{L}_\theta$*-majorized* if for each $x \in X$ with $\phi(x) \neq \emptyset$, there exists an $\mathcal{L}_\theta$-majorant (ϕ_x, N_x) of ϕ at x.

We would like to note that the class $\mathcal{L}_\theta$ and $\mathcal{L}_\theta$-majorized mappings in Definition 4.6.1 above are independent from class KF and KF-majorized (and thus class F and F-majorized) mappings introduced in Definition 4.2.2 (resp., Definition 4.3.2) in Section 2 (resp., in Section 3).

In this section, we shall deal mainly with either the case (I): $X = Y$ and is a non-empty convex subset of the topological vector space E and $\theta = I_X$, the identity mapping on X, or the case (II): $X = \Pi_{i \in I} X_i$ and $\theta = \pi_j : X \to X_j$ is the projection of X onto X_j and $Y = X_j$ is a non-empty convex subset of a topological vector space. In both cases (I) and (II), we shall write $\mathcal{L}$ in place of $\mathcal{L}_\theta$.

We shall need the following result which was first given by Kim [299] as an application of KKM principle, i.e., Theorem 2.2.8.

Lemma 4.6.2. Let X be a non-empty convex subset of a Hausdorff topological vector space E. Suppose $T : X \to 2^X$ is a set-valued mapping satisfying:

(1) $x \notin coT(x)$ for each $x \in X$;

(2) $T^{-1}(y)$ is compactly open in X for each $y \in X$; and

(3) there exists a non-empty compact subset D of X and a non-empty compact and convex subset X_0 of X such that $X_0 \cap coT(x) \neq \emptyset$ for each $x \in X \setminus D$.

Then T has a maximal element, i.e., there exists $\hat{x} \in X$ such that $T(\hat{x}) = \emptyset$.

Proof. If possible, suppose $T(x) \neq \emptyset$ and so that $coT(x) \neq \emptyset$ for each $x \in X$. It is clear that $(coT)^{-1}(y)$ is also compactly open in X for each $y \in X$ (e.g., see Lemma 5.1 of Yannelis [585]). Now we define a set-valued mapping $F : X \to 2^X$ by $F(x) = X \setminus (coT)^{-1}(x)$ for each $x \in X$. Then $F(x)$ is compactly closed in X for each $x \in X$. Now we claim that $\cap_{x \in X_0} F(x) \subset D$. Suppose the contrary, i.e., there exists a point $y \in \cap_{x \in X_0} F(x)$ such that $y \in X \setminus D$. It then will imply that $X_0 \cap (coT)(y) = \emptyset$, which is a contradiction. Thus we must have $\cap_{x \in X_0} F(x)$ is a closed subset of D so that $\cap_{x \in X_0} F(x)$ is compact. As $coT(x) \neq \emptyset$ for each $x \in X$, we have $X = \cup_{x \in X}(coT)^{-1}(x)$. Therefore

$$\cap_{x \in X} F(x) = \cap_{x \in X}(X \setminus (coT)^{-1}(x)) = \emptyset.$$

By Fan's Lemma, i.e., Theorem 2.2.8, it follows that F is not a KKM mapping. Therefore there exists a non-empty finite subset $A \in \mathcal{F}(X)$ and $w \in co(A)$ such that $w \notin \cup_{x \in A} F(x)$. Hence we have $w \in X \setminus \cap_{x \in A} F(x) = \cap_{x \in A}(coT)^{-1}(x)$, i.e., $x \in (coT(w)$ for each $x \in A$ which implies that $w \in co(A) \subset coT(w)$ as $coT(w)$ is convex. This contradicts assumtption (1). Thus there must exist at least one point $\hat{x} \in X$ such that $T(\hat{x}) = \emptyset$. □

As an application of Lemma 4.6.2, we have the following existence of maximal element for $\mathcal{L}$-majorized mapping in topological vector spaces without the assumption of paracompactness and compactness which is due to Kim [299].

Lemma 4.6.3. Let D be a non-empty compact subset of a non-empty convex set X in a Hausdorff topological vector space E and $y_0 \in D$ a fixed point. Let $\psi : X \to 2^X$ be an $\mathcal{L}$-majorized correspondence such that for each $x \in X \setminus D$, $y_0 \in co\psi(x)$. Then there exists a point $x_0 \in X$ such that $\psi(x_0) = \emptyset$.

Proof. Suppose for each $x \in X$, $\psi(x) \neq \emptyset$. As ψ is $\mathcal{L}$-majorized, there exists an $\mathcal{L}$-majorant ψ_x and an open neighborhood N_x of x in X such that

(i) $\psi(z) \subset \psi_x(z)$ for each $z \in X$;

(ii) $z \notin co\psi_x(z)$ for each $z \in n_x$; and

(iii) $\psi_x^{-1}(y)$ is open in X for each $y \in X$.

The family $\{N_x : x \in D\}$ is an open covering of D. As D is compact, there exists a finite subset $x_1, \cdots x_n \in X$ such that $\cup_{i=1}^n N_{x_i} \supset D$. For each $i = 1, \cdots, n$, denoted by $co\psi'_{x_i}$ the restriction of $co\psi_{x_i}$ on U. Then for each $i = 1, \cdots, n$ and each $y \in X$, it is easy to see that $(co\psi_{x_i})^{-1}(y)$ is open in X. We now define a set-valued mapping $\Psi : X \to 2^X$ by

$$\Psi(x) = \begin{cases} \cap_{i=1}^n co\psi'_{x_i}(x), & \text{if } x \in U; \\ \{y_0\}, & \text{if } x \notin U. \end{cases}$$

Then for each $z \in X$, we have $z \notin co\Psi(z)$ and moreover $(co\Psi)^{-1}(y)$ is open in X for each $y \in X$. Indeed, in the case $y \neq y_0$, we have

$$\begin{aligned} \Psi^{-1}(y) &= \{z \in X : y \in \Psi(z)\} \\ &= \{z \in U : y \in \Psi(z)\} = U \cap [\cap_{i=1}^n (co\psi_{x_i})^{-1}(y)]. \end{aligned}$$

In the case $y = y_0$, we can have

$$\begin{aligned} \Psi^{-1}(y_0) &= \{z \in X : y_0 \in \Psi(z)\} \\ &= [U \cap_{i=1}^n (co\psi_{x_i})^{-1}(y_0)] \cup (X \setminus U). \end{aligned}$$

By the assumption and condition (i), for each $z \in X \setminus D$, we have $y_0 \in co\psi(z) \subset co\psi_{x_i}(z)$ for each $i = 1, \cdots n$. Therefore we have $\Psi^{-1}(y_0) = \cap_{i=1}^n (co\psi_{x_i})^{-1}(y_0)$ which is open X. It follows that Ψ is of class $\mathcal{L}$. Moreover, we can show that $y_0 \in co\Psi(z)$ for each $z \in X \setminus D$. Thus all hypotheses of Lemma 4.6.2 are satisfied by letting $X_0 = \{y_0\}$. By Lemma 4.6.2, there exists $\hat{x} \in X$ such that $\Psi(\hat{x}) = \emptyset$. This implies that $\hat{x} \in U$. Since $\psi(z) \subset \Psi(z)$ for each $z \in U$ and thus $\psi(\hat{x}) = \emptyset$, which is a contradiction. Hence there must exist a maximal element $\hat{x} \in X$ such that $\psi(\hat{x}) = \emptyset$. The proof is complete. □

As applications of Lemma 4.6.3, we now give a complete proof for the existence of equilibria for qualitative games without assumption of paracompactness.

Theorem 4.6.4. Let $\Gamma = (X_i, P_i)_{i \in I}$ be a qualitative game such that for each $i \in I$, the following conditions are satisfied:

(a) D_i is a non-empty compact subset of a non-empty convex set X_i in a topological vector space E_i;

(b)$P_i : X(:= \Pi_{i \in I} X_i) \to 2^{X_i}$ is $\mathcal{L}$-majorized;

(c) the set $E^i = \{x \in X : P_i(x) \neq \emptyset\}$ is open in X; and

(d) there exists a point $y^i \in D_i$ such that for each $x \in X \setminus D$, $y^i \in coP_i(x)$, where $D = \Pi_{i \in I} D_i$.

Then there exists a point $x \in X$ such that $P_i(x_i) = \emptyset$ for all $i \in I$.

Proof. Let $D = \Pi_{i \in I} D_i$. Then D is non-empty compact as D_i is compact for each $i \in I$. For each $x \in X$, let $I(x) := \{i \in I : P_i(x) \neq \emptyset\}$ and we define a correspondence $P : X \to 2^X$ by

$$P(x) = \begin{cases} \cap_{i \in I(x)} P'_i(x), \text{ if } I(x) \neq \emptyset; \\ \emptyset, \text{ otherwsie}; \end{cases}$$

where $P_i'(x) := \Pi_{j\neq i, j\in I} X_j \otimes P_i(x)$ for each $x \in X$. Then we shall show that P is $\mathcal{L}$-majorized.

For each $x \in X$, it is clear $I(x) \neq \emptyset$ if and only if $P(x) \neq \emptyset$. Let $x \in X$ be such that $P(x) \neq \emptyset$. Then $P'_{i(x)}(x) \neq \emptyset$ for $i(x) \in I(x)$. Since $P_{i(x)}$ is of $\mathcal{L}$-majorized at x, there exist an open neighborhood $N(x)$ of x in X and a mapping $\phi_{i(x)} : X \to 2^{X_i}$ which is an $\mathcal{L}$-majorant $\phi_{i(x)}$ of $P_{i(x)}$ at x such that (i): $P_{i(x)}(z) \subset \phi_{i(x)}(z)$ for each $z \in X$ and so that $P'_{i(x)}(z) \subset \phi'_{i(x)}(z)$ for each $z \in X$, where $\phi'_{i(x)}(u) := \Pi_{j\in I, j\neq i(x)} X_j \otimes \phi_{i(x)}(u)$ for each $u \in X$; (ii): $z_{i(x)} := \pi_{i(x)}(z) \notin co\phi_{i(x)}(z)$ for each $z \in N(x)$ and (iii): $\phi^{-1}_{i(x)}(y)$ is open in X for each $y \in X$. Note that each subset of a topological vector space is regular, without loss of generality, we may assume that there exists a non-empty open neighborhood $O(x)$ of x such that $x \in O(x) \subset \overline{O(x)} \subset N(x)$. Then the properties (i) and (ii) are also satisfied by $\phi_{i(x)}(z)$ for each $z \in O(x)$ as $O(x) \subset N(x)$. By hypothesis (c), we may assume that $N(x) \subset E^i$. Thus $P_{i(x)}(z) \neq \emptyset$ for all $z \in N(x)$ and it follows that $i(x) \in \cap_{z\in N(x)} I(z)$. Therefore $P(z) = \cap_{j\in I(z)} P_j'(z) \subset P'_{i(x)}(z)$ for each $z \in N(x)$. In order to prove that P is $\mathcal{L}$-majorized, we define another mapping $\Phi_x : X \to 2^X$ by

$$\Phi_x(z) = \begin{cases} \phi'_{i(x)}(z), \text{ if } z \in \overline{O(x)}; \\ X, \text{ if } z \in X \setminus \overline{O(x)}; \end{cases}$$

where $\phi'_{i(x)}(z) := \Pi_{j\in I, j\neq i(x)} X_j \otimes \phi_{i(x)}(z)$ for each $z \in X$. We now claim that $(\Phi_x, O(x))$ is an $\mathcal{L}$-majorant of P at x. Note that for each $z \in X$, if $z \in \overline{O(x)}$, it follows that $P(z) = \cap_{i\in I(z)} P_i'(z) \subset \phi'_{i(x)}(z) = \Phi_x(z)$. In the case $z \in X \setminus \overline{O(x)}$, we always have that $P(z) = \cap_{i\in I(z)} P_i'(z) \subset X = \Phi_x(z)$ by the definition of Φ_x. Therefore $P(z) \subset \Phi_x(z)$ for all $z \in X$. Secondly, by the definition of Φ_x, it is clear that $z \notin co\Phi_x(z)$ for each $z \in O(x)$ as $z_{i(x)} = \pi_{i(x)}(z) \notin co\phi_{i(x)}(z)$ for each $z \in \overline{O(x)} \subset N(x)$ by (ii). Moreover for each $y \in X$, it is easy to verify that $\Phi_x^{-1}(y) = (\phi'_{i(x)})^{-1}(y) \cup (X \setminus \overline{O(x)})$ which is open in X as $(\phi'_{i(x)})^{-1}(y)$ is open by the openness of $\phi^{-1}_{i(x)}(y)$ from property (iii).

Thus we have shown that $(\Phi_x, O(x))$ is an $\mathcal{L}$-majorant of P at x, which means that P is of $\mathcal{L}$-majorized. By assumption (d) above, P satisfies all hypotheses of Lemma 4.6.3. By Lemma 4.6.3, it follows that there exists a point $x \in D$ such that $P(x) = \emptyset$ which implies that $P_i(x) = \emptyset$ for all $i \in I$ and thus we complete the proof. □

As an application of Theorem 4.6.4, we have the following existence theorem for generalized games (abstract economics) without compactness and paracomapct hypotheses which is different from the previous results in Sections 2, 3, 4 and 5 above.

Theorem 4.6.5. Let $G = (X_i, A_i, B_i, P_i)_{i\in I}$ be an abstract economy where I is a any (countable or uncountable) set such that for each $i \in I$

(i) X_i is a non-empty convex subset of a topological vector space E_i and D_i is a non-empty compact subset of X_i;

(ii) for each $x \in X = \Pi_{i\in I}X_i$, $A_i(x)$ is non-empty and $coA_i(x) \subset B_i(x)$;

(iii) the set $A_i^{-1}(y)$ is open in X for each $y \in X_i$;

(iv) the mapping $A_i \cap P_i : X \to 2^{X_i}$ is $\mathcal{L}$-majorized;

(v) the set $E^i = \{x \in X : A_i \cap P_i(x) \neq \emptyset\}$ is open in X; and

(vi) there exists $y^i \in D_i$ such that for each $x \in X \setminus D$, $y^i \in co(A_i \cap P_i)(x)$, where $D = \Pi_{i\in I}D_i$.

Then there exists $x \in X$ such that $x_i \in \overline{B_i}(x)$ and $A_i(x) \cap P_i(x) = \emptyset$ for each $i \in I$.

Proof. For each $i \in I$, let $F_i = \{x \in X : x_i \in \overline{B_i}(x)\}$, the set F_i is closed in X. We now define $Q_i : X \to 2^{X_i}$ by

$$Q_i(x) = \begin{cases} (A_i \cap P_i)(x_i), & \text{if } x \in F_i, \\ coA_i(x), & \text{if } x \notin F_i. \end{cases}$$

We shall show that the qualitative game $\Gamma = (X_i, Q_i)_{i\in I}$ satisfies the hypotheses of Theorem 4.6.4. First we have that for each $i \in I$, the set

$$\begin{aligned} \{x \in X : Q_i(x) \neq \emptyset\} &= \{x \in X \setminus F_i : Q_i(x) \neq \emptyset\} \cup \{x \in F_i : Q_i(x) \neq \emptyset\} \\ &= (X \setminus F_i) \cup \{x \in F_i : (A_i \cap P_i)(x) \neq \emptyset\} \\ &= (X \setminus F_i) \cup (F_i \cap E^i) \\ &= (X \setminus F_i) \cup E^i \end{aligned}$$

is open in X by (v).

Let $x \in X$ be such that $Q_i(x) \neq \emptyset$. We consider the following two cases. Case 1: $x \notin F_i$. Let $\Phi_x = coA_i$ and $N_x = X \setminus F_i$, then N_x is an open neighborhood of x in X such that (1): for each $z \in X$, $Q_i(z) \subset \Phi_x(z)$; (2) $\pi_i(z) \notin co\Phi_x(z)$ for each $z \in N_x$ and (3) $\Phi_x^{-1}(y)$ is open in X for all $y \in X_i$ by (iii) and Lemma 5.1 of Yannelis and Prabhakar [587]. Therefore Q_x is an $\mathcal{L}$-majorant of Q_i at x. Case 2: $x \in F_i$. Since $Q_i(x) = (A_i \cap P_i)(x)$ and $A_i \cap P_i$ is $\mathcal{L}$-majorized, there exists an open neighborhood N_x of x in X and a correspondence $\psi_x : X \to 2^{X_i}$ such that (1) $(A_i \cap P_i)(z) \subset \psi_x(z)$ for each $z \in X$; (2) $\pi_i(z) \notin co\psi_x(z)$ for each $z \in N_x$ and (3) $\psi_x^{-1}(y)$ is open in X for each $y \in X_i$. Define $\Phi_x : X \to 2^{X_i}$ by

$$\Phi_x(z) = \begin{cases} coA_i(z) \cap \psi_x(z), & \text{if } z \in F_i, \\ coA_i(z), & \text{if } z \notin F_i. \end{cases}$$

Note that as $(A_i \cap P_i)(z) \subset \psi_x(z)$ for each $z \in X$, we have $Q_i(z) \subset \Phi_x(z)$ for each $z \in X$. It is easy to see that $\pi_i(z) \notin co\Phi_x(z)$ for all $z \in N_x$. Moreover for each $y \in X_i$, the set

$$\begin{aligned}(\Phi_x)^{-1}(y) &= \{z \in X : y \in \Phi_x(z)\} \\ &= \{z \in F_i : y \in \Phi_x(z)\} \cup \{z \in X \setminus F_i : y \in \Phi_x(z)\} \\ &= \{z \in F_i : y \in coA_i(z) \cap \psi_x(z)\} \cup \{z \in X \setminus F_i : y \in coA_i(z)\} \\ &= [F_i \cap (coA_i)^{-1}(y) \cap \psi_x^{-1}(y)] \cup [(X \setminus F_i) \cap (coA_i)^{-1}(y)] \\ &= [\psi_x^{-1}(y) \cup (X \setminus F_i)] \cap (coA_i)^{-1}(y)\end{aligned}$$

is open in X. Therefore Φ_x is an $\mathcal{L}$-majorant of Q_i at x.

Moreover for each $i \in I$, by (vi) there exists $y^i \in D_i$ such that $y^i \in co(A_i(x) \cap P_i(x)) \subset coQ_i(x)$ for all $x \in X \setminus D$. Therefore the qualitative game $\Gamma = (X_i, Q_i)_{i \in I}$ satisfies all hypotheses of Theorem 4.6.4. By Theorem 4.6.4, there exists $x \in D$ such that $Q_i(x) = \emptyset$. Since for each $x \in X$, $A_i(x) \neq \emptyset$, therefore we have that $x_i \in \overline{B_i}(x)$ and $A_i(x) \cap P_i(x) = \emptyset$. □

Remark 4.6.6. (1) Theorem 4.6.5 improves Theorem 3 of Kim [299] in the following ways: (i) the set I is any (countable or uncountable) set instead of being finite; (ii) for each $i \in I$, the mapping $A_i \cap P_i$ is $\mathcal{L}$-majorized instead of being a class of $\mathcal{L}$. Therefore Theorem 4.6.5 not only answers the question raised by Kim [299](p.805) positively, but also answers the question posed by Yannelis and Prabhakar [587, p.243] in the affirmative with weaker hypotheses. The following result also shows that Theorem 3 of Ding et al. [141] is still true in topological vector spaces instead of locally convex topological vector spaces.

Theorem 4.6.7. Let $\Gamma = (X_i; P_i)_{i \in I}$ be a qualitative game such that for each $i \in I$;

(a) X_i is a non-empty convex subset of a topological vector space;

(b) $P_i : X = \Pi_{i \in I} X_i \to 2^{D_i}$ is $\mathcal{L}$-majorized, where D_i is a non-empty compact subset of X_i;

(c) the set $E^i = \{x \in X : P_i(x) \neq \emptyset\}$ is open in X; and

(d) there exists a non-empty subset F_i of D_i such that $F_i \cap P_i(x) \neq \emptyset$ for each $x \in E^i$ and $coF_i \subset D_i$.

Then Γ has an equilibrium point.

Proof. For each $i \in I$, let $K_i = cl_{D_i}(coF_i)$ and $K = \Pi_{i \in I} K_i$. Since D_i is non-empty compact and $coF_i \subset D$, so that K_i is compact and convex in X_i and K is a compact and convex subset in X. Define $P_i' : X \to 2^{K_i}$ by $P_i'(x) = P_i(x) \cap K_i$ for each $x \in X$. By (d), we have that $\{x \in X : P_i(x) \neq \emptyset\} = \{x \in X : P_i'(x) \neq \emptyset\}$. Note that P_i' is also $\mathcal{L}$-majorized since P_i is $\mathcal{L}$ majorized. Now for each $i \in I$, let P_i^K be the restriction of P_i' onto K. Then the qualitative game $(K_i; P_i^K)_{i \in I}$ satisfies all hypotheses of Theorem 4.6.4 in the compact case. By Theorem 4.6.4, there exists $x \in K$ such that $P_i^K(x) = \emptyset$ for all $i \in I$. Since $P_i^K(x) = P_i'(x) = P_i(x) \cap K_i = \emptyset$, by (d) again, we must have that $P_i(x) = \emptyset$ for all $i \in I$. □.

Theorem 4.6.7 shows that Theorem 3 of Ding et al. [141] holds in topological vector spaces. Since Theorem 4 of Ding et al. [141] was proved by their Theorem 4 which is a special case of Theorem 4.6.7 above, it is easy to see that Theorem 4 of Ding et al. [141] is still true in topological vector spaces instead of locally convex Hausdorff topological vector spaces by employing the same arguments in the proof of Theorem 4 in [141].

In what follows, we shall use the approximation method to obtain equilibrium existence theorems for generalized games. First we shall prove the following existence theorem of *approximate* equilibrium points for a one-person game, then another existence theorem of equilibrium points of generalized games with infinite (countable or uncountable) sets of players given, which in turn implies the existence of equilibrium points of generalized games in which the constraint correspondences are lower semicontionous instead of having open lower sections in the locally convex topological vector spaces.

Theorem 4.6.8. Let X be a non-empty convex subset of a topological vector space E and $A, B, P : X \to 2^X$ be such that:

(i) A is lower semicontinuous, and for each $x \in X$, $A(x)$ is non-empty and $coA(x) \subset B(x)$;

(ii) $A \cap P$ is $\mathcal{L}$-majorized; and

(iii) there exists a non-empty compact subset D of X and $y \in D$ such that $y \in co(A(x) \cap P(x))$ for each $x \in X \setminus D$.

Then for each open convex neighborhood V of zero in E, the one person game $(X; A, \overline{B_V}; P)$ has an equilibrium point in D, i.e., there exists a point $x_V \in K$ such that $x_V \in \overline{B_V}(x_V)$ and $A(x_V) \cap P(x_V) = \emptyset$ where $B_V(x) = (B(x) + V) \cap X$ for each $x \in X$.

Proof. Let V be an open convex neighborhood of zero in E and define $A_V, B_V : X \to 2^X$ by $A_V(x) = (A(x) + V) \cap X$ and $B_V(x) = (B(x) + V) \cap X$ for each $x \in X$. Since A is lower semicontinuous, A_V has an open graph in $X \times X$ by Lemma 4.1 of Chang [77, p.244] (or see Tulcea [561, p.7]). By (i), $A_V(x) \subset B_V(x)$ for each $x \in X$. Let $F_V = \{x \in X : x \notin \overline{B_V}(x)\}$, then F_V is open in X. Define $\Psi_V : X \to 2^X$ by

$$\Psi_V = \begin{cases} A(x) \cap P(x), & \text{if } x \notin F_V, \\ A_V(x), & \text{if } x \in F_V. \end{cases}$$

Suppose $x \in X$ is such that $\Psi_V(x) \neq \emptyset$.

Case 1: Suppose $x \in F_V$. Let $\Phi_x = A_V$ and $N_x = F_V$, then N_x is an open neighborhood of x in X such that (a) for each $z \in N_x$, $\Psi_V(z) = A_V(z) = \Phi_x(z)$, $z \notin \overline{B_V}(z)$ so that $z \notin A_V(z) = \Phi_x(z)$; (b) for each $y \in X$, $\Phi_x^{-1}(y) = A_V^{-1}(y)$ is open in X since A_V has an open graph in $X \times X$ and (c) for each $z \in X$, $\Psi_V(z) \subset \Phi_x(z)$. Thus Φ_x is an $\mathcal{L}$-majorant of Ψ_V at x.

Case 2: Suppose $x \notin F_V$. Then $\Psi_V(x) = A(x) \cap P(x) \neq \emptyset$. Since $A \cap P$ is $\mathcal{L}$-majorized by (ii), there exists an open neighborhood N_x of x in X and a correspondence $\Phi'_x : X \to 2^X$ such that (a) $\Psi_V(z) = A(z) \cap P(z) \subset \Phi'_x(z)$ for all $z \in X$; (b) $z \notin co\Phi'_x(z)$ for each $z \in N_x$ and (c) for each $y \in X$, $(\Phi'_x)^{-1}(y)$ is open in X. Define $\Phi_x : X \to 2^X$ by

$$\Phi_x = \begin{cases} A_V(z) \cap \Phi'_x(z), & \text{if } z \notin F_V, \\ A_V(z), & \text{if } z \in F_V. \end{cases}$$

Note that (i) for each $z \in X$, clearly $\Psi_V(z) \subset \Phi_x(z)$; (ii) it is easy to see that $z \notin co\Phi_x(z)$ for each $z \in N(x)$ and (iii) for each $y \in X$, $\Phi_x^{-1}(y) = [F_V \cup (\Phi'_x)^{-1}y)] \cap A_V^{-1}(y)$ is open in X. Hence Φ_x is an $\mathcal{L}$-majorant of Ψ_V at x.

Therefore Ψ_V is an $\mathcal{L}$-majorized correspondence. Moreover by (iii), there exist a non-empty compact subset D of X and $y \in D$ such that for each $x \in X \setminus D$, $y \in co(A(x) \cap P(x))$ so that $y \in co(A(x) \cap P(x)) \subset co\Psi_V(x)$. Thus by Lemma 4.6.3, there exists $\hat{x} \in D$ such that $\Psi_V(\hat{x}) = \emptyset$. Since $A(\hat{x}) \neq \emptyset$ by (i), we must have $\hat{x} \in \overline{B_V}(\hat{x})$ and $A(\hat{x}) \cap P(\hat{x}) = \emptyset$. □

The following is an existence theorem for the *approximate* equilibrium points of a generalized game:

Theorem 4.6.9. Let I be an any (countable or uncountable) set. For each $i \in I$, let X_i be a non-empty convex subset of a topological vector space E_i and $A_i, B_i, P_i : X = \Pi_{j \in I} X_j \to 2^{X_i}$ be such that:

(a) A_i is lower semicontinuous, and for each $x \in X$, $A_i(x)$ is non-empty and $coA_i(x) \subset B_i(x)$;

(b) $A_i \cap P_i$ is $\mathcal{L}$-majorized; and

(c) the set $E^i = \{x \in X : (A_i \cap P_i)(x) \neq \emptyset\}$ is open in X.

Suppose that there exists a non-empty compact subset D_0 of X and $y^0 \in D_0$ such that $y_i^0 \in co(A_i(x) \cap P_i(x))$ for all $x \in X \setminus D_0$ and for all $i \in I$.

Then given any $V = \Pi_{i \in I} V_i$ where for each $i \in I$, V_i is an open convex neighborhood of zero in E_i, the generalized game $\Gamma_V = (X_i; A_i, B_{V_i}; P_i)_{i \in I}$ has an equilibrium point in D_0; i.e. there exists a point $x_V = (x_{V_i})_{i \in I} \in D_0$ such that $x_{V_i} \in \overline{B_{V_i}}(x_V)$ and $A_i(x_V) \cap P_i(x_V) = \emptyset$ for each $i \in I$, where $B_{V_i} = (B_i(x) + V_i) \cap X_i$ for each $x \in X$ and each $i \in I$.

Proof. Let $V = \Pi_{i \in I} V_i$ be given where for each $i \in I$, V_i is an open convex neighborhood of zero in E_i. Fix any $i \in I$ and define $A_{V_i}, B_{V_i} : X \to 2^{X_i}$ by $A_{V_i}(x) = (coA_i(x) + V_i) \cap X_i$ and $B_{V_i}(x) = (B_i(x) + V_i) \cap X_i$ for each $x \in X$. By (a), A_i is lower semicontinuous so that coA_i is also lower semicontinuous by Proposition 2.6 of Michael [371, p.366], it follows from Lemma 4.1 of Chang [77, p.244] or from Tulcea [561, p.7] that A_{V_i} has an open graph in $X \times X$. Now let $F_{V_i} = \{x \in X : x_i \notin \overline{B_{V_i}}(x)\}$, then F_{V_i} is open in X. Define the mapping

$Q_{V_i} : X \to 2^{X_i}$ by

$$Q_{V_i}(x) = \begin{cases} (A_i \cap P_i)(x), & \text{if } x \notin F_{V_i}, \\ A_i(x), & \text{if } x \in F_{V_i}. \end{cases}$$

We shall prove that the qualitative game $\mathcal{T} = (X_i, Q_{V_i})_{i\in I}$ satisfies all conditions of Theorem 4.6.4. First we note that for each $i \in I$, the set

$$\begin{aligned} \{x \in X : Q_{V_i}(x) \neq \emptyset\} &= F_{V_i} \cup \{x \in X \setminus F_{V_i} : A_i(x) \cap P_i(x) \neq \emptyset\} \\ &= F_{V_i} \cup ((X \setminus F_{V_i}) \cap E^i) = F_{V_i} \cup E^i \end{aligned}$$

is open in X by (c). Let $x \in X$ be such that $Q_{V_i}(x) \neq \emptyset$. We consider the following two cases.

Case 1: $x \in F_{V_i}$. Let $\Psi_x = A_{V_i}$ and $N_x = F_{V_i}$, then N_x is an open neighborhood of x in X such that (i) $Q_{V_i}(z) \subset \Psi_x(z)$ for each $z \in X$; and by (b), $\pi_i(z) \notin co\Psi_x(z)$ for each $z \in N_x$; (ii) $co\Psi_x(z) \subset X_i$ for each $z \in X$ by (b) and (iii) $\Psi_x^{-1}(y) = A_{V_i}^{-1}(y)$ is open in X for all $y \in X_i$ since A_{V_i} has an open graph. Therefore, Ψ_x is an $\mathcal{L}$-majorant of Q_{V_i} at x.

Case 2: $x \notin F_{V_i}$. Since $Q_{V_i}(x) = (A_i \cap P_i)(x) \neq \emptyset$ and $A_i \cap P_i$ is $\mathcal{L}$-majorized, there exists an open neighborhood N_x of x in X and a correspondence $\phi_x : X \to 2^{X_i}$ such that (i) for each $z \in X$, $(A_i \cap P_i)(z) \subset \phi_x(z)$; (ii) $\pi_i(z) \notin co\phi_x(z)$ for each $z \in N_x$; and (iii) $co\phi_x(z) \subset X_i$ for each $z \in X$ and (iv) $\phi_x^{-1}(y)$ is open in X for each $y \in X_i$. Define $\Psi_x : X \to 2^{X_i}$ by

$$\Psi_x(z) = \begin{cases} A_{V_i}(z) \cap \phi_x(z), & \text{if } z \notin F_{V_i}, \\ A_{V_i}(z), & \text{if } z \in F_{V_i}. \end{cases}$$

Note that as $(A_i \cap P_i)(z) \subset \phi_x(z)$ and $A_i(z) \subset A_{V_i}(z)$ for each $z \in N_x$, we have $Q_{V_i}(z) \subset \Psi_x(z)$ and $co\Psi_x(z) \subset X_i$. It is easy to see that $\pi_i(z) \notin co\Psi_x(z)$ for all $z \in X$. Moreover, for any $y \in X_i$, the set

$$\begin{aligned} \Psi_x^{-1}(y) &= \{z \in X : y \in \Psi_x(z)\} \\ &= \{z \in X \setminus F_{V_i} : y \in \Psi_x(z)\} \cup \{z \in F_{V_i} : y \in \Psi_x(z)\} \\ &= \{z \in X \setminus F_{V_i} : y \in A_{V_i}(z) \cap \phi_x(z)\} \cup \{z \in F_{V_i} : y \in A_{V_i}(z)\} \\ &= [(X \setminus F_{V_i}) \cap A_{V_i}^{-1}(y) \cap \phi_x^{-1}(y)] \cup [F_{V_i} \cap A_{V_i}^{-1}(y)] \\ &= [\phi_x^{-1}(y) \cup F_{V_i}] \cap A_{V_i}^{-1}(y) \end{aligned}$$

is open in X. Therefore, Ψ_x is an $\mathcal{L}$-majorant of Q_{V_i} at point x.

Hence Q_{V_i} is an $\mathcal{L}$-majorized correspondence. Now by the assumption, there exists a non-empty compact subset D_0 of X and a point $y^0 \in D_0$ such that for each $x \in X \setminus D_0$, with $y_i^0 \in co(A_i(x) \cap P_i(x))$ for all $i \in I$. Note that if $x \notin F_{V_i}$, then $y_i \in co(A_i(x) \cap P_i(x)) \subset coQ_{V_i}(x)$ and if $x \in F_{V_i}$, then $\{x \in X : Q_{V_i}(x) \neq \emptyset\} = F_{V_i} \cup \{x \in X \setminus F_{V_i} : (A_i \cap P_i)(x) \neq \emptyset\} = F_{V_i} \cup E^i$ is open in X by condition (c). Therefore all hypotheses of Theorem 4.6.4 are satisfied, so that by Theorem 4.6.4, there exists a point $x_V = (x_{V_i})_{i\in I} \in D_0$ such that $Q_{V_i}(x_V) = \emptyset$ for all $i \in I$. Since for each $i \in I$, $A_i(x)$ is non-empty, we must have $x_{V_i} \in \overline{B_{V_i}}(x_V)$ and $A_i(x_V) \cap P_i(x_V) = \emptyset$. □

By Theorem 4.6.9 and Lemma 4.6.3, we shall present the existence theorem of equilibrium of generalized games in which the constraint correspondences are lower semicontinuous instead of having open lower sections in the locally convex topological vector spaces.

Theorem 4.6.10. Let $G = (X_i; A_i, B_i; P_i)_{i\in I}$ be a generalized game such that $X = \Pi_{i\in I} X_i$. Suppose the following conditions are satisfied:

(a) for each $i \in I$, X_i is a non-empty convex subset of a locally convex topological vector space E_i;

(b) for each $i \in I$, $A_i : X \to 2^{X_i}$ is lower semicontinuous such that for each $x \in X$, $A_i(x)$ is non-empty and $coA_i(x) \subset B_i(x)$;

(c) for each $i \in I$, $A_i \cap P_i$ is $\mathcal{L}$-majorized;

(d) for each $i \in I$, the set $E^i = \{x \in X : (A_i \cap P_i)(x) \neq \emptyset\}$ is open in X; and

(e) there exist a non-empty compact subset D_0 of X and a point $y^0 \in D_0$ such that for each $x \in X \setminus D_0$ with $y_i^0 \in co(A_i(x) \cap P_i(x))$ for all $i \in I$.

Then $\mathcal{G}$ has an equilibrium point in D_0, i.e. there exists a point $\hat{x} = (\hat{x}_i)_{i\in I} \in D_0$ such that for each $i \in I$, $\hat{x}_i \in \overline{B_i}(\hat{x})$ and $A_i(\hat{x}) \cap P_i(\hat{x}) = \emptyset$.

Proof. For each $i \in I$, let $\mathcal{B}_i$ be the collection of all open convex neighborhoods of zero in E_i and $\mathcal{B} = \Pi_{i\in I}\mathcal{B}_i$. Given any $V \in \mathcal{B}$, let $V = \Pi_{j\in I} V_j$, where $V_j \in \mathcal{B}_j$ for each $j \in I$. By Theorem 4.6.9, there exists a $\hat{x}_V \in D_0$ such that $\hat{x}_{V_i} \in \overline{B_{V_i}}(\hat{x}_V)$ and $A_i(\hat{x}_V) \cap P_i(\hat{x}_V) = \emptyset$ for each $i \in I$, where $B_{V_i}(x) = (B_i(x) + V_i) \cap X_i$ for each $x \in X$. It follows that the set $Q_V := \{x \in D_0 : x_i \in \overline{B_{V_i}}(x) \text{ and } A_i(x) \cap P_i(x) = \emptyset\}$ is a non-empty closed subset of D_0 by (d).

Now we want to prove $\{Q_V\}_{V\in\mathcal{B}}$ has the finite intersection property. Let $\{V_1, \cdots, V_n\}$ be any finite subset of $\mathcal{B}$. For each $i = 1, \cdots, n$, let $V_i = \Pi_{j\in I} V_{ij}$ where $V_{ij} \in \mathcal{B}_i$ for each $j \in I$; let $V = \Pi_{j\in I}(\cap_{i=1}^n V_{ij})$, then $Q_V \neq \emptyset$. Clearly $Q_V \subset \cap_{i=1}^n Q_{V_i}$ so that $\cap_{i=1}^n Q_{V_i} \neq \emptyset$. Therefore the family $\{Q_V : V \in \mathcal{B}\}$ has the finite intersection property. Since D_0 is compact, $\cap_{V\in\mathcal{B}} Q_V \neq \emptyset$. Now take any $\hat{x} \in \cap_{V\in\mathcal{B}} Q_V$, then for each $i \in I$, $\hat{x}_i \in \overline{B_{V_i}}(\hat{x})$ for each $V_i \in \mathcal{B}_i$ and $A_i(\hat{x}) \cap P_i(\hat{x}) = \emptyset$. By Lemma 4.6.3, we also have for each $i \in I$, $\hat{x}_i \in \overline{B_i}(\hat{x})$. □

Corollary 4.6.11. Let $\mathcal{G} = (X_i; A_i, B_i; P_i)_{i\in I}$ be a generalized game such that $X = \Pi_{i\in I} X_i$. Suppose the following conditions are satisfied:

(a) for each $i \in I$, X_i is a non-empty convex subset of a locally convex topological vector space E_i;

(b) for each $i \in I$ and for each $x \in X$, $A_i(x)$ is non-empty and $coA_i(x) \subset B_i(x)$;

(c) for each $i \in I$, A_i and P_i have open lower sections;

(d) for each $i \in I$, $A_i \cap P_i$ is $\mathcal{L}$-majorized; and

(e) there exist a non-empty compact subset D_0 of X and a point $y^0 \in D_0$ such that for each $x \in X \setminus D_0$ with $y_i^0 \in co(A_i(x) \cap P_i(x))$ for all $i \in I$.

Then $\mathcal{G}$ has an equilibrium point in D_0, i.e., there exists point $\hat{x} \in D_0$ such that for each $i \in I$, $\hat{x}_i \in \overline{B_i}(\hat{x})$ and $A_i(\hat{x}) \cap P_i(\hat{x}) = \emptyset$.

Proof. By (c), the mapping $A_i : X \to 2^{X_i}$ is lower semicontinuous and the set $E^i = \{x \in X : A_i(x) \cap P_i(x) \neq \emptyset\}$ is open in X. Therefore all the hypotheses in Theorem 4.6.10 are satisfied, so that the conclusion follows. □

Corollary 4.6.11 (and hence also Theorem 4.6.10) improves Theorem 6.1 of Yannelis and Prabhakar [587] in the following ways: (i) the index I need not be countable, (ii) for each $i \in I$, the set X_i need not metrizable and (iii) for each $i \in I$, $A_i \cap P_i$ need not be of $\mathcal{L}$ class.

Corollary 4.6.12. Let $G = (X_i, A_i, B_i, P_i)_{i \in I}$ be a generalized game such that $X = \Pi_{i \in I} X_i$. Suppose the following conditions are satisfied:

(a) for each $i \in I$, X_i is a non-empty convex subset of a locally convex topological vector space;

(b) for each $i \in I$ and for each $x \in X$, $A_i(x)$ is non-empty and $coA_i(x) \subset B_i(x)$;

(c) for each $i \in I$, A_i has an open graph in $X \times X_i$ (respectively, is lower semicontinuous) and P_i is lower semicontinuous (respectively, has an open graph in $X \times X_i$);

(d) for each $i \in I$, $A_i \cap P_i$ is $\mathcal{L}$-majorized; and

(e) there exist a non-empty compact convex subset D_0 of X and a point $y^0 \in D_0$ such that for each $x \in X \setminus D_0$ with $y_i^0 \in co(A_i(x) \cap P_i(x))$ for all $i \in I$.

Then $\mathcal{G}$ has an equilibrium point in D_0, i.e., there exists point $\hat{x} \in D_0$ such that for each $i \in I$, $\hat{x}_i \in \overline{B_i}(\hat{x})$ and $A_i(\hat{x}) \cap P_i(\hat{x}) = \emptyset$

Proof. For each $i \in I$, since A_i has an open graph in $X \times X_i$ (respectively, is lower semicontinuous) and P_i is lower semicontinuous (respectively, has open graph in $X \times X_i$), the mapping $A_i \cap P_i : X \to 2^{X_i}$ is also lower semicontinuous by Lemma 4.2 of Yannelis [586], so that the set $E^i = \{x \in X : A_i(x) \cap P_i(x) \neq \emptyset\}$ is open in X. Therefore all conditions of Theorem 4.6.10 are satisfied and the conclusion follows. □

Corollary 4.6.12 (and hence also Theorem 4.6.10) generalizes Theorems 3.1 and 4.1 of Chang [77], and the exisiting Theorem of Shafer and Sonnenschein [459].

4.7 Equilibria of $\mathcal{U}$-Majorized Mappings

The objective of this section is to give some existence theorems for maximal elements and equilibria in qualitative games without the compactness (or paracompactness) assumption on the domain of the preferences which are majorized by upper semicontinuous correspondences instead of being majorized by correspondences which have lower open sections. Our intention is to merely illustrate a certain technique that we think will be of use in various problems of mathematical economics. Many other results of the type proved here may be proved under more general conditions. By following Tan and Yuan [508], we have the following definition.

Definition 4.7.1. Let X be a topological space, Y a non-empty subset of a vector space E, let $\theta : X \to E$ be a mapping and $\phi : X \to 2^Y \cup \{\emptyset\}$ a correspondence. Then (1) ϕ is said to be of class $\mathcal{U}_\theta$ if (a) for each $x \in X$, $\theta(x) \notin \phi(x)$ and (b) ϕ is upper semicontinuous with closed and convex values in Y; (2) ϕ_x is a $\mathcal{U}_\theta$-majorant of ϕ at x if there is an open neighborhood $N(x)$ of x in X and $\phi_x : N(x) \to 2^Y$ such that (a) for each $z \in N(x)$, $\phi(z) \subset \phi_x(z)$ and $\theta(z) \notin \phi_x(z)$ and (b) ϕ_x is upper semicontinuous with closed and convex values; (3) ϕ is said to be $\mathcal{U}_\theta$-majorized if for each $x \in X$ with $\phi(x) \neq \emptyset$, there exists a $\mathcal{U}_\theta$-majorant ϕ_x of ϕ at x.

We remark that when $X = Y$ and $\theta = I_X$, the identity mapping on X, our notions of a $\mathcal{U}_\theta$-majorant of ϕ at x and an $\mathcal{U}_\theta$-majorized correspondence are generalizations of upper semicontinuous correspondences which are irreflexive (i.e., $x \notin \phi(x)$ for all $x \in X$) and have closed convex values. Here we shall deal mainly with either the case (I): $X = Y$ and is a non-empty convex subset of the topological vector space E and $\theta = I_X$, the identity mapping on X, or the case (II): $X = \Pi_{i \in I} X_i$ and $\theta = \pi_j : X \to X_j$ is the projection of X onto X_j and $Y = X_j$ is a non-empty convex subset of a topological vector space. In both cases (I) and (II), we shall write $\mathcal{U}$ in place of $\mathcal{U}_\theta$.

We shall need the following result.

Theorem 4.7.2. Let X be a topological space and Y a normal space. If $F, G : X \to 2^Y \cup \{\emptyset\}$ have closed values and are upper semicontinuous at $x \in X$, then $F \cap G$ is also upper semicontinuous at x.

Proof. If $F(x) \cap G(x) \neq \emptyset$, the conclusion follows from Hildenbrand [229, Proposition B.III.2, p.23-23] (also see Klein and Thompson [317, Theorem 7.3.10, p.86]). If $F(x) \cap G(x) = \emptyset$, since Y is normal, it is easy to see that there exists an open neighborhood N of x in X such that $F(z) \cap G(z) = \emptyset$ for all $z \in \mathbb{N}$; thus $F \cap G$ is also upper semicontinuous at x. □

We remark here that in Theorem 4.7.2 above, we do not require $F(x) \cap$

$G(x) \neq \emptyset$ for each $x \in X$.

We shall also need the following result which generalizes and extends Lemma 6.1 of Yannelis and Prabhakar [587].

Lemma 4.7.3. Let X and Y be two topological spaces and A be a closed (respectively, open) subset of X. Suppose $F_1 : X \to 2^Y \cup \{\emptyset\}$, $F_2 : A \to 2^Y \cup \{\emptyset\}$ are lower semicontinuous (respectively, upper semicontinuous) such that $F_2(x) \subset F_1(x)$ for all $x \in A$. Then the mapping $F : X \to 2^Y \cup \{\emptyset\}$ defined by

$$F(x) = \begin{cases} F_1(x), & \text{if } x \notin A; \\ F_2(x), & \text{if } x \in A \end{cases}$$

is also lower semicontinuous (respectively, upper semicontinuous).

Proof. Let U be any closed (respectively, open) subset of Y. Clearly

$$\begin{aligned} \{x \in X : F(x) \subset U\} &= \{x \in A : F_2(x) \subset U\} \cup \{x \in X \setminus A : F_1 \subset U\} \\ &\subset \{x \in A : F_2(x) \subset U\} \cup \{x \in X : F_1(x) \subset U\}. \end{aligned}$$

Conversely, if $x \in A$ and $F_2(x) \subset U$, then $F(x) = F_2(x) \subset U$ and if $x \in A$ and $F_1(x) \subset U$, then $F(x) = F_2(x) \subset F_1(x) \subset U$. If $x \in X \setminus A$, then $F(x) = F_1(x) \subset U$. This shows that we have $\{x \in X : F(x) \subset U\} \supset \{x \in A : F_2(x) \subset U\} \cup \{x \in X : F_1(x) \subset U\}$. Hence $\{x \in X : F(x) \subset U\} = \{x \in A : F_2(x) \subset U\} \cup \{x \in X : F_1(x) \subset U\}$. Since A and U are closed (respectively, open) and F_1 and F_2 are lower semicontinuous (respectively, upper semicontinuous), the set $\{x \in X : F(x) \subset U\}$ is also closed (respectively, open). Therefore, F is lower semicontinuous (respectively, upper semicontinuous). □

We also need the following result.

Lemma 4.7.4. Let X be a paracompact space and Y be a non-empty normal subset of a topological vector space E. Let $\theta : X \to E$ and $P : X \to 2^Y \cup \{\emptyset\}$ be $\mathcal{U}$-majorized. Then there exists a correspondence $\Psi : X \to 2^Y \cup \{\emptyset\}$ of class $\mathcal{U}$ such that $P(x) \subset \Psi(x)$ for each $x \in X$.

Proof. Since P is $\mathcal{U}$-majorized, for each $x \in X$, let $N(x)$ be an open neighborhood of x in X and $\psi_x : N(x) \to 2^Y \cup \{\emptyset\}$ be such that (1) for each $z \in N(x)$, $P(z) \subset \psi_x(z)$ and $\theta(z) \notin \psi_x(z)$ and (2) ψ_x is upper semicontinuous with closed and convex values. Since X is paracompact and $X = \cup_{x \in X} N(x)$, by Theorem VIII.1.4 of Dugundji [147, p.162], the open covering $\{N(x)\}$ of X has an open precise neighborhood-finite refinement $\{N'(x)\}$. For each $x \in X$, define $\psi'_x : X \to 2^Y \cup \{\emptyset\}$ by

$$\psi'_x(z) = \begin{cases} \psi_x(z), & \text{if } z \in N'(x); \\ Y, & \text{if } z \notin N'(x), \end{cases}$$

then ψ'_x is also upper semicontinuous on X by Theorem 4.7.2 above and is such that $P(z) \subset \psi'_x(z)$ for each $z \in X$.

Now define $\Psi : X \to 2^Y \cup \{\emptyset\}$ by $\Psi(z) = \cap_{x\in X}\psi'_x(z)$ for each $z \in X$. Clearly, Ψ has closed and convex values and $P(z) \subset \Psi(z)$ for each $z \in X$. Let $z \in X$ be given, then $z \in N'(x)$ for some $x \in X$ so that $\psi'_x(z) = \psi_x(z)$ and hence $\Psi(z) \subset \psi_x(z)$; as $\theta(z) \notin \psi_x(z)$, we must also have that $\theta(z) \notin \Psi(z)$. Thus $\theta(z) \notin \Psi(z)$ for all $z \in X$.

Now we shall show that Ψ is upper semicontinuous. For any given $u \in X$, there exists an open neighborhood M_u of u in X such that the set $\{x \in X : M_u \cap N(x) \neq \emptyset\}$ is finite, say $= \{x(u,1), \cdots, x(u,n(u))\}$. Thus we have that

$$\Psi(w) = \cap_{x\in X}\psi'_x(w) = \cap_{i=1}^{n(u)}\psi'_{x(u,i)}(w) \qquad \text{for all } w \in M_u.$$

For $i = 1, \cdots, n(u)$, since each $\psi'_{x(u,i)}$ is upper semicontinuous on X and hence on M_u with closed values and Y is normal, by Theorem 4.7.2 above again, $\Psi : M_u \to 2^Y$ is also upper semicontinuous at u. Since M_u is open, $\Psi : X \to 2^Y$ is also upper semicontinuous at u. Hence Ψ is of class $\mathcal{U}$. □

We now prove the following theorem concerning the existence of a maximal element for $\mathcal{U}$-majorized correspondences.

Theorem 4.7.5. Let X be a non-empty convex subset of a locally convex topological vector space and D a non-empty compact subset of X. Let $P : X \to 2^D \cup \{\emptyset\}$ be $\mathcal{U}$-majorized (i.e., $\mathcal{U}_{I_X}$-majorized). Then there exists a point $x \in coD$ such that $P(x) = \emptyset$.

Proof. Suppose the contrary, i.e., for all $x \in coD$, $P(x) \neq \emptyset$. Then for each $x \in coD$, $P(x) \neq \emptyset$ and coD is also paracompact by Lemma 1 of Ding et al. [141, p.206] (see also Lassonde [336, p.49]). Now applying Lemma 4.7.4, there exists a correspondence $\Psi : coD \to 2^D$ of class $\mathcal{U}$ such that for each $x \in coD$, $P(x) \subset \Psi(x)$. Since Ψ is upper semicontinuous with non-empty closed and convex values, by the fixed point Theorem of Himmelberg [233, Theorem 2, p.206], there exists $x \in coD$ such that $x \in \Psi(x)$. This contradicts that Ψ is of class $\mathcal{U}$. Hence the conclusion must hold. □

In what follows, we shall give some applications of Lemma 4.7.4 and Theorem 4.7.5. First we have the following:

Theorem 4.7.6. Let X be a non-empty convex subset of a locally convex topological vector space and D be a non-empty compact subset of X. Let $P : X \to 2^D$ be $\mathcal{U}$-majorized and $A : X \to 2^D$ be upper semicontinuous with closed and convex values. Then there exists a point $\hat{x} \in coD$ such that either $\hat{x} \in A(\hat{x})$ and $P(\hat{x}) = \emptyset$ or $\hat{x} \notin A(\hat{x})$ and $A(\hat{x}) \cap P(\hat{x}) = \emptyset$.

Proof. Let $F = \{x \in X : x \in A(x)\}$. We first note that F is closed in X

since A is upper semicontinuous with closed values. Define $\phi : X \to 2^D$ by

$$\phi(x) = \begin{cases} P(x), & \text{if } x \in F, \\ A(x) \cap P(x), & \text{if } x \notin F. \end{cases}$$

If $x \notin F$ and $A(x) \cap P(x) \neq \emptyset$, then $X \setminus F$ is an open neighborhood of x in X and since P is $\mathcal{U}$-majorized, there exists an open neighborhood $N(x)$ of x in X and a mapping $\psi_x : N(x) \to 2^D$ such that (1) for each $z \in N(x)$, $P(z) \subset \psi_x(z)$ and $z \notin \psi_x(z)$ and (2) ψ_x is upper semicontinuous with closed and convex values.

Without loss of generality, we may assume that $N(x) \subset X \setminus F$. We now define the mapping $\Psi_x : X \to 2^D$ by $\Psi_x(z) = A(z) \cap \psi_x(z)$ for each $z \in N(x)$. Then again by Lemma 4.7.3 (note that D is compact so that D is normal), we have (1) Ψ_x is upper semicontinuous with closed and convex values and (2) for each $z \in N(x)$, $z \notin \Psi_x(z)$. Thus Ψ_x is a $\mathcal{U}$-majorant of ϕ at x.

Now suppose that $x \in F$ and $P(x) \neq \emptyset$; then by assumption there exists an open neighborhood $N(x)$ of x in X and $\psi_x : N(x) \to 2^D$ such that (a) $P(z) \subset \psi_x(z)$ and $z \notin \psi_x(z)$ for each $z \in N(x)$ and (b) ψ_x is upper semicontinuous with closed and convex values. Define $\psi'_x : N(x) \to 2^D$ by

$$\psi'_x(z) = \begin{cases} \psi_x(z), & \text{if } z \in N(x) \cap F, \\ A(x) \cap \psi_x(z), & \text{if } z \notin N(x) \setminus F, \end{cases}$$

then (i) for each $z \in N(x)$, it is easy to see that $\phi(z) \subset \psi'_x(z)$ and $z \notin \psi'_x(z)$; and (ii) the mapping $A \cap \psi_x : N(x) \setminus F \to 2^D$ defined by $(A \cap \psi_x)(z) = A(z) \cap \psi_x(z)$ for each $z \in N(x) \setminus F$ is upper semicontinuous with closed and convex values by Lemma 4.7.3. It follows that the mapping ψ'_x is also upper semicontinuous with closed and convex values by Lemma 4.7.3 since $N(x) \setminus F$ is open in $N(x)$. This shows that ψ'_x is a $\mathcal{U}$-majorant of ϕ at x.

Therefore ϕ is $\mathcal{U}$-majorized. By Theorem 4.7.5, there exists a point $\hat{x} \in coD \subset X$ such that $\phi(\hat{x}) = \emptyset$. By the definition of ϕ, either $P(\hat{x}) = \emptyset$ and $\hat{x} \in A(\hat{x})$ or $A(\hat{x}) \cap P(\hat{x}) = \emptyset$ and $\hat{x} \notin A(\hat{x})$. □

The following is an equilibrium existence theorem for a qualitative game.

Theorem 4.7.7. Let $\Gamma = (X_i, P_i)_{i \in I}$ be a qualitative game such that for each $i \in I$:

(a) X_i is a non-empty convex subset of a locally convex topological vector space E_i and D_i is a non-empty compact subset of X_i;

(b) the set $E^i = \{x \in X : P_i(x) \neq \emptyset\}$ is open in X;

(c) $P_i : E^i \to 2^{D_i} \cup \{\emptyset\}$ is $\mathcal{U}$-majorized; and

(d) there exists a non-empty compact and convex subset F_i of D_i such that $F_i \cap P_i(x) \neq \emptyset$ for each $x \in E^i$.

Then there exists a point $x \in X$ such that $P_i(x_i) = \emptyset$ for all $i \in I$.

Proof. Since D_i is a non-empty compact subset of X_i for each $i \in I$, the set $D = \Pi_{i\in I} D_i$ is also a non-empty compact subset of X. Now for each $x \in X$, let $I(x) = \{i \in I : P_i(x) \neq \emptyset\}$. Define a correspondence $P : X \to 2^D \cup \{\emptyset\}$ by

$$P(x) = \begin{cases} \cap_{i\in I(x)} P_i'(x), & \text{if } I(x) \neq \emptyset, \\ \emptyset, & \text{if } I(x) = \emptyset, \end{cases}$$

where $P_i'(x) = \Pi_{j\neq i, j\in I} F_j \times P_i(x)$ for each $x \in X$.

Then by condition (d) and definition of P, for each $x \in X$ with $I(x) \neq \emptyset$, $P(x) \neq \emptyset$. Let $x \in X$ be such that $P(x) \neq \emptyset$. Fix an $i \in I(x)$. By assumption (c), there exists an open neighborhood $N(x)$ of x in E^i and $\phi_i : N(x) \to 2^{D_i}$ such that (i) for each $z \in N(x)$, $P_i(z) \subset \phi_i(z)$ and $\pi_i(z) \notin \phi_i(z)$ and (ii) ϕ_i is upper semicontinuous with closed and convex values. Note that by (b), $N(x)$ is also an open neighborhood of x in X and for each $z \in N(x)$, $P_i(z) \neq \emptyset$ so that $i \in I(z)$ for each $z \in N(x)$. Now we define $\Phi_x : N(x) \to 2^D$ by $\Phi_x(z) = \Pi_{j\neq i, j\in I} F_j \otimes \phi_i(z)$ for each $z \in N(x)$. We observe that (1) for each $z \in N(x)$, $P(z) \subset P_i'(z) \subset \Phi_x(z)$ and $z \notin \Phi_x(z)$; (2) Φ_x has closed and convex values and (3) since $\Pi_{j\neq i, j\in I} F_j$ and $\phi_x(z)$ are compact for each $z \in N(x)$, it is easy to see that Φ_x is also upper semicontinuous. Therefore, Φ_x is a $\mathcal{U}$-majorant of P at x. Thus P is $\mathcal{U}$-majorized. Now by Theorem 4.7.5, there exists a point $x \in coD \subset X$ such that $P(x) = \emptyset$ which implies that $P_i(x) = \emptyset$ for all $i \in I$. □

Theorem 4.7.8. Let $\Gamma = (X_i; A_i, B_i; P_i)_{i\in I}$ be a generalized game where I is any (countable or uncountable) set of players such that for each $i \in I$:

(i) X_i is a non-empty compact and convex subset of a locally convex topological vector space E_i;

(ii) for each $x \in X$, $A_i(x)$ is non-empty, $A_i(x) \subset B_i(x)$ and $B_i(x)$ is convex;

(iii) the set $E^i = \{x \in X : A_i(x) \cap P_i(x) \neq \emptyset\}$ is paracompact (which is satisfied if X_i is metrizable) and open in X; and

(iv) the mapping $A_i \cap P_i : X \to 2^{X_i} \cup \{\emptyset\}$ is $\mathcal{U}$-majorized on E^i.

Then Γ has an equilibrium point, i.e., there exists a point $x \in X$ such that $\pi_i(x) \in \overline{B_i}(x)$ and $A_i(x) \cap P_i(x) = \emptyset$ for all $i \in I$.

Proof. Let $I_0 = \{i \in I : E^i \neq \emptyset\}$. Suppose $I_0 = \emptyset$. Then for each $i \in I$ and for each $x \in X$, $A_i(x) \cap P_i(x) = \emptyset$. Define $B : X \to 2^X \cup \{\emptyset\}$ by $B(x) = \Pi_{i\in I} \overline{B_i}(x)$ for each $x \in X$. Since each $\overline{B_i}$ has a closed graph, it is easy to see that B also has a closed graph. Since X is compact, B is upper semicontinuous. By the Fan-Glicksberg fixed point theorem, there exists $\hat{x} \in X$ such that $\hat{x} \in B(\hat{x})$, i.e., $\pi_i(\hat{x}) \in \overline{B_i}(\hat{x})$ for all $i \in I$. Thus we may assume without loss of generality that $I_0 \neq \emptyset$.

Case 1: For each $i \in I_0$, by (iv) and Lemma 4.7.4 (note that the set X_i, being compact Haudorff, is normal), there exists a mapping $\psi_i : E^i \to 2^{X_i} \cup \{\emptyset\}$ which is upper semicontinuous with closed and convex values such that $A_i(x) \cap P_i(x) \subset \psi_i(x)$ and $\pi_i(x) \notin \psi_i(x)$ for each $x \in E^i$. Since $\overline{B_i} : X \to 2^{X_i} \cup \{\emptyset\}$ is upper semicontinuous with closed and convex values, the correspondence $\psi_i \cap \overline{B_i} : E^i \to 2^{X_i} \setminus \{\emptyset\}$ is also upper semicontinuous with non-empty closed and convex values by (ii) and Proposition B.III.2 of Hildenbrand [229, p.23-24]. Define a correspondence $\phi_i : X \to 2^{X_i} \setminus \{\emptyset\}$ by

$$\phi_i(x) = \begin{cases} \overline{B_i}(x), & \text{if } x \notin E^i, \\ (\psi_i \cap \overline{B_i})(x), & \text{if } x \in E^i. \end{cases}$$

Then Lemma 4.7.3 implies that ϕ_i is upper semicontinuous with non-empty closed and convex values.

Case 2: For each $i \in I \setminus I_0$, we define a correspondence $\phi_i : X \to 2^{X_i} \cup \{\emptyset\}$ by $\phi_i(x) = \overline{B_i}(x)$ for each $x \in X$. Then ϕ_i is upper semicontinuous with non-empty compact and convex values.

Now define the correspondence $\Psi : X \to 2^X$ by $\Psi(x) = \Pi_{i \in I} \phi_i(x)$ for each $x \in X$. Then Ψ is also upper semicontinuous with non-empty compact and convex values. By the Fan-Glicksberg fixed point theorem again, there exists a point $x \in X$ such that $x \in \Psi(x)$. If there exists $i \in I_0$ such that $x \in E^i$, then $\pi_i(x) \in \phi_i(x) = \overline{B_i}(x) \cap \psi_i(x) \subset \psi_i(x)$ which is a contradiction; it follows that $x \notin E^i$ for all $i \in I_0$. Hence we must have $\pi_i(x) \in \overline{B_i}(x)$ and $A_i(x) \cap P_i(x) = \emptyset$ for all $i \in I$. $\square$

It seems natural to replace condition (iii) of Theorem 4.7.8 by condition '*the set $E^i = \{x \in X : A_i(x) \cap P_i(x) \neq \emptyset\}$ is closed in X*'; however the following simple example shows that this can not be done:

Example 4.7.9. Let $X = [0,1]$ and define $A, B, P : X \to 2^X \cup \{\emptyset\}$ by

$$B(x) = A(x) = \begin{cases} [1/2, 1], & \text{if } x \in [0, 1/2), \\ [0, 1], & \text{if } x = 1/2, \\ [0, 1/2], & \text{if } x \in (1/2, 1]. \end{cases}$$

and

$$P(x) = \begin{cases} \{x/4\}, & \text{if } x \in [\frac{1}{2}, 1], \\ \emptyset, & \text{if } x \in [0, 1/2). \end{cases}$$

It is easy to see that A and P are both upper semicontinuous with closed and convex values and $x \notin P(x)$ for each $x \in X$; thus $A \cap P$ is $\mathcal{U}$-majorized. Note that the subset $E = \{x \in X : A(x) \cap P(x) \neq \emptyset\} = [1/2, 1]$ is closed in $[0,1]$ and A, B and P satisfy the hypotheses (i), (ii), and (iv) but not (iii) of Theorem 4.7.8. However, at the unique fixed point $1/2$ of the correspondence A, we have $A(\frac{1}{2}) \cap P(\frac{1}{2}) = [0,1] \cap \{1/8\} \neq \emptyset$. Thus the generalized game $([0,1]; A, B; P)$ has no equilibrium point.

4.8 Equilibria of Ψ-Condensing Mappings

In this section, we shall prove some existence theorems for maximal elements for Ψ-condensing correspondences which are either KF_C-majorized or $\mathcal{U}$-majorized and whose domain are non-compact sets in locally convex topological vector spaces. As an application, we obtain an existence theorem for equilibrium points for a one-person game from which an existence theorem for N-person games is derived. Finally, we give an existence theorem equilibria of generalized games with a countable or uncountable set of players such that the intersection of constraint and preference correspondences are $\mathcal{U}$-majorized and constraint correspondences are Ψ-condensing.

The object of this section is to present a method for proving the existence of maximal elements and equilibria of generalized games which enables one to remove altogether the compactness (or paracompactness) assumption on the domain (and/or codomain) of the preference and constraint correspondences. This is done by strengthening the assumptions on the preference or constraint correspondences. The basic idea underlying the method may be explained as follows.

Let X be a non-empty subset of a locally convex topological vector space E. We introduce a function $\Psi : 2^X \to \mathbb{R}$ which assigns to each relatively compact subset D of X the value zero, and call it a measure of non-compactness. Intuitively, $\Psi(D)$ measures how far a set is from being relatively compact. The larger the value $\Psi(D)$ the *more non-compact* a set D is in X. Now a set-valued mapping $T : X \to 2^E \cup \{\emptyset\}$ is said to be Ψ-condensing if for each subset D that is not relatively compact, we have $\Psi(T(D)) < \Psi(D)$.

It should be noted that if $T : X \to 2^E \cup \{\emptyset\}$ is a compact mapping (i.e., if $T(X)$ is contained in a compact subset K of E), then the mapping T is automatically Ψ-condensing for any measure of non-compactness.

Definition 4.8.1. Let X be a non-empty set. A mapping $P : X \to 2^X \cup \{\emptyset\}$ is said to be acyclic (e.g., see Bergstrom [42, p.403]) if for each $n \in \mathbb{N}$ with $x_{i+1} \in P(x_i)$ for $i = 1, 2, \cdots n-1$ implies that $x_1 \notin P(x_n)$.

Let C denote a lattice with a least element 0. We now also recall some definitions introduced by Fitzpatick and Petryshyn [175].

Definition 4.8.2. Let X be a locally convex topological vector space. Then a mapping $\Psi : 2^X \to C$ is called a measure of non-compactness provided that the following conditions hold for any $A, B \in 2^X$:

(1) $\Psi(A) = 0$ if and only if A is precompact (i.e., it is relative compact);
(2) $\Psi(\overline{co}A) = \Psi(A)$, where $\overline{co}A$ denotes the closed convex hull of A; and
(3) $\Psi(A \cup B) = max\{\Psi(A), \Psi(B)\}$.

It follows from (3) that if $A \subset B$, then $\Psi(A) \leq \Psi(B)$. The above notion is a generalization of the set-measure of non-compactness of Kuratowski [331] and the ball-measure of non-compactness of Sadovskii [450] defined either in terms of a family of seminorms when X is a locally convex topological vector space by Gohberg et al. [200] or in terms of a single norm when X is a Banach space.

Definition 4.8.3. Let $\Psi : 2^X \to C$ be a measure of non-compactness of X and $D \subset X$. A mapping $T : D \to 2^X$ is called Ψ-condensing provided that if $\Omega \subset D$ and $\Psi(T(\Omega)) \geq \Psi(\Omega)$, then Ω is relatively compact.

Note that if $T : D \to 2^X$ is a compact mapping (i.e. $T(D)$ is precompact), then T is Ψ-condensing for any measure of non-compactness Ψ. Various Ψ-condensing mappings which are not compact have been considered by Borisovich et al. [49], Gohberg et al. [200] and Furi and Vignoli [187]. Moreover, when the measure of non-compactness Ψ is either the set-measure of non-compactness or ball-measure of non-compactness, Ψ-condensing mappings are called condensing mappings, e.g., see Nussbaum [405].

Throughout the rest of this section, E denotes a locally convex topological vector space, D denotes a non-empty closed convex subset of E, C denotes a lattice with a least element 0 and $\Psi : 2^E \to C$ denotes a measure of non-compactness.

Lemma 4.8.4. If $T : D \to 2^D \cup \{\emptyset\}$ is Ψ-condensing, then there exists a non-empty compact convex subset K of D such that $T(x) \subset K$ for each $x \in K$.

Proof. Let x_0 be an element of D and consider the family $\mathcal{F}$ of all closed convex subsets C of D such that $x_0 \in C$ and $T(x) \subset C$ for each $x \in C$. Clearly $\mathcal{F}$ is non-empty. Let $C_0 = \cap_{C \in \mathcal{F}} C$. Then C_0 is a non-empty closed and convex subset of D and $x_0 \in C_0$. If $x \in C_0$, then $T(x) \subset C$ for all $C \in \mathcal{F}$ so that $T(x) \subset C_0$.

Now we shall prove that C_0 is also compact. Let $C_1 = \overline{co}(\{x_0\} \cup T(C_0))$. Then $C_1 \subset C_0$, which implies that $T(C_1) \subset T(C_0) \subset C_1$. Thus $C_1 \in \mathcal{F}$ and hence $C_0 \subset C_1$. Therefore $C_0 = C_1$. Hence,

$$\begin{aligned} \Psi(C_0) &= \Psi(C_1) = \Psi(\overline{co}(\{x_0\} \cup T(C_0))) = \Psi(\{x_0\} \cup T(C_0)) \\ &= max\{\Psi(\{x_0\}), \Psi(T(C_0))\} = \Psi(T(C_0)) \end{aligned}$$

so $\Psi(C_0) \leq \psi(T(C_0))$ which implies that C_0 is compact. □

Theorem 4.8.5. Suppose that $T : D \to 2^D$ satisfies the following conditions:

(1) T is Ψ-condensing and $T(x)$ is non-empty convex for each $x \in D$; and

(2) for each $y \in D$, the set $T^{-1}(y) = \{x \in D : y \in T(x)\}$ is compactly open in D.

Then T has a fixed point in D.

Proof. Since T is Ψ-condensing, by Lemma 4.8.4, there exists a non-empty compact convex subset K of D such that $T : K \to 2^K$. For each $y \in K$, $T^{-1}(y)$ is also open in K by (2). Now by the Browder-Fan fixed point theorem in Section 3, there exists $x \in K$ such that $x \in T(x)$. □

Theorem 4.8.6. Suppose that $T : D \to 2^D$ is upper semicontinuous and Ψ-condensing such that $T(x)$ closed and convex for each $x \in D$. Then T has a fixed point.

Proof. Since T is a Ψ-condensing, by Lemma 4.8.4, there exists a non-empty compact convex subset K of D such that $T : K \to 2^K$ is also upper semicontinuous with non-empty closed and convex values. Now by the Fan-Glicksberg fixed point theorem in Fan [156] or Glicksberg [196], there exists $x \in K$ such that $x \in T(x)$. □

Theorem 4.8.5 and Theorem 4.8.6 generalize mant well-known fixed point theorems in locally convex topological vector spaces, e.g., see Dugundji and Granas [149] and Reich [433], Smart [491], Istratescu [264] and Zeidler [613].

We now prove the following theorem on the existence of a maximal element which generalizes the corresponding result of Toussaint [560, Theorem 2.4] and of Yannelis and Prabhakar [587, Corollary 5.1].

Theorem 4.8.7. Let $T : D \to 2^D \cup \{\emptyset\}$ be Ψ-condensing such that either (i) T is $\mathcal{U}$-majorized or (ii) T is KF_C-majorized. Then there exists a point $x^* \in D$ such that $T(x^*) = \emptyset$.

Proof. Suppose $T(x) \neq \emptyset$ for all $x \in D$. By Lemma 4.8.4, there exists a non-empty compact convex subset K of D such that $T : K \to 2^K$. If assumption (i) holds, then by Lemma 4.7.4 in section 7, there exists an upper semicontinuous mapping $S : K \to 2^K$ such that for each $x \in K$, $S(x)$ is non-empty closed and convex, $T(x) \subset S(x)$ and $x \notin S(x)$. But then by the classical Fan-Glicksberg fixed point theorem ([156] or [196]), there exists a point $x \in K$ such that $x \in S(x)$ which is a contradiction. Now if assumption (ii) holds, since T is KF_C-majorized, by Theorem 4.3.10 (see also Lemma 2 of Ding and Tan [138]), there exists a mapping $S : K \to 2^K$ such that (a) $T(x) \subset S(x)$ for each $x \in K$, (b) $S^{-1}(y)$ is open in K for each $y \in K$ and (c) $x \notin coS(x)$ for each $x \in K$. By Lemma 5.1 of Yannelis and Prabhakar [587], $(coS)^{-1}(y)$ is also open in K for each $y \in K$. Then by the Browder-Fan fixed point theorem in Section 3 (see also Browder [59]), there exists a point $x \in K$ such that $x \in coS(x)$ which is a contradiction. Therefore the conclusion must hold. □

We now also have the following extension of a theorem of Bergstrom [42] to a locally convex topological vector space and a non-compact setting.

Theorem 4.8.8. Suppose $P : D \to 2^D \cup \{\emptyset\}$ satisfies the following conditions:

(i) P is Ψ-condensing;
(ii) $P^{-1}(y)$ is compactly open for each $y \in D$; and
(iii) P is acyclic.

Then there exists $x^* \in D$ such that $P(x^*) = \emptyset$.

Proof. By Lemma 4.8.4, there exists a non-empty compact convex subset K of D such that $P : K \to 2^K \cup \{\emptyset\}$. Clearly, the restriction $P|_K$ of P to K is also acyclic and has open inverse images. Therefore the conclusion follows from Bergstrom's theorem [42, p.403]. □.

As an application of Theorem 4.8.7, we shall first prove the following existence theorem for equilibrium points for a one-person game.

Theorem 4.8.9. Let $A, B, P : D \to 2^D \cup \{\emptyset\}$ be such that:
(i) $A \cap P$ is KF_C-majorized,
(ii) for each $x \in D$, $A(x)$ is non-empty and $coA(x) \subset B(x)$ and for each $y \in D$, $A^{-1}(y)$ is compactly open in D; and
(iii) B is Ψ-condensing.

Then there exists $x \in D$ such that $x \in \overline{B}(x)$ and $A(x) \cap P(x) = \emptyset$; that is, the one person game $(D; A, B; P)$ has an equilibrium point.

Proof. Let $F = \{x \in D : x \in \overline{B}(x)\}$, then F is closed in D. Define $\Lambda : D \to 2^D \cup \{\emptyset\}$ by

$$\Lambda(x) = \begin{cases} A(x) \cap P(x), & \text{if } x \in F, \\ A(x), & \text{if } x \notin F. \end{cases}$$

Suppose $\Lambda(x) \neq \emptyset$. If $x \notin F$, then $D \setminus F$ is an open neighborhood of x in D such that for each $z \in D \setminus F$, $z \notin \overline{B}(z)$. Now define $\Phi_x : D \to 2^D \cup \{\emptyset\}$ by $\Phi_x(z) = A(z)$ for each $z \in D$ and $N_x = D \setminus F$, then N_x is an open neighborhood of x in D such that

(i) $\Lambda(z) \subset \Phi_x(z)$ and $z \notin co\Phi_x(z)$ for each $z \in N_x$, and

(ii) $\Phi_x^{-1}(y) = A^{-1}(y)$ is compactly open in D.

Therefore Φ_x is a KF_C-majorant of Λ at x. On the other hand, if $x \in F$, then $\Lambda(x) = A(x) \cap P(x) \neq \emptyset$. Since $A \cap P$ is KF_C-majorized, there exists an open neighborhood N_x of x in D and a correspondence $\Phi_x : D \to 2^D \cup \{\emptyset\}$ such that $\Lambda(z) = A(z) \cap P(z) \subset \Phi_x(z)$ and $z \notin co\Phi_x(z)$ for each $z \in N_x$, and $\Phi_x^{-1}(y)$ is compactly open in D for each $y \in D$. Define the mapping $\Phi'_x : D \to 2^D \cup \{\emptyset\}$ by

$$\Phi'_x(z) = \begin{cases} A(z) \cap \Phi_x(z), & \text{if } z \in F, \\ A(z), & \text{if } z \notin F. \end{cases}$$

Note that $\Lambda(z) \subset \Phi'_x(z)$ for each $z \in N_x$. It is easy to see that $z \notin co\Phi'_x(z)$ for all $z \in D$. Moreover, for any $y \in D$, the set $(\Phi'_x)^{-1}(y) = [\Phi_x^{-1}(y) \cup (D \setminus F)] \cap A^{-1}(y)$

is compactly open in D. It follows that Φ'_x is a KF_C-majorant of Λ at the point x. Hence Λ is a KF_C-majorized correspondence.

Since $\Lambda(x) \subset A(x) \subset B(x)$ for each $x \in D$, and B is Ψ-condensing by condition (ii), Λ is also Ψ-condensing .

Now by Lemma 4.8.4, there exists a non-empty compact convex subset K of X such that $\Lambda : K \to 2^K \cup \{\emptyset\}$. Clearly, the restriction $\Lambda|_K$ of Λ to K satisfies all hypotheses of Theorem 4.8.7. By Theorem 4.8.7, there exists $x \in K \subset D$ such that $\Lambda(x) = \emptyset$. Since $A(x) \neq \emptyset$, we must have $x \in \overline{B}(x)$ and $A(x) \cap P(x) = \emptyset$. □

Let I be a finite set and X_i be a topological space, and $X = \Pi_{i \in I} X_i$. For a given correspondence $A_i : X \to 2^{X_i}$, define a mapping $A'_i : X \to 2^X$ by $A'_i(x) = \{y \in X : y_i \in A_i(x)\} = \pi_i^{-1}(A_i(x))$ for each $x \in X$, where $\pi_i : X \to X_i$ is the projection. Then it is easy to see that A'_i is of class KF_C if and only if for each $x \in X$, $x_i \notin A_i(x)$ and $A_i^{-1}(y)$ is compactly open for each $y \in X$.

From Theorem 4.8.9, we shall now derive another existence theorem for equilibrium points for N-person games, where $N > 1$.

Theorem 4.8.10. Let $(X_i; A_i, B_i; P_i)_{i=1}^N$ be an N-person game. Suppose for each $i = 1, 2, \cdots, N$:

(i) X_i is a non-empty closed convex subset of a locally convex topological vector space E_i;

(ii) for each $x \in X$, $A_i(x)$ is non-empty and $coA_i(x) \subset B_i(x)$;

(iii) for each $y \in X_i$, $A_i^{-1}(y)$ is compactly open in X;

(iv) $A'_i \cap P'_i : X \to 2^X \cup \{\emptyset\}$ is of class KF_C, where $A'_i(x) = \pi_i^{-1}(A_i(x))$ and $P'_i(x) = \pi_i^{-1}(P_i(x))$; and

(v) the mapping $B : X \to 2^X$ defined by $B(x) = \Pi_{i=1}^N \overline{B_i}(x)$ for each $x \in X$ is Ψ-condensing (where $E = \Pi_{i=1}^N E_i$).

Then there exists $x \in X$ such that for each $i = 1, 2, \cdots, N$, $x \in \overline{B_i}(x)$ and $A_i(x) \cap P_i(x) = \emptyset$; that is, the N-person game $(X_i; A_i, B_i; P_i)_{i=1}^N$ has an equilibrium point.

Proof. For each $x \in X$, let $I(x) = \{i \in I : A_i(x) \cap P_i(x) \neq \emptyset\}$. Define the correspondences $A, B, P : X \to 2^X \cup \{\emptyset\}$ by $A(x) = \Pi_{i=1}^N A_i(x)$, $B(x) = \Pi_{i=1}^N \overline{B_i}(x)$ and

$$P(x) = \begin{cases} \cap_{i \in I(x)} P'_i(x), & \text{if } I(x) \neq \emptyset, \\ \emptyset, & \text{if } I(x) = \emptyset \end{cases}$$

for each $x \in X$. Since $B : X \to 2^X \cup \{\emptyset\}$ is Ψ-condensing, by Lemma 4.8.4, there exists a non-empty compact and convex susbet K of X such that $B(x) \subset K$ for each $x \in K$. By (ii), it follows that the A is also a self-mapping on K.

Condition (iii) implies that $A^{-1}(y) \cap K = \cap_{i=1}^N (A_i^{-1}(y_i) \cap K)$ is open in K for each $y \in K$. In order to apply Theorem 4.8.9, it remains to prove that the correspondence $A \cap P : K \to 2^K \cup \{\emptyset\}$ is KF_C-majorized.

Suppose $x \in K$ and $(A \cap P)(x) \neq \emptyset$; then $I(x) \neq \emptyset$. Choose any $y \in A(x) \cap P(x)$, then $x \in \cap_{i\in I} A_i^{-1}(y_i) \cap \cap_{i\in I(x)} P_i^{-1}(y_i) \cap K \subset \cap_{i\in I(x)}((A'_i \cap P'_i)^{-1}(y) \cap K) := N_x$ which is an open neighborhood of x in K by (iv). Note that if $z \in N_x$, then $I(x) \subset I(z)$: since for each $i \in I(x)$, $y \in (A'_i \cap P'_i)(z)$, we have $y_i \in A_i(z) \cap P_i(z)$ so that $i \in I(z)$. Now fix any $i_0 \in I(x)$. Define $P_K, \psi_x : K \to 2^K \cup \{\emptyset\}$ by $P_K(z) = P(z) \cap K$ and $\psi_x(z) = A'_{i_0}(z) \cap P'_{i_0}(z) \cap K$ for each $z \in K$, then by (iv) again, ψ_x is of class KF_C (in fact, KF) and for each $z \in N_z$, since $A(z) \subset B(z) \subset K$,

$$\begin{aligned} A(z) \cap P(z) \quad &= A(z) \cap P_K(z) = \Pi_{i\in I} A_i(z) \cap \cap_{i\in I(x)} P'_i(z) \cap K \\ &\subset \pi_{i_0}^{-1}(A_{i_0}(z)) \cap \pi_{i_0}^{-1}(P_i(z)) \cap K = A'_{i_0}(z) \cap P'_{i_0}(z) \cap K \\ &= \psi_x(z) \end{aligned}$$

Thus ψ_x is a KF_C-majorized of $A \cap P$ at x. Thus the mappings $A, P_K, B : K \to 2^K \cup \{\emptyset\}$ satisfy all the hypotheses of Theorem 4.8.9. By Theorem 4.8.9, there exists $x_0 \in K$ such that $x_0 \in \overline{B}(x_0)$ and $A(x_0) \cap P_K(x_0) = \emptyset$. Since $x_0 \in K$ and $A(x_0) \subset K$, it follows that $A(x_0) \cap P(x_0) = A(x_0) \cap P(x_0) \cap K = A(x_0) \cap P_K(x_0) = \emptyset$. Note that $B = \Pi_{i\in I}\overline{B_i}$ has a closed graph, we have $\overline{B}(x) = B(x)$ for each $x \in K$ and also $I(x_0) = \emptyset$. Hence the conclusion follows. □

The following is an existence theorem for equilibrium points of generalized games in which the intersection of constraint and preference correspondences are $\mathcal{U}$-majorized and constraint correspondences are Ψ-condensing. We emphasize that the set of players need not be finite.

Theorem 4.8.11. Let $\mathcal{G} = (X_i; A_i, B_i, P_i)_{i\in I}$ be a generalized game where I is any (countable or uncountable) set of players such that:

(a) for each $i \in I$, X_i is a non-empty closed convex subset of a locally convex topological vector space E_i;

(b) for each $i \in I$ and for each $x \in X$, $A_i(x)$ is non-empty and $coA_i(x) \subset \overline{B_i}(x)$;

(c) for each $i \in I$, the set $E^i = \{x \in X : (A_i \cap P_i)(x) \neq \emptyset\}$ is paracompact and open in X;

(d) for each $i \in I$, $A_i \cap P_i$ is $\mathcal{U}$-majorized; and

(e) the correspondence $B : X \to 2^X$ defined by $B(x) = \Pi_{i\in I}\overline{B_i}(x)$ for each $x \in X$ is Ψ-condensing (where $E = \Pi_{i\in I} E_i$).

Then $\mathcal{G}$ has an equilibrium point in X, i.e., there exists a point $\hat{x} = (\hat{x}_i)_{i\in I} \in X$ such that for each $i \in I$, $\hat{x}_i \in \overline{B_i}(\hat{x})$ and $A_i(\hat{x}) \cap P_i(\hat{x}) = \emptyset$.

Proof. Without loss of generality, we may assume that $\overline{B_i}(x)$ is convex for each $x \in X$ (otherwise replace $\overline{B_i}$ by $\overline{coA_i}$). Since the correspondence $B : X \to 2^X \cup \{\emptyset\}$ is Ψ-condensing, by Lemma 4.8.4, there exists a non-empty compact and convex subset K of X such that $B : K \to 2^K$.

Let $E_K^i = K \cap E^i$ for each $i \in I$. Let $I_0 = \{i \in I : E_K^i \neq \emptyset\}$. If $I_0 = \emptyset$, then $E_K^i = \emptyset$ for all $i \in I$ so that $(A_i \cap P_i)(x) = \emptyset$ for all $x \in K$. On the other hand, since B has a closed graph, B is upper semicontinuous on K. Also, B has closed and convex values. Thus by the Fan-Glicksberg fixed point theorem, there exists $x \in K$ such that $x \in B(x)$. It follows that $x_i \in \overline{B_i}(x)$ for all $i \in I$ and hence x is an equilibrium point of $\mathcal{G}$. Therefore we may assume that $I_0 \neq \emptyset$.

For each $i \in I$, let $K_i = \pi_i(K)$; note that each K_i is compact and convex and that $\overline{B_i}(x) \subset K_i$ for each $x \in K$.

Case 1: Suppose $i \in I_0$. Note that E_K^i is paracompact (e.g., see Theorem VIII.2.4 of Dugundji [147, p.165]) and open in K. By (d) and Lemma 4.7.4, there exists an upper semicontinuous mapping $\psi_i : E_K^i \to 2^{K_i}$ such that for each $x \in E_K^i$, (i) $\psi_i(x)$ is closed and convex, (ii) $\pi_i(x) \notin \psi_i(x)$ and (iii) $A_i(x) \cap P_i(x) \subset \psi_i(x)$. Since $\overline{B_i} : K \to 2^{K_i}$ is also upper semicontinuous with closed and convex values, the mapping $\psi_i \cap \overline{B_i} : E_K^i \to 2^{K_i}$ is also upper semicontinuous with closed and convex values by Theorem 7.3.10 of Klein and Thompson [317]. Define a correspondence $\phi_i : K \to 2^{K_i}$ by

$$\phi_i(x) = \begin{cases} \overline{B_i}(x), & \text{if } x \notin E_K^i, \\ (\psi_i \cap \overline{B_i})(x), & \text{if } x \in E_K^i. \end{cases}$$

Then Lemma 4.7.3 implies that ϕ_i is upper semicontinuous with non-empty closed and convex values.

Case 2: Suppose $i \in I \setminus I_0$. Define a correspondence $\phi : K \to 2^{K_i}$ by $\phi_i = \overline{B_i}(x)$ for each $x \in K$. Then ϕ is upper semicontinuous with compact and convex values.

Finally we define a correspondence $\Phi : K \to 2^{\Pi_{i \in I} K_i}$ by $\Phi(x) = \Pi_{i \in I} \phi_i(x)$ for each $x \in K$. Then Φ is also upper semicontinuous and has non-empty compact and convex values. Since $\Phi(x) \subset B(x) \subset K$ for each $x \in K$, $\Phi : K \to 2^K$ is in fact a self-mapping on K. Now the Fan-Glicksberg fixed point theorem again implies that there exists a point $x \in K$ such that $x \in \Phi(x)$. It follows that $\pi_i(x) \in \overline{B_i}(x)$ for all $i \in I$. If there exists $i \in I_0$ such that $x \in E_K^i$, then $\pi_i(x) \in \phi_i(x) = \overline{B_i}(x) \cap \psi_i(x) \subset \psi_i(x)$ which contradicts (ii). Therefore $x \notin E_K^i$ for all $i \in I_0$. Hence we also have $A_i(x) \cap P_i(x) = \emptyset$ for all $i \in I$. □

4.9 Abstract Economics in Frechet Spaces

So far, we have given a number of existence theorems for equilibria of N-person games, qualitative games and generalized games in H-spaces, topological vcetor space and locally convex topological spaces. In this section, the underlying spaces of generalized games are Frechet spaces and we shall obtain existence theorems for generalized games by Michael's selection theorems in [371].

Theorem 4.9.1. Let $\mathcal{G} = (X_i; A_i; P_i)_{i\in I}$ be a generalized game and $X = \Pi_{i\in I}X_i$ be paracompact, where I is any (countable or uncountable) set. Suppose that for each $i \in I$, the following conditions are satisfied:

(i) X_i is a non-empty closed and convex subset of a Frechet space E_i;

(ii) A_i is lower semicontinuous with non-empty closed convex values;

(iii) the mapping $A : X \to 2^X$ defined by $A(x) = \Pi_{i\in I}A_i(x)$ is Ψ-condensing for each $x \in X = \Pi_{i\in I}X_i$, where C is a lattice with a least element 0 and $\Psi : 2^{\Pi_{j\in I}E_j} \to C$ is a measure of non-compactness;

(iv) for each $x \in X$, $\pi_i(x) \notin A_i(x) \cap P_i(x)$;

(v) the set $U_i := \{x \in X : A_i(x) \cap P_i(x) \neq \emptyset\}$ is closed in X; and

(vi) the mapping $A_i \cap P_i$ is lower semicontinuous on U_i such that for each $x \in U_i$, $A_i(x) \cap P_i(x)$ is closed and convex.

Then there exists $x^* \in X$ such that for each $i \in I$, $\pi_i(x^*) \in A_i(x^*)$ and $A_i(x^*) \cap P_i(x^*) = \emptyset$.

Proof. Fix an $i \in I$. Define $F_i : X \to 2^{X_i}$ by

$$F_i(x) = \begin{cases} A_i(x) \cap P_i(x), & \text{if } x \in U_i; \\ A_i(x), & \text{if } x \notin U_i. \end{cases}$$

By Lemma 4.7.3, F_i is lower semicontinuous with non-empty closed and convex values. Then by Michael's selection theorem [371, Theorem 3.2''] and the Remark of Aubin [10, p.551]), there exists a continuous mapping $f_i : X \to X_i$ such that $f_i(x) \in F_i(x)$ for each $x \in X$.

Now define $f : X \to X$ by $f(x) = \{f_i(x)\}_{i\in I}$ for each $x \in X$. Then f is continuous and $f(x) \in F(x) = \Pi_{i\in I}F_i(x) \subset \Pi_{i\in I}A_i(x)$. Since A is Ψ-condensing, it follows that f is also Ψ-condensing. Since $X = \Pi_{i\in I}X_i$ is a non-empty closed and convex subset of the locally convex topological vector space $\Pi_{i\in I}E_i$, f satisfies all hypotheses of Theorem 4.8.6. By Theorem 4.8.6, there exists $x^* \in X$ such that $f(x^*) = x^*$. Note that for each $i \in I$, if $x_i^* \in U_i$, then $\pi_i(x^*) = f_i(x^*) \in A_i(x^*) \cap P_i(x^*)$ which contradicts (iv). Hence for each $i \in I$, we must have $\pi_i(x^*) \notin U_i$ and thus $\pi_i(x^*) \in A_i(x^*)$ and $A_i(x^*) \cap P_i(x^*) = \emptyset$. □

We also have the following result.

Theorem 4.9.2. Let $\mathcal{G} = (X_i; A_i; P_i)_{i\in I}$ be a generalized game, where I is any set. Suppose for each $i \in I$, the following conditions are satisfied:

(i) X_i is a non-empty closed and convex subset of a Frechet space E_i;

(ii) A_i is upper semicontinuous with non-empty closed convex values;

(iii) the mapping $A : X = \Pi_{i\in I}X_i \to 2^X$ defined by $A(x) = \Pi_{i\in I}A_i(x)$ for each $x \in X$ is Ψ-condensing, where C is a lattice with a least element 0 and $\Psi : 2^{\Pi_{j\in I}E_j} \to C$ is a measure of non-compactness;

(iv) the set $U_i = \{x \in X : A_i(x) \cap P_i(x) \neq \emptyset\}$ is paracompact and open in X; and

(v) the mapping $A_i \cap P_i$ is lower semicontinuous on U_i such that for each $x \in U_i$, $A_i(x) \cap P_i(x)$ is closed and convex.

Then there exists $x^* \in X$ such that for each $i \in I$, either $\pi_i(x^*) \in A_i(x^*) \cap P_i(x^*)$ or $\pi_i(x^*) \in A_i(x^*)$ and $A_i(x^*) \cap P_i(x^*) = \emptyset$.

Proof. Fix an $i \in I$. By (v), Theorem 3.2″ of Michael [371] and the Remark of Aubin [10, p.551], let $f_i : U_i \to X_i$ be a continuous function such that $f_i(x) \in A_i(x) \cap P_i(x)$ for each $x \in U_i$. Define $F_i : X \to 2^{X_i}$ by

$$F_i(x) = \begin{cases} \{f_i(x)\}, & \text{if } x \in U_i; \\ A_i(x), & \text{if } x \notin U_i, \end{cases}$$

then by (ii), (iv) and Lemma 4.7.3, F_i is upper semicontinuous with closed convex values.

Now we define $F : X \to 2^X$ by $F(x) = \Pi_{i\in I} F_i(x)$ for each $x \in X$. Then F is upper semicontinuous with closed convex values and $F(x) \subset A(x)$ for each $x \in X$. Since A is Ψ-condensing, F is also Ψ-condensing. Therefore by Theorem 2.6.3, there exists $x^* \in X$ such that $x^* \in F(x^*)$. It follows that for each $i \in I$, either $\pi_i(x^*) \in A_i(x^*) \cap P_i(x^*)$ or $\pi_i(x^*) \in A_i(x^*)$ and $A_i(x^*) \cap P_i(x^*) = \emptyset$. □

Note that if the set I is countable, the set $U_i = \{x \in X : A_i(x) \cap P_i(x) \neq \emptyset\}$ is a subset of a metrizable set $X = \Pi_{i\in I} X_i$ so that U_i is automatically paracompact for each $i \in I$.

By Theorem 3.1‴ of Michael [371, p.368] instead of his Theorem 3.2″ [371, p.367], the same argument used in proving Theorem 4.9.1 and Theorem 4.9.2 can likewise be used to prove the following:

Theorem 4.9.3. Let $\mathcal{G} = (X_i; A_i; P_i)_{i\in I}$ be a generalized game and $X = \Pi_{i\in I} X_i$ be paracompact, where I is any (countable or uncountable) set. Suppose that for each $i \in I$, the following conditions are satisfied:

(i) X_i is a non-empty closed and convex subset of a finite dimensional space E_i;

(ii) A_i is lower semicontinuous with non-empty convex values (but not necessarily closed);

(iii) the mapping $A : X \to 2^X$ defined by $A(x) = \Pi_{i\in I} A_i(x)$ is Ψ-condensing for each $x \in X = \Pi_{i\in I} X_i$, where C is a lattice with a least element 0 and $\Psi : 2^{\Pi_{j\in I} E_j} \to C$ is a measure of non-compactness;

(iv) for each $x \in X$, $\pi_i(x) \notin A_i(x) \cap P_i(x)$;

(v) the set $U_i := \{x \in X : A_i(x) \cap P_i(x) \neq \emptyset\}$ is closed in X; and

(vi) the mapping $A_i \cap P_i$ is lower semicontinuous on U_i such that for each $x \in U_i$, $A_i(x) \cap P_i(x)$ is convex (but not necessarily closed).

Then there exists $x^* \in X$ such that for each $i \in I$, $\pi_i(x^*) \in A_i(x^*)$ and $A_i(x^*) \cap P_i(x^*) = \emptyset$.

Theorem 4.9.4. Let $\mathcal{G} = (X_i, A_i, P_i)_{i\in I}$ be a generalized game where I is any set. Suppose for each $i \in I$, the following conditions are satisfied:

(i) X_i is a non-empty closed and convex subset of a finite dimensional space E_i;

(ii) A_i is upper semicontinuous with non-empty closed convex values;

(iii) the mapping $A : X = \Pi_{i\in I} X_i \to 2^X$ defined by $A(x) = \Pi_{i\in I} A_i(x)$ for each $x \in X$ is Ψ-condensing, where C is a lattice with a least element 0 and $\Psi : 2^{\Pi_{j\in I} E_j} \to C$ is a measure of non-compactness;

(iv) the set $U_i := \{x \in X : A_i(x) \cap P_i(x) \neq \emptyset\}$ is paracompact and open in X; and

(v) the mapping $A_i \cap P_i$ is lower semicontinuous on U_i such that for each $x \in U_i$, $A_i(x) \cap P_i(x)$ is convex (but not necessarily closed).

Then there exists $x^* \in X$ such that for each $i \in I$, either $\pi_i(x^*) \in A_i(x^*) \cap P_i(x^*)$ or $\pi_i(x^*) \in A_i(x^*)$ and $A_i(x^*) \cap P_i(x^*) = \emptyset$.

As an application of Theorem 4.9.3, we have the following.

Corollary 4.9.5. Let I be any set. For each $i \in I$, let X_i be a non-empty compact convex subset of a finite dimensional space E_i and $P_i : X = \Pi_{j\in I} X_j \to 2^{X_i} \cup \{\emptyset\}$ be lower semicontinuous on the set $U_i = \{x \in X : P_i(x) \neq \emptyset\}$ such that for each $x \in U_i$, $P_i(x)$ is convex. If for each $i \in I$, U_i is paracompact and is either open or closed in X, then there exists $x^* \in X$ such that for each $i \in I$, either $\pi_i(x^*) \in P_i(x^*)$ or $P_i(x^*) = \emptyset$.

Proof. For each $i \in I$, let $A_i : X \to 2^{X_i}$ be defined by $A_i(x) = X_i$ for each $x \in X$. Then A_i is continuous with closed convex values and A_i is also Ψ-condensing since $X = \Pi_{i\in I} X_i$ is compact. Therefore by Theorem 4.9.3, there exists $x^* \in X$ such that for each $i \in I$, either $\pi_i(x^*) \in P_i(x^*)$ or $P_i(x^*) = \emptyset$. □

Corollary 4.9.5 generalizes Theorem 1 of Barbolla [23] which in turn improves the fixed point theorem of Gale and Mas-Colell [191] and Florenzano [177] in the following ways: (1) the index set I need not be finite and (2) for each $i \in I$, U_i is either open or closed instead of U_i being open for all $i \in I$ or U_i being closed for all $i \in I$. We remark that our argument in proving Theorem 4.9.2 is different from that of Barbolla [23]. Finally we note that the results in this section improve or generalize the results of Mehta [368] in the case that the set I of agents (or players) is any (countable or uncountable) set instead of a countable set.

4.10 Abstract Economics

The purpose of this section is two-fold. We first obtain some sufficient conditions for the intersection of two lower semicontinuous mappings to be again lower semicontinuous and secondly, by applying Michael's selection theorem [371], a fixed point theorem is derived and is applied together with earlier results on intersection of lower semicontinuous maps to obtain existence theorems for equilibrium points of a generalized game and of a qualitative game.

4.10.1 Some Properties of Lower Semicontinuous Set-Valued Mappings

The aim of this section is to establish some properties for lower semicontinous set-valued amppings.

Let E be a vector space and $A \subset E$. We shall denote by $aff(A)$ the affine span of A. The subset A is said to be finite dimensional if A is contained in a finite dimensional subspace of E. If E is a topological vector space and $A \subset E$, $ri(A)$ denotes the relative interior of A in $aff(A)$.

First we observe that the proof of Proposition 1.1 of Marano [360, p.286] actually produced the following slightly strengthened version (where the original assumption that ϕ is lower semicontinuous is replaced by the weaker assumption that ϕ is lower semicontinuous at s_0).

Lemma 4.10.1. Let S be a topological space and $\phi : S \to 2^{\mathbb{R}^n}$ be a mapping with non-empty convex values. If there exists $s_0 \in S$ such that ϕ is lower semicontinuous at s_0 and $\theta_{\mathbb{R}}^n \in \text{int}\phi(s_0)$ where $0_{\mathbb{R}}^n$ is the zero vector in $\mathbb{R}^n$, then there exists a neighborhood U_0 of s_0 in S such that $0_{\mathbf{R}^n} \in \text{int}\phi(s)$ for all $s \in U_0$.

Since the translation of a mapping (resp., convex set, open set) preserves its lower semicontinuity at a point (resp., convexity, openness), Lemma 4.10.1 is equivalent to the following.

Lemma 4.10.2. Let S be a topological space and $\phi : S \to 2^{\mathbb{R}^n}$ be a mapping with convex values. Suppose ϕ is lower semicontinuous at some $s_0 \in S$ and $y_0 \in \text{int}\phi(x_0)$, then there exists a neighborhood U_0 of s_0 in S such that $y_0 \in \text{int}\phi(s)$ for all $s \in U_0$.

By slightly modifying the proof of Lemma 1 of Monteiro [373], we formulate the following property for lower semicontinuous mappings in finite dimensional spaces.

Lemma 4.10.3. Let S be a topological space and $\phi : S \to 2^{\mathbb{R}^n} \cup \{\emptyset\}$ be a mapping with convex values. Suppose ϕ is lower semicontinuous at some $x_0 \in S$ and $B(y_0, t) \subset \phi(x_0)$ where $B(y_0, t)$ is the open ball of positive radius t centered at y_0. Then for any number r with $0 < r < t$, there exists a neighborhood U_0 of x_0 such that $B(y_0, r) \subset \phi(x)$ for all $x \in U_0$. In particular, if ϕ is open-valed, then ϕ has an open graph.

Proof. By restricting ϕ to the set $\{x \in X : \phi(x) \cap \text{int}\phi(x_o) \neq \emptyset\}$ which by the lower semicontinuity of ϕ at x_0, is an open neighborhood of x_0 in S. We may assume without loss of generality that ϕ is non-empty valued. Choose any r' with $r < r' < t$, then we have $\overline{B(y_0, r')} \subset \overline{B(y_0, t)} \subset \phi(x_0)$ where $\overline{B(y_0, r')}$ is the closure of $B(y_0, r')$ in $\mathbb{R}^n$. Note that $\overline{B(y_0, r')}$ is compact. Choose any ϵ with $0 < \epsilon < r' - r$. Let $\{B(y_i, \epsilon/2) : i = 1, \cdots, m\}$ be a finite cover of $\overline{B(y_0, r')}$ with $y_i \in B(y_0, r')$ for all $i = 1, \cdots, m$. By Lemma 4.10.2 1′, there exists an open neighborhood U_1 of x_0 in S such that $y_0 \in \text{int}\phi(x)$ for all $x \in U_1$; in particular, $\text{int}\phi(x) \neq \emptyset$ for all $x \in U_1$. By lower semicontinuity of ϕ at x_0 again, there exists an open neighborhood U_0 of x_0 in S such that for each $x \in U_0$, $\phi(x) \cap B(y_i, \epsilon/2) \neq \emptyset$ for all $i = 1, \cdots, m$. By replacing U_0 by $U_1 \cap U_0$, we may also assume that $U_0 \subset U_1$. Now given any non-empty subsets A and B of $\mathbb{R}^n$, define $e(A, B) = \sup\{d(a, B); a \in A\}$, where d is the Euclidean metric on $\mathbb{R}^n$ and $d(a, B) = \inf\{d(a, b) : b \in B\}$. It is easy to see that for each $x \in U_0$,

$$e(B(y_0, r'), \phi(x)) \leq e(\cup_{i=1}^m B(y_i, \epsilon/2), \phi(x)) \leq \epsilon.$$

By the Lemma of [373], for each $x \in U_0$,

$$d(y_0,\ R^n \setminus B(y_0, r')) \leq d(y_0,\ R^n \setminus \phi(x)) + e(B(y_0, r'), \phi(x)),$$

so that

$$d(y_0,\ R^n \setminus \phi(x)) \geq d(y_0, R^n \setminus B(y_0, r')) - e(B(y_0, r'), \phi(x)) \geq r' - \epsilon > r.$$

It follows that $B(y_0, r) \subset \phi(x)$ for all $x \in U_0$. □

Remark 1.1 of Marano [360, p.287] shows that Lemma 4.10.3 does not hold if ϕ takes its values in an infinite dimensional Hilbert space.

Now let X and Y be topological spaces and $\psi,\ \phi : X \to 2^Y$ be lower semicontinuous. We observe that if graph ϕ is open in $X \times Y$, then it is shown in Yannelis [586, Lemma 4.2] that $\phi \cap \psi$ is also lower semicontinuous. Clearly, if graph ϕ is open in $X \times Y$, then ϕ is open-valued. The following example shows that (a) graph ϕ need not be open in $X \times Y$ even though ϕ is open-valued, and (b) even when $\phi(x) \cap \psi(x) \neq \emptyset$ for all $x \in X$, if condition that 'graph ϕ is open in $X \times Y$' is weakened and replaced by the condition ϕ *is open-valued*, then $\phi \cap \psi$ need not be lower semicontinuous:

Example 4.10.4. Let $F_1, F_2 : [0,1] \to 2^{[0,1]}$ be defined by

$$F_1(x) = \begin{cases} [0,1], & \text{if } x = 0; \\ (0,1] \setminus \{1/n : n = 2, 3, \cdots, \}, & \text{if } x \neq 0 \end{cases}$$

and

$$F_2(x) = \begin{cases} \{0\}, & \text{if } x = 0; \\ \{1/n : n = 1, 2, \cdots\}, & \text{if } x \neq 0\,. \end{cases}$$

Then it is easy to see that (i) F_1 and F_2 are lower semicontinuous, (ii) $graph F_1$ is not open in $[0,1] \times [0,1]$ and (iii) F_1 is open-valued such that for each $x \in [0,1]$,

$$F_1(x) \cap F_2(x) = \begin{cases} \{0\}, & \text{if } x = 0; \\ \{1\}, & \text{if } x \neq 0. \end{cases}$$

It follows that $F_1 \cap F_2$ is not lower semicontinuous.

Theorem 4.10.5. Let S be a topological space and $F_1,\ F_2 : S \to 2^{\mathbf{R}^n}$ be lower semicontinuous at $x_0 \in S$ such that F_1 is open and convex-valued. Then $F_1 \cap F_2$ is also lower semicontinuous at x_0.

Proof. Let U be an open subset of $\mathbb{R}^n$ such that $F_1(x_0) \cap F_2(x_0) \cap U \neq \emptyset$. Choose any $y_0 \in F_1(x_0) \cap F_2(x_0) \cap U$. Since $F_1(x_0) \cap U$ is open, y_0 is an interior point of $F_1(x_0) \cap U$. By Lemma 4.10.3, there is an open neighborhood V_1 of x_0 in S and $r > 0$ such that $B(y_0, r) \subset F_1(x)$ for all $x \in V_1$, where $B(y_0, r)$ is an open ball in $\mathbb{R}^n$ of radius r and centered at y_0. Since U is open in $\mathbb{R}^n$, we may assume without loss of generality that $B(y_0, r) \subset U$. Since F_2 is lower semicontinuous at x_0, there is an open neighborhood V_2 of x_0 in S such that $F_2(x) \cap B(y_0, r) \neq \emptyset$ for all $x \in V_2$. Let $V = V_1 \cap V_2$, then V is an open neighborhood of x_0 in S such that for each $x \in V$, $F_1(x) \cap F_2(x) \cap U \supset F_2(x) \cap B(y_0, r) \neq \emptyset$. Hence $F_1 \cap F_2$ is lower semicontinuous. □

We note that Theorem 4.10.5 is different from Theorem B of Lechicki and Spakowski [338, p.121] in the following ways: (a) $\text{int}(F_1 \cap F_2)(x_0)$ is not required to be non-empty, but (*b*) the mapping F_1 is required to have open values.

The following example from Lechicki and Spakowski [338, Example 3] shows that even if F_2 has closed (or open) convex values, the conclusion of Theorem 4.10.5 fails to hold if $\mathbb{R}^n$ is replaced by an infinite dimensional Banach space.

Example 4.10.6. Let $Y = l^\infty$, the Banach space of all bounded sequences $x = (x_n)_{n=1}^\infty$ of real numbers with $\|x\|_\infty = \sup_{n \in N} |x_n| < \infty$ and $S = [0,1]$. Define $G_1, G_2 : S \to 2^Y$ by

$$G_1(t) = \{x \in Y : x_1 \geq t, x_k \leq k - t \text{ for } k \geq 2\}$$

and

$$G_2(t) = \{x \in Y : x_1 \leq 1 - t, x_k \leq k(1 - x_1 - t) \text{ and } x_k \leq k + x_1/k - t/k \text{ for } k \geq 2\}.$$

Then G_1 and G_2 are both lower semicontinuous at 0 with closed convex values and $\text{int}_Y(G_1(0) \cap G_2(0)) \neq \emptyset$, but $G_1 \cap G_2$ is not lower semicontinuous at 0 (see Example 3 of [338]). Now we define $F_1, F_2 : S \to 2^Y$ by $F_1(t) = \text{int}_Y G_1(t)$ and $F_2(t) = \text{int}_Y G_2(t)$ (or $G_2(t)$) for each $t \in S$. Since

$$\overline{F_1(0) \cap F_2(0)} = \overline{(\text{int}_Y G(0)) \cap (\text{int}_Y G_2(0))} = \overline{(\text{int}_Y G_1(0)) \cap G_2(0)} = G_1(0) \cap G_2(0),$$

$F_1 \cap F_2$ is also not lower semicontinuous at $t = 0$ by Proposition 7.3.3 of Klein and Thompson [317] (or Proposition 2.3 of Michael [371, p.366]).

Theorem 4.10.7. Let S be a topological space and $F_1,\ F_2 : S \to 2^{\mathbb{R}^n}$ be lower semicontinuous at $x_0 \in S$ such that for each $x \in S$, $F_1(x)$ and $F_2(x)$ are both convex and $(\text{int}_{\mathbb{R}^n} F_1(x)) \cap F_2(x) \neq \emptyset$ or $F_1(x) \cap (\text{int}_{\mathbb{R}^n} F_2(x)) \neq \emptyset$ whenever $F_1(x) \cap F_2(x) \neq \emptyset$. Then $F_1 \cap F_2$ is also lower semicontinuous at $x_0 \in S$.

Proof. Without loss of generality, we may assume that $(\text{int}_{\mathbb{R}^n} F_1(x_0)) \cap F_2(x_0) \neq \emptyset$. Since $F_1(x_0)$ and $F_2(x_0)$ are convex, by Theorem of Dolecki [146, p.253],

$$\overline{(\text{int}^n_{\mathbb{R}} F_1(x_0)) \cap F_2(x_0)} = \overline{F_1(x_0) \cap F_2(x_0)} = \overline{F_1(x_0)} \cap \overline{F_2(x_0)}.$$

In order to prove $F_1 \cap F_2$ is lower semicontinuous at x_0, it suffices to prove that $(\text{int}^n_{\mathbb{R}} F_1) \cap F_2$ is lower semicontinuous at x_0. Now let U be any non-empty open subset of $\mathbb{R}^n$ such that $(\text{int}^n_{\mathbb{R}} F_1(x_0)) \cap F_2(x_0) \cap U \neq \emptyset$. Choose any $y_0 \in (\text{int}^n_{\mathbb{R}} F_1(x_0)) \cap F_2(x_0) \cap U$. Since $\text{int}^n_{\mathbb{R}} F_1(x_0) \cap U$ is open, y_0 is an interior point of $\text{int}^n_{\mathbb{R}} F_1(x_0) \cap U$. Since F_1 is lower semicontinuous at x_0 and $\text{int}^n_{\mathbb{R}} F(x_0) \neq \emptyset$, we have $\overline{F_1(x_0)} = \overline{\text{int}^n_{\mathbb{R}} F_1(x_0)}$. Therefore $\text{int}^n_{\mathbb{R}} F_1$ is also lower semicontinuous at x_0. By Lemma 4.10.3, there is an open neighborhood V_1 of x_0 in S and $r > 0$ such that $B(y_0, r) \subset \text{int}^n_{\mathbb{R}} F_1(x)$ for all $x \in V_1$, where $B(y_0, r)$ is an open ball in $\mathbb{R}^n$ of radius r and centered at y_0. Since U is open in $\mathbb{R}^n$, we may assume without loss of generality that $B(y_0, r) \subset U$. Since F_2 is lower semicontinuous at x_0, there is an open neighborhood V_2 of x_0 in S such that $F_2(x) \cap B(y_0, r) \neq \emptyset$ for all $x \in V_2$. Let $V = V_1 \cap V_2$, then V is an open neighborhood of x_0 in S such that for each $x \in V$, $(\text{int}^n_{\mathbb{R}} F_1(x)) \cap F_2(x) \cap U \supset F_2(x) \cap B(y_0, r) \neq \emptyset$. Hence $(\text{int}^n_{\mathbb{R}} F_1) \cap F_2$ is lower semicontinuous at x_0, so that $F_1 \cap F_2$ is also lower semicontinuous at x_0.
□

Theorem 4.10.7 generalizes a result of Obukhovskii [406] (see also Borisovich et al. [49, Corollary 1.3.10, p.725].

Theorem 4.10.8. Let S be a topological space, X be a non-empty subset of a finite dimensional topological vector space and $F_1, F_2 : S \to 2^{aff(X)}$ be lower semicontinuous at $x_0 \in S$. If for each $x \in S$, $F_1(x)$ is convex and open in $aff(X)$, then $F_1 \cap F_2$ is also lower semicontinuous at x_0.

Proof. Choose any $a_0 \in aff(X)$ and let $Y = aff(X) - a_0$, then Y is a finite dimensional topological vector space and is therefore topologically isomorphic to some $\mathbb{R}^n$. Define $\tilde{F}_1,\ \tilde{F}_2 : S \to 2^Y$ by $\tilde{F}_1(x) = F_1(x) - a_0$, $\tilde{F}_2(x) = F_2(x) - a_0$ for all $x \in S$. Then $\tilde{F}_1$ and $\tilde{F}_2$ are lower semicontinuous at x_0 such that for each $x \in S$, $\tilde{F}_1(x)$ is convex and is open in Y. By Theorem 4.10.5, $\tilde{F}_1 \cap \tilde{F}_2$ is lower semicontinuous at x_0. It follows that $F_1 \cap F_2$ is also lower semicontinuous at x_0.
□

By using the same proof as in Theorem 4.10.8, we have the following.

Theorem 4.10.9. Let S be a topological space, X be a non-empty subset of a finite dimensional topological vector space and $F_1, F_2 : S \to 2^{aff(X)}$ be lower semicontinuous at $x_0 \in S$. If for each $x \in S$, $F_1(x)$ and $F_2(x)$ are convex and $ri(F_1(x)) \cap F_2(x) \neq \emptyset$ or $F_1(x) \cap ri(F_2(x)) \neq \emptyset$ whenever $F_1(x) \cap F_2(x) \neq \emptyset$, then $F_1 \cap F_2$ is also lower semicontinuous at x_0.

Next we shall prove a fixed point theorem as follows.

Theorem 4.10.10. Let I be a non-empty countable set. For each $i \in I$, let C_i be a non-empty finite dimensional compact convex subset of a topological vector space E_i and $F_i : C := \Pi_{j\in I} C_j \to 2^{E_i}$ be lower semicontinuous such that:
(a) $F_i(x)$ is convex for each $x \in C$;
(b) $F_i(x) \cap ri(C_i) \neq \emptyset$ for each $x \in C$; and
(c) $F_i(C) \subset aff(C_i)$.

Then the mapping $F := \Pi_{i\in I} F_i$ has a fixed point in C.

Proof. Let $i \in I$ be arbitrarily fixed. Define $F_i' : C \to 2^{E_i}$ by $F_i'(x) = F_i(x) \cap ri(C_i)$ for each $x \in C$. Let A be an open subset of E_i such that $ri(C_i) = aff(C_i) \cap A$. If B is any open subset of E_i, then by (c),

$$\begin{aligned} \{x \in C : F_i'(x) \cap B \neq \emptyset\} &= \{x \in C : F_i(x) \cap ri(C_i) \cap B \neq \emptyset\} \\ &= \{x \in C : F_i(x) \cap A \cap B \neq \emptyset\} \end{aligned}$$

is open in C since F_i is lower semicontinuous and $A \cap B$ is open in E_i. This shows that F_i' is lower semicontinuous on C. Note that C is metrizable, being a countable product of finite dimensional sets; thus C is perfectly normal. Also for each $x \in C$, $F_i'(x)$ is a convex subset of C_i which is contained in some Euclidean space. Hence by Theorem 3.1$'''$ of Michael [371], there exists a continuous mapping $f_i : C \to C_i$ such that $f_i(x) \in F_i'(x)$ for all $x \in C$.

Now define $f : C \to C$ by $f := \Pi_{i\in I} f_i$, then f is continuous. Since C is a compact convex subset of some locally convex space (in fact, a countable product of Euclidean spaces), by the Schauder-Tychonoff fixed point theorem, there exists $x^* \in C$ such that $f(x^*) = x^*$. It follows that $x^* = \Pi_{i\in I} f_i(x^*) \in \Pi_{i\in I} F_i'(x^*) \subset \Pi_{i\in I} F_i(x^*) = F(x^*)$; that is, $F = \Pi_{i\in I} F_i$ has a fixed point in C. □

Corollary 4.10.11. Let X be a non-empty subset of a topological vector space E, C a non-empty finite dimensional compact convex subset of X and let $F : X \to 2^E$ be lower semicontinuous such that $F(C) \subset aff(C)$ and for each $x \in C$, $F(x)$ is convex and $F(x) \cap ri(C) \neq \emptyset$. Then F has a fixed point in X.

Corollary 4.10.11 improves Theorem 1 of Cubiotti [106] where F is assumed to have closed values.

We remark that in Theorem 4.10.10: (1) if $I = \{1, 2, \cdots, n\}$ is finite, then $ri(\Pi_{i=1}^n X_i) = \Pi_{i=1}^n ri(X_i)$; in this case, Theorem 4.10.10 can also be obtained from Corollary 4.10.11; and (2) if $I = \{1, 2, \cdots ,\}$ is infinite and the set $\{j \in I : X_j$ is not a singleton set $\}$ is infinite, then $ri(\Pi_{j=1}^{\infty} X_j) = \emptyset$; in this case, Theorem 4.10.10 can not be deduced from Corollary 4.10.11. Thus Theorem 4.10.10 is a true generalization of Theorem 1 of Cubiotti [106].

We shall also need the following affine version of Corollary 6.3.2 of Rockafeller [444, p.46].

Lemma 4.10.12. Let C be a non-empty convex subset of a finite dimensional topological vector space E. Then every open set in $aff(C)$ which meets clC also meets $ri(C)$.

Proof. Choose any $a_0 \in aff(C)$ and let $Y = aff(C) - a_0$, then Y is a finite dimensional topological vector space and is therefore topologically isomorphic to some $\mathbb{R}^n$. Let U be an open set in $aff(C)$ such that $U \cap cl_E(C) \neq \emptyset$. Then $U - a_0$ is open in Y and $\emptyset \neq (U - a_0) \cap (cl(C) - a_0) = (U - a_0) \cap cl(C - a_0)$. By Corollary 6.3.2 in [444, p.46], $(U - a_0) \cap ri(C - a_0) \neq \emptyset$. As $ri(C - a_0) = ri(C) - a_0$, it follows that $U \cap ri(C) \neq \emptyset$. □

4.10.2 Equilibria

The aim of this section is to establish the existence of equilibria for abstract economics in finite dimensional spaces.

We first have the following existence theorem for equilibrium points of generalized games.

Theorem 4.10.13. Let $\mathcal{G} = (X_i; A_i, B_i; P_i)_{i \in I}$ be a generalized game where I is countable. Suppose that for each $i \in I$, the following properties hold:

(1) X_i is a non-empty subset of a topological vector space E_i;

(2) $A_i, B_i : X = \Pi_{i \in I} X_i \to 2^{X_i}$ are lower semicontinuous such that $coA_i(x)$; $\subset B_i(x)$ for each $x \in X$;

(3) $P_i : X \to 2^{X_i} \cup \{\emptyset\}$ is lower semicontinuous on D_i, where $D_i = \{x \in X : P_i(x) \cap A_i(x) \neq \emptyset\}$ is closed in X;

(4) there exists a non-empty finite dimensional compact convex subset C_i of X_i such that (a) for each $x \in C := \Pi_{j \in I} C_j$, $P_i(x)$ is open in $aff(X_i)$, (b) for each $x \in C$, $coA_i(x) \cap ri(C_i) \neq \emptyset$, (c) for each $x \in C \cap D_i$, $coP_i(x) \cap coA_i(x) \cap C_i \neq \emptyset$ and (d) $coA_i(x) \subset aff(C_i)$; and

(5) for each $x \in C$, $\pi_i(x) \notin coP_i(x)$.

Then $\mathcal{G}$ has an equilibrium point x^* in C.

Proof. Fix an arbitrary $i \in I$. Define $F_i : C \to 2^{E_i}$ by

$$F_i(x) = \begin{cases} coP_i(x) \cap coA_i(x), & \text{if } x \in C \cap D_i, \\ coA_i(x), & \text{if } x \notin C \cap D_i \ . \end{cases}$$

Since A_i and P_i are lower semicontinuous, coA_i and coP_i are lower semicontinuous. Since for each $x \in C$, $coP_i(x)$ is convex and open in $aff(X_i)$ by (4)(a), the mapping $x \mapsto coA_i(x) \cap coP_i(x)$ is also lower semicontinuous on C by Theorem 4.10.8. Since $C \cap D_i$ is closed in C, F_i is lower semicontinuous by Lemma 4.7.3. Moreover, we have

(i) $F_i(x)$ is convex for each $x \in C$; and

(ii) $F_i(C) \subset coA_i(C) \subset aff(C_i)$ by condition (4)(d).

Now we shall show that $F_i(x) \cap ri(C_i) \neq \emptyset$ for each $x \in C$. Indeed, if $x \in C \setminus D_i$, then $F_i(x) \cap ri(C_i) = coA_i(x) \cap ri(C_i) \neq \emptyset$ by (4)(b). Suppose $x \in C \cap D_i$, then by (4)(c), $coP_i(x) \cap coA_i(x) \cap C_i \neq \emptyset$; as $coP_i(x) \cap aff(C_i)$ is open in $aff(C_i)$ by (4)(a), $coP_i(x) \cap ri(coA_i(x) \cap C_i) \neq \emptyset$ by Lemma 4.10.12. Also, it follows from $coA_i(x) \cap C_i \neq \emptyset$ that $ri(coA_i(x) \cap C_i) = ri(coA_i(x)) \cap ri(C_i)$ by Theorem 6.5 of Dolecki [146, p.47]. Thus

$$\begin{aligned}\emptyset &\neq coP_i(x) \cap ri(coA_i(x) \cap C_i)\\ &= coP_i(x) \cap ri(coA_i(x)) \cap ri(C_i)\\ &\subset coP_i(x) \cap coA_i(x) \cap ri(C_i) = F_i(x) \cap ri(C_i).\end{aligned}$$

This shows that all hypotheses of Theorem 4.10.10 are satisfied so that there exists $x^* \in C$ such that $x^* \in \Pi_{i\in I} F_i(x^*)$. By (5), we must have $x^* \notin D_i$ and $\pi_i(x^*) \in coA_i(x^*)$ for all $i \in I$. It follows that for each $i \in I$, $\pi_i(x^*) \in B_i(x^*)$ by (2) and $P_i(x^*) \cap A_i(x^*) = \emptyset$. Hence x^* is equilibrium point of $\mathcal{G}$. □

As a special case of Theorem 4.10.13, we have the following result.

Theorem 4.10.14. Let $\mathcal{G} = (X_i; A_i, B_i; P_i)_{i\in I}$ be a generalized game where I is countable. Suppose that for each $i \in I$, the following properties hold:

(1) X_i is a non-empty compact convex subset of a finite dimensional topological vector space;

(2) $A_i, B_i : X = \Pi_{i\in I} X_i \to 2^{X_i}$ are lower semicontinuous such that $coA_i(x) \subset B_i(x)$ for each $x \in X$;

(3) $P_i : X \to 2^{X_i} \cup \{\emptyset\}$ is lower semicontinuous on D_i, where $D_i = \{x \in X : P_i(x) \cap A_i(x) \neq \emptyset\}$ is closed in X; and

(4) For each $x \in X$, (a) $P_i(x)$ is open in $aff(X_i)$, (b) $coA_i(x) \cap ri(X_i) \neq \emptyset$ and (c) $\pi_i(x) \notin coP_i(x)$.

Then $\mathcal{G}$ has an equilibrium point x^* in X.

The following example shows that the assumption '$D_i = \{x \in X : A_i(x) \cap P_i(x) \neq \emptyset\}$ *is closed in* X' in condition (3) of Theorem 4.10.13 and Theorem 4.10.14 is necessary.

Example 4.10.15. Define $A : [0,1] \to 2^{[0,1]}$ by $A(x) := [0, 1 - x]$ for each $x \in [0,1]$. Clearly, A is continuous with closed convex values. Now define

$A_1 : [0,1] \to 2^{[0,1]}$ by

$$A_1(x) = \begin{cases} [0, 1-x], & \text{if } x \in (0,1], \\ \{1\}, & \text{if } x = 0, \end{cases}$$

is also lower semicontinuous by Lemma 4.7.3. We also define $A_2 : [0,1] \to 2^{[0,1]}$ by

$$A_2(x) = \begin{cases} co([0, 1-x] \cup \{1/2\}), & \text{if } x \in (0,1], \\ co(\{1\} \cup \{1/2\}), & \text{if } x = 0, \end{cases}$$

then A_2 is also lower semicontinuous and in fact

$$A_2(x) = \begin{cases} [0, 1-x], & \text{if } x \in (0, 1/2], \\ [0, 1/2], & \text{if } x \in (1/2, 1), \\ [1/2, 1], & \text{if } x = 0, \\ [0, 1/2], & \text{if } x = 1. \end{cases}$$

We now define $P : [0,1] \to 2^{[0,1]}$ by $P(x) = [0, x)$ for each $x \in [0,1]$ then for any $y \in [0,1]$, we have $P^{-1}(y) = (y, 1]$. Since $\{x \in [0,1], P(x) \cap A_2(x) \neq \emptyset\} = (0,1]$ and the fixed point set of A_2 is $(0, 1/2]$, but for each $x \in (0, 1/2]$, $A_2(x) \cap P(x) \neq \emptyset$.

Finally we shall derive the following existence theorem for an equilibrium point of a qualitative game.

Theorem 4.10.16. Let $\mathcal{G} = (X_i; P_i)_{i \in I}$ be a qualitative game where I is countable. Suppose that for each $i \in I$, the following properties hold:

(1) X_i is a non-empty subset of a topological vector space E_i;

(2) $P_i : X = \Pi_{j \in I} X_j \to 2^{X_i} \cup \{\emptyset\}$ is lower semicontinuous on D_i, where $D_i = \{x \in X : P_i(x) \neq \emptyset\}$ is closed in X; and

(3) there exists a non-empty finite dimensional compact convex subset C_i of X_i such that (a) for each $x \in C := \Pi_{j \in I} C_j$, $P_i(x) \subset aff(C_i)$; (b) for each $x \in C \cap D_i$, $coP_i(x) \cap ri(C_i) \neq \emptyset$; (c) for each $x \in C$, $\pi_i(x) \notin coP_i(x)$.

Then $\mathcal{G}$ has an equilibrium point x^* in C.

Proof. Fix an arbitrary $i \in I$. Define $F_i : C \to 2^{E_i}$ by

$$F_i(x) = \begin{cases} coP_i(x), & \text{if } x \in C \cap D_i; \\ aff(C_i), & \text{if} \in X \setminus (C \cap D_i) \,. \end{cases}$$

Then by (2), (3)(a) and Lemma 4.7.3, F_i is lower semicontinuous on C. Clearly $F_i(C) \subset aff(C_i)$ by (3)(a) and for each $x \in X$, $F_i(x)$ is convex. We shall now show that for each $x \in C$, $F_i(x) \cap ri(C_i) \neq \emptyset$. Suppose $x \in C \cap D_i$, then by (3)(b), $F_i(x) \cap ri(C_i) = coP_i(x) \cap ri(C_i) \neq \emptyset$. If $x \in X \setminus (C \cap D_i)$, then

$F_i(x) \cap ri(C_i) = aff(C_i) \cap ri(C_i) = ri(C_i) \neq \emptyset$ by Theorem 6.2 of Rockafellar [444, p.45].

Therefore all hypotheses of Theorem 4.10.10 are satisfied so that there exists $x^* \in C$ such that $x^* \in \Pi_{i \in I} F_i(x^*)$. By (3)(c), we must have $P_i(x^*) = \emptyset$ for all $i \in I$. Thus x^* is an equilibrium point of $\mathcal{G}$. □

By Theorem 4.10.16, we have the following result.

Corollary 4.10.17. Let $\mathcal{G} = (X_i; P_i)_{i \in I}$ be a qualitative game where I is countable. Suppose that for each $i \in I$, the following properties hold:

(1) X_i is a non-empty finite dimensional compact convex subset of a topological vector space E_i;

(2) $P_i : X = \Pi_{i \in I} X_i \to 2^{X_i} \cup \{\emptyset\}$ is lower semicontinuous on D_i, where $D_i = \{x \in X : P_i(x) \neq \emptyset\}$ is closed in X;

(3) For each $x \in X$, $coP_i(x) \cap ri(X_i) \neq \emptyset$; and

(4) For each $x \in X$, $\pi_i(x) \notin coP_i(x)$.

Then $\mathcal{G}$ has an equilibrium point $x^* \in X$.

Chapter 5

Variational Inequalities - Part I

The theory of variational inequalities provides very effective and powerful techniques for studying a wide class of problems arising in partial differential equations, mechanics, optimization and control problems, transportation and economic equilibria, operations research, contact problems in elasticity and other branches of mathematics and engineering sciences. In recent years, variational inequalities have been generalized and applied in various directions (e.g., see Aubin and Ekeland [15], Baiocchi and Capelo [22], Bensoussan and Lions [39], Browder [62], Chang [65], Cottle et al. [101], Cubiotti [105]-[110], Ding and Tarafdar [144]-[145], Noor et al. [390]-[403], Shih and Tan [465]-[472], Tan and Yuan [510], Harker and Pang [220], Hartman and Stampacchia [221], Kinderlehrer and Stampacchia [304], and many others.

The outline of Chapter 5 is as follows. We first reduce the existence problems for variational inequalities and generalized quasi-variational inequalities to the existence problem for equilibria of generalized games; in other words, the solutions of variational inequalities are nothing else, but are exactly equilibria of their equivalent model of generalized games. This simple fact enables us to consider the existence of solutions for non-compact variational inequalities and generalized quasi-variational inequalities in infinite dimensional topological vector spaces by the existence theorems for equilibria of non-compact generalized games established in Chapter 4.

More precisely, in Section 2, as applications of equilibria for non-compact generalized games in Chapter 4, the existence theorems for non-compact variational inequalities are given under various non-compact conditions in locally convex topological vector spaces. As applications, existence theorems for equilibria of constraint games are established. These results improve and unify many corresponding results in the literature.

In Section 3, as applications of variational inequalities in Section 2, two types of non-compact generalized quasi-variational inequalities are considered under various hypotheses in locally convex topological vector spaces.

In Section 4, the stability of solutions for two types of generalized quasi-variational inequalities is established. Our results show that almost all generalized quasi-variational inequalities are stable in the sense of the Baire category.

As many vector-valued optimization problems can be formulated as the existence of solutions for vector-valued variational inequalities, the purpose of Section 5 is to introduce a new kind of variational inequality which is called *generalized vector variational-like inequality* including several classical and well-known variational inequalities as special cases. As an application of the Knaster-Kuratowski-Mazurkiewicz principle again, we prove that there exist solutions for generalized vector-valued variational-like inequality under reasonable hypotheses. These results generalize corresponding results in the literature as special cases.

Finally, we study a new kind of generalized quasi-variational inclusion with non-compact set-valued mappings. The purpose of Section 6 is to give some iterative algorithms and to establish their convergence theory for algorithms under some suitable assumptions by using essentially properties of proximal mappings.

5.1 Introduction

Let $\mathbb{N}$ and $\mathbb{R}$ denote the set of all natural numbers and a real line, respectively. Let X be a non-empty subset of $\mathbb{R}^n$, where $n \in \mathbb{N}$. Suppose $F : X \to 2^X$ and $T : X \to 2^{\mathbb{R}^n}$ are two set-valued mappings. In 1982, for the purpose of study on the problems from operations research, mathematical programming and optimization theory, Chan and Pang [64] first introduced the following so-called *generalized quasi-variational inequality* problem $(GQVI(X; F, T))$ which is to find $(\hat{x}, \hat{u}) \in X \times \mathbb{R}^n$ such that

$$\hat{x} \in F(\hat{x}), \hat{u} \in T(\hat{x}) \text{ and } \sup_{y \in F(\hat{x})} \langle \hat{u}, \hat{x} - y \rangle \leq 0.$$

They established the following existence result.

Theorem 5.1.1 (Chan and Pang [64]). Let X be a non-empty compact convex subset of $\mathbb{R}^n$, where $n \in \mathbb{N}$. Suppose $F : X \to 2^X$ is continuous with non-empty compact and convex values and $T : X \to 2^{\mathbb{R}^n}$ is upper semicontinuous with non-empty compact and contractible values. Then $GQVI(X; F, T)$ has at least one solution.

In 1985, Shih and Tan [466] were the first to study the $GQVI(X; F; T)$ in infinite dimensional locally convex Hausdorff topological vector spaces as follows.

Theorem 5.1.2. Let E be a locally convex Hausdorff topological vector space, E^* be the dual space of E and X be a non-empty compact convex subset of E. Let $F : X \to 2^X$ be continuous such that for each $x \in X$, $A(x)$ is a non-empty closed convex subset of X, and $T : X \to 2^{E^*}$ be upper semicontinuous from the relative topology of X to the strong topology of E^* such that for each $x \in X$, $T(x)$ is a non-empty strongly compact subset of E^*. Then there exists a point $\hat{y} \in X$ such that

$$\begin{cases} \hat{y} \in F(\hat{y}) \text{ and} \\ \sup_{x \in F(\hat{y})} \inf_{w \in T(\hat{y})} Re\langle w, \hat{y} - x\rangle \leq 0. \end{cases}$$

Since then, there have been a number of generalizations of the existence theorems about GQVI(X; F; T), e.g., see Cubiotti [106], Ding and Tan [135], Harker and Pang [220], Kim [298], Shih and Tan [474] and Tian and Zhou [556] and references therein. These results have wide applications to problems in game theory and economics, mathematical programming (e.g., see Aubin [10], Aubin and Ekeland [15], Chan and Pang [64], Harker and Pang [220] and reference therein). Most existence theorems mentioned above, however, are obtained on compact sets in finite dimensional spaces or infinite dimensional locally convex Hausdorff topological vector spaces, and both F and T are either continuous or upper (lower) semicontinuous.

On the other hand, in economic and game applications, it is known that the choice space (or the space of feasible allocations) generally is not compact in any topology of the choice space (even though it is closed and bounded), a key situation in infinite dimensional topological vector spaces. Moreover, we note that there is essentially no existence theorem of solutions of generalized quasi-variational inequalities on non-compact sets in infinite dimensional spaces. This motivates our work in this chapter and the next one to give a series of existence theorems on generalized quasi-variational inequalities by relaxing the compactness conditions and continuity. By the existence theorems of generalized quasi-variational inequalities, the stability of solutions for two types of generalized quasi-variational inequalities are also established.

Throughout this chapter, all topological spaces are assumed to be Hausdorff for simplicity unless otherwise specified. Let E^* be the dual space of a topological vector space E, i.e., E^* consisting of all continuous linear functionals on E. The real part of pairing between E^* and E is denoted by $Re\langle w, x\rangle$ for each $w \in E^*$ and $x \in E$. Let X be a topological space, E a topological vector space and $F : X \to 2^E$ a set-valued mapping. We recall that the mapping F is said to be upper hemicontinuous if the mapping $x \mapsto \sup_{u \in F(x)} Re\langle p, u\rangle$ is upper semicontinuous (in the sense of real functions) for each $p \in E^*$. It is clear that each upper semicontinuous set-valued mapping with non-empty bounded values is upper hemicontinuous (e.g., see Section 3 of Chapter 1, and also Aubin and Ekeland [15]). We also recall that a topological vector space E is said to *have sufficiently many continuous linear functionals* if for each $x \neq y$ in E, there exists $p \in E^*$ such that $p(x - y) \neq 0$. By the Hahn-Banach theorem, it is well-known that each locally convex Hausdorff topological vector space E has sufficiently many continuous linear functionals. However there exists some topological vector space with sufficiently continuous linear functionals which is not necessarily a locally convex topological vector space (see, e.g., Rudin [449]). We now recall some definitions which will be used through this chapter.

Definition 5.1.3. Let X be a non-empty convex subset of a topological vector space and $\psi : X \times X \to \mathbb{R} \cup \{-\infty, +\infty\}$ an extended real-valued function. We recall that (e.g., see Zhou and Chen [622]):

(1) ψ is said to be 0-*diagonal quasi-concave in* y if for each $A \in \mathcal{F}(X)$ and $x_0 \in co(A)$, we have $\min_{y \in A} \psi(x, y) \leq 0$; and

(2) ψ is said to be 0-*diagonal concave in* y if for each $A := \{y_1, \cdots y_n\} \in \mathcal{F}(X)$ and $\alpha_i \in (0, 1]$ for $i = 1, \cdots, n$ with $\Sigma_{i=1}^n \alpha_i = 1$ and $x_0 := \Sigma_{i=1}^n \alpha_i y_i$, we have $\Sigma_{i=1}^n \alpha_i \psi(x_0, y_i) \leq 0$.

Remark 5.1.4. Let $\psi(x, x) \leq 0$ for each $x \in X$. It is clear that each $\psi(x, y)$ which is concave in y (for each $x \in X$) is 0-diagonal concave in y; and if $\psi(x, y)$ is 0-diagonal concave, then $\psi(x, y)$ is also 0-diagonal quasi-concave in y, but the converse does not hold by the Remark 2.2 of Zhou and Chen [622, p.215].

Definition 5.1.5. Let X be a non-empty subset of a topological vector space E and $T : X \to 2^{E^*}$ a set-valued mapping with non-empty values. Then T is said to be

(1) *monotone* (e.g., see Mosco [375]) if for each $x, y \in X$, $u \in T(x)$ and $v \in T(y)$, $Re\langle x - y,\, u - v\rangle \geq 0$;

(2) *pseudo-monotone* (e.g., see Saigal [451, p.263]) if for each $x, y \in X$, $u \in T(x)$ and $v \in T(y)$, $Re\langle x - y,\, v\rangle \geq 0$ implies $Re\langle x - y,\, u\rangle \geq 0$; and

(3) *quasi-monotone* if for each $x, y \in X$,

$$\inf_{v \in T(y)} Re\langle x - y,\, v\rangle > 0 \text{ implies } \inf_{u \in T(x)} Re\langle x - y,\, u\rangle > 0.$$

From the definition, it is clear that each monotone mapping is pseudo-monotone, and Proposition 3.2 of Karamardian and Schaible [283, p.39] also shows that each quasi-monotone set-valued mapping is a set-valued generalization of each single-valued mapping which is pseudo-monotone.

Let X be a topological space and $f : X \to \mathbb{R}$ is said to be *compactly lower semicontinuous* if f is lower semicontinuous on each non-empty compact subset C of X.

5.2 Variational Inequalities in Locally Convex Spaces

In this section, as applications of equilibria for abstract economics in Chapter 4, we establish the existence of solutions for the system of generalized quasi-variational inequalities in locally convex topological vector spaces. These results improve and unify many corresponding results in the literature (e.g., see Aubin [10], Aubin and Ekeland [15], and references therein).

As consequents of Theorem 4.5.3 (resp., Theorem 4.5.2) and Theorem 4.5.14 (resp., Theorem 4.5.13), we have the following existence theorems for non-compact quasi-variational inequalities in locally convex topological vector spaces.

Theorem 5.2.1. For each $i = 1, 2, \cdots, N$, let E_i be a locally convex topological vector space and $X = \cup_{j=1}^{\infty} C_{i,j}$ where $\{C_{i,j}\}_{j=1}^{\infty}$ is an increasing sequence of non-empty compact convex subsets of E_i. Let $X = \Pi_{i=1}^{N} X_i$. Suppose the following conditions are satisfied:

(i) for each $i = 1, 2, \cdots, N$, $A_i : X \to 2^{X_i}$ is lower semicontinuous with closed graph and convex values;

(ii) for each $i = 1, 2, \cdots, N$, $\psi_i : X \times X_i \to \mathbb{R} \cup \{-\infty, +\infty\}$ is such that for each $y_i \in X_i$, $x \mapsto \psi_i(x, y_i)$ is lower semicontinuous on X;

(iii) for each $i = 1, 2, \cdots, N$ and for each fixed $x \in X$, $x_i \notin co\{y_i \in X_i : \psi_i(x, y_i) > 0\}$);

(iv) for each $i = 1, 2, \cdots, N$, the set $\{x \in X : \sup_{y_i \in A_i(x)} \psi_i(x, y_i) > 0\}$ is open in X; and

(v) for each sequence $(y_n)_{n=1}^{\infty}$ in X which is escaping from X relative to $\{C_n\}_{n=1}^{\infty}$, where $C_n = \Pi_{i \in I} C_{i,n}$, there exists $n \in \mathbb{N}$ and $x_n \in C_n$ such that $\pi_i(x_n) \in A_i(y_n) \cap \{z \in X_i : \psi_i(y_n, z) > 0\}$ for each $i = 1, 2, \cdots, N$.

Then there exists $x^* \in X$ such that for each $i = 1, 2, \cdots, N$,

$$x_i^* \in A_i(x^*) \text{ and } \sup_{y_i \in A_i(x^*)} \psi_i(x^*, y_i) \leq 0.$$

Proof. For each $i = 1, 2, \cdots, N$, define $P_i : X \to 2^{X_i} \cup \{\emptyset\}$ by $P_i(x) = \{y_i \in X_i : \psi_i(x, y_i) > 0\}$ for each $x \in X$. First (ii) implies that for each $i = 1, 2, \cdots, N$, P_i has open lower sections in X so that by (iii), P_i is of class KF; it follows that $A_i \cap P_i$ is KF-majorized. Condition (iv) implies that for each $i = 1, \cdots, N$, the set $\{x \in X : A_i(x) \cap P_i(x) \neq \emptyset\}$ is open in X. Therefore $\mathcal{G} = (X_i; A_i; P_i)_{i=1}^N$ satisfies all the hypotheses of Theorem 4.5.3 with $A_i = B_i$ for each $i = 1, 2, \cdots, N$. By Theorem 4.5.3, there exists an $x^* \in X$ such that for each $i = 1, 2, \cdots, N$, $x_i^* \in A_i(x^*)$ and $A_i(x^*) \cap P_i(x^*) = \emptyset$. Since $\{x \in X : A_i(x) \cap P_i(x) \neq \emptyset\} = \{x \in X : \sup_{y_i \in A_i(x)} \psi_i(x, y_i) > 0\}$, it follows that for each $i = 1, 2, \cdots, N$, $x_i^* \in A_i(x^*)$ and $\sup_{y_i \in A_i(x^*)} \psi_i(x^*, y_i) \leq 0$. □.

A proof similar to that of Theorem 5.2.1, with Theorem 4.5.2 being applied instead of Theorem 4.5.3 gives the following result and is thus omitted.

Theorem 5.2.2. Let I be any index set. For each $i \in I$, let X_i be a non-empty convex subset of a locally convex topological vector space E_i. Let $X = \Pi_{i \in I} X_i$ be paracompact. Suppose that:

(i) for each $i \in I$, $A_i : X \to 2^{X_i}$ is a lower semicontinuous correspondence with closed graph and convex values;

(ii) for each $i \in I$, $\psi_i : X \times X_i \to \mathbb{R} \cup \{-\infty, +\infty\}$ is such that for each $y_i \in X_i$, $x \mapsto \psi_i(x, y_i)$ is lower semicontinuous on X;

(iii) for each $i \in I$ and for each $x \in X$, $x_i \notin co(\{y_i \in X_i : \psi_i(x, y_i) > 0\})$;

(iv) for each $i \in I$ and for each $y_i \in X_i$, the set $\{x \in X : \sup_{y_i \in A_i(x)} \psi_i(x, y_i) > 0\}$ is open in X; and

(v) there exists a non-empty compact convex subset X_0 of X and a non-empty compact subset K of X such that for each $y \in X \setminus K$, there exists $x = (x_i)_{i \in I} \in co(X_0 \cup \{y\})$ with $x_i \in co(A_i(y) \cap \{z_i \in X_i : \psi_i(y, z_i) > 0\})$ for all $i \in I$.

Then there exists $x^* \in K$ such that for each $i \in I$,

$$x_i^* \in A_i(x^*) \text{ and } \sup_{y_i \in A_i(x^*)} \psi_i(x^*, y_i) \leq 0.$$

Now by Theorem 4.5.14 instead of Theorem 4.5.14, we also have the following results for upper semicontinuous correspondences.

Theorem 5.2.3. For each $i = 1, 2, \cdots, N$, let E_i be a locally convex topological vector space and let $X_i = \cup_{j=1}^{\infty} C_{ij}$ be closed in E_i and have property (K), where $\{C_{i,j}\}_{j=1}^{\infty}$ is an increasing sequence of non-empty compact convex subsets of E_i. Let $X = \Pi_{i=1}^N X_i$. Suppose the following conditions are satisfied:

(i) for each $i = 1, 2, \cdots, N$, $A_i : X \to 2^{X_i}$ is compact and upper semicontinuous with compact and convex values;

(ii) for each $i = 1, 2, \cdots, N$, $\psi_i : X \times X_i \to \mathbb{R} \cup \{-\infty, +\infty\}$ is such that for each $y_i \in X_i$, $x \mapsto \psi_i(x, y_i)$ is lower semicontinuous on X;

(iii) for each $i = 1, 2, \cdots, N$ and for each $x \in X$, $x_i \notin co(\{y_i \in X_i : \psi_i(x, y_i) > 0\})$;

(iv) for each $i = 1, 2, \cdots, N$, the set $\{x \in X : \sup_{y_i \in A_i(x)} \psi_i(x, y_i) > 0\}$ is open in X; and

(v) for each sequence $(y_n)_{n=1}^{\infty}$ in X which is escaping from X relative to $\{C_n\}_{n=1}^{\infty}$, where $C_n = \Pi_{i\in I}C_{i,n}$, there exist $n \in \mathbb{N}$ and $x_n \in C_n$ such that $\pi_i(x_n) \in A_i(y_n) \cap \{z \in X_i : \psi_i(y, z) > 0\}$ for all $i = 1, 2, \cdots, N$.

Then there exists $x^* \in X$ such that for each $i = 1, 2, \cdots, N$,

$$x_i^* \in A_i(x^*) \text{ and } \sup_{y_i \in A_i(x^*)} \psi_i(x^*, y_i) \leq 0.$$

Proof. For each $i = 1, 2, \cdots, N$, define $P_i : X \to 2^{X_i} \cup \{\emptyset\}$ by $P_i(x) = \{y_i \in X_i : \psi_i(x, y_i) > 0\}$ for each $x \in X$, then we shall show that $\mathcal{G} = (X_i, A_i, P_i)_{i=1}^N$ satisfies all the hypotheses of Theorem 4.5.14. First, (ii) implies that for each $i = 1, 2, \cdots, N$, P_i has open lower sections in X so that by (iii), P_i is of class KF and that P_i is lower semicontinuous. Condition (iv) implies that for each $i = 1, 2, \cdots, N$, the set $\{x \in X : A_i(x) \cap P_i(x) \neq \emptyset\}$ is open in X. Therefore all hypotheses of Theorem 4.5.14 are satisfied with $A_i = B_i$ for each $i = 1, 2, \cdots, N$. By Theorem 4.5.14, there exists an $x^* \in K$ such that for each $i = 1, 2, \cdots, N$, $A_i(x^*) \cap P_i(x^*) = \emptyset$ and $x_i^* \in A_i(x^*)$. Since $\{x \in X : A_i(x) \cap P_i(x) \neq \emptyset\} = \{x \in X : \sup_{y_i \in A_i(x)} \psi_i(x, y_i) > 0\}$, it follows that for each $i = 1, 2, \cdots, N$, $x_i^* \in A_i(x^*)$ and $\sup_{y_i \in A_i(x^*)} \psi_i(x^*, y_i) \leq 0$. □.

Similarly, by applying Theorem 4.5.13 instead of Theorem 4.5.14, we have the following variational inequalities.

Theorem 5.2.4. Let I be any index set. For each $i \in I$, let X_i be a non-empty closed convex subset of a locally convex topological vector space E_i and X_i have property (K). Let $X = \Pi_{i\in I}X_i$ be paracompact. Suppose that:

(i) for each $i \in I$, $A_i : X \to 2^{X_i}$ is compact and upper semicontinuous with compact convex values;

(ii) for each $i \in I$, $\psi_i : X \times X_i \to \mathbb{R} \cup \{-\infty, +\infty\}$ is such that for each $y_i \in X_i$, $x \mapsto \psi_i(x, y_i)$ is lower semicontinuous on X;

(iii) for each $i \in I$ and for each $x \in X$, $x_i \notin co(\{y_i \in X_i : \psi_i(x, y_i) > 0\})$;

(iv) for each $i \in I$, the set $\{x \in X : \sup_{y_i \in A_i(x)} \psi_i(x, y_i) > 0\}$ is open in X for each fixed $y_i \in X_i$; and

(v) there exists a non-empty compact and convex subset X_0 of X and a non-empty compact subset K of X such that for each $y \in X \setminus K$, there exists $x = (x_i)_{i\in I} \in co(X_0 \cup \{y\})$ with $x_i \in co(A_i(y) \cap \{z_i \in X_i : \psi_i(y, z_i) > 0\})$ for all $i \in I$.

Then there exists $x^* \in K$ such that for each $i \in I$,

$$x_i^* \in A_i(x^*) \text{ and } \sup_{y_i \in A_i(x^*)} \psi_i(x^*, y_i) \leq 0.$$

Letting $I = \{1\}$ in Theorem 5.2.1 and Theorem 5.2.3, we have the following existence results for quasi-variational inequalities in non-compact settings.

Corollary 5.2.5. Let $X = \cup_{i=1}^{\infty} C_i$ where $\{C_i\}_{i=1}^{\infty}$ is an increasing sequence of non-empty compact convex subsets of a locally convex topological vector space E. Suppose the following conditions are satisfied:

(i) $A : X \to 2^X$ is lower semicontinuous with closed graph and convex values;

(ii) $\psi : X \times X \to \mathbb{R} \cup \{-\infty, +\infty\}$ is such that for each $y \in X$, $x \mapsto \psi(x, y)$ is lower semicontinuous on X;

(iii) for each $x \in X$, $x \notin co(\{y \in X : \psi(x, y) > 0\})$;

(iv) the set $\{x \in X : \sup_{y \in A(x)} \psi(x, y) > 0\}$ is open in X; and

(v) for each sequence $(y_n)_{n=1}^{\infty}$ in X which is escaping from X relative to $\{C_n\}_{n=1}^{\infty}$, there exist $n \in N$ and $x_n \in C_n$ such that $x_n \in (A(y_n) \cap \{z \in X : \psi(y_n, z) > 0\})$.

Then there exists $x^* \in X$ such that $x^* \in A(x^*)$ and $\sup_{y \in A(x^*)} \psi(x^*, y) \leq 0$.

Corollary 5.2.6. Let E be a locally convex topological vector space, let $X = \cup_{i=1}^{\infty} C_i$ be closed in E and have property (K), where $\{C_i\}_{i=1}^{\infty}$ is an increasing sequence of non-empty compact convex subsets of E. Suppose the following conditions are satisfied:

(i) $A : X \to 2^X$ is compact and upper semicontinuous and with compact convex values;

(ii) $\psi : X \times X \to \mathbb{R} \cup \{-\infty, +\infty\}$ is such that for each $y \in X$, $x \mapsto \psi(x, y)$ is lower semicontinuous;

(iii) for each $x \in X$, $x \notin co(\{y \in X : \psi(x, y) > 0\})$;

(iv) the set $\{x \in X : \sup_{y \in A(x)} \psi(x, y) > 0\}$ is open in X; and

(v) for each sequence $(y_n)_{n=1}^{\infty}$ in X which is escaping from X relative to $\{C_n\}_{n=1}^{\infty}$, there exists $n \in \mathbb{N}$ and $x_n \in C_n$ such that $x_n \in (A(y_n) \cap \{z \in X_i : \psi(y, z) > 0\})$.

Then there exists $x^* \in X$ such that $x^* \in A(x^*)$ and $\sup_{y \in A(x^*)} \psi(x^*, y) \leq 0$.

By Corollary 5.2.6, we have the following slight generalization of Theorem 3 of Tian and Zhou [556].

Corollary 5.2.7. Let X be a non-empty convex subset of a locally convex topological vector space E. Suppose that:

(i) $F : X \to 2^X$ has closed graph and convex values;

(ii) $\psi : X \times X \to \mathbb{R} \cup \{-\infty, +\infty\}$ is such that for each $y \in X$, $x \mapsto \psi(x, y)$ is lower semicontinuous on X for each fixed $y \in X$;

(iii) for each $x \in X$, $x \notin co(\{y \in X : \psi(x, y) > 0\}$;

(iv) there exists a non-empty compact convex subset $Z \subset X$ and a non-empty subset $C \subset Z$ such that

(iv_a) $F(C) \subset Z$;

(iv_b) $F(z) \cap Z \neq \emptyset$ for each $z \in Z$;

(iv_c) for each $x \in Z \setminus C$ there exists $y \in F(x) \cap Z$ with $\psi(x, y) > 0$; and

(iv_d) the set $\{z \in Z : \sup_{y \in F(x) \cap Z} \psi(x, y) \leq 0\}$ is closed in Z.

Then there exists $x^* \in F(x^*)$ such that $\sup_{y \in F(x^*)} \psi(x^*, y) \leq 0$.

Proof. Define $F_1 : Z \to 2^Z$ by $F_1(x) = F(x) \cap Z$ for each $x \in Z$. Then $F_1(x)$ is non-empty closed and convex for each $x \in Z$ by (i) and (iv_b). By (i), F has a closed graph so that F_1 is also closed. It follows that F_1 is upper semicontinuous. Now conditions (ii) and (iii) imply that F_1 and ψ satisfy all hypotheses of Corollary 5.2.6 with $F_1 = A$ and $X = Z$. By Corollary 5.2.6, there exists $x^* \in Z$ such that $x^* \in F_1(x^*)$ and $\sup_{y \in F_1(x^*) \cap Z} \psi(x^*, y) \leq 0$. By (iv_c), x^* must be in C. Therefore $F_1(x^*) = F(x^*) \cap Z = F(x^*)$ by $(v)_a$. Hence, $\sup_{y \in F(x^*)} \psi(x^*, y) \leq 0$. □

Corollary 5.2.7 also generalizes Theorem 3.1 of Zhou and Chen [622], Theorem 15.2.1 of Aubin [10] and Theorem 4 of Fan [167]. We conclude this section with an application of Corollary 5.2.5 to give an existence theorem for equilibria of constrained games.

Definition 5.2.8. Let $I = \{1, 2, \cdots, N\}$. Each agent i chooses a strategy x_i in a subset X_i of a locally convex topological vector space E_i. Denote by X the (Cartesian) product $\Pi_{j=1}^N X_j$ and X_{-i} the product $\Pi_{j \in I, j \neq i} X_j$. Denote by x and x_{-i} an element of X and X_{-i} respectively. Each agent i has a payoff (utility) function $u_i : X \to \mathbb{R} \cup \{-\infty, +\infty\}$. Given x_{-i} (the strategies of others), the choice of the ith agent is restricted to a non-empty compact and convex set $A_i(x_{-i}) \subset X_i$, the *feasible strategy set*; the ith agent chooses $x_i \in A_i(x_{-i})$ so as to minimize $u_i(x_{-i}, x_i)$ over $A_i(x_{-i})$, where (x_{-i}, x_i) is the point $y = (y_j)_{j \in I}$ such that $y_i = x_{-i}$ and $y_i = x_i$. The family $\mathcal{G} = (X_i; A_i; u_i)_{i=1}^N$ is then called a *constrained N-person game* and an equilibrium for $\mathcal{G}$ is an $x^* \in X$ such that $x_i^* \in A_i(x_{-i}^*)$ and $u_i(x^*) \leq u_i(x_{-i}^*, x_i)$ for all $x_i \in A_i(x_{-i}^*)$ (e.g., $u_i(x^*) = \inf_{x_i \in A_i(x_{-i}^*)} u_i(x_{-i}^*, x_i)$) for each $i = 1, 2, \cdots, N$.

Note that if $A_i(x_{-i}) = X_i$ for each $i = 1, 2, \cdots, N$, the constrained N-person game reduces to the conventional game $\mathcal{G} = (X_i; u_i)_{i \in I}$ and its equilibrium is called a Nash-equilibrium.

Theorem 5.2.9. Let $\mathcal{G} = (X_i; A_i; U_i)_{i=1}^N$ be a constrained game and $X = \Pi_{i=1}^N X_i = \cup_{j=1}^\infty C_j$ where $\{C_j\}_{j=1}^\infty$ is an increasing sequence of non-empty compact convex subsets of a locally convex topological vector space E. Suppose the following conditions are satisfied:

(i) the correspondence $A : X \to 2^X$ defined by $A(x) = \Pi_{i=1}^N A_i(x_{-i})$ for each $x = (x_{-i}, x_i) \in X$ is lower semicontinuous with closed graph and convex values;

(ii) the function $\psi : X \times X \to \mathbb{R} \cup \{-\infty, +\infty\}$ defined by

$$\psi(x, y) = \Sigma_{i=1}^N [u_i(x_{-i}, x_i) - u_i(x_{-i}, y_i)]$$

for each $(x, y) \in X \times X$ is such that for each $y \in X$, $x \mapsto \psi(x, y)$ is lower semicontinuous on X, where $x = (x_{-i}, x_i)$ and $y = (y_{-i}, y_i)$;

(iii) for each $x \in X$, $x \notin co(\{y \in X : \psi(x, y) > 0\})$;

(iv) the set $\{x \in X : \sup_{y \in A(x)} \psi(x, y) > 0\}$ is open in X; and

(v) for each sequence $(y_n)_{n=1}^{\infty}$ in X which is escaping from X relative to $\{C_n\}_{n=1}^{\infty}$, there exists $n \in \mathbb{N}$ and $x_n \in C_n$ such that $x_n \in A_i(y_n) \cap \{z \in X_i : \psi(y, z) > 0\})$.

Then there exists $x^* \in X$ such that for each $i = 1, 2, \cdots, N$,

$$x_i^* \in A_i(x_{-i}^*) \text{ and } u_i(x^*) \leq \inf_{x_i \in A(x_{-i}^*)} u_i(x_{-i}^*, x_i).$$

Proof. By (i)-(v), A and ψ satisfy all hypotheses of Corollary 5.2.5. By Corollary 5.2.5, there exists $x^* \in X$ such that $x^* \in A(x^*)$ and $\sup_{y \in A(x^*)} \psi(x^*, y) \leq 0$. For each $i \in I$, and $y_i \in A_i(x_{-i}^*)$, let $y = (x_{-i}^*, y_i)$. Then $y \in A(x^*)$ so that $(u_i(x^*) - u_i(x_{-i}^*, y_i)) = \Sigma_{i=1}^{N}[u_i(x^*) - u_i(x_{-i}^*, y_i)] = \psi(x^*, y) \leq \sup_{y \in A(x^*)} \psi(x^*, y) \leq 0$. Therefore $(u_i(x^*) - u_i(x_{-i}^*, y_i)) \leq 0$ for all $y_i \in A_i(x_{-i}^*)$. Hence x^* is an equilibrium point of the constrained game $\mathcal{G} = (X_i; A_i; u_i)_{i=1}^{N}$. □

Theorem 5.2.9 generalizes the corresponding results of Aubin [10, p.282-283] and Aubin and Ekeland [15, p.350-351] in the following ways: (i) the feasible correspondence A_i is lower (or upper) semicontinuous instead of continuous and (ii) the strategy set X_i need not be compact. More study of Nash-equilibria will be carried out in Chapter 8.

5.3 Generalized Quasi-Variational Inequalities

In this section, we shall first give some (non-compact) existence theorems of two types of generalized quasi-variational inequalities (in short, GQVI(X; F; T)) where, the domain X is a non-empty convex but not necessarily compact subset of a locally convex topological vector space E and the mapping F need not be continuous and the mapping T need not have any continuity property. These results generalize and improve many known results in the literature, e.g., Aubin [10], Aubin and Ekeland [15], Chan and Pang [64], Harker and Pang [220], Kim [298], Shih and Tan [466], [474], Tarafdar [531] and Tian and Zhou [556] and the references therein.

If E is a topological vector space, we shall denote by E^* the dual space of E. The dual pairing between E^* and E is denoted by $\langle w, x\rangle$ for $w \in E^*$ and $x \in E$ and $Re\langle w, x\rangle$ denotes the real part of $\langle w, x\rangle$. We reacll the following definition agagin.

Definition 5.3.1. Let X be a non-empty convex subset of a locally convex topological vector E. Then $T : X \to 2^{E^*}$ is said to be monotone (see Browder [60, p.79]) if for each $x, y \in X$ and for each $u \in T(x)$, $w \in T(y)$, $Re\langle w - u, y - x\rangle \geq 0$.

Suppose $F : X \to 2^X$, $T : X \to 2^{E^*}$ and $f : X \times X \to \mathbb{R} \cup \{-\infty, +\infty\}$. By applying Theorem 5.2.1 and Corollary 5.2.5, we shall prove the existence of a solution $\hat{x} \in X$ to the following generalized quasi-variational inequalities:

$$\text{(I)} \qquad \begin{cases} \hat{x} \in F(\hat{x}), \\ \sup_{u \in T(\hat{x})} Re\langle u, \hat{x} - y\rangle + f(\hat{x}, y) \leq 0 \text{ for all } y \in F(\hat{x}) \end{cases}$$

or the existence of solutions $\hat{x} \in X$ and $\hat{u} \in E^*$ to the following generalized quasi-variational inequalities:

$$\text{(II)} \qquad \begin{cases} \hat{x} \in F(\hat{x}) \text{ and } \hat{u} \in T(\hat{x}), \\ Re\langle \hat{u}, \hat{x} - y\rangle + f(\hat{x}, y) \leq 0 \text{ for all } y \in F(\hat{x}). \end{cases}$$

Now we recall some definitions (see, Zhou and Chen [622]).

Definition 5.3.2. Let X be a non-empty convex subset of a topological vector space. A function $\psi : X \times X \to \mathbb{R} \cup \{-\infty, +\infty\}$ is said to be

(1) γ-diagonally quasi-convex (resp., γ-diagonally quasi-concave) in y, in short γ-DQCX (respectively, γ-DQCV) in y, if for each $A \in \mathcal{F}(X)$ and each $y \in co(A)$, $\gamma \leq \max_{x \in A} \psi(x, y)$ (respectively, $\gamma \geq \inf_{x \in A} \psi(x, y)$); and

(2) γ-diagonally convex (respectively, γ-diagonally concave) in y, in short γ-DCX (respectively, γ-DCV) in y, if for each $A \in \mathcal{F}(X)$ and each $y \in co(A)$ with $y = \Sigma_{i=1}^m \lambda_i y_i$ (where $\lambda_i \geq 0$ for each $i = 1, 2, \cdots, n$ and $\Sigma_{i=1}^m \lambda_i = 1$), we have $\gamma \leq \Sigma_{i=1}^m \lambda_i \psi(y_i, y)$ (respectively, $\gamma \geq \Sigma_{i=1}^m \lambda_i \psi(y_i, y)$).

Definition 5.3.3. Let X and Y be two non-empty convex subsets of E, we also recall that a function $\psi : X \times Y \to \mathbb{R} \cup \{-\infty, +\infty\}$ is quasi-convex (respectively, quasi-concave) in y, if for each fixed $x \in X$, for each $A \in \mathcal{F}(Y)$ and each $y \in co(A)$, $\psi(x, y) \leq \max_{z \in A} \psi(x, z)$ (respectively, $\psi(x, y) \geq \inf_{z \in A} \psi(x, z)$).

Remark 5.3.4. It is easy to see that:
(i) if $\psi(x, y)$ is γ-DCX (respectively, γ-DCV) in y, then $\psi(x, y)$ is γ-DQCX (respectively, γ-DCV) in y;
(ii) if $\psi_i(x, y)$ for each $i = 1, 2, \cdots, N$ is γ-DCX (respectively, γ-DCV) in y, then $\psi(x, y) = \Sigma_{i=1}^m a_i(x) \psi_i(x, y)$ is still γ-DCX (respectively, γ-DCV) in y, where $a_i : X \to \mathbb{R}$ with $a_i(x) \geq 0$ and $\Sigma_{i=1}^m a_i(x) = 1$ for each $x \in X$; and
(iii) the function $\psi(x, y) : X \times X \to \mathbb{R} \cup \{-\infty, +\infty\}$ is 0-DQCV in y if and only if $x \notin co(\{y \in X : \psi(x, y) > 0\})$ for each $x \in X$.

Let K be a non-empty convex subset of a topological vector space E and $A : K \to E^*$ be not monotone. We define a function $f : K \times K \to \mathbb{R}$ by $f(x, y) = \langle A(y), x - y\rangle$ for each $(x, y) \in K \times K$. It is clear (e.g., see Zhou and Chen [622, p.216-217]) that for each fixed $x \in K$, the function $f(x, \cdot)$ is both DQCX and DQCV in y, but not quasi-convex or quasi-concave in y.

First we have the following existence result for the problem (I) where $T : X \to 2^{E^*}$ is a set-valued monotone mapping.

Theorem 5.3.5. Let $X = \cup_{i=1}^{\infty} C_i$ where $\{C_i\}_{i=1}^{\infty}$ is an increasing sequence of non-empty compact convex subsets of a locally convex topological vector space E. Suppose the following conditions are satisfied:

(i) $F : X \to 2^X$ is lower semicontinuous with closed graph and convex values;

(ii) $T : X \to 2^{E^*}$ is monotone such that for each one-dimensional flat $L \subset E$, $T|_{L \cap X}$ is lower semicontinuous when E^* is equipped with the weak*-topology $\sigma(E^*, E)$;

(iii) $f : X \times X \to \mathbb{R} \cup \{\infty, +\infty\}$ is such that for each $y \in X$, $x \mapsto f(x,y)$ is lower semicontinuous on X, and for each $x \in X$, $y \mapsto f(x,y)$ is concave and $f(x,x) = 0$ for each $x \in X$;

(iv) the set $\{x \in X : \sup_{y \in F(x)}[\sup_{u \in T(y)} Re\langle u, x - y\rangle + f(x,y)] > 0\}$ is open in X; and

(v) for each sequence $(y_n)_{n=1}^{\infty}$ in X which is escaping from X relative to $\{C_n\}_{n=1}^{\infty}$, there exists $n \in N$ and $x_n \in C_n$ such that $x_n \in F(y_n) \cap \{z \in X : \sup_{u \in T(z)} Re\langle u, y_n - z\rangle + f(y_n, z) > 0\}$.

Then there exists $\hat{x} \in X$ such that $\hat{x} \in F(\hat{x})$ and

$$\sup_{u \in T(\hat{x})} Re[\langle u, \hat{x} - y) + f(\hat{x}, y\rangle] \leq 0 \text{ for all } y \in F(\hat{x})$$

Proof. Define a function $\psi : X \times X \to \mathbb{R} \cup \{-\infty, +\infty\}$ by

$$\psi(x,y) = \sup_{u \in T(y)} Re\langle u, x - y\rangle + f(x,y)$$

for each $(x,y) \in X \times X$. Then for each $y \in X$, $x \mapsto \psi(x,y)$ is lower semicontinuous on X. Since T is monotone and by (iii), it is easy to see that for each fixed $x \in X$, $y \mapsto \psi(x,y)$ is 0-DCV by Proposition 3.2 of Zhou and Chen [622]. Thus all the hypotheses of Corollary 5.2.5 (hence also of Theorem 5.2.1) are satisfied so that there exists $\hat{x} \in K$ such that $\hat{x} \in F(\hat{x})$ and $\sup_{u \in T(y)}[\langle u, \hat{x} - y\rangle + f(\hat{x}, y)] \leq 0$ for all $y \in F(\hat{x})$. Since for each one-dimensional flat $L \subset E$, $T|_{L \cap X}$ is lower semicontinuous when E^* is equipped with the weak*-topology $\sigma(E^*, E)$, by the same argument as in the step 2 of Shih and Tan [466, p.338], we can show that

$$\sup_{y \in T(\hat{x})} [\langle u, \hat{x} - y\rangle + f(\hat{x}, y)] \leq 0$$

for all $y \in F(\hat{x})$. Thus the proof is complete. □

Now by Corollary 5.2.6 (hence also Theorem 5.2.3) and the same argument as in Theorem 5.3.5, we have the following result.

Theorem 5.3.6. Let $X = \cup_{i=1}^{\infty} C_i$ be closed and have property (K), where $\{C_i\}_{i=1}^{\infty}$ is an increasing sequence of non-empty compact convex subsets of a locally convex topological vector space E. Suppose the following conditions are satisfied:

(i) $F : X \to 2^X$ is compact upper semicontinuous with compact and convex values;

(ii) $T : X \to 2^{E^*}$ is monotone such that for each one-dimensional flat $L \subset E$, $T|_{L\cap X}$ is lower semicontinuous when E^* is equipped with the weak*-topology $\sigma(E^*, E)$;

(iii) $f : X \times X \to \mathbb{R} \cup \{\infty, +\infty\}$ is such that for each $y \in X$, $x \mapsto f(x, y)$ is lower semicontinuous on X and for each $x \in X$, $y \mapsto f(x, y)$ is concave and $f(x, x) = 0$ for each $x \in X$;

(iv) the set $\{x \in X : \sup_{y\in F(x)}[\sup_{u\in T(y)} Re\langle u, x - y\rangle + f(x, y)] > 0\}$ is open in X; and

(v) for each sequence $(y_n)_{n=1}^{\infty}$ in X which is escaping from X relative to $\{C_n\}_{n=1}^{\infty}$, there exists $n \in \mathbb{N}$ and $x_n \in C_n$ such that $x_n \in F(y_n) \cap \{z \in X : \sup_{u\in T(z)} Re\langle u, y_n - z\rangle + f(y_n, z) > 0\}$.

Then there exists $\hat{x} \in X$ such that $\hat{x} \in F(\hat{x})$ and

$$\sup_{u\in T(\hat{x})}[Re\langle u, \hat{x} - y\rangle + f(\hat{x}, y)] \leq 0 \text{ for all } y \in F(\hat{x}).$$

As an immediate consequence of Theorem 5.3.6, we have the following result.

Corollary 5.3.7. Let X be a non-empty compact convex subset of a locally convex topological vector space E and let $F : X \to 2^X$ be upper semicontinuous with closed and convex values. If $T : X \to 2^{E^*}$ is monotone such that for each one-dimensional flat $L \subset E$, $T|_{L\cap X}$ is lower semicontinuous when E^* is equipped with the weak*-topology $\sigma(E^*, E)$. Suppose $f : X \times X \to \mathbb{R} \cup \{-\infty, +\infty\}$ is such that for each $y \in X$, $x \mapsto f(x, y)$ is lower semicontinuous on X and for each $x \in X$, $y \mapsto f(x, y)$ is concave and $f(x, x) = 0$ for each $x \in X$. Suppose further that the set $\{x \in X : \sup_{y\in F(x)}[\sup_{u\in T(y)} Re\langle u, x - y\rangle + f(x, y)] > 0\}$ is open in X. Then there exists an $\hat{x} \in X$ such that $\hat{x} \in F(\hat{x})$ and $\sup_{u\in T(\hat{x})}[Re\langle u, \hat{x} - y\rangle + f(\hat{x}, y)] \leq 0$ for all $y \in F(\hat{x})$.

Theorem 5.3.6 generalizes Theorem 1 of Shih and Tan [466] to a non-compact setting. We note that our proofs of Theorem 5.3.5 and Theorem 5.3.6 depend on the existence theorems for equilibria of generalized games instead of the partition of unity arguments used by Aubin [10], Aubin and Ekeland [15], Shih and Tan [466] and Zhou and Chen [622]. When $T = 0$ in Corollary 5.2.2, Corollary 5.2.2 generalizes the Joly-Mosco Theorem (see Theorem 15.2.2 of Aubin [10]), and also gives the well-known Fan-Glicksberg fixed point theorem (see [156] and [196]).

Recall that for a topological vector space E, the strong topology on its dual space E^* is the topology on E^* generated by the family $\{U(B; \omega) : B$ is a non-empty bounded subset of E and $\omega > 0\}$ as a base for the neighborhood system at zero, where $U(B; \omega) := \{f \in E^* : \sup_{x\in B} |\langle f, x\rangle| < \omega\}$.

We now observe that in Theorem 5.3.5 and Theorem 5.3.6, the interaction between the correspondences T and F (namely, condition (v)) can be achieved by imposing additional continuity conditions on T and F.

Theorem 5.3.8. Let $X = \cup_{i=1}^{\infty} C_i$ be bounded, where $\{C_i\}_{i=1}^{\infty}$ is an increasing sequence of non-empty compact convex subsets of a locally convex topological vector space E. Suppose that $F : X \to 2^X$ is continuous with closed and convex values and $T : X \to 2^{E^*}$ is monotone such that T is lower semicontinuous when E^* is equipped with the strong topology. Suppose that

(i) $f : X \times X \to \mathbb{R} \cup \{\infty, +\infty\}$ is lower semicontinuous such that for each $x \in X$, $y \mapsto f(x, y)$ is concave and $f(x, x) = 0$ for each $x \in X$;

(ii) for each sequence $(y_n)_{n=1}^{\infty}$ in X which is escaping from X relative to $\{C_n\}_{n=1}^{\infty}$, there exist $n \in N$ and $x_n \in C_n$ such that $x_n \in F(y_n) \cap \{z \in X : \sup_{u \in T(z)} Re\langle u, y_n - z\rangle + f(y_n, z) > 0\}$.

Then there exists $\hat{x} \in X$ such that $\hat{x} \in F(\hat{x})$ and

$$\sup_{u \in T(\hat{x})}[Re\langle u, \hat{x} - y\rangle + f(\hat{x}, y)] \leq 0 \text{ for all } y \in F(\hat{x}).$$

Proof. By Theorem 5.3.5, we need only show that the set

$$\Sigma := \{x \in X : \sup_{y \in F(x)} [\sup_{u \in T(y)} Re\langle u, x - y\rangle + f(x, y)] > 0\}$$

is open in X.

Since X is bounded and $f(\cdot, \cdot)$ is lower semicontinuous, the function $(u, x, y) \mapsto Re\langle u, x - y\rangle + f(x, y)$ is lower semicontinuous from $E^* \times X \times X$ to $\mathbb{R}$. Therefore $(x, y) \mapsto \sup_{u \in T(y)}[Re\langle u, x - y\rangle + f(x, y)]$ is also lower semicontinuous by the lower semicontinuity of $T(\cdot)$ and Proposition III-19 of Aubin and Ekeland [15]. Since F is lower semicontinuous, $x \mapsto \sup_{y \in F(x)} \sup_{u \in T(y)}[Re\langle u, x - y\rangle + f(x, y)]$ is lower semicontinuous by Proposition III-19 of Aubin [10] again. Thus the set $\Sigma := \{x \in X : \sup_{y \in F(x)} \sup_{u \in T(y)}[Re\langle u, x - y\rangle + f(x, y)] > 0\}$ is open in X. □

Theorem 5.3.8 also generalizes Theorem 2 of Shih and Tan [466] to a non-compact setting.

Now we consider the existence of solutions of the problems (I) and (II) where the correspondence $T : X \to 2^{E^*}$ need not be monotone. But, first we recall and state Kneser's minimax theorem [318] (see also Aubin [10, p.40-41]) as follows.

Theorem 5.3.9 (Kneser Minimax Inequality [318]). Let X be a non-empty convex set in a vector space and let Y be a non-empty compact convex subset of a topological vector space. Suppose that f is a real-valued function on $X \times Y$ such that for each fixed $x \in X$, $f(x, y)$ is lower semicontinuous and convex on Y, and for each fixed $y \in Y$, $f(x, y)$ is concave on X. Then

$$\min_{y \in Y} \sup_{x \in X} f(x, y) = \sup_{x \in X} \min_{y \in Y} f(x, y).$$

We then have the following existence result of variational inequalities.

Theorem 5.3.10. Let $X = \cup_{i=1}^{\infty} C_i$ where $\{C_i\}_{i=1}^{\infty}$ is an increasing sequence of non-empty compact convex subsets of a locally convex topological vector space E. Suppose that

(i) $F : X \to 2^X$ is lower semicontinuous with closed graph and compact and convex values;

(ii) $T : X \to 2^{E^*}$ has compact convex values such that for each fixed $y \in X$, $x \mapsto \inf_{u \in T(x)} Re\langle u, x - y\rangle$ is lower semicontinuous;

(iii) $f : X \times X \to \mathbb{R} \cup \{-\infty, +\infty\}$ is such that for each $y \in X$, $x \mapsto f(x, y)$ is lower semicontinuous on X and for each $x \in X$, $y \mapsto f(x, y)$ is 0-diagonally concave;

(iv) the set $\{x \in X : \sup_{y \in F(x)}[\inf_{u \in T(x)} Re\langle u, x - y\rangle + f(x, y)] > 0\}$ is open in X; and

(v) for each sequence $(y_n)_{n=1}^{\infty}$ in X which is escaping from X relative to $\{C_n\}_{n=1}^{\infty}$, there exists $n \in \mathbb{N}$ and $x_n \in C_n$ such that $x_n \in F(y_n) \cap \{z \in X : \inf_{u \in T(y_n)} Re\langle u, y_n - z\rangle + f(y_n, z) > 0\}$.

Then there exists $\hat{x} \in X$ such that $\hat{x} \in F(\hat{x})$ and

$$\sup_{y \in F(\hat{x})} \inf_{u \in T(\hat{x})}[\langle u, \hat{x} - y\rangle + f(\hat{x}, y)] \leq 0.$$

If in addition, for each $x \in X$, $y \mapsto f(x, y)$ is concave, then there exists $\hat{u} \in T(\hat{x})$ such that $\sup_{y \in F(\hat{x})}[\langle \hat{u}, \hat{x} - y\rangle + f(\hat{x}, y)] \leq 0$.

Proof. Define the functional $\psi : X \times X \to \mathbb{R} \cup \{-\infty, +\infty\}$ by

$$\psi(x, y) = \inf_{u \in T(x)}[Re\langle u, x - y\rangle + f(x, y)]$$

for each $(x, y) \in X \times X$. Then we have:

(1) for each fixed $y \in X$, $x \mapsto \psi(x, y)$ is lower semicontinuous on X and $x \notin co(\{y \in X : \psi(x, y) > 0\})$ for each $x \in X$ by (iv);

(2) the set $\{x \in X : \sup_{y \in F(x)} \psi(x, y) > 0\}$ is open in X; and

(3) for each sequence $(y_n)_{n=1}^{\infty}$ in X which is escaping from X relative to $\{C_n\}_{n=1}^{\infty}$, there exists $n \in N$ and $x_n \in C_n$ such that $x_n \in co(\{F(y_n) \cap \{z \in X : \psi(y_n, z) > 0\})$.

Therefore F and ψ satisfy all conditions of Corollary 5.2.5. By Corollary 5.2.5, there exists an $\hat{x} \in K$ such that $\hat{x} \in F(\hat{x})$ and $\psi(\hat{x}, y) \leq 0$ for all $y \in F(\hat{x})$.

If in addition, for each $x \in X$, $y \mapsto f(x, y)$ is concave, define the function $f_1 : F(\hat{x}) \times T(\hat{x}) \to \mathbb{R} \cup \{-\infty, +\infty\}$ by

$$f_1(x, y) = Re\langle x, \hat{x} - y\rangle + f(\hat{x}, y)$$

for each $(x, y) \in F(\hat{x}) \times T(\hat{x})$. Then for each $y \in X$, $x \mapsto f_1(x, y)$ is lower semicontinuous and for each $x \in X$, $y \mapsto f_1(x, y)$ is concave. By Kneser minimax Theorem 5.3.9,

$$\inf_{u \in T(\hat{x})} \sup_{y \in F(\hat{x})} [Re\langle u, \hat{x} - y\rangle + f(\hat{x}, y)] = \sup_{y \in F(\hat{x})} \inf_{u \in T(\hat{x})} [Re\langle u, \hat{x} - y\rangle + f(\hat{x}, y)] \leq 0,$$

so that there exists $\hat{u} \in T(\hat{x})$ such that $\sup_{y \in F(\hat{x})}[Re\langle \hat{u}, \hat{x} - y\rangle + f(\hat{x}, y)] \leq 0$. □

By the same arguments used in Theorem 5.3.10 and by applying Corollary 5.2.6 instead of Corollary 5.2.5, we have the following result whose proof is omitted:

Theorem 5.3.11. Let $X = \cup_{i=1}^{\infty} C_i$ be closed and have property (K), where $\{C_i\}_{i=1}^{\infty}$ is an increasing sequence of non-empty compact convex subsets of a locally convex topological vector space E. Suppose that:

(i) $F : X \to 2^X$ is compact and upper semicontinuous with compact and convex values;

(ii) $T : X \to 2^{E^*}$ has compact and convex values such that for each $y \in X$, $x \mapsto \inf_{u \in T(x)} Re\langle u, x - y\rangle$ is lower semicontinuous;

(iii) $f : X \times X \to \mathbb{R} \cup \{-\infty, +\infty\}$ is such that for each $y \in X$, $x \mapsto f(x, y)$ is lower semicontinuous on X for each $y \in X$ and for each $x \in X$, $y \mapsto f(x, y)$ is 0-diagonally concave;

(iv) the set $\{x \in X : \sup_{y \in F(x)}[\inf_{u \in T(x)} Re\langle u, x - y\rangle + f(x, y)] > 0\}$ is open in X; and

(vi) for each sequence $(y_n)_{n=1}^{\infty}$ in X which is escaping from X relative to $\{C_n\}_{n=1}^{\infty}$, there exists $n \in N$ and $x_n \in C_n$ such that $x_n \in F(y_n) \cap \{z \in X : \inf_{u \in T(y_n)} Re\langle u, y_n - z\rangle + f(y_n, z) > 0\}$.

Then there exists $\hat{x} \in X$ such that

$$\hat{x} \in F(\hat{x}) \text{ and } \sup_{y \in F(\hat{x})} \inf_{u \in T(\hat{x})}[\langle u, \hat{x} - y\rangle + f(\hat{x}, y)] \leq 0.$$

If in addition, for each $x \in X$, $y \mapsto f(x, y)$ is concave, then there exists $\hat{u} \in T(\hat{x})$ such that $\sup_{y \in F(\hat{x})} Re[\langle \hat{u}, \hat{x} - y\rangle + f(\hat{x}, y)] \leq 0$.

If X is a bounded subset of a locally convex topological vector space E and $T : X \to 2^{E^*}$ has compact and convex values and is upper semicontinuous when E^* is equipped with the strong topology, then the function $g : X \times X \to \mathbb{R} \cup \{-\infty, +\infty\}$ defined by $g(x, y) = \inf_{u \in T(x)}\langle u, x - y\rangle$ has the property that for each $y \in X$, $x \mapsto g(x, y)$ is lower semicontinuous on X by Lemma 2 of Kim and Tan [303]. Thus, Theorem 5.3.11 generalizes Theorem 3 of Shih and Tan [466] and we have the following.

Corollary 5.3.12. Let $X = \cup_{i=1}^{\infty} C_i$ be bounded, where $\{C_i\}_{i=1}^{\infty}$ is an increasing sequence of non-empty compact convex subsets of a locally convex topological vector space E. Suppose that:

(i) $F : X \to 2^X$ is lower semicontinuous with closed graph and compact and convex values;

(ii) $T : X \to 2^{E^*}$ has (strongly) compact convex values and is upper semicontinuous when E^* is equipped with the strong topology;

(iii) $f : X \times X \to \mathbb{R} \cup \{-\infty, +\infty\}$ is such that for each $y \in X$, $x \mapsto f(x, y)$ is lower semicontinuous and for each $x \in X$, $y \mapsto f(x, y)$ is 0-diagonally concave;

(iv) the set $\{x \in X : \sup_{y\in F(x)}[\inf_{u\in T(x)} Re\langle u, x-y\rangle + f(x,y)] > 0\}$ is open in X; and

(v) for each sequence $(y_n)_{n=1}^{\infty}$ in X which is escaping from X relative to $\{C_n\}_{n=1}^{\infty}$, there exists $n \in \mathbb{N}$ and $x_n \in C_n$ such that $x_n \in F(y_n) \cap \{z \in X : \inf_{u\in T(y_n)} Re\langle u, y_n - z\rangle + f(y_n, z) > 0\}$.

Then there exists $\hat{x} \in X$ such that $\hat{x} \in F(\hat{x})$ and

$$\sup_{y\in F(\hat{x})} \inf_{u\in T(\hat{x})}[\langle u, \hat{x}-y\rangle + f(\hat{x},y)] \leq 0.$$

If in addition, for each $x \in X$, $y \mapsto f(x,y)$ is concave, then there exists $\hat{u} \in T(\hat{x})$ such that $\sup_{y\in F(\hat{x})}[\langle \hat{u}, \hat{x}-y\rangle + f(\hat{x},y)] \leq 0$.

Proof. Define $\psi : X \times X \to \mathbb{R} \cup \{-\infty, +\infty\}$ by

$$\psi(x,y) = \inf_{u\in T(x)}[Re\langle u, x-y\rangle + f(x,y)],$$

for each $(x,y) \in X \times X$. Since X is bounded, by Lemma 2 of Kim and Tan [303], for each $y \in X$, $x \mapsto \inf_{u\in T(x)}\langle u, x-y\rangle$ is lower semicontinuous. Therefore ψ and F satisfy all the hypotheses of Theorem 5.3.10. Thus the conclusion follows from Theorem 5.3.10. □

In the above proof, if we apply Theorem 5.3.11 instead of Theorem 5.3.10, we have:

Corollary 5.3.13. Let $X = \cup_{i=1}^{\infty} C_i$ be bounded, closed and have property (K), where $\{C_i\}_{i=1}^{\infty}$ is an increasing sequence of non-empty compact convex subsets of a locally convex topological vector space E_i. Suppose that:

(i) $F : X \to 2^X$ is compact and upper semicontinuous with compact and convex values;

(ii) $T : X \to 2^{E^*}$ has (strongly) compact convex values and is upper semicontinuous when E^* is equipped with the strong topology;

(iii) $f : X \times X \to \mathbb{R} \cup \{-\infty, +\infty\}$ is such that for each $y \in X$, $x \mapsto f(x,y)$ is lower semicontinuous and for each $x \in X$, $y \mapsto f(x,y)$ is 0-diagonally concave;

(iv) the set $\{x \in X : \sup_{y\in F(x)}[\inf_{u\in T(x)} Re\langle u, x-y\rangle + f(x,y)] > 0\}$ is open in X; and

(v) for each sequence $(y_n)_{n=1}^{\infty}$ in X which is escaping from X relative to $\{C_n\}_{n=1}^{\infty}$, there exists $n \in \mathbb{N}$ and $x_n \in C_n$ such that $x_n \in F(y_n) \cap \{z \in X : \inf_{u\in T(y_n)} Re\langle u, y_n - z\rangle + f(y_n, z) > 0\}$.

Then there exists $\hat{x} \in X$ such that $\hat{x} \in F(\hat{x})$ and

$$\sup_{y\in F(\hat{x})} \inf_{u\in T(\hat{x})}[\langle u, \hat{x}-y\rangle + f(\hat{x},y)] \leq 0.$$

If in addition, for each $x \in X$, $y \mapsto f(x,y)$ is concave, then there exists $\hat{u} \in T(\hat{x})$ such that $\sup_{y\in F(\hat{x})}[Re\langle \hat{u}, \hat{x}-y\rangle + f(\hat{x},y)] \leq 0$.

Proof. Define $\psi : X \times X \to \mathbb{R} \cup \{-\infty, +\infty\}$ by

$$\psi(x,y) = \inf_{u\in T(x)}[Re\langle u, x-y\rangle + f(x,y)]$$

for each $(x, y) \in X \times X$. Since X is bounded, by Lemma 2 of Kim and Tan [303], for each $y \in X$, $x \mapsto \inf_{u \in T(x)} \langle u, x - y \rangle$ is lower semicontinuous. Therefore ψ and F satisfy all hypotheses of Theorem 5.3.11. Thus the conclusion follows from Theorem 5.3.11. □

Now if we impose a continuity condition on the correspondence F, then we have the following result.

Theorem 5.3.14. Let $X = \cup_{i=1}^{\infty} C_i$ be bounded, where $\{C_i\}_{i=1}^{\infty}$ is an increasing sequence of non-empty compact convex subsets of a locally convex topological vector space E. Suppose that:

(i) $F : X \to 2^X$ is continuous with non-empty compact and convex values;

(ii) $T : X \to 2^{E^*}$ has (strongly) compact convex values and is upper semicontinuous when E^* is equipped with the strong topology;

(iii) $f : X \times X \to \mathbb{R} \cup \{-\infty, +\infty\}$ is lower semicontinuous and for each $x \in X$, $y \mapsto f(x, y)$ is 0-diagonally concave; and

(iv) for each sequence $(y_n)_{n=1}^{\infty}$ in X which is escaping from X relative to $\{C_n\}_{n=1}^{\infty}$, there exists $n \in N$ and $x_n \in C_n$ such that $x_n \in F(y_n) \cap \{z \in X : \inf_{u \in T(y_n)} Re\langle u, y_n - z \rangle + f(y_n, z) > 0\}$.

Then there exists $\hat{x} \in X$ such that $\hat{x} \in F(\hat{x})$ and

$$\sup_{y \in F(\hat{x})} [\inf_{u \in T(\hat{x})} Re\langle u, \hat{x} - y \rangle + f(\hat{x}, y)] \leq 0.$$

If in addition, for each $x \in X$, $y \mapsto f(x, y)$ is concave, then there exists $\hat{u} \in T(\hat{x})$ such that $\sup_{y \in F(\hat{x})} [Re\langle \hat{u}, \hat{x} - y \rangle + f(\hat{x}, y)] \leq 0$.

Proof. Define the function $\psi_1 : X \times X \times E^* \to \mathbb{R} \cup \{-\infty, +\infty\}$ by $\psi_1(x, y, u) = Re\langle u, x - y \rangle$ for each $(x, y, u) \in X \times X \times E^*$. Since X is a bounded subset of the locally convex topological vector space E and E^* is equipped with the strong topology, ψ_1 is continuous. Since $T : X \to 2^{E^*}$ is upper semicontinuous with (strong) compact and convex values, by Theorem 1 of Aubin [10, p.67], the function $\psi_2 : X \times X \to \mathbb{R} \cup \{-\infty, +\infty\}$ defined by $\psi_2(x, y) = \inf_{u \in T(x)} \langle u, x - y \rangle$ is also lower semicontinuous on $X \times X$ so that $(x, y) \mapsto \inf_{u \in T(x)} Re\langle u, x - y \rangle + f(x, y)$ is lower semicontinuous by (iii). As $F : X \to 2^X$ is lower semicontinuous, by Theorem 2 of Aubin [10, p.69], the function $x \mapsto \sup_{y \in F(x)} Re \inf_{u \in T(x)} [\langle u, x - y \rangle + f(x, y)]$ is lower semicontinuous from X to $\mathbb{R} \cup \{-\infty, +\infty\}$. It follows that the set $\Sigma = \{x \in X : \sup_{y \in F(x)} \inf_{u \in T(x)} Re[\langle u, x - y \rangle + f(x, y)] > 0\}$ is open in X. Thus F, T and f satisfy all hypotheses of Corollary 5.3.12. Thus the conclusion follows from Corollary 5.3.12. □

Remark 5.3.15. In Theorems 5.3.5, 5.3.6, 5.3.8, 5.3.10, 5.3.11 and 5.3.14, we assume that the correspondence $T : X \to 2^{E^*}$ satisfies some kind of continuity. In fact, under appropriate conditions, the existence theorems for solutions of the problems (I) and (II) still hold without assuming any continuity of T, for more details, see Ricceri [436] and the references therein. We also note that

Corollary 5.3.13 (and hence also Theorem 5.3.11) generalizes the Theorem of Kim [298] which in turn improves Theorem 4 of Shih and Tan [466]. For the applications of quasi-variational inequalities and generalized quasi-variational inequalities to game theory and economics theory, we refer to Aubin [10], Aubin and Ekeland [15], Border [48] and references therein.

5.4 Stability of Quasi-Variational Inequalities

The aim of this section is to establish the generic stability theory of quasi-variational inequalities. We first show that almost all quasi-variational are stable from viewpoint of Baire category theory; in another word, each quasi-variational nequality problem can be approximated arbitrarily by a quasi-variational inequality which is stable in the sense that the small change of the objective function results in the small change of its solution set. Thus thus we might say that almost all of quasi-variational inequality problems are stable. Secondly, we show that each variational inequality problem has, at least, one connected component of its solutions which is stable though in general its solution set may not have a good behavior (i.e., not stable). Our results show that if a variational inequality problem has only one connected solution set, it is then always stable without the assumption of the objective function is either Lipschitz or differentiable.

5.4.1 Generic Stability of Quasi-Variational Inequalities

In this section, the general stability theory of quasi-variational inequalities will be established. In particular, we shall study the stability of the following set

$$S_A(D,S,T) := \{y \in D : y \in S(y) \text{ and } \sup_{x\in S(y)} \sup_{w\in T(y)} Re\langle w, y-x\rangle \le 0\}$$

(resp., the set

$$S_B(D,S,T) := \{y \in D : y \in S(y) \text{ and } \sup_{x\in S(y)} \inf_{w\in T(y)} Re\langle w, y-x\rangle \le 0\}),$$

where D is a non-empty compact convex subset of a complete convex subset X of the normed space E, $S : X \to 2^X$ is a continuous set-valued mapping with non-empty compact convex values and $T : X \to 2^{E^*}$ is monotone and lower semicontinuous with non-empty bounded closed values (respectively, $T : X \to 2^{E^*}$ is upper semicontinuous with non-empty compact values).

Throughout this section, we assume that (i) E denotes a normed space with norm $\|\cdot\|$; (ii) E^* denotes the dual space of E with the norm $\|\cdot\|^*$; (iii) X denotes a non-empty complete convex subset of E; (iv) $K(X)$ denotes the family of all non-empty compact subsets of X and (v) h (respectively, h^*) denotes the Hausdorff metric defined on the family $bc(E)$ (respectively, $bc(E^*)$) of all bounded closed subsets of E (respectively, E^*) which is induced by the norm $\|\cdot\|$ (respectively, $\|\cdot\|^*$). Note that $K(X)$ is a complete metric space when equipped with the Hausdorff metric h (e.g., see [6]). If $A \subset E$ is non-empty, $x \in E$ and $\delta > 0$, let $U(\delta, x) = \{y \in E : \|x - y\| < \delta\}$ and $U(\delta, A) = \{y \in E : \|y - a\| < \delta$ for some $a \in A\}$.

As a special case of Theorem 5.3.8, we have the following generalized quasi-variational inequalities (GQVI) which is Theorem 2 of Shih and Tan [466]:

Theorem 5.4.1. Let E be a locally convex topological vector space, E^* be the dual space of E and X be a non-empty compact convex subset of E. Let $S : X \to 2^X$ be continuous such that for each $x \in X$, $S(x)$ is a closed convex subset of X and let $T : X \to 2^{E^*}$ be monotone such that T is lower semicontinuous from the relative topology of X to the strong topology of E^*. Then there exists a point $\hat{y} \in X$ such that

$$GQVI(A) \qquad \begin{cases} \hat{y} \in S(\hat{y}) \text{ and} \\ \sup_{x \in S(\hat{y})} \sup_{w \in T(\hat{y})} Re\langle w, \hat{y} - x\rangle \leq 0. \end{cases}$$

By Theorem 5.3.14, we also have the following generalized quasi-variational inequalities which is Theorem 4 of Shih and Tan [466]:

Theorem 5.4.2. Let E be a locally convex topological vector space, E^* the dual space of E and let X be a non-empty compact convex subset of E. Let $S : X \to 2^X$ be continuous such that for each $x \in X$, $S(x)$ is a closed convex subset of X, and $T : X \to 2^{E^*}$ be upper semicontinuous from the relative topology of X to the strong topology of E^* such that for each $x \in X$, $T(x)$ is a strongly compact subset of E^*. Then there exists a point $\hat{y} \in X$ such that

$$GQVI(B) \qquad \begin{cases} \hat{y} \in S(\hat{y}) \text{ and} \\ \sup_{x \in S(\hat{y})} \inf_{w \in T(\hat{y})} Re\langle w, \hat{y} - x\rangle \leq 0. \end{cases}$$

We call such a point $\hat{y}$ in Theorem 5.4.1 (resp., Theorem 5.4.2) a solution of the generalized quasi-variational inequality (A) (respectively, (B)), in short, GQVI(A) (respectively, GQVI(B)) for (S, T) in X and denote by $S_A(X, S, T)$ (respectively, $S_B(X, S, T)$) the set of all solutions of the generalized quasi-variational inequality (A) (respectively, (B)) for (S, T) in X. Thus under the stated conditions above, the associated set $S_A(X, S, T)$ (respectively, $S_B(X, S, T)$) is non-empty.

We shall study the stability of the solution set $S_A(D, S, T)$ of of $GQVI(A)$ and the solution set $S_B(D, S, T)$ of $GQVI(B)$ with D, S and T varying.

Let $C(X) = \{S : X \to K(X) : S$ is continuous on $X\}$, $L(X) = \{T : X \to 2^{E^*} : T$ is lower semicontinuous with bounded closed values $\}$, and $U(X) = \{T :$

$X \to K(E^*)$: T is upper semicontinuous }. For each $S_1, S_2 \in C(X)$, define $d_1(S_1, S_2) = \sup_{x \in X} h(S_1(x), S_2(x))$. For each $T_1, T_2 \in L(X)$ (respectively, T_1, $T_2 \in U(X)$), define $d_2(T_1, T_2) = \sup_{x \in X} h^*(T_1(x), T_2(x))$. It can be routinely checked that $(C(X), d_1)$, $(L(X), d_2)$ and $(U(X), d_2)$ are complete metric spaces. For each $u_1 = (D_1, S_1, T_1)$, $u_2 = (D_2, S_2, T_2)$ in $Y_L := K(X) \times C(X) \times L(X)$ (respectively, $Y_U = K(X) \times C(X) \times U(X)$), define $\rho(u_1, u_2) = h(D_1, D_2) + d_1(S_1, S_2) + d_2(T_1, T_2)$. Then (Y, ρ) is a complete metric space for $Y = Y_L$ or $Y = Y_U$.

As a special case of Theorem 5.3.8, we have the following generalized quasi-variational inequalities (GQVI) which is Theorem 2 of Shih and Tan [466] (see also Theorem 2 of Shih and Tan [474] with $f \equiv 0$):

Theorem 5.4.1. Let E be a locally convex topological vector space, E^* be the dual space of E and X be a non-empty compact convex subset of E. Let $S : X \to 2^X$ be continuous such that for each $x \in X$, $S(x)$ is a closed convex subset of X and let $T : X \to 2^{E^*}$ be monotone such that T is lower semicontinuous from the relative topology of X to the strong topology of E^*. Then there exists a point $\hat{y} \in X$ such that

$$GQVI(A) \qquad \begin{cases} \hat{y} \in S(\hat{y}) \text{ and} \\ \sup_{x \in S(\hat{y})} \sup_{w \in T(\hat{y})} Re\langle w, \hat{y} - x \rangle \leq 0. \end{cases}$$

By Theorem 5.3.14, we also have the following generalized quasi-variational inequalities which is Theorem 4 of Shih and Tan [466]:

Theorem 5.4.2. Let E be a locally convex topological vector space, E^* the dual space of E and let X be a non-empty compact convex subset of E. Let $S : X \to 2^X$ be continuous such that for each $x \in X$, $S(x)$ is a closed convex subset of X, and $T : X \to 2^{E^*}$ be upper semicontinuous from the relative topology of X to the strong topology of E^* such that for each $x \in X$, $T(x)$ is a strongly compact subset of E^*. Then there exists a point $\hat{y} \in X$ such that

$$GQVI(B) \qquad \begin{cases} \hat{y} \in S(\hat{y}) \text{ and} \\ \sup_{x \in S(\hat{y})} \inf_{w \in T(\hat{y})} Re\langle w, \hat{y} - x \rangle \leq 0. \end{cases}$$

We call such a point $\hat{y}$ in Theorem 5.4.1 (resp., Theorem 5.4.2) a solution of the generalized quasi-variational inequality (A) (respectively, (B)), in short, GQVI(A) (respectively, GQVI(B)) for (S, T) in X and denote by $S_A(X, S, T)$ (respectively, $S_B(X, S, T)$) the set of all solutions of the generalized quasi-variational inequality (A) (respectively, (B)) for (S, T) in X. Thus under the stated conditions above, the associated set $S_A(X, S, T)$ (respectively, $S_B(X, S, T)$) is non-empty.

Lemma 5.4.3. Let $\{A_n\}_{n=1}^{\infty}$ be a sequence in $K(X)$ which converges to $A \in K(X)$. Then every sequence $(x_n)_{n=1}^{\infty}$ in X with $x_n \in A_n$ for each $n \in \mathbb{N}$, has a subsequence which converges to a point in A.

Proof. Since A_n and A are compact and $A_n \to A$, by A.5.1 (ii) of Mas-Colell [363, p.10], $\cup_{n=1}^{\infty} A_n \cup A$ is compact. Since $x_n \in A_n \subset \cup_{n=1}^{\infty} A_n \cup A$, the sequence $\{x_n\}$ has a subsequence $\{x_{n_i}\}$ which converges to $x \in \cup_{n=1}^{\infty} A_n \cup A$. Now by Lemma 1(2) of Yu [594, p.231], $x \in A$. □

Let $M_L = \{(D, S, T) \in Y_L$: there exists $y \in D$ such that $y \in S(y)$ and $\sup_{x\in S(y)} \sup_{w\in T(y)} Re\langle w, y - x\rangle \leq 0 \}$

and

$M_U = \{(D, S, T) \in Y_U$: there exists $y \in D$ such that $y \in S(y)$ and $\sup_{x\in S(y)} \inf_{w\in T(y)} Re\langle w, y - x\rangle \leq 0\}$.

Define $S_A : M_L \to 2^X$ (respectively, $S_B : M_U \to 2^X$) by

$$S_A(u) = \{y \in D : y \in S(y) \text{ and } \sup_{x\in S(y)} \sup_{w\in T(y)} Re\langle w, y - x\rangle \leq 0\}$$

for each $u = (D, S, T) \in M_L$ (respectively,

$$S_B(u) = \{y \in D : y \in S(y) \text{ and } \sup_{x\in S(y)} \inf_{w\in T(y)} Re\langle w, y - x\rangle \leq 0\}$$

for each $u = (D, S, T) \in M_U$).

Lemma 5.4.4.

(i) The space M_L is closed in Y_L.

(ii) The space M_U is closed in Y_U.

Proof. Let $((D_n, S_n, T_n))_{n=1}^{\infty}$ be a sequence in M_L (respectively, M_U) such that $(D_n, S_n, T_n) \to (D, S, T) \in Y_L$ (respectively, Y_U). For each $n \in \mathbb{N}$, let $y_n \in D_n$ be such that

(1) $y_n \in S_n(y_n)$ and

(2) $\sup_{x\in S_n(y_n)} \sup_{w\in T_n(y_n)} Re\langle w, y_n - x\rangle \leq 0$ (respectively, $\sup_{x\in S_n(y_n)} \inf_{w\in T_n(y_n)} Re\langle w, y_n - x\rangle \leq 0$).

Since $D_n \to D$, by Lemma 5.4.3, without loss of generality, we may assume that $y_n \to y_0$ as $n \to \infty$ and $y_0 \in D$.

Now we shall show that

(i) $y_0 \in S(y_0)$ and

(ii) $\sup_{x\in S(y_0)} \sup_{w\in T(y_0)} Re\langle w, y_0 - x\rangle \leq 0$ (respectively, $\sup_{x\in S(y_0)} \inf_{w\in T(y_0)} Re\langle w, y_0 - x\rangle \leq 0\}$).

(i) Suppose that $y_0 \notin S(y_0)$, then there exists $a > 0$ such that $U(a, S(y_0)) \cap U(a, y_0) = \emptyset$. Since $S_n \to S$ as $n \to \infty$ there exists $n_1 \in \mathbb{N}$ such that for each

$n \in \mathbb{N}$ with $n \geq n_1$, $S_n(u) \subset U(a/2, S(u))$ for all $u \in X$. Since $y_n \to y_0$, by the upper semicontinuity of S, there exists a positive integer $n_2 \geq n_1$ such that $y_n \in U(a, y_0)$ and $S(y_n) \subset U(a/2, S(y_0))$ for all $n \in \mathbb{N}$ with $n \geq n_2$. Now for any integer $n \geq n_2$, $S_n(y_n) \subset U(a, S(y_0))$ and $y_n \in U(a, y_0)$. Note that $y_n \in S_n(y_n)$ which contradicts that $U(a, S(y_0)) \cap U(a, y_0) = \emptyset$. Therefore we must have $y_0 \in S(y_0)$.

(ii) Suppose that

$$\sup_{x \in S(y_0)} \sup_{w \in T(y_0)} Re\langle w, y_0 - x\rangle > 0$$

(respectively,

$$\sup_{x \in S(y_0)} \inf_{w \in T(y_0)} Re\langle w, y_0 - x\rangle > 0),$$

then there exist $x_0 \in S(y_0)$ and $\epsilon > 0$ such that $\sup_{w \in T(y_0)} Re\langle w, y_0 - x_0\rangle > \epsilon > 0$ (respectively, $\inf_{w \in T(y_0)} Re\langle w, y_0 - x_0\rangle > \epsilon > 0$). As T is lower semicontinuous (respectively, upper semicontinuous) and the mapping $(w, y, x) \mapsto Re\langle w, y - x\rangle$ is continuous from $E^* \times X \times X$ to $\mathbb{R}$, the mapping $(y, x) \mapsto \sup_{w \in T(y)} Re\langle w, y - x\rangle$ (respectively, $(y, x) \mapsto \inf_{w \in T(y)} Re\langle w, y - x\rangle$) is lower semicontinuous by Proposition 19 (respectively, Proposition 21) of Aubin and Ekeland [15, p.118] (respectively, [15, p.119]). Since $y_n \to y_0$ as $n \to \infty$, there exist $\delta > 0$ and $n_0 \in \mathbb{N}$ such that for each integer $n \geq n_0$, $y_n \in U(\delta, y_0)$ and for each $(y', x') \in U(\delta, y_0) \times U(\delta, x_0)$, $\sup_{w \in T(y')} Re\langle w, y' - x'\rangle > \epsilon > 0$ (respectively, $\inf_{w \in T(y')} Re\langle w, y' - x'\rangle > \epsilon > 0$). In particular, for each integer $n \geq n_0$ and $x' \in U(\delta, x_0)$,

$$\sup_{w \in T(y_n)} Re\langle w, y_n - x'\rangle > \epsilon > 0 \text{ (respectively, } \inf_{w \in T(y_n)} Re\langle w, y_n - x'\rangle > \epsilon > 0). \tag{5.4.1}$$

Because $S_n \to S$ as $n \to \infty$, there exists an integer $n_1 \geq n_0$ such that $S(u) \subset U(\delta/4, S_n(u))$ for all $u \in X$. Since $S(y_n) \to S(y_0)$ and $x_0 \in S(y_0)$, there exists an integer $n_2 \geq n_1$ such that $S(y_n) \cap U(\delta/4, x_0) \neq \emptyset$ for all integers $n \geq n_2$. Therefore for all $n \in N$ with $n \geq n_2$, $S(y_n) \cap U(\delta/4, x_0) \neq \emptyset$ and $S(u) \subset U(\delta/4, S_n(u))$ for all $u \in X$ which imply that

$$S_n(y_n) \cap U(\delta/2, x_0) \neq \emptyset. \tag{5.4.2}$$

Now by (5.4.1) and (5.4.2), for each integer $n \geq n_2$, we have

$$\sup_{w \in T(y_n)} Re\langle w, y_n - x'\rangle > \epsilon > 0 \text{ (respectively, } \inf_{w \in T(y_n)} Re\langle w, y_n - x'\rangle > \epsilon > 0) \tag{5.4.3}$$

for all $x' \in S_n(y_n) \cap U(\delta/2, x_0)$.

Since $T_n \to T$ as $n \to \infty$, there exists an integer $n_3 \geq n_2$ such that for each $n \geq n_3$, $d_2(T_n, T) < \frac{\epsilon}{2p}$, where $p = \|x_0\| + \delta + \sup\{\|y_n\| : n \geq 1\}$. Fix an

$n \geq n_3$. Let $w \in T(y_n)$ (respectively, $w \in T_n(y_n)$) be arbitrarily fixed. Since $h^*(T_n(y_n), T(y_n)) < \frac{\epsilon}{2p}$, there exists $w' \in T_n(y_n)$ (respectively, $w' \in T(y_n)$) with $\|w - w'\|^* < \frac{\epsilon}{2p}$. By (5.4.2), choose any $x' \in S_n(y_n) \cap U(\delta/2, x_0)$. Then

$$|Re\langle w' - w, y_n - x'\rangle| \leq \|w' - w\|^*(\|y_n\| + \|x' - x_0\| + \|x_0\|) < \frac{\epsilon}{2},$$

so that

$$\sup_{z\in T(y_n)} Re\langle z, y_n - x'\rangle \geq Re\langle w', y_n - x'\rangle \geq Re\langle w, y_n - x'\rangle - \epsilon/2$$

(respectively,

$$Re\langle w, y_n - x'\rangle \geq Re\langle w', y_n - x'\rangle - \epsilon/2 \geq \inf_{w\in T(y_n)} Re\langle w, y_n - x'\rangle - \epsilon/2).$$

Since $w \in T(y_n)$ (respectively, $w \in T_n(y_n)$) is arbitrary, we have by (5.4.3),

$$\sup_{z\in T_n(y_n)} Re\langle z, y_n - x'\rangle \geq \sup_{w\in T(y_n)} Re\langle w, y_n - x'\rangle - \epsilon/2 > \epsilon - \epsilon/2 = \epsilon/2 > 0$$

(respectively,

$$\inf_{w\in T_n(y_n)} Re\langle w, y_n - x'\rangle \geq \inf_{w\in T(y_n)} Re\langle w, y - x'\rangle > \epsilon - \epsilon/2 = \epsilon/2 > 0).$$

This contradicts (2) as $x' \in S_n(y_n)$. Hence (ii) must hold.

Therefore $y_0 \in S(u)$. Thus $(D, S, T) \in M_L$ (respectively, $(D, S, T) \in M_U$) so that M_L (respectively, M_U) is closed in Y_L (respectively, Y_U). □

For convenience, we recall and again state Theorem 2 of Fort [178, p.101] as following Lemma 5.4.5 (see also Lemma 2.5.10) below:

Lemma 5.4.5. Suppose W is a metric space, Z is a topological space and $S : Z \to K(W)$ is upper semicontinuous. Then S is continuous at points of a residual set in Z.

Lemma 5.4.6.

(i) $S_A(u) \in K(X)$ for each $u \in M_L$.

(ii) $S_B(u) \in K(X)$ for each $u \in M_U$.

Proof. Suppose $u = (D, S, T) \in M$ (respectively, $u = (D, S, T) \in M_U$). Since $S_A(u) \subset D$ (respectively, $S_B(u) \subset D$), it is sufficient to prove that $S_A(u)$ (respectively, $S_B(u)$) is closed in D. Let $(y_n)_{n=1}^{\infty}$ be a sequence in $S_A(u)$ (respectively, $S_B(u)$) which converges to a point $y_0 \in D$. By the definition of S_A (respectively, S_B), we have (i) $y_n \in S(y_n)$ and (ii) $\sup_{x\in S(y_n)} \sup_{w\in T(y_n)} Re\langle w, y_n - x\rangle \leq 0$ (respectively, $\sup_{x\in S(y_n)} \inf_{w\in T(y_n)} Re\langle w, y_n - x\rangle \leq 0$). Since S is upper semicontinuous on D with compact values and D is compact, $y_0 \in S(y_0)$. Note that T is lower semicontinuous (respectively, upper semicontintinuous) and the mapping $(w, y, x) \mapsto Re\langle w, y - x\rangle$ is continuous from $E^* \times X \times X$ to R, it follows that the mapping $(y, x) \mapsto \sup_{w\in T(y)} Re\langle w, y - x\rangle$ (respectively, $(y, x) \mapsto \inf_{w\in T(y)} Re\langle w, y - x\rangle$) is lower semicontinuous from $X \times X$ to $\mathbb{R}$ by Proposition 19 (respectively, Proposition 21) of Aubin and Ekeland [15, p.118] (respectively [15, p.119]). Since S is also lower semicontinuous, the mapping $y \mapsto \sup_{x\in S(y)} \sup_{w\in T(y)} Re\langle w, y - x\rangle$ (respectively, $y \mapsto \sup_{x\in S(y)} \inf_{w\in T(y)} Re\langle w, y - x\rangle$) is also lower semicontinuous by Proposition 19 of Aubin and Ekeland [15, p.118] again. Thus

$$\sup_{x\in S(y_0)} \sup_{w\in T(y_0)} Re\langle w, y_0 - x\rangle \leq \liminf_{n\to\infty} \sup_{x\in S(y_n)} \sup_{w\in T(y_n)} Re\langle w, y_n - x\rangle \leq 0$$

(respectively,

$$\sup_{x\in S(y_0)} \inf_{w\in T(y_0)} Re\langle w, y_0 - x\rangle \leq \liminf_{n\to\infty} \sup_{x\in S(y_n)} \inf_{w\in T(y_n)} Re\langle w, y_n - x\rangle \leq 0).$$

Hence $y_0 \in S_A(u)$ (respectively, $y_0 \in S_B(u)$). Therefore $S_A(u)$ (respectively, $S_B(u)$) is closed in D. □

Lemma 5.4.7. The correspondences $S_A : M_L \to K(X)$ (respectively, $S_B : M_U \to K(X)$) is upper semicontinuous.

Proof. Suppose that S_A (respectively, S_B) were not upper semicontinuous at some point $u = (D, S, T) \in M_L$ (respectively, M_U), then there exists an open subset G of X with $G \supset S_A(u)$ (respectively, $G \supset S_B(u)$) and a sequence $\{u_n\}_{n\in\mathbb{N}}$ in M_L (respectively, M_U) with $u_n \to u \in M_L$ (respectively, M_U) such that for each $n \in \mathbb{N}$, there exists $y_n \in S_A(u_n)$ (respectively, $y_n \in S_B(u_n)$) with $y_n \notin G$. Let $u_n = (D_n, S_n, T_n)$, then $D_n \to D$; $S_n \to S$ and $T_n \to T$. Since $y_n \in D_n$, for each $n \in \mathbb{N}$, by Lemma 5.4.3, we may assume without loss of generality that $y_n \to y_0 \in D$. Note that $y_n \notin G$ for all $n \in \mathbb{N}$ so that $y_0 \notin G \supset S_A(u)$ (respectively, $y_0 \notin G \supset S_B(u)$). Now the same argument as in the proof of Lemma 5.4.4 shows that (1) $y_0 \in S(y_0)$ and (2) $\sup_{x\in S(y_0)} \sup_{w\in T(y_0)} Re\langle w, y_0 - x\rangle \leq 0$ (respectively, $\sup_{x\in S(y_0)} \inf_{w\in T(y_0)} Re\langle w, y_0 - x\rangle \leq 0$). Therefore $y_0 \in S_A(u)$ (respectively, $y_0 \in S_B(u)$). This contradicts the fact that $y_0 \notin G \supset S_A(u)$ (respectively, $y_0 \notin G \supset S_B(u)$). Therefore S_A (respectively, S_B) is upper semicontinuous. □

In what follows, let $M = M_L$ or M_U and $S = S_A$ or S_B, respectively. Now let M_1 be an arbitrarily fixed non-empty closed subset of M.

Definition 5.4.8. If $u \in M_1$, then (i) a point $y \in S(u)$ is essential relative to M_1 if for each open neighborhood $N(y)$ of y in X, there exists an open neighborhood $O(u)$ of u in M_1 such that $S(u') \cap N(y) \neq \emptyset$ for each $u' \in O(u)$ and (ii) u is essential relative to M_1 if every $y \in S(u)$ is essential relative to M_1.

Theorem 5.4.9. (i) S is lower semicontinuous at $u \in M_1$ if and only if u is essential relative to M_1.

(ii) S is continuous at $u \in M_1$ if and only if u is essential relative to M_1.

Proof. (i) S is lower semicontinuous at $u \in M_1$ if and only if each $y \in S(u)$ is essential relative to M_1 if and only if u is essential relative to M_1.

(ii) This follows from (i) and Lemma 5.4.7. □

Theorem 5.4.10. If $u \in M_1$ is such that $S(u)$ is a singleton set, then u is essential relative to M_1.

Proof. Suppose $S(u) = \{x\}$. Let G be any open set in X such that $S(u) \cap G \neq \emptyset$, then $x \in G$ so that $S(u) \subset G$. Since S is upper semicontinuous at u by Lemma 5.4.7, there is an open neighborhood $O(u)$ of u in M such that $S(u') \subset G$ for each $u' \in O(u)$; in particular, $G \cap S(u') \neq \emptyset$ for each $u' \in O(u)$. Thus S is lower semicontinuous at u. By Theorem 5.4.9 (ii), u is essential relative to M. □

Theorem 5.4.11. There exists a dense G_δ subset Q_1 of M_1 such that u is essential relative to M_1.

Proof. Note that S is upper semicontinuous with compact and convex values, by Lemma 5.4.6 and Lemma 5.4.7. By Lemma 5.4.5, there exists a dense G_δ subset Q_1 of M_1 such that S is lower semicontinuous at each $u \in Q_1$. By Theorem 5.4.9 (ii), u is essential relative to M_1 for each $u \in Q_1$. □

Remark 5.4.12. Note that the mapping $S : M_1 \to K(X)$ is continuous at $u = (D, S, T) \in M_1$ if and only if for each $\epsilon > 0$, there is $\delta > 0$ such that $h(S(u), S(u')) < \epsilon$ whenever $u' \in M_1$ and $\rho(u, u') < \delta$; i.e., the solution set $S(u)$ of u is stable in M_1: $S(u')$ is close to $S(u)$ whenever $u' \in M_1$ is close to u. Theorem 5.4.9 (ii) implies that if $u \in M_1$, then u is essential relative to M_1 if and only if the solution set $S(u)$ is stable in M_1. Theorem 5.4.11 implies that there exists a dense G_δ subset Q_1 of M_1 such that for each $u = (D, S, T) \in Q_1$, the solution set $S(u)$ of the $GQVI$ is stable in M_1. In particular, most (in the sense of the Baire category) u in M_1 have stable solution set $S(u)$.

Now let

$$\begin{aligned}
CK(X) &= \{A \in K(X) : A \text{ is convex }\},\\
CC(X) &= \{S \in C(X) : S(x) \in CK(X) \text{ for each } x \in X\},\\
ML(X) &= \{T \in L(X) : T \text{ is monotone }\},\\
M'_L &= \{X\} \times CC(X) \times ML(X),\\
M'_U &= \{X\} \times CC(X) \times U(X).
\end{aligned}$$

It is easy to see that $CC(X)$ is closed in $K(X)$ so that M'_U is closed in M_U.

Also $ML(X)$ is closed in $L(X)$ so that M'_L is also closed in M_L. The following is an application of the results obtained in this section.

Theorem 5.4.13. Let X be a non-empty compact and convex subset of the normed space E. Then there exists a dense G_δ subset Q' of M'_L (respectively, M'_U) such that u is essential relative to M'_L (respective, M'_U) for each $u \in Q'$. Thus most (in the sense of the Baire category) of the solutions of the $GQVI(A)$ (respectively, $GQVI(B)$) for (S,T) in X are stable in M'_L (respectively, M'_U).

Let $M''_L = \{X\} \times \{O_X\} \times ML(X)$ (respectively, $M''_U = \{X\} \times \{O_X\} \times U(X)$), where $O_X(x) = X$ for all $x \in X$. Clearly, M''_L (respectively, M''_U) is a closed subset of M'_L (respectively, M'_U) if X is compact. The following deals with the stability of Hartman-Stampacchia variational inequalities in Hartman and Stampacchia [221]:

Theorem 5.4.14. Let X be a non-empty compact and convex subset of the normed space E. Then there exists a dense G_δ subset Q'' of M''_L (resp., M''_U) such that u is essential relative to M''_L (resp., M''_U) for each $u \in Q''$.

5.4.2 Essentially Connected Components

The aim of this section is to establish the existence of essentially connected components for Hartman-Stampacchia type variational inequalities for both set-valued and single-valued mappings in normed spaces. Our results show that each variational inequality problem has, at least, one connected component of its solutions which is stable though in general its solution set may not have a good behavior (i.e., not stable). Thus if a variational inequality problem has only one connected solution set, it must be stable. Here we don't need to require the objective mapping to be either Lipschitz or differential.

Let D be a non-empty compact convex subset of a (complete) convex subset X of a normed space E and $T : X \to 2^{E^*}$ be a set-valued mapping with non-empty compact values. We recall that a variational inequality problem (in short, $VI(T,D)$) is is to find $y \in D$ such that

$$\sup_{x \in X} \inf_{w \in T(y)} Re\langle w, y - x\rangle \leq 0. \tag{5.4.4}$$

If (5.4.4) has a solution $\hat{y} \in D$ and $T(\hat{y})$ is compact and convex, it follows by Kneser minimax inequality in [318] that

$$\inf_{w \in T(\hat{y})} \sup_{x \in X} \langle w, \hat{y} - x\rangle = \sup_{x \in X} \inf_{w \in T(\hat{y})} \langle w, \hat{y} - x\rangle.$$

Note that the mapping $w \mapsto \sup_{x \in X}\langle w, \hat{y} - x\rangle$ is lower semicontinuous, and thus there exists $w_0 \in T(\hat{y})$ such that

$$Re\langle w_0, \hat{y} - x\rangle \leq 0$$

for all $x \in X$.

Throughout this section we assume that (i): E denotes a normed space with norm $\|\cdot\|$; (ii): E^* denotes the dual space of E with the norm $\|\cdot\|^*$; (iii): X denotes a non-empty convex (not necessarily compact) subset of E; (iv): $K(X)$ (resp., $CK(X)$) denotes the family of all non-empty compact (resp., compact and convex) subsets of X and (v): h^* denotes the Hausdorff metric defined on the family $bc(E^*)$ of all bounded closed subsets of E^* which is induced by the norm $\|\cdot\|^*$. Note that $K(X)$ (resp., $CK(X)$) is a complete metric space when equipped with the Hausdorff metric h by Theorem 4.3.9 of Klein and Thompson [317]. If $A \subset E$ is non-empty, $x \in E$ and $\delta > 0$, let $U(\delta, x) = \{y \in E : \|x - y\| < \delta\}$ and $U(\delta, A) = \{y \in E : \|y - a\| < \delta$ for some $a \in A\}$.

Let $U(X) := \{T : X \to K(E^*) : T$ is upper semicontinuous $\}$. For each $T_1, T_2 \in U(X)$, define $d_2(T_1, T_2) = \sup_{x \in X} h^*(T_1(x), T_2(x))$. It can be routinely checked that $(U(X), d_2)$ is a complete metric space.

Set $Y := U(X) \times K(X)$ and for each $u_1 = (T_1, D_1)$ and $u_2 = (T_2, D_2)$ in Y, we define a metric ρ in Y by $\rho(u_1, u_2) := h(D_1, D_2) + d_2(T_1, T_2)$. Then it is easy to see that (Y, ρ) is a complete metric space. We also denote by $Y_1 := U(X) \times CK(X)$. It is clear that Y_1 is a subclass of Y.

Let $M := \{(T, D) \in Y :$ there exists $y \in D$ such that $\sup_{x \in D} \inf_{w \in T(y)} Re\langle w, y - x\rangle \leq 0\}$ and define a solution mapping $S : M \to 2^X$ by

$$S(u) := \{y \in D(\subset X) : \sup_{x \in D} \inf_{w \in T(y)} Re\langle w, y - x\rangle \leq 0\};$$

for each $u = (T, D) \in M$. Then we have that $S(u) \neq \emptyset$ for each $u \in M$.

The principle aim of this section is to study the stability of connected components of solutions set for variational inequality problem $VI(T, D)$ with the perturbations of both D and T varying, where T may be either a set-valued mapping or a single-valued mapping.

Theorem 5.4.2 shows that for each $u = (T, D) \in Y_1$, its solution set $S(u) \neq \emptyset$ and indeed Y_1 is a closed subset of M.

For various existence theory of variational inequalities and its numerous applications, please see Noor's survey paper [401], or Hyers et al. [244] and related references therein.

In this section, we shall establish the upper semicontinuity of of solution set mapping for variational inequalities $S : Y \to 2^X$.

We first need the following fact.

Lemma 5.4.15. The space M is closed in Y.

Proof. It is a special case of Lemma 5.4.4. □

Lemma 5.4.16. The set $S(u) \in K(X)$ for each $u \in M$.

Proof. Indeed, it is a special case of Lemma 5.4.6. □

Lemma 5.4.17. The solution mapping $S : M \to K(X)$ is upper semicontinuous.

Proof. It is a spceial case of Lemma 5.4.7. □

Let M_1 be an arbitrarily fixed non-empty closed subset of M. Corresponding to Definition 5.4.8, we recall the following notion (see also Tan et al. [519]).

Definition 5.4.18. Let $u \in M_1$ be any given variational problem. Then (i): a solution $y \in S(u)$ for a variational problem u is said to be *essential* relative to M_1 if for each open neighborhood $N(y)$ of y in X, there exists an open neighborhood $O(u)$ of u in M_1 such that $S(u') \cap N(y) \neq \emptyset$ for each $u' \in O(u)$; (ii): the variational inequality problem u is *essential* relative to M_1 if every $y \in S(u)$ is essential relative to M_1; and (iii): the variational inequality problem u is *weakly essential* relative to M_1 if there exists one solution $y \in S(u)$ which is essential relative to M_1.

Remark 5.4.19. Note that the mapping $S : M_1 \to K(X)$ is continuous at $u = (T, D) \in M_1$ if and only if for each $\epsilon > 0$, there is $\delta > 0$ such that $h(S(u), S(u')) < \epsilon$ whenever $u' \in M_1$ and $\rho(u, u') < \delta$; i.e., the solution set $S(u)$ of u is stable in M_1 in the sense that : $S(u')$ is close to $S(u)$ whenever $u' \in M_1$ is close to u.

Now by introducing the concept of essentially components of solution set for variational inequalities, we will establish the existence theory of essentially connected components for generally variational inequalities of upper semicontinuous set-valued mappings.

For each variational inequality problem $u = VI(D, T) \in Y$, the component of a solution $y \in S(u)$ is the union of all connected subsets of $S(u)$ which contain the point y. From Engelking [154], we know that components are connected closed subsets of $S(u)$ and thus they are also compact as $S(u)$ is compact. It is also easy to see that the components of two distinct points of $S(u)$ either coincide or are disjoint, so that all components constitute a decomposition of $S(u)$ into connected pairwise disjoint compact subsets, i.e.,

$$S(u) = \bigcup_{\alpha \in \Lambda} S_\alpha(u)$$

where Λ is an index set, for any $\alpha \in \Lambda$, $S_\alpha(u)$ a nonempty connected compact and for any $\alpha, \beta \in \Lambda (\alpha \neq \beta)$, $S_\alpha(u) \cap S_\beta(u) = \emptyset$.

In order to study the existence of essentially connected components of generally variational problems, we also introduce the following notion.

Definition 5.4.20. For each variational inequality problem $u := VI(T, D) \in Y$, suppose the set $S(u) = \bigcup_{\alpha\in\Lambda} S_\alpha(u)$. Then a component $S_\alpha(u)$ for some $\alpha \in \Lambda$, is said to be an *essential component* of u if for each open set O containing $S_\alpha(u)$, there exists $\delta > 0$ such that for any other variational problem $u' = VI(D', T') \in Y$ with $\rho(u, u') = \sup_{x\in K} \rho(T(x), T'(x)) + h(D, D') < \delta$, we have that $S(u') \cap O \neq \emptyset$.

Remark 5.4.21. Definition 5.4.20 says that even though we could not expect the continuity for all solutions of any given variational inequality problem $u = VI(T, D)$, however, there is a case that some of solutions enjoy the continuous stability. In the rest part of this section, we will show that the existence of such nice solutions for each general variational problem $VI(T, D)$ of upper semicontinuous set-valued mappings.

We recall that for given non-empty subsets A and B of a metric space E, the Hausdorff metric h between A and B is defined by $h(A, B) := \inf\{\epsilon : A \subset O(B, \epsilon) \text{ and } B \subset O(A, \epsilon)\}$. Then we have the following simple fact.

Now we consider the stability of those variational inequality problems from the subclass Y_1 of Y as described in the beginning of this section.

Theorem 5.4.22. Let $y = VI(T, D) \in Y_1$ be a given variational inequality problem (thus the solution set is non-empty). Then there exists at least one essentially connected component of $S(y)$.

Proof. For any given $y = VI(T, D) \in Y_1$, suppose that the solution set $S(y)$ of the variational problem $VI(T, D)$ is decomposed as follows:

$$S(y) = \bigcup_{\alpha\in\Lambda} S_\alpha(y)$$

where Λ is an index set, for any $\alpha \in \Lambda$, $S_\alpha(y)$ is a connected compact and for any $\alpha, \beta \in \Lambda(\alpha \neq \beta)$, $S_\alpha(x) \cap S_\beta(x) = \emptyset$. We shall prove that there exists at least one essential component of $S(y)$. Let us suppose otherwise there is no any essential connected component. Then for any $\alpha \in \Lambda$, there exists an open set $O_\alpha \supset S_\alpha(y)$ such that for any $\varepsilon > 0$, there is $y_\alpha \in Y_1$ with $\rho(y, y_\alpha) < \varepsilon$ such that $S(y_\alpha) \cap O_\alpha = \emptyset$. As $S(y)$ is compact, there exist two open and finite coverings $\{V_i\}_{i=1}^n$ and $\{W_i\}_{i=1}^n$ which satisfy the following conditions (e.g., see Kinoshita [311]): (1) $\overline{W_i} \subset V_i$; (2) $V_i \cap V_j = \emptyset$ for each $i \neq j$; and (3) V_i contains at least one $S_{\alpha_i}(y)$ with $O_{\alpha_i} \supset V_i \supset S_{\alpha_i}(y)$. By Lemma 5.4.17, the solution mapping S is upper semicontinuous at y and $\bigcup_{i=1}^n W_i \supset S(y)$ and $\bigcup_{i=1}^n W_i$ is open, then there exists a $\delta > 0$ such that $\bigcup_{i=1}^n W_i \supset S(y')$ for any $y' \in Y_1'$ with $\rho(y, y') < \delta$. Thus there is $y_{\alpha_i} \in Y_1'$ with $\rho(y, y_{\alpha_i}) < \delta$ such that $S(y_{\alpha_i}) \cap O_{\alpha_i} = \emptyset$.

Let $y = VI(T, D)$ and $y_{\alpha_i} = VI(T_{\alpha_i}, D)$, where $i = 1, \cdots, n$. We define a set-valued mapping $T^* : X \to 2^{\mathbb{E}^*}$ by

$$T^*(x) = \begin{cases} T(x), & \text{if } x \in X \setminus \bigcup_{i=1}^n V_i, \\ T_i^*(x), & \text{if } x \in V_i \text{ for some } i = 1, 2, \cdots, n \end{cases}$$

where $T_i^* : X \to 2^{E^*}$ is a set-valued mapping defined by

$$T_i^*(x) := \lambda_i(x)T(x) + \mu_i(x)T_{\alpha_i}(x)$$

for each $x \in X$, and λ_i and μ_i are two functions from X to $\mathbb{R}$ defined by

$$\lambda_i(x) := \frac{d(x, \overline{W}_i)}{d(x, \overline{W}_i) + d(x, X \setminus \bigcup_{i=1}^n V_i)}$$

and

$$\mu_i(x) := \frac{d(x, X \setminus \bigcup_{i=1}^n V_i)}{d(x, \overline{W}_i) + d(x, X \setminus \bigcup_{i=1}^n V_i)}$$

for each $x \in X$. It is clear that λ_i and μ_i are continuous. Note that T and T_{α_i} are upper semicontinuous with compact values, it follows by Theorems 7.3.11, 7.3.14 and 7.3.15 of Klein and Thompson [317] that T_i^* is upper semicontinuous. Secondly, for each given $x \in X$, if $x \in X \setminus \cup_{i=1}^n V_i$, then $T(x) = T_i^*(x)$ if $x \in X \setminus \cup_{i=1}^n V_i$ for all $i = 1, 2, \cdots, n$; and $T_i^*(x) := T_{\alpha_i}(x)$ if $x \in \overline{W}_i$. Next we show that T^* is upper semicontinuous on X. Without loss of generality, let G be any given open subset of E^* and suppose there exists $x_0 \in X$ such that $T^*(x_0) \subset G$. If there exists $i \in \{1, 2, \cdots, n\}$ such that $x_0 \in V_i$, then $T^*(x_0) = T_i^*(x_0)$. By the upper semicontinuity of T_i^*, there exists a non-empty open neighbourhood $N(x_0)$ of x_0 in V_i such that $T^*(u) = T_i^*(u) \subset G$. As V_i is open in X, we may assume that $N(x_0)$ is also an open neighbrhood of X and thus T^* is upper semicontinuous at $x_0 \in V_i$ by the definition of upper semicontinuity. In the case $x_0 \in X \setminus \cup_{i=1}^n V_i$, by the facts that $T^*(x_0) = T_i^*(x_0) = T(x)$ for all $i = 1, 2, \cdots, n$, and the upper semicontinuity of T and T_i, it follows that for each $i = 1, 2, \cdots, n$, there exists a non-empty open neighbourhood $V_i(x_0)$ of x_0 in X such that $T_i^*(u) \subset G$ for all $u \in V_i(x_0)$ (as $T_i^*(x_0) \subset G$). Since T is also upper semicontinuous, there exists a non-empty open neighbourhood $V_0(x_0)$ of x_0 in X such that $T(u) \subset G$ for each $u \in V_0(x_0)$. Let $N(x_0) := \cap_{i=0}^n V_i(x_0)$. Then $N(x_0)$ is a non-empty open neighborhood of x_0 in X. Moreover, it is easy to see that $T^*(u) \subset G$ for each $u \in N(x_0)$. Indeed, if $u \in N(x_0) \cap (X \setminus \cup_{i=1}^n V_i)$, then $T^*(u) = T(u) \subset G$; in the case $u \in N(x_0) \cap V_i$ for some $i = 1, \cdots, n$, then $T^*(u) = T_i^*(u) \subset G$. Therefore we have shown that for each open neighborhood G of E^* with $T^*(x_0) \subset G$ for some $x_0 \in X$, there exists a non-empty open neighborhood $N(x_0)$ of x_0 such that $T^*(u) \subset G$ for each $u \in N(x_0)$. This means T^* is upper semicontinuous. Thus $y^* = VI(T^*, D) \in Y_1'$, so that $S(y^*) \neq \emptyset$. Note that $\rho(y, y_{\alpha_i}) < \delta$ for $i = 1, 2, \cdots, n$, it follows by Lemma 2.8.20 that

$$h(T(x), \lambda_i(x)T(x) + \mu_i(x)T_{\alpha_i}(x)) \leq h(T(x), T_{\alpha_i}(x)).$$

Therefore $\rho(y, y^*) < \delta$ and $S(y^*) \subset \bigcup_{i=1}^n W_i$. Note that for any $x_0 \in S(y^*)$, there is an index i_0 such that $x_0 \in W_{i_0}$, and hence $x_0 \in W_{i_0} \subset \overline{W}_{i_0} \subset O_{\alpha_{i_0}}$. Therefore $T^*(x_0) = T_{\alpha_{i_0}}(x_0)$ and $x_0 \in S(y_{\alpha_{i_0}})$. Thus contradicts our assumption that $S(y_{\alpha_{i_0}}) \cap O_{\alpha_{i_0}} = \emptyset$. Hence there exists at least one essentially connected component of $S(y)$. This completes the proof.□

As an application of Theorem 5.4.22, we give a sufficient condition that $y = VI(T, D) \in Y_1$ is weakly essential or essential.

Theorem 5.4.23. If $y = VI(T, D) \in Y_1$ is such that the solution set $S(y)$ of variational inequality VI(T, D) is a totally disconnected set, then $VI(T, D)$ is weakly essential. In particular, if the solution set $S(y)$ of the variational inequality problem $VI(T, D)$ is either a singleton or a connected set, then the problem $VI(T, D)$ is essential.

Proof. Since $S(y)$ is a totally disconnected set, then $S(y) = \cup_{\alpha \in \Lambda} S_\alpha(\phi)$, where $S_\alpha(\phi)$ is a singleton set for each $\alpha \in \Lambda$. By Theorem 5.4.22, there exists $S_{\alpha_0}(\phi) = \{x_0\}$, which is an essential component of $S(y)$. It is clear that x_0 is essential and thus it is weakly essential. In the case, $S(y)$ is connected, it must be essential by Theorem 5.4.22 and hence the variational inequality $VI(T, D)$ is essential and the proof is complete. □

Remark 5.4.24. Let M_1 be a non-empty closed subset of Y_1 ($\subset M$). As the mapping $S : M_1 \to K(X)$ is continuous at $u = (T, D) \in M_1$ if and only if for each $\epsilon > 0$, there is $\delta > 0$ such that $h(S(u), S(u')) < \epsilon$ whenever $u' \in M_1$ and $\rho(u, u') < \delta$; i.e., the solution set $S(u)$ of u is continuous at $u \in M_1$ in the sense that: $S(u')$ is close to $S(u)$ whenever $u' \in M_1$ is close to u. Thus the essential property of variational inequality $VI(T, D)$ characterizes the stability of its solution set. Theorems 5.4.22 and 5.4.23 tell us that each variational inequality problem has, at least, one connected component of its solutions which is stable though in general its solution set may not have a good behavior (i.e., not stable). If we know that the solution set of a nonlinearly variational inequality problem $VI(T, D)$ is connected, Theorem 5.4.23 tells us that it must be stable. Here we don't need to require the objective mapping T to be either Lipschitz or differential.

5.5 Generalized Vector-Valued Variational-Like Inequalities

The purpose of this section is to introduce a new kind of variational inequality, a 'generalized vector variational-like inequality' which includes several classical and well-known variational inequalities as special cases. As an application of the

Knaster-Kuratowski-Mazurkiewicz principle as extended by Fan [157], we prove that there exist solutions for our generalized vector variational-like inequality under reasonable hypotheses. These results generalize corresponding results given by Chen et al. [80]-[84], Giannessi [195], Harker and Pang [220], Hartman and Stampacchia [221], Isac [252], Lee et al. [339], Noor [394], Saigal [451], Siddiqi et al. [479] and Yang [582].

Vector-valued variational inequalities were first introduced and studied by Giannessi [195] in a finite dimensional Euclidean space. Later Chen et al. [80]-[81] and Yang [581]-[582] have intensively studied some kinds of vector variational inequalities, vector quasi-variational inequalities and vector complementarity problems in Banach spaces. Recently, Lee et al. [339] and Siddiqi et al. [479] established some existence theorems for solutions of some kinds of vector-valued inequality equations and vector variational inequalities in Banach space and in Hausdorff topological vector spaces.

The purpose of this section is to study the existence problem for solutions of more general vector-generalized variational-like inequalities involving set-valued η-pseudo-monotone mappings in Banach space.

Definition 5.5.1. Let Y be an ordered Banach space, C be a closed pointed convex cone of Y with $intC \neq \emptyset$ and A a nonempty subset of Y. By using C we can define a partial order "$\leq$" as follows:

$$x \leq y \text{ if and only if } y - x \in C.$$

Following Yu [601], a point $z_0 \in A$ is called a *vector maximal point* of A if the set $\{z \in A : z_0 \leq z, z \neq z_0\} = \emptyset$, which is equivalent to

$$A \cap (z_0 + C) = \{z_0\}.$$

We denote the set of all such vector maximal points of A by Max$A$.

It should be noted that from the definition above we have:

$$\text{Max}(A) \subset A, \text{Max}(\alpha A) = \alpha \text{Max}(A)$$

for all $\alpha > 0$ and

$$\text{Max}(A + a) = \text{Max}(A + a)$$

for all $a \in Y$.

We first need the following simple fact which was first given by Luc [351].

Lemma 5.5.2 (Luc [351]). Let A be a compact set in Y. Then

$$\text{Max}(A) \neq \emptyset.$$

In the sequel we assume that X is a Banach space, Y is an ordered Banach space, K is a nonempty closed convex subset of X, $L(X, Y)$ is the space of all continuous linear operators from X into Y and $C : K \longrightarrow 2^Y$ is a set-valued mapping such that for each $x \in K$, $C(x)$ is a closed pointed convex cone in Y with $intC(x) \neq \emptyset$. We suppose that $\eta : K \times K \longrightarrow X$ is a continuous mapping and $T : X \to 2^{L(X,Y)}$ is a set-valued mapping.

We consider the following generalized vector-valued variational-like inequality (GVVLI) problem of finding $x_* \in K$ such that

$$\text{Max}\langle T(x_*), \eta(y, x_*)\rangle \subset Y \setminus (-intC(x_*)) \qquad (GVVLI)$$

for all $y \in K$, where

$$\langle T(x_*), \eta(y, x_*)\rangle = \bigcup_{u \in T(x_*)} \langle u, \eta(y, x_*)\rangle,$$

$$\text{Max}\langle T(x_*), \eta(y, x_*)\rangle = \sup_{u \in T(x_*)} \langle u, \eta(y, x_*)\rangle,$$

and $\langle u, \eta(y, x_*)\rangle$ denotes the evaluation of linear operator u from X into Y at $\eta(y, x_*)$ and $\text{Max}(A)$ denotes the set of all vector maximal points of A.

Throughout this section we shall call the above problem (GVVLI) a *generalized vector variational-like inequality.* We will study the existence of solutions for (GVVLI) using KKM theory developed in Chapter 2.

Some special cases.

(I): If $T : X \to L(X, Y)$ is a single-valued mapping, and $\eta(y, x) = y - g(x)$, where $g : K \to K$ is a single-valued mapping, then the problem (GVVLI) reduces to finding $x_* \in K$ such that

$$\langle T(x_*), y - g(x_*)\rangle \notin -intC(x_*)$$

for all $y \in K$. This is considered in Siddiqi et al. [479].

(II): If $T : X \to L(X, Y)$ is a single-valued mapping, and $\eta(y, x) = y - x$, then the problem (GVVLI) reduces to finding $x_* \in K$ such that

$$\langle T(x_*), y - x_*\rangle \notin -intC(x_*)$$

for all $y \in K$. This is known as a vector-valued variational inequality and was considered by Chen [81].

(III): If $C(x) \equiv C$ for all $x \in K$, where C is a closed convex cone in Y with $intC \neq \emptyset$, and $\eta(y, x) = y - x$, then the problem (GVVLI) reduces to finding $x_* \in K$ that satisfies that for each $y \in K$, there exists $u_* \in T(x_*)$ such that

$$\langle u_*, y - x_*\rangle \notin -intC.$$

This is considered in Lee et al. [339].

(IV): If $C(x) \equiv C$ for all $x \in K$ and C is the same as in (II), then the problem (III) reduces to finding $x_* \in K$ such that

$$\langle T(x_*), y - x_* \rangle \notin -intC$$

for all $y \in K$. This is considered in Chen et al. [83]-[84].

(V): If $Y = \mathbb{R}$, $L(X, Y) = X^*$ (which is the dual of X), $C(x) \equiv \mathbb{R}^+$ for all $x \in K$, then the problem (GVVLI) reduces to finding $x_* \in K$ such that

$$\sup_{u \in T(x_*)} \langle u, \eta(y, x_*) \rangle \geq 0$$

for all $y \in K$. This is considered in Cottle and Yao [103], Isac [252], and by Noor [394] in some special cases.

(VI): If $Y = \mathbb{R}$, $X = \mathbb{R}^n$, $C(x) \equiv \mathbb{R}+$ for all $x \in K \subset \mathbb{R}^n$, $L(X, Y) = \mathbb{R}^n$ and $\eta(y, x) = y - x$, then the problem (IV) collapses to finding $x_* \in K$ such that

$$\langle T(x_*), y - x_* \rangle \geq 0$$

for all $y \in K$. This is due to Hartman and Stampacchia [221].

The special cases (I)-(VI) show that the generalized vector variational-like inequalities (GVVLI) provide a more general and unified setting for the study of the above class of problems.

Let D be a subset of a topological vector space E. Once again, we recall that a set-valued mapping $T : D \to 2^E$ is called a KKM mapping if for each nonempty finite subset $\{x_1, \cdots, x_n\} \subset D$, we have

$$co\{x_1, \cdots, x_n\} \subset \bigcup_{i=1}^{n} T(x_i).$$

Definition 5.5.3. Let E and F be two topological spaces, and $T : E \to 2^F$ a set-valued mapping. T is said to be closed , if for any $x \in E$, $\{x_n\} \subset E$ with $x_n \to x$ and any $\{y_n\} \subset F$ with $y_n \in T(x_n)$ and $y_n \to y$, then we have $y \in T(x)$.

Definition 5.5.4. Let X and Y be two normed spaces, $T : X \to 2^{L(X,Y)}$ a set-valued mapping. Let K be a nonempty subset of X and $C : K \to 2^Y$ a mapping such that for each $x \in K$, $C(x)$ is a closed pointed convex cone in Y with $intC(x) \neq \emptyset$ and $C(x) \neq Y$, and let $\eta(x, y) : K \times K \to X$ be a mapping. Then

(1) T is said to be *η-pseudo-monotone*, , if for any $x, y \in X$, $u \in T(x)$, and $v \in T(y)$, we have that
$\langle u, \eta(y, x) \rangle \notin -intC(x)$ implies that $\langle v, \eta(y, x) \rangle \notin -intC(x)$.

(2) T is said to be *V-hemicontinuous*, , if for any $x,\, y \in X$, $\alpha > 0$ we have $\langle T(x+\alpha y),\, z\rangle \to \langle T(x),\, z\rangle$ as $\alpha \to 0^+$, for each $z \in X$. Here $\langle T(u),\, z\rangle = \bigcup_{w\in T(u)}\langle w,\, z\rangle$.

Remark 5.5.5. If $\eta : K\times X \to X$, is given by $\eta(x,\, y) = x-y$ for all $x \in K$ and $y \in X$ then the η-pseudo-monotonicity of T coincides with the ordinary pseudo-monotonicity of the set-valued mapping T (see Saigal [451]).

Theorem 5.5.6. Let X and Y be two Banach spaces, $K \subset X$ a nonempty closed convex subset, $C : K \to 2^Y$, $T : X \to 2^{L(X,Y)}$ and $\eta : K \times K \to X$ be three mappings satisfying the following conditions:

(i) for each $x \in K$, $C(x)$ is a pointed closed convex cone in Y with $intC(x) \neq \emptyset$ and $C(x) \neq Y$;

(ii) the mapping $S : X \to 2^Y$ defined by

$$S(x) = Y \setminus (-intC(x))$$

is closed;

(iii) T is a η-pseudo-monotone nonempty compact-valued and V-hemicontinuous mapping;

(iv) η is continuous, affine and $\eta(x,\, x) = \theta$ for all $x \in K$, where θ is the zero in X; and

(v) there exists a nonempty compact subset D of X satisfying for each finite subset N of K there exists a nonempty compact convex subset L_N of K such that $N \subset L_N$ and for each $x \in L_N \setminus D$ there exists $y \in L_N$ such that

$$\mathrm{Max}\langle T(y),\, \eta(y,\, x)\rangle \cap (-intC(x)) \neq \emptyset.$$

Then the generalized vector variational-like inequality (GVVLI) has a solution in K, i.e., there exists $x_* \in K$ such that

$$\mathrm{Max}\langle T(x_*),\, \eta(y,\, x_*)\rangle \subset Y \setminus (-intC(x_*))$$

for all $y \in K$.

Proof. Define two set-valued mappings F and $G : K \to 2^K$ by

$$F(y) = \{x \in K : \mathrm{Max}\langle T(x),\, \eta(y,x)\rangle \subset Y \setminus (-intC(x))\}$$

and

$$G(y) = \{x \in K : \mathrm{Max}\langle T(y),\, \eta(y,x)\rangle \subset Y \setminus (-intC(x))\}$$

for each $y \in K$. Since T is compact-valued and $\langle \cdot,\, \cdot\rangle$ is continuous, $\langle T(x),\, \eta(y,x)\rangle$ and $\langle T(y),\, \eta(y,x)\rangle$ are both compact subsets in Y. By Lemma 5.5.2, we know that for each $y \in K$, $F(y)$ and $G(y)$ are both well-defined.

(1): By conditions (iv), for each $y \in K$, $y \in F(y)$ and $y \in G(y)$, so that for each $y \in K$, $F(y) \neq \emptyset$ and $G(y) \neq \emptyset$.
Next we prove that $F(y) \subset G(y)$ for all $y \in K$. In fact, if $x \in F(y)$, then we have

$$\text{Max}\langle T(x), \eta(y, x)\rangle \subset Y \setminus (-intC(x)).$$

Since $\langle T(x), \eta(y, x)\rangle$ is a compact set in Y, by Lemma 5.5.2 there exists $u \in T(x)$ such that

$$\langle u, \eta(y, x)\rangle \notin -intC(x).$$

By the η-pseudo-monotonicity of T for each $v \in T(y)$ we have

$$\langle v, \eta(y, x)\rangle \notin -intC(x),$$

and thus

$$\text{Max}\langle T(y), \eta(y, x)\rangle \subset Y \setminus (-intC(x)).$$

This implies that $x \in G(y)$. Therefore $F(y) \subset G(y)$ for all $y \in K$.

(2): Next we prove that $F : K \to 2^K$ is a KKM mapping.

Suppose on the contrary that F is not a KKM mapping. Then there exists a finite set $\{y_1, \cdots, y_n\} \subset K$ such that

$$co\{y_1, \cdots, y_n\} \not\subset \bigcup_{i=1}^{n} F(y_i).$$

Therefore there exists $x = \sum_{i=1}^{n} \alpha_i y_i \in co\{y_1, \cdots, y_n\}$, but $x \notin F(y_i)$ for $i = 1, 2, \cdots, n$, where $\alpha_i \geq 0$, $i = 1, \cdots, n$ and $\sum_{i=1}^{n} \alpha_i = 1$. Since $x \notin F(y_i)$ for $i = 1, \cdots, n$, we have that

$$\text{Max}\langle T(x), \eta(y_i, x)\rangle \not\subset Y \setminus (-intC(x)), \quad i = 1, 2, \cdots, n.$$

Therefore there exist $u_i \in T(x)$, $i = 1, 2, \cdots, n$ such that

$$\langle u_i, \eta(y_i, x)\rangle \in -intC(x), \quad i = 1, \cdots, n.$$

Since $intC(x)$ is convex, we have

$$\sum_{i=1}^{n} \alpha_i \langle u_i, \eta(y_i, x)\rangle \in -intC(x).$$

Again since

$$\begin{aligned}\sum_{i=1}^{n}\alpha_i\langle u_i,\, \eta(y_i,\, x)\rangle &\in \sum_{i=1}^{n}\alpha_i\langle T(x),\, \eta(y_i,\, x)\rangle \\ &= \langle T(x),\, \eta(\sum_{i=1}^{n}\alpha_i y_i,\, x)\rangle \text{ (by condition (iv))} \\ &= \langle T(x),\, \eta(x,\, x)\rangle \\ &= \langle T(x),\, \theta\rangle \\ &= \theta,\end{aligned}$$

and thus

$$\sum_{i=1}^{n}\alpha_i\langle u_i,\, \eta(y_i,\, x)\rangle = \theta \in -intC(x).$$

This contradicts the assumption that $C(x)$ is a pointed convex cone.

This implies that F is a KKM-mapping, and thus G is also a KKM mapping.

(3): Next we prove that for each $y \in K$, $G(y)$ is a closed subset in K. Let $\{x_n\} \subset G(y)$ be a sequence such that $x_n \to x \in K$. Therefore we have

$$\text{Max}\langle T(y),\, \eta(y,\, x_n)\rangle \subset Y \setminus (-intC(x_n)) = S(x_n), \quad i = 1,\, 2,\, \cdots .$$

By condition (ii), S is closed. From

$$\text{Max}\langle T(y),\, \eta(y,\, x_n)\rangle \to \text{Max}\langle T(y),\, \eta(y,\, x)\rangle(\text{ as } n \to \infty),$$

we have

$$\text{Max}\langle T(y),\, \eta(y,\, x)\rangle \subset S(x),$$

i.e, $\text{Max}\langle T(y),\, \eta(y,\, x)\rangle \subset Y\setminus(-intC(x))$. This implies that $x \in G(y)$, i.e, $G(y)$ is closed for each $y \in K$.

(4): Now we prove that

$$\bigcap_{y\in K} G(y) \neq \emptyset.$$

In fact, by condition (v) for each finite set $N \subset K$, there exists a nonempty compact convex subset $L_N \subset K$ such that $N \subset L_N$ and for each $x \in L_N \setminus D$ there exists $y \in L_N$ such that $x \notin G(y)$. Hence we have

$$L_N \cap \{G(y) :\, y \in L_N\} \subset D.$$

By Theorem 2.2.5 (precisely, see Theorem 1 of Park [422, p.27]), it follows that

$$\bigcap_{y\in K} G(y) \neq \emptyset. \tag{5.5.1}$$

(5): Finally we prove that

$$\bigcap_{y\in K} F(y) = \bigcap_{y\in K} G(y). \tag{5.5.2}$$

Since $F(y) \subset G(y)$ for all $y \in K$, we know that

$$\bigcap_{y\in K} F(y) \subset \bigcap_{y\in K} G(y).$$

To prove the conclusion, it is sufficient to prove

$$\bigcap_{y\in K} G(y) \subset \bigcap_{y\in K} F(y).$$

In fact, if $x \in \bigcap_{y\in K} G(y)$, then for each $y \in K$, $x \in G(y)$, i.e., we have

$$\text{Max}\langle T(y), \eta(y, x)\rangle \subset Y \setminus (-intC(x)) \text{ for all } y \in K.$$

Since K is convex, for any $\alpha \in (0, 1)$, $y_\alpha = \alpha y + (1-\alpha)x \in K$. Hence we have

$$\begin{aligned}
&\text{Max}\langle T(y_\alpha), \eta(y_\alpha, x)\rangle \\
&\quad = \text{Max}\langle T(y_\alpha), \eta(\alpha y + (1-\alpha)x, x\rangle \\
&\quad = \text{Max}\langle T(y_\alpha), \alpha\eta(y, x) + (1-\alpha)\eta(x, x)\rangle \text{ (by condition (iv))} \\
&\quad = \text{Max}\langle T(y_\alpha), \alpha\eta(y, x)\rangle \text{ (since } \eta(x, x) = 0) \\
&\quad \subset Y \setminus (-intC(x)).
\end{aligned}$$

This implies that for each $\alpha \in (0, 1)$,

$$\text{Max}\langle T(y_\alpha), \alpha\eta(y, x)\rangle \cap (-intC(x)) = \emptyset.$$

Dividing α, we also have

$$\text{Max}\langle T(y_\alpha), \eta(y, x)\rangle \cap (-intC(x)) = \emptyset.$$

Letting $\alpha \to 0^+$ and using the V-hemicontinuity of T we have

$$\begin{aligned}
\lim_{\alpha\to 0^+} \text{Max}\langle T(y_\alpha), \eta(y, x)\rangle &= \text{Max}\langle T(x), \eta(y, x)\rangle \\
&\subset Y \setminus (-intC(x)),
\end{aligned}$$

i.e, $x \in F(y)$ for all $y \in K$. Therefore $x \in \bigcap_{y\in K} F(y)$, and so that $\cap_{y\in K} F(y) = \cap_{y\in K} G(y)$ is true.

Combining the cases of (5.5.1) and (5.5.2) above, we know that

$$\bigcap_{y\in K} F(y) = \bigcap_{y\in K} G(y) \neq \emptyset.$$

Taking $x_* \in \bigcap_{y \in K} F(y)$, we have

$$\text{Max}\langle T(x_*),\, \eta(y,\, x_*)\rangle \subset Y \setminus (-intC(x_*)) \text{ for all } y \in K.$$

This completes the proof of Theorem 5.5.6.

Theorem 5.5.7. In Theorem 5.5.6, if condition (v) is replaced by the following condition (v)′:

(v) ′ there exist a nonempty compact subset D of K and a $y_0 \in D$ such that

$$\text{Max}\langle T(y_0),\, \eta(y_0,\, x)\rangle \bigcap (-intC(x)) \neq \emptyset \text{ for all } x \in K \setminus D,$$

then the conclusion of Theorem 5.5.6 remains true.

Proof. It suffices to show that (v)′ implies (v) in Theorem 5.5.6. In fact, for any finite subset $N \subset K$, let

$$L_N = co(\{y_0\} \cup N \cup D) \subset K.$$

By condition (v)′ for any $x \in L_N \setminus D \subset K \setminus D$ there exists $y_0 \in L_N$ such that

$$\text{Max}\langle T(y_0),\, \eta(y_0,\, x)\rangle \cap (-intC(x)) \neq \emptyset.$$

Hence condition (v) holds.

Remark 5.5.8. (1) Yang [582, Theorem 2] is a special case of Theorem 2.2 with T being a single-valued mapping, $C(x) \equiv C$ for all $x \in K$, where C is a closed convex cone in Y with $intC \neq \emptyset$ and $\eta(y,\, x) = y - x$.

(2) Theorem 5.5.6 extends and improves the main result of Siddiqi et al. [479, Theorem 2.1].

(3) Chen and Yang [84, Theorem 2.1] is also a special case of Theorem 5.5.6 with $C(x) \equiv C$, where C is the same as in Remark 5.5.5, K being a nonempty compact convex subset of X and $\eta(y,\, x) = y - x$.

(4) Lee et al. [339, Theorem 2.1] is a special case of Theorem 5.5.6 with X being a reflexive Banach space, $C(x) \equiv C$ for all $x \in K$, where C is the same as above and $\eta(y,\, x) = y - x$.

(5) Cottle and Yao [103, Theorem 2.1] is a special case of Theorem 5.5.7 with $Y = X^*$ (the dual space of X), K being a nonempty compact convex subset of X, $T : X \to X^*$ a single-valued mapping and $\eta(y,\, x) = y - x$.

(6) Theorem 5.5.6 also generalizes the corresponding results in Isac [252] and Noor [394].

Finally, we give an existence theorem for solutions of the generalized vector-valued variational-like inequalities (GVVLI). We first have the following result.

Theorem 5.5.9. Let X be a Banach space, $K \subset X$ a nonempty closed convex subset, $T : K \to 2^{X^*}$ a nonempty compact valued mapping and $\eta : K \times K \to X$ a continuous affine mapping with $\eta(x, x) = \theta$ for all $x \in K$. Assume that the following conditions are satisfied:

(i) for any $x, y \in X$, $u \in T(x)$, $v \in T(y)$, $\langle u, \eta(y, x)\rangle \geq 0$ implies $\langle v, \eta(y, x)\rangle \geq 0$;

(ii) T is hemi-continuous, i.e, for any $x, y \in X$, $\alpha > 0$, we have

$$\lim_{\alpha \to 0}\langle T(x + \alpha y), z\rangle = \langle T(x), z\rangle$$

for each $z \in X$, where $\langle T(w), z\rangle = \{\langle u, z\rangle : u \in T(w)\}$; and

(iii) there exists a nonempty compact subset $D \subset X$ such that for each finite subset N of K, there exists a nonempty compact convex subset L_N of K such that $N \subset L_N$ and for each $x \in L_N \setminus D$ there exists $y \in L_N$ such that

$$\sup_{v \in T(y)} \langle v, \eta(y, x)\rangle < 0.$$

Then there exists $x_* \in K$ such that

$$\sup_{u \in T(x_*)} \langle u, \eta(y, x_*)\rangle \geq 0$$

for all $y \in K$, where $\langle \cdot, \cdot\rangle$ denotes the pairing of X and X^*.

In addition, if $T(x_*)$ is a convex set in X^*, then there exists a point $u_* \in T(x_*)$ such that $\langle u_*, \eta(y, x_*)\rangle \geq 0$ for all $y \in K$.

Proof. Taking $Y = \mathbb{R}$, $C(x) \equiv \mathbb{R}^+$ for all $x \in X$, and $L(X, Y) = X^*$ in Theorem 5.5.6, then the conclusion of Theorem 5.5.9 follows immediately from Theorem 5.5.6.

Finally, if $T(x_*)$ is a convex set in X^*, we define a function $\varphi : K \times T(x_*) \to \mathbb{R}$ by

$$\varphi(y, u) = \langle u, \eta(y, x_*)\rangle$$

for each $(y, u) \in K \times T(x_*)$. Then for each $y \in K$, $u \mapsto \varphi(y, u)$ is linear and continuous and for each $u \in T(x^*)$, $y \mapsto \varphi(y, u)$ is affine. By Kneser minimax theorem above we have

$$\sup_{u \in T(x_*)} \min_{y \in K}\langle u, \eta(y, x_*)\rangle = \inf_{y \in K} \sup_{u \in T(x_*)} \langle u, \eta(y, x_*)\rangle \geq 0.$$

Since $T(x_*)$ is a compact set and $u \mapsto \inf_{y \in K}\langle u, \eta(y, x_*)\rangle$ is a continuous function on $T(x_*)$, there exists $u_* \in T(x_*)$ such that

$$\inf_{y \in K}\langle u_*, \eta(y, x_*)\rangle = \sup_{u \in T(x_*)} \inf_{y \in K}\langle u, \eta(y, x_*)\rangle \geq 0.$$

Therefore we have

$$\langle u_*, \eta(y, x_*)\rangle \geq 0$$

for all $y \in K$. This completes the proof.

Remark 5.5.10. Theorem 5.5.9 is a scalar-valued variational-like inequality, which generalizes the corresponding results of Cottle and Yao [103], Isac [252] and Noor [394].

5.6 Algorithms of Generalized Quasi-Variational Inclusion

In this section, a kind of generalized quasi-variational inclusion with non-compact set-valued mappings is introduced which includes many variational and quasi-variational inequalities studied by Chang [65], Hartman and Stampacchia [221], Hassouni and Moudafi [228], Huang and Xu [243], Isac [252], Jou and Yao [278], Noor et al. [403], Pitonyak, Shi and Shillor [431], Siddiqi and Ansari [478], Shi [463], Zeng [616] and others as special cases. Using essentially properties of proximal mappings, the equivalence between variational inclusion and fixed points of corresponding mappings is given. This fact allows us to design two algorithms which, in turn are used to prove the existence of solutions for set-valued variational inclusion. Finally convergence theory of two algorithms are also established. Our results unify and improve corresponding algorithms and results for solving variational inequality (resp., inclusion) problems in recent literature.

Recently, the iterative algorithms for finding approximate solutions of new classes of variational inequalities or quasi-variational inequalities have been considered and studied by many people. The purpose of this section is to extend ideas of Hassouni and Moudafi [228] to more general problems. By using essentially properties of proximal mappings, the equivalence between variational inclusion and some kind of subgradient inclusion is established. This equivalence is then used to establish two iterative algorithms for solving generalized quasi-variational inclusion. Furthermore, the convergence theory of algorithms are also established under some suitable assumptions.

Throughout this section, let H be an Hilbert space, whose norm and inner product are denoted by $\|\cdot\|$ and $\langle\cdot,\cdot\rangle$, respectively. We always assume that K is a non-empty closed convex set in H and also we denote by 2^H and $CB(H)$ the families of all non-empty subsets and all non-empty bounded and closed subsets of H, respectively.

For the sake of convenience for our discussion, we first recall and give some definitions and notions. Let $T, V : H \to 2^H$ be two set-valued mappings and $g, A, m : H \to H$ be three given single-valued mappings. Suppose $\psi : H \to \mathbb{R} \cup \{+\infty\}$ is a proper convex lower semicontinuous function and $\partial\psi$ is the subdifferential of ψ. Let us consider the following:

The Generalized Set-valued Quasi-Variational Inclusion (GSQVI):

Find $u \in H$, $x \in T(u)$, $y \in V(u)$ such that

$$\begin{cases} g(u) \in dom\partial\psi + m(u) \\ \langle x - Ay, v + m(u) - g(u)\rangle \geq \psi(g(u) - m(u)) - \psi(v) \end{cases} \qquad (GSQVI)$$

for all $v \in H$. The problem (GSQVI) has potential applications in mechanics, physics, differential equations, and pure and applied sciences.

Some special cases.

(I): If T is a single-valued mapping, $V \equiv I$, the identity mapping and $m(u) \equiv 0$ for all $u \in H$, then the problem (GSQVI) is equivalent to finding $u \in H$, $g(u) \in dom\partial\psi$ such that

$$\langle T(u) - Au, v - g(y)\rangle \geq \psi(g(u)) - \psi(v)$$

for all $v \in H$, which is known as the varitaional inclusion problem which was introduced and studied by Hassouni and Moudafi [228].

(II): If $\psi \equiv \delta_K$, the indicator functions of closed convex set K in H defined by

$$\psi(x) = \begin{cases} 0, & x \in K; \\ +\infty, & x \notin K. \end{cases}$$

Then problem (GSQVI) is equivalent to finding $u \in H$, $x \in T(u)$, $y \in V(u)$, $g(u) \in K(u) := K + m(u)$ for each $u \in H$ such that

$$\langle x - Ay, v + m(u) - g(u)\rangle \geq 0$$

for all $v \in K$, i.e.,

$$\langle x - Ay, v - g(u)\rangle \geq 0 \qquad (5.6.1)$$

for all $v \in K(u)$, which is known as the generalized set-valued strongly nonlinear quasi-variational inequality problem which was studied in a recent paper of Noor et al. [403].

(III): If $A \equiv I$, from (5.6.1) we can obtain the variational inequalities considered by Zeng [616]. Especially, if $m(u) \equiv 0$ for all $u \in H$, from (5.6.1) we can obtain the following variational inequality

$$\langle x - Ay, v - g(u)\rangle \geq 0$$

for all $v \in K$ which is studied by Zeng [616].

(IV): If T is single-valued, and $V \equiv I$, the identity mapping, then the problem (5.6.1) is equivalent to finding $u \in H$ such that $g(u) \in K(u)$ and

$$\langle Tu - Au, v - g(u) \rangle \geq 0$$

for all $v \in K(u)$, which is known as the generalized strongly nonlinear quasi-variational inequality introduced and studied by Jou and Yao [278], Siddiqi and Ansari [476] and Huang and Hu [243].

(V): If for each $u \in H$, $K(u)$ is a closed convex cone in H and $K^*(u)$ is the polar cone of $K(u)$ defined by

$$K^*(u) = \{w \in H : \langle w, v \rangle \geq 0 \text{ for all } v \in K(u)\}$$

then Problem (5.6.1) is equivalent to finding $u \in H$, $x \in T(u)$, $y \in V(u)$ and $g(u) \in K(u)$ such that $x - Ay \in K^*(u)$ and

$$\langle x - Ay, g(u) \rangle = 0.$$

The problem above appears to be a new one and we shall call it '*the generalized set-valued strongly nonlinear quasi-complementarity problem*'.

(VI): If $A \equiv 0$, $g \equiv I$, the identity mapping, $m(u) \equiv 0$ and T is a single-valued mapping, then problem (5.6.1) collapses to finding $u \in K$ such that

$$\langle Tu, v - u \rangle \geq 0$$

for all $v \in K$, which is known as the classical variational inequality problem which is originally due to Hartman and Stampacchia [221].

For appropriate suitable choice of the operators T, V, A, g and the convex set K, we can obtain various classes of variational inequalities, variational inclusion and complementarity problems given in the literature as special cases of our variational inequality (GSQVI). Thus our generalized set-variational inclusion problem (GSQVI) is the most general and unifying one.

In this section, we first establish the following equivalence between the existence of solutions for variational inclusion problem (GSQVI) and the existence of fixed points for corresponding set-valued mappings, and then two algorithms for solving the problem (GSQVI) are proposed by the motivation of this equivalence.

Lemma 5.6.1. Let $\psi : H \to \mathbb{R} \cup \{+\infty\}$ be a proper convex lower semi-continuous functions and $\rho > 0$. Then we have the following:

(1) The mapping $J_\rho^\psi := (I + \rho\partial\psi)^{-1} : H \to H$ is nonexpansive, i.e.,

$$\|J_\rho^\psi(u) - J_\rho^\psi(v)\| \leq \|u - v\|$$

for all $u, v \in H$, where I is the identity mapping on H.

(2) Especially, if $\psi = \delta_K$, where K is a closed convex subset of H and δ_K is the indicator function of K, then $J_\rho^\psi(u) = P_K(u)$ for all $\rho > 0$ and $u \in H$, where P_K is the projective operator of H onto K.

Proof. The conclusion (1) can be found from Minty [372]. Now we only prove the conclusion (2) (see also Rockafellar [444]).

In fact, for any given $u \in H$, denoting $x := J_\rho^\psi(u)$, then we have $u \in x + \rho\partial\psi(x)$. By the definition of $\partial\psi$, we have

$$u - x \in \rho\partial\psi(x) = \rho\{p \in H : \langle p, y - x\rangle \leq 0 \text{ for all } y \in K\}$$

and so $\langle \rho^{-1}(u - x), y - x\rangle \leq 0$ for all $y \in K$, $\rho > 0$, and thus

$$\langle x - u, y - x\rangle \geq 0$$

for all $y \in K$. Hence we have $x = P_K(u)$, and it follows that

$$P_K(u) = J_\rho^\psi(u)$$

for all $\rho > 0$ and $u \in H$. This complete the proof of Lemma 5.6.1. □

Theorem 5.6.2. The problem (GSQVI) has a solution if and only if for some $\rho > 0$, the mapping $F : H \to H$ define by

$$\begin{aligned} F(u) &= \cup_{x\in T(u)} \cup_{y\in V(u)} [u - g(u) + m(u) \\ &+ J_\rho^\psi(g(u) - \rho(x - Ay) - m(u))] \end{aligned}$$

for each $u \in H$, has a fixed point in H.

Proof. If (u^*, x^*, y^*) is a solution of (GSQVI), hence $u^* \in H$, $x^* \in T(u^*)$, $y^* \in V(u^*)$ such that $g(u^*) \in dom\partial\psi + m(u^*)$ and

$$\langle x^* - Ay^*, v + m(u^*) - g(u^*)\rangle \geq \psi(g(u^*) - m(u^*)) - \psi(v)$$

for all $v \in H$. By the definition of $\partial\psi$ and the preceding inequality above, we have

$$Ay^* - x^* \in \partial\psi(g(u^*) - m(u^*)).$$

Hence for any given $\rho > 0$, we have

$$g(u^*) - m(u^*) - \rho(x^* - Ay^*) \in g(u^*) - m(u^*) + \rho\partial\psi(g(u^*) - m(u^*)).$$

It follows from the definition of J_ρ^ψ that

$$g(u^*) - m(u^*) = J_\rho^\psi(g(u^*) - m(u^*) - \rho(x^* - Ay^*)),$$

and so,

$$\begin{aligned} u^* &= u^* - g(u^*) + m(u^*) + J_\rho^\psi(g(u^*) - \rho(x^* - Ay^*) - m(u^*)) \\ &\in \cup_{x\in T(u^*)} \cup_{y\in V(u^*)} [u^* - g(u^*) + m(u^*) + J_\rho^\psi(g(u^*) - \rho(x - Ay) - m(u^*))] \\ &= F(u^*). \end{aligned}$$

This implies that u^* is a fixed point of F in H.

Conversely, if $u^* \in H$ is a fixed point of F, i.e., $u^* \in F(u^*)$. By the definition of F, there exists $x^* \in T(u^*)$ and $y^* \in V(u^*)$ such that

$$u^* = u^* - g(u^*) + m(u^*) + J_\rho^\psi(g(u^*) - \rho(x^* - Ay^*) - m(u^*)),$$

i.e.,

$$g(u^*) - m(u^*) = J_\rho^\psi(g(u^*) - m(u^*) - \rho(x^* - Ay^*)).$$

Therefore we have

$$g(u^*) - m(u^*) - \rho(x^* - Ay^*) \in g(u^*) - m(u^*) + \rho\partial\psi(g(u^*) - m(u^*)).$$

Because of $\rho > 0$, we have

$$Ay^* - x^* \in \partial\psi(g(u^*) - m(u^*)).$$

This implies that $g(u^*) \in dom\partial\psi(u^*) + m(u^*)$ and by the definition of $\partial\psi$, we have

$$\langle x^* - Ay^*, v + m(u^*) - g(u^*)\rangle \geq \psi(g(u^*) - m(u^*)) - \psi(v)$$

for all $v \in H$. Thus (u^*, x^*, y^*) is a solution of the problem (GSQVI) and the proof of Theorem 5.6.2 is completed. □

Next we consider the iterative algorithms of solutions for the variational inclusion problem (GSQVI).

Algorithms 5.6.3. Let $T, V : H \to CB(H)$ be set-valued mappings and $m, g, A : H \to H$ be single-valued mappings such that for each $u \in H$, $G(u) \subset g(H)$, where $G : H \to 2^H$ is a set-valued mapping defined by

$$G(u) = \cup_{x\in T(u)} \cup_{y\in V(u)} [m(u) + J_\rho^\psi(g(u) - \rho(x - Ay) - m(u))]$$

for each $u \in H$. Now for each given $u_0 \in H$, take $x_0 \in T(u_0)$, $y_0 \in V(u_0)$. Letting

$$w_0 = m(u_0) + J_\rho^\psi(g(u_0) - \rho(x_0 - Ay_0) - m(u_0)) \in G(u_0) \subset g(H),$$

then there exists $u_1 \in H$ such that $w_0 = g(u_1)$. Since $x_0 \in T(u_0) \in BC(H)$, $y_0 \in V(x_0) \in BC(H)$, it follows (e.g., see Nalder [378]) that for each given $\epsilon > 0$, there exists$x_1 \in T(u_1)$ and $y_1 \in V(u_1)$ such that

$$\|x_0 - x_1\| \leq (1 + \epsilon)\hat{H}(T(u_0), T(u_1))$$

and

$$\|y_0 - y_1\| \leq (1+\epsilon)\hat{H}(V(u_0), V(u_1)).$$

where $\hat{H}$ is the Hausdorff metric on $CB(H)$. Letting

$$w_1 = m(u_1) + J_\rho^\psi(g(u_1) - \rho(x_1 - Ay_1) - m(u_1)) \in G(u_1) \subset g(H),$$

there exists $u_2 \in H$ such that $w_1 = g(u_2)$. Inductively, we can obtain three sequences $\{u_n\}$, $\{x_n\}$ and $\{y_n\}$ such that for each $n = 1, 2\cdots$,

$$x_n \in T(u_n), \quad \|x_n - x_{n+1}\| \leq (1 + \frac{\epsilon}{2^n})\hat{H}(T(u_n), T(u_{n+1}))$$

$$y_n \in V(u_n), \quad \|y_n - y_{n+1}\| \leq (1 + \frac{\epsilon}{2^n})\hat{H}(V(u_n), V(u_{n+1}))$$

and

$$g(u_{n+1}) = m(u_n) + J_\rho^\psi(g(u_n) - \rho(x_n - Ay_n) - m(u_n))$$

where $\rho > 0$ is a constant.

Remark 5.6.4. Algorithm 3.1 given by Siddiqi and Ansari [478] is a special case of Algorithm 5.6.3 with $\psi = \delta_K$, $V \equiv I$ and T being a single-valued mapping.

Algorithm 5.6.5. Let $T, V : H \to 2^H$ be set-valued mappings and $m, g, A : H \to H$ be single-valued mappings. For any given $u_0 \in H$, we define iterative sequences $\{x_n\}$, $\{y_n\}$ and $\{u_n\} \subset H$ as follows:

$$x_n \in T(u_n), \qquad y_n \in V(u_n)$$

and

$$u_{n+1} = u_n - g(u_n) + m(u_n) + J_\rho^{\psi_n}(g(u_n) - \rho(x_n - Ay_n) - m(u_n)) + e_n$$

for each $n \in \mathbb{N}$, where e_n is an error and $\{\psi_n\}$ is the sequence approximating ψ and $\rho > 0$ is a constant.

Remark 5.6.6. If $m \equiv 0$, T is a single-valued mapping and $V \equiv I$, then Algorithm 5.6.5 is reduced to Algorithm 2.1 of Hassouni and Moudafi [228].

In this section, we establish the convergence theory for iterative algorithms for solving variational inclusion problem (GSQVI) proposed above for Lipschitz continuous set-valued mappings T and V. Our results unify and improve related results and algorithms for solving the variational inequality (resp., inclusion) problem (GSQVI) given by recent literature.

We first need the following defintions and notions.

Definition 5.6.7. A mapping $F : H \to H$ is said to be

(1) λ-Lipschitz continuous if

$$\|F(u) - F(v)\| \leq \lambda \|u - v\|$$

for all $u, v \in H$, where $\lambda > 0$ is a constant; and

(2) μ-strongly monotone if

$$\langle F(u) - F(v), u - v \rangle \geq \mu \|u - v\|^2$$

for all $u, v \in H$ and $\mu > 0$ is a constant.

Definition 5.6.8. Let $g : H \to H$ be a single-valued mapping and $T : H \to CB(H)$ be a set-valued mapping. Then T is said to be

(1) α-$\hat{H}$-Lipschitz continuous with respect to g if

$$\hat{H}(T(u), T(v)) \leq \alpha \|g(u) - g(v)\|$$

for all $u, v \in H$, where $\alpha > 0$ is a constant, and $\hat{H}$ is the Hausdorff metric on $CB(H)$; and

(2) β-strongly monotone with respect to g if

$$\langle x - y, g(u) - g(v) \rangle \geq \beta \|g(u) - g(v)\|^2$$

for all $u, v \in H$, $x \in T(u)$ and $y \in T(v)$, where $\beta > 0$ is a constant.

Remark 5.6.9. We would like to note a simple fact that the Lipschitz continuity of a set-valued mapping, in general, does not imply the single-valued property of that mapping. On the other hand, a set-valued mapping with both Lipschitz continuity and strongly monotone property may reduce back to the single-valued mapping subject to the choice of coefficients α and β in the definition above. In order to give the general results, in what follows we state two convergence results (i.e., Theorems 5.6.10 and 5.6.12) in the form of set-valued mappings (though they may reduce back to single-valued cases due to certain values for the coefficients α or β).

Theorem 5.6.10. Let $g : H \to H$ be a single-valued mapping and $T : H \to CB(H)$ an α-strongly monotone and β-$\hat{H}$-Lipschitz continuous mapping with respect to g. Suppose $V : H \to CB(H)$ is a λ-$\hat{H}$-Lipschitz continuous mapping with respect to g; $m : H \to H$ is a μ-Lipschitz continuous mapping with respect to g and $A : H \to H$ is a γ-Lipschitz continuous mapping with $0 < \gamma \leq 1$. Suppose further that $g(H)$ is closed and there exists a constant $\rho > 0$ such that

$$\begin{cases} \lambda(1 - 2\mu) + 2\sqrt{(\beta^2 - \lambda^2)\mu(1 - \mu)} < \alpha, \\ \rho < \lambda + 2\mu < 1, \text{ and} \\ \lambda < \alpha \leq \beta \end{cases} \tag{5.6.2}$$

and

$$\begin{cases} |\rho - 7\dfrac{\alpha - \lambda(1-2\mu)}{\beta^2 - \lambda^2}| \\ < \dfrac{\sqrt{[\alpha - \lambda(1-2\mu)^2 - 4(\beta^2 - \lambda^2)\mu(1-\mu)]}}{\beta^2}. \end{cases} \tag{5.6.3}$$

Then the sequences $\{g(u_n)\}$, $\{x_n\}$ and $\{y_n\}$ generated by Algorithm (5.6.5) strongly converge to $g(u^*)$, x^* and y^*, respectively and (u^*, x^*, y^*) is a solution of the generalized set-valued quasi-variational inclusion problem (GSQVI).

Proof. From the Lemma 5.6.1 and Algorithm 5.6.3, we have that

$$\begin{aligned} \|g(u_{n+1}) - g(u_n)\| &\le \|m(u_n) - m(u_{n-1})\| + \|g(u_n) - g(u_{n-1}) \\ &\quad - \rho(x_n - x_{n-1}) + \rho(Ay_n - Ay_{n-1}) - (m(u_n) - m(u_{n-1})\| \\ &\le 2\|m(u_n) - m(u_{n-1})\| + \|g(u_n) - g(u_{n-1}) - \rho(x_n - x_{n-1})\| \\ &\quad + \rho\|Ay_n - Ay_{n-1}\| \\ &\le 2\|m(u_n) - m(u_{n-1})\| + \|g(u_n) - g(u_{n-1}) - \rho(x_n - x_{n-1})\| \\ &\quad + \rho\gamma\|y_n - y_{n-1}\| \text{ (as } A \text{ is } \gamma\text{-Lipschitz with } \gamma \le 1) \\ &\le 2\|m(u_n) - m(u_{n-1})\| + \|g(u_n) - g(u_{n-1}) - \rho(x_n - x_{n-1})\| \\ &\quad + \rho(1 + \frac{\epsilon}{2^{n-1}})\hat{H}(V(u_n), V(u_{n-1})). \end{aligned} \tag{5.6.4}$$

Since V is λ-$\hat{H}$-Lipschitz continuous with respect to g, we know that

$$\hat{H}(V(u_n), V(u_{n-1})) \le \lambda\|g(u_n) - g(u_{n-1})\|. \tag{5.6.5}$$

Again since T is α-strongly monotone and β-$\hat{H}$- Lipschitz continuous with respect to g, we have

$$\begin{aligned} \|g(u_n) - g(u_{n-1}) - \rho(x_n - x_{n_1})\|^2 &= \|g(u_n) - g(u_{n-1})\|^2 \\ &\quad -2\rho\langle x_n - x_{n-1}, g(u_n) - g(u_{n-1})\rangle + \rho^2\|x_n - x_{n-1}\|^2 \\ &\le \|g(u_n) - g(u_{n-1})\|^2 - 2\rho\alpha\|g(u_n) - g(u_{n-1})\|^2 \\ &\quad +\rho^2[(1 + \frac{\epsilon}{2^{n-1}})\hat{H}(T(u_n), T(u_{n-1}))]^2 \\ &\le (1 - 2\rho\alpha + \rho^2\beta^2(1 + \frac{\epsilon}{2^{n-1}})^2)\|g(u_n) - g(u_{n-1}\|^2. \end{aligned} \tag{5.6.6}$$

Since m is μ-Lipschitz continuous with respect to g, from inequalities above, we have

$$\|g(u_{n+1}) - g(u_n)\| \le \theta_n\|g(u_n) - g(u_{n-1})\| \tag{5.6.7}$$

where

$$\theta_n = 2\mu + \sqrt{1 - 2\rho\alpha + \rho^2\beta^2(1 + \frac{\epsilon}{2^{n-1}})^2} + \rho(1 + \frac{\epsilon}{2^{n-1}})\lambda.$$

Denoted by $\theta := \lim_{n\to\infty}\theta_n$, it is easy to see that

$$\theta = 2\mu + \sqrt{1 - 2\rho\alpha + \rho^2\beta^2} + \rho\lambda.$$

By conditions (5.6.2) and (5.6.3), we know that $\theta < 1$. Note that $\theta_n \searrow \theta$ (i.e, θ_n converges to θ from the right side), there exists n_0 such that for any $n \geq n_0$, $\theta_n < 1$. Therefore it follows from (5.6.7) that

$$\|g(u_{n+1}) - g(u_n)\| \leq \theta_{n_0}\|g(u_n) - g(u_{n-1})\| \tag{5.6.8}$$

for all $n \geq n_0$ This implies that $\{g(u_n)\}$ is a Cauchy sequence in $g(H)$. As $g(H)$ is closed, there exists an $u^* \in H$ such that $g(u_n) \to g(u^*)$.

Furthermore, by using Algorithm 5.6.3, the β-$\hat{H}$-Lipschitz continuity of T with respect to g and the λ-$\hat{H}$-Lipschitz continuity of V with respect to g, we have

$$\begin{aligned} \|x_n - x_{n+1}\| &\leq (1 + \frac{\epsilon}{2^n})\hat{H}(T(u_n), T(u_{n+1})) \\ &\leq (+\frac{\epsilon}{2^n})\beta\|g(u_n) - g(u_{n+1})\| \end{aligned}$$

$$\begin{aligned} \|y_n - y_{n+1}\| &\leq (1 + \frac{\epsilon}{2^n})\hat{H}(V(u_n), V(u_{n+1})) \\ &\leq (+\frac{\epsilon}{2^n})\lambda\|g(u_n) - g(u_{n+1})\|. \end{aligned}$$

Therefore $\{x_n\}$ and $\{y_n\}$ are both Cauchy sequences in H. Let $x_n \to x^*$ and $y_n \to y^*$. Denote

$$w^* := m(u^*) + J_\rho^\psi(g(u^*) - \rho(x^* - Ay^*) - m(u^*)).$$

Since m is μ-Lipschitz continuous with respect to g, we have

$$\begin{aligned} \|g(u_{n+1}) &- w^*\| \leq 2\|m(u_n) - m(u^*)\| + \|g(u_n) - g(u^*)\| \\ &+ \rho\|x_n - x^*\| + \rho\|Ay_n - Ay^*\| \\ &\leq (2\mu + 1)\|g(u_n) - g(u^*)\| \\ &+ \rho\{\|x_n - x^*\| + \|Ay_n - Ay^*\|\} \\ &\to 0 \end{aligned}$$

as $n \to \infty$ and thus

$$g(u^*) = w^* = m(u^*) + J_\rho^\psi(g(u^*) - \rho(x^* - Ay^*) - m(u^*)). \tag{5.6.9}$$

Now it is easy to see that $g(u^*) \in dom\partial\psi(u^*) + m(u^*)$. On the other hand, we also have

$$\begin{aligned} d(x^*, T(u^*)) &\leq \|x^* - x_n\| + d(x_n, T(u^*)) \\ &\leq \|x^* - x_n\| + \hat{H}(T(u_n), T(u^*)) \\ &\leq \|x^* - x_n\| + \beta\|g(u_n) - g(u^*)\| \to 0. \end{aligned}$$

Since $T(u^*) \in CB(H)$, $x^* \in T(u^*)$. Similarly, we can also prove that $y^* \in V(u^*)$. Therefore from (5.6.9) we have

$$u^* \in F(u^*) = \cup_{x\in T(u^*)}\cup_{y\in V(u^*)}[u^*-g(u^*)+m(u^*)+J_\rho^\psi(g(u^*)-\rho(x-Ay)-m(u^*))].$$

By Theorem 5.6.2, we know that (u^*, x^*, y^*) is a solution of the problem (GSQVI) and thus the proof of Theorem 5.6.10 is complete. □

Next we consider the convergence of the iterative sequences $\{u_n\}$, $\{x_n\}$ and $\{y_n\}$ generated by Algorithm 5.6.5 and the existence of solutions for the problem (GSQVI).

Definition 5.6.11. Let $T : H \to 2^H$ be a set-valued mapping. Then T is said to be

(1) α-strongly monotone if

$$\langle x - y, u - v\rangle \geq \alpha\|u - v\|^2$$

for all $u, v \in H$, $x \in T(u)$ and $y \in T(v)$, where α is a positive constant; and

(2) β-Lipschitz continuous if

$$\delta(T(u), T(v)) \leq \beta\|u - v\|$$

for all $u, v \in H$, where β is a positive constant and $\delta(A, B) := \sup\{\|a - b\| : a \in A, b \in B\}$ for each $A, B \in 2^H$.

By Remark 5.6.9, the different choice of coefficients β and or γ for set-valued mappings in Theorem 5.6.12 below, may result in the single-valued properties of set-valued mappings, nevertheless we still give its general version in the form of set-valued mappings.

Theorem 5.6.12. Let $T : H \to 2^H$ be an α-strongly monotone and β-Lipschitz continuous set-valued mapping; the mapping $V : H \to 2^H$ be a γ-Lipschitz continuous mapping and the mapping $A : H \to H$ be η-Lipschitz continuous with $\eta \leq 1$ and $m : H \to H$ be μ-Lipschitz continuous. Suppose $g : H \to H$ is a λ-strongly monotone and σ-Lipschitz continuous mapping. If there exists a constant $\rho > 0$ such that

$$\begin{cases} k := 2(\mu + \sqrt{1 - 2\lambda + \sigma^2} < 1 \\ \gamma(1 - k)+\sqrt{(\beta^2 - \gamma^2)k(2 - k)} < \alpha \end{cases} \tag{5.6.10}$$

and

$$|\rho - \frac{\alpha + \gamma(k - 1)}{\beta^2 - \gamma^2}| < \frac{\sqrt{[\alpha + \gamma(k - 1)]^2 - (\beta^2 - \gamma^2)k(2 - k)}}{\beta^2 - \gamma^2} \tag{5.6.11}$$

then the generalized set-valued quasi-varitaional inclusion problem (GSQVI) has a solution (u^*, x^*, y^*).

In addition, if for each $v \in H$,

$$\lim_{n\to\infty} \|J^{\psi}_{\rho_n}(v) - J^{\psi}_{\rho}(v)\| = 0 \text{ and } \lim_{n\to\infty} \|e_n\| = 0,$$

then the iterative sequences $\{u_n\}$, $\{x_n\}$ and $\{y_n\}$ generalized by Algorithms 5.6.5 converge strongly to u^*, x^* and y^*, respectively.

Proof. (I): First we prove that problem (GSQVI) has a solution (u^*, x^*, y^*). For the purpose, by Theorem 5.6.2, it is sufficient to prove that the mapping F defined in Theorem 5.6.2 has a fixed point $u^* \in H$. For any $u, v \in H$, $a \in F(u)$ and $b \in F(v)$, there exists $x_1 \in T(u)$, $y_1 \in V(u)$, $x_2 \in T(v)$ and $y_2 \in V(v)$ such that

$$a = u - g(u) + m(u) + J^{\psi}_{\rho}(g(u) - \rho(x_1 - Ay_1) - m(u)),$$

and

$$b = v - g(v) + m(v) + J^{\psi}_{\rho}(g(v) - \rho(x_2 - Ay_2) - m(v)).$$

By Lemma 5.6.1 (1), we have

$$\begin{aligned}\|a &- b\| \le \|u - v - (g(u) - g(v))\| + \|m(u) - m(v)\| \\ &+ \|J^{\psi}_{\rho}(g(u) - \rho(x_1 - Ay_1) - m(u)) \\ &- J^{\psi}_{\rho}(g(v) - \rho(x_2 - Ay_2) - m(u))\| \\ &\le 2\|u - v - (g(u) - g(v))\| + 2\|m(u) - m(v)\| \\ &+ \|u - v - \rho(x_1 - x_2)\| + \rho\|Ay_1 - Ay_2\|.\end{aligned}$$

By the assumption, A is η-Lipschitz continuous with $\eta \le 1$, from the preceding expression we have

$$\begin{aligned}\|a - b\| \le& 2\|u - v - (g(u) - g(v))\| + 2\|m(u) - m(v))\| \\ &+\|u - v - \rho(x_1 - x_2)\| + \rho\|y_1 - y_2\| \\ &\le 2\|u - v - (g(u) - g(v))\| + 2\|m(u) - m(v))\| \\ &+\|u - v - \rho(x_1 - x_2)\| + \rho\delta(V(u), V(v)),\end{aligned} \tag{5.6.12}$$

where $\delta((V(u), V(v)) = \sup\{\|x - y\| : x \in V(u), y \in V(v)\}$. Since g is λ-strongly monotone and σ-Lipschitz continuous, we have

$$\begin{aligned}\|u - v - (g(u) - g(v))\|^2 =& \|u - v\|^2 - 2\langle u - v, g(u) - g(v)\rangle \\ &+\|g(u) - g(v)\|^2 \\ &\le(1 - 2\lambda + \delta^2)\|u - v\|^2.\end{aligned} \tag{5.6.13}$$

Similarly, we also have

$$\|u - v - \rho(x_1 - x_2)\|^2 \le (1 - 2\rho\alpha + \rho^2\beta^2)\|u - v\|^2. \tag{5.6.14}$$

Again, since A and m are γ- and μ-Lipschitz continuous, respectively, we have

$$\delta(V(u), V(v)) \leq \gamma\|u - v\|, \tag{5.6.15}$$

and

$$\|m(u) - m(v) \leq \mu\|u - v\|. \tag{5.6.16}$$

Combining inequalities (5.6.12)-(5.6.16) above, we have

$$\delta(F(u), F(v)) \leq \theta\|u - v\| \tag{5.6.17}$$

where

$$\theta = 2[\mu + \sqrt{1 - 2\lambda + \sigma^2}] + \sqrt{1 - 2\rho\alpha + \rho^2\beta^2} + \rho\gamma.$$

By assumptions (5.6.10) and (5.6.11) of Theorem 5.6.12, it is easy to prove that $\theta < 1$. By Theorem 3.1 of Siddiqi and Ansari [476], F has a fixed point $u^* \in H$. Hence by Theorem 5.6.2, there exists $x^* \in T(u^*)$, and $y^* \in V(u^*)$ such that (u^*, x^*, y^*) is a solution of the generalized set-valued variational inclusion problem (GSQVI).

(II) Finally, we consider the convergence of the iterative sequences $\{u_n\}$, $\{x_n\}$ and $\{y_n\}$ generated by the Algorithm 5.6.5.

Since (u^*, x^*, y^*) is a solution of (GSQVI), we have

$$u^* = u^* - g(u^*) + m(u^*) + J_\rho^\psi(g(u^*) - \rho(x^* - Ay^*) - m(u^*)).$$

Denoting

$$h(u^*) := g(u^*) - \rho(x^* - Ay^*) - m(u^*)$$

and

$$h(u_n) := g(u_n) - \rho(x_n - Ay_n) - m(u_n);$$

then we have

$$\begin{aligned}\|u_{n+1} - u^*\| \leq \|u_n - u^* - (g(u_n) - g(u^*))\| + \|m(u_n) - m(u^*)\| \\ + \|J_\rho^{\psi_n}(h(u_n)) - J_\rho^\psi(h(u^*))\| + \|e_n\|.\end{aligned} \tag{5.6.18}$$

By Lemma 5.6.1, we have

$$\begin{aligned}\|J_\rho^{\psi_n}(h(u_n)) - J_\rho^\psi(h(u^*))\| \leq \|J_\rho^{\psi_n}(h(u_n)) \\ -J_\rho^{\psi_n}(h(u^*))\| + \|J_\rho^{\psi_n}(h(u^*) - J_\rho^\psi(h(u^*))\| \\ \leq \|h(u_n) - h(u^*)\| + \|J_\rho^{\psi_n}(h(u^*) - J_\rho^\psi(h(u^*))\| \\ \leq \|u_n - u^* - (g(u_n) - g(u^*))\| + \|m(u_n) - m(u^*)\| \\ +\|u_n - u^* - \rho(x_n - x^*)\| + \rho\|Ay_n - Ay^*\| + \|c_n\|\end{aligned} \tag{5.6.19}$$

where

$$c_n = \|J_\rho^{\psi_n}(h(u^*)) - J_\rho^\psi(h^*))\|.$$

By substituting (5.6.19) into (5.6.18) and noting the η-Lipschitz continuity of A with $\eta \leq 1$, we have

$$\begin{aligned}\|u_{n+1} &- u^*\| \leq 2\|u_n - u^* - (g(u_n) - g(u^*))\| \\ &+ 2\|m(u_n) - m(u^*) + \|u_n - u^* - \rho(x_n - x^*)\| \\ &+ \rho\|y_n - y^*\| + c_n + \|e_n\|.\end{aligned}$$

By the same argument as stated in the first section, we can also prove that

$$\|u_n - u^* - (g(u_n) - g(u^*))\|^2 \leq (1 - 2\lambda + \sigma^2)\|u_n - u^*\|,$$

$$\|u_n - u^* - \rho(x_n - x^*)\| \leq (1 - 2\rho\alpha + \rho^2\beta^2)\|u_n - u^*\|,$$

$$\|y_n - y^*\| \leq \delta(V(u_n), V(u^*)) \leq \gamma\|u_n - u^*\|,$$

and

$$\|m(u_n) - m(u^*)\| \leq \mu\|u_n - u^*\|.$$

Therefore we have

$$\|u_{n+1} - u^*\| \leq \theta\|u_n - u^*\| + c_n + \|e_n\|,$$

where

$$\theta = 2(\mu + \sqrt{1 - 2\lambda + \delta^2}) + \sqrt{1 - 2\rho\alpha + \rho^2\beta^2} + \rho\gamma < 1.$$

Denoting

$$\epsilon_n = c_n + \|e_n\| = \|J_\rho^{\psi_n}(h(u^*)) - J_\rho^{\psi}(h(u^*))\| + \|e_n\|,$$

then we have

$$\|u_{n+1} - u^*\| \leq \theta\|u_n - u^*\| + \epsilon_n,$$

and so we have

$$\|u_{n+1} - u^*\| \leq \theta^{n+1}\|u_0 - u^*\| + \Sigma_{j=0}^n \theta^j \epsilon_{n-j}.$$

By the assumption of Theorem, $\lim_{n\to\infty} \epsilon_n = 0$. Hence the sequence $\{u_n\}$ strongly converges to u^* (e.g., see, Ortega and Rheinboldt [407, p.338]). Again since $x_n \in T(x_n)$, $x^* \in T(u^*)$ and T is β-Lipschitz continuous, we have

$$\|x_n - x^*\| \leq \delta(T(u_n), T(u^*)) \leq \beta\|u_n - u^*\| \to 0.$$

This implies that $\lim_{n\to\infty} x_n = x^*$. Similarly we can prove that $\lim_{n\to\infty} y_n = y^*$. Thus the proof of Theorem 5.6.12 is completed. □

Remark 5.6.13. The aim of this section was to design theoretic algorithms and then to establish their convergence theory for solving solutions of set-valued variational inclusion problem (GSQVI) (they might reduce back to single-valued cases in some situaions) under reasonable assumptions. So far, we do not give any numerical experiments by using these algorithms for the study of nonlinear mathematical programming and or complementarity problems.

In this section, we studied the general existence of solutions for variational inequalities by using results from generalized games that were established in Chapter 4. However, there are many other kinds of variational inequalities which have not been included here. For more information in those directions, interested readers are referred to some ideas proposed by Harker and Pang [220], Isac [252] and references therein.

Chapter 6

Variational Inequalities - Part II

6.1 Introduction

This chapter is a continuation of Chapter 5. As an application of one-person game established in Chapter 4, we first establish several non-compact generalized quasi-variational inequalities in locally convex topological spaces in Section 2 and then the classical results of $(GQVI)$ in finite dimensional spaces have been improved and generalized in several aspects.

In Section 3, it is our purpose to establish some existence results for $(GQVI)$ without upper semicontinuity as a direct application of fixed point theorems. As special cases, a number of classical results in finite dimensional spaces have been extended and unified.

In Section 4, as an application of the Ky Fan minimax inequality in its non-compact version, we study a general existence theorem of solution for variational inequalities in locally convex topological vector spaces. This result improves and/or unifies the corresponding results in the literature as special cases.

In Section 5, as applications of the celebrated Ky Fan minimax principle again, we first establish a general existence theorem for generalized quasi-variational inequalities in topological vector spaces with sufficiently many continuous linear functionals. This result includes a number of existence theorems for generalized quasi-variational inequalities as special cases. As applications, several existence theorems of quasi-variational inequalities for set-valued mappings, which have either upper semicontinuous or monotone properties are derived, and one of these existence results is applied to give a new non-compact generalization of the Fan-Glicksberg fixed point theorem for set-valued mappings in topological vector spaces. Also we provide a simple example which shows that the lower semicontinuity assumption is essential for the guarantee of existence of solutions for generalized quasi-variational inequalities. Our results unify a number of existence theorems for generalized variational inequalities which have been given in the literature of the last decade.

In Section 6, by using the concept called *quasi-monotone operator* for set-valued mappings, which is a generalization of monotone mapping for set-valued mapping, the general quasi-variational inequalities have been established. Then we obtain the corresponding non-compact $(GQVI)$ results when underlying spaces are reflexive Banach spaces in both Sections 7 and 8.

In Section 9, as applications of variational inequalities established in Banach spaces and locally convex topological vector spaces, fixed point theorems for non-expansive set-valued mappings in Hilbert spaces are established. Moreover the general existence of solutions for minimization problems have been given as applications of generalized quasi-variational inequalities.

Finally in Sections 10 and 11, by establishing implicitly quasi-variational inequalities, the general complementarity problems have been investigated.

6.2 Non-Compact Generalized Quasi-Variational Inequalities

In this section, as applications of existence of equilibria for generalized games, we give a new approach to prove several existence theorems of solutions for non-compact generalized quasi-variational inequalities in both locally convex spaces and finite dimensional spaces. These results generalize or improve most of Chan and Pang [64], Harker and Pang [220], Kim [298], Shih and Tan [466], Tarafdar [531], Tian and Zhou [556].

In 1985, Shih and Tan [466] were the first to study the $GQVI(X; A; B)$ in infinite dimensional locally convex Hausdorff topological vector spaces. One of their main results [466, Theorem 4] was later generalized in [474, Theorem 11], which is stated as a special case as follows (see also Kim [298]).

Theorem 6.2.1. Let E be a locally convex Hausdorff topological vector space, E^* be the dual space of E and X be a non-empty compact convex subset of E. Let $A : X \to 2^X$ be continuous such that for each $x \in X$, $A(x)$ is a non-empty closed convex subset of X, and $B : X \to 2^{E^*}$ be upper semicontinuous from the relative topology of X to the strong topology of E^* such that for each $x \in X$, $B(x)$ is a non-empty strongly compact subset of E^*. Then there exists a point $\hat{y} \in X$ such that

$$\begin{cases} \hat{y} \in A(\hat{y}) \text{ and} \\ \inf_{x \in A(\hat{y})} \sup_{w \in B(\hat{y})} Re\langle w, \hat{y} - x \rangle \leq 0. \end{cases}$$

Since then, there have been a number of generalizations of the existence theorems about $GQVI(X; A; B)$. Most existence theorems mentioned above, however, are obtained on compact sets in finite dimensional spaces or infinite dimensional locally convex Hausdorff topological vector spaces, and both A and B are either continuous or upper (lower) semicontinuous. On the other hand, in economic and game applications, it is known that the choice space (or the space of feasible allocations) generally is not compact in any topology of the choice space (even though it is closed and bounded), a key situation in infinite dimensional topological vector spaces. Moreover, we note that there are essentially no existence theorems of solutions of generalized quasi-variational inequalities on non-compact sets in infinite dimensional spaces. This motivates our work to give a series of existence theorems on generalized quasi-variational inequalities by relaxing the compactness conditions and continuity. In this section, as applications of existence theorems of non-compact generalized games, we shall give some non-compact existence theorems of $GQVI(X; A; B)$; i.e., the domain X is a nonempty convex subset of infinite (or finite) dimensional locally convex Hausdorff topological vector space E instead of compact convex; secondly, the mappings A may not be continuous and B may not have the continuity property.

For the convenience of our study, we first state the following result which is Lemma 4.10.2.

Lemma 6.2.2. Let S be a topological space and $\phi : S \to 2^{\mathbb{R}^n}$ be a mapping with convex values. Suppose ϕ is lower semicontinuous at some $x_0 \in S$ and $B(y_0, t) \subset \phi(x_0)$, where $B(y_0, t)$ is the open ball of positive radius t centered at y_0. Then for any number r with $0 < r < t$, there exists a neighborhood U_0 of x_0 such that $B(y_0, r) \subset \phi(x)$ for all $x \in U_0$.

Lemma 6.2.3. Let X be a non-empty subset of a topological vector space E. Suppose that the mapping $B : X \to 2^{E^*}$ has non-empty compact values and the set $\{x \in X : \inf_{u \in B(x)} Re\langle u, w \rangle > 0\}$ is open in X for each $w \in X - X$. Then the mapping $P : X \to 2^X$ defined by $P(x) = \{y \in X : \inf_{u \in B(x)} Re\langle u, x - y \rangle > 0\}$ for each $x \in X$ is lower semicontinuous with open convex values. In particular, if E is a finite dimensional space, the graph of P is open.

Proof. For each $x \in X$, since $B(x)$ is non-empty compact it follows that the function $\psi : E \to \mathbb{R} \cup \{-\infty, +\infty\}$ defined by $\psi(y) = \inf_{u \in B(x)} Re\langle u, x - y \rangle$

for each $y \in E$ is continuous. Therefore $P(x)$ is open. Now we wish to prove that P is lower semicontinuous. For each non-empty open subset U in X, suppose $y \in P(x) \cap U$, this means $\inf_{u \in B(x)}\langle u, x-y\rangle > 0$. Let $N(x) = \{x' \in X : \inf_{u \in B(x')}\langle u, x-y\rangle > 0\}$. Then $N(x)$ is a non-empty open neighborhood of x in X by the hypothesis and $V(x) := N(x) \cap [(x-y)+U]$ is a non-empty open neighborhood of x. For any $x' \in V(x)$, there exists $y' \in U$ is such that $y' = x' - (x-y)$, and $\inf_{u \in B(x')}\langle u, x'-y'\rangle = \inf_{u \in B(x')}\langle u, x-y\rangle > 0$. Hence for each $x' \in V(x)$, there exists $y' \in U$ such that $y' \in P(x')$; that is, $P(x') \cap U \neq \emptyset$ for all $x' \in V(x)$. Thus P is lower semicontinuous with open values. In particular, if E is a finite dimensional space $\mathbb{R}^n$, where $n \in \mathbb{N}$, $P(x)$ is lower semicontinuous with open and convex values. So that the graph of P is open in $X \times \mathbb{R}^n$ by Lemma 6.2.2. □

Remark 6.2.4. In Lemma 6.2.3, if the topological vector space E is a Frechet space (complete metrizable space) and the mapping $B : X \to 2^{E^*}$ has non-empty bounded values. Then for each fixed $x \in X$, the function $\psi_x : E \to \mathbb{R} \cup \{-\infty, +\infty\}$ defined by $\psi_x(y) = \inf_{u \in B(x)} Re\langle u, x-y\rangle$ for each $y \in E$ is concave and upper semicontinuous. Note that the domain of ψ_x is the set $D := \{y \in E : -\infty < \psi_x(y) < +\infty\} = E$, by Proposition 4 of Aubin [10, p.93], the function ψ is continuous. Following the proof of Lemma 6.2.3, it is clear that the conclusions of Lemma 6.2.3 still hold if the underlying space E is a Frechet space and provided that the mapping B has non-empty bounded values instead of having non-empty compact values.

In order to give the existence of solutions for non-compact generalized quasi-variational inequalities in locally convex topological vector spaces, we need the following two lemmas which are special cases of Theorem 4.5.2 and Theorem 4.5.6 of Section 5 in Chapter 4 (see also Theorem 3.3 and Theorem 5.4 of Tan and Yuan in [512] and [513], respectively) and hence their proofs are omitted here.

Lemma 6.2.5. Let X be a non-empty paracompact convex subset of a locally convex topological vector space E and $A, P : X \to 2^X$ such that

(i) A is lower semicontinuous with closed graph, and $A(x)$ is non-empty convex for each $x \in X$;
(ii) $A \cap P$ is L-majorized;
(iii) the set $D = \{x \in X : (A \cap P)(x) \neq \emptyset\}$ is open in X; and
(iv) there exists a non-empty compact convex subset X_0 of X and a non-empty compact subset K of X such that for each $y \in X \setminus K$, with $co(X_0 \cup \{y\}) \cap co(A(y) \cap P(y)) \neq \emptyset$.

Then there exists a point $\hat{x} \in K$ is such that $\hat{x} \in A(x)$ and $A(\hat{x}) \cap P(\hat{x}) = \emptyset$.

Lemma 6.2.6. Let X be a non-empty paracompact closed convex subset of a locally convex topological vector space E such that X has the property K.

Suppose that $A, P : X \to 2^X$ satisfy that

(i) $A : X \to 2^X$ is compact and upper semicontinuous with non-empty compact convex values;

(ii) $P : X \to 2^X$ is lower semicontinuous and L-majorized;

(iii) $D := \{x \in X : (A \cap P)(x) \neq \emptyset\}$ is open in X; and

(iv) there exists a non-empty compact convex subset X_0 of X and a non-empty compact subset K of X such that for each $x \in X \setminus K$ there is an $y \in co(X_0 \cup \{x\})$ with $y \in co(A(x) \cap P(x))$.

Then there exists $\bar{x} \in K$ such that $\bar{x} \in A(\bar{x})$ and $A(\bar{x}) \cap P(\bar{x}) = \emptyset$.

As applications of Lemma 6.2.2 to Lemma 6.2.6, we shall now give several existence theorems for non-compact generalized quasi-variational inequalities in locally convex spaces and finite dimensional spaces. We first have:

Theorem 6.2.7. Let X be a non-empty papacompact convex subset of a locally convex topology vector space E. Suppose that there exist a non-empty compact subset K and a non-empty compact convex subset X_0 of X such that:

(i) the mapping $A : X \to 2^X$ is lower semicontinuous with closed graph, and $A(x)$ is non-empty convex for each $x \in X$;

(ii) the mapping $B : X \to 2^{E^*} \setminus \{\emptyset\}$ has compact values such that the set $\{x \in X : \inf_{u \in B(x)} Re\langle u, x - y\rangle > 0\}$ is open in X for each fixed $y \in X$, and $B(x)$ is convex for $x \in K$ with $x \in A(x)$;

(iii) the set $\{x \in X \sup_{y \in A(x)} \inf_{u \in B(x)} Re\langle u, x - y\rangle > 0\}$ is open in X; and

(iv) for each $x \in X \setminus K$, there exists $y \in A(x) \cap co(X_0 \cup \{x\})$ with $\inf_{u \in B(x)} \langle u, x - y\rangle > 0$.

Then there exist $\hat{x} \in K$ with $\hat{x} \in A(\hat{x})$ and $\hat{u} \in B(\hat{x})$ such that $\sup_{y \in A(\hat{x})} Re\langle \hat{u}, \hat{x} - y\rangle \leq 0$.

Proof. Define a mapping $P : X \to 2^X$ by $P(x) = \{y \in X : \inf_{u \in B(x)} Re\langle u, x - y\rangle > 0\}$ for each $x \in X$. Then $x \notin P(x)$ and $P(x)$ is open and convex for each $x \in X$, since for each fixed $x \in X$, $y \mapsto \inf_{u \in B(x)} Re\langle u, x - y\rangle$ is continuous. For each $y \in X$, $P^{-1}(y) = \{x \in X : \inf_{u \in B(x)} Re\langle u, x - y\rangle > 0\}$ is open in X by (ii). Therefore P is of class L, so that the mapping $A \cap P$ is L-majorized. Note that $D = \{x \in X : A(x) \cap P(x) \neq \emptyset\} = \{x \in X : \sup_{y \in A(x)} \inf_{u \in B(x)} Re\langle u, x - y\rangle > 0\}$ is open by (iii). Therefore all conditions of Lemma 6.2.5 are satisfied. By Lemma 6.2.5, there exists $\hat{x} \in K$ such that $\hat{x} \in A(\hat{x})$ and $A(\hat{x}) \cap P(\hat{x}) = \emptyset$. By the definition of P, $\sup_{y \in A(\hat{x})} \inf_{u \in B(\hat{x})} Re\langle u, \hat{x} - y\rangle \leq 0$. Note that $B(\hat{x})$ is a non-empty compact and convex subset by hypothesis (ii), using Kneser minimax Theorem (e.g., see Theorem 5.3.9 above), there exists $\hat{u} \in B(\hat{x})$ such that

$$\sup_{y \in A(\hat{x})} Re\langle \hat{z}, \hat{x} - y\rangle = \inf_{u \in B(\hat{x})} \sup_{y \in A(\hat{x})} Re\langle u, \hat{x} - y\rangle = \sup_{y \in A(\hat{x})} \inf_{u \in B(\hat{x})} Re\langle z, \hat{x} - y\rangle \leq 0$$

which completes the proof. □

In Theorem 6.2.7, if X is a convex subset of a finite dimensional space $\mathbb{R}^n$, Lemma 6.2.3 shows that condition (iii) of Theorem 6.2. is satisfied automatically. Therefore we have the following existence theorem for non-compact generalized quasi-variational inequalities in finite dimensional spaces.

Theorem 6.2.8. Let X be a non-empty convex subset of a finite dimensional space $\mathbb{R}^n$, where $n \in \mathbb{N}$. Suppose that there exists a non-empty compact subset K and a non-empty compact convex subset X_0 of X such that:

(i) the mapping $A : X \to 2^X$ is lower semicontinuous with closed graph, and $A(x)$ is non-empty convex for each $x \in X$;

(ii) the mapping $B : X \to 2^{\mathbb{R}^n} \setminus \{\emptyset\}$ has compact values and $B(x)$ is compact convex for $x \in K$ with $x \in A(x)$;

(iii) for each $y \in X - X$, the set $\{x \in X : \inf_{u \in B(x)} Re\langle u, y\rangle \leq 0\}$ is closed; and

(iv) for each $x \in X \setminus K$, there exists $y \in A(x) \cap co(X_0 \cup \{x\})$ with $\inf_{u \in B(x)}\langle u, x - y\rangle > 0$.

Then there exist $\hat{x} \in K$ with $\hat{x} \in A(\hat{x})$ and $\hat{u} \in B(\hat{x})$ such that $\sup_{y \in A(\hat{x})} Re\langle \hat{u}, \hat{x} - y\rangle \leq 0$.

Proof. Define a mapping $P : X \to 2^X$ by $P(x) = \{y \in X : \inf_{u \in B(x)} Re\langle u, x - y\rangle > 0\}$ for each $x \in X$. Then by Lemma 6.2.3, $P(x)$ has an open graph in $X \times \mathbb{R}^n$. Therefore $P^{-1}(y) = \{x \in X : y \in P(x)\} = \{x \in X : \inf_{u \in B(x)} Re\langle u, x - y\rangle > 0\}$ is open in X. Moreover, since A is lower semicontinuous so that the intersect mapping $A \cap P : X \to 2^{\mathbb{R}^n}$ defined by $(A \cap P)(x) = A(x) \cap P(x)$ for each $x \in X$ is lower semicontinuous (e.g., see Theorem 4.10.4 above). Thus the set $\{x \in X : A(x) \cap P(x) \neq \emptyset\} = \{x \in X : (A \cap P)(x) \cap X \neq \emptyset\}$ which is open in X by the definition of lower semicontinuity. Note that the set $D := \{x \in X : A(x) \cap P(x) \neq \emptyset\} = \{x \in X : \sup_{y \in A(x)} \inf_{u \in B(x)} Re\langle u, x - y\rangle > 0\}$. Therefore D is open in X. Thus all hypotheses of Theorem 6.2.7 are satisfied. By Theorem 6.2.7, the conclusion follows. □

In Theorems 6.2.7 and 6.2.8, if the topological vector space E is a Frechet space, and condition (ii) is replaced by condition (ii)′: *the mapping* $B : X \to 2^{E^*} \setminus \{\emptyset\}$ *has bounded values and* $B(x)$ *is compact and convex for each* $x \in K$ *with* $x \in A(x)$, then the conclusions of Theorems 6.2.7 and 6.2.9 hold by the Remark 6.2.4 after Lemma 6.2.3.

Note that in Theorem 6.2.8, our conditions (ii) and (iii) are exactly conditions (ii) and (i) of Theorem 1 of Cubiotti [107, p.82]. Here the domain set X does not need to be closed and our condition (i) is slightly weaker than condition (iii) of Theorem 1 of Cubiotti [107, p.82], but our non-compact condition (iv) seems to be stronger than the hypothesis (iv) of Theorem 1 in [107, p.82]. Moreover, both arguments are different.

In Theorem 6.2.8 for given mapping $B : X \to 2^{E^*} \setminus \{\emptyset\}$, we only need condition that the set $\{x \in X : \inf_{u\in B(x)} Re\langle u, w\rangle > 0\}$ is open for each $w \in X - X$. This condition was first introduced by Ricceri [436], and may be satisfied without any upper (or lower) semicontinuity for the mapping B. Therefore Theorem 6.2.8 refers to the existence of solutions for non-compact generalized quasi-variational inequalities without continuity in infinite (or finite) dimensional spaces.

In what follows, we shall consider the existence of solutions for generalized quasi-variational inequalities under condition that the mapping B is upper semicontinuous.

Theorem 6.2.9. Let X be a non-empty paracompact and bounded convex subset of a locally convex topological vector space E. Suppose that the following condition are satisfied:

(i) the mapping $A : X \to 2^X \setminus \{\emptyset\}$ is lower semicontinuous with a closed graph;

(ii) the mapping $B : X \to 2^{E^*} \setminus \{\emptyset\}$ is upper semicontinuous with non-empty (strong) compact values, and $B(x)$ is compact convex for each $x \in X$ with $x \in A(x)$; and

(iii) there exists a non-empty compact convex subset X_0 of X and a non-empty compact subset K of X such that for each $x \in X \setminus K$, one has

$$\sup_{y\in A(x)\cap co(X_0\cup\{x\})} \inf_{u\in B(x)} Re\langle u, x-y\rangle > 0.$$

Then there exists $\hat{x} \in K$ with $\hat{x} \in A(\hat{x})$ and $\hat{u} \in B(\hat{x})$ such that $\sup_{y\in A(\hat{x})} Re\langle \hat{u}, \hat{x} - y\rangle \le 0$.

Proof. Since X is a bounded subset of a locally convex topological vector space E, and E^* is equipped with the strongly topology, the function $\psi_1 : E^* \times X \times X \to \mathbb{R} \cup \{-\infty, +\infty\}$, defined by $\psi_1(u, x, y) = Re\langle u, x-y\rangle$ for each $(u, x, y) \in E^* \times X \times X$ is continuous. Since B is upper semicontinuous with non-empty strongly compact values, by Theorem 1 of Aubin [10, p.67], the function $\psi_2 : X\times X \to \mathbb{R}\cup\{-\infty, +\infty\}$ defined by $\psi_2(x, y) = \inf_{u\in B(x)} Re\langle u, x-y\rangle$ is lower semicontinuous. Thus for each fixed $y \in X$, the set $\{x \in X : \inf_{u\in B(x)} Re\langle u, x - y\rangle > 0\}$ is open in X. As the mapping $A : X \to 2^X\setminus\{\emptyset\}$ is lower semicontinuous, by proposition III-19 of Aubin and Ekeland [15, p.118] the function $\psi_3 : X \to \mathbb{R} \cup \{-\infty, +\infty\}$ defined by $\psi_3(x) = \sup_{y\in A(x)} \inf_{u\in B(x)} Re\langle u, x - y\rangle$ is lower semicontinuous, so that $\{x \in X : \sup_{y\in A(x)} \inf_{u\in B(x)} Re\langle u, x - y\rangle > 0\}$ is open in X.

Define a mapping $P : X \to 2^X$ by $P(x) = \{y \in X : \inf_{u\in B(x)} Re\langle u, x - y\rangle > 0\}$ for each $x \in X$. Then $P(x)$ is convex and $x \notin P(x)$ for each $x \in X$. Note that $P^{-1}(y) = \{x \in X : \inf_{u\in B(x)} Re\langle u, x - y\rangle > 0\}$ is open in X for each $y \in X$. Therefore P is lower semicontinuous so that P is of class L. Let $D =$

$\{x \in X : A(x) \cap P(x) \neq \emptyset\}$. Note that $D = \{x \in X : \inf_{u \in B(x)} Re\langle u, x-y\rangle > 0\}$ is open in X. Therefore A and P satisfy all conditions of Theorem 6.2.7. By Theorem 6.2.7, there exists $\hat{x} \in K$ with $\hat{x} \in A(\hat{x})$ and $\hat{u} \in B(\hat{x})$ such that $\sup_{y \in A(\hat{x})} Re\langle \hat{u}, \hat{x} - y\rangle \leq 0$. □

Theorem 6.2.9 is a non-compact generalization of Theorem due to Kim [298], which in turn improves Theorem 4 of Shih and Tan [466, p.341] and Theorem 2 in [474, p.69-70] (with $M = 0$).

As consequences of Theorems 6.2.8 and 6.2.9, we have the following:

Corollary 6.2.10. Let X be a non-empty bounded closed and convex subset of a finite dimensional space $\mathbb{R}^n$, where $n \in \mathbb{N}$. Suppose that:

(i) the mapping $A : X \to 2^X \setminus \{\emptyset\}$ is continuous with closed values;

(ii) the mapping $B : X \to 2^{\mathbb{R}^n} \setminus \{\emptyset\}$ has compact values and $B(x)$ is compact and convex for $x \in X$ with $x \in A(x)$; and

(iii) for each $y \in X - X$, the set $\{x \in X : \inf_{u \in B(x)} Re\langle u, y\rangle \leq 0\}$ is closed.

Then there exist $\hat{x} \in X$ with $\hat{x} \in A(\hat{x})$ and $\hat{u} \in B(\hat{x})$ such that $\sup_{y \in A(\hat{x})} Re\langle \hat{u}, \hat{x} - y\rangle \leq 0$.

As noted above, Corollary 6.2.10 is true if condition (ii) is replaced by condition (ii)′: the mapping B has non-empty bounded values and $B(x)$ is compact convex for each $x \in X$ with $x \in A(x)$.

Corollary 6.2.11. Let X be a non-empty compact and convex subset of a locally convex topological vector space E. Suppose that the following condition are satisfied:

(i) the mapping $A : X \to 2^X \setminus \{\emptyset\}$ is continuous with closed values;

(ii) the mapping $B : X \to 2^{E^*} \setminus \{\emptyset\}$ is upper semicontinuous with non-empty (strong) compact values, and $B(x)$ is compact convex for each $x \in X$ with $x \in A(x)$.

Then there exist $\hat{x} \in K$ with $\hat{x} \in A(\hat{x})$ and $\hat{u} \in B(\hat{x})$ such that $\sup_{y \in A(\hat{x})} Re\langle \hat{u}, \hat{x} - y\rangle \leq 0$.

Finally, we shall give the existence theorems of solutions for non-compact generalized quasi-variational inequalities in which the mapping A is upper semicontinuous instead of being lower semicontinuous.

Theorem 6.2.12. Let X be a non-empty paracompact closed convex subset of a locally convex topological vector space E such that X has the property K. Suppose that:

(i) the mapping $A : X \to 2^X \setminus \{\emptyset\}$ is compact and upper semicontinuous with non-empty compact convex values;

(ii) the mapping $B : X \to 2^{E^*} \setminus \{\emptyset\}$ is such that $\{x \in X : \inf_{u \in B(x)} Re\langle u, x - y\rangle > 0\}$ is open in X, and $B(x)$ is compact convex for each $x \in K$ with $x \in A(x)$;

(iii) the set $\{x \in X : \sup_{y\in A(x)} \inf_{u\in B(x)} Re\langle u, x-y\rangle > 0\}$ is open in X; and

(iv) there exists a non-empty compact convex subset X_0 of X and a non-empty compact subset K of X such that for each $x \in X \setminus K$, one has

$$\sup_{y\in A(x)\cap co(X_0\cup\{x\})} \inf_{u\in B(x)} Re\langle u, x-y\rangle > 0.$$

Then there exist $\hat{x} \in K$ with $\hat{x} \in A(\hat{x})$ and $\hat{u} \in B(\hat{x})$ such that $\sup_{y\in A(\hat{x})} Re\langle \hat{u}, \hat{x}-y\rangle \le 0$.

Proof. Define a mapping $P : X \to 2^X$ by $P(x) = \{y \in X : \inf_{u\in B(x)} Re\langle u, x-y\rangle > 0\}$ for each $x \in X$. Then $P(x)$ is convex and $x \notin P(x)$ for each $x \in X$. Note that $P^{-1}(y) = \{x \in X : \inf_{u\in B(x)} Re\langle u, x-y\rangle > 0\}$ is open in X by (ii) for each $y \in X$. Therefore P is of class L (in fact P is lower semicontinuous). Let $D = \{x \in X : A(x) \cap P(x) \neq \emptyset\}$. Since $D = \{x \in X : \sup_{y\in A(x)} \inf_{u\in B(x)} Re\langle u, x-y\rangle > 0\}$ is open in X by (iii), so that A and P satisfy all conditions of Lemma 6.2.6. By Lemma 6.2.6, there exists $\hat{x} \in K$ such that $\hat{x} \in A(\hat{x})$ and $A(\hat{x}) \cap P(\hat{x}) = \emptyset$, i.e., $\hat{x} \in A(\hat{x})$ and $\sup_{y\in A(\hat{x})} \inf_{u\in B(\hat{x})} Re\langle u, \hat{x}-y\rangle \le 0$. Since $B(\hat{x})$ is non-empty compact and convex by (ii), employing Kneser minimax theorem, there exists $\hat{u} \in B(\hat{x})$ such that $\sup_{y\in A(\hat{x})} Re\langle \hat{z}, \hat{x}-y\rangle = \inf_{u\in B(\hat{x})} \sup_{y\in A(\hat{x})} Re\langle u, \hat{x}-y\rangle = \sup_{y\in A(\hat{x})} \inf_{u\in B(\hat{x})} Re\langle u, \hat{x}-y\rangle \le 0$. □

We shall now observe that in Theorem 6.2.12, the interaction between the mappings A and B (namely, condition (iii)) can be achieved by imposing additional continuity on the mappings A and B.

Theorem 6.2.13. Let X be a non-empty paracompact closed convex subset of a locally convex topological vector space E such that X has the property K. Suppose that:

(i) the mapping $A : X \to 2^X$ is compact and upper semicontinuous with non-empty compact convex values;

(ii) the mapping $B : X \to 2^{E^*} \setminus \{\emptyset\}$ is such that for each $y \in X$, $x \mapsto \inf_{u\in B(x)} Re\langle u, x-y\rangle$ is lower semicontinuous, and $B(x)$ is compact convex with $x \in A(x)$;

(iii) the set $\{x \in X : \sup_{y\in A(x)} \inf_{u\in B(x)} Re\langle u, x-y\rangle > 0\}$ is open in X; and

(iv) there exists a non-empty compact convex subset X_0 of X and a non-empty compact subset K of X such that for each $x \in X \setminus K$, one has

$$\sup_{y\in A(x)\cap co(X_0\cup\{x\})} \inf_{u\in B(x)} Re\langle u, x-y\rangle > 0.$$

Then there exist $\hat{x} \in K$ with $\hat{x} \in A(\hat{x})$ and $\hat{u} \in B(\hat{x})$ such that $\sup_{y\in A(\hat{x})} Re\langle \hat{u}, \hat{x}-y\rangle \le 0$.

Proof. By (ii), for each $y \in E$, the mapping $x \mapsto \inf_{u\in B(x)} Re\langle u, x-y\rangle$ is lower semicontinuous, so that the set $\{x \in X : \inf_{u\in B(x)} Re\langle u, x-y\rangle > 0\}$ is open in X for each fixed $y \in E$. Therefore all hypotheses of Theorem 6.2.12 are satisfied. By Theorem 6.2.12, there exists $\hat{x} \in K$ with $\hat{x} \in A(\hat{x})$ and $\hat{u} \in B(\hat{x})$ such that $\sup_{y \in A(\hat{x})} Re\langle \hat{u}, \hat{x} - y\rangle \leq 0$. □

We note that when X is bounded and the mapping B is upper semicontinuous in Theorem 6.2.13, condition (ii) of Theorem 6.2.13 is satisfied automatically. Thus Theorem 6.2.13 is a non-compact generalization of Theorem 3 of Shih and Tan [466, p.340]. Also if X is a compact subset, then Theorem 7 of Tian and Zhou [556] is an immediately consequence of Theorem 6.2.13.

We also remark that Corollary 6.2.10 improves Theorem 6.1 of Harker and Pang [220] (see also Chan and Pang [64] in the following ways: (i) the set X need not compact and (ii) the mapping B does not need to be upper semicontinuous.

6.3 Generalized Quasi-Variational Inequalities without Upper Semicontinuity

It is our purpose in this section to establish the existence of generalzied quasi-variational inequalities in finte dimensional spaces without upper semicontinuity for set-valued mappings. As applications, the classical quasi-variational inequalities have been unified and generalized.

We now recall some notions and definitions. Given a nonempty closed convex subset X of $\mathbb{R}^n$ and two set-valued mappings $\Gamma : X \to 2^X$ and $\Phi : X \to 2^{\mathbb{R}^n}$, the *generalized quasi-variational inequality* problem associated with X, Γ and Φ (in short, $GQVI(X, \Gamma, \Phi)$) is to find $(\hat{x}, \hat{z}) \in X \times \mathbb{R}^n$ such that

$$\hat{x} \in \Gamma(\hat{x}), \quad \hat{z} \in \Phi(\hat{x}) \quad \text{and} \quad \langle \hat{z}, \hat{x} - y\rangle \leq 0 \quad \text{for all} \quad y \in \Gamma(\hat{x}),$$

where $\langle \cdot, \cdot \rangle$ is the usual scalar product of $\mathbb{R}^n$ which induces the Euclidean norm $\|\cdot\|$. Such a problem, which was introduced by Chan and Pang in [64], extends at one time the *generalized variational inequality* problem ($\Gamma(x) \equiv X$) and the *quasi-variational inequality* problem (where Φ is single-valued), both being generalizations of the classical variational inequality problem. The reader is referred to [220] for an excellent and detailed treatment of the basic facts in both theory and applications of finite dimensional variational inequalities and for a very detailed reference. For other applications of the problem $GQV(X, \Gamma, \Phi)$, we refer, for instance to Cubiotti [108].

In the paper [64], Chan and Pang also established their classical existence result [64, Corollary 3.1], which remains undoubtedly one of the basic facts in the theory of variational inequalities. For comparsion, we state Chan and Pang's result as the following (e.g., see also Theorem 5.1.1).

Theorem 6.3.1. Let $X \subseteq \mathbb{R}^n$ be nonempty, compact and convex, $\Gamma : X \to 2^X$ both lower and upper semicontinuous with nonempty closed convex values, $\Phi : X \to 2^{\mathbb{R}^n}$ upper semicontinuous with nonempty compact convex values.

Then $GQVI(X, \Gamma, \Phi)$ admits a solution.

Recently, a general existence result for the problem $GQVI(X, \Gamma, \Phi)$ was established by Cubiotti [107], which improves Theorem 6.3.1 above in some directions. In particular, the set-valued mappings Φ was allowed to lie in a larger class than the one of upper semicontinuous set-valued mappings. We now state result as follows.

Theorem 6.3.2 (Cubiotti [107, Theorem 1]). Let $X \subseteq \mathbb{R}^n$ be closed and convex, $\Gamma : X \to 2^X$ and $\Phi : X \to 2^{\mathbb{R}^n}$ two set-valued mappings, $K \subseteq X$ a nonempty compact set. Assume that:

(i) the set $\Phi(x)$ is nonempty and compact for each $x \in X$, and convex for each $x \in X$, with $x \in \Gamma(x)$;

(ii) for each $y \in X - X$, the set $\{x \in X : \inf_{z \in \Phi(x)} \langle z, y \rangle \leq 0\}$ is closed;

(iii) Γ is a lower semicontinuous set-valued mappings with closed graph and nonempty convex values;

(iv) $\Gamma(x) \cap K \neq$ for all $x \in X$; and

(v) for each $x \in X \setminus K$, with $x \in \Gamma(x)$, one has

$$\sup_{y \in \Gamma(x) \cap K} \inf_{z \in \Phi(x)} \langle z, x - y \rangle > 0.$$

Then $GQVI(X, \Gamma, \Phi)$ has at least one solution.

We recall that when the set X is compact, assumption (iii) above means exactly that Γ is both lower and upper semicontinuous with nonempty closed convex values. Moreover, it is easy to realize that if the compact-valued set-valued mappings Φ is upper semicontinuous, then it satisfies assumption (ii) of Theorem Theorem 6.3.2, while the converse is not necessarily true. Therefore, Theorem 6.3.2 strictly contains Theorem 6.3.1. We refer to [592] for the basic properties and a nice characterization of the set-valued mappings satisfying assumptions (i) and (ii) of Theorem 6.3.2.

The aim of this section is to establish the following result (see also Cubiotti and Yuan [111]).

Theorem 6.3.3. Theorem 6.3.2 (hence, Theorem 6.3.1) above remain true if we replace assumption (iii) by the following more general assumption:

(iii)′ Γ is lower semicontinuous with nonempty convex values and the set $\{x \in X : x \in \Gamma(x)\}$ is closed.

Let S and V be two metric spaces, $\Psi : S \to 2^V$ a set-valued mapping. We say that Ψ is *upper semicontinuous* in S if for each closed $\Omega \subseteq V$ the set $\Psi^-(\Omega) = \{s \in S : \Psi(s) \cap \Omega \neq\}$ is closed in S. We say that Ψ is *lower semicontinuous* in S if for each open $\Omega \subseteq V$ the set $\Psi^-(\Omega)$ is open in S. We recall, in particular, that if for each $v \in V$ the set $\Psi^-(\{v\})$ is open in S, then Ψ is lower semicontinuous. In the sequel we shall write $\Psi^-(v)$ instead of $\Psi^-(\{v\})$. The graph of Ψ is the set $\{(s, v) \in S \times V : v \in \Psi(s)\}$. We briefly recall some basic implications (for more details on set-valued mappings, the reader is referred to Aubin and Frankowski [16], Beer [28] or Klein and Thompson in [317]):

(a) if Ψ has closed graph, then each set $\Psi(s)$ is closed;

(b) if Ψ has closed graph and V is compact, then Ψ is upper semicontinuous;

(c) if Ψ is upper semicontinuous with nonempty closed values, then the graph of Ψ is closed; and

(d) if $V = S$ and the graph of Ψ is closed, then the set $\{s \in S : s \in \Psi(s)\}$ is also closed.

If $A \subseteq \mathbb{R}^n$, we denote by $\overline{A}$ the closure of A. Also, we denote by ri(A) the relative interior of A (that is, the interior of A in its affine hull), while span(A) will denote the linear subspace of $\mathbb{R}^n$ spanned by A. We recall that each nonempty convex subset of $\mathbb{R}^n$ has nonempty relative interior.

If $x \in \mathbb{R}^n$ and $r > 0$, we put $B(x, r) = \{v \in \mathbb{R}^n : \|x - v\| < r\}$, $B_r = B(0_{\mathbb{R}^n}, r)$ and $B_r^c = \{v \in \mathbb{R}^n : \|v\| \leq r\}$. The following fact follows at once from Theorem 4.10.6 in Chapter 4 and thus its proof is omitted here.

Lemma 6.3.4. Let D be a metric space and $\Gamma : D \to 2^{\mathbb{R}^n}$ a set-valued mapping, where $r > 0$ is a constant. Assume that:

(i) Γ is lower semicontinuous with convex values; and

(ii) $\Gamma(x) \cap B_r \neq \emptyset$ for all $x \in D$.

Then the set-valued mapping $x \to \Gamma(x) \cap B_r^c$ is lower semicontinuous in D.

Before we prove Theorem 6.3.3, we first have the following result (see also Theorem 3.2 of Cubiotti and Yuan [111]).

Theorem 6.3.5. Let X be a nonempty compact and convex subset of $\mathbb{R}^n$, $\Gamma : X \to 2^X$ and $\Phi : X \to 2^{\mathbb{R}^n}$ two set-valued mappings. Assume that:

(i) the set $\Phi(x)$ is nonempty and compact for each $x \in X$, and convex for each $x \in X$, with $x \in \Gamma(x)$;

(ii) for each $w \in X - X$, the set $\{x \in X : \inf_{z \in \Phi(x)} \langle z, w \rangle \leq 0\}$ is closed;

(iii) the set-valued mapping Γ is lower semicontinuous with nonempty convex values; and

(iv) the set $\{x \in X : x \in \Gamma(x)\}$ is closed.

Then $GQVI(X, \Gamma, \Phi)$ admits a solution.

Proof. Define $F : X \to 2^X$ by

$$F(x) = \{y \in \Gamma(x) : \inf_{z\in\Phi(x)} \langle z, x - y\rangle > 0\}$$

for each $x \in H$. We shall prove that F is lower semicontinuous. We first define a set-valued mapping $F_1 : X \to 2^X$ by $F_1(x) = \{y \in X : \inf_{z\in\Phi(x)}\langle z, x-y\rangle > 0\}$ for each $x \in X$. Then F_1 has convex values and its graph is open by Lemma 4.10.2. Note that Γ is lower semicontinuous and $F(x) = F_1(x) \cap \Gamma(x)$ for each $x \in X$. It is easy to see that F is lower semicontinous by the definition of the lower semicontinuity. Now, observe that by Michael [371, Theorem 3.1‴] there exists a continuous $g : X \to X$ such that $g(x) \in \Gamma(x)$ for all $x \in X$. By Brouwer's classical fixed point theorem, see, for instance, see Aubin [13, Theorem 7.1]), there exists $\overline{x} \in X$ such that $\overline{x} = g(\overline{x}) \in \Gamma(\overline{x})$. Therefore, the set $H := \{x \in X : x \in \Gamma(x)\}$ is nonempty. Consider the set-valued mapping $G : X \to 2^X$ defined by putting

$$G(x) = \begin{cases} F(x), & \text{if } x \in H, \\ \Gamma(x), & \text{if } x \in X \setminus H. \end{cases}$$

We claim that there exists $x \in H$ such that $F(x) =$. Arguing by contradiction, assume that $F(x) \neq$ for all $x \in H$. Thus, the set-valued mapping G has nonempty convex values. Moreover, since H is closed by assumption, F and Γ are lower semicontinuous and $F(x) \subseteq \Gamma(x)$ for all $x \in X$, it is easily seen (see, for instance, [384, Lemma 2.3]) that the set-valued mapping G is lower semicontinuous. Again by [371, Theorem 3.1‴] and Brouwer's fixed point theorem, there exists $x^* \in X$ such that $x^* \in G(x^*)$. Since $G(x) \subseteq \Gamma(x)$ for all $x \in X$, we get $x^* \in H \cap F(x^*)$. In particular, we have $\inf_{z\in\Phi(x^*)}\langle z, x^* - x^*\rangle > 0$, which is absurd. Consequently, there exists $\hat{x} \in H$ such that $F(\hat{x}) = \emptyset$. That is, $\hat{x} \in \Gamma(\hat{x})$ and $\inf_{z\in\Phi(\hat{x})}\langle z, \hat{x} - y\rangle \leq 0$ for all $y \in \Gamma(\hat{x})$. By Kneser minimax inequality, we get

$$\inf_{z\in\Phi(\hat{x})} \sup_{y\in\Gamma(\hat{x})} \langle z, \hat{x} - y\rangle \leq 0.$$

Since the function $z \mapsto \sup_{y\in\Gamma(\hat{x})}\langle z, \hat{x} - y\rangle$ is lower semicontinuous (see [153, p.10]) and the set $\Phi(\hat{x})$ is compact, there exists $\hat{z} \in \Phi(\hat{x})$ such that

$$\sup_{y\in\Gamma(\hat{x})} \langle \hat{z}, \hat{x} - y\rangle = \inf_{z\in\Phi(\hat{x})} \sup_{y\in\Gamma(\hat{x})} \langle z, \hat{x} - y\rangle \leq 0.$$

The proof is complete. □

Now we have the following result.

Theorem 6.3.6. Let $X \subseteq \mathbb{R}^n$ be closed and convex, $\Gamma : X \to 2^X$ and $\Phi : X \to 2^{\mathbb{R}^n}$ two set-valued mappings. Assume that:

(i) the set $\Phi(x)$ is nonempty and compact for each $x \in X$, and convex for each $x \in X$, with $x \in \Gamma(x)$;

(ii) for each $w \in X - X$, the set $\{x \in X : \inf_{z\in\Phi(x)}\langle z, w\rangle \leq 0\}$ is closed; and

(iii) the set-valued mapping Γ is lower semicontinuous with convex values and the set $\{x \in X : x \in \Gamma(x)\}$ is closed.

Moreover, assume that there exists $r > 0$ such that the following conditions hold:

(iv) $X \cap B_r \neq \emptyset$ and $\Gamma(x) \cap B_r \neq \emptyset$ for all $x \in X \cap B_r^c$; and

(v) for each $x \in X$, with $x \in \Gamma(x)$ and $\|x\| = r$, and each $z \in \Phi(x)$, there exists $y \in \Gamma(x)$, with $\|y\| < r$, such that $\langle z, x - y\rangle \geq 0$.

Then there exists $(\hat{x}, \hat{z}) \in X \times \mathbb{R}^n$, with $\|\hat{x}\| \leq r$, which solves $GQVI(X, \Gamma, \Phi)$.

Proof. Put $X_r = X \cap B_r^c$. By Lemma 6.3.4, the set-valued mapping $\Gamma_r : X_r \to 2^{X_r}$ defined by putting $\Gamma_r(x) = \Gamma(x) \cap B_r^c$ is lower semicontinuous in X_r with nonempty convex values. Moreover, the set $\{x \in X_r : x \in \Gamma_r(x)\} = \{x \in X : x \in \Gamma(x)\} \cap B_r^c$ is closed by (iii). Finally, we observe that for each $w \in X_r - X_r$ the set $\{x \in X_r : \inf_{z\in\Phi(x)}\langle z, w\rangle \leq 0\}$ is closed. Therefore, by Theorem 6.3.5, there exists $\hat{x} \in \Gamma(\hat{x})$, with $\|\hat{x}\| \leq r$, and $\hat{z} \in \Phi(\hat{x})$ such that

$$\langle \hat{z}, \hat{x} - y\rangle \leq 0 \quad \text{for all} \quad y \in \Gamma(\hat{x}) \cap B_r^c. \tag{6.3.1}$$

We claim that $(\hat{x}, \hat{z})$ solves $GQVI(X, \Gamma, \Phi)$. To see this, let $v \in \Gamma(\hat{x})$. We distinguish two cases.

(a) $\|\hat{x}\| < r$. In this case, there exists $\lambda \in (0, 1)$ such that $v_\lambda := (1 - \lambda)\hat{x} + \lambda v \in \Gamma(\hat{x}) \cap B_r$. By (6.3.1), we get

$$0 \geq \langle \hat{z}, \hat{x} - v_\lambda\rangle = \lambda\langle \hat{z}, \hat{x} - v\rangle,$$

hence $\langle \hat{z}, \hat{x} - v\rangle \leq 0$, as desired.

(b) $\|\hat{x}\| = r$. By assumption (v) there exists $u \in \Gamma(\hat{x}) \cap B_r$ such that $\langle \hat{z}, \hat{x} - u\rangle \geq 0$. By (6.3.1), we get

$$\langle \hat{z}, \hat{x} - u\rangle = 0. \tag{6.3.2}$$

Let $b \in (0, 1)$ be such that $x_b := (1 - b)u + bv \in \Gamma(\hat{x}) \cap B_r$. By both (6.3.1) and (6.3.2) we get

$$0 \geq \langle \hat{z}, \hat{x} - x_b\rangle = b\langle \hat{z}, \hat{x} - v\rangle,$$

hence $\langle \hat{z}, \hat{x} - v\rangle \leq 0$, as desired. The proof is complete. □

As a special case of Theorem 6.3.6, we have the following existence theorem:

Theorem 6.3.7. Let $X \subseteq \mathbb{R}^n$ be closed and convex, $K \subseteq X$ a nonempty compact set, $\Gamma : X \to 2^X$ and $\Phi : X \to 2^{\mathbb{R}^n}$ two set-valued mappings. Assume that:

(i) the set $\Phi(x)$ is nonempty and compact for each $x \in X$, and convex for each $x \in X$, with $x \in \Gamma(x)$;

(ii) for each $w \in X - X$, the set $\{x \in X : \inf_{z \in \Phi(x)} \langle z, w \rangle \leq 0\}$ is closed;

(iii) the set-valued mapping Γ is lower semicontinuous with convex values and the set $\{x \in X : x \in \Gamma(x)\}$ is closed;

(iv) $\Gamma(x) \cap K \neq$ for all $x \in X$; and

(v) for each $x \in X \setminus K$, with $x \in \Gamma(x)$, and each $z \in \Phi(x)$, there exists $y \in \Gamma(x) \cap K$ such that $\langle z, x - y \rangle \geq 0$.

Then $GQVI(X, \Gamma, \Phi)$ admits a solution.

Remark 6.3.8. (i). Theorem 6.3.7 follows at once from Theorem 6.3.6 by choosing any $r > 0$ in such a way that $K \subseteq B_r$. Moreover, it is worth noticing that the coercivity condition (v) of Theorem 6.3.7 is weaker than the corresponding assumption (v) in Theorem 6.3.2.

(ii). It is easy to construct an example of a situation where Theorem 6.3.5 (hence, Theorems 6.3.6 and 6.3.7) applies but Theorems 6.3.1 and 6.3.2 do not. To see this, take $X = [0, 1]$, $\Phi(x) = \{x\}$ and

$$\Gamma(x) = \begin{cases} [0, 1], & \text{if } x \neq \frac{1}{2}; \\ \{\frac{1}{2}\} & \text{if } x = \frac{1}{2}. \end{cases}$$

In fact, such Γ is lower semicontinuous with nonempty closed convex values, but the graph of Γ is not closed, hence Γ is not upper semicontinuous. However, we have $\{x \in X : x \in \Gamma(x)\} = [0, 1]$, hence Theorem 6.3.5 applies. We also note that the interior of $\Gamma(\frac{1}{2})$ is empty, hence [109, Theorem 2.1] cannot be applied.

6.4 Non-Compact Variational Inequalities in Locally Convex Spaces

The aim of this section is to establish a general existence theorem of $GVIP(F, K)$ for quasi-monotone set-valued mappings in locally convex Hausdorff topological vector spaces as an application of Ky-Fan minimax inequalities in its non-compact version.

Throughout this section, all topological vector spaces (resp., locally convex topological vector spaces) are assumed to be *Hausdorff* unless specified. Let K be a nonempty convex subset of E and $F : K \to 2^{E^*}$ a set-valued mapping with non-empty values. The *generalized variational inequality problem* $GVIP(F,\ K))$ is to find $\hat{y} \in K$ and $\hat{v} \in F(\hat{y})$ such that

$$Re\langle x - \hat{y},\ \hat{v} \rangle \geq 0 \quad \text{for all} \quad x \in K.$$

The aim of this section is to establish the following existence theorem of $GVIP(F, K)$ for quasi-monotone set-valued mappings in locally convex Hausdorff topological vector spaces.

Theorem 6.4.1. Let K be nonempty convex subset of a locally convex Hausdorff topological vector space E with $E^* \neq \emptyset$ (hence, E need not to be of second category). Let $F K \to 2^{E^*}$ be a quasi-monotone mapping such that F is upper semicontinuous from the line segments in K to the weak* topology of E^* and each $F(x)$ is non-empty compact in the strong topology on E^*. Suppose that there exists a non-empty weakly compact convex subset K_0 of K and a non-empty weakly compact subset D of K such that for each $y \in K \setminus D$, there exists $x \in \text{co}(K_0 \cup \{y\})$ with

$$Re\langle y - x, u\rangle > 0 \quad \text{for all} \quad u \in F(x).$$

Then there exists a point $\hat{y} \in K$ such that

$$\inf_{v \in F(\hat{y})} Re\langle \hat{y} - x, v\rangle \leq 0 \quad \text{for all} \quad x \in K.$$

If, in addition, $F(\hat{y})$ is also convex, then there exists $\hat{v} \in F(\hat{y})$ such that

$$Re\langle x - \hat{y}, \hat{v}\rangle \geq 0 \quad \text{for all} \quad x \in K,$$

that is, $(\hat{y}, \hat{v})$ is a solution of $GVIP(F, K)$.

Remark 6.4.2. Theorem 6.4.1 shows that Theorem A of Ding [134] still holds without the assumption that E is of secondary category, thus Theorem 6.4.6 includes Theorem A of Ding [134] as a special case.

In order to prove Theorem 6.4.6, we first recall the following generalization of the Ky Fan minimax principle which is Theorem 2.5.3 in Chapter 2.

Lemma 6.4.3. Let K be a non-empty convex subset of a Hausdorff topological vector space E and $\psi : X \times X \to \mathbb{R} \cup \{-\infty, +\infty\}$ be an extended real-valued function such that

(1) for each fixed $x \in K$, $y \mapsto \psi(x, y)$ is compactly lower semicontinuous;
(2) for each $A \in \mathcal{F}(X)$ and $y_0 \in co(A)$, $\min_{x \in A} \psi(x, y_0) \leq 0$; and
(3) there exists a non-empty compact and convex subset K_0 of K and a non-empty compact subset D of K such that for each $y \in K \setminus D$, there exists $x \in co(K_0 \cup \{x\})$ with $\psi(x, y) > 0$.

Then there exists $\hat{y} \in D$ such that $\sup_{x \in X} \psi(x, \hat{y}) \leq 0$.

We also need the following result which was originally proved by Tan and Yuan [510, Lemma 8]; and for completeness, we include its proof here.

Lemma 6.4.4. Let K be a non-empty convex subset of a topological vector space E and $T : K \to 2^{E^*}$ be upper semicontinuous from the line segments in K

to the weak* topology of E^* such that each $T(x)$ is non-empty weak* compact. If $\hat{y} \in K$, then the inequality

$$\inf_{u \in T(x)} Re\langle u, \hat{y} - x\rangle \le 0 \text{ for all } x \in K$$

implies the inequality

$$\inf_{w \in T(\hat{y})} Re\langle w, \hat{y} - x\rangle \le 0 \text{ for all } x \in K.$$

Proof. Let $x \in K$ be arbitrarily fixed and $z_t = tx + (1-t)\hat{y} = \hat{y} - t(\hat{y} - x)$ for each $t \in [0,1]$. Since K is convex, $z_t \in K$ for all $t \in [0,1]$. Thus for all $t \in (0,1]$,

$$t \cdot \inf_{u \in T(z_t)} Re\langle u, \hat{y} - x\rangle = \inf_{u \in T(z_t)} Re\langle u, \hat{y} - z_t\rangle \le 0$$

so that

$$\inf_{u \in T(z_t)} Re\langle u, \hat{y} - x\rangle \le 0$$

for all $t \in (0,1]$. If $\inf_{w \in T(\hat{y})} Re\langle w, \hat{y} - x\rangle > 0$, let $G := \{w \in E^* : Re\langle w, \hat{y} - x\rangle > 0\}$. Then G is a weak*-open set in E^* such that $T(\hat{y}) \subset G$. As $z_t \to \hat{y}$ when $t \to 0^+$, by the upper semicontinuity of T on $\{z_t : t \in [0,1]\}$, there exists $t_0 \in (0,1]$ such that $T(x_t) \subset G$ for all $t \in (0, t_0)$. Note that $T(z_t)$ is weak* compact, $\inf_{u \in T(z_t)} Re\langle u, \hat{y} - x\rangle > 0$ for all $t \in (0, t_0)$ which contradicts above inequality. Thus we must have $\inf_{w \in T(\hat{y})} Re\langle w, \hat{y} - x\rangle \le 0$ and the proof is complete. □

Lemma 6.4.5. Let K be a non-empty convex subset of a topological vector space E and $f : K \to 2^{E^*}$ a set-valued quasi-monotone mapping with non-empty values. Then the mapping $\psi : K \times K \to \mathbb{R} \cup \{\infty, +\infty\}$ defined by

$$\psi(x, y) := \inf_{u \in f(y)} Re\langle u, x - y\rangle$$

for each $(x, y) \in K \times K$ is 0-diagonal concave in y for each fixed $x \in K$.

Proof. The conclusion follows straightforward from the definitions of quasi-monotone and 0-diagonal concave mappings. □

By combining Lemmas 6.4.3, 6.4.4 and 6.4.5, we can now prove Theorem 6.4.6 above, i.e., we have the following:

Remark 6.4.6. The proof of Theorem 6.4.1 above: Let K be nonempty convex subset of a locally convex Hausdorff topological vector space E. Let $F\, K \to 2^{E^*}$ be a quasi-monotone mapping with non-empty values such that F is u.s.c. from the line segments in K to the weak* topology of E^* and each $F(x)$ is compact in the strong topology on E^*. Suppose that there exist nonempty weakly compact convex subset K_0 of K and nonempty weakly compact subset D of K such that for each $y \in K \setminus D$, there is a $x \in \mathrm{co}(K_0 \cup \{y\})$ with

$$Re\langle y - x,\, u\rangle > 0 \quad \text{for all} \quad u \in F(x).$$

Then there exists a point $\hat{y} \in K$ such that

$$\inf_{v \in F(\hat{y})} Re\langle \hat{y} - x,\, v\rangle \leq 0 \quad \text{for all} \quad x \in K.$$

If, in addition, $F(\hat{y})$ is also convex, then there exists $\hat{v} \in F(\hat{y})$ such that

$$Re\langle x - \hat{y},\, \hat{v}\rangle \geq 0 \quad \text{for all} \quad x \in K,$$

that is, $(\hat{y},\, \hat{v})$ is a solution of $GVIP(F,\, K)$.

Proof. Define a mapping $\varphi : K \times K \to \mathbb{R}$ by

$$\varphi(x, y) = \inf_{u \in F(x)} Re\langle u, y - x\rangle$$

for each $(x, y) \in K \times K$. Then we have

(a) for each fixed $y \in K$, $\varphi(x, y)$ is 0-diagonal concave in x by Lemma 6.4.5 as F is quasi-monotone; and

(b) for each $x \in K$, the function $y \mapsto \varphi(x,\, y)$ is weakly lower semicontinuous on each weakly compact subset A of K.

Note that E is locally convex, and for each $x \in K$ and for any nonempty weakly compact subset A of K, $A - x$ is weakly bounded and hence, by Theorem 3.18 of Rudin [449], $A - x$ is strongly bounded in E. Now we only need to prove that for each fixed $x \in K$ and for any real number $\lambda \in \mathbb{R}$ the set $A_\lambda = \{y \in A : \inf_{u \in F(x)} Re\langle y - x,\, u\rangle \leq \lambda\}$ is weakly closed. Let $\{y_\alpha\}_{\alpha \in \Gamma} \subset A_\lambda$ be a net, $y_0 \in A$ and $y_\alpha \rightharpoonup y_0$. Since $F(x)$ is strong compact, for each $\alpha \in \Gamma$, there exists an $u_\alpha \in F(x)$ such that

$$Re\langle y_\alpha - x,\, u_\alpha\rangle = \inf_{u \in F(x)} Re\langle y_\alpha - x,\, u\rangle \leq \lambda,$$

and there exists a subnet $\{u_\beta\}$ of $\{u_\alpha\}_{\alpha \in \Gamma}$ such that $u_\beta \to u_0 \in F(x)$. Let $B = \{y_\alpha - y : \alpha \in \Gamma\}$. Then B is strongly bounded. For any given $\varepsilon > 0$, let $W(B, \varepsilon) = \{u \in E^* : \sup_{y \in B} |\langle y,\, u - u_0\rangle\,| < \frac{\varepsilon}{3}\}$, then $W = W(B, \varepsilon)$ is a strongly open neighborhood of u_0 in E^*. Let $U = \{y \in A : |\langle y - y_0,\, u\rangle| < \frac{\varepsilon}{3}\}$. It follows that U is a weakly open neighborhood of y_0 in A. Choose $\beta_0 \in \Gamma$ such that for all $\beta \geq \beta_0$, $u_\beta \in W$ and $y_\beta \in U$. It follows that

$$\begin{aligned} Re\langle y_\beta - x,\,, u_\beta\rangle &- Re\langle y_0 - x,\, u_0\rangle| \leq |\langle y_\beta - y_0,\, u_\beta - u_0\rangle| \\ &+ |\langle y_0 - x,\, u_\beta - u_0\rangle| \\ &\quad +|\langle y_\beta - y_0,\, u_0\rangle| \\ &< \varepsilon/3 + \varepsilon/3 + \varepsilon/3 = \varepsilon. \end{aligned}$$

and so $\lim_{\beta\in\Gamma} Re\langle y_\beta - x,\, u_\beta\rangle = Re\langle y_0 - x,\, u_0\rangle$.
Hence, we have

$$\inf_{u\in F(x)} Re\langle y_0 - x,\, u\rangle \le Re\langle y_0 - x,\, u_0\rangle = \lim_{\beta\in\Gamma} Re\langle y_\beta - x,\, u_\beta\rangle \le \lambda,$$

i.e., $y_0 \in A_\lambda$ and A_λ is weakly closed in A. This shows that for each $x \in K$, $y \mapsto \varphi(x,\, y)$ is weakly lower semicontinuous on each weakly compact subset A of K.

(c) there exists a nonempty weakly compact convex subset K_0 of K and a nonempty weakly compact subset D of K such that for each $y \in K \setminus D$, there exists $x \in \mathrm{co}(K_0 \cup \{y\})$ with

$$Re\langle y - x,\, u\rangle > 0 \quad \text{for all} \quad u \in F(x).$$

Since $F(x)$ is strongly compact, therefore $\varphi(x,\, y) = \inf_{u\in F(x)} Re\langle y - x,\, u\rangle > 0$.

Now equip E with the weak topology and we see that all hypotheses of Lemma 6.4.3 hold. Hence it follows that there exists an $\hat{y} \in K$ such that

$$\varphi(x, \hat{y}) = \inf_{u\in F(x)} Re\langle \hat{y} - x,\, u\rangle \le 0 \quad \text{for all} \quad x \in K.$$

By Lemma 6.4.4, we have that

$$\varphi(x,\, \hat{y}) = \inf_{u\in F(\hat{y})} Re\langle \hat{y} - x,\, u\rangle \le 0 \quad \text{for all} \quad x \in K.$$

Finally, if $F(\hat{y})$ is convex, define $g:\, K \times F(\hat{y}) \to \mathbb{R}$ by

$$g(x,\, v) = Re\langle \hat{y} - x,\, v\rangle.$$

Then for each $x \in K$, $v \mapsto g(x,\, v)$ is weak* continuous and affine and for each $v \in F(\hat{y})$, $x \mapsto g(x,\, y)$ is also affine. By Kneser minimax theorem in [318], i.e., see Theorem 5.3.9, it follows that

$$\min_{v\in F(\hat{y})} \sup_{x\in K} Re\langle \hat{y} - x,\, v\rangle = \sup_{x\in K} \min_{v\in F(\hat{y})} Re\langle \hat{y} - x,\, v\rangle \le 0.$$

Since $F(\hat{y})$ is norm compact, there exists $\hat{v} \in F(\hat{y})$ such that

$$Re\langle \hat{y} - x,\, \hat{v}\rangle \le 0 \quad \text{for all} \quad x \in K$$

and so

$$Re\langle x - \hat{y},\, \hat{v}\rangle \ge 0 \quad \text{for all} \quad x \in K.$$

i.e., $(\hat{y},\, \hat{v})$ is a solution of $GVIP(F, K)$. □

As a special case of Theorem 6.4.6, we have the following:

Corollary 6.4.7. Let K be a non-empty compact and convex subset of a locally convex Hausdorff topological vector space E. Let $FK \to 2^{E^*}$ be a quasi-monotone mapping such that F is upper semicontinuous from the line segments in K to the weak* topology of E^* and each $F(x)$ is non-empty compact in the strong topology on E^*. Then there exists a point $\hat{y} \in K$ such that

$$\inf_{v \in F(\hat{y})} Re\langle \hat{y} - x, v\rangle \leq 0 \quad \text{for all} \quad x \in K.$$

If, in addition, $F(\hat{y})$ is also convex, then there exists $\hat{v} \in F(\hat{y})$ such that

$$Re\langle x - \hat{y}, \hat{v}\rangle \geq 0 \quad \text{for all} \quad x \in K,$$

that is, $(\hat{y}, \hat{v})$ is a solution of $GVIP(F, K)$.

Remark 6.4.8. We would like to note that Theorem 6.4.6 not only shows that Theorem A of Ding [134] is true without the assumption that E is of the secondary category, but also shows that the conclusion of Ding's result in [134] holds in locally convex topological vector spaces. By following the same ideas used in the literature such as Aubin [10], Cottle and Yao [103], Ding and Tarafdar [145], Harker and Pang [220], Karamardian [282], Saigal [451], Siddiqi and Ansari [476], Tan and Yuan [510] and so on, we can also prove a number of existence theorems of $GVIP(F, K)$ and $GCP(F, K)$ in which underlying spaces are not of secondary category, we omit fully details here.

Finally we wish to discuss the difference between our Theorem 6.4.6 and Theorem 1 of Cubiotti [107] (which is an extension of Theorem 2.1 of Cubiotti [105]) which also concerns the existence of solutions for $GQVI$ in finite dimensional spaces. First we note that Cubiotti's coercive condition (iii) of Theorem 1 in [107] is different from ours in Theorem 6.4.6. Secondly our set-valued mapping F in Theorem 6.4.6 is quasi-monotone and upper semicontinuous (thus it may not be lower semicontinious); however the corresponding set-valued mapping Γ of Theorem 1 in [107] is lower semicontinuous and its graph is closed (hence it may not be quasi-monotone nor upper semicontinuous). Furthermore the underlying space E of our Theorem 6.4.6 is a locally convex topological vector space which may be finite or infinite-dimensional, but the underlying space of Theorem 1 (see also Theorem 2.4 in [105] is finite dimensional. Thus our Theorem 6.4.6 and Theorem 1 of Cubiotti [107] (see also Theorem 2.4 in [105]) are independent each other.

6.5 General Quasi-Variational Inequalities and Applications

In this section, as applications of the celebrated Ky Fan minimax principle again, we first establish a general existence theorem of generalized quasi-variational inequalities in topological vector spaces with sufficient many continuous linear functionals. This result includes a number of existence theorems for generalized quasi-variational inequalities as special cases. As applications, several existence theorems of quasi-variational inequalities for set-valued mappings,which have either upper semicontinuous or monotone properties are derived, and one of these existence results is applied to give a new non-compact generalization of the Fan-Glicksberg fixed point theorem for set-valued mappings in topological vector spaces. Also we provide a simple example which shows that the lower semicontinuity assumption is essential for the guarantee of existence of solutions for generalized quasi-variational inequalities. Our results unify a number of existence theorems for generalized variational inequalities which have been given in the literature.

Now we have the following general quasi-variational variational inequality without lower semicontinuity assumption in topological vector spaces.

Theorem 6.5.1. Let X be a non-empty paracompact compact convex subset of a topological vector space E which has sufficiently many continuous linear functionals. Suppose $F : X \to 2^X$ is upper hemicontinuous with non-empty compact and convex values and $\psi : X \times X \to \mathbb{R}$ is a real-valued function such that:

(1) for each fixed $y \in X$, $x \mapsto \psi(x, y)$ is compactly lower semicontinuous;

(2) for each fixed $x \in X$, $y \mapsto \psi(x, y)$ is 0-diagonally concave in y;

(3) there exists a non-empty compact and convex subset X_0 of X and a non-empty compact subset K of X such that for each $x \in X \setminus K$, there exists $y \in F(x) \cap co(X_0 \cup \{x\})$ such that $\psi(x, y) > 0$; and

(4) the set $\{x \in X : \sup_{y \in F(x)} \psi(x, y) \leq 0\}$ is closed in X.

Then there exists $x \in K$ such that $x \in F(x)$ and $\sup_{y \in F(x)} \psi(x, y) \leq 0$.

Proof. We shall prove the conclusion as application of the Ky Fan minimax principle and modify arguments used by Aubin and Ekeland [15] (see also Shih and Tan [466] and Yuan [604]). Suppose the assertion of Theorem 6.5.1 were false. Then for each $x \in X$, either $x \notin F(x)$ or there exists $u \in F(x)$ such that $\psi(x, u) > 0$. In the case $x \notin F(x)$, note that E has sufficiently many continuous linear functionals, by Hahn-Banach theorem (e.g., see Theorem 3.4 of Rudin in [449, p.58]), there exists $p \in E^*$ such that $Re\langle p, x\rangle - \sup_{z \in F(x)}\langle p, z\rangle > 0$. Let $V(p) := \{x \in X : Re\langle p, x\rangle - \sup_{z \in F(x)}\langle p, z\rangle > 0\}$. As F is upper hemicontinuous, it is clear that $F(p)$ is an open neighborhood of x in X. In the case that there exists $u \in F(x)$ such that $\psi(x, u) > 0$, then $\sup_{y \in F(x)} \psi(x, y) > 0$. Let $\Sigma := \{x \in X : \sup_{y \in F(x)} \psi(x, y) > 0\}$. Then Σ is open by the condition (5). For the convenience of our discussion, in what follows, let $V_0' := \Sigma$ and $V_p := V(p)$ for each $p \in E^*$. Then we have that $X = V_0' \cup \Sigma \cup_{p \in E^*} V_p$. By a fact that (e.g., see Theorem VIII.1.4 of Dugundji [147, p.162]) each open cover of a paracompact space has an open precise neighborhood-finite refinement, and note that X is paracompact, without loss of generality, let $\{V_0, V'_{p_i} : i \in I\}$ be an open neighborhood-finite refinement of the open cover $\{V'_0, V_p : p \in E^*\}$ of X, where V_{p_i} (resp., V_0) is a non-empty open subset of V_{p_i} (resp., V'_0) for $p_i \in E^*$. In the following, we shall denote V'_{p_i} by V_i for the convenience of our discussion for each $i \in I$. Now assume $\{\beta_0, \beta_i : i \in I\}$ be a continuous partition of unity on X subordinate to the open cover $\{V_0, V_i : i \in I\}$. Then β_i (resp., β_0) is a continuous non-negative real-valued function on X such that β_i (resp., β_0) vanishes on $X \setminus V_i$ (resp., $X \setminus V_0$) for each $i \in I$. Moreover for each $x \in X$, there exists an open neighborhood $O(x)$ and $n(x) \in \mathbb{N}$ such that $\Sigma_{i=0}^{n(x)} \beta_i(z) = 1$ for each $z \in O(x)$. Now we define a real-valued mapping $\psi : X \times X \to \mathbb{R}$ by

$$\phi(x, y) := \beta_0(x)\psi(x, y) + \Sigma_{i\in I}\beta_i(x)Re\langle p_i, x - y\rangle$$

for each $(x, y) \in X \times X$. Then we have that

(a): for each fixed $x \in X$, $y \mapsto \phi(x, y)$ is also 0-diagonal concave;

(b): for each fixed $y \in X$, $x \mapsto \phi(x, y)$ is compactly lower semicontinuous; and

(c): for each $x \in X \setminus K$, there exists $\hat{y} \in co(X_0 \cup \{x\})$ such that $\phi(x, \hat{y}) > 0$. Indeed, as $\{V_0, V_i : i \in I\}$ is an open cover of X, either $x \in V_0$ or $x \in V_i$ for some $i \in I$. If $x \in V_0$, it is clear that $\beta_0(x) > 0$. By our assumption (3), there exists $\hat{y} \in F(x) \cap co(X_0 \cup \{x\})$ such that $\psi(x, \hat{y}) > 0$, so that $\beta_0(x)\psi(x, \hat{y}) > 0$. In the case $x \in V_i$, where $i \in I$. Let $N(x) := \{i \in I : x \in V_i\}$. Then $N(x)$ is a finite positive number by the finite open intersection property of the open cover $\{V_0, V_i : i \in I\}$, and $\beta_i(x)$ vanishes on $X \setminus V_i$ for each $i \in N(x)$. From the definition of V_i for each $i \in N(x)$, it follows that

$$Re\langle p_i, x\rangle > \sup_{z\in F(x)} Re\langle p_i, z\rangle \geq Re\langle p_i, u\rangle$$

for all $u \in F(x)$. Note that $\hat{y} \in F(x) \cap co(X_0 \cup \{x\})$, we also have $Re\langle p_i, x\rangle > Re\langle p_i, \hat{y}_i\rangle$, i.e., $Re\langle p_i, x - \hat{y}\rangle > 0$. Therefore

$$\phi(x, \hat{y}) = \beta_0(x)\psi(x, \hat{y}) + \Sigma_{i\in I}\beta_i(x)Re\langle p_i, x - \hat{u}\rangle > 0,$$

i.e., we have proved that for each $x \in X \setminus K$, there exists $\hat{y} \in co(X_0 \cup \{x\})$ such that $\phi(x, \hat{y}) > 0$.

Hence ϕ satisfies all hypotheses of the Ky Fan minimax principle, there exists $\hat{x} \in X$ such that

$$\sup_{y\in X} \phi(\hat{x}, y) \leq 0. \tag{6.5.1}$$

On the other hand, as we know that $X \subset V_0 \cup_{i\in I} V_i$ and $\{\beta_i, i = 0 \text{ and } i \in I\}$ is a partition of unity, there exists at least one index $i \in \{0, j : j \in I\}$ such that $\beta_i(\hat{x}) > 0$. If $\hat{x} \in V_0$, there exists $\hat{u} \in F(\hat{x})$ such that $\psi(\hat{x}, \hat{u}) > 0$. In the case $\hat{x} \in V_i$ for some $i \in I$, we have $Re\langle p_i, \hat{x}\rangle > \sup_{z\in F(\hat{x})} Re\langle p_i, z\rangle \geq$

$Re\langle p_i, u\rangle$ for all $u \in F(\hat{x})$. As $\hat{u} \in F(\hat{x})$, it follows that $\beta_i(\hat{x})Re\langle p_i, \hat{x} - \hat{u}\rangle > 0$. Hence we have that

$$\phi(\hat{x}, \hat{u}) = \beta_0(\hat{x})\psi(\hat{x}, \hat{u}) + \Sigma_{i\in I}\beta_i(\hat{x})Re\langle p_i, \hat{x} - \hat{u}\rangle > 0$$

which contradicts our inequality (6.5.1) above. Therefore the conclusion must hold and we complete the proof. □

We would like to note that Theorem 6.5.1 includes a number of existence theorems for quasi-variational inequalities in recent literature as a special case (e.g., see references [1-8], [20-27], [29-31] and [35]). For example, as a special case of Theorem 6.5.1, we have the following result.

Corollary 6.5.2. Let X be a non-empty compact convex subset of a topological vector space E which has sufficiently many continuous linear functionals. Suppose $F : X \to 2^X$ is upper semicontinuous with non-empty compact and convex values and $f : X \times X \to \mathbb{R}$ is a real-valued function such that:

(1) for each fixed $y \in X$, $x \mapsto \psi(x, y)$ is lower semicontinuous;
(2) for each fixed $x \in X$, $y \mapsto \psi(x, y)$ is 0-diagonal concave; and
(3) the set $\{x \in X : \sup_{y\in F(x)} \psi(x, y) \leq 0\}$ is closed in X.

Then there exists $x \in X$ such that $x \in F(x)$ and $\sup_{y\in F(x)} \psi(x, y) \leq 0$.

Proof. Note that each compact space is paracompact and each set-valued self-mapping with compact values in a compact space is upper semicontinuous if and only if it is upper hemicontinuous. Thus the conclusion follows from Theorem 6.5.1. □

Remark 6.5.3. From the proof of Theorem 6.5.1, in order to guarantee the existence of solutions for Theorem 6.5.1, we assume the following condition:

$$\text{The set } \{x \in X : \sup_{y\in F(x)} \psi(x, y) \leq 0\} \text{ is closed in } X. \qquad (6.5.2)$$

It is natural to ask that if Theorem 6.5.1 still holds without the the condition (6.5.2). However, the following simple example from Yuan [604, p.125] shows that there is no more any existence of solutions for Theorem 6.5.1 if we withdraw condition (6.5.2) for the problem $GQVI(X; F, \psi)$:

Example 6.5.4. In Theorem 6.5.1, let $X := [0, 1]$ and $F : X \to 2^X$ and $\psi : X \times X \to \mathbb{R}$ be defined by $\psi(x, y) := x - y$ for each $(x, y) \in X \times X$ and

$$F(x) := \begin{cases} [\frac{1}{2}, 1], & \text{if } x \in [0, \frac{1}{2}), \\ [0, 1], & \text{if } x = \frac{1}{2}, \\ [0, \frac{1}{2}], & \text{if } x \in (\frac{1}{2}, 1]. \end{cases}$$

It is clear that F is upper semicontinuous with non-empty compact convex values, but F is not lower semicontinuous. Note that F has only one fixed point $\{\frac{1}{2}\}$. Obviously, $\{\frac{1}{2}\}$ is not a solution of Theorem 6.5.1 and thus there is no existence of solutions for the problem $GQVI(X; F, \psi)$. On the other hand, we note that X, F and ψ satisfy all hypotheses of Theorem 6.5.1 except condition (4) as '*the set* $\{x \in X : \sup_{y\in F(\hat{x})} \psi(x,y) \leq 0\} = [0, \frac{1}{2})$' is only open in X.

Note that condition (6.5.2) above is automatically satisfied if F is lower semicontinuous and ψ is assumed to be jointly lower semicontinuous in Theorem 6.5.1 as $x \mapsto \sup_{y\in F(x)} \psi(x,y)$ is lower semicontinuous at this circumstance (e.g., by Proposition III-19 of Aubin and Ekeland [15, p.118]). Thus we might say condition (6.5.2) involves implicitly lower semicontinuity of the mapping F.

We should also mention that the Example 6.5.4 indicates that in order to guarantee the existence of solutions for general quasi-variational inequalities, the lower semicontinuity condition is essential in some sense, and this fact was also remarked by Cubiotti [110].

Without the assumption of the lower semicontinuity for the set-valued mappings F in our Theorem 6.5.1, we could not expect the existence of solutions for the problem $GQVI(X; F, T)$ in general, and thus we may say that the lower semicontinuity assumption is essential for the guarantee of the existence of solutions of generalized quasi-variational inequalities.

As another consequence of Theorem 6.5.1, we have the following result which contains Theorem of Kim [298] as a special case (the latter one also improves Theorem 4 of Shih and Tan [466]).

Theorem 6.5.5. Let X be a non-empty paracompact and bounded convex subset of a a topological vector space E which has sufficiently many continuous linear functionals. Suppose $F : X \to 2^X$ is continuous with non-empty compact and convex values and $T : X \to 2^{E^*}$ is upper semicontinuous with non-empty compact values from the relative topology of X to the strong topology of E^*. Furthermore assume that there exists a non-empty compact and convex subset X_0 of X and a non-empty compact subset K of X such that for each $x \in X \setminus K$, there exists $y \in co(X_0 \cup \{x\})$ such that $\inf_{u\in T(x)} Re\langle u, x-y\rangle > 0$. Then there exists $\hat{x} \in K$ such that

$$\hat{x} \in F(\hat{x}) \text{ and } \sup_{y\in F(\hat{x})} \inf_{u\in T(\hat{x})} Re\langle u, \hat{x}-y\rangle \leq 0.$$

Moreover if $T(\hat{x})$ is convex, there exists $\hat{u} \in T(\hat{x})$ such that $\sup_{y\in F(\hat{x})} Re\langle \hat{u}, \hat{x}-y\rangle \leq 0$, and thus we have $\mathrm{Re}\langle \hat{u}, \hat{x}\rangle = \inf_{y\in F(\hat{x})} \mathrm{Re}\langle \hat{u}, y\rangle$.

Proof. Define a mapping $\psi : X \times X \to \mathbb{R}$ by

$$\psi(x,y) := \inf_{u\in T(x)} Re\langle u, x-y\rangle$$

for each $(x, y) \in X \times X$. Then we have that:

(a): as X is bounded and T is upper semicontinuous with compact values, it follows that ψ jointly lower semicontinuous by III-Proposition 21 of Aubin and Ekeland [15, p.119], so that $x \mapsto \psi(x, y)$ is compactly lower semicontinuous for each fixed $y \in X$;

(b): for each fixed $x \in X$, $y \mapsto \psi(x, y)$ is concave so that $\psi(x, y)$ is 0-diagonal concave in y;

(c): there exists a non-empty compact and convex subset X_0 of X and a non-empty compact subset K of X such that for each $x \in X \setminus K$, there exists $y \in F(x) \cap co(X_0 \cup \{x\})$ such that $\psi(x, y) > 0$ and

(d): by the continuity of F and upper semicontinuityt of T, we have that the set $\{x \in X : \sup_{y \in F(x)} \inf_{u \in T(x)} \text{Re}\langle u, x - y\rangle \leq 0\}$ is closed in X as $x \mapsto \sup_{y \in F(x)} \psi(x, y)$ is lower semicontoinuous (by III-Proposition 19 of Aubin and Ekeland [15, p.118]).

Thus all hypotheses of Theorem 6.5.1 are satisfied. By Theorem 6.5.1, there exists $\hat{x} \in K$ such that $\hat{x} \in F(\hat{x})$ and $\psi(\hat{x}, y) \leq 0$, which implies that

$$\sup_{y \in F(\hat{x})} \inf_{u \in T(\hat{x})} Re\langle u, \hat{x} - y\rangle \leq 0.$$

In the case $T(\hat{x})$ is convex, by applying Kneser minimax theorem, we have that

$$\inf_{u \in T(\hat{x})} \sup_{y \in F(\hat{x})} Re\langle u, \hat{x} - y\rangle = \sup_{y \in F(\hat{x})} \inf_{u \in T(\hat{x})} Re\langle u, \hat{x} - y\rangle \leq 0.$$

Note that $u \mapsto \sup_{y \in F(\hat{x})} Re\langle u, \hat{x} - y\rangle$ is lower semicontinuous as $F(\hat{x})$ is compact and $T(\hat{x})$ also is compact, there exists $\hat{u} \in T(\hat{x})$ such that

$$\sup_{y \in F(\hat{x})} Re\langle \hat{u}, \hat{x} - y\rangle = \inf_{u \in T(\hat{x})} \sup_{y \in F(\hat{x})} Re\langle u, \hat{x} - y\rangle \leq 0.$$

Note that $\hat{x} \in F(\hat{x})$, it is obvious that $\text{Re}\langle \hat{u}, \hat{x}\rangle = \inf_{y \in F(\hat{x})} \text{Re}\langle \hat{u}, y\rangle$ and hence we complete the proof. □

Indeed, as another application of Theorem 6.5.1, we can study implicitly quasi-variational inequalities which was first introduced and studied by Mosco [375].

Theorem 6.5.6. Let X be a non-empty compact convex subset of a topological vector space E and Y be a non-empty convex subset of a topological vector space W. Let $G : X \to 2^X$ be upper hemicontinuous with non-empty compact and convex values and $F : X \to 2^Y$ a set-valued mapping with non-empty values. Suppose $f : X \times X \times Y \to \mathbb{R} \cup \{-\infty, +\infty\}$ and $\psi : X \to \mathbb{R} \cup \{-\infty, +\infty\}$ are two extended-valued functions such that

(1) $u \mapsto \psi(u) + \sup_{y \in F(x)} f(x, u, y)$ is diagonal convex; and $\sup_{y \in F(x)} f(x, x, y) \geq 0$ for each $x \in X$;

(2) for each fixed $u \in X$, $x \mapsto \psi(x) - \sup_{y\in F(x)} f(x,u,y)$ is compactly lower semicontinuous;

(3) the set

$$\{x \in X : g(x,x) - \inf_{u\in G(x)} [\sup_{y\in F(x)} f(x,u,y) + g(x,u)] \leq 0\}$$

is closed in X; and

(4) the set

$$\{x \in X : \psi(x) - \inf_{u\in G(x)} [\psi(u) + \sup_{y\in F(x)} f(x,u,y)] \leq 0\}$$

is closed in X.

Then there exists $\hat{x} \in X$ with $\hat{x} \in G(\hat{x})$ such that

$$\psi(\hat{x}) \leq \psi(u) + \sup_{y\in F(x)} f(\hat{x},u,y)$$

for all $u \in G(\hat{x})$.

Proof. Define a mapping $g : X \times X \to \mathbb{R} \cup \{-\infty, +\infty\}$ by

$$g(x,u) := \psi(x) - \psi(u) - \sup_{y\in F(x)} f(x,u,y)$$

for each $(x,u) \in X \times X$. Then it is clear to see that g satisfies all hypotheses of Theorem 6.5.1. By Theorem 6.5.1, there exists $\hat{x} \in X$ such that $\hat{x} \in G(\hat{x})$ and $\sup_{u\in G(\hat{x})} g(\hat{x},u) \leq 0$. By the definition of the function g, it follows that $\psi(\hat{x}) - \psi(u) - \sup_{y\in F(\hat{x})} f(\hat{x},u,y) \leq 0$ for each $u \in G(\hat{x})$. Thus

$$\psi(\hat{x}) \leq \inf_{u\in G(\hat{x})} \psi(u) + \inf_{u\in G(\hat{x})} \sup_{y\in F(\hat{x})} f(\hat{x},u,y)$$

and the proof is complete. □

Moreover, as an application of Theorem 6.5.1 again, we have the following result.

Theorem 6.5.7. Let X be a non-empty compact convex subset of a topological vector space E and Y a non-empty closed and convex subsets of a topological vector space W. Let $G : X \to 2^X$ be an upper hemicontinuous semicontinuous set-valued mapping with non-empty closed and convex values and $F : X \to 2^Y$ be a set-valued mapping with non-empty values. Suppose $f : X \times X \times Y \to \mathbb{R} \cup \{-\infty, +\infty\}$ and $g : X \times X \to \mathbb{R} \cup \{-\infty, +\infty\}$ are such that:

(1) for each fixed $u \in X$, $x \mapsto \sup_{y\in F(x)} f(x,u,y) + g(x,u)$ is compactly upper semicontinuous; and $x \mapsto g(x,x)$ is compactly lower semicontinuous;

(2) $u \mapsto \sup_{y\in F(x)} f(x,u,y) + g(x,u)$ is diagonal convex; and $\sup_{y\in F(x)} f(x,x,y) \geq 0$ for all $x \in X$;

(3) for each fixed $x \in X$, $y \mapsto f(x,u,y) + g(x,u)$ is compactly upper semicontinuous and diagonal quasi-concave;

(4) for each fixed $(x,y) \in X \times Y$, $u \mapsto f(x,u,y) + g(x,u)$ is compactly lower semicontinuous; and

(5) the set $\{x \in X : g(x,x) - \inf_{u\in G(x)}[\sup_{y\in F(x)} f(x,u,y) + g(x,u)] \leq 0\}$ is closed in X.

Then there exists $(\hat{x},\hat{y}) \in X \times Y$ such that $\hat{x} \in G(\hat{x})$ and $\hat{y} \in F(\hat{x})$ and

$$g(\hat{x},\hat{x}) \leq f(\hat{x},u,\hat{y}) + g(\hat{x},u)$$

for each $u \in G(\hat{x})$.

Proof. Define a mapping $h : X \times X \to \mathbb{R} \cup \{-\infty,+\infty\}$ by

$$h(x,u) := g(x,x) - \sup_{y\in F(x)} f(x,u,y) - g(x,u)$$

for each $(x,y) \in X \times X$. Then we have:

(a) for each fixed $u \in X$, $x \mapsto h(x,u)$ is compactly lower semicontinuous on X by condition (1); and

(b) for each fixed $x \in X$, $u \mapsto h(x,u)$ is 0-diagonal concave by the hypothesis (2).

Thus G and h satisfy all hypotheses of Theorem 6.5.1. By Theorem 6.5.1, there exists $\hat{x} \in X$ such that $\hat{x} \in G(\hat{x})$ and $\sup_{u\in G(\hat{x})} h(\hat{x},u) \leq 0$. By the definition of h, we have that

$$\sup_{u\in G(\hat{x})} [g(\hat{x},\hat{x}) - \sup_{y\in F(\hat{x})} f(\hat{x},u,y) - g(\hat{x},u)] \leq 0$$

which implies that

$$g(\hat{x},\hat{x}) \leq \inf_{u\in G(\hat{x})} \sup_{y\in F(\hat{x})} [f(\hat{x},u,y) + g(\hat{x},u)].$$

We note that for each fixed $y \in X$, the mapping $u \mapsto f(\hat{x},u,y) + g(\hat{x},u)$ is compactly lower semicontinuous and quasi-concave by condition (4) and for each fixed $u \in X$, $y \mapsto f(\hat{x},u,y) + g(\hat{x},u)$ is compactly upper semicontinuous and diagonal quasi-convex. Thus the Neumann-Sion type minimax theorem (e.g., see Theorem 2.14 of Zhou and Chen [622] or Yuan [604]) implies that

$$\sup_{y\in F(\hat{x})} \inf_{u\in G(\hat{x})} [f(\hat{x},u,y) + g(\hat{x},u)] = \inf_{u\in G(\hat{x})} \sup_{y\in F(\hat{x})} [f(\hat{x},u,y) + g(\hat{x},u)].$$

As $y \mapsto \inf_{u\in G(\hat{x})}[f(\hat{x},u,y) + g(\hat{x},u)]$is upper semicontinuous and $F(\hat{x})$ is compact, there exists $\hat{y} \in F(\hat{x})$ such that

$$\inf_{u\in G(\hat{x})} [f(\hat{x},u,\hat{y}) + g(\hat{x},u)] = \sup_{y\in F(\hat{x})} \inf_{u\in G(\hat{x})} [f(\hat{x},u,y) + g(\hat{x},u)]$$

and thus the conclusion follows. □

Finally, we give some applications of quasi-variational inequalities established in this section. More applications with details will be carried out later in this chapter.

In what follows, we will give some applications of our generalized quasi-variational inequalities above to Fan's alternative theorem which concerns best approximation and fixed point theorems for set-valued mappings, the existence of solutions for the Lions-Stampacchia variational inequality, and Gale-Nakaido-Debreu (GND) model in economics and generalized complementarity problems in Banach spaces, which are not necessarily reflexive.

As an application of Theorem 6.5.7, we have the following alternative result which concerns Fan's best approximation theorem (e.g., see Fan [162]):

(A): Fan's alternative theorem.

Theorem 6.5.8. Let X be a non-empty compact convex subset of a topological vector space E which has sufficiently many continuous functionals. Suppose $F : X \to 2^E$ is a continuous set-valued mapping with non-empty compact and convex values. Then we have

(1) either F has a fixed point in X;
or there exists $\hat{x} \in X$, $\delta > 0$ and a continuous semi-norm $p \in E^*$ such that

$$\sup_{y\in F(\hat{x})} [p(u-y) - p(\hat{x}-y)] \geq 0$$

for all $u \in X$ and $\inf_{y\in F(\hat{x})} p(\hat{x}-y) \geq \delta$.

Proof. Suppose that F has no fixed point in X. Then for each $x \in X$, we have $0 \notin x - F(x)$. By the virtue of Hahn-Banach theorem (e.g., see Theorem 3.4 of Rudin [449, p.58]), there exists $\delta_x > 0$ and a linear functional $p_x \in E^*$ such that

$$\inf_{u\in F(x)} |p_x(x-u)| > \delta_x.$$

Noting that F is upper semicontinuous with compact values, so that $x \to \inf_{u\in F(x)} |p_x(x-u)|$ is lower semicontinuous and thus there exists an open neighborhood $N(x)$ of x such that $\inf_{u\in F(z)} |p_x(z-u)| > \delta_x$ for all $z \in N(x)$. As X is compact, there exists a non-empty finite subset $\{x_1, \cdots, x_n\}$ of X such that $X = \cup_{i=1}^n N(x_i)$. Let $p := \max\{|p_{x_i}| : i = 1, \cdots, n\}$ and $\delta := \min\{\delta_{x_i} : i = 1, \cdots, n\}$. Then p is a continuous semi-norm on E and for each $x \in X$, we have $\inf_{u\in F(x)} p(x-u) > \delta$. Now define a function $f : X \times X \times X \to \mathbb{R} \cup \{-\infty, +\infty\}$ by

$$f(x, u, y) := p(u-y) - p(x-y)$$

for each $(x, u, y) \in X \times X \times X$; and let $E(x) := X$ for each $x \in X$ in Theorem 6.5.7. As F is upper semicontinuous, it is clear that for each fixed $u \in X$, $x \mapsto \sup_{y \in F(x)} f(x, u, y)$ is upper semicontinuous. Moreover it is not difficult to verify that the mapping $u \mapsto \sup_{y \in F(x)} f(x, u, y)$ is 0-diagonal convex. Thus all hypotheses of Theorem 6.5.7 are satisfied. By Theorem 6.5.7, there exists $\hat{x} \in X$ such that

$$\sup_{y \in F(\hat{x})} [p(u - y) - p(\hat{x} - y)] \geq 0$$

for all $u \in X$ and the proof is complete. □.

For some other best approximation results of Fan's alternative theorems in locally convex topological spaces for upper semicontinuous set-valued mappings, more results have been studied in Section 2.10 of Chapter 2.

(B): Lions-Stampacchia variational inequality. See, Bensoussan and Lions [39], or Kinderlehrer and Stampacchia [304] and many others.

Theorem 6.5.9. Let X be a non-empty closed, bounded and convex subset of a reflexive Banach space E and $G : X \to 2^X$ be continuous with non-empty closed andconvex values. Suppose $a : E \times E \to \mathbb{R}$ is a continuous bilinear for with $a(u, u) \geq 0$ for all $u \in E$ and x^* is any fixed point in E^*. Then there exists $\hat{x} \in X$ such that $\hat{x} \in G(\hat{x})$ and $a(\hat{x}, \hat{x} - u) \leq x^*(\hat{x} - u)$ for all $u \in G(\hat{x})$.

Proof. Let X and E be equipped with weak topology and F be the identity mapping in Theorem 6.5.7. We define $f : X \times X \to \mathbb{R} \cup \{-\infty, +\infty\}$ by $f(x, u) := a(x, x - u) - x^*(x - u)$ for each $(x, u) \in X \times X$. Then the conclusion follows from Theorem 6.5.7 and the proof is complete. □

(C): Gale-Nikaido-Debreu lemma . See Aubin and Ekeland [15], Gwinner [212], or Chapter 9 in this book.

Theorem 6.5.10. Let $(X, Y, \langle \cdot, \cdot \rangle)$ be a dual system of real topological vector spaces E and W. Let the bilinear form $\langle \cdot, \cdot \rangle$ be compactly continuous in $E \times W$. Suppose C and D are non-empty compact and convex subsets of E and W, respectively. Let P be the the convex cone generated by C, i.e., $P = \cup_{r \in \mathbb{R}_+} \{rC\}$ and $F : C \to 2^D$ be set-valued continuous with non-empty closed and convex values. Then there exists $\hat{x} \in C$ such that $F(\hat{x}) \cap P^* \neq \emptyset$, where $P^* := \{y \in W : \langle y, p \rangle \geq 0 \text{ for all } p \in P\}$.

Proof. In Theorem 6.5.7, let $G(x) := C$ for each $x \in X$ and $g := 0$. Define $f : X \times Y \to \mathbb{R}$ by $f(x, u, y) := \langle u, x \rangle$ for each $(x, u, y) \in X \times X \times Y$. By Theorem 6.5.7, there exists $\hat{x} \in C$ and $\hat{y} \in F(\hat{x})$ such that $\langle u, \hat{y} \rangle \geq 0$ for all $u \in C$. Thus $\langle z, \hat{y} \rangle \geq 0$ for all $z \in P$, which implies that $\hat{y} \in P^*$. Therefore $\hat{y} \in F(\hat{x}) \cap P^* \neq \emptyset$ and the proof is complete. □

(D): Generalized complementarity problem. See Cottle et al. [102], Harker and Pang [220], or Isac [253].

Let C be a non-empty subset of a topological vector space E and P a cone of E. Suppose $F : C \to 2^{E^*}$ is a set-valued mapping, $\psi : X \to E$ and $\phi : C \to E^*$ are two functions. The generalized implicit complementarity problem (in short, GICP) is to find $\hat{x} \in C$ such that there exists $\hat{y} \in F(\hat{x})$ satisfying

$$\psi(\hat{x}) \in P, \qquad \phi(\hat{x}) + \hat{y} \in P \text{ and } \langle \phi(\hat{x}) + \hat{y}, \psi(\hat{x}) \rangle = 0. \qquad (GICP)$$

For the simlicity of our discussion in this section, we shall only condider the existence of solutions for generalized complementarity problems in Euclidean spaces.

Definition 6.5.11. Let X be a subset of an Euclidean space $\mathbb{R}^n$, where $n \in \mathbb{N}$. We recall that C is said to be locally compact if there exists a fundamental system $\mathcal{V}$ of neighborhoods of zero such that $U \cap C$ is compact for each $U \in \mathcal{V}$.

We also need the following result.

Lemma 6.5.12. Let X be a cone in a topological vector space E, $T : X \to 2^{E^*}$, $\hat{y} \in X$ and $\hat{w} \in T(\hat{y})$. Then the following statements are equivalent:

(1) $Re\langle \hat{w}, \hat{y} - x \rangle \leq 0$ for all $x \in X$; and
(2) $Re\langle \hat{w}, \hat{y} \rangle = 0$ and $\hat{w} \in X^*$.

Proof. The conclusion follows by the fact that X is a cone. □

Now we have the following result.

Theorem 6.5.13. Let P be a locally compact convex cone of an Euclidean space $\mathbb{R}^n$, where $n \in \mathbb{N}$, and $F : P \to 2^{\mathbb{R}^n}$ be upper semicontinuous with non-empty compact and convex values. Suppose $\psi : P \to P$ and $\phi : P \to 2^{\mathbb{R}^n}$ are continuous such that:
(1) the mapping $u \mapsto \langle y, \psi(u) \rangle$ is convex for each fixed $y \in P^*$;
(2) there exists $\hat{u} \in P$ and $r > 0$ with $\|u\| \leq r > 0$ such that

$$\langle \phi(x) + y, \hat{u} - \psi(x) \rangle \leq 0 \text{ for all } x \in P \text{ with } \|x\| = r \text{ and } y \in F(x).$$

Then $GICP$ has a solution.

Proof. Let $B_r(P) := \{x \in P : \|x\| \leq r\}$ for the given $r \in \mathbb{R}$. Then $B_r(P)$ is compact and convex. Define $f : P \times P \times \mathbb{R}^n \to \mathbb{R}$ by

$$f(x, u, y) := \langle \phi(x) + y, u - \psi(x) \rangle$$

for each $(x, u, y) \in P \times P \times \mathbb{R}^n$ and setting $G(x) := B_r(P)$, $Y := D := \mathbb{R}^n$ and $g := 0$ for each $x \in P$ in Theorem 6.5.7. By Theorem 6.5.7, there exists $\hat{x} \in B_r(P)$ and $\hat{y} \in F(\hat{x})$ such that

$$\langle \phi(\hat{x}) - \hat{y}, u - \psi(\hat{x}) \rangle \geq 0 \text{ for all } u \in P.$$

We shall now prove that

$$\langle \phi(\hat{x}) + \hat{y}, u - \psi(\hat{x}) \rangle \geq 0 \text{ for all } u \in P.$$

In the case $\|\hat{x}\| = r$, it follows that

$$\langle \phi(\hat{x}) + \hat{y}, \hat{u} - \psi(\hat{x}) \rangle = 0.$$

For each $u \in P$, let $z_t := tu + (1-t)\hat{u}$ for $t \in (0,1)$. Then $z_t \in B_r(P)$ and we have

$$0 \leq \langle \phi(\hat{x}) + \hat{y}, z - \psi(\hat{x}) \rangle \leq t\langle \phi(\hat{x}) + \hat{y}, u - \psi(\hat{x}) \rangle,$$

which implies that

$$\langle \phi(\hat{x}) + \hat{y}, u - \psi(\hat{x}) \rangle \geq 0$$

for all $u \in P$.

If $\|\hat{x}\| < r$, let $z_t := tu + (1-t)\hat{x}$ for each $t \in (0,1)$. Similarly, we can prove the inequality above. Now by Lemma 6.5.12, it follows that

$$\langle \phi(\hat{x}) + \hat{y}, \psi(\hat{x}) \rangle = 0$$

and $\phi(\hat{x}) + \hat{y} \in P^*$. Thus the proof is complete.□

We would like to point out that a result similar to Theorem 6.5.6 has been given by Fu [186] under different hypotheses. For more details in complementarity problems, the reader is referred to Isac's book [253], or Hyers et al [244].

Before concluding this section, we wish to remark that some applications of variational inequalities to the study of best approximation problems, mathematical economics and fixed point theorems can be found in Yuan [609] and related references. For the generalized variational inequalities without upper semicontinous assumption, the interested reader is referred to Cubiotti [109], Cubiotti and Yuan [111], and references therein.

6.6 Generalized Quasi-Variational Inequalities of Monotone and Quasi-Monotone Mappings

In this section, as applications of generalized quasi-variational inequalities in last section, we shall study the existence theorems of $GQVI(X; F, T)$ when T is a monotone mapping. Before we prove the existence theorems of $GQVI(X; F, T)$, we state the following result which is a special case of Lemma 2 of Tan and Yuan [510].

Lemma 6.6.1. Let X be a non-empty convex subset of a topological vector space E and $T : X \to 2^{E^*}$ lower semicontinuous with non-empty values from the line segments in X to the weak* topology on E^*. If $\hat{y} \in X$, then the inequality

$$\sup_{u \in T(x)} Re\langle u, \hat{y} - x \rangle \leq 0 \text{ for all } x \in X$$

implies

$$\sup_{w \in T(\hat{y})} Re\langle w, \hat{y} - x \rangle \leq 0 \text{ for all } x \in X.$$

Now we have the following existence result.

Theorem 6.6.2. Let X be a non-empty paracompact convex subset of a topological vector space E which has sufficiently many continuous linear functionals. Suppose $F : X \to 2^X$ is upper hemicontinuous with non-empty compact and convex values and the monotone mapping $T : X \to 2^{E^*}$ is lower semicontinuous with non-empty weak* compact values from the line segment of X to the weak* topology of E^*, and the set $\{x \in X : \sup_{y \in F(x)} \sup_{u \in T(x)} \mathrm{Re}\langle u, x - r \rangle \leq 0\}$ is closed in X. Furthermore assume that there exists a non-empty compact and convex subset K of X and a non-empty compact subset K of X such that for each $x \in X \setminus K$, there exists $y \in co(X_0 \cup \{x\})$ such that $\sup_{u \in T(y)} Re\langle u, x - y \rangle >$ 0. Then there exists $\hat{x} \in K$ such that

$$\hat{x} \in F(\hat{x}) \text{ and } \sup_{y \in F(\hat{x})} \sup_{u \in T(\hat{x})} Re\langle u, \hat{x} - y \rangle \leq 0.$$

Proof. Define a mapping $\psi : X \times X \to \mathbb{R}$ by

$$\psi(x, y) := \sup_{u \in T(y)} Re\langle u, x - y \rangle$$

for each $(x, y) \in X \times X$. Then we have that:

(a): for each fixed $y \in X$, $x \mapsto \psi(x, y)$ is compactly lower semicontinuous;

(b): for each fixed $x \in X$, $\psi(x, y)$ is 0-diagonal concave in y as T is monotone (e.g., see Proposition 3.2 of Zhou and Chen [622, p.222]);

(c): there exists a non-empty compact and convex subset X_0 of X and a non-empty compact subset K of X such that for each $x \in X \setminus K$, there exists $y \in F(x) \cap co(X_0 \cup \{x\})$ such that $\psi(x, y) > 0$; and

(d): the set $\{x \in X : \sup_{y \in F(x)} \psi(x, y) \leq 0\}$ is closed in X by our assumption.

Thus ψ satisfies all hypotheses of Theorem 6.5.1. By Theorem 6.5.1, there exists $\hat{x} \in K$ such that $\hat{x} \in F(\hat{x})$ and $\sup_{u \in T(y)} \psi(\hat{x}, u) \leq 0$ for all $y \in F(\hat{x})$, i.e.,

$$\sup_{y \in F(\hat{x})} \sup_{u \in T(y)} Re\langle u, \hat{x} - y \rangle \leq 0.$$

By Lemma 6.6.1, it follows that,

$$\sup_{y \in F(\hat{x})} \sup_{u \in T(\hat{x})} Re\langle u, \hat{x} - y \rangle \leq 0,$$

and we complete the proof. □

In what follows, we shall show that the conclusion of Theorem 6.6.2 still holds when T is a monotone operator with a weaker upper semicontinuity.

The following result was originally given by Tan and Yuan [510, Lemma 8], for completeness we include its proof here.

Lemma 6.6.3. Let X be a non-empty convex subset of a topological vector space E and $T : X \to 2^{E^*}$ be upper hemicontinuous from the line segments in X to the weak* topology of E^* such that each $T(x)$ is weak* compact. If $\hat{y} \in X$, then the inequality

$$\inf_{u \in T(x)} Re\langle u, \hat{y} - x \rangle \leq 0 \text{ for all } x \in X$$

implies the inequality

$$\inf_{w \in T(\hat{y})} Re\langle w, \hat{y} - x \rangle \leq 0 \text{ for all } x \in X.$$

Proof. Let $x \in X$ be arbitrarily fixed and $z_t = tx + (1-t)\hat{y} = \hat{y} - t(\hat{y} - x)$ for each $t \in [0,1]$. Since X is convex, $z_t \in X$ for all $t \in [0,1]$. Thus for all $t \in (0,1]$,

$$t \cdot \inf_{u \in T(z_t)} Re\langle u, \hat{y} - x \rangle = \inf_{u \in T(z_t)} Re\langle u, \hat{y} - z_t \rangle \leq 0$$

so that

$$\inf_{u \in T(z_t)} Re\langle u, \hat{y} - x \rangle \leq 0$$

for all $t \in (0,1]$. If $\inf_{w \in T(\hat{y})} Re\langle w, \hat{y} - x \rangle > 0$, let $G := \{w \in E^* : Re\langle w, \hat{y} - x \rangle > 0\}$. Then G is a weak*-open set in E^* such that $T(\hat{y}) \subset G$. As $z_t \to \hat{y}$ when $t \to 0^+$, by the upper semicontinuity of T on $\{z_t : t \in [0,1]\}$, there exists $t_0 \in (0,1]$ such that $T(x_t) \subset G$ for all $t \in (0, t_0)$. Note that $T(z_t)$ is weak* compact, $\inf_{u \in T(z_t)} Re\langle u, \hat{y} - x \rangle > 0$ for all $t \in (0, t_0)$ which contradicts the inequality above. Thus we must have $\inf_{w \in T(\hat{y})} Re\langle w, \hat{y} - x \rangle \leq 0$ and complete the proof. □

As an application of both Theorem 6.5.1 and Lemma 6.6.3, we have the following:

Theorem 6.6.4. Let X be a non-empty paracompact convex subset of a topological vector space E which has sufficiently many continuous linear functionals. Suppose $F : X \to 2^X$ is upper semicontinuous with non-empty compact and convex values and $T : X \to 2^{E^*}$ is monotone and upper semicontinuous from the line segments in X to the weak* topology of E^* such that each $T(x)$ is non-empty weak* compact, and the set $\{x \in X : \sup_{y\in F(x)} \inf_{u\in T(y)} \mathrm{Re}\langle u, x - y\rangle \le 0\}$ is closed in X. Furthermore assume that there exists a non-empty compact and convex subset X_0 of X and a non-empty compact subset K of X such that for each $x \in X \setminus K$, there exists $y \in co(X_0 \cup \{x\})$ such that $\inf_{u\in T(y)} Re\langle u, x - y\rangle > 0$. Then there exists $\hat{x} \in K$ such that $\hat{x} \in F(\hat{x})$ and $\sup_{y\in F(\hat{x})} \inf_{u\in T(\hat{x})} Re\langle u, \hat{x} - y\rangle \le 0$. Moreover if $T(\hat{x})$ is convex, there exists $\hat{u} \in T(\hat{x})$ such that

$$\sup_{y\in F(\hat{x})} Re\langle \hat{u}, \hat{x} - y\rangle \le 0$$

and thus $\mathrm{Re}\langle \hat{u}, \hat{x}\rangle = \inf_{y\in F(\hat{x})} \mathrm{Re}\langle \hat{u}, y\rangle$.

Proof. Define a mapping $\psi : X \times X \to \mathbb{R}$ by

$$\psi(x, y) := \inf_{u\in T(y)} Re\langle u, x - y\rangle$$

for each $(x, y) \in X \times X$. In order to apply Theorem 6.6.2, we shall show that $\psi(x, y)$ is 0-diagonal concave in y and compactly lower semicontinuous in x. For each fixed $y \in X$, the continuity of $x \mapsto \psi(x, y)$ is obvious, so that $x \mapsto \psi(x, y)$ is compactly lower semicontinous for each fixed $y \in X$. We need only prove $\psi(x, y)$ is 0-diagonal concave in y for each fixed $x \in X$. Suppose $y_0 := \Sigma_{i=1}^n \lambda_i y_i$, where $\alpha_i \in (0, 1]$ with $\Sigma_{i=1}^n \lambda_i = 1$ and $n \in \mathbb{N}$. Then $y_0 \in X$. Note that $T(y_i)$ is weak* compact, there exists $u_i \in T(y_i)$ such that $\inf_{u\in T(y_i)} Re\langle u, y_0 - y_i\rangle = Re\langle u_i, y_0 - y_i\rangle$. Let u_0 be any fixed element in $T(y_0)$. Then $Re\langle u_i, y_0 - y_i\rangle \le Re\langle u_0, y_0 - y_i\rangle$ by the monotonicity of T for each $i = 1, \cdots, n$. Thus,

$$\begin{aligned}
\Sigma_{i=1}^n \lambda_i \psi(y_0, y_i) &= \Sigma_{i=1}^n \lambda_i \inf_{u\in T(y_i)} Re\langle u, y_0 - y_i\rangle \\
&= \Sigma_{i=1}^n \lambda_i Re\langle u_i, y_0 - y_i\rangle \\
&\le \Sigma_{i=1}^n Re\langle u_0, y_0 - y_i\rangle \\
&= Re\langle u_0, y_0 - y_0\rangle = 0,
\end{aligned}$$

which implies that $\Sigma_{i=1}^n \lambda_i \psi(y_0, y_i) \le 0$, i.e., $\psi(x, y)$ is 0-diagonal in y. Moreover, by the hypothesis of Theorem 6.6.4, there exists a non-empty compact and convex subset X_0 of X and a non-empty compact subset K of X such that for each $x \in X \setminus K$, there exists $y \in F(x) \cap co(X_0 \cup \{x\})$ such that $\psi(x, y) > 0$. Thus ψ satisfies all hypotheses of Theorem 6.5.1. By Theorem 6.5.1, there exists $\hat{x} \in K$ such that $\hat{x} \in F(\hat{x})$ and $\sup_{u\in T(y)} \psi(\hat{x}, u) \le 0$ for all $y \in F(\hat{x})$, i.e.,

$$\sup_{y\in F(\hat{x})} \inf_{u\in T(y)} Re\langle u, \hat{x} - y\rangle \le 0.$$

As T is upper semicontinuous from the line segment of X to E^*, Lemma 6.6.3 implies that

$$\sup_{y\in F(\hat{x})} \inf_{u\in T(\hat{x})} Re\langle u, \hat{x} - y\rangle \le 0.$$

If $T(\hat{x})$ is convex, note that $T(\hat{x})$ is weak* compact and convex, Kneser minimax theorem shows that,

$$\inf_{u\in T(\hat{x})} \sup_{y\in F(\hat{x})} Re\langle u, \hat{x} - y\rangle = \sup_{y\in F(\hat{x})} \inf_{u\in T(\hat{x})} Re\langle u, \hat{x} - y\rangle \leq 0.$$

Note that $u \mapsto \sup_{y\in F(\hat{x})} Re\langle u, \hat{x} - y\rangle$ is weak* lower semicontinuous by the fact that $F(\hat{x})$ is compact and $T(\hat{x})$ is weak* compact, there exists $\hat{u} \in T(\hat{x})$ such that

$$\sup_{y\in F(\hat{x})} Re\langle \hat{u}, \hat{x} - y\rangle = \inf_{u\in T(\hat{x})} \sup_{y\in F(\hat{x})} Re\langle u, \hat{x} - y\rangle \leq 0.$$

Finally the above inequality implies that $\mathrm{Re}\langle \hat{u}, \hat{x}\rangle = \inf_{y\in F(\hat{x})} \mathrm{Re}\langle \hat{u}, y\rangle$ and we complete the proof. □

As compact versions of Theorems 6.5.5, 6.6.2 and 6.6.4, we have the following consequences.

Corollary 6.6.5. Let X be a non-empty compact and convex subset of a topological vector space E which has sufficiently many continuous linear functionals. Suppose $F : X \to 2^X$ is continuous with non-empty compact and convex values and $T : X \to 2^{E^*}$ is upper semicontinuous with non-empty compact values from the relative topology of X to the strong topology of E^*. Then there exists $\hat{x} \in X$ such that

$$\hat{x} \in F(\hat{x}) \text{ and } \sup_{y\in F(\hat{x})} \inf_{u\in T(\hat{x})} Re\langle u, \hat{x} - y\rangle \leq 0.$$

Moreover if $T(\hat{x})$ is convex, there exists $\hat{u} \in T(\hat{x})$ such that $\sup_{y\in F(\hat{x})} Re\langle \hat{u}, \hat{x} - y\rangle \leq 0$ and also $\mathrm{Re}\langle \hat{u}, \hat{x}\rangle = \inf_{y\in F(\hat{x})} \mathrm{Re}\langle \hat{u}, y\rangle$.

Corollary 6.6.6. Let X be a non-empty compact convex subset of a topological vector space E which has sufficiently many continuous linear functionals. Suppose $F : X \to 2^X$ is continuous with non-empty compact and convex values and the monotone mapping $T : X \to 2^{E^*}$ is lower semicontinuous with non-empty weak* compact values from X to the weak* topology of E^*. Then there exists $\hat{x} \in X$ such that

$$\hat{x} \in F(\hat{x}) \text{ and } \sup_{y\in F(\hat{x})} \sup_{u\in T(\hat{x})} Re\langle u, \hat{x} - y\rangle \leq 0.$$

Proof. Note that F is continuous and T is lower semicontinuous. It follows that $x \mapsto \sup_{y\in F(x)} \sup_{u\in T(x)} \mathrm{Re}\langle u, x - r\rangle$ is lower semicontinuous by III-Propositions 19 and 21 of Aubin and Ekeland [15, p.118 and p.119]. Thus the conclusion follows from Theorem 6.6.2. □

Corollary 6.6.7. Let X be a non-empty compact convex subset of a topological vector space E which has sufficiently many continuous linear functionals.

Suppose $F : X \to 2^X$ is upper semicontinuous with non-empty compact and convex values and $T : X \to 2^{E^*}$ is monotone and upper semicontinuous from the line segments in X to the weak* topology of E^* such that each $T(x)$ is non-empty weak* compact, and the set $\{x \in X : \sup_{y\in F(x)} \inf_{u\in T(y)} \text{Re}\langle u, x-y\rangle \le 0\}$ is closed in X. Then there exists $\hat{x} \in X$ such that

$$\hat{x} \in F(\hat{x}) \text{ and } \sup_{y\in F(\hat{x})} \inf_{u\in T(\hat{x})} Re\langle u, \hat{x} - y\rangle \le 0.$$

Moreover if $T(\hat{x})$ is convex, there exists $\hat{u} \in T(\hat{x})$ such that $\sup_{y\in F(\hat{x})} Re\langle \hat{u}, \hat{x} - y\rangle \le 0$ and thus we also have

$$\text{Re}\langle \hat{u}, \hat{x}\rangle = \inf_{y\in F(\hat{x})} \text{Re}\langle \hat{u}, y\rangle.$$

Finally as another application of Theorem 6.5.1, we have the following new generalization of the Fan-Glicksberg fixed point theorem for upper semicontinuous mappings in topological vector spaces, which have sufficiently many continuous linear functionals.

Theorem 6.6.8. Let X be a non-empty paracompact convex subset of a topological vector space E which has sufficiently many continuous linear functionals. Suppose $F : X \to 2^X$ is upper hemicontinuous with non-empty compact and convex values. Moreover assume that there exists a non-empty compact and convex subset X_0 of X and a non-empty compact subset K of X such that for each $x \in X\backslash K$ there exists $y \in F(x)\cap co(X_0 \cup \{x\})$ such that for each $x \in X\backslash K$, there exists $y \in F(x) \cap co(X_0 \cup \{x\})$ with $\inf_{u\in Z_0}\langle u, x-y\rangle > 0$; and the set $\{x \in X : \sup_{y\in F(x)} \inf_{u\in Z_0} \text{Re}\langle u, x-y\rangle \le 0\}$ is closed in X. Then there exists $x \in K$ such that $x \in F(x)$, i.e., F has a fixed point in X.

Proof. Let $\psi(x,y) := \inf_{u\in Z_0}\langle u, x-y\rangle$ for each $(x,y) \in X \times X$ in Theorem 6.5.1, then the conclusion follows and we complete the proof.□

It is clear that Theorem 3.6 reduces back to the original version of the Fan-Glicksberg fixed point theorem when X is compact and E is a locally convex Hausdorff topological vector space.

In what follows, we shall study the existence of solutions for quasi-monotone mappings. But we first need the following result.

Lemma 6.6.9. Let X be a convex subset of a topological vector space E and $T : X \to 2^*$ a set-valued quasi-monotone with non-empty values. Then the mapping $\psi : X \times X \to \mathbb{R} \cup \{-\infty, +\infty\}$ defined by

$$\psi(x,y) := \inf_{u\in T(y)} Re\langle u, x-y\rangle$$

for each $(x,y) \in X \times X$ is 0-diagonal concave in y for each fixed $x \in X$.

Proof. If it were not true, there exists $n \in \mathbb{N}$, $y_i \in X$ and $\lambda_i \in (0,1]$ for $i = 1, \cdots, n$ with $\Sigma_{i=1}^n \lambda_i = 1$ such that $\psi(x_0, y_i) > 0$, where $x_0 = \Sigma_{i=1}^n \lambda_i y_i$; i.e., $\inf_{u \in T(y_i)} Re\langle u, x_0 - y_i \rangle > 0$ for each $i = 1, \cdots, n$. As T is quasi-monotone, it follows that

$$\inf_{u \in T(x_0)} Re\langle u, x_0 - y_i \rangle > 0$$

for each $i = 1, \cdots, n$. Now multiplying both sides of the inequality above by λ_i and summing them for $i = 1, \cdots, n$, we then have

$$0 = \inf_{u \in T(x_0)} Re\langle u, x_0 - x_0 \rangle = \Sigma_{i=1}^n \lambda_i \inf_{u \in T(x_0)} Re\langle u, x_0 - y_i \rangle > 0$$

which is not true. Thus the conclusion must hold. $\square$.

As applications of Theorem 6.5.1, Lemmas 6.6.3 and 6.6.9, we have the following:

Theorem 6.6.10. Let X be a non-empty paracompact convex subset of a topological vector space E which has sufficiently many continuous linear functionals. Suppose $F : X \to 2^X$ is upper hemicontinuous with non-empty compact and convex values and $T : X \to 2^{E^*}$ is quasi-monotone and upper semicontinuous from the line segments in X to the $\delta(E^*, E)$ topology of E^* such that each $T(x)$ is non-empty $\delta(E^*, E)$ compact, and such that the set

$$\{x \in X : \sup_{y \in F(x)} \inf_{u \in T(y)} \mathrm{Re}\langle u, x - y \rangle \leq 0\}$$

is closed. Furthermore assume that there exists a non-empty compact and convex subset X_0 of X and a non-empty compact subset K of X such that for each $x \in X \backslash K$, there exists $y \in co(X_0 \cup \{x\})$ such that $\inf_{u \in T(y)} Re\langle u, x - y \rangle > 0$. Then there exists $\hat{x} \in K$ such that $\hat{x} \in F(\hat{x})$ and $\sup_{y \in F(\hat{x})} \inf_{u \in T(\hat{x})} Re\langle u, \hat{x} - y \rangle \leq 0$.

Moreover if $T(\hat{x})$ is convex, there exists $\hat{u} \in T(\hat{x})$ such that

$$\sup_{y \in F(\hat{x})} Re\langle \hat{u}, \hat{x} - y \rangle \leq 0$$

or equivalently,

$$Re\langle \hat{u}, \hat{x} \rangle = \inf_{y \in F(\hat{x})} Re\langle \hat{u}, y \rangle.$$

Proof. Define a mapping $\psi : X \times X \to \mathbb{R}$ by

$$\psi(x, y) := \inf_{u \in T(y)} Re\langle u, x - y \rangle$$

for each $(x, y) \in X \times X$. As T is quasi-monotone, Lemma 6.6.3 implies that $\psi(x, y)$ is 0-diagonal concave in y for each fixed $x \in X$. Moreover, it is obvious that $x \mapsto \psi(x,y)$ is compactly lower semicontinuous for each fixed $y \in X$. By our hypotheses, there exists a non-empty compact and convex subset X_0 of X and a non-empty compact subset K of X such that for each $x \in X \setminus K$, there exists $y \in F(x) \cap co(X_0 \cup \{x\})$ such that $\psi(x,y) > 0$. Thus ψ satisfies all hypotheses of Theorem 6.5.1. By Theorem 6.5.1, there exists $\hat{x} \in K$ such that $\hat{x} \in F(\hat{x})$ and $\sup_{u\in T(y)} \psi(\hat{x}, u) \leq 0$ for all $y \in F(\hat{x})$, i.e.,

$$\sup_{y\in F(\hat{x})} \inf_{u\in T(y)} Re\langle u, \hat{x} - y\rangle \leq 0.$$

As T is upper semicontinuous from the line segment of X to E^*, Lemma 6.6.1 implies that

$$\sup_{y\in F(\hat{x})} \inf_{u\in T(\hat{x})} Re\langle u, \hat{x} - y\rangle \leq 0.$$

If $T(\hat{x})$ is convex, note that $T(\hat{x})$ is $\delta(E^*, E)$ compact and convex, Kneser minimax theorem in [318] tells us,

$$\inf_{u\in T(\hat{x})} \sup_{y\in F(\hat{x})} Re\langle u, \hat{x} - y\rangle = \sup_{y\in F(\hat{x})} \inf_{u\in T(\hat{x})} Re\langle u, \hat{x} - y\rangle \leq 0.$$

Note that $u \mapsto \sup_{y\in F(\hat{x})} \mathrm{Re}\langle u, \hat{x} - y\rangle$ is $\delta(E^*, E)$ lower semicontinuous as $F(\hat{x})$ is compact and $t(\hat{x})$ is $\delta(E^*, E)$ compact, there exists $\hat{u} \in T(\hat{x})$ such that

$$\sup_{y\in F(\hat{x})} Re\langle \hat{u}, \hat{x} - y\rangle = \inf_{u\in T(\hat{x})} \sup_{y\in F(\hat{x})} Re\langle u, \hat{x} - y\rangle \leq 0. \tag{6.6.1}$$

It is clear that (6.6.1) implies that

$$Re\langle \hat{u}, \hat{x}\rangle = \inf_{y\in F(\hat{x})} Re\langle \hat{u}, y\rangle$$

as $\hat{x} \in F(\hat{x})$ and thus we complete the proof.□

Remark 6.6.11. We would like to note that the condition '*the set* $\{x \in X : \sup_{y\in F(x)} \inf_{u\in T(y)} Re\langle u, x-y\rangle \leq 0\}$ *is closed*' in Theorem 6.6.10 is automatically satisfied if F is a continuous set-valued mapping and T is upper semicontinuous.

As a compact version of Theorem 6.6.10, we have the following result.

Corollary 6.6.12. Let X be a non-empty compact convex subset of a topological vector space E which has sufficiently many continuous linear functionals. Suppose $F : X \to 2^X$ is continuous with non-empty compact and convex values and the quasi-monotone mapping $T : X \to 2^{E^*}$ is quasi-monotone and upper semicontinuous with non-empty $\delta(E^*, E)$ compact values from (the line segment) of X to the $\delta(E^*, E)$ topology of E^*. Then there exists $\hat{x} \in X$ such that

$$\hat{x} \in F(\hat{x}) \text{ and } \sup_{y\in F(\hat{x})} \inf_{u\in T(\hat{x})} Re\langle u, \hat{x} - y\rangle \leq 0.$$

If $\hat{T}(\hat{x})$ is convex, there exists $\hat{u} \in T(\hat{x})$ such that

$$Re\langle \hat{u}, \hat{x}\rangle = \inf_{y\in F(\hat{x})} Re\langle u, y\rangle.$$

6.7 Variational Inequalities in Banach Spaces

In this section, in order to give applications of variational inequalities to fixed point theorems and minimization problems, we first establish existence theorems of solutions of variational inequalities for monotone and quasi-monotone set-valued mappings in either topological vector spaces or reflexive Banach spaces by applying Ky Fan minimax inequalities. These results are then used to study the existence of solutions for generalized complementarity problems, and some fixed point theorems for set-valued quasi-contractive mappings, which are generalizations of pseudo-contractive nonexpansive mappings (which, in turn are generalizations of nonexpansive mappings). Finally existence theorems on minimizers of minimization problem for the sum of convex lower semicontinuous functions are established. These results include the corresponding results in the literature as special cases.

This section consists of two parts. In the first part, by applying the Ky Fan minimax principle [164] and Kneser's minimax theorem [318], the existence theorems of solutions of variational inequalities for monotone and quasi-monotone set-valued mapping, in either topological vector spaces or reflexive Banach spaces are established. In the second part, these results are used to study the existence of solutions for generalized complementarity problems, and some fixed point theorems for set-valued quasi-contractive (see definition below) and set-valued nonexpansive mappings are obtained. Finally existence theorems on minimizers of minimization problems for the sum of convex lower semicontinuous functions are established. These results generalize and unify the corresponding results of literature as special cases.

Definition 6.7.1. If X is a non-empty subset of E. Then a mapping $T : X \to 2^{E^*}$ is said to be

(1) *monotone* on X if for each $x, y \in X$, $u \in T(x)$ and $w \in T(y)$, $Re\langle w - u, y - x\rangle \geq 0$;

(2) *semi-monotone* (e.g., see Bae et al. [20]) on X if for each $x, y \in X$, $u \in T(x)$ and $w \in T(y)$,

$$\inf_{u\in T(x)} Re\langle u, y - x\rangle \leq \inf_{w\in T(y)} Re\langle w, y - x\rangle;$$

(3) *pseudo-monotone* (see Definition 3.4 of Saigal [451, p.263]) if for each $x, y \in K$, $u \in T(x)$ and $v \in T(y)$, $Re\langle x - y,\, v\rangle \geq 0$ implies $Re\langle x - y,\, u\rangle \geq 0$; and

(4) *quasi-monotone* if for each $x, y \in K$, $\inf_{v\in T(y)} Re\langle x - y,\, v\rangle > 0$ implies $\inf_{u\in T(x)} Re\langle x - y,\, u\rangle > 0$.

As we mentioned in Chapter 5, it is clear that each monotone mapping is pseudo-monotone, and Definition (4) above is a set-valued generalization of the

Definition (3) for a single-valued mapping. Also each semi-monotone mapping is quasi-monotone. But the converse is not necessarily true (e.g., see Karamardian and Schaible [283, Example] and references therein).

Definition 6.7.2. Let E be a vector space and $y \in E$. The set

$$I_X(y) := \{x \in E : x = y + r(u - y) \text{ for some } u \in X \text{ and } r > 0\}$$

is called the inward set (e.g., see Halpern and Bergman [218]) of y with respect to X.

If E is a topological vector space and $X \subset E$, we denote by either $\overline{X}$ or $cl_E X$ (in short, clX) the closure of X in E. If E is a normed space with norm $\|\cdot\|$, A is a non-empty subset of E and $x \in E$, we denote by $d(x, A) := \inf\{\|x - y\| : y \in A\}$ the distance from x to A. Let $CB(E) := \{A \subset E : A \text{ is non-empty closed and bounded}\}$. Then the Hausdorff metric D on $CB(E)$ induced by the norm is defined by

$$\begin{aligned} D(A_1, A_2) &= \inf\{r > 0 : A_1 \subset B_r(A_2) \textit{ and } A_2 \subset B_r(A_1)\} \\ &= \max\{\sup_{x \in A_1} d(x, A_2), \sup_{y \in A_2} d(y, A_1)\} \end{aligned}$$

where $B_r(A) := \{x \in E : d(x, A) > r\}$ for any $A \in 2^E$ and $r > 0$.

Definition 6.7.3. (1) A mapping $T : X \to CB(E)$ is said to be *nonexpansive* on X if for each $x, y \in X$, we have that $D(T(x), T(y)) \leq \|x - y\|$.

(2) A mapping $T : X \to 2^E$ is said to be *pseudo-contractive* (e.g., see Bae et al. [20]) on X if for each $x, y \in X$ and $w \in T(y)$, there exists $u \in T(x)$ such that $\|x - y\| \leq \|(1 + r)(x - y) - r(u - w)\|$ for all $r > 0$.

We note that Definition 6.7.3(2) is a set-valued generalization of pseudo-contractive (single-valued) mappings as defined by Browder [57].

The concept of *quasi-monotone mappings* as given above motivates us to introduce the following definition.

Definition 6.7.4. Let X be a non-empty subset of a normed space $(E, \|\cdot\|)$ and $T : X \to 2^E$ a set-valued mapping with non-empty values. Then T is said to be *quasi-contractive* if the mapping $I - T$ is a quasi-monotone mapping.

By the definition of a quasi-contractive mapping we shall see that (e.g., Lemma 6.9.1 in section 9 below) each pseudo-contractive mapping is quasi-contractive; but the converse is not necessarily true. Thus the quasi-monotone mapping is a generalization of a pseudo-monotone mapping which, in turn, is a generalization of nonexpansive mapping.

In this section, we first state some lemmas which have their own interests and will be used to study the existence of solutions of variational inequalities and complementarity problems in the sequel below.

Lemma 6.7.5. Let X be a non-empty convex subset of a topological vector space E and $T : X \to 2^{E^*}$ a set-valued quasi-monotone with non-empty values. Then the mapping $\psi : X \times X \to [-\infty, +\infty]$ defined by

$$\psi(x,y) := \inf_{u \in T(y)} Re\langle u, x-y\rangle$$

for each $(x,y) \in X \times X$ is 0-diagonal concave in y for each fixed $x \in X$.

Proof. If it were not true, there would exist $n \in \mathbb{N}$, $y_i \in X$ and $\lambda_i \in (0,1]$ for $i = 1, \cdots, n$ with $\Sigma_{i=1}^n \lambda_i = 1$ such that $\psi(x_0, y_i) > 0$, where $x_0 = \Sigma_{i=1}^n \lambda_i y_i$; i.e., $\inf_{u \in T(y_i)} Re\langle u, x_0 - y_i\rangle > 0$ for each $i = 1, \cdots, n$. As T is quasi-monotone, it follows that

$$\inf_{u \in T(x_0)} Re\langle u, x_0 - y_i\rangle > 0$$

for each $i = 1, \cdots, n$. Now multiplying both sides of the inequality above by λ_i and summing them for $i = 1, \cdots, n$, we then have

$$0 = \inf_{u \in T(x_0)} Re\langle u, x_0 - x_0\rangle = \Sigma_{i=1}^n \lambda_i \inf_{u \in T(x_0)} Re\langle u, x_0 - y_i\rangle > 0$$

which is not true. Thus the conclusion must hold. □.

Lemma 6.7.6. Let X be a convex subset of a topological vector space E and $T : X \to 2^{E^*}$ a set-valued monotone with non-empty values. Then the mapping $\psi : X \times X \to [-\infty, +\infty]$ defined by

$$\psi(x,y) := \inf_{u \in T(y)} Re\langle u, x-y\rangle$$

and

$$\psi(x,y) := \sup_{u \in T(y)} Re\langle u, x-y\rangle)$$

for each $(x,y) \in X \times X$ is also 0-diagonal concave in y for each fixed $x \in X$.

Proof. The conclusions follow immediately from the definitions of 0-diagonal concave and monotone of mappings. □

Lemma 6.7.7. Let X be a non-empty convex subset of a topological vector space E and $T : X \to 2^{E^*}$ a set-valued mapping with non-empty values. Assume the mapping $\psi : X \times X \to [-\infty, +\infty]$ defined by

$$\psi(x,y) := \inf_{u \in T(y)} Re\langle u, x-y\rangle$$

for each $(x,y) \in X \times X$ is 0-diagonal concave in y for each fixed $x \in X$. Then we have that for each $x, y \in X$ with $\inf_{u \in T(y)} Re\langle u, x-y\rangle > 0$, it implies that $\inf_{u \in T(x)} Re\langle u, y-x\rangle \leq 0$.

Proof. If the conclusion were not true, i.e., there exists $x, y \in X$ with $\inf_{u\in T(y)} Re\langle u, x-y\rangle > 0$, but $\inf_{u\in T(x)} Re\langle u, y-x\rangle > 0$. Let $x_0 = \frac{1}{2}(x+y)$. Then we have that $\psi(x_0, x) > 0$ and $\psi(x_0, y) > 0$, which contradict that $\psi(x, y)$ is 0-diagonal concave in y. Thus the conclusion is true. □

We shall need the following Kneser's minimax theorem [318] (see also Theorem 5.3.9 in the last chapter):

Theorem 6.7.8 (Kneser Minimax Inequality). Let X be a non-empty convex set in a vector space V and Y a non-empty compact convex subset of a Hausdorff topological vector space E. Suppose that f is a real-valued function on $X \times Y$ such that for each fixed $x \in X$, $f(x, y)$ is lower semicontinuous and convex on Y, and for each fixed $y \in Y$, $f(x, y)$ is concave on X. Then

$$\min_{y\in Y}\sup_{x\in X} f(x,y) = \sup_{x\in X}\min_{y\in Y} f(x,y).$$

The following result is a special case of Lemma 3 of Ding and Tan [135] and we state it as:

Lemma 6.7.9. Let X be a non-empty convex subset of a topological vector space E, $h : X \to [-\infty, +\infty]$ a convex function and $T : X \to 2^{E^*}$ lower semicontinuous from the line segments of X to the weak topology on E^*. If $\hat{y} \in X$, then the inequality

$$\sup_{u\in T(x)} Re\langle u, \hat{y} - x\rangle \le h(x) - h(\hat{y}) \text{ for all } x \in X$$

implies the inequality

$$\sup_{w\in T(\hat{y})} Re\langle w, \hat{y} - x\rangle \le h(x) - h(\hat{y}) \text{ for all } x \in X.$$

The same proofs of Lemmas 1 and 2 of Shih and Tan [471] can be modified to obtain the following slight improvement of Lemma 2 [471] and the proof is thus omitted.

Lemma 6.7.10. Let X be a non-empty convex subset of a topological vector space E, $h : X \to [-\infty, +\infty]$ a convex and lower semicontinuous function and $T : X \to 2^{E^*}$ a set-valued mapping such that each $T(x)$ is a $\sigma(E^*, E)$ compact subset of E^* and T is upper semicontinuous from the line segments of X to the weak topology of E^*. If $\hat{y} \in X$, then the inequality

$$\sup_{u\in T(x)} Re\langle u, \hat{y} - x\rangle \le h(x) - h(\hat{y}) \text{ for all } x \in X$$

implies the inequality

$$\inf_{w\in T(\hat{y})} Re\langle w, \hat{y} - x\rangle \le h(x) - h(\hat{y}) \text{ for all } x \in X.$$

Now it is time for us to establish variational inequalities of monotone mappings in Banach spaces.

As an application of the Ky Fan minimax inequality, we first have the following existence theorem of solution of variational inequalities of set-valued monotone mappings in topological vector spaces.

Theorem 6.7.11. Let X be a non-empty convex subset of a Hausdorff topological vector space E, $h : X \to [-\infty, +\infty]$ a convex and lower semicontinuous function and $T : X \to 2^{E^*}$ set-valued monotone with non-empty values. Suppose there exists a non-empty compact and convex subset X_0 of X and a non-empty compact subset K of X such that for each $y \in X \setminus K$, there exists $x \in co(X_0 \cup \{y\})$ such that

$$\sup_{u \in T(x)} Re\langle u, y - x\rangle + h(y) - h(x) > 0.$$

Then there exists $\hat{y} \in K$ such that

$$\sup_{u \in T(x)} Re\langle u, \hat{y} - x\rangle \leq h(x) - h(\hat{y}) \text{ for all } x \in X.$$

Proof. Define $\Phi : X \times X \to [-\infty, +\infty]$ by

$$\Phi(x, y) := \sup_{u \in T(x)} Re\langle u, y - x\rangle + h(y) - h(x)$$

for each $(x, y) \in X \times X$. Then Φ satisfies all hypotheses of the Ky Fan minimax inequality. By Fan's minimax inequality, there exists $\hat{y} \in K$ such that $\Phi(x, \hat{y}) \leq 0$ for all $x \in X$. By the definition of Φ, it follows that

$$\sup_{x \in X} \sup_{u \in T(x)} Re\langle u, \hat{y} - x\rangle \leq h(x) - h(\hat{y}).$$

Thus the proof is completed. $\square$

As an application of Theorem 6.7.11, we have the following result which is essentially Lemma 4 of Tan and Yuan [510].

Theorem 6.7.12. Let X be a non-empty closed convex subset of a reflexive Banach space $(E, \|\cdot\|)$. Let $h : X \to [-\infty, +\infty]$ be a convex and lower semicontinuous function and $T : X \to 2^{E^*}$ set-valued monotone with non-empty values. Assume that the following condition is satisfied:

(E) For each sequence $(y_n)_{n=1}^{\infty}$ in X with $\|y_n\| \to \infty$ as $n \to \infty$, there exists a sequence $(x_n)_{n=1}^{\infty}$ in X with $\|x_n\| \leq \|y_n\|$ for all $n = 1, 2, \cdots$ such that

$$\overline{\lim}_{n\to\infty}\{\sup_{u \in T(x_n)} Re\langle u, y_n - x_n\rangle + h(y_n) - h(x_n)\} > 0.$$

Then there exists $\hat{y} \in X$ such that

$$\sup_{u \in T(x)} Re\langle u, \hat{y} - x \rangle \leq h(x) - h(\hat{y}) \text{ for all } x \in X.$$

Proof. Define $\Phi : X \times X \to [-\infty, +\infty]$ by

$$\Phi(x, y) := \sup_{u \in T(x)} Re\langle u, y - x \rangle + h(y) - h(x)$$

for each $(x, y) \in X \times X$. Then we have

(1) As T is monotone, $\Phi(x, y)$ is 0-diagonal concave in y for each $x \in X$ by Lemma 6.7.6; and

(2) Note that h is convex and lower semicontinuous respect to the norm topology of X, h is lower semicontinuous with respect to the weak topology of X, it follows that for each $x \in X$, the function $y \to \Phi(x, y)$ is weakly lower semicontinuous.

Now let $X_N := \{x \in X : \|x\| \leq N\}$ for each $N \in \mathbb{N}$. We may assume that $X_N \neq \emptyset$ for all $N \geq N_0$. Note that for each $N \geq N_0$, X_N is weakly compact and convex since E is reflexive; equip X_N with the weak topology. By applying Theorem 6.7.11, there exists $\hat{y}_N \in X_N$ such that

$$\Phi(x, \hat{y}_N) \leq 0 \text{ for all } x \in X_N.$$

Suppose $\|\hat{y}_N\| \to \infty$ as $N \to \infty$, then by the assumption (E), there exists a sequence $(x_N)_{N \geq N_0}$ in X with $\|x_N\| \leq \|\hat{y}_N\|$ for all $N \geq N_0$ such that

$$\overline{\lim}_{N \to \infty} \sup_{u \in T(x_N)} Re\langle u, \hat{y}_N - x_N \rangle + h(\hat{y}_N) - h(x_N) > 0.$$

But, for each $N \geq N_0$, $\|x_N\| \leq \|\hat{y}_N\| \leq N$ implies $x_N \in X_N$ so that by the inequality above, $\Phi(x_N, \hat{y}_N) \leq 0$ for all $N \geq N_0$; i.e.,

$$\sup_{u \in T(x_N)} Re\langle u, \hat{y}_N - x_N \rangle + h(\hat{y}_N) - h(x_N) \leq 0 \text{ for all } N \geq N_0$$

which contradicts (2). Therefore we must have $\|\hat{y}_N\| \not\to \infty$ as $N \to \infty$. It follows that there exists a positive integer $M > N_0$ and a subsequence $(\hat{y}_{N(i)})_{i=1}^{\infty}$ of $(\hat{y}_N)_{N \geq N_0}$ such that $\|\hat{y}_{N(i)}\| \leq M$ for all $i = 1, 2, \cdots$. Thus $(\hat{y}_{N(i)})_{i=1}^{\infty}$ is a sequence in the weakly compact set X_M so that by Eberlein-Smulian Theorem (e.g., see Dunford and Schwartz [150, p.430]), there exists another subsequence $(\hat{y}_{N(i(j))})_{j=1}^{\infty}$ of $(\hat{y}_{N(i)})_{i=1}^{\infty}$ and $\hat{y} \in X_M$ such that $(\hat{y}_{N(i(j))})_{j=1}^{\infty}$ converges weakly to $\hat{y}$.

Now let $x \in X$ be given. Choose any positive integer $M' \geq M$ with $x \in X_{M'}$. Take any $j_0 \in \mathbb{N}$ with $N(i(j_0)) \geq M'$; then for all $j \geq j_0$, $x \in X_{M'} \subset X_{N(i(j_0))}$

so that by (1), $\Phi(x, \hat{y}_{N(i(j))}) \leq 0$. Since $(\hat{y}_{N(i(j))})_{j=1}^{\infty}$ converges weakly to $\hat{y}$ and $y \mapsto \Phi(x, y)$ is weakly lower semicontinuous by (2), we must have

$$\Phi(x, \hat{y}) \leq \underline{\lim}_{j \to \infty} \Phi(x, \hat{y}_{N(i(j))}) \leq 0.$$

Hence

$$\sup_{u \in T(x)} Re\langle u, \hat{y} - x \rangle \leq h(x) - h(\hat{y}) \text{ for all } x \in X. \qquad \square$$

Remark 6.7.13. We would like to note that our coercive condition (E) of Theorem 6.7.12 is quite general. For instance, our condition (E) includes the following coercive condition which is often used in the literature.

The Coercive Condition (D)**:** Let X be a non-empty subset of a Banach (more general, normed) space $(E, \|\cdot\|)$ and $T : X \to 2^{E^*}$ a set-valued mapping with non-empty values. Suppose there exists $x_0 \in X$ such that

$$\lim_{\|y\| \to \infty, y \in X} \inf_{w \in T(y)} Re\langle w, y - x_0 \rangle > 0,$$

or more strongly, we have

$$\lim_{\|y\| \to \infty, y \in X} \inf_{w \in T(y)} Re \frac{\langle w, y - x_0 \rangle}{\|y\|} = \infty.$$

If X is bounded, condition (E) of Theorem 6.7.12 is automatically satisfied. For example, we have the following:

Corollary 6.7.14. Let X be a non-empty bounded closed convex subset of a reflexive Banach space $(E, \|\cdot\|)$. Let $h : X \to [-\infty, +\infty]$ a convex and lower semicontinuous function and $T : X \to 2^{E^*}$ set-valued monotone with non-empty values. Then there exists $\hat{y} \in X$ such that

$$\sup_{u \in T(x)} Re\langle u, \hat{y} - x \rangle \leq h(x) - h(\hat{y}) \text{ for all } x \in X.$$

Proof. As X is bounded, condition (E) of Theorem 6.7.12 is satisfied and thus the conclusion follows from Theorem 6.7.12. $\square$

An another application of Theorem 6.7.11, we have the following result.

Theorem 6.7.15. Let X be a non-empty convex subset of a Hausdorff topological vector space E, $h : X \to [-\infty, +\infty]$ a convex and lower semicontinuous function and $T : X \to 2^{E^*}$ set-valued monotone with non-empty values. Suppose there exists a non-empty compact and convex subset X_0 of X and a non-empty compact subset K of X such that for each $y \in X \setminus K$, there exists $x \in co(X_0 \cup \{y\})$ such that

$$\sup_{u \in T(x)} Re\langle u, y - x \rangle + h(y) - h(x) > 0.$$

Then we have the following:

(I): If T is lower semicontinuous from the line segments of X to the weak topology of E^*, then there exists $\hat{y} \in K$ such that

$$\sup_{w \in T(\hat{y})} Re\langle w, \hat{y} - x\rangle \leq h(x) - h(\hat{y}) \text{ for all } x \in X.$$

(II): If T is upper semicontinuous from the line segments of X to the weak topology of E^* and each $T(x)$ is weakly compact, then there exists $\hat{y} \in K$ such that

$$\inf_{w \in T(\hat{y})} Re\langle w, \hat{y} - x\rangle \leq h(x) - h(\hat{y}) \text{ for all } x \in X.$$

If, in addition, $T(\hat{y})$ is also convex, then there exists $\hat{w} \in T(\hat{y})$ such that

$$Re\langle \hat{w}, \hat{y} - x\rangle \leq h(x) - h(\hat{y}) \text{ for all } x \in X.$$

Proof. By Theorem 6.7.11, there exists $\hat{y} \in K$ such that

$$\sup_{u \in T(x)} Re\langle u, \hat{y} - x\rangle \leq h(x) - h(\hat{y}) \text{ for all } x \in X.$$

(I): If T is lower semicontinuous from the line segments of X to the weak topology of E^*, then by the inequality above and Lemma 6.7.9,

$$\sup_{w \in T(\hat{y})} Re\langle w, \hat{y} - x\rangle \leq h(x) - h(\hat{y}) \text{ for all } x \in X.$$

(II): If T is upper semicontinuous from the line segments of X to the weak topology of E^* and each $T(x)$ is weakly compact, by the inequality above again and Lemma 6.7.10, we have

$$\inf_{w \in T(\hat{y})} Re\langle w, \hat{y} - x\rangle \leq h(x) - h(\hat{y})$$

for all $x \in X$.

If, in addition, $T(\hat{y})$ is also convex, define $f : X \times T(\hat{y}) \to [-\infty, +\infty]$ by

$$f(x, w) = Re\langle w, \hat{y} - x\rangle + h(\hat{y}) - h(x).$$

Note that for each fixed $x \in X$, $w \mapsto f(x, w)$ is weakly lower semicontinuous and affine and for each fixed $w \in T(\hat{y})$, $x \mapsto f(x, w)$ is concave. Thus by Kneser's minimax theorem, it follows that

$$\min_{w \in T(\hat{y})} \{\sup_{x \in X} Re\langle w, \hat{y} - x\rangle + h(\hat{y}) - h(x)\}$$

$$= \sup_{x \in X} \inf_{w \in T(\hat{y})} Re\langle w, \hat{y} - x\rangle + h(\hat{y}) - h(x)\}.$$

Since $T(\hat{y})$ is weakly compact, there exists $\hat{w} \in T(\hat{y})$ such that by the inequality above and the inequality of (II),

$$\begin{aligned}\sup_{x\in X}\{Re \quad \langle & \hat{w}, \hat{y} - x\rangle + h(\hat{y}) - h(x)\} \\ &= \min_{w\in T(\hat{y})} \sup_{x\in X}\{Re\langle w, \hat{y} - x\rangle + h(\hat{y}) - h(x)\} \leq 0,\end{aligned}$$

that is, $Re\langle \hat{w}, \hat{y} - x\rangle \leq h(x) - h(\hat{y})$ for all $x \in X$. □

Remark 6.7.16. We would like to note that Theorem 6.7.15 generalizes Theorem 2 of Yen [591, p.479-480] in several aspects. By following exactly the same proof of Theorem 6.7.15 and as an application of Theorem 6.7.12 instead of Theorem 6.7.11, we also have the following non-compact version of variational inequalities in reflexive Banach space under another coercive condition (see also Theorem 1 of Tan and Yuan [510]).

Theorem 6.7.17. Let X be a non-empty closed convex subset of a reflexive Banach space $(E, \|\cdot\|)$, $h : X \to [-\infty, +\infty]$ a convex and lower semicontinuous function and $T : X \to 2^{E^*}$ a set-valued monotone mapping. Assume that the following condition is satisfied:

(1)"(E)" For each sequence $(y_n)_{n=1}^{\infty}$ in X with $\|y_n\| \to \infty$ as $n \to \infty$, there exists a sequence $(x_n)_{n=1}^{\infty}$ in X with $\|x_n\| \leq \|y_n\|$ for all $n = 1, 2, \cdots$ such that

$$\overline{\lim}_{n\to\infty}\{\sup_{u\in T(x_n)} Re\langle u, y_n - x_n\rangle + h(y_n) - h(x_n)\} > 0.$$

(I): If T is lower semicontinuous from the line segments of X to the weak topology of E′, then there exists $\hat{y} \in X$ such that

$$\sup_{w\in T(\hat{y})} Re\langle w, \hat{y} - x\rangle \leq h(x) - h(\hat{y}) \text{ for all } x \in X.$$

(II): If T is upper semicontinuous from the line segments of X to the weak topology of E^* and each $T(x)$ is weakly compact, then there exists $\hat{y} \in X$ such that

$$\inf_{w\in T(\hat{y})} Re\langle w, \hat{y} - x\rangle \leq h(x) - h(\hat{y}) \text{ for all } x \in X.$$

If, in addition, $T(\hat{y})$ is also convex, then there exists $\hat{w} \in T(\hat{y})$ such that

$$Re\langle \hat{w}, \hat{y} - x\rangle \leq h(x) - h(\hat{y}) \text{ for all } x \in X.$$

Proof. By following the same proof used in the Proof of Theorem 6.7.15 and applying Theorem 6.7.12 instead of Theorem 6.7.11, we obtain the Theorem 6.7.17. □

Theorem 6.7.18. Let X be a non-empty closed convex subset of a reflexive Banach space $(E, \|\cdot\|)$, $h : X \to [-\infty, +\infty]$ a convex and lower semicontinuous function and $T : X \to 2^{E^*}$ a set-valued monotone mapping with non-empty values. Assume that the following condition is satisfied:

$(E)_\infty$ For each sequence $(y_n)_{n=1}^\infty$ in X with $\|y_n\| \to \infty$ as $n \to \infty$, there exists a sequence $(x_n)_{n=1}^\infty$ in X with $\|x_n\| \le \|y_n\|$ for all $n = 1, 2, \cdots$ such that

$$\overline{\lim}_{n\to\infty}\{\sup_{u\in T(x_n)} Re\langle u, y_n - x_n\rangle + h(y_n) - h(x_n)\}/\|y_n\| = \infty.$$

(I): If T is lower semicontinuous from line segments in X to the weak topology of E^*, then for each given $w_0 \in E^*$, there exists $\hat{y} \in X$ such that

$$\sup_{w\in T(\hat{y})} Re\langle w - w_0, \hat{y} - x\rangle \le h(x) - h(\hat{y}) \text{ for all } x \in X.$$

(II): If T is upper semicontinuous from line segments in X to the weak topology of E^* and each $T(x)$ is weakly compact and convex, then for each given $w_0 \in E^*$, there exists $\hat{y} \in X$ and $\hat{w} \in T(\hat{y})$ such that

$$Re\langle \hat{w} - w_0, \hat{y} - x\rangle \le h(x) - h(\hat{y}) \text{ for all } x \in X.$$

Proof. Let $w_0 \in E^*$ be given and we define $T^* : X \to 2^{E^*}$ by

$$T^*(y) := T(y) - w_0$$

for each $y \in X$. By condition $(E)_\infty$, for each sequence $(y_n)_{n=1}^\infty$ in X with $\|y_n\| \to \infty$ as $n \to \infty$, there exists a sequence $(x_n)_{n=1}^\infty$ in X with $\|x_n\| \le \|y_n\|$ for all $n = 1, 2, \cdots$ such that

$$\begin{aligned}
\overline{\lim}_{n\to\infty} \quad & \{\sup_{u\in T^*(x_n)} Re\langle u, y_n - x_n\rangle + h(y_n) - h(x_n)\}/\|y_n\| \\
&= \overline{\lim}_{n\to\infty}\{\sup_{u\in T(x_n)} Re\langle u - w_0, y_n - x_n\rangle + h(y_n) - h(x_n)\}/\|y_n\| \\
&\ge \overline{\lim}_{n\to\infty}\{\sup_{u\in T(x_n)} Re\langle u, y_n - x_n\rangle + h(y_n) - h(x_n)\}/\|y_n\| - 2\|w_0\| \\
&= \infty
\end{aligned}$$

since $|Re\langle w_0, y_n - x_n\rangle|/\|y_n\| \le \|w_0\| + \|w_0\|\|x_n\|/\|y_n\| \le 2\|w_0\|$.
It follows that

$$\overline{\lim}_{n\to\infty}\{\sup_{u\in T^*(x_n)} Re\langle w_0, y_n - x_n\rangle + h(y_n) - h(x_n)\} = \infty > 0$$

and hence the conclusion follows from Theorem 6.7.17 and we complete the proof. □

We note that condition (E) in Theorems 6.7.12 and 6.7.17 and the assumption that $(E)_\infty$ of Theorem 6.7.18 are automatically satisfied if X is bounded.

Also Theorem 4.1 of Chang and Zhang [76] is a special case of Theorem 6.7.18 under case (II). We also note that Theorems 6.7.15 and 6.7.17 are very closely related to but not comparable to Theorems 1 and 2 of Shih and Tan [471] and Theorems 3 and 4 of Shih and Tan [472].

Recall that subset X of a vector space E is said to be a cone if X is a non-empty convex set such that $\alpha X \subset X$ for all $\alpha \geq 0$. If X is a cone in a topological vector space E, X^* will denote the dual cone of X in E^*, i.e.,

$$X^* = \{y \in E^* : Re\langle y, x\rangle \geq 0 \text{ for all } x \in X\}.$$

The same proof of Lemma 2 of Shih and Tan [467] can be modified to obtain the following results and is thus omitted.

Lemma 6.7.19. Let X be a cone in a topological vector space E. Suppose that $T : X \to 2^{E^*}$ is a set-valued mapping with non-empty values and $\hat{y} \in X$. Then the following statements are equivalent:

(a) $\sup_{w\in T(\hat{y})} Re\langle w, \hat{y} - x\rangle \leq 0$ for all $x \in X$;
(b) $Re\langle w, \hat{y}\rangle = 0$ for all $w \in T(\hat{y})$ and $T(\hat{y}) \subset X^*$.

Lemma 6.7.20. Let X be a cone in a topological vector space E, $T : X \to 2^{E^*}$, $\hat{y} \in X$ and $\hat{w} \in T(\hat{y})$. Then the following statements are equivalent:
(a) $Re\langle \hat{w}, \hat{y} - x\rangle \leq 0$ for all $x \in X$;
(b) $Re\langle \hat{w}, \hat{y}\rangle = 0$ and $\hat{w} \in X^*$.

When E is real, Lemma 6.7.20 was also obtained by Fang (e.g., see Chan and Pang [64, p.213]).

In view of Lemmas 6.7.19 and 6.7.20, by taking $h \equiv 0$ in Theorem 6.7.15, we have the following theorem on the generalized complementarity problem.

Theorem 6.7.21. Let X be a closed cone in a reflexive Banach space $(E, \|\cdot\|)$ and $T : X \to 2^{E^*}$ monotone with non-empty values. Assume that the following condition is satisfied:

$(E)_0$ For each sequence $(y_n)_{n=1}^{\infty}$ in X with $\|y_n\| \to \infty$ as $n \to \infty$, there exists a sequence $(x_n)_{n=1}^{\infty}$ in X with $\|x_n\| \leq \|y_n\|$ for all $n = 1, 2, \cdots$ such that

$$\overline{\lim}_{n\to\infty} \sup_{u\in T(x_n)} Re\langle u, y_n - x_n\rangle > 0.$$

(I): If T is lower semicontinuous from the line segments in X to the weak topology of E^*, then there exists $\hat{y} \in X$ such that Re$\langle w, \hat{y}\rangle = 0$ for all $w \in T(\hat{y})$ and $T(\hat{y}) \subset X^*$.

(II): If T is upper semicontinuous from the line segments in X to the weak topology of E^* and each $T(x)$ is closed bounded and convex, then there exists $\hat{y} \in X$ and $\hat{w} \in T(\hat{y})$ such that $Re\langle \hat{w}, \hat{y}\rangle = 0$ and $\hat{w} \in X^*$.

Remark 6.7.22. Before we conclude this section, we would like to remark that if X is bounded, condition (E) (resp., $(E)_\infty$ or $(E)_0$) of Theorem 6.7.15 (resp., Theorem 6.7.18 or Theorem 6.7.21) is automatically satisfied. Thus our Theorems 6.7.15, 6.7.18 and 6.7.21 are non-compact versions of the corresponding variational inequalities known in the literature.

6.8 Variational Inequalities of Quasi-Monotone Mappings in Banach Spaces

In this section, we shall first establish the existence theorems of solutions for quasi-monotone mappings in reflexive Banach spaces and then these existence results will be used to derive fixed point theorems for set-valued mappings in Hilbert spaces.

We now state the following simple fact (see, also Lemma 1 of Bae et al. [20]):

Lemma 6.8.1. Let E^* be the dual space of a Hausdorff topological vector space and E. Let A be a non-empty bounded subset of E and C a non-empty(strongly) compact subset of E^*. Define $f : A \to [-\infty, +\infty]$ by

$$f(x) = \min_{u \in C} Re\langle u, x\rangle \text{ for all } x \in A.$$

Then f is weakly continuous on A.

Now we have the following result.

Lemma 6.8.2. Let X be a non-empty convex subset of a Hausdorff topological vector space E, $h : X \to [-\infty, +\infty]$ a convex function and $T : X \to 2^{E^*}$ upper semicontinuous from the line segments of X to the $\sigma(E^*, E)$ topology on E^* such that each $T(x)$ is $\sigma(E^*, E)$ compact. If $\hat{y} \in X$, then the inequality

$$\inf_{u \in T(x)} Re\langle u, \hat{y} - x\rangle \leq h(x) - h(\hat{y}) \text{ for all } x \in X$$

implies the inequality

$$\inf_{w \in T(\hat{y})} Re\langle w, \hat{y} - x\rangle \leq h(x) - h(\hat{y}) \text{ for all } x \in X.$$

Proof. Let $x \in X$ be arbitrarily fixed. For each $t \in [0, 1]$, let $z_t = tx + (1-t)\hat{y} = \hat{y} - t(\hat{y} - x)$. Since X is convex, $z_t \in X$ for all $t \in [0, 1]$. Thus for all $t \in (0, 1]$,

$$t \cdot \inf_{u \in T(z_t)} Re\langle u, \hat{y} - x\rangle = \inf_{u \in T(z_t)} Re\langle u, \hat{y} - z_t\rangle$$

$\le h(z_t) - h(\hat{y}) \le t \cdot h(x) + (1-t) \cdot h(\hat{y}) - h(\hat{y}) = t \cdot (h(x) - h(\hat{y}))$ so that $\inf_{u \in T(z_t)} Re\langle u, \hat{y} - x\rangle \le h(x) - h(\hat{y})$ for all $t \in (0.1]$.

If $\inf_{w \in T(\hat{y})} Re\langle w, \hat{y} - x\rangle > h(x) - h(\hat{y})$, let $G = \{w \in E^* : Re\langle w, \hat{y} - x\rangle + h(\hat{y}) - h(x) > 0\}$; then G is a $\sigma(E^*, E)$-open set in E^* such that $T(\hat{y}) \subset G$. As $z_t \to \hat{y}$ as $t \to 0^+$, by upper semicontinuity of T on $\{z_t : t \in [0,1]\}$, there exists $t_0 \in (0,1]$ such that $T(x_t) \subset G$ for all $t \in (0, t_0)$. As $T(z_t)$ is $\sigma(E^*, E)$ compact, $\inf_{u \in T(z_t)} Re\langle u, \hat{y} - x\rangle + h(\hat{y}) - h(x) > 0$ for all $t \in (0, t_0)$ which contradicts the inequality above. Thus we must have $\inf_{w \in T(\hat{y})} Re\langle w, \hat{y} - x\rangle \le h(x) - h(\hat{y})$. □

Theorem 6.8.3. Let X be a non-empty closed and convex subset of a reflexive Banach space $(E, \|\cdot\|)$, $h : X \to [-\infty, +\infty]$ a convex and lower semicontinuous function and $T : X \to 2^{E^*}$ quasi-monotone such that each $T(x)$ is compact in the norm topology on E^*. Assume that the following condition is satisfied:

$(E)^*$ for each sequence $(y_n)_{n=1}^{\infty}$ in X with $\|y_n\| \to \infty$ as $n \to \infty$, there exists a sequence $(x_n)_{n=1}^{\infty}$ in X with $\|x_n\| \le \|y_n\|$ for all $n = 1, 2, \cdots$ such that

$$\overline{\lim}_{n \to \infty} \{ \inf_{u \in T(x_n)} Re\langle u, y_n - x_n\rangle + h(y_n) - h(x_n)\} > 0.$$

Then there exists $\hat{y} \in X$ such that

$$\inf_{u \in T(x)} Re\langle u, \hat{y} - x\rangle \le h(x) - h(\hat{y}) \text{ for all } x \in X.$$

Proof. Define $\Phi : X \times X \to [-\infty, +\infty]$ by

$$\Phi(x, y) = \inf_{u \in T(x)} Re\langle u, y - x\rangle + h(y) - h(x)$$

for each $(x, y) \in X \times X$. Then we have

(1) for each fixed $y \in X$, $\Phi(x, y)$ is 0-diagonal concave in x by Lemma 6.7.5 as T is quasi-monotone; and

(2) since h is convex and lower semicontinuous with respect to the norm topology on X, h is also lower semicontinuous with respect to the weak topology on X. It follows that for each $x \in X$, the function $y \mapsto \Phi(x, y)$ is weakly lower semicontinuous on A for each non-empty norm-bounded subset A of X by Lemma 6.8.1.

Then the same proof of Theorem 6.7.12 with necessary modifications (all "$\sup_{u \in T(x)}$" and all "$\sup_{u \in T(x_N)}$" being replaced by "$\inf_{u \in T(x)}$" and "$\inf_{u \in T(x_N)}$" respectively), we see that there exists $\hat{y} \in X$ such that

$$\inf_{u \in T(x)} Re\langle u, \hat{y} - x\rangle \le h(x) - h(\hat{y}) \text{ for all } x \in X. \qquad \square$$

Theorem 6.8.4. Let X be a non-empty closed and convex subset of a reflexive Banach space $(E, \|\cdot\|)$, $h : X \to [-\infty, +\infty]$ be a convex and lower semicontinuous function and $T : X \to 2^{E^*}$ be quasi-monotone and upper semicontinuous from the line segments of X to the weak topology on E^* such that each $T(x)$ is compact in the norm topology on E^*. Assume that the following condition is satisfied:

$(E)^*$ for each sequence $(y_n)_{n=1}^{\infty}$ in X with $\|y_n\| \to \infty$ as $n \to \infty$, there exists a sequence $(x_n)_{n=1}^{\infty}$ in X with $\|x_n\| \leq \|y_n\|$ for all $n = 1, 2, \cdots$ such that

$$\overline{\lim}_{n\to\infty}\{\inf_{u\in T(x_n)} Re\langle u, y_n - x_n\rangle + h(y_n) - h(x_n)\} > 0.$$

Then there exists $\hat{y} \in X$ such that

$$\inf_{w\in T(\hat{y})} Re\langle w, \hat{y} - x\rangle \leq h(x) - h(\hat{y}) \text{ for all } x \in X.$$

If, in addition, $T(\hat{y})$ is also convex, then there exists $\hat{w} \in T(\hat{y})$ such that

$$Re\langle \hat{w}, \hat{y} - x\rangle \leq h(x) - h(\hat{y}) \text{ for all } x \in X.$$

Proof. By Lemma 6.8.3, there exists $\hat{y} \in X$ such that

$$\inf_{u\in T(x)} Re\langle u, \hat{y} - x\rangle \leq h(x) - h(\hat{y}) \text{ for all } x \in X.$$

By Lemma 6.8.2, we have

$$\inf_{w\in T(\hat{y})} Re\langle w, \hat{y} - x\rangle \leq h(x) - h(\hat{y}) \text{ for all } x \in X.$$

If, in addition, $T(\hat{y})$ is also convex, by Kneser's minimax theorem, and using the same argument as in the proof of Theorem 6.7.17, there exists $\hat{w} \in T(\hat{y})$ such that

$$Re\langle \hat{w}, \hat{y} - x\rangle \leq h(x) - h(\hat{y}) \text{ for all } x \in X. \qquad \square$$

In the proof of Theorem 6.7.18, if we replace all "$\sup_{u\in T^*(x_n)}$" and all "$\sup_{u\in T(x_n)}$" by "$\inf_{u\in T^*(x_n)}$" and "$\inf_{u\in T(x_n)}$" respectively, we have the following application of Theorem 6.8.4:

Theorem 6.8.5. Let X be a non-empty closed and convex subset of a reflexive Banach space $(E, \|\cdot\|)$, $h : X \to [-\infty, +\infty]$ a convex and lower semicontinuous function and $T : X \to 2^{E^*}$ quasi-monotone and upper semicontinuous from the line segments of X to the weak topology on E^* such that each $T(x)$ is convex and compact in the norm topology on E^*. Assume that the following condition is satisfied:

$(E)^*_\infty$ For each sequence $(y_n)_{n=1}^\infty$ in X with $\|y_n\| \to \infty$ as $n \to \infty$, there exists a sequence $(x_n)_{n=1}^\infty$ in X with $\|x_n\| \le \|y_n\|$ for all $n = 1, 2, \cdots$ such that

$$\overline{\lim}_{n\to\infty}\{\inf_{u\in T(x_n)} Re\langle u, y_n - x_n\rangle + h(y_n) - h(x_n)\}/\|y_n\| = \infty.$$

Then for each given $w_0 \in E^*$, there exists $\hat{y} \in X$ and $\hat{w} \in T(\hat{y})$ such that

$$Re\langle \hat{w} - w_0, \hat{y} - x\rangle \le h(x) - h(\hat{y}) \text{ for all } x \in X.$$

By Lemma 6.7.20 and by taking $h \equiv 0$ in Theorem 6.8.4, we have the following theorem on the generalized complementarity problem.

Theorem 6.8.6. Let X be a non-empty closed and convex subset of a reflexive Banach space $(E, \|\cdot\|)$ and $T : X \to 2^{E^*}$ quasi-monotone and upper semicontinuous from the line segments in X to the weak topology on E^* such that each $T(x)$ is convex and compact in the norm topology on E^*. Assume that the following condition is satisfied:

$(E)^*_0$ For each sequence $(y_n)_{n=1}^\infty$ in X with $\|y_n\| \to \infty$ as $n \to \infty$, there exists a sequence $(x_n)_{n=1}^\infty$ in X with $\|x_n\| \le \|y_n\|$ for all $n = 1, 2, \cdots$ such that

$$\overline{\lim}_{n\to\infty}\{\inf_{u\in T(x_n)} Re\langle u, y_n - x_n\rangle\} > 0.$$

Then there exists $\hat{y} \in X$ and $\hat{w} \in T(\hat{y})$ such that

$$Re\langle \hat{w}, \hat{y}\rangle = 0 \quad \textit{and} \quad \hat{w} \in X^*.$$

6.9 Applications to Fixed Points and Minimization Problems

The purpose of this section is to apply variational inequalities to study fixed point theorems and minimization problems . Before we prove the fixed point theorem for quasi-contractive set-valued mappings in Hilbert spaces, we first recall the following result which is the Proposition of Bae et al. [20]; we shall include its proof for the sake of completeness.

Lemma 6.9.1. Let X be a non-empty subset of a Hilbert space H. Then we have

(a) If $T : X \to CB(H)$ is nonexpansive and for each $x \in X$, $T(x)$ is weakly compact, then T is pseudo-contractive on X; and
(b) If $T : X \to 2^H$ is pseudo-contractive on X, then $I - T$ is semi-monotone on X where $I(x) := x$ for all $x \in X$.

Proof. (a) Let $x, y \in X$ and $w \in T(y)$. Since $T(y)$ is weakly compact, there exists $u \in T(x)$ such that $\|w-u\| = d(w, T(x))$. But we see that $d(w, T(x)) \leq D(T(x), T(y)) \leq \|x-y\|$, it follows that $\|w-u\| \leq \|x-y\|$. Thus for all $r > 0$,

$$\|(1+r)(x+y) - r(u-w)\| \geq (1+r)\|x-y\| - r\|u-w\| \geq \|x-y\|.$$

Therefore T is pseudo-contractive on X.

(b) Since T is pseudo-contractive on X, for each $x, y \in X$ and for each $w \in T(x)$, there exists $u \in T(x)$ such that for all $r > 0$,

$$\begin{aligned} \|x-y\|^2 &\leq \|(1+r)(x-y) - r(u-w)\|^2 \\ &= \|(x-y) + r((x-y) - (u-w))\|^2 \\ &= \|x-y\|^2 + 2r \cdot Re\langle (x-u) - (y-w), x-y\rangle \\ &+ r^2 \cdot \|(x-y) - (u-w)\|^2. \end{aligned}$$

Therefore we have

$$2Re\langle (x-u) - (y-w), x-y)\rangle + r \cdot \|(x-y) - (u-w)\|^2 \geq 0.$$

By taking $r \to 0^+$, we have $Re\langle (x-u) - (y-w), x-y\rangle \geq 0$ so that

$$Re\langle y-w, y-x\rangle \geq Re\langle x-u, y-x\rangle \geq \inf_{u \in T(x)} Re\langle x-u, y-x\rangle.$$

Thus $\inf_{w \in T(y)} Re\langle y-w, y-x\rangle \geq \inf_{u \in T(x)} Re\langle x-u, y-x\rangle$ and $I - T$ is a semi-monotone mapping on X. □

Remark 6.9.2. Lemma 6.9.1 shows that the class of quasi-contractive mappings includes the class of pseudo-contractive mappings as a special case.

Now as an application of Theorem 6.8.4, we have the following fixed point theorem in Hilbert spaces.

Theorem 6.9.3. Let X be a non-empty closed convex subset of a Hilbert space H and $T : X \to 2^H$ be quasi-contractive and upper semicontinuous from the line segments in X to the weak topology on H such that each $T(x)$ is compact in the norm topology on H. Assume that the following condition is satisfied:

$(E)^+$ For each sequence $(y_n)_{n=1}^{\infty}$ in X with $\|y_n\| \to \infty$ as $n \to \infty$, there exists a sequence $(x_n)_{n=1}^{\infty}$ in X with $\|x_n\| \leq \|y_n\|$ for all $n = 1, 2, \cdots$ such that

$$\overline{\lim}_{n\to\infty}\{\inf_{u \in T(x_n)} Re\langle x_n - u, y_n - x_n\rangle\} > 0.$$

Then there exists $\hat{y} \in X$ such that

$$\inf_{w \in T(\hat{y})} Re\langle \hat{y} - w, \hat{y} - x\rangle \leq 0 \; for \; all \; x \in X.$$

If $T(\hat{y})$ is also convex, then there exists $\hat{w} \in T(\hat{y})$ such that

$$Re\langle \hat{y} - \hat{w}, \hat{y} - x\rangle \leq 0 \text{ for all } x \in \overline{I_X(\hat{y})},$$

and if, in addition, either $\hat{y}$ is an interior point of X in H or $P(\hat{y}) \in \overline{I_X(\hat{y})}$, where $P(\hat{y})$ is the projection of $\hat{y}$ on $T(\hat{y})$, then $\hat{y}$ is a fixed point of T, i.e., $\hat{y} \in T(\hat{y})$.

Proof. As T is quasi-monotone, by Theorem 6.8.4 with $h = 0$, there exists $\hat{y} \in X$ such that

$$\inf_{w \in T^*(\hat{y})} Re\langle w, \hat{y} - x\rangle \leq 0 \text{ for all } x \in X;$$

this is,

$$\inf_{w \in T(\hat{y})} Re\langle \hat{y} - w, \hat{y} - x\rangle \leq 0 \text{ for all } x \in X.$$

If $T(\hat{y})$ is also convex, then by Theorem 6.8.4 again, there exists $\hat{w} \in T(\hat{y})$ such that

$$(1) \qquad Re\langle \hat{y} - \hat{w}, \hat{y} - x\rangle \leq 0 \text{ for all } x \in X.$$

If $x \in I_X(\hat{y})$, then $x = \hat{y} + r(u - \hat{y})$ for some $u \in X$ and $r > 0$. Thus $\hat{y} - x = r(\hat{y} - u)$ so that by (1),

$$Re\langle \hat{y} - \hat{w}, \hat{y} - x\rangle = r \cdot Re\langle \hat{y} - \hat{w}, \hat{y} - u\rangle \leq 0.$$

It follows that

$$(2) \qquad Re\langle \hat{y} - \hat{w}, \hat{y} - x\rangle \leq 0 \text{ for all } x \in \overline{I_X(\hat{y})}.$$

Now, if $\hat{y}$ is an interior point of X in H, then (2) implies that $\hat{y} = \hat{w} \in T(\hat{y})$. Next suppose $P(\hat{y}) \in \overline{I_X(\hat{y})}$. Since $P(\hat{y})$ is the projection of $\hat{y}$ on $T(\hat{y})$, we must have, by Theorem I.2.3 of Kinderlehrer and Stampacchia [304], $P(\hat{y}) \in T(\hat{y})$ and $Re\langle P(\hat{y}) - \hat{y}, w - P(\hat{y})\rangle \geq 0$ for all $w \in T(\hat{y})$. Since $\hat{w} \in T(\hat{y})$, by (2) we have

$$\begin{aligned} 0 &\leq Re \langle P(\hat{y}) - \hat{y}, \hat{w} - P(\hat{y})\rangle \\ &= Re\langle P(\hat{y}) - \hat{y}, \hat{w} - \hat{y} + \hat{y} - P(\hat{y})\rangle \\ &= Re\langle P(\hat{y}) - \hat{y}, \hat{w} - \hat{y}\rangle - \|\hat{y} - P(\hat{y})\|^2 \\ &= Re\langle \hat{y} - \hat{w}, \hat{y} - P(\hat{y})\rangle - \|\hat{y} - P(\hat{y})\|^2 \leq -\|P(\hat{y}) - \hat{y}\|^2 \end{aligned}$$

so that $\|P(\hat{y}) - \hat{y}\|^2 \leq 0$ and hence $\hat{y} = P(\hat{y}) \in T(\hat{y})$. □

As each nonexpansive set-valued mapping with weakly compact values is pseudo-contractive, and each pseudo-contractive set-valued mapping is a quasi-contractive mapping, we have the following fixed point theorems in Hilbert spaces.

Theorem 6.9.4. Let X be a non-empty closed convex subset of a Hilbert space H and $T : X \to 2^H$ be pseudo-contractive and upper semicontinuous (for example, T is nonexpansive) such that each $T(x)$ is compact convex and $P(y) \in \overline{I_X(y)}$ for each $y \in \partial X$ where $P(y)$ is the projection of y on $T(y)$ and ∂X is the boundary of X in H. Assume that the following condition is satisfied:

$(E)^+$ For each sequence $(y_n)_{n=1}^\infty$ in X with $\|y_n\| \to \infty$ as $n \to \infty$, there exists a sequence $(x_n)_{n=1}^\infty$ in X with $\|x_n\| \leq \|y_n\|$ for all $n = 1, 2, \cdots$ such that

$$\overline{\lim}_{n\to\infty}\{\inf_{u \in T(x_n)} Re\langle x_n - u, y_n - x_n\rangle\} > 0.$$

Then T has a fixed point in X.

We remark that except that the set X is required to be closed in H, the above result is a generalization of Theorem 8 of Tan [504, p.561] to set-valued and non-self mappings.

As an special case of Theorem 6.9.4, we have the following fixed point theorem for nonexpansive set-valued mappings in Hilbert spaces.

Corollary 6.9.5. Let X be a non-empty bounded closed convex subset of a Hilbert space H and $T : X \to 2^H$ be pseudo-contractive and upper semicontinuous (for example, T is nonexpansive) such that each $T(x)$ is compact convex and $P(y) \in \overline{I_X(y)}$ for each $y \in \partial X$ where $P(y)$ is the projection of y on $T(y)$ and ∂X is the boundary of X in H. Then T has a fixed point in X.

Proof. As X is bounded, condition $(E)^+$ of Theorem 6.9.4 is satisfied and thus the conclusion folllows. □

We also note that condition $(E)^*$ in Lemma 6.8.3 and Theorem 6.8.4, condition $(E)^*_\infty$ in Theorem 6.8.5 and $(E)^+$ of Theorems 6.9.3 and 6.9.4 are automatically satisfied if the set X is bounded.

Finally we remark that given any increasing sequence $(N_n)_{n=1}^\infty$ of positive integers, the conclusions of Theorems 6.7.12, 6.8.3 and Theorems 6.7.15 to Theorem 6.9.4 remain valid if we replace the phrase "$\cdots$ there exists a sequence $(x_n)_{n=1}^\infty$ in X with $\|x_n\| \leq \|y_n\|$ for all $n = 1, 2, \cdots$" in conditions (E), $(E)_\infty$, $(E)_0$, $(E)^*$, $(E)^*_\infty$, $(E)^*_0$ and $(E)^+$ by the phrase "$\cdots$ there exists a sequence $(x_n)_{n=1}^\infty$ in X with $\|x_n\| \leq N_n$ for all $n = 1, 2, \cdots$".

We would like to note that our results in this section improve the corresponding results of Aubin and Ekeland [15], Bae et al. [20], Browder [62], Chan and Pang [64], Chang and Zhang [76], Ding and Tan [135]-[136], Ding and Tarafdar [144], Shih and Tan [466]-[467], Tan [504], Tan and Yuan [510], Yen [591], Zhou and Chen [622], and many others. Finally we also note that as applications of variational inequalities, some existence theorems of fixed points for quasi-contractive set-valued mappings have been established. For an updated study of the subject for fixed point theory in both metric spaces and topological spaces, the interested reader is referred to Goebel and Kirk [198], or a recent book written by Jaworowski et al. [266]. For some other study of variational inequalities and topological fixed point theorems as applications of Ky Fan type minimax inequalities in general topological spaces (such as generalized convex spaces and so on), we refer them to Park's paper [412] and references therein.

Before we conclude this section, as an application of variational inequalities that were established in Section 7 above, we shall consider the existence of solutions for the following minimization problem:

$$\inf_{x \in E} f(x) \tag{6.9.1}$$

where f is the sum of two extended real-valued functions $g, h : E \to (-\infty, +\infty]$, and E is a normed space. Before we prove the existence of solutions for (6.9.1), we recall the following definition of subdifferential (e.g., see Aubin and Ekeland [15, 187]):

Definition 6.9.6. Let X be a non-empty convex subset of a topological vector space E. Suppose $f : X \to (-\infty, +\infty]$ is a function with a non-empty domain. If $x_0 \in Domf$, the '*subdifferential* $\partial f(x_0)$ *of* f *at* x_o' is the subset (which may be empty) of E^* defined by

$$\partial f(x_0) = \{p \in E^* : f(x_0) - f(x) \leq \langle p, x_0 - x\rangle \text{ for all } x \in X\}.$$

The element $p \in \partial f(x_0)$ are also called subgradients.

The following simple Proposition shows that the existence of solutions of variational inequalities guarantees sufficiently the existence of the minimizers for the minimization problem (6.9.1).

Proposition 6.9.7. Let X be a non-empty convex subset of a Hausdorff topological vector space E. Suppose $f = g + h$ is the sum of a convex function g and a subdifferential function h defined on a non-empty convex subset X, i.e., $g, h : X \to (-\infty, +\infty]$. Then a point $\hat{x} \in X$ minimizes f if the there exists $p \in \partial h(\hat{x})$ such that

$$\sup_{x \in X}[\langle p, \hat{x} - x\rangle + g(\hat{x}) - g(x)] \leq 0.$$

Proof. Suppose there exists $p \in \partial h(\hat{x})$ such that $\langle p, \hat{x} - x\rangle + g(\hat{x}) - g(x) \leq 0$ for all $x \in X$. Then we have that

$$f(\hat{x}) - f(x) = h(\hat{x}) - h(x) + g(\hat{x}) - g(x) \leq \langle p, \hat{x} - x\rangle + g(\hat{x}) - g(x) \leq 0$$

for all $x \in X$. Thus $\hat{x}$ minimizes f and the proof is completed. □

Now we have the following general existence theorem which guarantee the existence of minimizers for the minimization problem (6.9.1).

Theorem 6.9.8. Let X be a non-empty convex subset of a Hausdorff topological vector space E and $f = g + h$ be the sum of two extended-valued functions $g, h : E \to (-\infty, +\infty]$, where g is a convex and lower semicontinuous function and the subdifferential mapping of h which is defined by $T(x) = \partial h(x)$ for each $x \in X$ is upper semicontinuous from the line of X to the weak topology of E^* with non-empty weakly compact and convex values. Suppose there exists a non-empty compact and convex subset X_0 of X and a non-empty compact subset K of X such that for each $y \in X \setminus K$, there exists $x \in co(X_0 \cup \{y\}$ such that

$$\sup_{u \in \partial h(x)} \mathrm{Re}\langle u, y - x\rangle + h(y) - h(x) > 0.$$

Then there exists a minimizer x_0 which is a solution of Problem (6.9.1), i.e., $f(x_0) = \inf_{x \in E} f(x) = \inf_{x \in E}[g(x) + h(x)]$.

Proof. By our hypotheses, g and T satisfy all conditions of Theorem 6.7.15 (II). By Theorem 6.7.15 (II), there exists $\hat{x} \in K$ and $p \in T(\hat{x})$ such that

$$\sup_{x \in X} \mathrm{Re}\langle p, \hat{x} - x\rangle \leq g(x) - g(\hat{x}).$$

Then Proposition 6.9.7 implies that $\hat{x}$ is a solution of the minimization problem (6.9.1) and thus the proof is completed. □

In what follows, we shall give some sufficient conditions for the function h such that h's subdifferential mapping satisfies all conditions posed in Proposition 6.9.7.

Theorem 6.9.9. Let E be a normed space and $f = g + h$ be the sum of two extended-valued functions $g, h : E \to (-\infty, +\infty]$, which are both convex and lower semicontinuous and the domain $Domh(:= \{x \in E : -\infty < h(x) < +\infty\}) = E$. Suppose there exists a non-empty compact and convex subset X_0 of E and a non-empty compact subset K of E such that for each $y \in E \setminus K$, there exists $x \in co(X_0 \cup \{y\}$ such that

$$\sup_{u \in \partial h(x)} \mathrm{Re}\langle u, y - x\rangle + h(y) - h(x) > 0.$$

Then there exists a minimizer x_0 which is a solution of Problem (6.9.1), i.e., $f(x_0) = \inf_{x \in E} f(x) = \inf_{x \in E}[g(x) + h(x)]$.

Proof. By the hypothesis, we have that $Domh = E$. As h is convex and lower semicontinuous, Theorem 17 of Aubin and Ekeland [15, p.199] implies that $\partial h(x)$ is a non-empty, bounded closed convex of E^* (and indeed weakly compact) for each $x \in X$. Now define $T : X \to 2^{E^*}$ by $T(x) := \partial h(x)$ for each $x \in X$. By the definition of subdifferential, it is clear to see that T is monotone and upper hemicontinuous on X (e.g., see Theorem 17 of Aubin and Ekeland [15, p.200] again). As $T(x)$ is weakly compact for each $x \in X$, it follows that T is also upper semicontinuous by Theorem 10 of Aubin and Ekeland [15, p.128]. Thus the conclusion follows from Theorem 6.9.8. Indeed, by applying the varitaional inequality Theorem 6.7., there exists $\hat{x} \in X$ and $p \in T(\hat{x})$ such that

$$\sup_{x \in X}[\langle p, \hat{x} - x\rangle + g(\hat{x}) - g(x)] \leq 0.$$

Therefore Proposition 6.9.7 above shows that $\hat{x}$ is a minimizer of the minimization problem (6.9.1) and the proof is complete. □

By the same proof of Theorem 6.9.9 and as an application of Theorem 6.7.17 instead of Theorem 6.7.15, we have the following existence of minimizer for the minimization problem (6.9.1) and its proof is omitted here.

Theorem 6.9.10. Let E be a reflexive Banach and $f = g + h$ be the sum of two extended-valued functions $g, h : E \to (-\infty, +\infty]$, which are both convex and lower semicontinuous and the domain $Domh = E$. Assume that the following condition (E) is satisfied:

(E) For each sequence $(y_n)_{n=1}^{\infty}$ in E with $\|y_n\| \to \infty$ as $n \to \infty$, there exists a sequence $(x_n)_{n=1}^{\infty}$ in E with $\|x_n\| \leq \|y_n\|$ for all $n = 1, 2, \cdots$ such that

$$\overline{\lim}_{n\to\infty}\{\sup_{u\in\partial h(x_n)} Re\langle u, y_n - x_n\rangle + h(y_n) - h(x_n)\} > 0.$$

Then there exists a minimizer x_0 which is a solution of the Problem (6.9.1), i.e., $f(x_0) = \inf_{x\in E} f(x) = \inf_{x\in E}[g(x) + h(x)]$.

6.10 Implicitly Quasi-Variational Inequalities

In this section, some new existence theorems for solutions of quasi-variational inequalities are given as applications of either generalized games or fixed point theorems. These results are then applied to obtain the existence of solutions for generalized complementarity problems. Finally the quasi-variational inequality of the obstacle type in reflexive Banach spaces is also studied. Our results include most of the corresponding results in the literature as special cases.

The aim of this section is two-fold. We first prove some new existence theorems for quasi-variational inequalities as applications of either generalized games or fixed point theorems. Secondly, these variational inequalities are then applied to discuss the existence of solutions for generalized complementarity problems. Finally the quasi-variational inequality of the obstacle type in reflexive Banach spaces is also obtained.

Let X and C be non-empty convex subsets of a locally convex topological vector space E; $A : X \to 2^X$ a set-valued mapping with non-empty compact and convex values, $F : X \to 2^C$ a set-valued mapping with non-empty compact and convex values and $\theta : X \times C \to E^*$ a single-valued mapping. We shall study the following generalized quasi-variational inequality $(GQVI)$:

To find $x_0 \in A(x_0)$ and $y_0 \in F(x_0)$ such that

$$\inf_{u \in A(x_0)} Re\langle \theta(x_0, y_0), u - x_0 \rangle \geq 0.$$

The study of the generalized quasi-variational inequalities $(GQVI)$ was first introduced and discussed by Chan and Pang in 1982 for the purpose of applications in the study of operation research, mathematical programming, and optimization theory in 1982. Since then, there have been a number of generalizations of the existence theorems about $(GQVI)$, e.g., see Aubin and Ekeland [15], Cottle et al. [101], Cubiotti [108], Harker and Pang [220] and many others. These results have wide applications to problems in games theory, economics and mathematical programming. For more details, the interested reader is referred to Cottle et al.'s survey book [102], Isac [253] and Hyers et al. [244]. We wish to note that the above problem $(GQVI)$ which will be discussed in this section is of a general form, and it includes most of generalized quasi-variational inequalities in the literature as special cases.

In order to give our first existence theorem of $(GQVI)$ problem, we state the following result which is a special case of Lemma 6.2.6 (see also Theorem 4.5.13).

Lemma 6.10.1. Let X be a non-empty paracompact closed convex subset of a locally convex topological vector space E such that X has the property K. Suppose that $A, P : X \to 2^X$ satisfy the following:

(i) $A : X \to 2^X$ is compact and upper semicontinuous with non-empty compact convex values;

(ii) $P : X \to 2^X$ has open inverse values, i.e., the set $P^{-1}(y) := \{x \in X : y \in P(x)\}$ is open in X for each $y \in X$;

(iii) the set $\{x \in X : (A \cap P)(x) \neq \emptyset\}$ is open in X; and

(iv) there exists a non-empty compact convex subset X_0 of X and a non-empty compact subset K of X such that for each $x \in X \setminus K$ there is an $y \in co(X_0 \cup \{x\})$ with $y \in co(A(x) \cap P(x))$.

Then there exists $\bar{x} \in K$ such that $\bar{x} \in A(\bar{x})$ and $A(\bar{x}) \cap P(\bar{x}) = \emptyset$.

Now we have the following implicitly quasi-variational inequality.

Theorem 6.10.2. Let X and C be non-empty convex subsets of a locally convex topological vector space E and X be bounded and paracompact with the property (K). Let $A : X \to 2^X$ be continuous with non-empty compact and convex values, $F : X \to 2^C$ be upper semicontinuous with non-empty compact and convex values and $\theta : X \times C \to E^*$ be continuous. Suppose there exists a non-empty compact and convex subset X_0 of X and a non-empty compact subset K of X such that:

(i) for each fixed $x \in X$, $y \mapsto \theta(x, y)$ is concave on Y; and
(ii) for each $x \in X \setminus K$, $\inf_{u \in A(x) \cap X_0} \sup_{y \in F(x)} Re\langle \theta(x, y), u - x \rangle < 0$.

Then there exist $x_0 \in A(x_0)$ and $y_0 \in F(x_0)$ such that

$$\inf_{u \in A(x_0)} Re\langle \theta(x_0, y_0), u - x_0 \rangle \geq 0.$$

Proof. Define a mapping $P : X \to 2^X$ by

$$P(x) = \{u \in X : \sup_{y \in F(x)} Re\langle \theta(x, y), u - x \rangle < 0\}$$

for each $x \in X$. Then we have the following:

(a) $x \notin P(x)$ for each $x \in X$; and
(b) $P(x)$ is convex for each $x \in X$. Indeed, let $u_1, u_2 \in P(x)$. In order to prove that $P(x)$ is convex, it suffices to show that $u_0 = \lambda u_1 + (1 - \lambda) u_2 \in P(x)$ for each $\lambda \in (0, 1)$. Note that for each $y \in F(x)$, $Re\langle \theta(x, y), u_0 - x \rangle = \lambda Re\langle \theta(x, y), u_1 - x \rangle + (1 - \lambda) Re\langle \theta(x, y), u_2 - x \rangle$, so that we have

$$\sup_{y \in F(x)} Re\langle \theta(x, y), u_0 - x \rangle \leq \sup_{u \in F(x)} \lambda Re\langle \theta(x, y), u_1 - x \rangle + \sup_{y \in F(x)} (1 - \lambda) Re\langle \theta(x, y), u_2 - x \rangle,$$

which implies that $\sup_{y \in F(x)} Re\langle \theta(x, y), u_0 - x \rangle < 0$. Therefore $u_0 \in P(x)$.

Since X is bounded, the function $(x, y, u) \mapsto Re\langle \theta(x, y), u - x \rangle$ from $X \times C \times X$ to $\mathbb{R}$ is continuous, and by the upper semicontinuity of F with compact values, the function $\phi : X \times X \to \mathbb{R} \cup \{-\infty, +\infty\}$ defined by

$$\phi(x, u) = \sup_{y \in F(x)} Re\langle \theta(x, y), u - x \rangle$$

for each $(x, u) \in X \times C$ is upper semicontinuous by Proposition 21 of Aubin and Ekeland [15, p.118]. Therefore we have

(c) the set $P^{-1}(u) = \{x \in X : \sup_{y \in F(x)} Re\langle \theta(x, y), u - x \rangle < 0\}$ is open in X for each $u \in X$.

As A is lower semicontinuous, the mapping $\psi : X \to \mathbb{R} \cup \{-\infty, +\infty\}$ defined by

$$\psi(x) = \inf_{u \in A(x)} \sup_{y \in F(x)} Re\langle \theta(x, y), u - x \rangle$$

for each $x \in X$ is upper semicontinuous by Proposition 19 of Aubin and Ekeland in [15, p.118] again. Thus the set

$$\{x \in X : \inf_{u \in A(x)} \sup_{y \in F(x)} Re\langle \theta(x, y), u - x \rangle < 0\}$$

is open in X. Note that

$$\{x \in X : A(x) \cap P(x) \neq \emptyset\} = \{x \in X : \inf_{u \in A(x)} \sup_{y \in F(x)} Re\langle \theta(x, y), u - x \rangle < 0\}.$$

Hence, we have

(d) the set $\{x \in X : A(x) \cap P(x) \neq \emptyset\}$ is open in X.

Moreover, by (ii), we also have

(e) for each $x \in X \setminus K$, $A(x) \cap P(x) \cap X_0 \neq \emptyset$.

Thus the triple $(X; A; P)$ satisfies all hypotheses of Lemma 6.10.1. By Lemma 6.10.1, there exists $x_0 \in X$ such that $A(x_0) \cap P(x_0) = \emptyset$. By the definition of P, it implies that

$$\inf_{u \in A(x_0)} \sup_{y \in F(x_0)} Re\langle \theta(x_0, y), u - x_0 \rangle \geq 0.$$

Note that $y \mapsto \theta(x, y)$ is concave by (i), the Kneser minimax theorem implies that

$$\inf_{u \in A(x_0)} \sup_{y \in F(x_0)} Re\langle \theta(x_0, y), u - x_0 \rangle = \sup_{y \in F(x_0)} \inf_{u \in A(x_0)} Re\langle \theta(x_0, y), u - x_0 \rangle.$$

Note that $y \mapsto \inf_{u \in A(x_0)} Re\langle \theta(x_0, y), u - x_0 \rangle$ is upper semicontinuous, thus there exists $y_0 \in F(x_0)$ such that

$$\inf_{u \in A(x_0)} Re\langle \theta(x_0, y_0), u - x_0 \rangle = \sup_{y \in F(x_0)} \inf_{u \in A(x_0)} Re\langle \theta(x_0, y_0), u - x_0 \rangle \geq 0,$$

and thus the conclusion follows. □

When the locally convex topological vector space E is of the second category, the following result shows that the conclusion of Theorem 6.10.2 still holds without the assumption that X is bounded, i.e., we have the following result.

Theorem 6.10.3. Let X and C be non-empty convex subsets of a locally convex topological vector space E and X be paracompact with the property (K). Let $A : X \to 2^X$ be continuous with non-empty compact and convex values, $F : X \to 2^C$ be upper semicontinuous with non-empty compact and convex values and $\theta : X \times C \to E^*$ be continuous. Suppose there exists a non-empty compact and convex subset X_0 of X and a non-empty compact subset K of X such that:

(i) for each fixed $x \in X$, $y \mapsto \theta(x, y)$ is concave on Y;

(ii) for each $x \in X \setminus K$, $\inf_{u \in A(x) \cap X_0} \sup_{y \in F(x)} Re\langle \theta(x, y), u - x \rangle < 0$; and

(iii) the space E is of the second category.

Then there exists $x_0 \in A(x_0)$ and $y_0 \in F(x_0)$ such that

$$\inf_{u \in A(x_0)} Re\langle \theta(x_0, y_0), u - x_0 \rangle \geq 0.$$

Proof. We shall follow the notations introduced in the proof of Theorem 6.10.2 to prove our conclusion. By the assumptions, and examining the arguments of Theorem 6.10.2, it is clear that (a) and (b) of Theorem 6.10.2 hold.

Note that E is of the second category, by Banach-Steinhaus Theorem II-2.5 of Rudin [449], the function $(x, y, u) \mapsto Re\langle\theta(x, y), u - x\rangle$ from $X \times C \times X$ to $\mathbb{R}$ is continuous. Now by the same argument used in Theorem 6.10.2, it is easy to verify that the following are still true:

(c) the set $P^{-1}(u) = \{x \in X : \sup_{y \in F(x)} Re\langle\theta(x, y), u - x\rangle < 0\}$ is open in X for each $u \in X$;
(d) the set $\{x \in X : A(x) \cap P(x) \neq \emptyset\}$ is open in X; and
(e) for each $x \in X \setminus K$, $A(x) \cap P(x) \cap X_0 \neq \emptyset$.

Then by applying the Kneser minimax theorem, i.e., Lemma 5.3.A above, the conclusion follows. □

As a compact case of Theorem 6.10.2, we have the following result.

Corollary 6.10.4. Let X be a non-empty compact and convex subset of a locally convex topological vector space E and C be a non-empty convex subset of E. Let $A : X \to 2^X$ be continuous with non-empty closed and convex values, $F : X \to 2^C$ be upper semicontinuous with non-empty compact and convex values and $\theta : X \times C \to E^*$ be continuous such that $y \mapsto \theta(x, y)$ is concave on C for each fixed $x \in X$. Then there exist $x_0 \in A(x_0)$ and $y_0 \in F(x_0)$ such that

$$\inf_{u \in A(x_0)} Re\langle\theta(x_0, y_0), u - x_0\rangle \geq 0.$$

Proof. As X is compact, X is paracompact and has the property (K). Thus all hypotheses of Theorem 6.10.2 are satisfied. By Theorem 6.10.2, the conclusion follows. □

By the concave assumption of the mapping θ in its second variable on C, Theorems 6.10.2, 6.10.3 and Corollary 6.10.4 guarantee the existence of solutions for $(GQVI)$. In what follows, as a direct application of the Himmelberg's fixed point theorem in [233], we shall show that the conclusions of Theorems 6.10.2 and 6.10.3 still hold without either concave or convex assumption for the function θ if both mappings A and F are assumed to be *completely compact* in the sense that each closure of each image set for both A and F are compact.

More formally, we have the following definition.

Definition 6.10.5. Let X and Y be two topological spaces and $F : X \to 2^Y$ be a set-valued mapping. Then F is said to be *completely compact* if the closure of the set $F(X) := \cup_{x \in X} F(x)$ is compact in Y.

We now recall the following fixed point theorem of the Himmelberg in [233, Theorem 2]) which is a generalization of the well known the Fan-Glicksberg fixed point theorem.

Lemma 6.10.6 (Himmelberg Fixed Point Theorem). Let T be a non-empty convex subset of a locally convex topological vector space E. Let

$F : T \to 2^T$ be upper semicontinuous with non-empty closed and convex values and $F(T)$ is contained in some compact subset C of T. Then F has a fixed point.

Now as an application of the Himmelberg fixed point theorem above, we have:

Theorem 6.10.7. Let X and C be non-empty convex subsets of a locally convex topological vector space E. Let $A : X \to 2^X$ be continuous with non-empty closed and convex values, $F : X \to 2^C$ be upper semicontinuous with non-empty closed and convex values and $\theta : X \times C \to E^*$ be continuous. Suppose that:

(i) both A and F are completely compact; and

(ii) either the space E is of second category or X is bounded.

Then there exists $x_0 \in A(x_0)$ and $y_0 \in F(x_0)$ such that

$$\inf_{u \in A(x_0)} Re\langle \theta(x_0, y_0), u - x_0 \rangle \geq 0.$$

Proof. Since A and F are completely compact, there exist non-empty compact subsets X_0 and C_0 of X and C, respectively such that $A(X) := \cup_{z \in X} A(z) \subset X_0$ and $F(X) := \cup_{z \in X} F(z) \subset C_0$. Now define $J : X \times C \to 2^X$ by

$$J(x, y) = \{u \in A(x) : Re\langle \theta(x, y), u - x \rangle = \inf_{s \in A(x)} Re\langle \theta(x, y), s - x \rangle\}$$

for each $(x, y) \in X \times C$. By condition (ii), the mapping $(x, y, u) \mapsto Re\langle \theta(x, y), u - x \rangle$ is continuous from $X \times C \times X$ to $\mathbb{R} \cup \{-\infty, +\infty\}$. Then $J(x, y) \neq \emptyset$ by the upper semicontinuity of A and $A(x)$ is non-empty compact for each $(x, y) \in X \times C$. Moreover the continuity of θ and A implies that the graph of J is also closed, i.e., graph J is closed in $X \times C \times X$. As $J(x, y) \subset A(x)$, we have that $J(X \times C) \subset A(X) \subset X_0$. Hence J is upper semicontinuous with non-empty closed and convex values.

Now define $G : X \times C \to 2^{X \times C}$ by

$$G(x, y) = (J(x, y), F(x))$$

for each $(x, y) \in X \times C$. Then G is upper semicontinuous with non-empty closed and convex values and in fact

$$\begin{aligned} G(X \times C) &= \cup_{(x,y) \in X \times C} (J(x, y), F(x)) \subset (J(X \times C), F(X)) \\ &\subset A(X) \times F(X) \subset X_0 \times C_0. \end{aligned}$$

Note that $X_0 \times C_0$ is non-empty and compact in $X \times C$, G is also completely compact and thus G satisfies all hypotheses of the Himmelberg fixed point theorem above. By Lemma 6.10.6, there exists $(x_0, y_0) \in X \times C$ such that $(x_0, y_0) \in G(x_0, y_0)$. By the definition of G, the conclusion follows. $\square$.

Note that each compact mapping may not be completely compact, hence both Theorems 6.10.2 (resp., 6.10.3) and 6.10.7 are independent to each other. As an immediately consequence of Theorem 6.10.7, we have the following result which includes Corollary 6.10.4 above as a special case:

Corollary 6.10.8. Let X be a non-empty compact and convex subset of a locally convex topological vector space E and C be a non-empty convex subset of E. Let $A : X \to 2^X$ be continuous with non-empty closed and convex values, $F : X \to 2^C$ be upper semicontinuous with non-empty compact convex values and $\theta : X \times C \to E^*$ be a single-valued continuous mapping.

Then there exists $x_0 \in A(x_0)$ and $y_0 \in F(x_0)$ such that

$$\inf_{u \in A(x_0)} Re\langle \theta(x_0, y_0), u - x_0 \rangle \geq 0.$$

Remark 6.10.9. Corollary 2.6 is an infinite dimensional generalization of Theorem 2.1 of Yao [589]. We also note that Theorems 6.10.2 and 6.10.3 include the corresponding results of Aubin and Ekeland [15], Chan and Pang [64], Tan and Shih [466], Tarafdar and Yuan [545] as special cases.

6.11 Applications to Complementarity Problems

In this section, as applications of our variational inequalities in section 10, we shall discuss the existence of solutions for generalized complementarity problem (GCP).

Definition 6.11.1. Let f be a mapping from $\mathbb{R}^n$ into itself. The '*classical nonlinear complementarity problem*' (in short, (GCP)) is to find a vector $x \in \mathbb{R}^n$ such that

$$x \geq 0, f(x) \geq 0 \text{ and } \langle x, f(x) \rangle = 0.$$

Since Cottle [98] first introduced and studied the existence of solutions for the complementarity problems, various extensions of the nonlinear complementarity problems have been proposed and investigated by many mathematicians in recent years. A number of related references can be found in the book of Cottle et al. [102], [253], Hyers et al. [244], and the survey paper [220]. For example, in [411], Parida and Sen considered and studied the following generalized complementarity problem: to find $x \in \mathbb{R}^n$ and $y \in F(x)$ such that

$$x \in K, \theta(x, y) \in K^* \text{ and } \langle x, \theta(x, y) \rangle = 0, \qquad (GCP')$$

where K is a closed convex cone in $\mathbb{R}^n$, $K^* = \{y \in R^n : \langle y, x \rangle \geq 0 \text{ for all } x \in K\}$, which is the polar cone of K, F is a set-valued mapping from K into a non-empty subset C of $\mathbb{R}^n$ and $\theta : K \times C \to \mathbb{R}^n$. More recently, Yao [589] studied the above (GCP') problem by allowing K to be a set-valued mapping. It is our purpose in this section to study the following generalized complementarity problem (GCP) which includes the above (GCP') as a special case. That is,

to find $x \in X$, $z \in M(x)$ and $y \in F(x)$ such that

$$x - z \in L(x),\ \theta(x,y) \in L(x)^* \text{ and } \langle \theta(x,y), x - m(x) \rangle = 0 \qquad (GCP)$$

where K and C is a non-empty subset of a locally convex topological vector space E, both $M, L : K \to 2^K$ are set-valued mappings and $L(x)$ is a non-empty cone for each $x \in K$, $A : X \to K$ is defined by $A(x) := M(x) + L(x)$ for each $x \in X$; $F : X \to 2^C$ is a set-valued mapping and $\theta : X \times C \to E^*$ is a single-valued mapping. We note that (GCP) reduces to the problem studied by Yao [589] if M is single-valued and $E := \mathbb{R}^n$ for some $n \in \mathbb{N}$, where the latter case in turn reduces to the problem considered by Chan and Pang [64] if $\theta(x,y) = y$ and $X = C = \mathbb{R}^n$. If we further assume that $L(x) = \mathbb{R}^n_+$ for each $x \in \mathbb{R}^n$ and F is a single-valued mapping, then the (GCP) problem above reduces to the problem studied by Noor and Zarae in [404]. Thus our (GCP) problem can be regarded as a more general and unifying problem which includes many complementarity problems considered in the literature as special cases.

In order to apply our varitaional inequalities, we first state the following result.

Lemma 6.11.2. The existence of solutions for $(GQVI)$ implies the existence of solutions for the complementarity problem (GCP). In particular, if the mapping M is a single-valued mapping, then both existence of solutions for $(GQVI)$ and (GCP) are equivalent.

Proof. Suppose $(GQVI)$ has a solution, i.e., there exists $x_0 \in X$ and $y_0 \in F(x_0)$ such that

$$x_0 \in A(x)_0,\ \langle \theta(x_0, y_0), u - x_0 \rangle \geq 0 \text{ for all } u \in A(x_0). \qquad (6.11.1)$$

Then we claim that x_0 is also a solution of (GCP). As $x_0 \in A(x_0)$, there exists $z_0 \in M(x_0)$ such that $x_0 - z_0 \in L(x_0)$. Since $L(x_0)$ is a cone, $z_0 \in A(x_0)$ and $2x_0 - z_0 \in A(x_0)$. It follows that $\langle \theta(x_0, y_0), z_0 - x_0 \rangle = 0$. Moreover, for each $w \in L(x_0)$, we have $z_0 + w \in A(x_0)$ and $\langle \theta(x_0, y_0), w \rangle = \langle \theta(x_0, y_0), z_0 + w - x_0 \rangle \geq 0$ from (6.11.1). Thus $\theta(x_0, y_0) \in L(x_0)^*$. Thus (GCP) has a solution.

In particular, if M is a single-valued mapping, it is easy to verify that the solution for (GCP) is also a solution of $(GQVI)$. Indeed, suppose x is a solution of (GCP), as M is a single-valued mapping, there exists $y \in F(x)$ such that

$$\theta(x,y) \in L(x)^*, \text{ and } \langle \theta(x,y), x - M(x) \rangle = 0.$$

For each $u \in A(x)$, $u = M(x) + z$ for some $z \in L(x)$. Since $\theta(x,y) \in L(x)^*$, we have

$$\theta(x,y), u - x \rangle = \langle \theta(x,y), z \rangle \geq 0.$$

Hence $(GQVI)$ has a solution and both statements are are equivalent. $\square$

Lemma 6.11.2 states that both existence solutions for (GCP) and $(GQVI)$ are equivalent in certain circumstances. This kind of nice relationship between quasi-variational inequalities and complemenratity problems has been studied by many authors, for example, see Chan and Pang [64], Cottle [98], Harker and Pang [220], Isac [253], Karamardin [282], Parida and Sen [411], Yao [589], and so on.

Lemma 6.11.3. Let X be a non-empty convex subset of a topological vector space E and D any non-empty subset of X and $x \in int_X D$. Then for each $u \in D$, there exists a $\lambda_0 \in (0,1)$ such that $\lambda x + (1-\lambda)u \in D$ for all $\lambda \in [0,1]$ with $\lambda \geq \lambda_0$.

Proof. For each $u \in X$, note that $x \in int_X D$ and X is a convex subset of a topological vector space E, by the continuity of the function $\lambda \to \lambda x + (1-\lambda)u$ at x, the conclusion follows. $\square$

Lemma 6.11.4. The following two statements are equivalent.

(i) A vector $x \in X$ with $x \in A(x)$ and a vector $y \in F(x)$ such that

$$\langle \theta(x,y), u - x\rangle \geq 0 \text{ for all } u \in A(x).$$

(ii) There exists a non-empty subset D of X such that $x \in int_{A(x)} D$ and $y \in F(x)$ such that

$$\langle \theta(x,y), u - x\rangle \geq 0 \text{ for all } u \in A(x) \cap D.$$

Proof. It is clear that (i) implies (ii). Now we show that (ii) implies (i). For any $u \in A(x)$, as $x \in int_{A(x)} D$, by Lemma 6.11.3, there exists $\lambda \in (0,1)$ such that $\lambda x + (1-\lambda)u \in D$ so that $\lambda x + (1-\lambda)u \in A(x) \cap D$ as $A(x)$ is convex. Then

$$0 \leq \langle \theta(x,y), \lambda x + (1-\lambda)u - x\rangle = (1-\lambda)\langle \theta(x,y), u - x\rangle,$$

so that $\langle \theta(x,y), u - x\rangle \geq 0$ for all $u \in A(x)$ and the conclusion follows. $\square$

Now we have the following:

Theorem 6.11.5. Let X be a non-empty convex subset of a locally convex vector topological space E and D a non-empty compact and convex subset of X such that the following conditions are satisfied:

(i) F is continuous with non-empty compact and convex values;

(ii) $A_1 : D \to 2^D$ defined by $A_1(x) = A(x) \cap D$ is upper semicontinuous with non-empty values, where the mapping is as defined before in (GCP); and

(iii) for each $x \in D$ and for each $z \in A(x) \setminus D$, there exists $u \in A(x) \cap D$ such that

$$\sup_{y \in F(x)} \langle \theta(x,y), u - z\rangle \leq 0.$$

Then (GCP) has a solution.

Proof. By Theorem 6.10.7, there exists $x \in D$ with $x \in A(x) \cap D$ and $y \in F(x)$ such that $\langle \theta(x,y), v - x \rangle \geq 0$ for all $v \in A(X) \cap D$. Now for each $z \in A(x) \setminus D$, by (iii), there exists $u \in A(x) \cap D$ such that $\langle \theta(x,y), z - u \rangle \geq 0$. On the other hand, we have $\langle \theta(x,y), u - x \rangle \geq 0$. By adding the last two inequalities, we have $\langle \theta(x,y), u - x \rangle \geq 0$ for all $u \in A(x)$. The conclusion then follows from Lemma 6.11.2. □

Let E be a topological vector space with topology τ. We shall denote by $E^* = (E, \tau)^*$ the topological dual of (E, τ) and $W := W(E, E^*)$ the weak topology of E. If X is a non-empty subset of E, we shall denote by (X, W) the set X equipped with relative weak topology W.

We also have the following existence theorem for a (GCP) problem when the underlying space E is a reflexive Banach space.

Theorem 6.11.6. Let X be a non-empty convex subset of a reflexive Banach space E and D a non-empty closed and bounded convex subset of X such that the following conditions are satisfied:

(i) $F : (X, W) \to 2^{(E,W)}$ is continuous with non-empty closed, bounded and convex values;

(ii) $A_1 : (D, W) \to 2^{(D,W)}$ defined by $A_1(x) := A(x) \cap D$ is upper semicontinuous with non-empty closed and convex values, where A is as defined as before; and

(iii) for each $x \in D$ with $x \in \partial_{A(x)} D$, there exists $u \in int_{A(x)} D$ such that

$$\sup_{y \in F(x)} \langle \theta(x,y), u - x \rangle \leq 0.$$

Then (GCP) has a solution.

Proof. As E is a reflexive Banach space and D is closed convex and bounded, the set (D, W) equipped with the relative weak topology W is W-compact. By Theorem 6.10.7, there exists $x \in D$ with $x \in A(x) \cap D$ and $y \in F(x)$ such that $\langle \theta(x,y), v - x \rangle \geq 0$ for all $v \in A(x) \cap D$. If $x \in int_{A(x)} D$, then the result follows from Lemma 6.11.4; otherwise $x \in \partial_{A(x)} D$. By (iii), there exists $u \in int_{A(x)} D$ such that $\langle \theta(x,y), u - z \rangle \leq 0$ for all $y \in F(x)$. It follows that $\langle \theta(x,y), u - x \rangle = 0$. Now for each $v \in A(x)$, choose $\lambda \in (0,1)$ such that $\lambda u + (1 - \lambda) v \in A(x) \cap D$. We have

$$0 \leq \langle \theta(x,y), \lambda u + (1 - \lambda) v - x \rangle = (1 - \lambda) \langle \theta(x,y), v - x \rangle.$$

Hence $\langle \theta(x,y), v - x \rangle \geq 0$ for each $v \in A(x)$ and the conclusion follows by Lemma 6.11.2. □

Remark 6.11.7. We note that both Theorems 6.11.5 and 6.11.6 improve Theorems 3.4 and 3.3 of Yao [589] in the following ways: (i) X is a subset of a locally convex vector topological space or reflexive Banach space instead of being a finite dimensional space; (ii) the mapping M is a set-valued mapping instead of being a single-valued mapping and (iii) F is a upper semicontinuous instead of being upper continuous and uniformly compact in the sense of Hogan [234].

In what follows, we assume that E is a reflexive Banach space unless otherwise specified, and

1: both X and C are non-empty convex subsets of a reflexive Banach space E and its dual space E^* respectively,

2: $M, L : X \to 2^X$ are set-valued mappings such that $L(x)$ is a non-empty cone for each $x \in X$;

3: $F : X \to 2^C$ is a set-valued mapping;

4: $A : X \to 2^X$ is defined by $A(x) = M(x) + L(x)$ for each $x \in X$;

5: for each $r > 0$, $A_r : B_r \to 2^{B_r}$ is defined by $A_r(x) := A(x) \cap B_r$ for each $x \in B_r$, where $B_r := \{x \in E : \|x\| \leq r\}$ and $C_r := \{x \in E : \|x\| = r\}$; and

6: $\theta : X \times C \to E$ is a continuous single-valued mapping.

We now have the following result.

Theorem 6.11.8. Let X be a non-empty cone of a reflexive Banach space E. Suppose that:

(i) $F : (X, W) \to 2^C$ is upper semicontinuous with non-empty closed and convex values;

(ii) there exists $x_0 \in \cap_{x \in X} A(x)$ such that

$$\limsup_{x \in A(x) \text{ and } \|x\| \to \infty} \sup_{y \in F(x)} \langle \theta(x, y), x_0 - x \rangle < 0,$$

(iii) there exists an $r_0 > 0$ such that the mapping $A_{r_0} : (X, W) \to 2^{(B_{r_0}, W)}$ is continuous with non-empty values for all $r \geq r_0$.

Then (GCP) has a solution.

Proof. By (ii), there exists an $r_1 > 0$ such that for any $x \in A(x)$ with $\|x\| \geq r_1$, we have

$$\sup_{y \in F(x)} \langle \theta(x, y), x_0 - y \rangle < 0.$$

Let $r \geq \max\{r_0, r_1, \|x_0\|\}$. As E is a reflexive Banach space, B_r is weakly compact. Then by Theorem 6.10.7 again, there exists $x \in A(x) \cap B_r(x)$ and $y \in F(x)$ such that $\langle \theta(x, y), u - x \rangle \geq 0$ for all $u \in A(x) \cap B_r$. It follows from inequality (2) above that $x \in A(x) \cap intB_r = int_{A(x)} B_r$. Consequently, the result follows from Lemmas 6.11.4 and 6.11.2. □

In the case where L is constant with $L(x) = X_0$ for all $x \in X$, condition that A_r is continuous on $X \cap B_r$ for large r is automatically satisfied provided the mapping M is continuous. As a special case of Theorem 6.11.8, we have:

Corollary 6.11.9. Let X be a non-empty cone of a finite dimensional Euclidean space E (so that E is a reflexive Banach space). Suppose that:

(i) $F : X \to 2^C$ is upper semicontinuous with non-empty closed and convex values;

(ii) $M : X \to 2^X$ is continuous;

(iii) there exists $u_0 \in X$ such that $u_0 - M(x) \subset X$ for all $x \in X$; and

(iv)

$$\limsup_{x \in A(x) \text{ and } ||x|| \to \infty} \sup_{y \in F(x)} \langle \theta(x, y), u_0 - x \rangle < 0.$$

Then (GCP) has a solution.

Proof. Note that in each finite dimensional Euclidean space, both strong (i.e., nomed) topology and weak topology coincide. By condition (iii), we have $u_0 \in \cap_{x \in X} A(x)$ and moreover the mapping A_r is also continuous for sufficiently large r (e.g., see Theorem 4.2 of Chan and Pang [64]. Hence the result follows from Theorem 6.11.6. □

With respect to the study of Yao [589], we can give an application of the complementarity problem to the obstacle problem in its infinite dimensional settings.

Let X be a closed convex cone in a reflexive Banach space E and $\leq_X$ be the partial order induced by X, i.e., $x \leq_X y$ if and only if $x - y \in X$ for all $x, y \in E$. Let f and m be functions from E into itself. The *quasi-variational inequality problem of obstacle type* (OVI) is to find $x^* \in E$ such that

$$x^* \geq_X m(x*), \langle f(x), x - x^* \rangle \geq 0 \text{ for all } x \leq_X m(x^*). \qquad (OVI)$$

Now let $A : E \to 2^E$ be a set-valued mapping defined by $A(x) = m(x) + X$ for each $x \in E$. It is easy to see that the (OVI) problem above is equivalent to finding $x^* \in m(x^*) + X$ such that

$$f(x^*) \in X^*, \langle f(x^*), x^* - m(x^*) \rangle = 0.$$

As an application of Corollary 6.11.9, we have the following existence theorem for (OVI) problem in reflexive Banach spaces.

Theorem 6.11.10. Let X be a closed non-empty convex cone in a reflexive Banach space E. Let f and m be continuous functions from (E, W) to (E, W) and $A(x) = m(x) + X$ for each $x \in X$. Suppose there exists a vector $u_0 \in E$ such that $u_0 - m(x) \in X$ for all $x \in E$ and

$$\limsup_{x \in A(x) \text{ and } ||x|| \to \infty} \langle f(x), u_0 - x \rangle < 0.$$

Then the (OVI) problem has a solution.

Proof. The conclusion follows by Corollary 6.11.9. □

Theorem 6.11.10 is an infinite dimensional version of Theorem 3.8 of Yao [589]. Moreover our Theorems 6.11.5, 6.11.6 and 6.11.8 improve most of the existence theorems for complementarity problems given by Cottle [98], Cottle et al. [102], Harker and Pang [220], Isac [253], Karamardian [282], Noor and Zarae [404], Parida and Sen [411], Saigal [451].

Finally, we also wish to point out that one of the special cases, (i.e., the space E is a Hilbert space and A is a single-valued mapping) of the (GCP) problem that we study in this section was studied as mathematical models for mechanical problems, and the post-critical equilibrium state of a thin elastic plate subjected to unilateral conditions. For more details we refer the interested reader to Cottle and Yao [103], Isac and Thera [257] and references therein.

6.12 Stability Results of Nonlinear Complementarity Problems

The aim of this part is to develop the generally generic stability theory for nonlinear complementarity problems in the setting of infinitey dimensional Banach spaces. We first show that each nonlinear complementarity problem can be approximated arbitrarily by a nonlinear complementarity problem which is stable in the sense that the small change of the objective function results in the small change of its solution set; and thus we might say that almost all of complementarity problems are stable from viewpoint of Baire category. Secondly, we show that each nonlinear complementarity problem has, at least, one connected component of its solutions which is stable though in general its solution set may not have a good behavior (i.e., not stable). Our results show that if a complementarity problem has only one connected solution set, it is then always stable without the assumption of the objective function is either Lipschitz or differentiable.

6.12.1 Generic Stability of Nonlinear Complementarity Problems

We know that the complementarity theory is dedicated to the study of complementarity problems - which is fundamental to the study of many optimization problems and the analysis and computation of equilibria in the physical and economical sense. It is well known that the complementarity theory has also many and remarkable applications in Engineering, Elasticity, Mechanics, Game Theory etc. The *solution set* of a complementarity problem can be *empty* or *non-empty*, *stable* or *unstable*. In this section, our principle aim is to study the stability of solutions for nonlinear complementarity problems without the traditional assumptions such as *Lipschitz* or *differential* conditions on the objective functions by introducing a new concept called *essential solution* which reflects the stability of solutions for complementarity problems.

We recall that a generally nonlinear complementarity problem (in short, NCP) (in $\mathbb{R}^n$) is described as follows: Let $f : \mathbb{R}^n \to \mathbb{R}^n$ be a continuous mapping and $K \subset \mathbb{R}^n$ be an acute convex closed cone with its vertex at the origin. Denoted by K^* the dual cone for K, i.e., $K^* = \{y \in \mathbb{R}^n : \langle x, y\rangle \geq 0\}$. The generally nonlinear complementarity problem (denoted by $NCP(f, K)$) is to find a vector $x \in K$ such that

$$f(x) \in K^* \text{ and } \langle x, f(x)\rangle = 0$$

where $\langle \cdot, \rangle$ denotes the inner product of $\mathbb{R}^n$.

By introducing a notion called *exceptional family*, the characteristic for the existence of solutions for generally nonlinear problems has been established recently by Bulavski et al. [63], Hyers et al. [244], Isac et al. [255]. As applications, many general existence theorems of complementarity problems have been given which unify and improve the corresponding existence theory of complementarity problems in the literature. For more details, see books of Hyers et al. [244], Isac [253] and related references therein.

Throughout this section, let K be a non-empty closed convex cone in a Banach space E and $f : K \to E^*$ a continuous mapping unless specified. We denote by $SNCP(f, K, D)$ the solution set which is contained in a non-empty subset D of K for the (nonlinear) complementarity problem $NCP(f, K)$, i.e.,

$$SNCP(f, K, D) := \{x \in D(\subset K) : f(x) \in K^* \text{ and } \langle x, f(x)\rangle = 0\}.$$

We first note that as a special case of Theorem 6.11.5 (see also Theorem 4.3.2 of Isac [253, p.116]), the following existence result for generally nonlinear complementarity problem holds.

Lemma 6.12.1. Let K be a non-empty closed and convex cone of a Banach space E and D be a non-empty compact subset of K. Suppose $f : K \to E^*$ is a continuous mapping such that for each $x \in K \setminus D$, there exists $y \in D$ such that $\langle x - y, f(x)\rangle > 0$. Then all solutions of $SNCP(f, K)$ are contained in D, i.e., $SNCP(f, K, D)$ is a non-empty and closed subset of D.

The stability study of solutions of nonlinear complementarity problem is an important topic in complementarity theory. In this section, our aim is to develop the generic stability of solutions for generally nonlinear complementarity problem $NCP(f, K)$ as introduced above under the setting of infinite dimensional Baanch spaces.

Let $K(E)$ be the space of all non-empty compact subsets of a metric space (E, d) (e.g, the Banach space $(E, \|\cdot\|)$) equipped with the Hausdorff metric h which is induced by the metric d (resp., the norm $\|\cdot\|$). For any $\epsilon > 0$, $x_0 \in E$ and $A \in K(E)$, let $U(\epsilon, A) = \{x \in E : d(u, x) < \epsilon \text{ for some } u \in A\}$ and $O(x_0, \epsilon) = \{x \in E : d(x_0, y) < \epsilon\}$.

Let Y be a topological space. We also recall that a subset $Q \subset Y$ is called a residual set if it is a countable intersection of open dense subsets of Y.

Let K be a non-empty closed and convex cone of a Banach space $(E, \|\cdot\|)$ and set $C := \{f : K \to E^* : \text{and } f \text{ is continuous }\}$ and define a metric ρ in C by $\rho(f, f') := \sup_{x \in K} \|f(x) - f'(x)\|^*$ for each $f, f' \in C$, where $\|\cdot\|^*$ denotes the norm of the dual space E^*. Clearly, ρ is a metric on C and we have the following fact.

Lemma 6.12.2. The metric space (C, ρ) is complete.

Denoted by $K(K)$ the collection of all non-empty compact subsets of the cone K in $(E, \|\cdot\|)$. Then we know that $K(K)$ is a complete metric space endowed with the Hausdorff metric h (induced from the norm $\|\cdot\|$ of E).

Let $Y := C \times K(K)$ and we define a metric d on Y by $d(y, y') := \rho(f, f') + h(A, A')$ for each $y = (f, A)$ and $y' = (f', A') \in Y$. Then it is clear that (Y, d) is also a complete metric space. Let $M := \{y = (f, A) \in Y\text{: such that } NCP(f, K) \text{ has solutions in } A, \text{ i.e., } SNCP(f, K, A) \neq \emptyset\}$. Then we have the following result.

Lemma 6.12.3. The space (M, d) is complete.

Proof. Since $M \subset Y$ and Y is complete, it is sufficient to prove that M is closed in Y. Let $\{y_n\}_{n=1}^{\infty}$ be a sequence in M and $y_n \to y \in Y$. Let $y_n := (f_n, A_n)$, $n = 1, 2, \cdots$ and $y = (f, A)$. Then $f_n \to f$ and $A_n \to A$. For each $n = 1, 2, \cdots$, since $y_n \in M$, there is $x_n \in A_n$ such that $f_n(x_n) \in K^*$ and $\langle x_n, f_n(x_n)\rangle = 0$. Since A_n and A are compact and $A_n \to A$, by A.5.1 (ii) of Mas-Colell [363, p.10], $\cup_{n=1}^{\infty} A_n \cup A$ is compact. Since $x_n \in A_n \subset \cup_{n=1}^{\infty} A_n \cup A$, without loss of generality we may assume that $x_n \to x \in \cup_{n=1}^{\infty} A_n \cup A$. If $x \notin A$, since A is compact, there is $a > 0$ such that $U(a, A) \cap O(x, a) = \emptyset$. Since $A_n \to A$ and $x_n \to x$, there is N_1 such that $A_n \subset U(a, A)$ and $x_n \in O(x, a)$ for all $n \geq N_1$, which contradicts the assumption that $x_n \in A_n$. Hence we must have $x \in A$. By the fact that K^* is closed, the continuity of f and $\lim_{n\to\infty} f_n = f$, it is easy to see that $f(x) \in K^*$ and $\langle x, f(x)\rangle = 0$. Therefore $y = (f, A) \in M$ so that M is closed in Y. $\square$

For each $y = (f, A) \in M$, we denote by $S(y)$ the solution set of the nonlinear complementarity problem $NCP(f, K)$ in A, i.e., $S(y) = SNCP(f, K, A)$. Then we have that $S(y) \neq \emptyset$. In what follows, we shall also use $NCP(f, K, A)$ to denote the generally nonlinear complementarity problem $NCP(f, K)$ associated with solutions in a subset A of the closed and convex cone K.

Lemma 6.12.4. The set $S(y) \in K(K)$ for each $y \in M$.

Proof. Let $y = (f, A) \in M$ be given. Since that $x_n \to x \in \cup_{n=1}^{\infty} A_n \cup A$. If $x \notin A$, since A is compact, there is $a > 0$ such that $U(a, A) \cap O(x, a) = \emptyset$. Since $A_n \to A$ and $x_n \to x$, there is N_1 such that $A_n \subset U(a, A)$ and $x_n \in O(x, a)$ for all $n \geq N_1$, which contradicts the assumption that $x_n \in A_n$. Hence we must have $x \in A$. Secondly, by the continuity of f, we must also have that $f(x) \in K^*$ and $\langle x, f(x) \rangle = 0$. Therefore $y = (f, A) \in M$ so that M is closed in Y. This completes the proof. □

By Lemma 6.12.4, the mapping $y \mapsto S(y)$ defines a solution mapping $S : M \to K(K)$ and indeed we have the following upper semicontinuity of the mapping S.

Lemma 6.12.5. The solution mapping $S : M \to K(K)$ is upper semicontinuous on M.

Proof. Suppose S is not upper semicontinuous at $y \in M$, then there exist $\epsilon_0 > 0$ and a sequence $\{y_n\}_{n=1}^{\infty}$ in M with $y_n \to y$ such that for each $n = 1, 2, \cdots$, there exists $x_n \in S(y_n)$ with $x_n \notin U(\epsilon_0, S(y))$. Let $y_n = (f_n, A_n)$ and $y = (f, A)$, then $f_n \to f$ and $A_n \to A$. Since $x_n \in A_n \subset \cup_{n=1}^{\infty} A_n \cup A$ and $\cup_{n=1}^{\infty} A_n \cup A$ is compact, without loss of generality, we may assume that $x_n \to x \in \cup_{n=1}^{\infty} A_n \cup A$. Note that we must have $x \notin U(\epsilon_0, S(y))$. Now the same argument as in the proof of Lemma 6.12.3 shows that $x \in A$, $f(x) \in K^*$ and $\langle x, f(x) \rangle = 0$, so that $x \in S(y)$. This contradicts that $x \notin U(\epsilon_0, S(y))$. Therefore S must be upper semicontinuous. This completes the proof. □

In order to study the stability of solution set for nonlinear complementarity problems, we now introduce the following notions.

Definition 6.12.6. Let M_1 be a non-empty closed subset of M (then M_1 is also complete as so is the space M). If $y = (f, A) \in M_1$, then a solution point x in $S(y)$ is said to be an *essential solution* of the nonlinear complementarity problem $NCP(f, K, A)$ with respect to M_1 provided that for any $\epsilon > 0$, there is $\delta > 0$ such that for any $y' = NCP(f', A') \in M_1$ with $d(y, y') = \rho(f, f') + h(A, A') < \delta$, there exists a solution $x' \in S(y')$ for the complementarity problem $NCP(f', K, A')$ with $d(x, x') < \epsilon$. The complementarity problem $NCP(f, K, A)$ is said to be *essential* (with respect to M_1) if every $x \in S(y)$ is an essential solution of y with respect to M_1.

Remark 6.12.7. Definition 6.12.6 says that if the complementarity problem $NCP(f, K, A)$ is essential, then for any $\epsilon > 0$, there exists a $\delta > 0$ such that for any other complementarity problem $NCP(f', K, A')$ with $\sup_{x \in X}\{d(f(x), f'(x)) : x \in X\} + h(A, A') < \delta$, having at least on solution x' in the ϵ-neighborhood of the solution set of $NCP(f, K, A)$; or equivalently to saying, the solution mapping S is continuous at the complementarity problem $NCP(f, K, A)$ when it is essential. Therefore the essential property of $NCP(f, K, A)$ characterizes the continuous property of its solution set in A.

Now we have the following characteristic of stability for solution set of complementarity problems.

Theorem 6.12.8. The solution mapping S is lower semicontinuous at $y \in M_1$ if and only if y is essential with respect to M_1.

Proof. Suppose S is lower semicontinuous at $y \in M_1$. Then for any $\epsilon > 0$, there is $\delta > 0$ such that for any $y' \in M_1$ with $d(y, y') < \delta$, we have $S(y) \subset U(\epsilon, S(y'))$ so that for any $x \in S(y)$, there is $x' \in S(y')$ with $d(x, x') < \epsilon$. Thus every $x \in S(y)$ is an essential solution of $y = NCP(f, K)$ with respect to M_1 and hence y $(= NCP(f, k))$ is essential with respect to M_1.

Conversely, suppose that y is essential with respect to M_1. If S were not lower semicontinuous at $y \in M_1$, then there exist $\epsilon_0 > 0$ and a sequence $\{y_n\}_{n=1}^{\infty}$ in M with $y_n \to y$ such that for each $n = 1, 2, \cdots$, there is $x_n \in S(y)$ with $x_n \notin U(\epsilon_0, S(y_n))$. Since $S(y)$ is compact, we may assume that $x_n \to x \in S(y)$. Since x is an essential solution of the nonlinear complementarity problem $y = NCP(f, K)$ with respect to M_1, $y_n \to y$ and $x_n \to x$, there is N such that $d(x_n, x) < \epsilon_0/2$ and $x \in U(\epsilon_0/2, S(y_n))$ for all $n \geq N$. Hence $x_n \in O(x, \epsilon_0/2) \subset U(\epsilon_0, S(y_n))$ for all $n \geq N$ which contradicts the assumption that $x_n \notin U(\epsilon_0, S(y_n))$ for all $n = 1, 2, \cdots$. Hence S must be lower semicontinuous at y. □

The following theorem says that each complementarity problem can be arbitrarily approximated by an essential complementarity problem.

Theorem 6.12.9. The set of essential points with respect to M_1 is a dense residual set in M_1. In particular, every point in M_1 can be arbitrarily approximated by an essential point in M_1.

Proof. By Lemma 6.12.4 and Lemma 6.12.5, $S : M \to K(K)$ is an usco mapping. Since M_1 is complete, it follows by Lemma 2.8.2 that the set of points where S is lower semicontinuous is a dense residual set in M_1. By Theorem 6.12.8, the set of essential points in M_1 is a dense residual set in M_1. □

By combining Lemma 6.12.5, Theorems 6.12.8 and 6.12.9, we have the following result.

Theorem 6.12.10. The solution mapping S is continuous at $y \in M_1$ if and only if y is essential with respect to M_1. Moreover, the set of points at which S is continuous is a dense residual set in M_1.

We remark that S is continuous at $y \in M_1$, if and only if for each $\epsilon > 0$, there is $\delta > 0$ such that $h(S(y), S(y')) < \epsilon$ for each $y' \in M$ with $d(y, y') < \delta$. Theorem 6.12.8 implies that if $y = (f, g, A) \in M_1$, then y is essential with respect to M_1 if and only if its set $S(y)$ of solution points is stable: $S(y')$ is close to $S(y)$ whenever y' is close to y.

We now give a sufficient condition that $y \in M_1$ is essential with respect to M_1:

Theorem 6.12.11. If $y \in M_1$ is such that $S(y)$ is a singleton set, then y is essential with respect to M_1.

Proof. Suppose $S(y) = \{x\}$. By Lemma 6.12.5, the mapping S is upper semicontinuous at y. Thus for any $\epsilon > 0$, there is $\delta > 0$ such that for each $y' \in M_1$, $d(y, y') < \delta$ implies $S(y') \subset U(\epsilon, S(y)) = O(x, \epsilon)$ so that $S(y) = \{x\} \subset U(\epsilon, S(y'))$. This shows that S is also lower semicontinuous at y. By Theorem 6.12.8, y is essential with respect to M_1. □

In this section, the generic stability results, mainly Theorems 6.12.9 and 6.12.10 tell us that though not all solution sets of complementarity problems have good behavior, however there always exists some complementarity problem with stable solutions to approximate arbitrarily each of them, this indicates that almost all of complementarity problems are stable in the sense of Baire category theory.

6.12.2 Existence of Essentially Connected Components

As we have seen in last section, in general not all solutions of complementarity problems are stable though there exist complementarity problems with essential solutions to approximate them arbitrarily. In this section, however we will show that there exists at least one connected component of solution set for each complementarity problem, which is stable by introducing the concept of essential components of solution set for a class of complementarity problems which satisfy so-called *strong Karamardian's condition* (whose definition will be given below).

Suppose x is a solution of the complementarity problem $y = NCP(f, K, A) \in Y$, then the component of the solution $x \in SNCP(f, K, A)$ is the union of all connected subsets of $S(y)$ which contain the point x. From Engelking [154], we know that components are connected closed subsets of $S(y)$ and thus they are also compact as $S(y)$ is compact. It is also easy to see that the components of two distinct points of $S(y)$ either coincide or are disjoint, so that all components constitute a decomposition of $S(y)$ into connected pairwise disjoint compact subsets, i.e.,

$$S(y) = \bigcup_{\alpha \in \Lambda} S_\alpha(y)$$

where Λ is an index set, for any $\alpha \in \Lambda$, $S_\alpha(y)$ is a nonempty connected compact and for any $\alpha, \beta \in \Lambda (\alpha \neq \beta)$, $S_\alpha(y) \cap S_\beta(y) = \emptyset$.

In order to study the existence of essentially connected components of generally complementarity problems, we first introduce the following definition.

Definition 6.12.12. For each complementarity problem $y = (f, K, A) \in Y$, suppose the set $S(y) = \bigcup_{\alpha\in\Lambda} S_\alpha(y)$. Then a component $S_\alpha(y)$ for some $\alpha \in \Lambda$, is said to be an *essential component* of y if for each open set O containing $S_\alpha(y)$, there exists $\delta > 0$ such that for any other complementarity problem $y' = NCP(f', K, A') \in Y$ with $\rho(y, y') = \sup_{x\in K} \rho(f(x), f'(x)) + h(A, A') < \delta$, we have that $S(y') \cap O \neq \emptyset$.

Remark 6.12.13. Definition 6.12.12 above means that even though we could not expect the continuity for all solutions set of a given complementarity problem $y = NCP(f, K, A)$, however, there is a case that maybe some component of its solution set enjoins the continuous stability. In the rest part of this section, we will show that the existence of such nice component for each generally complementarity problem $NCP(f, K, A)$.

In order to establish the general existence of essentailly connected components of soultion set for complementarity problems, we introduce some kind of Karamardian's condition (see Karamardian [282] and also Isac [253, p.116-117] and related references), which is called the *Strong Karamardian Condition* as follows:

The Strong Karamardian Condition: A continuous function $f : K \to E^*$ is said to satisfy *Strong Karamardian's condition* on K if there exists a compact subset D of K such that for each $x \in K \setminus D$, $\langle x - y, f(x)\rangle > 0$ for all $y \in D$.

Let Y_1 be a collection of all complementarity problems satisfying the strong Karamardian's condition, i.e.,

$$Y_1 := C_1(X) \times \{A \in K(K) : A \supset D\},$$

wheere C_1 is the collection of all continuous mappings from K to E^* which satisfy the strong Karamardian's condition above. It is clear that $Y_1 \subset Y$ and the solution set $\emptyset \neq SNCP(y) \subset D$ for each $y = NCP(f, K, A) \in Y_1$ by Lemma 6.12.1, and thus Y_1 is also a subset of M.

Now we have the following general existence result of essentially connected components of solution set for any complementarity problem from the class Y_1.

Theorem 6.12.14. Let $y = NCP(f, K, A)$ be a given complementarity problem in Y_1. Then there exists at least one essentially connected component of the solution set $S(y)$.

Proof. For any given $y = NCP(f, K, A) \in Y_1$, suppose that the solution set

$S(y)$ of the complementarity problem $NCP(f, K, A)$ is decomposed as follows:

$$S(y) = \bigcup_{\alpha \in \Lambda} S_\alpha(y)$$

where Λ is an index set, for any $\alpha \in \Lambda$, $S_\alpha(y)$ is a connected compact and for any $\alpha, \beta \in \Lambda(\alpha \neq \beta)$, $S_\alpha(x) \cap S_\beta(x) = \emptyset$. We shall prove that there exists at least one essential component of $S(y)$. Let us suppose otherwise there is no any essential connected component. Then for any $\alpha \in \Lambda$, there exists an open set $O_\alpha \supset S_\alpha(y)$ such that for any $\varepsilon > 0$, there is $y_\alpha \in Y$ with $\rho(y, y_\alpha) < \varepsilon$ such that $S(y_\alpha) \cap O_\alpha = \emptyset$. As $S(y)$ is compact, there exist two open and finite coverings $\{V_i\}_{i=1}^n$ and $\{W_i\}_{i=1}^n$ which satisfy the following conditions:

(1) $\overline{W}_i \subset V_i$; (2) $V_i \cap V_j = \emptyset$ for each $i \neq j$; and (3) V_i contains at least one $S_{\alpha_i}(y)$ with $O_{\alpha_i} \supset V_i \supset S_{\alpha_i}(y)$.

Then it follows by Lemma 6.12.5 that the solution mapping S is upper semi-continuous at y and $\bigcup_{i=1}^n W_i \supset S(y)$ and $\bigcup_{i=1}^n W_i$ is open, then there exists a $\delta > 0$ such that $\bigcup_{i=1}^n W_i \supset S(y')$ for any $y' \in Y$ with $\rho(y, y') < \delta$. Thus there exists $y_{\alpha_i} \in Y$ with $\rho(y, y_{\alpha_i}) < \delta$ such that $S(y_{\alpha_i}) \cap O_{\alpha_i} = \emptyset$.

Let $y = NCP(f, K, A)$ and $y_{\alpha_i} = NCP((f_{\alpha_i}, K, A)$, where $i = 1, 2, ..., n$. We define a mapping $f^* : K \to E^*$ by

$$f^*(x) = \begin{cases} f(x), & \text{if } x \in X \setminus \bigcup_{i=1}^n V_i, \\ f_{\alpha_i}(x), & \text{if } x \in \overline{W}_i \\ \lambda_i(x) f(x) + \mu_i(x) f_{\alpha_i}(x), & \text{if } x \in V_i \setminus \overline{W}_i. \end{cases}$$

where

$$\lambda_i(x) = \frac{d(x, \overline{W}_i)}{d(x, \overline{W}_i) + d(x, X \setminus \bigcup_{i=1}^n V_i)}$$

and

$$\mu_i(x) = \frac{d(x, X \setminus \bigcup_{i=1}^n V_i)}{d(x, \overline{W}_i) + d(x, X \setminus \bigcup_{i=1}^n V_i)}.$$

By the definition of f^* and Theorem 5.4.22 (or Theorem 2.8.21), it is easy to verify that f^* is continuous and f^* also satisfies the strong Karamardian condition, thus $y^* = NCP(f^*, K, A) \in Y_1$. Therefore $S(y^*) \neq \emptyset$. Note that $\rho(y, y_{\alpha_i}) < \delta$ for $i = 1, 2, \cdots, n$, it follows by Lemma 2.8.20 that

$$h(f(x), \lambda_i(x) f(x) + \mu_i(x) f_{\alpha_i}(x)) \leq h(f(x), f_{\alpha_i}(x)).$$

Therefore $\rho(y, y^*) < \delta$ and $S(y^*) \subset \bigcup_{i=1}^n W_i$. Note that for any $x_0 \in S(y^*)$, there is an index i_0 such that $x_0 \in W_{i_0}$, and hence $x_0 \in W_{i_0} \subset \overline{W}_{i_0} \subset O_{\alpha_{i_0}}$. Therefore $f^*(x_0) = f_{\alpha_{i_0}}(x_0)$, $g^*(x_0) = g_{\alpha_{i_0}}(x_0)$ and $x_0 \in S(y_{\alpha_{i_0}})$. This contradicts our assumption that $S(y_{\alpha_{i_0}}) \cap O_{\alpha_{i_0}} = \emptyset$. Hence there exists at least one essentially connected component of $S(y)$. This completes the proof.□

Theorem 6.12.15. If the complementarity problem $y = NCP(f, K, A) \in Y_1$ is such that the solution set $S(y)$ of complementarity problem NCP(f, K, A) is either totally disconnected set, then $NCP(f, K, A)$ is weakly essential. In particular, if the solution set of $NCP(f, K, A)$ is either a singleton, or it is connected, then the problem $NCP(f, K, A)$ is essential.

Proof. Since $S(y)$ is a totally disconnected set, then $S(y) = \cup_{\alpha \in \Lambda} S_\alpha(\phi)$, where $S_\alpha(\phi)$ is a singleton set for each $\alpha \in \Lambda$. By Theorem 6.12.14, there exists $S_{\alpha_0}(\phi) = \{x_0\}$, which is an essential component of $S(y)$. It is clear that x_0 is essential and thus it is weakly essential. In the case, the solution set $S(y)$ is a singleton set or it is connected, then y is is essential by Theorem 6.12.14 and thus the complementarity problem $NCP(f, K, A)$ is essential and the proof is complete. □

Theorem 6.12.14 tells us that each nonlinear complementarity problem has, at least, one connected component of its solutions which is stable though in general its solution set may not have a good behavior (i.e., not stable). Theorem 6.12.15 tells us that if a complementarity problem has only one connected solution set, it must be stable, Here we don't need to require the objective function f to be either Lipschitz or differential.

Remark 6.12.16. In both chapters 5 and 6, we have developed some existence theory for both variational inequality and related complementarity problems, On the other hand, we know that for any given NCP, the existence of its solution is not always assured (see Kojima [324], Isac [253] and references there). Because of this aspect, many people investigated this issue and proposed a variety of existence theorems. These include the existence conditions developed by Eaves, Kojima, Karamardian and many others. Only recently after the new notion, '*exceptional family of elements*', introduced by Kalashnikov [281] and Isac et al. [255], has some characterisation been given, but only in finite dimensional Euclidean spaces (see also [255]). The notion is a new variant of a similar unpublished one under the name '*opposite radial sequence*' introduced initially by Isac in early 1980's and is more general than the notion of '*exceptional sequences*' due to Smith [493]. So far, in infinite dimensional Hilbert spaces characterisations have been developed recently, but only in some very special cases such as k-set fields of mappings [254].

Definition 6.12.17. Let K be a closed convex cone of the Euclidean space R^n and $f : \mathbb{R}^n \to \mathbb{R}^n$ be a vector-valued function. The family of elements $\{x^r\}_{r>0}$ is an '*exceptional family*' of elements for f with respect to K if for each $r > 0$, there exists a real number $\mu_r > 0$ such that $u_r = f(x^r) + u_r x^r$ and the following conditions are satisfied:

(a) $u_r \in K^*$;
(b) $\langle u_r, x^r \rangle = 0$; and
(c) $\lim_{r\to\infty} \|x^r\| = \infty$, where K^* is the dual cone of K, i.e., $K^* = \{y \in R^n :$

$\langle x, y\rangle \geq 0$ for all $x \in K\}$.

As an application of topological degree we can prove the following general alternative: For any continuous function $f : \mathbb{R}^n \to \mathbb{R}^n$, there exists either a solution for the $NCP(f, K)$, or an exceptional family of elements for f with respect to K. For its proof, see Hyers et al. [244] and Isac et al. [255]. The concept of an exceptional family of elements for a continuous function opens a new research direction in complementarity theory. It seems to be important as it allows us to characterise the existence of solutions. For example, when a continuous function from R^n into itself is without exceptional family of elements, then the explicit NCP associated to this function has a solution. By combining this with topological degree theory, one of the recent development (e.g, see Hyers et al. [244]) states that: '*for any continuous mapping* $f : R^n_+ \to R^n$, *there exists either a solution of the complementarity problem or an exceptional family of elements for* f' (*unfortunately, it is only for finite-dimensional spaces so far*). Hence, an important issue is now, how to find tests or an algorithm which can be used to decide when a given function is without exceptional families of elements. On the other hand, the study of exceptional families in infinite dimensional spaces is another important topic as many useful problems in applications are in the setting of infinite dimensional spaces, as we have said above (see also books of [244], [253] and the references therein). How to extend the notion of 'exceptional family' from existing finite spaces to infinite dimensional spaces is still an open question. So far, only some very special cases in the setting of infinite dimensional Hilbert spaces are developed for such as k-set fields of nonlinear mappings by Isac recently in [254]. Many questions are still unanswered. The general theory is not yet developed. By the connection of NCP and variational inequlity problems (VIP), following the basic idea of exceptional family in the study of nonlinear complementarity problems above, the new notion of 'exceptional family' for VIP over polyhedral convex stes in finite-dimensional spaces has been recently developed by Zhao [617] (see also [618] and references therein). This concept provides a new way to invistage variational inequality problems (VIP) since it enables us to characterize the existence of solutions for variational inequality theory. As applications, some general existence theory for VIP in finite dimensional spaces has been developed, and whose sufficient conditions for the guarantee of VIPs' solutions are weaker than the most of the well-known existence conditions in existing literature. However, the same situation as the study of nonlinear complementarity problems, how to extend the concept of exceptional family for VIP to infinite dimensional space is still an open problem. It is worthwhile to develop this new and useful theory in the framework of infinite dimensional spaces.

Finally we also note that by using the same idea used in this section for the class of complementarity problems satisfying the strong Karamardian condition, the general existence results of essentially connected components of solutions for complementarity problems which satisfy such as *coercive*, or *weakly coercive condition*, or some other kinds of coercive conditios (see the book of Hypers et al. [244, p.63], or Zeidler's book [614, p.472] for their definitions) can be also established; and thus we omit all of their deatils here. Secondly, we can also establish the existence theory of essentially connected components of solutions sets for a class of nonlinear complementarity problem $NCP(f, K)$ in which the objective function f is not an *exceptional family* when its solutions set is contained in a compact set.

Chapter 7

Applications to Multi-Objective Optimization

7.1 Introduction

It is well-known in games theory that the existence of equilibria of scalar games for two persons is equivalent to the following minimax theorem:

$$\inf_{x\in X}\sup_{y\in Y} f(x,y) = \sup_{y\in Y}\inf_{x\in X} f(x,y)$$

where, A and B are nonempty subsets of topological vector spaces and f is a real-valued function on $A \times B$. It is clear that the real-valued function f above has a saddle point $(x^*, y^*) \in A \times B$ if and only if the minimax theorem above holds. However, if the specified game has multiple non-comparable criteria, the game is called a *multi-criteria game*. In this case, the payoff function f takes its value in a vector space (which is infinite dimensional in most cases) instead of being in the real line. Moreover, the set of all equilibrium optimal strategy pairs does not always have interchange ability condition, which holds for scalar games. This observation motivates us to study the general existence of saddle points and minimax theorem for vector-valued functions in topological vector spaces, and indeed this problem has been studied in the past (e.g., see Blackwell [47]; Chen [80], Corley [97]; Luc [351]-[352]; Takahashi [502]; Tan et al. [514], Tanaka [523]-[527]; Yu [602]-[601] and others). On the other hand, minimax theorems for a vector-valued mapping and some generalizations for saddle point concepts have been explored actively (e.g., see Ferro [172]; Nieuwenhuis [386]; Simons [485]; Sion [490]; Tanaka [523] and many others).

A multi-criteria two-person zero-sum game, or say, a two-person vector-valued zero-sum game is a quadruple-tuple

$$\Gamma = (X, Y; -f, f),$$

where X and Y are nonempty sets and f is a mapping $f : X \times Y \to Z$, where Z is an ordered topological vector space with an ordering defined by a convex cone C. The set X (resp., Y) is the set of strategies of player 1 (resp., player 2) and the mapping $-f$ (resp., f) is the payoff function of this player. When player 1 and player 2 choose a strategy $x \in X$ and a strategy $y \in Y$, respectively, the payoffs with respect to player 1 and player 2 are given by the vectors $-f(x, y)$ and $f(x, y)$, respectively. Each player chooses a strategy in order to increase his/her payoff and wants to find a strategy maximizing his payoff.

As applications of the Browder-Fan fixed point theorem developed in Chapter 2, the aim of this chapter is to first give extensive study for the existence of saddle points (resp., vector-valued saddle points) mappings (resp., vector-valued mappings) defined in underlying spaces which may be topological vector spaces, locally convex spaces, Banach spaces and even H-spaces. Then the stability of vector-valued saddle points and minimax theorems will be established in Section 6. Our stability results show that the majority of C-saddle points are stable in the sense of the Baire category. Finally vector-valued minimax inequalities were be established in Section 7.

7.2 ε-Saddle Points in Topological Vector Spaces

In this section, as applications of the Browder-Fan fixed point theorem, new ϵ-saddle point saddle point theorems are obtained in topological vector spaces which generalize the corresponding results in the literature as special cases. We first recall the following notion.

Definition 7.2.1. Let X, Y be Hausdorff topological spaces and f be a real-valued function on the product space $X \times Y$. If $\epsilon > 0$, a point $(x_\epsilon^*, y_\epsilon^*) \in X \times Y$ is said to be an ϵ-saddle point of f if for each $(x, y) \in X \times Y$, we have

$$f(x, y_\epsilon^*) - \epsilon < f(x_\epsilon^*, y_\epsilon^*) < f(x_\epsilon^*, y) + \epsilon.$$

Also a point $(x^*, y^*) \in X \times Y$ is said to be a saddle point of f if for each $(x, y) \in X \times Y$, we have

$$f(x, y^*) \leq f(x^*, y^*) \leq f(x^*, y).$$

We shall denote by $\mathbb{R}$ and $\mathbb{N}$ the set of all real numbers and the set of all natural numbers respectively. If X is a convex subset of a vector space and $f : X \to \mathbb{R}$, then f is said to be quasi-concave if for each $t \in \mathbb{R}$, the set $\{x \in X : f(x) > t\}$ is convex. f is said to be quasi-convex if $-f$ is quasi-concave. f is said to be quasi-convex if $-f$ is quasi-concave. If Y is a compact Hausdorff space, $C(Y)$ denotes the Banach space of all continuous real-valued functions on Y with supremum norm.

We begin with the following result.

Theorem 7.2.2. Let X be a nonempty convex subset of a Hausdorff topological vector space, Y be a nonempty compact convex subset of a Hausdorff topological vector space. Suppose that $f : X \times Y \to \mathbb{R}$ satisfies the following conditions:

(i) for each $(x, y) \in X \times Y$, $\inf_{v \in Y} f(x, v) > -\infty$ and $sup_{u \in X} f(u, y) < +\infty$;

(ii) for each fixed $y \in Y$, $x \mapsto f(x, y)$ is quasi-concave and $x \mapsto f(x, y) - \inf_{v \in Y} f(x, v)$ is upper semicontinuous; and

(iii) for each fixed $x \in X$, $y \mapsto f(x, y)$ is quasi-convex and $y \mapsto f(x, y) - sup_{u \in X} f(u, y)$ is lower semicontinuous.

Then f has an ϵ-saddle point $(x_\epsilon^*, y_\epsilon^*) \in X \times Y$ for each $\epsilon > 0$.

Proof. Let $\epsilon > 0$ be given. For each $(x, y) \in X \times Y$, define

$$T(x) = \{y \in Y : f(x, y) - \inf_{v \in Y} f(x, v) < \epsilon\},$$

and

$$S(y) = \{x \in X : f(x, y) - sup_{u \in X} f(u, y) > -\epsilon\},$$

then by (i), (ii) and (iii), $T(x)$ and $S(y)$ are nonempty and convex. For each $(x, y) \in X \times Y$, the sets

$$T^{-1}(y) = \{x \in X : f(x, y) - \inf_{v \in Y} f(x, v) < \epsilon\},$$

$$S^{-1}(x) = \{y \in Y : f(x, y) - sup_{u \in X} f(u, y) > -\epsilon\}$$

are open by (ii) and (iii).

Thus, by the Browder-Fan fixed point theorem in Chapter 2, there exists $(x_\epsilon^*, y_\epsilon^*) \in X \times Y$ such that $y_\epsilon^* \in T(x_\epsilon^*)$ and $x_\epsilon^* \in S(y_\epsilon^*)$, i.e.,

$$f(x, y_\epsilon^*) - \epsilon < f(x_\epsilon^*, y_\epsilon^*) < f(x_\epsilon^*, y) + \epsilon$$

for each $(x,y) \in X \times Y$. □

Theorem 7.2.3. Let X be a nonempty convex subset of a Hausdorff topological vector space, Y be a nonempty compact convex subset of a Hausdorff topological vector space. suppose that $f : X \times Y \to \mathbb{R}$ satisfies the following conditions:

(i) for each fixed $y \in Y$, $\sup_{u\in X} f(u,y) < +\infty$;

(ii) for each fixed $y \in Y$, $x \mapsto f(x,y)$ is quasi-concave and $x \mapsto f(x,y) - \inf_{v\in Y} f(x,v)$ is upper semicontinuous;

(iii) for each fixed $x \in X$, $y \mapsto f(x,y)$ is quasi-convex and lower semicontinuous and $y \mapsto f(x,y) - sup_{u\in X} f(u,y)$ is lower semicontinuous; and

(iv) for each sequence $\{(x_k, y_k)\}_{k\in \mathbb{N}}$ in $X \times Y$ where for each $k \in \mathbb{N}$, (x_k, y_k) is an ϵ_k-saddle point of f and $\epsilon_k \to 0^+$, there exists a subsequence $\{x_{n_k}\}_{k\in\mathbb{N}}$ and $x^* \in X$ such that for each $y \in Y$,

$$\overline{\lim}_{k\to\infty} f(x_{n_k}, y) \le f(x^*, y).$$

Then f has saddle point $(x^*, y^*) \in X \times Y$.

Proof. For each fixed $x \in X$, since Y is compact and $y \mapsto f(x,y)$ is lower semicontinuous by (ii), we must have $\inf_{v\in Y} f(x,v) > -\infty$.

For each $k \in \mathbb{N}$, by Theorem 7.2.2, there exists $(x_k^*, y_k^*) \in X \times Y$ such that for each $(x,y) \in X \times Y$,

$$f(x, y_k^*) - \epsilon_k < f(x_k^*, y_k^*) < f(x_k^*, y) + \epsilon_k.$$

By (iv), there exists a subsequence $\{x_{n_k}^*\}_{k\in\mathbb{N}}$ of $\{x_k^*\}_{k\in\mathbb{N}}$ and $x^* \in X$ such that for each $y \in Y$,

$$\overline{\lim}_{k\to\infty} f(x_{n_k}^*, y) \le f(x^*, y).$$

Since Y is compact, there exists a subnet $\{y_\alpha\}_{\alpha\in\Gamma}$ of $\{y_{n_k}^*\}_{k\in\mathbb{N}}$ and $y^* \in Y$ such that $y_\alpha^* \to y^*$. We shall prove that (x^*, y^*) is a saddle point of f.

For each $y \in Y$ and each $\alpha \in \Gamma$, by the last second inequality,

$$\begin{aligned} f(x^*,y^*) &= f(x^*,y^*) - f(x_\alpha^*, y_\alpha^*) + f(x_\alpha^*, y_\alpha^*) \\ &< f(x^*,y^*) - f(x_\alpha^*, y_\alpha^*) + f(x_\alpha^*, y) + \epsilon_\alpha \\ &= [f(x^*,y^*) - f(x^*, y_\alpha^*)] + [f(x^*, y_\alpha^*) - f(x_\alpha^*, y_\alpha^*)] + f(x_\alpha^*, y) + \epsilon_\alpha \\ &< [f(x^*,y^*) - f(x^*, y_\alpha^*)] + f(x_\alpha^*, y) + 2\epsilon_\alpha. \end{aligned}$$

By (iii) and (iv), it follows that

$$\begin{aligned} f(x^*,y^*) &\le \overline{\lim}_\alpha [f(x^*,y^*) - f(x^*, y_\alpha^*)] + \overline{\lim}_\alpha f(x_\alpha^*, y) \\ &\le f(x^*,y^*) - \underline{\lim}_\alpha f(x^*, y_\alpha^*) + f(x^*, y) \le f(x^*, y). \end{aligned}$$

Next, for each $x \in X$ and each $\alpha \in \Gamma$, by the inequality above again,

$$\begin{aligned} f(x^*, y^*) &= f(x^*, y^*) - f(x_\alpha^*, y_\alpha^*) + f(x_\alpha^*, y_\alpha^*) \\ &> f(x^*, y^*) - f(x_\alpha^*, y_\alpha^*) + f(x, y_\alpha^*) - \epsilon_\alpha \\ &= [f(x^*, y^*) - f(x_\alpha^*, y^*)] + [f(x_\alpha^*, y^*) - f(x_\alpha^*, y_\alpha^*)] + f(x, y_\alpha^*) - \epsilon_\alpha \\ &> [f(x^*, y^*) - f(x_\alpha^*, y^*)] + f(x, y_\alpha^*) - 2\epsilon_\alpha. \end{aligned}$$

By (iii) and (iv), it follows that

$$\begin{aligned} f(x^*, y^*) &\geq \underline{\lim}_\alpha [f(x^*, y^*) - f(x_\alpha^*, y^*)] + \underline{\lim}_\alpha f(x, y_\alpha^*) \\ &\geq f(x^*, y^*) - \overline{\lim}_\alpha f(x_\alpha^*, y^*) \geq f(x, y^*). \end{aligned}$$

Therefore $(x^*, y^*) \in X \times Y$ is a saddle point of f. □

Lemma 7.2.4. Let X be a nonempty convex subset of a Hausdorff topologicalvector space, Y be a nonempty compact convex subset of a Hausdorff topological vector space. Suppose that $f : X \times Y \to \mathbb{R}$ satisfies the following conditions:

(i) $f(x, y)$ is continuous on $X \times Y$;

(ii) for each fixed $y \in Y$, $x \mapsto f(x, y)$ is quasi-concave;

(iii) for each fixed $x \in X$, $y \mapsto f(x, y)$ is quasi-convex;

(iv) $\sup_{x \in X} \min_{y \in Y} f(x, y) < \infty$;

(v) the family $\{f(x, \cdot) : x \in X\}$ of real-valued functions on Y is equicontinuous and closed in the Banach space $C(Y)$; and

(vi) there exists a sequence $\{(x_k, y_k)\}_{k \in \mathbb{N}}$ in $X \times Y$ where for each $k \in \mathbb{N}$, (x_k, y_k) is an ϵ_k-saddle point of f on $X \times Y$ and $\epsilon_k \to 0^+$.

Then $\sup\{|f(x_k, y)| : k \in \mathbb{N}, y \in Y\} < +\infty$.

Proof. Since Y is compact, by (i), (ii), (iii) and (iv), the beginning of Proof. in the Theorem 3 of Komyia [326] shows that there exists a number M_1 such that $f(x, y) \leq M_1$ for all $x \in X$ and all $y \in Y$. It follows that the mapping $y \mapsto \sup_{x \in X} f(x, y)$ is real-valued and lower semicontinuous. By compactness of Y again, there exists a number M_2 such that $\sup_{x \in X} f(x, y) \geq M_2$ for all $y \in Y$.

Since $\epsilon_k \to 0^+$, we may suppose that $\epsilon_k < 1$ for each $k \in \mathbb{N}$. For each $k \in \mathbb{N}$ and each $y \in Y$, by (vi),

$$M_1 \geq f(x_k, y) \geq f(x_k, y_k) - \epsilon_k \geq \sup_{x \in X} f(x, y_k) - 2\epsilon_k \geq M_2 - 2$$

so that

$$\sup\{|f(x_k, y)| : k \in \mathbb{N}, y \in Y\} \leq max(|M_1|, |M_2 - 2|). \ \square$$

As an application of Theorem 7.2.3, we have the following result which is Theorem 3 of Komyia [326].

Corollary 7.2.5. Let X be a nonempty convex subset of a Hausdorff topological vector space, Y be a nonempty compact convex subset of a Hausdorff topological vector space. Suppose that $f : X \times Y \to \mathbb{R}$ satisfies the following conditions:

(i) $f(x,y)$ is continuous on $X \times Y$;

(ii) for each fixed $y \in Y$, $x \mapsto f(x,y)$ is quasi-concave;

(iii) for each fixed $x \in X$, $y \mapsto f(x,y)$ is quasi-convex;

(iv) $\sup_{x\in X} min_{y\in Y} f(x,y) < +\infty$; and

(v) the family $\{f(x,\cdot) : x \in X\}$ of real-valued functions on Y is equicontinuous and closed in the Banach space $C(Y)$.

Then f has a saddle point $(x^*, y^*) \in X \times Y$.

Proof. For each fixed $y \in Y$, by Corollary 2 of Aubin and Cellina [14, p.53], $x \mapsto \sup_{v\in Y}[-f(x,v)] = -\inf_{v\in Y} f(x,v)$ is upper semicontinuous so that by (i), $x \mapsto f(x,y) - \inf_{v\in Y} f(x,v)$ is upper semicontinuous.

For each fixed $y \in Y$ and for any $\eta > 0$, by (v), there exists a neighborhood V of y such that

$$f(u,y') - f(u,y) < \eta$$

for any $u \in X$ and $y' \in V$; thus

$$\sup_{u\in X} f(u,y') \le \sup_{u\in X} f(u,y) + \eta.$$

It follows that $y \mapsto \sup_{u\in X} f(u,y)$ is upper semicontinuous so that by (i), $y \mapsto f(x,y) - \sup_{u\in X} f(u,y)$ is also lower semicontinuous.

Now if $\{(x_k, y_k)\}_{k\in\mathbb{N}}$ is a sequence in $X \times Y$, where for each $k \in \mathbb{N}$, (x_k, y_k) is an ϵ_k-saddle point of f and $\epsilon_k \to 0^+$, by Lemma 7.2.4 and Ascoli's Theorem (e. g., see [449, p.369]), there exists a subsequence $\{f(x_{n_k},\cdot)\}_{k\in\mathbb{N}}$ of $\{f(x_k,\cdot)\}_{k\in\mathbb{N}}$ and $x^* \in X$ such that $\{f(x_{n_k},\cdot)\}_{k\in\mathbb{N}}$ converges uniformly to $f(x^*,\cdot)$ on Y. The conclusion now follows from Theorem 7.2.3. □

In order to obtain another new saddle point theorem, we recall the concept of an escaping sequence which was first introduced by Border in [48, p.34] as follows.

Let E be a Hausdorff topological vector space and Y be a subset of E such that $Y = \cup_{n=1}^{\infty} K_n$ where $\{K_n\}_{n\in\mathbb{N}}$ is an increasing sequence of nonempty compact sets. Then a sequence $\{y_n\}_{n\in\mathbb{N}}$ in Y is said to be escaping from Y (relative to $\{K_n\}_{n\in\mathbb{N}}$) if for each $n \in \mathbb{N}$, there exists a positive integer M such that $y_k \notin K_n$ for all $k \ge M$ (see also Definition 2.3.1).

Theorem 7.2.6. Let X be a nonempty convex subset of a Hausdorff topological vector space, Y be a nonempty convex subset of a Hausdorff topological vector space such that $Y = \cup_{n=1}^{\infty} K_n$ where $\{K_n\}_{n\in\mathbb{N}}$ is an increasing sequence of nonempty compact convex subsets of Y. Suppose that $f : X \times Y \to \mathbb{R}$ satisfies the following conditions:

(i) $f(x, y)$ is continuous on $X \times Y$;

(ii) for each fixed $y \in Y$, $x \mapsto f(x, y)$ is quasi-concave;

(iii) for each fixed $x \in X$, $y \mapsto f(x, y)$ is quasi-convex;

(iv) for each $n \in \mathbb{N}$, $\sup_{x \in X} min_{y \in K_n} f(x, y) < +\infty$;

(v) for each $n \in \mathbb{N}$, the family $\{f(x, \cdot) : x \in X\}$ of real-valued functions on K_n is equi-continuous and closed in the Banach space $C(K_n)$; and

(vi) for each sequence $\{(x_n, y_n)\}_{n \in \mathbb{N}}$ in $X \times Y$ where (x_n, y_n) is a saddle point of f on $X \times K_n$ for each $n \in \mathbb{N}$ which is escaping from $X \times Y$ relative to $\{X \times K_n\}_{n \in \mathbb{N}}$, there exists $n_0 \in \mathbb{N}$ and $x'_{n_0} \in X$ (or $y'_{n_0} \in K_{n_0}$) such that

$$f(x'_{n_0}, y_{n_0}) > f(x_{n_0}, y_{n_0}) \quad (\text{ or } f(x_{n_0}, y_{n_0}) > f(x_{n_0}, y'_{n_0})).$$

The f has a saddle point $(x^*, y^*) \in X \times Y$.

Proof. By Corollary 7.2.5, for each $n \in \mathbb{N}$, there exists a saddle point $(x_n, y_n) \in X \times K_n$ of f on $X \times K_n$. Suppose that the sequence $\{(x_n, y_n)\}_{n \in \mathbb{N}}$ in $X \times Y$ were escaping from $X \times Y$ relative to $\{X \times K_n\}_{n \in \mathbb{N}}$, then by (vi) there exists $n_0 \in \mathbb{N}$ and $x'_{n_0} \in X$ (or $y'_{n_0} \in K_{n_0}$) such that

$$f(x'_{n_0}, y_{n_0}) > f(x_{n_0}, y_{n_0}) \quad (\text{ or } f(x_{n_0}, y_{n_0} > f(x_{n_0}, y'_{n_0}))$$

which contradicts the fact that $(x_{n_0}, y_{n_0}) \in X \times K_{n_0}$ is a saddle point of f on $X \times K_{n_0}$. Therefore the sequence $\{(x_n, y_n)\} \in X \times K_n\}_{n \in \mathbb{N}}$ is not escaping from $X \times Y$ relative to $\{X \times K_n\}_{n \in \mathbb{N}}$; thus some subsequence $\{(x_{n_k}, y_{n_k})\}_{k \in \mathbb{N}}$ of $\{(x_n, y_n)\}_{n \in \mathbb{N}}$ must lie entirely in some $X \times K_s$, where $s \in \mathbb{N}$. By (v), Lemma 7.2.4 and Ascoli's Theorem again, there exists a subsequence $\{f(x_{n_k}, \cdot)\}_{k \in \mathbb{N}}$ of $\{f(x_n, \cdot)\}_{n \in \mathbb{N}}$ and $x^* \in X$ such that $\{f(x_{n_k}, \cdot)\}_{k \in \mathbb{N}}$ converges uniformly to $f(x^*, \cdot)$ on K_s. Since K_s is compact, there exists a subnet $\{y_\alpha\}_{\alpha \in \Gamma}$ of $\{y_{n_k}\}_{k \in \mathbb{N}}$ and $y^* \in K_s$ such that $y_\alpha \to y^*$.

For each $y \in Y$, there exists $s_1 > s$ such that $y \in K_{s_1}$. Let $\alpha_0 \in \Gamma$ be such that $\alpha_0 \geq s_1$, then for any $\alpha \geq \alpha_0$, $y \in K_\alpha$ and

$$\begin{aligned} f(x^*, y^*) &= f(x^*, y^*) - f(x^*_\alpha, y^*_\alpha) + f(x^*_\alpha, y^*_\alpha) \\ &\leq f(x^*, y^*) - f(x^*_\alpha, y^*_\alpha) + f(x^*_\alpha, y) \\ &= [f(x^*, y^*) - f(x^*, y^*_\alpha)] + [f(x^*, y^*_\alpha) - f(x^*_\alpha, y^*_\alpha)] + f(x^*_\alpha, y) \end{aligned}$$

so that by (i) and inequality above, we have $f(x^*, y^*) \leq f(x^*, y)$.

For each $x \in X$,

$$\begin{aligned} f(x^*, y^*) &= f(x^*, y^*) - f(x^*_\alpha, y^*_\alpha) + f(x^*_\alpha, y^*_\alpha) \\ &\geq f(x^*, y^*) - f(x^*_\alpha, y^*_\alpha) + f(x, y^*_\alpha) \\ &= [f(x^*, y^*) - f(x^*, y^*_\alpha)] + [f(x^*, y^*_\alpha) - f(x^*_\alpha, y^*_\alpha)] + f(x, y^*_\alpha) \end{aligned}$$

so that by (i) and the inequality above again, we also have $f(x^*, y^*) \geq f(x, y^*)$. Therefore (x^*, y^*) is a saddle point of f. □

7.3 Saddle Points beyond Topological Vector Spaces

The purpose of this section is to prove a new topological saddle point theorem, which in turn includes classical saddle point theorems such as Sion's minimax theorem [490] and others in topological vector spaces as special cases without using the Browder-Fan fixed point theorem.

Let X and Y be nonempty sets and $f : X \times Y \to \mathbb{R} \cup \{-\infty, +\infty\}$ be an extended real-valued function. A point $(x^*, y^*) \in X \times Y$ is said to be a saddle point of f in $X \times Y$ if

$$f(x, y^*) \leq f(x^*, y^*) \leq f(x^*, y)$$

for all $(x, y) \in X \times Y$.

Now consider a two-person zero-sum game G_f generated by the function f. This means that the first player selects a point x from X and the second player selects a point y from Y, and a point as a result of this choice such that the second pays the first one the amount $f(x, y)$. A point $(x^*, y^*) \in X \times Y$ is said to be a solution of the game G_f if and only if it is a saddle point of f in $X \times Y$.

The first saddle point theorem was proved by von Neumann [567]. Sion [490] proved a very general saddle point theorem for a function which is quasi-concave and upper semicontinuous in its first variable and quasi-convex and lower semicontinuous in its second variable in topological vector spaces.

In this section, we shall prove some new saddle point theorems without any linear structure by an elementary proof that mainly depends the connectedness of topological spaces. This method has been used by many authors, see also Joo and Stachó [276], Simons [485] and related references therein. Thus our result includes Sion's saddle point theorem and related minimax inequalities in topological vector spaces as special cases.

Let X and Y be two Hausdorff topological spaces. Let f be a real-valued function defined in $X \times Y$ such that for each fixed $x \in X$, $y \mapsto f(x, y)$ is lower semicontinuous and for each fixed $y \in Y$, $x \mapsto f(x, y)$ is upper semicontinuous. If X and Y are both compact, we know that $\min_{y \in Y} \max_{x \in X} f(x, y)$ and $\max_{x \in X} \min_{y \in Y} f(x, y)$ both exist. Denote by

$$\underline{v} := \max_{x \in X} \min_{y \in Y} f(x, y)$$

and

$$\overline{v} := \min_{y \in Y} \max_{x \in X} f(x, y),$$

then it is clear that $\underline{v} \leq \overline{v}$. Moreover it is well known and easily proves that $\underline{v} = \overline{v}$ if and only if f has a saddle point.

Let $\mathbb{N}$ and $\mathbb{R}$ denote the set of all positive integers and the set of all real numbers. We then have the following topological saddle point theorem by an elementary proof which mainly depends the connectedness of topological spaces.

Theorem 7.3.1. Let X and Y be compact Hausdorff topological spaces (indeed, Y is not necessarily compact). Then the function $f : X \times Y \to \mathbb{R} \cup \{-\infty, +\infty\}$ has a saddle point if the following conditions are satisfied:

(1) for each fixed $x \in X$, $y \mapsto f(x, y)$ is lower semicontinuous;

(2) for each fixed $y \in Y$, $x \mapsto f(x, y)$ is upper semicontinuous;

(3) for each $c \in \mathbb{R}$, $m \in \mathbb{N}$ and $y_i \in Y$, where $i = 1, \cdots, m$, the set

$$\cap_{i=1}^{m} \{x \in X : f(x, y_i) \geq c\}$$

is either connected or empty; and

(4) for each $y_1, y_2 \in Y$, there exists a function $S_{y_1,y_2} : [0,1] \to Y$ such that $S(0) = y_1$, $S(1) = y_2$ and for each $x \in X$, $n \in \mathbb{N}$, $j = 1, \cdots, 2^{n-1}$,

$$f(x, S(\frac{2j-1}{2^n})) \leq \max\{f(x, S(\frac{j-1}{2^{n-1}})), f(x, S(\frac{j}{2^{n-1}}))\}$$

and for any $\xi \in [0,1]$,

$$\liminf_{\lambda \in T, \lambda \to \xi} f(x, S(\lambda)) \geq f(x, S(\xi)),$$

where $T := \{\frac{j}{2^n} : n \in \mathbb{N}; j = 0, 1, \cdots, 2^n\}$.

Proof. It suffices to prove that $\underline{v} = \overline{v}$. We set

$$\mathcal{F} := \{F : F = \{x \in X : f(x, y) \geq c\}, \text{ where } y \in Y \text{ and } c < \overline{v}\}.$$

Let us show that the family $\mathcal{F}$ has the finite intersection properties, i.e.,

$$\cap_{i=1}^{n} \{x \in X : f(x, y_i) \geq c_i\} \neq \emptyset$$

for each $n \in \mathbb{N}$, where $y_i \in Y$, $c_i < \overline{v}$ and $i = 1, \cdots, n$. We prove above formula by induction on $n \in \mathbb{N}$. For $n = 1$, if there exists $y \in Y$ and $c_0 < \overline{v}$ such that $\{x \in X : f(x, y_0) \geq c_0\} = \emptyset$, then $f(x, y_0) < c_0$ for any $x \in X$ and

$$\overline{v} = \min_{y \in Y} \max_{x \in X} f(x, y) \leq \max_{x \in X} f(x, y_0) \leq c_0 < \overline{v}$$

which is a contradiction.

For a given $n \in \mathbb{N}$, we suppose that for any $y_i \in Y$ and $c_i < \overline{v}$, where $i = 1, \cdots, n$, it holds $\cap_{i=1}^{n} \{x \in X : f(x, y_i) \geq c_i\} \neq \emptyset$, but by contradiction we suppose that there exist $y_i \in Y$, $i = 1, \cdots, n+1$ such that $\cap_{i=1}^{n+1} \{x \in X : f(x, y_i) \geq c_i\} = \emptyset$. Without loss of the generality, we may assume that $c_1 \geq c_2 \geq \cdots \geq c_{n+1}$. Then

$$\cap_{i=1}^{n+1} \{x \in X : f(x, y_i) \geq c_1\} \subset \cap_{i=1}^{n+1} \{x \in X : f(x, y_i) \geq c_i\} = \emptyset.$$

By condition (4), there exists a function $S : [0,1] \to Y$ such that $S(0) = y_1$ and $S(1) = y_2$. For any $\lambda \in [0,1]$, define

$$I(\lambda) := \cap_{i=3}^{n+1}\{x \in X : f(x,y_i) \geq c_1\} \cap \{x \in X : f(x, S(\lambda)) \geq c_1\}.$$

In the previous definition it is understood that $\cap_{i=3}^{n+1}\{x \in X : f(x,y_i) \geq c_1\} = X$ if $n = 1$. By the induction assumption, $I(\lambda)$ is nonempty and

$$I(0) \cap I(1) \subset \cap_{i=1}^{n+1}\{x \in X : f(x,y_i) \geq c_1\} = \emptyset.$$

On the other hand, from conditions (2) and (3), $I(\lambda)$ is closed (hence compact) and connected. Now we construct two sequences $\{u_k\}_{k\in\mathbb{N}}$ and $\{w_k\}_{k\in\mathbb{N}}$ in T as follows: Let $u_1 := 0$ and $w_1 := 1$. Suppose that we have defined $u_h \in T$, $w_h \in T$ with $1 \leq h \leq k$ such that $u_h < w_h$, $u_h \in T_h := \{\frac{i-1}{2^{h-1}} : i = 1, \cdots, 2^{h-1}\}$, $w_h = u_h + \frac{1}{2^{h-1}}$, $u_h \leq u_{h+1}$, $w_{h+1} \leq w_h$, $I(u_h) \cap I(w_h) = \emptyset$ and $I(\frac{u_h+w_h}{2}) \subset I(u_h) \cup I(w_h)$. Now we define u_{k+1} and w_{k+1} as follows: by the induction hypotheses and connectedness properties we have either $I(\frac{u_k+w_k}{2}) \subset I(u_k)$ or $I(\frac{u_k+w_k}{2}) \subset I(w_k)$. If $I(\frac{u_k+w_k}{2}) \subset I(u_k)$, let $u_{k+1} := \frac{u_k+w_k}{2} \in T$ and $w_{k+1} := w_k$. Otherwise set $u_{k+1} := u_k$ and $w_{k+1} := \frac{u_k+w_k}{2} \in T$. Repeating this proceeding, we obtain two sequences $\{u_k\}_{k\in\mathbb{N}}$ and $\{w_k\}_{k\in N}$ such that

$$I(u_1) \supset I(u_2) \supset I(u_3) \supset \cdots,$$

and

$$I(w_1) \supset I(w_2) \supset I(w_3) \supset \cdots,$$

with $u_n \to \xi$ and $w_n \to \xi \in [0,1]$. Denote $y^* := S(\xi) \in Y$ and select $\overline{v} > c^* > c_1$ and by the induction assumption, we have

$$x^* \in \cap_{i=3}^{n+1}\{x \in X : f(x,y_i) \geq c_1\} \cap \{x \in X : f(x,y^*) \geq c^*\}.$$

Then

$$f(x^*,y^*) \geq c^* > c_1, \quad f(x^*,y_i) \geq c_1, \text{ for } i = 3, \cdots, n+1.$$

From condition (4), we have

$$\liminf_{n\to\infty} f(x^*, S(u_k)) \geq f(x^*, S(\xi)) = f(x^*,y^*) > c_1$$

and

$$\liminf_{n\to\infty} f(x^*, S(w_k)) > c_1.$$

Therefore there exists u_p and w_q in Y such that $f(x^*, S(u_p)) > c_1$ and $f(x^*, S(w_q)) > c_1$, so that $x^* \in I(u_p) \cap I(w_q)$. We may assume that $p \leq q$. Since $I(w_p) \supset I(w_q)$, so that $x^* \in I(w_p)$, which contradicts with $I(u_p) \cap I(w_p) = \emptyset$. Therefore $\mathcal{F}$ has the finite intersection property and $\cap_{F\in\mathcal{F}} F \neq \emptyset$ as F is compact for each $F \in \mathcal{F}$.

Remark 7.3.2. The proof of Theorem 7.3.1 actually tells us that following minimax inequality holds

$$\max_{x\in X}\inf_{y\in Y} f(x,y) = \inf_{y\in Y}\max_{x\in X} f(x,y)$$

when f satisfies hypotheses (1), (2), (3) and (4) of Theorem 7.3.1, where X is a compact Hausdorff space but Y may be any Hausdorff topological space which is not necessarily compact.

In particular, we have the following minimax inequality theorem as an application of Theorem 7.3.1.

Theorem 7.3.3. Let X be a compact Hausdorff topological space and Y a nonempty convex subset of a Hausdorff topological vector space F. Suppose $f : X \times Y \to \mathbb{R} \cup \{-\infty, +\infty\}$ is a function such that the following conditions are satisfied:

(1) for each fixed $x \in X$, $y \mapsto f(x,y)$ is lower semicontinuous;
(2) for each fixed $y \in Y$, $x \mapsto f(x,y)$ is upper semicontinuous;
(3) for each $c \in \mathbb{R}$, $m \in \mathbb{N}$ and $y_i \in Y$, where $i = 1, \cdots, m$, the set

$$\cap_{i=1}^m \{x \in X : f(x, y_i) \geq c\}$$

is either connected or empty; and
(4) for each fixed $x \in X$, $y \mapsto f(x,y)$ is quasi-convex.

Then we have

$$\max_{x\in X}\min_{y\in Y} f(x,y) = \min_{y\in Y}\max_{x\in X} f(x,y).$$

In particular, if Y is compact, then f has a saddle point in $X \times Y$.

Proof. It suffices to verify that f satisfies all hypotheses of Theorem 7.3.1. For each y_1, $y_2 \in Y$, define $S : [0,1] \to Y$ by $S(\lambda) = (1-\lambda)y_1 + \lambda y_2$. For each $x \in X$, by (1), we have

$$\begin{aligned} f(x, \quad & (1 - \frac{2j-1}{2^n} y_1 + \frac{2j-1}{2^n} y_2)) \\ & \leq \max\{f(x, (1 - \frac{j-1}{2^{n-1}})y_1 + \frac{j-1}{2^{n-1}} y_2), f(x, (1 - \frac{j}{2^{n-1}})y_1 + \frac{j}{2^{n-1}} y_2)\}. \end{aligned}$$

As S is continuous, for each $\xi \in [0,1]$, we also have

$$\liminf_{\lambda\in T \text{ and } \lambda\to\xi} f(x, S(\lambda)) \geq f(x, S(\xi)).$$

Thus all hypotheses of Theorem 7.3.1 are satisfies. By Theorem 7.3.1, it follows that

$$\max_{x\in X}\min_{y\in Y} f(x,y) = \min_{y\in Y}\max_{x\in X} f(x,y).$$

Thus f has a saddle point in $X \times Y$ and the proof of Theorem 7.3.3 is complete.

Remark 7.3.4. As minimax inequality plays a very important role in the study of optimization and other subjects, the existence of topological minimax inequalities has been extensively studied by many authors in recently years (e.g., see Simons [485]) However, all topological minimax theorems from [353] and related references therein are independent from our results in this section.

As an immediately consequence of Theorem 7.3.3, we have the following saddle point theorem in topological vector space which includes Sion's classical minimax inequality in [490] as a special case.

Corollary 7.3.5. Let X be a nonempty compact and convex subsets of a Hausdorff topological vector spaces E and Y be a nonempty convex subset of a Hausdorff topological space F, respectively. Suppose $f : X \times Y \to \mathbb{R}$ satisfies that:

(1) for each fixed $x \in X$, $y \mapsto f(x,y)$ is lower semicontinuous and quasi-convex; and

(2) for each fixed $y \in Y$, $x \mapsto f(x,y)$ is upper semicontinuous and quasi-concave.

Then we have

$$\max_{x\in X}\min_{y\in Y} f(x,y) = \min_{y\in Y}\max_{x\in X} f(x,y).$$

Moreover if Y is compact, then f has a saddle point in $X \times Y$.

Proof. For each fixed $m \in \mathbb{N}$, each $y_i \in Y$, where $i = 1, \cdots, m$ and for each fixed $c \in \mathbb{R}$, if $\cap_{i=1}^m \{x \in X : f(x,y_i) \geq c\} \neq \emptyset$, then by (2), the set $\{x \in X : f(x,y_i) \geq c\}$ is convex for $i = 1, \cdots, m$. Therefore $\cap_{i=1}^m \{x \in X : f(x,y_i) \geq c\}$ is nonempty convex and hence nonempty connected. Thus all hypotheses of Theorem 7.3.3 are satisfied. By Theorem 7.3.2, the conclusion follows. □

In this note, we establish a topological minimax result (i.e., Theorem 7.3.1), which can allow us to consider the existence of saddle points without the traditional linear structure. Also our results are new and independent from all results in the literature (see [353]).

Finally, we wish to note that for the application of saddle point theorem and its applications to mathematical economics and game theory, the interested reader can find more details from Aubin and Ekeland [15], Border [48], Ichiishi [245] and references therein. Furthermore a very comprehensive bibliography for the study of topological versions of saddle points and minimax theorems can be also found from Simons [485].

7.4 Vector-Valued Saddle Points in Topological Vector Spaces

In this section, as applications of the Browder-Fan fixed point theorems in topologoical vector spaces, we prove some new existence theorems for loose saddle points and for saddle points of set-valued maps or vector-valued functions. These theorems generalize the corresponding results in the literature.

Let A and B be nonempty sets, f a real-valued function on $A \times B$. A point $(x^*, y^*) \in A \times B$ is called a saddle point of f in $A \times B$ if

$$f(x^*, y) \leq f(x^*, y^*) \leq f(x, y^*) \tag{7.4.1}$$

for all $(x, y) \in A \times B$.

The aim of the present section is to establish some new existence theorems for loose saddle points and for saddle points of vector-valued functions or set-valued maps under some weaker conditions, in particular, the underlying spaces need not be locally convex. These theorems generalize the corresponding results in the literature in several ways.

Throughout this section, all Hausdorff topological vector spaces are assumed to be real unless specified. Denote by $\mathbb{N}$ and $\mathbb{R}$ the set of all natural numbers and the set of all real numbers, respectively. Let A be a nonempty set. If B is another nonempty set, then a mapping $F : A \to 2^B$ is called a set-valued mapping from A to B.

Let Z be a vector space. If C is a convex cone in Z (i.e., C is convex and $tC \subset C$ for all $t \geq 0$), then the partial order $\geq$ on Z generated by C is defined by $z_1 \geq z_2$ if and only if $z_1 - z_2 \in C$. Let $\geq$ be a partial order on Z. Then for a, $b \in Z$, $a \leq b$ if and only if $b \geq a$ and $a < b$ (resp., $a > b$) if and only if $a \leq b$ (resp., $a \geq b$) and $a \neq b$. There is an one-to-one correspondence between vector orderings of a real ordered vector space space S and convex cones in S, we assume that each real ordered vector space has a convex cone C which induces the ordering as defined above.

Let Z be an ordered topological vector space with an ordering defined by a convex cone C. The convex cone C is assumed to be *pointed*, i.e., , $C \cap (-C) = \{0\}$, and hence the ordering is anti-symmetric and $0 \in C$. Moreover, we assume that the convex cone C to be *solid*, , i.e., its topological interior $intC$ is nonempty, and hence $C^0 := (intC) \cup \{0\}$ is a pointed convex cone and induces another (anti-symmetric) vector ordering $\leq_{C^0}$, which is weaker than $\leq_C$ in E. We also remark that the orderings $\leq_C$ and $\leq_{C^0}$ are two directed anti-symmetric partial orderings. Here an ordering $\leq$ is directed in the sense that for each given $x, y \in E$, there exists $z \in E$ such that $x \leq z$ and $y \leq z$.

Let A be a nonempty subset of Z. Then a point $a_0 \in A$ is said to be a *maximal* (resp., *minimal*) *point* of A if $a \in A$ and $a \geq a_0$ (resp., $a_0 \geq a$) implies $a = a_0$. The set of maximal (resp., minimal) points of A is denoted by $max\ A$ (resp., $min\ A$). We shall adopt the concept of '*cone extreme point*' which was proposed by Yu [601]. It is clear that an element z_0 of A is a C-minimal point of A if and only if the set $\{z \in A : z \leq_C z_0, z \neq z_0\} = \emptyset$; which is equivalent to $A \cap (z_0 - C) = \{z_0\}$. We denote the set of all such C-minimal (resp, C-maximal) points by Min$A$ (resp., $MaxA$). Also C^0-minimal and C^0-maximal points of A are defined similarly. In this section, the set of maximal (resp., minimal) points of A is denote by max A (resp., min A).

We would also like to point out that if the $intC \neq \emptyset$, there exists a nonzero continuous linear functional ψ such that $\psi(x) \geq 0$ for all $x \in C$, i.e., $C^* \setminus \{0\} \neq \emptyset$, where $C^* := \{\psi \in E^* : \psi(x) \geq 0$ for all $x \in C\}$, which is also called the dual cone of C.

Let X be a nonempty convex subset of a vector space, Z a Hausdorff topological vector space with a partial order $\geq$ and F a set-valued mapping from X to Z. Then (i) F is compact-valued if $F(x)$ is compact for all $x \in X$; (ii) F is convex (resp., concave) if for every $x_1, x_2 \in X$, $\lambda \in [0,1]$, $y_1 \in F(x_1)$ and $y_2 \in F(x_2)$, there exists $y_3 \in F(\lambda x_1 + (1-\lambda)x_2)$ such that

$$y_3 \leq \lambda y_1 + (1-\lambda)y_2 \ \ (\text{resp., } y_3 \geq \lambda y_1 + (1-\lambda)y_2);$$

(iii) F is quasi-convex (resp., quasi-concave) if for any $a \in Z$, the set $\{x \in X :$ there is $y \in F(x)$ with $y \leq a$ (resp., $y \geq a$)$\}$ is convex. If F is single-valued and the order of Z is generated by a convex cone $C \subset Z$, the convexity (resp., concavity) and quasi-convexity (resp., quasi-concavity) coincide with that of C-convexity (resp., C-concavity) and C-quasi-convexity (resp., C-quasi-concavity). . If $g : X \to Z$ is a function, then g is said to be properly quasi-C-convex [525, p.327] if either

$$g(\lambda x_1 + (1-\lambda)x_2) \leq g(x_1) \ \text{ or } \ g(\lambda x_1 + (1-\lambda)x_2) \leq g(x_2)$$

for every $x_1, x_2 \in X$ and $\lambda \in [0,1]$. Also g is said to be properly quasi-C-concave (e.g., see Tanaka [525]) if $-g$ is properly quasi-C-convex. If $p : Z \to \mathbb{R}$ is a function, then p is monotonic (resp., strictly monotonic) if $p(a) \geq p(b)$ (resp., $p(a) > p(b)$) whenever $a \geq b$ (resp., $a > b$).

Lemma 7.4.1. Let X be a nonempty convex subset of a vector space, Z a Hausdorff topological vector space, $\geq$ a partial order on Z which is generated by a convex cone $C \subset Z$ and F a set-valued mapping from X to Z. If F is convex (resp., concave), then F is quasi-convex (resp., quasi-concave).

Proof. For any $a \in Z$, we need to prove that the set $L = \{x \in X$: there is $y \in F(x)$ with $y \leq a$ (resp., $y \geq a$)$\}$ is convex. For any $x_1, x_2 \in L$ and

$\lambda \in [0,1]$, there exists $y_1 \in F(x_1)$ and $y_2 \in F(x_2)$ such that $y_1 \leq a$ and $y_2 \leq a$ (resp., $y_1 \geq a$ and $y_2 \geq a$). Since F is convex (resp., concave), there exists $y_3 \in F(\lambda x_1 + (1-\lambda)x_2)$ such that

$$y_3 \leq \lambda y_1 + (1-\lambda)y_2 \text{ (resp., } y_3 \geq \lambda y_1 + (1-\lambda)y_2).$$

Since $\geq$ is generated by a convex cone,

$$\lambda y_1 + (1-\lambda)y_2 \leq a \text{ (resp., } \lambda y_1 + (1-\lambda)y_2 \geq a),$$

so that $y_3 \leq a$ (resp., $y_3 \geq a$). Hence

$$\lambda x_1 + (1-\lambda)x_2 \in L$$

and L is convex. □

Lemma 7.4.2. Let X be a nonempty convex subset of a vector space, Z a Hausdorff topological vector space and $\geq$ a partial order on Z. Let F be a set-valued mapping from $A \times B$ to Z and p a real-valued function on Z. If F is convex (resp., concave) and p is monotonic and convex (resp., concave), then the composite set-valued map pF is convex (resp., concave).

Proof. If F and p are convex, Proposition 2.2 of Luc and Vargas [353] shows that pF is also convex. If F and p are concave, then the same proof of Proposition 2.2 of Luc and Vargas [353] can be modified to show that pF is also concave. □

Let X, Y and Z be Hausdorff topological vector spaces, $\geq$ a partial order on Z, $A \subset X$, $B \subset Y$ and F a set-valued mapping from $A \times B$ to Z. For $x \in A$ and $y \in B$, let $F(A,y) = \cup_{a \in A} F(a,y)$ and $F(x,B) = \cup_{b \in B} F(x,b)$. Then a point $(x^*,y^*) \in A \times B$ is said to be (1) a loose saddle point of F in $A \times B$ if $F(x^*,y^*) \cap \min F(A,y^*) \neq \emptyset$ and $F(x^*,y^*) \cap \max F(x^*,B) \neq \emptyset$; (2) a saddle point of F in $A \times B$ if $F(x^*,y^*) \cap \min F(A,y^*) \cap \max F(x^*,B) \neq \emptyset$.

Note that every saddle point is a loose saddle point; the converse is in general not true. To see this, take $X = Y = E = \mathbb{R}$, $A = B = [0,1]$ and define $F : X \times Y \to 2^{\mathbb{R}}$ by $F(x,y) = [0,1]$ for each $(x,y) \in X \times Y$. In this case, every point $(x,y) \in A \times B$ is a loose saddle point of F but F has no saddle point.

When F is single-valued, every loose saddle point is also a saddle point. Let $p : X \to \mathbb{R}$ be strictly monotonic, then any loose saddle point (saddle point) of pF in $A \times B$ is also a loose saddle point (saddle point) of F in $A \times B$, see Lemma 3.2 of Luc and Vargas [353].

We also need the following two simple facts:

Lemma 7.4.3. Let X, Y, Z be three Hausdorff topological vector spaces, A and B nonempty subsets of X and Y, respectively. Suppose that A is compact and the set-valued mapping F from $A \times B$ to Z is upper semicontinuous such that for each fixed $x \in A$, $y \to F(x, y)$ is lower semicontinuous. Then the set-valued mapping T from B to Z, defined by $T(y) = \cup_{x \in A} F(x, y)$ for each $y \in B$, is continuous.

Proof. Let $y \in B$ and G be an open set in Z with $G \cap T(y) \neq \emptyset$, then there is $x \in A$ such that $G \cap F(x, y) \neq \emptyset$. Since $y \to F(x, y)$ is lower semicontinuous at y, there exists an open neighborhood $O(y)$ of y in B such that for each $y' \in O(y)$, $G \cap F(x, y') \neq \emptyset$ so that $G \cap T(y') \neq \emptyset$. Thus T is lower semicontinuous.

Now let $y \in B$ and G be any open set in Z with $G \supset T(y)$, then $G \supset F(x, y)$ for all $x \in A$. Since F is upper semicontinuous on $A \times B$, there exists two open neighborhoods $O(x)$ of x in A and $O_x(y)$ of y in B such that $G \supset F(x', y')$ for any $x' \in O(x)$ and $y' \in O_x(y)$. Since $\cup_{x \in A} O(x) = A$ and A is compact, there exists $\{x_1, \cdots x_n\} \subset A$ such that $\cup_{i=1}^n O(x_i) = A$. Set $O(y) = \cap_{i=1}^n O_{x_i}(y)$, then $O(y)$ is an open neighborhood of y in B. If $y' \in O(y)$ and $x' \in A$, there is $i \in \{1, \cdots, n\}$ such that $x' \in O(x_i)$; since $y' \in O_{x_i}(y)$, $G \supset F(x', y')$. Thus $G \supset T(y')$ for each $y' \in O(y)$. Hence T is also upper semicontinuous. □

Lemma 7.4.4. Let Z be a Hausdorff topological space and Φ be a set-valued mapping from Z to $\mathbb{R}$ which is compact-valued and continuous. Then the functions $p, q : Z \to \mathbb{R}$, defined by $p(z) = \max \Phi(z)$ and $q(z) = \min \Phi(z)$ are continuous.

Proof. Fix an arbitrary $z \in Z$. For any $\epsilon > 0$, since $(-\infty, p(z) + \epsilon) \supset \Phi(z)$ and Φ is upper semicontinuous at z, there exists an open neighborhood $O_1(z)$ of z in Z such that $(-\infty, p(z) + \epsilon) \supset \Phi(z')$ for any $z' \in O_1(z)$, hence $p(z') = \max \Phi(z') < p(z) + \epsilon$ for any $z' \in O_1(z)$. Since $(p(z) - \epsilon, p(z) + \epsilon) \cap \Phi(z) \neq \emptyset$ and Φ is lower semicontinuous at z, there exists an open neighborhood $O_2(z)$ of z in Z such that $(p(z) - \epsilon, p(z) + \epsilon) \cap \Phi(z') \neq \emptyset$ for any $z' \in O_2(z)$, hence $p(z') = \max \Phi(z') > p(z) - \epsilon$ for any $z' \in O_2(z)$. Set $O(z) = O_1(z) \cap O_2(z)$, then $O(z)$ is an open neighborhood of z in Z such that $|p(z) - p(z')| < \epsilon$ for any $z' \in O(z)$ so that p is continuous at z. Therefore p is continuous. Similarly, the function q is also continuous. □

Theorem 7.4.5. Let A and B be two nonempty compact convex subsets of two Hausdorff topological vector spaces X and Y respectively, Z a Hausdorff topological vector space with a partial order $\geq$. Suppose that the set-valued mapping F from $A \times B$ to Z satisfies the following conditions:

(i) F is compact-valued and upper semicontinuous such that for each fixed $x \in A$, $y \to F(x, y)$ is lower semicontinuous on B and for each fixed $y \in B$, $x \to F(x, y)$ is lower semicontinuous on A; and

(ii) there exists a function $p : Z \to \mathbb{R}$ which is strictly monotonic and continuous such that for each fixed $x \in A$, $y \to pF(x, y) = \{p(z) : z \in F(x, y)\}$ is quasi-concave and for each fixed $y \in B$, $x \to pF(x, y)$ is quasi-convex.

Then F has a loose saddle point.

Proof. Note that for each $(x, y) \in A \times B$, by (i) and Theorem 7.4.2 of Klein and Thompson [317], the sets $F(A, y)$ and $F(x, B)$ are compact. Since p is continuous, the sets $pF(A, y)$ and $pF(x, B)$ are also compact. Fix an arbitrary $n \in \mathbb{N}$ and define the set-valued mapping W_n from $A \times B$ to $A \times B$ by

$$W_n(x,y) = \left\{ (u,v) \in A \times B : \begin{array}{c} \min pF(A,y) > \min pF(u,y) - 1/n \\ \max pF(x,B) < \max pF(x,v) + 1/n \end{array} \right\} \quad (7.4.2)$$

for each $(x, y) \in A \times B$. First we shall show that W_n is convex-valued for each $n \in \mathbb{N}$. Indeed, suppose $(x, y) \in A \times B$, (u_1, v_1), $(u_2, v_2) \in W_n(x, y)$ and $\lambda \in [0, 1]$ and let $a = \max\,(\min pF(u_1, y),\ \min pF(u_2, y))$. Since $u \to pF(u, y)$ is quasi-convex, for each $a \in \mathbb{N}$ the set $S = \{u \in A :$ there exists $s \in pF(u, y)$ with $s \leq a\}$ is convex. From this, for each $u_1, u_2 \in S$ and $\lambda \in [0.1]$, it follows that min $pF(\lambda u_1 + (1-\lambda)u_2, y) \leq a$ so that $\min pF(A, y) > \min pF(\lambda u_1 + (1-\lambda)u_2, y) - 1/n$. Similarly, $\max pF(x, B) < \max pF(x, \lambda v_1 + (1-\lambda)v_2) + 1/n$. This shows that $\lambda(u_1, v_1) + (1-\lambda)(u_2, v_2) \in W_n(x, y)$. Hence $W_n(x, y)$ is convex. Next we shall show that for each $(u, v) \in A \times B$, the set

$$W_n^{-1}(u,v) = \left\{ (x,y) \in A \times B : \begin{array}{c} \min pF(A,y) > \min pF(u,y) - \dfrac{1}{n} \\ \max pF(x,B) < \max pF(x,v) + \dfrac{1}{n} \end{array} \right\} \quad (7.4.3)$$

is open in $A \times B$. Indeed, by (i) and Lemma 7.4.3, the set-valued maps $y \to F(A, y)$ and $x \to F(x, B)$ are continuous. By Theorem 7.3.11 of Klein and Thompson [317], the set-valued maps $y \to pF(A, y)$, $y \to pF(u, y)$, $x \to pF(x, B)$ and $x \to pF(x, v)$ are also continuous. By Lemma 7.4.4, the functions $y \to \min\, pF(A, y)$, $y \to \min\, pF(u, y)$, $x \to \max\, pF(x, B)$ and $x \to \max pF(x, v)$ are continuous. Hence $W_n^{-1}(u, v)$ is open in $A \times B$. Now by the Browder-Fan fixed point Theorem 2.3.19; or see Theorem 1 of Browder [59]), there exists $(x_n^*, y_n^*) \in A \times B$ such that $(x_n^*, y_n^*) \in W_n(x_n^*, y_n^*)$, i.e.,

$$\begin{aligned} \min pF(A, y_n^*) &> \min pF(x_n^*, y_n^*) - 1/n, \\ \max pF(x_n^*, B) &< \max pF(x_n^*, y_n^*) + 1/n. \end{aligned} \quad (7.4.4)$$

Let $s_n^* \in F(x_n^*, y_n^*)$ and $t_n^* \in F(x_n^*, y_n^*)$ be such that $p(s_n^*) = \min pF(x_n^*, y_n^*)$ and $p(t_n^*) = \max pF(x_n^*, y_n^*)$.

Since A and B are compact and F is upper semicontinuous on $A \times B$, $F(A, B) = \cup_{(x,y)\in A\times B} F(x, y)$ is compact by Theorem 7.4.2 [317]. Since $x_n^* \in A$, $y_n^* \in B$, $s_n^* \in F(A, B)$ and $t_n^* \in F(A, B)$, there exists a subnet $\{n(\alpha)\}_{\alpha\in\Gamma}$ of $\{n\}_{n=1}^{\infty}$ and $x^* \in A$, $y^* \in B$ and s^*, $t^* \in F(A, B)$ such that $x_{n(\alpha)}^* \to x^*$, $y_{n(\alpha)}^* \to y^*$, $s_{n(\alpha)}^* \to s^*$ and $t_{n(\alpha)}^* \to t^*$. Note that $n(\alpha) \to \infty$ and for each $\alpha \in \Gamma$,

$$\min pF(A, y_{n(\alpha)}^*) > p(s_{n(\alpha)}^*) - 1/n(\alpha), \quad (7.4.5a)$$

$$\max pF(x^*_{n(\alpha)}, B) < p(t^*_{n(\alpha)}) + 1/n(\alpha). \tag{7.4.5b}$$

By (7.4.5a), (7.4.5b) and the continuity of the functions of $y \to \min pF(A, y)$, $x \to \max pF(x, B)$ and p, we have

$$\min pF(A, y^*) \geq p(s^*), \tag{7.4.6a}$$

$$\max pF(x^*, B) \leq p(t^*). \tag{7.4.6b}$$

Since $s^* \in F(x^*, y^*) \subset F(A, y^*)$ and $t^* \in F(x^*, y^*) \subset F(x^*, B)$, (7.4.6a) and (7.4.6b) imply that

$$\min pF(A, y^*) = p(s^*), \tag{7.4.7a}$$

$$\max pF(x^*, B) = p(t^*) \tag{7.4.7b}$$

so that $pF(x^*, y^*) \cap \min pF(A, y^*) \neq \emptyset$ and $pF(x^*, y^*) \cap \max pF(x^*, B) \neq \emptyset$. Therefore $(x^*, y^*) \in A \times B$ is a loose saddle point of pF in $A \times B$, and hence it is also a loose saddle point of F in $A \times B$. □

Theorem 7.4.5 generalizes Theorem 3.3 of Luc and Vargas [353] in the following ways: (i) X and Y need not be locally convex (in the proof of Theorem 3.3 of Luc and Vargas [353], even though the authors did not assume the underlying space to be locally convex, they applied a fixed point theorem of Browder [62] in this section. The actual fixed point theorem was due to Fan [156] or Glicksberg [196], which requires the space to be locally convex) and (ii) the continuity condition of F is weakened. We remark that our proof is different from that of Luc and Vargas [353].

The following example from Tan et al. [521] shows that Theorem 7.4.5 is a true generalization of Theorem 3.3 of Luc and Vargas [353].

Example 7.4.6. Let $A = B = [0, 1]$. Let P and Q be the points $(1, 0)$ and $(0, 1)$ in $A \times B$ respectively. Define a set-valued mapping F from $A \times B$ to $A \times B$ by $F(0, 0) = \overline{PQ}$ and $F(x, y) =$ the line segment joining P to the point $(\frac{xy}{x^2+y^2}, 1 - \frac{xy}{x^2+y^2})$ if $(x, y) \neq (0, 0)$. Then it can be shown that F is compact-(and convex-) valued and is upper semicontinuous such that for each fixed $x \in A$, $y \to F(x, y)$ is lower semicontinuous on B and for each fixed $y \in B$, $x \to F(x, y)$ is lower semicontinuous on A. Now let $U = \{(x, y) \in A \times B : 0 \leq x < 1/4,\ 3/4 < y \leq 1\}$, then U is open in $A \times B$ such that $U \cap F(0, 0) \neq \emptyset$. However, for any δ with $0 < \delta < 1$, we see that $F(\frac{1}{2}\delta, \frac{1}{2}\delta) =$ the line segment joining P and the point $(1/2, 1/2)$ so that $F(\frac{1}{2}\delta, \frac{1}{2}\delta) \cap U = \emptyset$. Therefore F is not lower semicontinuous at $(0, 0)$. Thus Theorem 3.3 of Luc and Vargas [353] is not applicable. Define $p : \mathbb{R}^2 \to \mathbb{R}$ by $p(x, y) = x + y$, then p is strictly monotonic and continuous (and linear). Note that $pF(x, y) = \{1\}$ for all $x, y \in [0, 1]$ so that for each fixed $x \in A$, $y \to pF(x, y)$ is (quasi-) concave and for each fixed $y \in B$, $x \to pF(x, y)$ is (quasi-) convex. Hence our Theorem 7.4.5 is applicable (in fact, every point in $A \times B$ is a saddle point of F).

Corollary 7.4.7. Let A and B be two nonempty compact convex subsets of two Hausdorff topological vector spaces X and Y respectively, Z a Hausdorff topological vector space with a partial order $\geq$ generated by a convex cone C in Z. Suppose that:

(i) the set-valued mapping F from $A \times B$ to Z satisfies condition (i) of Theorem 7.4.5;

(ii) there exists a real-valued continuous linear functional p on Z such that $p(z) > 0$ for all $z \in C \setminus \{0\}$;

(iii) for each fixed $x \in A$, the mapping $y \to F(x, y)$ is convex on B; and

(iv) for each fixed $y \in B$, the mapping $x \to F(x, y)$ is concave on A.

Then F has a loose saddle point in $A \times B$.

Proof. Note that (a) p is strictly monotonic, (b) for each fixed $x \in A$, the mapping $y \to pF(x, y)$ is convex on B and (c) for each fixed $y \in B$, the mapping $x \to pF(x, y)$ is concave on A. By Lemmas 7.4.1 and 7.4.2, we have: (i) for each fixed $x \in A$, the mapping $y \to pF(x, y)$ is quasi-convex on B and (ii) for each fixed $y \in B$, the mapping $x \to pF(x, y)$ is quasi-concave on A. The result now follows from Theorem 7.4.5. □

Corollary 7.4.8. Let A and B be two nonempty compact convex subsets of two Hausdorff topological vector spaces X and Y respectively, Z a Hausdorff topological vector space with a partial order $\geq$ generated by a convex cone C in Z such that $C \setminus \{0\}$ is open in Z. Suppose that (i) the set-valued mapping F from $A \times B$ to Z satisfies condition (i) of Theorem 7.4.11; (ii) for each fixed $x \in A$, the mapping $y \to F(x, y)$ is convex on B and (iii) for each fixed $y \in B$, the mapping $x \to F(x, y)$ is concave on A. Then F has a loose saddle point in $A \times B$.

Proof. By Theorem 3.4 of Rudin [449, p.58], there is a continuous linear functional p on Z such that $p(z) > 0$ for all $z \in C \setminus \{0\}$. The result now follows from Corollary 7.4.7. □

Corollary 7.4.7 and Corollary 7.4.8 generalize Corollary 3.4 and Corollary 3.5 of Luc and Vargas [353] respectively to topological vector spaces which are not necessarily locally convex.

Corollary 7.4.9. Let A and B be two nonempty compact convex subsets of two Hausdorff topological vector spaces X and Y respectively, Z a Hausdorff topological vector space with a partial order $\geq$ generated by a convex cone C in Z such that $C \setminus \{0\}$ is open in Z. Suppose that the single-valued mapping $f : A \times B \to Z$ is continuous. Then f has a saddle point in $A \times B$ if f satisfies any one of the following conditions:

(i) for each fixed $y \in B$, the mapping $x \to f(x, y)$ is properly quasi C-convex and for each fixed $x \in A$, the mapping $y \to f(x, y)$ is properly quasi C-concave;

(ii) for each fixed $y \in B$, the mapping $x \to f(x,y)$ is properly quasi C-convex and for each fixed $x \in A$, the mapping $y \to f(x,y)$ is C-concave;

(iii) for each fixed $y \in B$, the mapping $x \to f(x,y)$ is C-convex and for each fixed $x \in A$, the mapping $y \to f(x,y)$ is properly quasi C-concave;

(iv) for each fixed $y \in B$, the mapping $x \to f(x,y)$ is C-convex and for each fixed $x \in A$, the mapping $y \to f(x,y)$ is C-concave.

Proof. There is a continuous linear functional p on Z such that $p(z) > 0$ for all $z \in C \setminus \{0\}$. It is easy to prove that for each fixed $y \in B$, the mapping $x \to pf(x,y)$ is quasi-convex and for each fixed $x \in A$, the mapping $y \to pf(x,y)$ is quasi-concave. By Theorem 7.4.5, f has a loose saddle point in $A \times B$. Since f is single-valued, this loose saddle point is also a saddle point. □

Corollary 7.4.9 generalizes Lemma 3.3 of Tanaka [525] to topological vector spaces which are not necessarily locally convex.

Theorem 7.4.10. Let A and B be two nonempty convex subsets of two locally convex Hausdorff topological vector spaces X and Y respectively. Suppose that $A = \cup_{n=1}^{\infty} C_n$ and $B = \cup_{n=1}^{\infty} D_n$, where $\{C_n\}_{n=1}^{\infty}$ and $\{D_n\}_{n=1}^{\infty}$ are increasing sequences of nonempty compact convex sets in X and Y respectively. Assume that the function $f : A \times B \to \mathbb{R}$ satisfies the following conditions:

(i) for each fixed $y \in B$, the function $x \to f(x,y)$ is lower semi-continuous and quasi-convex;

(ii) for each fixed $x \in A$, the function $y \to f(x,y)$ is upper semicontinuous and quasi-concave; and

(iii) for each sequence $\{(x_n, y_n)\}_{n=1}^{\infty}$ in $A \times B$, where $(x_n, y_n) \in C_n \times D_n$ for each $n = 1, 2, \cdots$, which is escaping relative to $\{C_n \times D_n\}_{n=1}^{\infty}$, there exists $n_0 \in \mathbb{N}$ and $(x'_{n_0}, y'_{n_0}) \in C_{n_0} \times D_{n_0}$ such that $f(x_{n_0}, y'_{n_0}) > f(x_{n_0}, y_{n_0})$ or $f(x_{n_0}, y_{n_0}) > f(x'_{n_0}, y_{n_0})$.

Then f has a saddle point in $A \times B$.

Proof. First we note that $A \times B = \cup_{n=1}^{\infty}(C_n \times D_n)$. For each $n \in \mathbb{N}$, since C_n and D_n are compact convex subsets of A and B respectively and conditions (i) and (ii) hold, by Sion's minimax inequality (e.g., see Theorem 3.4 of Sion [490]), there exists $(x_n, y_n) \in C_n \times D_n$ such that

$$f(x_n, y) \le f(x_n, y_n) \le f(x, y_n) \tag{7.4.8}$$

for all $(x,y) \in C_n \times D_n$. Suppose that the sequence $\{(x_n, y_n)\}_{n=1}^{\infty}$ is escaping from $A \times B$ relative to $\{C_n \times D_n\}_{n=1}^{\infty}$, then by (iii), there exists $n_0 \in \mathbb{N}$ and $(x'_{n_0}, y'_{n_0}) \in C_{n_0} \times D_{n_0}$ such that

$$f(x_{n_0}, y'_{n_0}) > f(x_{n_0}, y_{n_0}) \quad \text{or} \quad f(x_{n_0}, y_{n_0}) > f(x'_{n_0}, y_{n_0})$$

which contradicts (7.4.8). Thus $\{(x_n, y_n)\}_{n=1}^{\infty}$ is not escaping from $A \times B$ relative to $\{C_n \times D_n\}_{n=1}^{\infty}$ so that there exists $m_1 \in \mathbb{N}$ and a subsequence $\{(x_{n_i}, y_{n_i})\}_{i=1}^{\infty}$

of $\{(x_n, y_n)\}_{n=1}^{\infty}$ such that $(x_{n_i}, y_{n_i}) \in C_{m_1} \times D_{m_1}$ for all $i = 1, 2, \cdots$. Since $C_{m_1} \times D_{m_1}$ is compact, there exists a subnet $\{(w_\alpha, z_\alpha)\}_{\alpha\in\Gamma}$ of $\{(x_{n_i}, y_{n_i})\}_{i=1}^{\infty}$ and $(x^*, y^*) \in C_{m_1} \times D_{m_1}$ such that $w_\alpha \to x^*$ and $z_\alpha \to y^*$. Let $w_\alpha = x_{n_{i(\alpha)}}$ and $z_\alpha = y_{n_{i(\alpha)}}$ where $i(\alpha) \to \infty$.

For any $y \in B$, since $B = \cup_{n=1}^{\infty} D_n$, there is $m_2 \geq m_1$ such that $y \in D_n$ for all $n \geq m_2$, so that $f(x_n, y) \leq f(x_n, y_n)$ by (7.4.8). Choose $\alpha_0 \in \Gamma$ such that $n_{i(\alpha)} \geq m_2$ for all $\alpha \geq \alpha_0$. It follows that for all $\alpha \geq \alpha_0$,

$$\begin{aligned} f(x^*, y^*) &= f(x^*, y^*) - f(x_{n_{i(\alpha)}}, y_{n_{i(\alpha)}}) + f(x_{n_{i(\alpha)}}, y_{n_{i(\alpha)}}) \\ &\geq f(x^*, y^*) - f(x_{n_{i(\alpha)}}, y_{n_{i(\alpha)}}) + f(x_{n_{i(\alpha)}}, y) \\ &= [f(x^*, y^*) - f(x^*, y_{n_{i(\alpha)}})] + [f(x^*, y_{n_{i(\alpha)}}) - f(x_{n_{i(\alpha)}}, y_{n_{i(\alpha)}})] + f(x_{n_{i(\alpha)}}, y) \\ &\geq [f(x^*, y^*) - f(x^*, y_{n_{i(\alpha)}})] + f(x_{n_{i(\alpha)}}, y). \end{aligned}$$

By (i) and (ii),

$$\begin{aligned} f(x^*, y^*) &\geq \liminf_\alpha [f(x^*, y^*) - f(x^*, y_{n_{i(\alpha)}})] + \liminf_\alpha f(x_{n_{i(\alpha)}}, y) \\ &\geq f(x^*, y^*) - \limsup_\alpha f(x^*, y_{n_{i(\alpha)}}) + f(x^*, y) \\ &\geq f(x^*, y). \end{aligned}$$

For any $x \in A$, since $A = \cup_{n=1}^{\infty} C_n$, there is $m_3 \geq m_1$ such that $x \in C_n$ for all $n \geq m_3$, so that $f(x_n, y_n) \leq f(x, y_n)$ by (8) again. Choose $\alpha_1 \in \Gamma$ such that $n_{i(\alpha)} \geq m_3$ for all $\alpha \geq \alpha_1$. It follows that for all $\alpha \geq \alpha_1$,

$$\begin{aligned} f(x^*, y^*) &= f(x^*, y^*) - f(x_{n_{i(\alpha)}}, y_{n_{i(\alpha)}}) + f(x_{n_{i(\alpha)}}, y_{n_{i(\alpha)}}) \\ &\leq f(x^*, y^*) - f(x_{n_{i(\alpha)}}, y_{n_{i(\alpha)}}) + f(x, y_{n_{i(\alpha)}}) \\ &= [f(x^*, y^*) - f(x_{n_{i(\alpha)}}, y^*)] + [f(x_{n_{i(\alpha)}}, y^*) - f(x_{n_{i(\alpha)}}, y_{n_{i(\alpha)}})] + f(x, y_{n_{i(\alpha)}}) \\ &\leq [f(x^*, y^*) - f(x_{n_{i(\alpha)}}, y^*)] + f(x, y_{n_{i(\alpha)}}). \end{aligned}$$

By (i) and (ii),

$$\begin{aligned} f(x^*, y^*) &\leq \limsup_\alpha [f(x^*, y^*) - f(x_{n_{i(\alpha)}}, y^*)] + \limsup_\alpha f(x, y_{n_{i(\alpha)}}) \\ &\leq f(x^*, y^*) - \liminf_\alpha f(x_{n_{i(\alpha)}}, y^*) + f(x, y^*) \\ &\leq f(x, y^*). \end{aligned}$$

Therefore (x^*, y^*) is a saddle point of f in $A \times B$. □

As an application of Theorem 7.4.10, we have the following existence theorems of saddle points in reflexive Banach spaces.

Theorem 7.4.11. Let A and B be two nonempty convex subsets of two reflexive Banach spaces X and Y respectively. Suppose that $A = \cup_{n=1}^{\infty} C_n$ and $B = \cup_{n=1}^{\infty} D_n$, where $\{C_n\}_{n=1}^{\infty}$ and $\{D_n\}_{n=1}^{\infty}$ are increasing sequences of nonempty bounded closed convex sets in X and Y respectively. Assume that the function $f : A \times B \to \mathbb{R}$ satisfies the following conditions:

(i) for each fixed $y \in B$, the function $x \to f(x, y)$ is weakly lower semicontinuous and quasi-convex;

(ii) for each fixed $x \in A$, the function $y \to f(x, y)$ is weakly upper semicontinuous and quasi-concave; and

(iii) for each sequence $\{(x_n, y_n)\}_{n=1}^{\infty}$ in $A \times B$, where $(x_n, y_n) \in C_n \times D_n$ for each $n = 1, 2, \cdots$, which is escaping relative to $\{C_n \times D_n\}_{n=1}^{\infty}$, there exists $n_0 \in \mathbb{N}$ and $(x'_{n_0}, y'_{n_0}) \in C_{n_0} \times D_{n_0}$ such that $f(x_{n_0}, y'_{n_0}) > f(x_{n_0}, y_{n_0})$ or $f(x_{n_0}, y_{n_0}) > f(x'_{n_0}, y_{n_0})$.

Then f has a saddle point in $A \times B$.

Proof. Let X and Y be equipped with weak topologies. Then the conclusion follows from Theorem 7.4.10. □

Theorem 7.4.12. Let A and B be two nonempty closed convex subsets of two reflexive Banach spaces X and Y respectively. Suppose that the function $f : A \times B \to \mathbb{R}$ satisfies the following conditions:

(i) for each fixed $y \in B$, the function $x \to f(x, y)$ is weakly lower semicontinuous and quasi-convex;

(ii) for each fixed $x \in A$, the function $y \to f(x, y)$ is weakly upper semicontiunous and quasi-concave; and

(iii) for each sequence $\{(x_n, y_n)\}_{n=1}^{\infty}$ in $A \times B$ with $\|x_n\|$, $\|y_n\| \leq n$ for all $n \geq m_0$ such that $\|x_n\| + \|y_n\| \to \infty$ as $n \to \infty$, there exists $n_0 \geq m_0$ and $(x, y) \in A \times B$ with $\|x\|$, $\|y\| \leq n_0$ such that $f(x_{n_0}, y) > f(x_{n_0}, y_{n_0})$ or $f(x_{n_0}, y_{n_0}) > f(x, y_{n_0})$.

Then f has a saddle point in $A \times B$.

Proof. We may assume that for some $a \in A$ and $b \in B$, $\|a\|$, $\|b\| \leq m_0$. For each $n \geq m_0$, define $C_n = \{x \in A : \|x\| \leq m_0 + n\}$ and $D_n = \{y \in B : \|y\| \leq m_0 + n\}$, then $\{C_n\}_{n=1}^{\infty}$, $\{D_n\}_{n=1}^{\infty}$ are increasing sequences of nonempty bounded closed and convex sets such that $A = \cup_{n=1}^{\infty} C_n$ and $B = \cup_{n=1}^{\infty} D_n$. If $\{(x_n, y_n)\}_{n=1}^{\infty}$ is any sequence in $A \times B$, where $(x_n, y_n) \in C_n \times D_n$ for each $n = 1, 2, \cdots$, which is escaping from $A \times B$ relative to $\{(C_n \times D_n)\}_{n=1}^{\infty}$, then $\|x_n\| + \|y_n\| \to \infty$ as $n \to \infty$ so that by (iii), there exists $n_0 \in \mathbb{N}$ and $(x, y) \in A \times B$ with $\|x\|$, $\|y\| \leq n_0$ such that $f(x_{n_0}, y) > f(x_{n_0}, y_{n_0})$ or $f(x_{n_0}, y_{n_0}) > f(x, y_{n_0})$. Moreover we must have $(x, y) \in C_{n_0} \times D_{n_0}$ so that condition (iii) of Theorem 7.4.11 is also satisfied. The conclusion now follows from Theorem 7.4.11. □

Theorem 7.4.13. Let A and B be two nonempty closed convex subsets of two reflexive Banach spaces X and Y respectively, Z a Hausdorff topological vector space with a partial order $\geq$ and $f : A \times B \to Z$ be a mapping. Suppose that there exists a function $p : Z \to \mathbb{R}$ which is strictly monotonic and continuous such that the following conditions are satisfied:

(i) for each fixed $y \in B$, the function $x \to pf(x,y)$ is weakly lower semicontinuous and quasi-convex and for each fixed $x \in A$, the function $y \to pf(x,y)$ is weakly upper semicontiunous and quasi-concave;

(ii) A is bounded or there exists $\overline{y} \in B$ such that $\lim_{\|x\|\to\infty} pf(x,\overline{y}) = +\infty$; and

(iii) B is bounded or there exists $\overline{x} \in A$ such that $\lim_{\|y\|\to\infty} pf(\overline{x},y) = -\infty$.

Then f has a saddle point in $A \times B$.

Proof. By Lemma 3.2 of Luc and Vargas [353], it suffices to show that pf has a saddle point in $A \times B$. If both A and B are bounded, then the result follows from Theorem 3.4 of Sion [490]). Thus we may assume that either A or B is unbounded. Fix any $\hat{x} \in A$ and $\hat{y} \in B$ and let $m_0 \in \mathbb{N}$ be such that $\max\{\|\overline{x}\|, \|\hat{x}\|, \|\overline{y}\|, \|\hat{y}\|\} \le m_0$. First we shall show that condition (iii) of Theorem 7.4.12 is satisfied for pf. If this were false, then there exists a sequence $\{(x_n,y_n)\}_{n=1}^{\infty}$ in $A \times B$ with $\|x_n\|$, $\|y_n\| \le n$for all $n \ge m_0$ and $\|x_n\| + \|y_n\| \to \infty$ as $n \to \infty$ such that for each $n \ge m_0$ and for each $(x,y) \in A \times B$ with $\|x\|$, $\|y\| \le n$,

$$pf(x_n,y) \le pf(x_n,y_n) \le pf(x,y_n). \tag{7.4.9}$$

Case 1. Suppose that A is bounded but B is unbounded. Note that $\|y_n\| \to \infty$ as $n \to \infty$. Since A is weakly compact by Eberlein-Smulian Theorem (e.g., see Dunford and Schwartz ([150, p.430]), we may assume without loss of generality that $x_n \rightharpoonup x_0$ for some $x_0 \in A$; here, "$\rightharpoonup$" denotes the weak convergence. Choose $n_0 \ge m_0$ such that $\|x_0\| \le n_0$. Then for any $n \ge n_0$, we have by (7.4.9), $pf(x_n,\hat{y}) \le pf(x_n,y_n) \le pf(\overline{x},y_n)$; it follows that $pf(x_0,\hat{y}) \le \liminf pf(x_n,\hat{y}) \le \liminf pf(\overline{x},y_n) = -\infty$ by (iii) since $\|y_n\| \to \infty$. This contradicts $pf(x_0,\hat{y}) > -\infty$.

Case 2. Suppose that B is bounded but A is unbounded. Note that $\|x_n\| \to \infty$ as $n \to \infty$. Since B is weakly compact, we may assume that $y_n \rightharpoonup y_0$ for some $y_0 \in B$. Choose $n_0 \ge m_0$ such that $\|y_0\| \le n_0$. Then for any $n \ge n_0$, we have by (9), $pf(x_n,\overline{y}) \le pf(x_n,y_n) \le pf(\hat{x},y_n)$; it follows that $pf(\hat{x},y_0) \ge \limsup pf(\hat{x},y_n) \ge \limsup pf(x_n,\overline{y}) = +\infty$ by (ii) since $\|x_n\| \to \infty$ as $n \to \infty$. This contradicts $pf(\hat{x},y_0) < \infty$.

Case 3. Suppose that both A and B are unbounded. If $\|x_n\| \nrightarrow \infty$ as $n \to \infty$ (resp., $\|y_n\| \nrightarrow \infty$ as $n \to \infty$)), then a subsequence $\{x_{n_i}\}_{i=1}^{\infty}$ of $\{x_n\}_{n=1}^{\infty}$ (resp., a subsequence $\{y_{n_i}\}_{i=1}^{\infty}$ of $\{y_n\}_{n=1}^{\infty}$) must be bounded. Since X (resp., Y) is reflexive and A (resp., B) is closed, an arguement similar to Case 2 (resp., Case 1) leads to a contradiction. Hence we may assume that $\|x_n\| \to \infty$ and $\|y_n\| \to \infty$ as $n \to \infty$. Since $pf(x_n,\overline{y}) \le pf(x_n,y_n) \le pf(\overline{x},y_n)$ for each $n \ge m_0$, by (ii) and (iii) we have

$$+\infty = \lim_{n\to\infty} pf(x_n,\overline{y}) \le \lim_{n\to\infty} pf(\overline{x},y_n) = -\infty$$

which is again a contradiction.

Hence condition (iii) of Theorem 7.4.12 must be satisfied by pf. Therefore pf has a saddle point in $A \times B$ by Theorem 7.4.12. □

Theorem 7.4.13 generalizes Corollary 3.3. of Tanaka [528] in the following ways: (i) p need not be linear, (ii) pf need not be "*convex − concave*" on $A \times B$ and (iii) the partial order $\geq$ need not be generated by a convex cone. We remark that in Corollary 3.3 of Luc and Vargas [353, p.135], the semicontinuity should be in the weak topology.

7.5 Vector-Valued Saddle Points in H-Spaces

In this section, from the motivation of multicritia games and multi-objective optimization in vector optimization theory, as applications of the Browder-Fan fixed point theorem's generalization in H-spaces, we first study the existence of C-saddle and weak C-saddle points for set-valued (and single-valued) vector-valued in H-spaces and Hausdorff topological vector spaces. Then, existence theorems of vector-valued minimax theorems will be established. Finally the stability of vector-valued saddle points will also be investigated and we prove that a majority of weak C-saddle points are stable in the sense of the Baire category. Our results generalize corresponding results in the literature.

Let Z be an ordered topological vector space with an ordering defined by a convex cone C. The convex cone C is assumed to be *pointed*, i.e., $C \cap (-C) = \{0\}$, and hence the ordering is anti-symmetric and $0 \in C$. Moreover, we assume that the convex cone C to be *solid*, i.e., its topological interior $intC$ is nonempty, and hence $C^0 := (intC) \cup \{0\}$ is a pointed convex cone and induces another (anti-symmetric) vector ordering $\leq_{C^0}$, which is weaker than $\leq_C$ in E. We also remark that the orderings $\leq_C$ and $\leq_{C^0}$ are two directed anti-symmetric partial orderings. Here an ordering $\leq$sense that for each given $x, y \in E$, there exists $z \in E$ such that $x \leq z$ and $y \leq z$.

Let A be a nonempty subset of Z. As defined in last section, a point $a_0 \in A$ is said to be a *maximal* (resp., *minimal*) *point* of A if $a \in A$ and $a \geq a_0$ (resp., $a_0 \geq a$) implies $a = a_0$. The set of maximal (resp., minimal) points of A is denoted by *max* A (resp., *min* A). It is clear that a element z_0 of A is a C-minimal point of A if and only if the set $\{z \in A : z \leq_C z_0, z \neq z_0\} = \emptyset$; which is equivalent to $A \cap (z_0 - C) = \{z_0\}$. We denote the set of all such C-minimal (resp, C-maximal) points by Min$A$ (resp., $MaxA$). Also C^0-minimal and C^0-maximal points of A defined similarly, and denoted by $\text{Min}_W A$ and $Max_W A$ respectively. These C^0-minimality and C^0-maximality are weaker concepts than C-minimality and C-maximality, respectively,. It should be noted that $\text{Min}A \subset \text{Min}_W A$ and $MaxA \subset Max_W A$. Moreover, $\text{Min}\alpha A = \alpha\text{Min}A$, $Max\alpha A = \alpha MaxA$, $\text{Min}_W \alpha A = \alpha\text{Min}_W A$, $Max_W \alpha A = \alpha Max_W A$ for any

$\alpha > 0$.

Let $F : X \times Y \to 2^Z$ be a vector-valued mapping on a product space $X \times Y$. We will also use the following symbols: $F(X,y) := \cup_{x\in X}\{F(x,y)\}$; $F(x,Y) := \cup_{y\in Y}\{F(x,y)\}$ and $F(X,Y) := \cup_{(x,y)\in X\times Y}\{F(x,y)\}$.

A multi-criteria two-person zero-sum game, or say, a two-person vector-valued zero-sum game (e.g., see Tanaka [524]) is a quadruple-tuple $\Gamma = (X, Y; -f, f)$, where X and Y are nonempty sets and f is a mapping $f : X \times Y \to Z$, where Z is an ordered topological vector space with an ordering defined by a convex cone C. The set X (resp., Y) is the set of strategies of player 1 (resp., player 2) and the mapping $-f$ (resp., f) is the payoff function of this player. When player 1 and player 2 choose a strategy $x \in X$ and a strategy $y \in Y$, respectively, the payoffs with respect to player 1 and player 2 are given by the vectors $-f(x,y)$ and $f(x,y)$, respectively. Each player chooses a strategy in order to increase his/her payoff and wants to find a strategy maximizing his payoff. That player 1 is regarded as minimizer and player 2 as maximizer. We now state the following definition (see also Tanaka [524]).

Definition 7.5.1. A strategy $x_0 \in X$ is said to be an optimal response strategy for the player 1 against a strategy $y \in Y$ of the player 2 if $f(x_0, y) \in \text{Min} f(X,y)$. Similarly, a strategy $y_0 \in Y$ is said to be one optimal response strategy for player 2 against a strategy $x \in X$ of player 1 if $f(x, y_0) \in \text{Max} f(x, Y)$.

Definition 7.5.2. Let X, Y and Z be Hausdorff topological vector spaces, $\geq$ a partial order on Z, $A \subset X$, $B \subset Y$ and F a set-valued mapping from $A \times B$ to Z. For $x \in A$ and $y \in B$, let $F(A,y) = \cup_{a\in A} F(a,y)$ and $F(x,B) = \cup_{b\in B} F(x,b)$. Then a point $(x^*, y^*) \in A \times B$ is said to be

(1) a *C-loose saddle point* (resp., *weak C-loose saddle point*) of F in $A \times B$ if $F(x^*,y^*) \cap \text{ Min } F(A,y^*) \neq \emptyset$ and $F(x^*,y^*) \cap \text{ Max } F(x^*,B) \neq \emptyset$ (resp., if $F(x^*,y^*) \cap \text{ Min}_W\ F(A,y^*) \neq \emptyset$ and $F(x^*,y^*) \cap \text{ Max}_W\ F(x^*,B) \neq \emptyset$); and

(2) a *C-saddle point* (resp., *weak C-saddle point*) of F in $A \times B$ if $F(x^*,y^*) \cap \text{ Min } F(A,y^*) \cap \ max\ F(x^*,B) \neq \emptyset$ (resp., $F(x^*,y^*) \cap \text{ Min}_W\ F(A,y^*) \cap \text{ Max}_W\ F(x^*,B) \neq \emptyset$).

Note that every saddle point is a loose saddle point; the converse is in general not true. To see this, take $X = Y = E = \mathbb{R}$, $A = B = [0,1]$ and define $F : X \times Y \to 2^{\mathbb{R}}$ by $F(x,y) = [0,1]$ for each $(x,y) \in X \times Y$. In this case, every point $(x,y) \in A \times B$ is a loose saddle point of F but F has no saddle point.

When F is single-valued, every loose saddle point is also a saddle point. Let $p : X \to \mathbb{R}$ be strictly monotonic, then any loose saddle point (saddle point) of pF in $A \times B$ is also a loose saddle point (saddle point) of F in $A \times B$, see Lemma 3.2 of Luc and Vargas [353].

We also need the following notion of semi-saddle points for a pair of functionals. This is also known as the Nash-equilibria (Nash [382] of a two-person nonzero-sum game in the game theory; e.g., see Tanaka [524] and references therein.

Definition 7.5.3. Let two functionals g_1 and g_2 be defined on a product space $X \times Y$. Then a point $(x_0, y_0) \in X \times Y$ is said to be

(1) a *semi-saddle point* of (g_1, g_2) is

$$g_1(x_0, y_0) \leq g_1(x, y_0) \text{ and } g_2(x_0, y_0) \geq g_2(x_0, y),$$

for all $(x, y) \in X \times Y$.

(2) a *strict semi-saddle point* of (g_1, g_2) if

$$g_1(x_0, y_0) < g_1(x, y_0) \text{ and } g_2(x_0, y_0) > g_2(x_0, y),$$

for any $x \in X$ and $y \in Y$ with $x \neq x_0$ and $y \neq y_0$.

Also we need the following definition which is essentially due to Jahn [272]:

Definition 7.5.4. Let z_0 be an element of a nonempty subset A of a topological vector space E. Then a functional $\psi : A \to \mathbb{R}$ is said to be

(1) *monotonically increasing* with respect to the lower (resp., upper) section of A at z_0 if

$$\psi(z) \leq \psi(z_0) \text{ (resp., } \psi(z) \geq \psi(z_0))$$

for any $z \in (\{z_0\} - C) \cap A$ (resp., $z \in (\{z_0\} + C) \cap A$);

(2) *strongly monotonically increasing* with respect to the lower (resp., upper) section on A at z_0 if

$$\psi(z) < \psi(z_0) \text{ for any } z \in (\{z_0\} + C_-) \cap A, z \neq z_0$$

(resp.,

$$\psi(z) > \psi(z_0) \text{ for any } z \in (\{z_0\} + C_+) \cap A, z \neq z_0); \text{ and}$$

(3) *strictly monotonically increasing* with respect to the lower (resp., upper) section on A at z_0 if

$$\psi(z) < \psi(z_0) \text{ (resp., } \psi(z) > \psi(z_0))$$

for any $z \in (\{z_0\} - intC) \cap A$ (resp., $z \in (\{z_0\} + intC) \cap A$).

Remark 7.5.5. We also like to mention that each functional $\psi \in C^*$ (resp., $\psi \in C^* \setminus \{0\}$) is monotonically (resp., strictly monotonically) increasing with respect to both the lower and upper sections on Z at any point $z \in Z$. As we assume that $intC \neq \emptyset$, it follows that $C^* \setminus \{0\} \neq \emptyset$.

If ψ is a nonzero linear functional on a vector space E, we associate with the hyperplane $H := \{x \in E : \psi(x) = c\}$ for some $c \in \mathbb{R}$. Then the sets $\{x \in E :$

$\psi(x) \leq c\}$, $\{x \in E : \psi(x) < c\}$, $\{x \in E : \psi(x) \geq c\}$ and $\{x \in E : \psi(x) > c\}$ are called the half-space determined by H. Now we recall the following definition which was first given by Tanaka [524, p.366]:

Definition 7.5.6. Let X be a topological space and Z be another topological vector space. A mapping $f : X \to Z$ is said to be *demicontinuous* on X if the set $f^{-1}(M) = \{x \in X : f(x) \in M\}$ is closed for each closed half-space $M \subset Z$.

Remark 7.5.7. It is clear that each ordinary continuous mapping is demicontinuous and indeed the fact that: X is a topological space and Z a topological vector space and $f : X \to Z$ is demicontinuous, then for each $\psi \in E^*$, the composite mapping $\psi \circ (f) : X \to \mathbb{R}$ is continuous, e.g., see Lemma 2.2 of Tanaka [524, p.366].

Definition 7.5.8. We recall again that a pair $(X, \Gamma_A\})$ is called an H-space (e.g., see Bardaro and Ceppitelli [24]) if X is a topological space and $\{\Gamma_A\}$ is a given family of nonempty contractible subsets of X, indexed by the finite subsets of X such that $A \subset B$ implies $\Gamma_A \subset \Gamma_B$.

Definition 7.5.9. Let $(X, \{\Gamma_A\})$ be an H-space. We then have the following:

(1) A subset D of X is said to be *H-convex*, if for every finite subset A of D, $\Gamma_A \subset D$.

(2) Let $F : X \to 2^{\mathbb{R}}$ be a set-valued mapping. Then F is said to be H-quasiconvex (resp., H-quasiconcave), if for any $\lambda \in \mathbb{R}$, the set

$$\{x \in X : \text{ there exists a } s \in F(x) \text{ such that } s < \lambda\}$$

(resp., $\{x \in X$: there exists a $s \in F(x)$ such that $s > \lambda\}$) is H-convex.

(3) If $f : X \to Z$ is a single-valued mapping, where Z is a topological vector space with a partial order '$\leq$' generated by a cone C which has a nonempty interior, i.e., $intC \neq \emptyset$, then f is said to be *strictly H-quasiconvex* if for each $z \in Z$, and $x_1, x_2 \in X$ with $x_1 \neq x_2$ such that $f(x_1) \leq z$ and $f(x_2) < z$, then there exists $x_0 \in \Gamma_{(x_1,x_2)}$ such that $f(x_0) < z$.

Definition 7.5.10. Let X be a nonempty convex subset of a vector space, Z a Hausdorff topological vector space with a partial order $\geq$ and F a set-valued mapping from X to Z. Then F is said to be

(1) *compact-valued if* $F(x)$ is compact for all $x \in X$;

(2) *convex* (resp., *concave*) if for every $x_1, x_2 \in X$, $\lambda \in [0,1]$, $y_1 \in F(x_1)$ and $y_2 \in F(x_2)$, there exists $y_3 \in F(\lambda x_1 + (1-\lambda)x_2)$ such that $y_3 \leq \lambda y_1 + (1-\lambda)y_2$ (resp., $y_3 \geq \lambda y_1 + (1-\lambda)y_2$); and

(3) *quasi-convex* (resp., *quasi-concave*) if for any $a \in Z$, the set $\{x \in X :$ there is $y \in F(x)$ with $y \leq a$ (resp., $y \geq a)\}$ is convex.

If F is single-valued and the order of Z is generated by a convex cone $C \subset Z$, the convexity (resp., concavity) and quasi-convexity (resp., quasi-concavity) coincide with that of C-convexity (resp., C-concavity) and C-quasi-convexity (resp., C-quasi-concavity) in Tanaka [525]-[524]. If $g : X \to Z$ is a function, then g is said to be properly quasi-C-convex (see Tanaka [525]) if either $g(\lambda x_1 + (1-\lambda)x_2) \leq g(x_1)$ or $g(\lambda x_1 + (1-\lambda)x_2) \leq g(x_2)$ for every $x_1, x_2 \in X$ and $\lambda \in [0,1]$. Also g is said to be properly quasi-C-concave (see Tanaka [32, p.327) if $-g$ is properly quasi-C-convex.

For the reader's convenience of comparison, we will recall the definition of various convexity concepts for single-valued mappings as follows.

Definition 7.5.11. Let X be a convex set in a real topological vector space E and Z an ordered topological vector space with a pointed convex cone C. A vector-valued mapping $f : X \to Z$ is said to be:

(1) C-*convex* on X if $f(\lambda x_1 + (1-\lambda)x_2) \leq_c \lambda f(x_1) + (1-\lambda)f(x_2)$ for all $x_1, x_2 \in X$ and $\lambda \in [0,1]$;

(2) *properly quasi* C-*convex* on X if either $f(\lambda x_1 + (1-\lambda)x_2) \leq_C f(x_1)$ or $f(\lambda x_1 + (1-\lambda)x_2) \leq_c f(x_2)$;

(3) Ferro's *quasi* C-*convex* on X if for each $z \in Z$, the set $A(z) := \{x \in X : f(x) \leq_c z\}$ is convex or empty;

(4) Luc's *quasi* C-*convex* on X for for each $x_1, x_2 \in X$ and $\lambda \in [0,1]$, $f(\lambda x_1 + (1-\lambda)x_2) \leq_c z$ for all $z \in C(f(x_1), f(x_2))$, where $C(f(x_1), f(x_2))$ is the set of upper bounds of $f(x_1)$ and $f(x_2)$, i.e.,

$$C(f(x_1), f(x_2)) := \{z \in Z : f(x_1) \leq Cz \text{ and } f(x_2) \leq_c z\};$$

(5) C-*convexlike* on X if for each $x_1, x_2 \in X$ and $\lambda \in [0,1]$, there exists $x \in X$ such that $f(x) \leq_c \lambda f(x_1) + (1-\lambda)f(x_2)$;

(6) *natural quasi* C-*convex* on X if $f(\lambda x_1 + (1-\lambda)x_2) \in co\{f(x_1), f(x_2)\} - C$; and

(7) $\cdots$ C-*concave* on X if $-f$ is $\cdots$ C-convex on X.

Remark 7.5.12. Let X be a convex set in a topological vector space E and Z an ordered vector topological vector space with a pointed convex cone C. Then it is clear that the followings are true (see Lemma 2.1 of Tanaka [523, p.363]):

(1) each C-convex mapping is natural quasi C-convex;

(2) each properly quasi C-convex mapping is natural quasi C-convex; (3) each natural quasi C-convex mapping is quasi C-convex; and

(4) each quasi C-convex mapping is C-convexlike whenever C is closed and f is continuous.

We also need the following result which is Theorem 2.4 of Tanaka [523]:

Lemma 7.5.13. Let X and Y be two nonempty convex subsets of topological vector spaces E and F, respectively. Let $f : X \times Y \to \mathbb{Z}$ be a vector-valued function defined on a product space $X \times Y$. Suppose ψ_1 and ψ_2 are two functionals from $f(X,Y) \to \mathbb{R}$ and $(x_0, y_0) \in X \times Y$ are given. Then we have the following:

(1) If both ψ_1 and ψ_2 are monotonically increasing with respect to both lower section and upper section on $f(X \times Y)$ at $f(x_0, y_0)$, respectively and the point $(x_0, y_0) \in X \times Y$ is a strict semi-saddle point of the mapping pair $(\psi_1 \circ f, \psi_2 \circ f)$, then $(x_0, y_0) \in X \times Y$ is a C-saddle point of f.

(2) If both ψ_1 and ψ_2 are strictly monotonically increasing with respect to both lower section and upper section on $f(X \times Y)$ at $f(x_0, y_0)$, respectively and the point $(x_0, y_0) \in X \times Y$ is a semi-saddle point of the mapping pair $(\psi_1 \circ f, \psi_2 \circ f)$, then $(x_0, y_0) \in X \times Y$ is a weak C-saddle point of f.

Lemma 7.5.14. Let X be a nonempty convex subset of a vector space Z which is a Hausdorff topological vector space, $\geq$ a partial order on Z which is generated by a convex cone $C \subset Z$ and F a set-valued mapping from X to Z. If F is convex (resp., concave), then F is quasi-convex (resp., quasi-concave).

Proof. For any $a \in Z$, we need to prove that the set $L = \{x \in X$: there is $y \in F(x)$ with $y \leq a$ (resp., $y \geq a)\}$ is convex. For any $x_1, x_2 \in L$ and $\lambda \in [0, 1]$, there exists $y_1 \in F(x_1)$ and $y_2 \in F(x_2)$ such that $y_1 \leq a$ and $y_2 \leq a$ (resp., $y_1 \geq a$ and $y_2 \geq a$). Since F is convex (resp., concave), there exists $y_3 \in F(\lambda x_1 + (1 - \lambda)x_2)$ such that $y_3 \leq \lambda y_1 + (1 - \lambda)y_2$ (resp., $y_3 \geq \lambda y_1 + (1-\lambda)y_2$). Since $\geq$ is generated by a convex cone, $\lambda y_1 + (1-\lambda)y_2 \leq a$ (resp., $\lambda y_1 + (1 - \lambda)y_2 \geq a$), so that $y_3 \leq a$ (resp., $y_3 \geq a$). Hence $\lambda x_1 + (1 - \lambda)x_2 \in L$ and L is convex. □

Lemma 7.5.15. Let X be a nonempty convex subset of a vector space, Z a Hausdorff topological vector space and $\geq$ a partial order on Z. Let F be a set-valued mapping from $A \times B$ to Z and p a real-valued function on Z. If F is convex (resp., concave) and p is monotonic and convex (resp., concave), then the composite set-valued map pF is convex (resp., concave).

Proof. If F and p are convex, Proposition 2.2 of Luc and Vargas [353] shows that pF is also convex. If F and p are concave, then the same proof of Proposition 2.2 in [353] can be modified to show that pF is also concave. □

We also need the generalization of the Browder-Fan type fixed point theorem in H-spaces for our study in this section. The following one is a special case of Theorem 2.3.15 (see also Theorem 2.3.16 in Chapter 2 or Tarafdar [536]):

Lemma 7.5.16 (Browder-Fan Fixed Point Theorem in H-Spaces). Let X be a compact topological space, $(Y, \{\Gamma_B\})$ be an H-space, and $T : X \to 2^Y$ be a set-valued mapping satisfying the following conditions:

(i) for each $x \in X$, $T(x)$ is nonempty H-convex; and
(ii) for each $y \in Y$, $T^{-1}(y)$ is open.

Then there exists a continuous function $h : X \to Y$ such that for any $x \in X$, $h(x) \in T(x)$.

In addition, if $X = Y$, and X is a compact H-space, and $T : X \to 2^X$ is a set-valued mapping satisfying conditions (i) and (ii), then T has a fixed point in X.

In this section, we shall study the existence of saddle points in H-sapces which may not have any linear structure.

We first need following two facts which are Lemmas 3.1 and 3.2 of Tan et al. [521] and thus we omit their proofs here.

Lemma 7.5.17. Let X, Y, Z be three Hausdorff topological vector spaces, A and B nonempty subsets of X and Y, respectively. Suppose that A is compact and the set-valued mapping F from $A \times B$ to Z is upper semicontinuous such that for each fixed $x \in A$, $y \to F(x,y)$ is lower semicontinuous. Then the set-valued mapping T from B to Z, defined by $T(y) = \cup_{x\in A} F(x,y)$ for each $y \in B$, is continuous (i.e., both lower and upper semicontinuous).

Lemma 7.5.18. Let Z be a Hausdorff topological space and Φ be a set-valued mapping from Z to $\mathbb{R}$ which is compact-valued and continuous. Then the functions $p, q : Z \to \mathbb{R}$, defined by $p(z) = max\ \Phi(z)$ and $q(z) = min\ \Phi(z)$ are continuous.

Now we have the following:

Theorem 7.5.19. Let $(X, \{\Gamma_A\})$ and $(Y, \{\Gamma_B\})$ be two compact Hausdorff H-spaces, Z be a Hausdorff topological vector space and C be a convex cone of Z. Suppose that $F : X \times Y \to 2^Z$ is a set-valued mapping and $\xi_1, \xi_2 : Z \to \mathbb{R}$ are two functions such that

(i) $\xi_1 F$ and $\xi_2 F$ are upper semicontinuous with nonempty compact values and for each fixed $x \in X$, $y \mapsto \xi_1 F(x,y)$ and $y \mapsto \xi_2 F(x,y)$ are lower semicontinuous; and for each fixed $y \in Y$, $x \mapsto \xi_1 F(x,y)$ and $x \mapsto \xi_2 F(x,y)$ are lower semicontinuous; and

(ii) the mapping $x \longmapsto \xi_1 F(x, y)$ is H-quasi-convex and the mapping $y \longmapsto \xi_2 F(x, y)$ is H-quasi-concave.

Then there exists $z^*, w^* \in F(x^*, y^*)$ such that

$$\xi_1(z^*) = \mathrm{Min}\xi_1 F(X, y^*) \text{ and } \xi_2(w^*) = \mathrm{Max}\xi_2 F(x^*, Y).$$

Proof. Define set-vauled mappings $S_n : Y \to 2^X$ and $T_n : X \to 2^Y$, $n =$

$1, 2, \cdots$ by

$$S_n(y) = \{u \in X : \min \xi_1 F(u, y) < \min \xi_1 F(X, y) + \frac{1}{n}\},$$
$$T_n(x) = \{v \in X : \max \xi_2 F(x, Y) - \frac{1}{n} < \max \xi_2 F(x, v)\}.$$

Since X and Y are compact, by condition (i), the set $\xi_1 F(X, y)$ and $\xi_2 F(x, Y)$ are both compact. Therefore for any $x \in X$ and for any $y \in Y$, the sets $S_n(y)$ and $T_n(x)$ are both nonempty.

By condition (ii), $S_n(y)$ and $T_n(x)$ are both H-convex for any $y \in Y$ and any $x \in X$, $n = 1, 2, \cdots$.

Next we prove that for any $u \in X$ and $v \in Y$, $S_n^{-1}(u)$ and $T_n^{-1}(v)$ are both open.

In fact, since

$$\begin{aligned} S_n^{-1}(u) &= \{y \in Y : u \in S_n(y)\} \\ &= \{y \in Y : \min \xi_1 F(u, y) - \min \xi F(X, y) < \frac{1}{n}\} \end{aligned}$$

and by condition (i), Lemmas 7.5.17 and 7.5.18, it follows that $y \mapsto \min \xi_1 F(u, y) - \min \xi_1 F(X, y)$ is continuous, so that the set $S_n^{-1}(u)$ is open for each $u \in X$. Similarly it follows that the set $T_n^{-1}(v)$ is open for each $v \in Y$. Therefore we have proved that for any $x \in X$ and $y \in Y$, $T_n(x)$ and $S_n(y)$ are both nonempty H-convex and $S_n^{-1}(u)$ and $T_n^{-1}(v)$ are both open for each $u \in X$ and $v \in Y$. Let D be any finite subset of $X \times Y$, and A be the projection of D onto X and B be the projection of D onto Y. Since $(X, \{\Gamma_A\})$ and $(Y, \{\Gamma_B\})$ are H-spaces, $(X \times Y, \{\Gamma_D\})$ is also an H-space. Now we define a set-valued mapping $W_n : X \times Y \to 2^{X \times Y}$ for each $n = 1, 2, \cdots$ by

$$W_n(x, y) = S_n(y) \times T_n(x).$$

Then for each $(x, y) \in X \times Y$, $W_n(x, y)$ is nonempty H-convex and for each $(u, v) \in X \times Y$, $W_n^{-1}(u, v) = T_n^{-1}(v) \times S_n^{-1}(u)$. Hence $W_n^{-1}(u, v)$ is open. By Browder-Fan fixed point theorem in H-spaces (e.g., see Lemma 7.5.16), for each $n = 1, 2, \cdots$, there exists a $(x_n, y_n) \in X \times Y$ such that

$$(x_n, y_n) \in W_n(x_n, y_n) = S_n(y_n) \times T_n(X_n).$$

Since $x_n \in S_n(y_n) \subset X$, and $y_n \in T_n(x_n) \subset Y$, we have

$$\min \xi_1 F(X, y_n) > \min \xi_1 F(x_n, y_n) - \frac{1}{n}, \tag{7.5.1}$$

and

$$\max \xi_2 F(x_n, Y) - \frac{1}{n} < \max \xi_2 F(x_n, y_n). \tag{7.5.2}$$

As $\xi_1 F$ and $\xi_2 F$ are compact-valued, there exists $s_n \in \xi_1 F(x_n, y_n)$ and $t_n \in \xi_2 F(x_n, y_n)$ such that

$$\begin{cases} s_n = \min \xi_1 F(x_n, y_n), \\ t_n = \max \xi_2 F(x_n, y_n). \end{cases} \tag{7.5.3}$$

¿From (7.5.1) and (7.5.2), we have

$$\begin{cases} \min \xi_1 F(X, y_n) > s_n - \dfrac{1}{n}, \\ \max \xi_2 F(x_n, Y) - \dfrac{1}{n} < t_n. \end{cases} \tag{7.5.4}$$

Since X and Y are compact, without loss of generality, we may assume that $x_n \to x^* \in X$ and $y_n \to y^* \in Y$. On the other hand, by condition (i) again, $\xi_1 F(X, Y)$ and $\xi_2 F(X, Y)$ are compact and the graph($\xi_1 F$) and graph($\xi_2 F$) of $\xi_1 F$ and $\xi_2 F$ are closed. Hence we may assume that $t_n \to t^* \in \mathbb{R}$, $s_n \to s^* \in \mathbb{R}$. Thus we have $s^* \in \xi_1 F(x^*, y^*)$ and $t^* \in \xi_2 F(x^*, y^*)$. Again by Lemmas 7.5.17 and 7.5.18, the mapping: $y \longmapsto \min \xi_2 F(X, y)$ and $x \longmapsto \max \xi_1 F(x, Y)$ are continuous. Hence from (7.5.3) and (7.5.4) we have $s^* \leq \min \xi_1 F(X, y^*)$ and $t^* \geq \max \xi_2 F(x^*, Y)$. Then it is clear $s^* = \min \xi_1 F(X, y^*)$ and $t^* = \max \xi_2 F(x^*, Y)$. Therefore there exists z^* and $w^* \in F(x^*, y^*)$ such that $s^* = \xi_1(z^*)$, $t^* = \xi_2(w^*)$ and

$$\xi_1(z^*) = \min \xi_1 F(X, y^*),$$
$$\xi_2(w^*) = \max \xi_2 F(x^*, Y).$$

Thus the proof is complete. □

As an application of Theorem 7.5.24, we have the following result.

Theorem 7.5.20. Let $(X, \{\Gamma_A\})$ and $(Y, \{\Gamma_B\})$ be two compact Hausdorff H-spaces, Z be a Hausdorff topological vector space and C be a convex cone of Z. Suppose that $F : X \times Y \to 2^Z$ is a set-valued mapping and $\xi_1, \xi_2 : Z \to \mathbb{R}$ are strongly increasing mappings such that:

(i) the mappings $(x, y) \mapsto \xi_1 F(x, y)$ and $(x, y) \mapsto \xi_2 F(x, y)$ are upper semicontinuous with nonempty compact values and for each fixed $x \in X$, $y \mapsto \xi_1 F(x, y)$ and $y \mapsto \xi_2 F(x, y)$ are lower semicontinuous; and for each fixed $y \in Y$, $x \mapsto \xi_1 F(x, y)$ and $x \mapsto \xi_2 F(x, y)$ are lower semicontinuous; and

(ii) the mapping $x \longmapsto \xi_1 F(x, y)$ is H-quasi-convex and the mapping $y \longmapsto \xi_2 F(x, y)$ is H-quasi-concave.

Then there exists $z^*, w^* \in F(x^*, y^*)$ such that

$$F(x^*, y^*) \cap \mathrm{Min} F(X, y^*) \neq \emptyset \text{ and } F(x^*, y^*) \cap \mathrm{Max} F(x^*, Y) \neq \emptyset$$

i.e., F has a loose saddle point.

Proof. By Theorem 7.5.19, there exists $(x^*, y^*) \in X \times Y$ and $z^*, w^* \in F(x^*, y^*)$ such that $\xi_1(z^*) = \min \xi_1 F(X, y^*)$ and $\xi_2(w^*) = \max \xi_2 F(x^*, Y)$. Since ξ_1 and ξ_2 are strongly increasing and by Lemma 7.5.13(2), it follows that $z^* \in \mathrm{Min}F(X, y^*)$ and $w^* \in \mathrm{Max}f(x^*, Y)$. Therefore we have $F(x^*, y^*) \cap \mathrm{Min}F(X, y^*) \neq \emptyset$ and $F(x^*, y^*) \cap \mathrm{Max}F(x^*, Y) \neq \emptyset$. The proof is complete. □

If we assume that $intC \neq \emptyset$, then we can still have the existence of loose weak saddle points by assuming that ξ_1 and ξ_2 are strictly increasing instead of being strongly increasing.

Theorem 7.5.21. Let $(X, \{\Gamma_A\})$ and $(Y, \{\Gamma_B\})$ be two compact Hausdorff H-spaces, Z be a Hausdorff topological vector space and C be a convex cone of Z such that $intC \neq \emptyset$. Suppose that $F : X \times Y \to 2^Z$ is a set-valued mapping and $\xi_1, \xi_2 : Z \to \mathbb{R}$ are strictly increasing mappings such that:

(i) $\xi_1 F$ and $\xi_2 F$ are upper semicontinuous with nonempty compact values and for each fixed $x \in X$, $y \mapsto \xi_1 F(x, y)$ and $y \mapsto \xi_2 F(x, y)$ are lower semicontinuous; and for each fixed $y \in Y$, $x \mapsto \xi_1 F(x, y)$ and $x \mapsto \xi_2 F(x, y)$ are lower semicontinuous; and

(ii) the mapping $x \longmapsto \xi_1 F(x, y)$ is H-quasi-convex and the mapping $y \longmapsto \xi_2 F(x, y)$ is H-quasi-concave. Then there exists a loose weak saddle point of F in $X \times Y$, i.e., there exists $(x^*, y^*) \in X \times Y$ such that $F(x^*, y^*) \cap \mathrm{Min}_W F(X, y^*) \neq \emptyset$ and $F(x^*, y^*) \cap \mathrm{Max}_W F(x^*, Y) \neq \emptyset$.

Proof. By Theorem 7.5.19, there exists $(x^*, y^*) \in X \times Y$ and $s^*, w^* \in F(x^*, y^*)$ such that $\xi_1(z^*) = \min \xi_1 F(X, y^*)$ and $\xi_2(w^*) = \max \xi_2 F(x^*, Y)$. We now claim that $s^* \in \mathrm{Min}_W F(X, y^*)$ and $w^* \in \mathrm{Max}_W F(x^*, Y)$. Suppose, if possible that $s^* \notin \mathrm{Min}_W F(X, y^*)$, then there exists $s \in F(X, y^*)$ with $s \neq s^*$ and $s - s^* \in intC$. As ξ_1 is strictly increasing, we have $\xi_1(s^*) - \xi_1(s) = \xi_1(s^* - s) > 0$, i.e, $\xi_1(s) < \xi_1(s^*) = \min \xi_1 F(X, y^*) \leq \xi_1(s)$. Thus we have $\xi_1(s) < \xi_1(s)$ which is impossible. By the same reason, we must have $w^* \in \mathrm{Max}_W F(x^*, Y)$. Thus $F(x^*, y^*) \cap \mathrm{Min}_W F(X, y^*) \neq \emptyset$ and $F(x^*, y^*) \cap \mathrm{Max}_W F(x^*, Y) \neq \emptyset$. Thus we complete the proof. □

As special cases of Theorems 7.5.2 and 7.5.21, we have the following result.

Theorem 7.5.22. Let X and Y be two nonempty convex and compact Hausdorff sets in two Hausdorff topological vector spaces and Z a Hausdorff topological vector space and C be a convex cone of Z. Suppose that $F : X \times Y \to 2^Z$ is a set-valued mapping and $\xi_1, \xi_2 : Z \to \mathbb{R}$ are strongly increasing continuous mappings such that:

(i) F is upper semicontinuous with nonempty compact values and for each fixed $x \in X$, $y \mapsto F(x, y)$ is lower semicontinuous; and for each fixed $y \in Y$, $x \mapsto F(x, y)$ is lower semicontinuous; and

(ii) the mapping $x \longmapsto \xi_1 F(x, y)$ is quasi-convex and the mapping $y \longmapsto \xi_2 F(x, y)$ is quasi-concave.

Then there exists $z^*, w^* \in F(x^*, y^*)$ such that

$$F(x^*, y^*) \cap \text{Min}F(X, y^*) \neq \emptyset \text{ and } F(x^*, y^*) \cap \text{Max}F(x^*, Y) \neq \emptyset.$$

Proof. As each compact and convex set in a topological vector space is H-convex and both ξ_1 and ξ_2 are continuous, it follows that all hypotheses of Theorem 7.5.20 are satisfied. By Theorem 7.5.20, the conclusion follows and we complete the proof. □

Theorem 7.5.23. Let X and Y be two nonempty compact and convex sets in two Hausdorff topological vector spaces, Z be a Hausdorff topological vector space and C be a convex cone of Z such that $intC \neq \emptyset$. Suppose that $F : X \times Y \to 2^Z$ is a set-valued mapping and $\xi_1, \xi_2 : Z \to \mathbb{R}$ are strictly increasing and continuous mappings such that:

(i) the mapping $(x, y) \mapsto F(x, y)$ is upper semicontinuous with nonempty compact values and for each fixed $x \in X$, $y \mapsto F(x, y)$ is lower semicontinuous; and for each fixed $y \in Y$, $x \mapsto F(x, y)$ is lower semicontinuous; and

(ii) the function $x \longmapsto \xi_1 F(x, y)$ is quasi-convex and the function $y \longmapsto \xi_2 F(x, y)$ is quasi-concave.

Then there exists a loose weak saddle point of F in $X \times Y$, i.e., there exists $(x^*, y^*) \in X \times Y$ such that $F(x^*, y^*) \cap \text{Min}_W F(X, y^*) \neq \emptyset$ and $F(x^*, y^*) \cap \text{Max}_W F(x^*, Y) \neq \emptyset$.

Proof. As an application of Theorem 7.5.21, it is clear the conclusion is true. □

As an application of Theorem 7.5.23, we have the following existence of weak saddle points in topological vector spaces.

Theorem 7.5.24. Let X and Y be two nonempty compact and convex sets in two Hausdorff topological vector spaces, Z be a Hausdorff topological vector space and C be a convex cone of Z such that $intC \neq \emptyset$. Suppose that $F : X \times Y \to 2^Z$ is a set-valued mapping such that:

(i) the mapping $(x, y) \mapsto F(x, y)$ is upper semicontinuous with nonempty compact values and for each fixed $x \in X$, $y \mapsto F(x, y)$ is lower semicontinuous; and for each fixed $y \in Y$, $x \mapsto F(x, y)$ is lower semicontinuous; and

(ii) the mapping $x \longmapsto F(x, y)$ is convex and the mapping $y \longmapsto F(x, y)$ is concave.

Then there exists a loose weak saddle point of F in $X \times Y$, i.e., there exists $(x^*, y^*) \in X \times Y$ such that $F(x^*, y^*) \cap \text{Min}_W F(X, y^*) \neq \emptyset$ and $F(x^*, y^*) \cap \text{Max}_W F(x^*, Y) \neq \emptyset$.

Proof. As $intC \neq \emptyset$, there exists $\xi \in E^*$ which is strictly increasing, i.e., $\xi(s) > 0$ for each $s \in intC$. By condition (ii) and the fact that ξ is linear and continuous, we have that the mapping $x \mapsto \xi F(x, y)$ is convex and $y \mapsto \xi F(x, y)$

is concave (e.g., see Lemma 2.2 of Tan et al. [521] or see Proposition 2.2 of Luc and Vargas [353]). Let $\xi_1 = \xi_2$ in Theorem 7.5.23. Then all hypotheses of Theorem 7.5.23 are satisfied and thus there exists $(x^*, y^*) \in X \times Y$ such that $F(x^*, y^*) \cap \text{Min}_W F(X, y^*) \neq \emptyset$ and $F(x^*, y^*) \cap \text{Max}_W F(x^*, Y) \neq \emptyset$. Thus the proof is complete. □

As a single-valued case of Theorem 7.5.23, we have the following existence of weak saddle points in topological vector spaces.

Theorem 7.5.25. Let X and Y be two nonempty compact and convex sets in two Hausdorff topological vector spaces, Z be a Hausdorff topological vector space and C be a convex cone of Z such that $intC \neq \emptyset$. Suppose that $f : X \times Y \to Z$ is a single-valued mapping such that:
(i) the mapping $(x, y) \mapsto f(x, y)$ is demicontinuous; and
(ii) the mapping $x \longmapsto f(x, y)$ is natural quasi-C-convex and the mapping $y \mapsto f(x, y)$ is natural quasi-C-concave.

Then f has a weak saddle point $(x^*, y^*) \in X \times Y$, i.e., there exists $(x^*, y^*) \in X \times Y$ such that $f(x^*, y^*) \cap \text{Min}_W f(X, y^*) \cap \text{Max}_W f(x^*, Y) \neq \emptyset$.

Proof. As $intC \neq \emptyset$, there exists $\xi \in E^*$ which is strictly increasing, i.e., $\xi(s) > 0$ for each $s \in intC$. As f is demicontinuous, we have $(x, y) \mapsto \xi f(x, y)$ is continuous by Remark 7.5.7 (e.g., see Lemma 2.2 of Tanaka [523]). Moreover by condition (ii) and the linearity of ξ, it is clear that $x \mapsto \xi f(x, y)$ is quasi-convex and $y \mapsto \xi f(x, y)$ is quasi-concave. Thus all hypotheses of Theorem 7.5.23 are satisfied. By Theorem 7.5.23, there exists $(x^*, y^*) \in X \times Y$ such that $f(x^*, y^*) \in \text{Min}_W f(X, y^*)$ and $f(x^*, y^*) \in f(x^*, Y)$. Thus (x^*, y^*) is a weak saddle point of f and we complete the proof. □

We would like to point out that Theorem 7.5.25 generalizes Theorem 3.1 of Tanaka [529] into topological vector spaces. Moreover our proof is different from his as he used Hartung's minimax inequality in locally convex space in [227] (see also Simons [481] or Sion [490]); however, our proof is basically on the use of the Browder-Fan fixed point theorem in topological vector spaces and Theorem 7.5.19.

As a special case of Theorem 7.5.25, we have the following corollary:

Corollary 7.5.26. Let X and Y be two nonempty compact and convex sets in two Hausdorff topological vector spaces, Z be a Hausdorff topological vector space and C be a closed pointed convex cone of Z such that $intC \neq \emptyset$. Suppose that $f : X \times Y \to Z$ is a single-valued mapping such that:
(i) f is demicontinuous; and
(ii) the mapping $x \longmapsto f(x, y)$ is C-convex and the mapping $y \mapsto f(x, y)$ is C-concave.
Then f has a weak saddle point $(x^*, y^*) \in X \times Y$, i.e., there exists $(x^*, y^*) \in$

$X \times Y$ such that $f(x^*, y^*) \cap \text{Min}_W f(X, y^*) \cap \text{Max}_W f(x^*, Y) \neq \emptyset$, i.e., $f(x^*, y^*) \in \text{Min}_W f(X, y^*) \cap \text{Max}_W f(x^*, Y)$.

Theorem 7.5.19 improves Theorem 3.3 of Tan et al. [521] which in turn includes Theorem 3.3 of Luc and Vargas [353] as a special case.

Indeed as an application of Theorem 7.5.19, we can also have the following existence of saddle points in topological vector spaces.

Theorem 7.5.27. Let X and Y be two nonempty compact and convex sets in two Hausdorff topological vector spaces, Z be a Hausdorff topological vector space and C be a convex cone of Z such that $intC \neq \emptyset$. Suppose that $f : X \times Y \to Z$ is a single-valued mapping such that:
(i) the mapping $(x, y) \mapsto f(x, y)$ is demicontinuous; and
(ii) the mapping $x \longmapsto f(x, y)$ is strictly natural quasi-C-convex and the mapping $y \mapsto f(x, y)$ is strictly natural quasi-C-concave.

Then f has a saddle point $(x^*, y^*) \in X \times Y$, i.e., there exists $(x^*, y^*) \in X \times Y$ such that $f(x^*, y^*) \in \text{Min} f(X, y^*) \cap \text{Max} f(x^*, Y) \neq \emptyset$.

Proof. As $intC \neq \emptyset$, there exists $\xi \in E^*$ which is strictly increasing, i.e., $\xi(s) > 0$ for each $s \in intC$. As f is demicontinuous, we have $(x, y) \mapsto \xi f(x, y)$ is continuous by Remark 7.5.7. Moreover by condition (ii) and the linearity of ξ, it is clear that $x \mapsto \xi f(x, y)$ is quasi-convex and $y \mapsto \xi f(x, y)$ is quasi-concave. Thus all hyoptheses of Theorem 7.5.19 are satisfied. By Theorem 7.5.19, there exists $(x^*, y^*) \in X \times Y$ such that $\xi f(x^*, y^*) = \text{Min}\xi f(X, y^*)$ and $\xi f(x^*, y^*) = \text{Max}\xi f(x^*, Y)$. Now we claim that $f(x^*, y^*) \in \text{Min} f(X, y^*)$ and $f(x^*, y^*) \in \text{Max} f(x^*, Y)$. If not, we may assume that $f(x^*, y^*) \notin \text{Min} f(X, y^*)$. Then there exists $x \in X$ with $x \neq x^*$ such that $f(x, y^*) < f(x^*, y^*)$. Then by the strictly quasi-convexity, there exists $\lambda \in (0, 1)$ such that

$$f(x_\lambda) \in f(x^*, y^*) - intC,$$

where $x_\lambda = \lambda x + (1 - \lambda)x^* \in X$. Then we have $\xi(f(x_\lambda)) < \xi(f(x^*, y^*))$. But we do know that $\xi(f(x^*, y^*)) \leq \xi(f(x_\lambda))$ which implies that $\xi(f(x_\lambda)) < \xi(f(x_\lambda))$. This is a contradiction. Thus we must have $f(x^*, y^*) \in \text{Min} f(X, y^*)$ and by the same reason, we also have $f(x^*, y^*) \in \text{Max} f(x^*, Y)$. Thus the proof is completed. □

By using the Brouwder-Fan fixed point theorem, we show that Theorem 4.1 of Tanaka [523] in locally convex spaces proved by using Tychonoff's fixed point theorem still holds in Hausdorff topological vector spaces. Moreover, we wish to note that as application of Theorem 7.5.19 and the same arguement used in the proof of Theorem 7.5.27, we do have the same existence result of loose saddle points in H-spaces, i.e., we have the following:

Theorem 7.5.28. Let $(X, \{\Gamma_A\})$ and $(Y, \{\Gamma_B\})$ be two compact Hausdorff H-spaces, Z be a Hausdorff topological vector space and C be a convex cone of Z such that $intC \neq \emptyset$. Suppose that $f : X \times Y \to Z$ is a single-valued mapping such that:

(i) the mapping $(x, y) \mapsto f(x, y)$ is demicontinuous; and

(ii) the mapping $x \longmapsto f(x, y)$ is strictly quasi-H-convex and the mapping $y \mapsto f(x, y)$ is strictly quasi-H-concave.

Then f has a saddle point $(x^*, y^*) \in X \times Y$, i.e., there exists $(x^*, y^*) \in X \times Y$ such that $f(x^*, y^*) \in \text{Min} f(X, y^*) \cap \text{Max} f(x^*, Y) \neq \emptyset$.

Proof. As $intC \neq \emptyset$, there exists $\xi \in E^*$ which is strictly increasing, i.e., $\xi(s) > 0$ for each $s \in intC$. As f is demicontinuous, we have $(x, y) \mapsto \xi f(x, y)$ is continuous by Remark 7.5.5. Moreover by condition (ii) and the linearity of ξ, it is clear that $x \mapsto \xi f(x, y)$ is quasi-H-convex and $y \mapsto \xi f(x, y)$ is quasi-H-concave. Thus all hypotheses of Theorem 7.5.19 are satisfied. By Theorem 7.5.19, there exists $(x^*, y^*) \in X \times Y$ such that $\xi f(x^*, y^*) = \text{Min} \xi f(X, y^*)$ and $\xi f(x^*, y^*) = \text{Max} \xi f(x^*, Y)$. Now we claim that $f(x^*, y^*) \in \text{Min} f(X, y^*)$ and $f(x^*, y^*) \in \text{Max} f(x^*, Y)$. If not, we may assume that $f(x^*, y^*) \notin \text{Min} f(X, y^*)$. Then there exists $x \in X$ with $x \neq x^*$ such that $f(x, y^*) < f(x^*, y^*)$. Then by the strictly quasi-convexity of f, there exists $\lambda \in (0, 1)$ such that

$$f(x_\lambda) \in f(x^*, y^*) - intC,$$

where $x_\lambda = \lambda x + (1 - \lambda)x^* \in X$. Then following the same arguements used in the proof of Theorem 7.5.27, we arrive at the conclusion that $f(x^*, y^*) \in Minf(X, y^*)$; similarly, we can prove that $f(x^*, y^*) \in Maxf(x^*, Y)$. The proof is complete. □

7.6 Generic Stability of Vector-Valued Saddle Points

In this section, we shall study the stability of C-saddle points for vector mappings. In order to do so, we first recall some notions and definitions. If X is a topological space, we shall denote by $K(X)$ and $CK(X)$ the space consisting of of all nonempty compact subsets of X and the space consisting of all nonempty compact and convex subsets of X, respectively, both endowed with the Vietoris topology (see, Klein and Thompson [317]).

Let Z be a topological space. We recall that a mapping $T : X \to 2^Z$ is said to be (1): upper (resp., lower) semicontinuous at $x \in X$, if for each open set G in Z with $G \supset T(x)$ (respectively, $G \cap T(x) \neq \emptyset$), there exists an open neighborhood $O(x)$ of x in X such that $G \supset T(x')$ (respectively, $G \cap T(x') \neq \emptyset$)

for each $x' \in O(x)$; (2): almost lower semicontinuous (in short, ALSC) at $x \in X$, if there exists $z \in T(x)$ such that for each open neighborhood $N(z)$ of z in Z, there exists an open neighborhood $O(x)$ of x in X with the property that $N(z) \cap T(x') \neq \emptyset$ for each $x' \in O(x)$; and (3): USCO if T is upper semicontinuous with nonempty compact values.

Definition 7.6.1. A space X is said to be (i): *Čech-complete* if it can be embedded as a G_δ subset of some compact Hausdorff space; and (ii): belong to the class $\mathcal{L}$ (see Kenderov [288]) if for each Čech-complete space Z, every USCO mapping $F : Z \to K(X)$ is almost lower semicontinuous on some dense G_δ subset of Z.

By Theorem 2 of Fort [178], each completely metrizable space belongs to $\mathcal{L}$ and each Banach space with its weak topology (which is non-metrizable if it is infinitely dimensional), is also in $\mathcal{L}$ (see Theorem 2 of Christensen [93]).

7.6.1 Stability Results in Compact Case

In this section, we always assume that X and Y are two nonempty compact convex subsets of topological vector space E and F, respectively and Z is a normed space with a partial order generated by a solid-pointed closed convex cone C.

Let $L(X \times Y, Z)$ be the family of all bounded real-valued vector-valued mappings defined on $X \times Y$ to Z in the sense that that for each $f \in L(X \times Y, Z)$, we have $\sup_{(x,y) \in X \times Y} \|f(x,y)\| < \infty$. For $f, g \in L(X \times Y, Z)$, define

$$\rho(f,g) = \sup_{(x,y) \in X \times Y} \|f(x,y) - g(x,y)\|.$$

Clearly, $(L(X \times Y, Z), \rho)$ is a complete metric space. Let $M = \{f \in L(X \times Y, Z) :$ f is C-convex in its first variable and f is C-concave in its second variable $\}$. Then for each $f \in M$, f has at least one weak C-saddle point (resp., C-saddle point) by Theorem 7.5.25.

As we assume that the solid convex cone C is closed, it is easy to verify that the M is a closed subset of $L(X \times Y, Z)$. Thus we have the following:

Lemma 7.6.2. (M, ρ) is a complete metric space.

Now for each $f \in M$, we define a solution mapping $S : L(X \times Y, Z) \to 2^{X \times Y}$ by $S(f) = \{(x,y) \in X \times Y : (x,y)$ is the C-saddle point of f in $X \times Y\}$. By the continuity of f, it is clear that $S(f)$ is closed and nonempty for each $f \in L(X \times Y, Z)$. Then we have the following:

Lemma 7.6.3. The mapping $S : M \to K(X) \times K(Y)$ is upper semicontinuous.

Proof. Since $X \times Y$ is compact, we only need to prove that the graph of S is closed in $M \times X \times Y$. For any net $\{(f_\alpha, x_\alpha, y_\alpha)\}_{\alpha\in\Gamma}$ in graph S with $(f_\alpha, x_\alpha, y_\alpha) \to (f_0, x_0, y_0) \in M \times X \times Y$, then $f_\alpha \to f_0$, $x_\alpha \to x_0$ and $y_\alpha \to y_0$ and $f_\alpha(x_\alpha, y_\alpha) \in Max_W f_\alpha(x_\alpha, Y) \cap \text{Min}_W f_\alpha(X, y_\alpha)$ for all $\alpha \in \Gamma$. In order to finish the proof, it suffices to show that $(x_0, y_0) \in S(f_0)$. If not, without loss of the generality, we may assume that $f_0(x_0, y_0) \notin \text{Min}_W f_0(X, y_0)$. Then there exists $z^* \in \text{Min}_W f_0(X, y_0)$ such that $z^* - f_0(x_0, y_0) \in -intC$. Thus there exists $x^* \in X$ such that $f_0(x^*, y_0) - f_0(x_0, y_0) \in -intC$. As $-intC$ is a nonempty open and convex set in Z, there exists a nonempty open and symmetric (or, balanced) neighborhood C_0 of zero in Z such that

$$f_0(x^*, y_0) - f_0(x_0, y_0) + 4C_0 \subset -intC.$$

As $f_\alpha \to f_0$ uniformly, there exists $\alpha_1 \in \Gamma$ such that for each $\alpha \geq \alpha_1$, we have

$$f_\alpha(x^*, y) - f_0(x^*, y) \in C_0 \text{ and } f_0(x, y) - f_\alpha(x, y) \in C_0$$

for all $(x, y) \in X \times Y$. Thus we have $f_\alpha(x^*, y_\alpha) - f_0(x^*, y_\alpha) \in C_0$ and $f_0(x_\alpha, y_\alpha) - f_\alpha(x_\alpha, y_\alpha) \in C_0$. Note that $x_\alpha \to x_0$ and $y_\alpha \to y_0$, there exists $\alpha_2 \in \Gamma$ such that for each $\alpha \geq \alpha_2$, we have $f_0(x^*, y_\alpha) - f_0(x^*, y_0) \in C_0$ and $f_0(x_0, y_0) - f_0(x_\alpha, y_\alpha) \in C_0$. Let $\alpha_3 = \max\{\alpha_1, \alpha_2\}$ and note that

$$\begin{aligned} f_\alpha(x^*, y_\alpha) &- f_\alpha(x_\alpha, y_\alpha) = f_\alpha(x^*, y_\alpha) - f_0(x^*, y_\alpha) \\ &+ f_0(x^*, y_\alpha) - f_0(x^*, y_0) + f_0(x_\alpha, y_\alpha) - f_\alpha(x_\alpha, y_\alpha) \\ &+ f_0(x_0, y_0) - f_0(x_\alpha, y_\alpha) + f_0(x^*, y_0) - f_0(x_0, y_0). \end{aligned}$$

It follows that for each $\alpha \geq \alpha_3$, we have

$$f_\alpha(x^*, y_\alpha) - f_\alpha(x_\alpha, y_\alpha) \in f_0(x^*, y_0) - f_0(x_0, y_0) + 4C_0 \subset -intC.$$

Thus $f_\alpha(x^*, y_\alpha) - f_\alpha(x_\alpha, y_\alpha) \in -intC$ which implies that $f_\alpha(x_\alpha, y_\alpha) \notin Minf(X, y_\alpha)$. This contradicts our assumption that $(f_\alpha, x_\alpha, y_\alpha) \in graphS$. Therefore the graph of S must be closed (thus S is upper semicontinuous) we complete the proof. □

Definition 7.6.4. For each $f \in M$, then (1) a point $(x, y) \in S(f)$ is said to be *essential* relative to M if for each open neighborhood $N((x, y))$ of (x, y) in $X \times Y$, there exists an open neighborhood $O(f)$ of f in M such that $S(f') \cap N(x, y) \neq \emptyset$ for each $f' \in O(f)$; (2) f is said to be *weakly essential* relative to M if there exists $(x, y) \in S(f)$ which is essential relative to M; and (3) f is said to be *essential* relative to M if every $(x, y) \in S(f)$ is essential relative to M.

We also need the following result which is Theorem 2 of Fort [178] (see also Lemma 2.4.3).

Lemma 7.6.5. If X is metrizable, Z is a Baire space and $S : Z \to K(X)$ is an USCO mapping, then the set of points where S is lower semicontinuous is a dense G_δ set in Z.

Now we have the following result.

Theorem 7.6.6. (i) S is almost lower semicontinuous at $f \in M$ if and only if f is weakly essential relative to M.

(ii) S is lower semicontinuous at $f \in M$ if and only if f is essential relative to M.

(iii) S is continuous at $f \in M$ if and only if f is essential relative to M.

Proof. (i) S is almost lower semicontinuous at $f \in M$ if and only if there exists $y \in S(f)$ such that y is essential relative to M if and only if f is weakly essential relative to M.

(ii) S is lower semicontinuous at $f \in M$ if and only if for each $(x, y) \in S(f)$ is essential relative to M if and only if f is essential relative to M.

(iii) This follows from (ii) and Lemma 7.6.2 and thus we complete the proof of Theorem 7.6.6. $\square$

Remark 7.6.7. If $X \times Y$ is metrizable by a metric d, then the Vietoris topology on $K(X \times Y)$ coincidences with the topology generated by the Hausdorff metric D induced by d (e.g., see Corollary 4.2.3 of Klein and Thompson [317]. Then S is continuous at $f \in M$ if and only if for each $\epsilon > 0$, there is $\delta > 0$ such that for each $g \in M$, $h(S(f), S(g)) < \epsilon$ whenever $\rho(f, g) < \delta$; i.e., $S(f)$ is stable (or say, continuous) in the sense that "*$F(g)$ is close to $F(f)$ whenever g is close to f*". Thus Theorem 5.4 (iii) shows that $S(f)$ is stable if and only if f is essential relative to M.

We shall give a sufficient condition that $f \in M$ is essential relative to M:

Theorem 7.6.8. If $f \in M$ is such that $S(f)$ is singleton set, then f is essential relative to M.

Proof. Suppose $S(f) = \{(x, y)\}$. Let G be any open set in $X \times Y$ such that $S(f) \cap G \neq \emptyset$, then $(x, y) \in G$ and $S(f) \subset G$. As S is upper semicontinuous at f by Lemma 7.6.3, there is an open neighborhood $O(f)$ of f in M such that $S(f') \subset G$ for each $f' \in O(f)$; in particular, we have $G \cap S(f') \neq \emptyset$ for each $f' \in O(f)$. Thus S is lower semicontinuous at f. By Theorem 7.6.6(ii), f is essential relative to M. $\square$

Theorem 7.6.9. (i) Suppose that $X \times Y$ belongs to class $\mathcal{L}$. Then there exists a dense G_δ subsets Q of M such that f is weakly essential relative to M for each $f \in Q$.

(ii) Suppose that X is metrizable. Then there exists a dense G_δ subset Q of M such that f is essential relative to M for each $f \in Q$.

Proof. (i) Since M is a complete metric space, M is Čech-complete. By Lemma 7.6.3, the mapping $S : M \to K(X \times Y)$ is upper semicontinuous. Since $X \times Y$ belongs to class $\mathcal{L}$, S is almost lower semicontinuous on some dense G_δ subset Q of M. By Theorem 7.6.6 (i), f is weakly essential relative to M for each $f \in Q$. (ii) By Lemma 7.6.3 and Lemma 7.6.5, S is lower semicontinuous on some dense G_δ subset Q of M. By Theorem 7.6.6 (ii), f is essential relative to M for each $f \in Q$. □

We note that if we define $M = \{f \in L(X \times Y, Z)$: f has at weak C- saddle point in $X \times Y\}$ and

$$S(f) = \{(x, y) \in X \times Y : f(x, y) \in Min_W f(X, y) \cap Max_W f(x, Y)\}$$

for each $f \in L(X \times Y, Z)$. By the uniform continuity of f, then all results in this section remain valid. In this section, Theorem 7.6.6 shows that the majority of weak C-saddle points are stable in the sense of the Baire category, where we only consider the property of the weak C-saddle points when f changes in $L(X \times Y, Z)$. It is a natural question to ask what will happen for the solution set of weak C-saddle points of $(A, B, f) \in K(X) \times K(Y) \times L(X \times Y, Z)$ if (A, B, f) has perturbation. We will answer this situation in next section.

7.6.2 Stability Results in Non-Compact Case

The purpose of this section is to investigate the property of a solution set of weak C-saddle points for each $(A, B, f) \in K(X) \times K(Y) \times L(X \times Y, Z)$ when (A, B, f) has perturbation in a general case. We prove that the analogy of Theorem 7.6.4 still holds when each $(A, B, f) \in K(X) \times K(Y) \times L(X \times Y, Z)$ varies with Hausdorff metric, which is induced by the metric of X, Y and ρ.

Let X and Y be two nonempty topological spaces, Z a normed linear space and $L(X \times Y, Z)$ is the the space of all bounded mappings defined on $X \times Y$ with the metric ρ defined by $\rho(f_1, f_2) := \sup_{(x,y) \in X \times Y} |f_1(x, y) - f_2(x, y)|$ for each $f_1, f_2 \in L(X \times Y, Z)$. Then it is clear that $L(X \times Y, Z), \rho)$ is a complete metric space.

Let $W = K(X) \times K(Y) \times L(X \times Y, Z)$. Now for each $u = (A, B, f) \in W$, we consider the following existence and stability problem of having $(x, y) \in A \times B$ such that

$$f(x,y) \in Min_W f(A, y) \cap Max_W f(x, B),$$

i.e., (x, y) is a weak C-saddle point of f on $A \times B$.

Before we study the stability of the set $S(u)$ of weak C-saddle points of f in $A \times B$ for each $u = (A, B, f) \in W$, we shall state several lemmas which will be used to prove our main results later. The proof of the following result is routine and is hence omitted.

Lemma 7.6.10. $(L(X \times Y, Z), \rho)$ is a complete metric space.

Lemma 7.6.11. Suppose X is a nonempty subset of a Hausdorff topological vector space. If $\{A_\alpha\}_{\alpha\in\Gamma}$ is a net of compact and convex sets in $K(X)$ which converges to $A \in K(X)$ in the Vietoris topology, then A is also convex.

Proof. It is Lemma 3.2 of Tan et al. [515]. □

Lemma 7.6.12. Let $\{A_\alpha\}_{\alpha\in\Gamma}$ be a equence of sets in $K(X)$ which converges to $A \in K(X)$ in the Vietoris topology. Then any net $\{x_\alpha\}_{\alpha\in\Gamma}$ with $x_\alpha \in A_\alpha$ for each $\alpha \in \Gamma$ has a cluster point in A.

Proof. It is Lemma 3.3. of Tan et al. [515]. □

Now define a subspace M of W by $M := \{(A, B, f) \in W$: there exists $(x, y) \in A \times B$ such that $f(x, y) \in Min_W f(A, y) \cap Max_W f(x, B)\}$. Then define a mapping $S : M \to 2^{X\times Y}$ by

$$S(u) := \{(x, y) \in A \times B : f(x, y) \in Min_W f(A, y) \cap Max_W f(x, B)\}$$

for each $u = (A, B, f) \in M$. Then we have the following:

Lemma 7.6.13. The set M is a closed subspace of W.

Proof. Suppose that $\{(A_\alpha, B_\alpha, f_\alpha)\}_{\alpha\in\Gamma}$ is a net in M such that $(A_\alpha, B_\alpha, f_\alpha) \to (A, B, f) \in W$. For each $\alpha \in \Gamma$, there exists $x_\alpha \in A_\alpha$ and $y_\alpha \in B_\alpha$ such that $f_\alpha(x_\alpha, y_\alpha) \in Min_W f_\alpha(A_\alpha, y_\alpha) \cap Max_W f_\alpha(x_\alpha, B_\alpha)$. Since $A_\alpha \to A$ and $B_\alpha \to B$ in the Vietoris topology, the net $\{x_\alpha\}_{\alpha\in\Gamma}$ and $\{y_\alpha\}_{\alpha\in\Gamma}$ have cluster points $x_0 \in A$ and $y_0 \in B$ by Lemma 7.6.12. Now we shall show that $f(x_0, y_0) \in Min_W f(A, y_0) \cap Max_W f(x_0, B)$. Suppose it were not true. Without loss of generality, we may assume that $f(x_0, y_0) \notin Max_W f(A, y_0)$. Thus there exists $x^* \in A$ such that $f(x^*, y_0) - f(x_0, y_0) \in -intC$. As f is continuous, there exists a symmetric (or, balanced) open neighborhood C_0 of zero in Z, an open neighborhood $N(x^*)$ of x^*, an open neighborhood $N(x_0)$ of x_0 and an open neighborhood $N(y_0)$ of y_0 in X and Y, respectively, such that

$$f(x', y) - f(x, y) - (f(x^*, y_0) - f(x_0, y_0)) \in C_0$$

for all $(x', x, y) \in N(x^*) \times N(x_0) \times N(y_0)$. Note that $f_\alpha \to f$ uniformly, there exists $\alpha_0 \in \Gamma$ such that for each $\alpha \geq \alpha_0$, we have $f_\alpha(x, y) - f(x, y) \in C_0$ for all $(x, y) \in X \times Y$. Note that

$$\begin{aligned} f_\alpha(x', y) - f_\alpha(x, y) &= (f_\alpha(x', y) - f(x', y)) \\ &+ (f(x, y) - f_\alpha(x, y)) + (f(x', y)) - f(x, y)) \end{aligned}$$

and

$$\begin{aligned} f_\alpha(x^*, y_0) - f_\alpha(x_0, y_0) &= f_\alpha(x^*, y_0) - f(x^*, y_0) \\ &- (f_\alpha(x_0, y_0) - f(x_0, y_0)) + (f(x^*, y_0) - f(x_0, y_0)). \end{aligned}$$

Thus we have

$$f_\alpha(x', y) - f_\alpha(x, y) - (f_\alpha(x^*, y_0) - f_\alpha(x_0, y_0)) \in 5C_0$$

for each $\alpha \geq \alpha_0$ and $(x', x, y) \in N(x^*) \times N(x_0) \times N(y_0)$. As $x^* \in A$, $A_\alpha \to A$, $B_\alpha \to B$, $x_\alpha \to x_0$ and $y_\alpha \to y_0$, there exists $\alpha_1 \geq \alpha_0$ such that for each $\alpha \geq \alpha_1$, we have $N(x^*) \cap A_\alpha \neq \emptyset$, $x_\alpha \in A_\alpha$ and $y_\alpha \in B_\alpha$. Now for each $\alpha \geq \alpha_1$, choose $x^{*'} \in N(x^*) \cap A_\alpha$, we have

$$f_\alpha(x^{*'}, y_\alpha) - f_\alpha(x_\alpha, y_\alpha) - (f_\alpha(x^*, y_0) - f_\alpha(x_0, y_0)) \in 5C_0.$$

As

$$\begin{aligned} f_\alpha(x^*, y_0) \quad &- \quad f_\alpha(x_0, y_0) = f_\alpha(x^*, y_0) - f(x^*, y_0) + f(x^*, y_0) \\ &- \quad [f_\alpha(x_0, y_0) - f(x_0, y_0)] - f(x_0, y_0), \end{aligned}$$

it follows that

$$f_\alpha(x^{*'}, y_\alpha) - f_\alpha(x_\alpha, y_\alpha) \in 7C_0 + [f(x^*, y_0) - f(x_0, y_0)].$$

Note that $f(x^*, y_0) - f(x_0, y_0) + 7C_0 \subset -intC$. we have $f_\alpha(x^{*'}, y_\alpha) - f_\alpha(x_\alpha, y_\alpha) \in -intC$ for each $\alpha \geq \alpha_1$, which implies that $f_\alpha(x_\alpha, y_\alpha) \notin Minf_\alpha(A_\alpha, y_\alpha)$. This contradicts our assumption that

$$f_\alpha(x_\alpha, y_\alpha) \in \mathrm{Min}_W f_\alpha(A_\alpha, y_\alpha) \cap \mathrm{Max}_W f_\alpha(x_\alpha, B_\alpha)$$

thus we must have

$$f(x_0, y_0) \in \mathrm{Min}_W f(, y_0) \cap \mathrm{Max}_W f(x_0, B),$$

and the proof is complete. □

Lemma 7.6.14. If X and Y are Čech-complete, then $L(X \times Y, Z)$ and hence M is Čech-complete.

Proof. The space $L(X \times Y, Z)$ is Čech-complete as it is a complete metric space by Lemma 7.6.11. As X and Y are Čech-complete, it follows that $K(X)$ and $K(Y)$ are also Čech-complete by Lemma 2.2 of Beer [27]. Therefore the product space $K(X) \times K(Y) \times L(X \times Y, Z)$ is Čech-complete by Theorem 3.9.8 of Engelking [154]. By our Lemma 7.6.13 and Theorem 3.9.6 of Engelking [154], M is also Čech-complete. □

Lemma 7.6.15. For the mapping $S : M \to 2^{X \times Y}$, we have that $S(u) \in K(X) \times K(Y)$ for each $u \in M$.

Proof. Since $S(u) \subset A \times B$, for each $u = (A, B, f) \in M$, it is sufficient to prove that $S(u)$ is closed in $A \times B$. Let $\{x_\alpha, (y_\alpha\}_{\alpha \in \Gamma}$ be a net in $S(u)$

which converges to a point $(x_0, y_0) \in A \times B$. By the definition of S, we have $f(x_\alpha, y_\alpha) \in Min_W f(x_\alpha, B) \cap Max_W f(A, y_\alpha)$ for each $\alpha \in \Gamma$. Now by the uniform continuity of f, we can prove that $f(x_0, y_0) \in Max_W f(A, y_0) \cap Min_W f(x_0, B)$ and thus $(x_0, y_0) \in S(u)$. Thus $S(u)$ is a closed subset of $A \times B$. □.

Lemma 7.6.16. The mapping $S : M \to K(X) \times K(Y)$ is upper semicontinuous.

Proof. If possible, suppose that S is not upper semicontinuous at some point $u = (A, B, f) \in M$, then there exists an open subset G of $X \times Y$ with $G \supset S(u)$ and a net $\{u_\alpha\}_{\alpha \in \Gamma}$ in M with $u_\alpha \to u \in M$ such that for each $\alpha \in \Gamma$, there exists $(x_\alpha, y_\alpha) \in S(u_\alpha)$ with $(x_\alpha, y_\alpha) \notin G$. Denote $u_\alpha = (A_\alpha, B_\alpha, f_\alpha)$ and $u = (A, B, f)$. Then we have $f_\alpha \to f$, $A_\alpha \to A$ and $B_\alpha \to B$. Since $x_\alpha \in A_\alpha$ and $y_\alpha \in B_\alpha$ for each $\alpha \in \Gamma$, by Lemma 7.6.12, as $(x_\alpha, y_\alpha) \notin G$ for each $\alpha \in \Gamma$, it follows that $(x_0, y_0) \notin G$. Therefore $f(x_0, y_0) \notin \mathrm{Min}_W f(A, y_0) \cap \mathrm{Max}_W f(x_0, B)$. Now by following the same arguement as given in the proof of Lemma 7.6.13, we can show that there exists $\alpha_1 \in \Gamma$ such that for for each $\alpha \geq \alpha_1$, we have $(x_\alpha, y_\alpha) \notin S(u_\alpha)$. However, we do know that $(x_\alpha, y_\alpha) \in S(u_\alpha)$. This is a contradiction. Thus S must be upper semicontinuous and the proof is complete. □

Now let M_1 be a nonempty closed subset of M. Corresponding to Definition 7.6.4, we also have the following notion.

Definition 7.6.17. For each $u \in M_1$, we define that

(1) a point $(x, y) \in S(u)$ is said to be *essential* relative to M_1 if each open neighborhood $N(x, y)$ of $(x, y) \in X \times Y$, there exists an open neighborhood $O(u)$ of u in M_1 such that $S(u') \cap N(x, y) \neq \emptyset$ for each $u' \in O(u)$;

(3) u is said to be *weakly essential* relative to M_1 if there exists $(x, y) \in S(u)$ which is essential relative to M_1; and

(4) u is said to be *essential* relative to M_1 if every $(x, y) \in S(u)$ is essential relative to M_1.

Theorem 7.6.18. (i) S is almost lower semicontinuous at $u \in M_1$ if and only if u is weakly essential relative to M_1.

(ii) S is lower semicontinuous at $u \in M_1$ if and only if u is essential relative to M_1.

(iii) S is continuous at $u \in M_1$ if and only if u is essential relative to M_1.

Proof. (i) S is almost lower semicontinuous at $u \in M_1$ if and only if there exists $(x, y)y \in S(u)$ such that (x, y) is essential relative to M_1 if and only if u is weakly essential relative to M_1.

(ii) S is lower semicontinuous at $u \in M_1$ if and only if each $(x, y) \in S(u)$ is essential relative to M_1 if and only if u is essential relative to M_1.

(iii) This follows from (ii) and Lemma 7.6.16. □

A proof analogous to that of Theorem 7.6.8 gives us the following result and its proof is thus omitted.

Theorem 7.6.19. If $u \in M_1$ is such that $S(u)$ is a singleton set, then u is essential relative to M_1.

Theorem 7.6.20. (i) Let $X \times Y$ be Čech-complete and belong to the class $\mathcal{L}$. Then there exists a dense G_δ subset Q of M_1 such that u is weakly essential relative to M_1 for each $u \in Q$.

(ii) Let $X \times Y$ be completely metrizable. Then there exists a dense G_δ subset Q of M_1 such that u is essential relative to M_1 for each $u \in Q$.

Proof. Note that S is USCO by Lemma 7.6.15 and Lemma 7.6.16.

(i) Since $X \times Y$ is Čech-complete, Lemma 7.6.14 implies that M is also also Čech-complete. Since M_1 is closed in M_1, M_1 is also Čech-complete by Theorem 3.9.6 of Engelking [154]. As $X \times Y \in \mathcal{L}$, there is a dense G_δ subset Q of M_1 such that S is almost lower semicontinuous at each $u \in Q$. By Theorem 7.6.6(i), u is weakly essential relative to M_1 for each $u \in Q$.

(ii) By Lemma 7.6.5, there exists a dense G_δ subset Q of M_1 such that S is lower semicontinuous at each $u \in Q$. By Theorem 7.6.18(ii), u is essential relative to M_1 for each $u \in Q$. □

Remark 7.6.21. If $X \times Y$ is a complete metric space with metric d, then $K(X) \times K(Y)$ is a complete metric space when equipped with the Hausdorff metric h induced by d. By Corollary 4.2.3 of Klein and Thompson [317, p.41], the Vietoris topology on $K(X) \times K(Y)$ coincides with the topology induced by the Hausdorff metric h. By Lemma 7.6.11, it follows that $M := K(X) \times K(Y) \times L(X \times Y, Z)$ and hence M and M_1 are also complete metric spaces when equipped with the metric D defined by

$$D(u, u') = \rho(f, f') + h((A, B), (A', B'))$$

for $u = (A, B, f)$ and $u' = (A', B', f')$. We note then, the mapping $S : M_1 \to K(X) \times K(Y)$ is continuous at $u = (A, B, f) \in M_1$ if and only if for each $\epsilon > 0$, there is $\delta > 0$ such that $h(S(u), S(u')) < \epsilon$ whenever $u' \in M_1$ and $D(u, u') < \delta$; i.e., the solution set $S(u)$ is stable in the sense that "*$S(u')$ is close to $S(u)$, whenever u' is close to u for all $u' \in M_1$*". Thus Theorem 7.6.18(iii) shows that if $u \in M_1$, then u is essential relative to M_1 if and only if the solution set $S(u)$ is stable.

Remark 7.6.22. Let X and Y be nonempty closed and convex subsets of Frechet spaces E_1 and E_2 equipped with a translation invariant metric d_1 and d_2. We denote by $CK(X) = \{A \in K(X) : A \text{ is convex}\}$; $CK(Y) := \{B \in K(Y) : B \text{ is convex}\}$; $CL(X \times Y, Z) := \{f \in L(X \times Y, Z) : f$ is C-convex in its first variable and f is C-concave in its second variable$\}$; and

$M' := CK(X) \times CK(Y) \times CL(X \times Y, Z)$. Then as an application of results obtained in this section, we have the following:

Theorem 7.6.23. (i) M' is a nonempty closed subset of M.

(ii) There exists a dense G_δ subset Q of M' such that u is essential relative to M' for each $u \in Q$.

Proof. (i) Clearly M' is nonempty. If $u = (A, B, f) \in M'$, then f has at least one weak C-saddle point by Corollary 7.5.26, i.e., there exists $(x, y) \in A \times B$ such that $f(x, y) \in \text{Min}_W f(A, y) \cap \text{Max}_W f(x, B)$. Thus $u \in M$ so that $M' \subset M$. Now if $\{(A_n, B_n, f_n)\}_{n=1}^{\infty}$ is a sequence in M' such that $(A_n, B_n, f_n) \to (A, B, f) \in M$, then $f_n \to f$ on $X \times Y$, and f is also continuous. By Lemma 7.6.11, both A and B are convex. Thus $(A, B, f) \in M'$ so that M' is closed in M.

Now (ii) follows from (i) and Theorem 7.6.19(ii). □

Remark 7.6.24. In this section, we have established the stability of weak C-saddle points of vector-valued mappings in vector spaces. Remark 7.6.22 (resp., Remark 7.6.7) tells us that the majority (in the sense of the Baire category) of weak C-saddle points are stable. However, we do not have similar stability results for the vector minimax inequalities in vector spaces due to the complexity of structures of the sets $\text{Min} \cup_{x \in X} \text{Max}_W F(x, Y)$ and $\text{Max} \cup_{y \in Y} \text{Min}_W F(X, y)$. However, we still expect that similar stability results should be valid by assuming reasonable assumptions.

7.7 Existence of Vector-Valued Minimax Inequalities

In previous sections, we studied the existence of C-saddle and weak C-saddle points in H-space and topological vector spaces. In this section, we shall investigate the general existence of vector minimax theorems. Before we prove the existence of minimax theorems, we need to introduce and recall some notations and definitions; e.g., see Takana [502] and references therein.

Let Z be a ordered topological vector space with a *solid*-pointed convex cone C and $F : X \times Y \to 2^Z$ a vector-valued mapping defined on a product space $X \times Y$. Then the following subsets of Z

$$\text{Min} \cup_{x \in X} \text{Max}_W F(x, Y) \text{ and } \text{Max} \cup_{y \in Y} \text{Min}_W F(X, y)$$

are called the *set of all minimax values* of F and the *set of all maximin values* of F, respectively. For convenience, we shall denote the set of all minimax (resp., maximin) values of F by

$$Minimax_{x \in X, y \in Y} F(x, y)$$

(resp.,

$$Maximin_{y \in Y, x \in X} F(x, y))$$

throughout this section.

Next we shall need one of the fundamental properties concerning the existence of minimax theorems. This property is represented by the following relations and called the *domination property* in vector optimization (e.g., see Luc [351] and related references).

The Domination Property (DP): Min$A \neq \emptyset$ and $A \subset$ Min$A + C$, where A is a nonempty subset of an ordered topological vector space Z with a pointed convex cone C.

It is well known that to study the hypotheses under which the property (DP) holds is one of the most important topics in vector optimization theory.

We also need the following definition which was given by Luc in [352] and we state it as:

Definition 7.7.1. A subset A is said to be C-complete if there are no covers of the form $\{z_\alpha - clC)^C\}_{\alpha \in \Gamma}$, where $\{z_\alpha\}_{\alpha \in \Gamma}$ is a net in A such that $z_\beta \leq_C z_\alpha$ and $z_\alpha \neq\leq_C z_\beta$ for each α, β in Γ with $\alpha < \beta$.

Remark 7.7.2. We recall that a set A is said to be C-compact if $(z - clC) \cap A$ is compact for each $z \in A$; and A is said to be C-semicompact if any cover of the form $\{z_\alpha - clC)^C\}_{\alpha \in \Gamma}$ has a finite subcover (see Corley [97]). Then it is obvious that each compact set is C-compact, and hence C-semicompact, and hence C-complete with whatever the convex cone C is (e.g., see Remark 5.1 of Tanaka [524] or Lemma 2.2 of Luc [352].

One of the general conditions that ensures condition (DP) was given by Luc is stated here as the following lemma (e.g., see Theorem 2.6 and Corollary 2.12 of Luc [352]):

Lemma 7.7.3. If a convex cone C of an ordered topological vector space E is corrected, i.e.,

$$clC + (C \setminus L) \subset C$$

where $L = C \cap (-C)$ and A is a nonempty and C-complete, then the property (DP) for A holds.

By Remark 7.7.2 and Lemma 7.7.3, we have the following existence result (cf., Lemma 5.2 and Lemma 5.3 of Tanaka [523]).

Lemma 7.7.4. Let Z be a topological vector space with a partial order $\leq$ generated by a solid-point cone C and A a nonempty subset of Z. Then we have the following

(1) if the convex cone C is correct (i.e., $clC + (C \setminus \{0\}) \subset C$) and A is C-complete (resp., A is $-C$-complete), then

$$\mathrm{Min}A \neq \emptyset \text{ and } A \subset \mathrm{Min}A + C$$

(resp.,

$$\mathrm{Max}A \neq \emptyset \text{ and } A \subset \mathrm{Max}A - C);$$

(2) if A is compact, then we have

$$\mathrm{Min}_W A \neq \emptyset \text{ and } A \subset \mathrm{Min}_W A + C^0$$

and

$$\mathrm{Max}_W A \neq \emptyset \text{ and } A \subset \mathrm{Max}_W A - C^0.$$

Proof. We note that the part (1) follows by Lemma 7.7.3. We only need to prove part (2). If A is compact, we have by Remark 7.7.2 that A is both C-complete and $-C$-complete. As C is convex and $intC \neq \emptyset$, we have $cl(intC) = clC$ and $intC = int(clC)$ (e.g., see Holmes [235, p.59]). then it follows (e.g.,see Remark 2.3 of Tanaka and Kuroiwa [530, p.96] that $clC^0 + (C^0 \setminus \{0\}) = clC^0 + intC = int(C + C^0) = int(C + intC \cup \{0\}) \subset C \subset C^0$. Thus we show that C^0 is correct. Then by part (1), we have $\mathrm{Min}_W A \neq \emptyset$ and $A \subset \mathrm{Min}_W A + C^0$ and $\mathrm{Max}_W A \neq \emptyset$ and $A \subset \mathrm{Max}_W A - C^0$. Thus the proof is complete. □

In what follows, we shall say that $F : X \times Y \to 2^Z$ holds the vector minimax inequality if there exists
$z_1 \in Minimax_{x \in X, y \in Y} F(x, y)$ and $z_2 \in Maximin_{y \in Y, x \in X} F(x, y)$
such that $z_1 \leq_C z_2$.

In order to establish our existence theorems of minimax inequalities in H-spaces, we first need:

Theorem 7.7.5. Let X and Y be two compact Hausdorff topological spaces, Z be a Hausdorff topological vector space and C be a closed pointed convex cone such that $intC \neq \emptyset$. Suppose $F : X \times Y \to 2^Z$ is upper semicontinuous with nonempty compact-valued; and for each fixed $x \in X$, $y \mapsto F(x, y)$ is lower semicontinuous and for each fixed $y \in Y$, $x \mapsto F(x, y)$ is lower semicontinuous. Then we have

$$\emptyset \neq \mathrm{Max}_W F(x, Y) \subset \mathrm{Min}_W \bigcup_{u \in X} \mathrm{Max}_W F(u, Y) + C^0,$$

for all $x \in X$, and

$$\emptyset \neq \mathrm{Min}_W F(X, y) \subset \mathrm{Max}_W \bigcup_{v \in Y} \mathrm{Min}_W F(X, v) - C^0$$

for all $y \in Y$.

Proof. By Lemma 7.7.4, for each $x \in X$ and $y \in Y$, we have

$$\mathrm{Max}_W F(x, Y) \neq \emptyset \text{ and } \mathrm{Min}_W F(X, y) \neq \emptyset.$$

By Lemma 7.5.17, it follows that the set-valued mapping $y \longmapsto F(X, y)$ is continuous. In order to prove that the set-valued mapping $y \longmapsto \mathrm{Min}_W F(X, y)$ is upper semicontinuous, it suffices to show that the mapping $y \mapsto \mathrm{Min}_W F(X, y)$ is closed. In fact, let $\{y_\alpha\}$ be a net in Y such that $y_\alpha \to y \in Y$, and $\{z_\alpha\}$ a net in Z such that $z_\alpha \in \mathrm{Min}F(X, y_\alpha)$ and $z_\alpha \to z \in Z$. Since $z_\alpha \in \mathrm{Min}_W F(X, y_\alpha)$, we have $z_\alpha \in F(X, y_\alpha)$. By the assumptions that X is compact and $F(X, \cdot)$ is continuous with nonempty compact values, it follows that F has nonempty closed values and the graph of $F(X, \cdot)$ is closed, so that $z \in F(X, y)$. Now we prove that $z \in \mathrm{Min}F(X, y)$. Suppose the contrary, i.e., $z \notin \mathrm{Min}_W F(X, y)$, then there exists $z^* \in F(X, y)$ such that

$$z^* - z \in -intC.$$

Since $y_\alpha \to y$, $z^* \in F(X, y)$ and the set-valued mapping $y \longmapsto F(X, y)$ is lower semicontinuous, there exists $z^*_\alpha \in F(X, y_\alpha)$ such that $z^*_\alpha \to z^*$. Therefore from (4.3) we have

$$z^*_\alpha - z_\alpha \in -intC$$

for sufficiently large α. i.e., $z_\alpha \notin \mathrm{Min}_W F(X, y_\alpha)$ for sufficiently large α. This contradicts that $z_\alpha \in \mathrm{Min}_W F(X, y_\alpha)$ for all α. Hence $z \in \mathrm{Min}_W F(X, y)$, i.e., the set-valued mapping $y \longmapsto \mathrm{Min}_W F(X, y)$ is closed, which implies that $y \mapsto \mathrm{Min}_W F(X, y)$ is upper semicontinuous. as Y is compact. Hence the set $\cup_{y \in Y} \mathrm{Min}_W F(X, y)$ is compact. Now by Lemma 7.7.4(2) again, we have:

$$\bigcup_{y \in Y} \mathrm{Min}_W F(X, y) \subset \mathrm{Max}_W\{\bigcup_{v \in Y} \mathrm{Min}_W F(X, v)\} - C^0,$$

i.e.,

$$\mathrm{Min}_W F(X, y) \subset \mathrm{Max}_W\{\bigcup_{v \in Y} \mathrm{Min}_W F(X, v)\} - C^0$$

for all $y \in Y$. By the same way, we can prove that (4.1) is true. This completes the proof.

Example 7.5.1 also shows that Theorem 7.7.5 is indeed a generalization of Theorem 2.1 in Tanaka [523].

Theorem 7.7.6. Let X and Y be two compact Hausdorff topological spaces, Z be a Hausdorff topological vector space and C be a closed pointed convex cone such that $intC \neq \emptyset$ and C is correct. Suppose $F : X \times Y \to 2^Z$ is upper semicontinuous with nonempty compact-valued; and for each fixed $x \in X$,

$y \mapsto F(x,y)$ is lower semicontinuous and for each fixed $y \in Y$, $x \mapsto F(x,y)$ is lower semicontinuous. Then we have

$$\emptyset \neq \mathrm{Max}_W F(x, Y) \subset \mathrm{Min} \bigcup_{u \in X} \mathrm{Max}_W F(u, Y) + C,$$

for all $x \in X$, and

$$\emptyset \neq \mathrm{Min}_W F(X, y) \subset \mathrm{Max} \bigcup_{v \in Y} \mathrm{Min}_W F(X, v) - C$$

for all $y \in Y$.

Proof. By the proof of Theorem 7.7.5, we first have $\mathrm{Max}_W F(x,Y) \neq \emptyset$, $\mathrm{Min}_W F(X,y) \neq \emptyset$ and $\cup_{x \in X} \mathrm{Max}_W F(x,Y)$ and $\cup_{y \in Y} \mathrm{Min}_W F(X,y)$ are compact. Then by Lemma 7.7.4(1), it follows that

$$\emptyset \neq \mathrm{Max}_W F(x,Y) \subset \mathrm{Min} \cup_{u \in X} \mathrm{Max}_W F(u,Y) + C$$

and

$$\emptyset \neq \mathrm{Max}_W F(x,Y) \subset \mathrm{Max} \cup_{v \in Y} \mathrm{Max}_W F(X,v) - C$$

for each $(x,y) \in X \times Y$. Thus we complete the proof. □

As applications of Theorem 7.7.5 and Theorem 7.7.6, we have the following vector minimax inequalities.

Theorem 7.7.7. Let $(X, \{\Gamma_A\})$ and $(Y, \{\Gamma_B\})$ be two compact Hausdorff H-spaces, Z be a Hausdorff topological vector spaceand C be a convex cone of Z such that $intC \neq \emptyset$. Suppose that $F : X \times Y \to 2^Z$ is a set-valued mapping and $\xi_1, \xi_2 : Z \to \mathbb{R}$ are strictly increasing continuous mappings such that:

(i) Suppose $F : X \times Y \to 2^Z$ is upper semicontinuous with nonempty compact-valued; and for each fixed $x \in X$, $y \mapsto F(x,y)$ is lower semicontinuous and for each fixed $y \in Y$, $x \mapsto F(x,y)$ is lower semicontinuous; and

(ii) the mapping $x \longmapsto \xi F(x, y)$ is H-quasiconvex and the set-valued mapping $y \longmapsto \xi F(x, y)$ is H-quasiconcave.

Then there exists $(\overline{x}, \overline{y}) \in X \times Y$ such that

$$F(\overline{x}, \overline{y}) \cap \{(\mathrm{Min}_W \bigcup_{x \in X} \mathrm{Max}_W F(x, Y)) + C^0\} \neq \emptyset$$

and

$$F(\overline{x}, \overline{y}) \cap \{(\mathrm{Max}_W \bigcup_{y \in Y} \mathrm{Min}_W F(X, y)) - C^0\} \neq \emptyset.$$

Moreover, if F is a single-valued mapping, then there exists

$$z_1 \in \mathrm{Min}_W \bigcup_{x \in X} \mathrm{Max}_W F(x, Y)$$

and

$$z_2 \in \mathrm{Max}_W \bigcup_{y\in Y} \mathrm{Min}_W F(X, y)$$

such that $z_2 > z_1$.

Proof. By Theorem 7.5.21, F has a loose saddle point $(\overline{x}, \overline{y}) \in X \times Y$, so that $F(\overline{x},\overline{y}) \cap \mathrm{Min}_W F(X,\overline{y}) \neq \emptyset$ and $F(\overline{x},\overline{y}) \cap \mathrm{Max}_W F(\overline{x}, Y) \neq \emptyset$. By Theorem 7.7.5, it follows that

$$F(\overline{x}, \overline{y}) \cap \{(\mathrm{Min}_W \bigcup_{x\in X} \mathrm{Max}_W F(x, Y)) + C^0\} \neq \emptyset$$

and

$$F(\overline{x}, \overline{y}) \cap \{(\mathrm{Max}_W \bigcup_{y\in Y} \mathrm{Min}_W F(X, y)) - C^0\} \neq \emptyset.$$

If F is single-valued, it follows that

$$F(\overline{x}, \overline{y}) \in \{\mathrm{Min}_W \bigcup_{x\in X} \mathrm{Max}_W F(x, Y) + C^0\} \cap \{\mathrm{Max}_W \bigcup_{y\in Y} \mathrm{Min}_W F(X, y) - C^0\}.$$

Therefore there exists

$$z_1 \in \mathrm{Min}_W \bigcup_{x\in X} \mathrm{Max}_W F(x, Y),$$

$$z_2 \in \mathrm{Max}_W \bigcup_{y\in Y} \mathrm{Min}_W F(X, y),$$

such that

$$F(\overline{x}, \overline{y}) \in z_1 + C^0 \text{ and } F(\overline{x}, \overline{y}) \in z_2 - C^0,$$

i.e., $F(\overline{x}, \overline{y}) - z_1 > 0$, and $z_2 - F(\overline{x}, \overline{y}) > 0$. Therefore we have

$$z_2 - z_1 > 0, \text{ i.e., } z_2 > z_1.$$

This completes the proof.

Theorem 7.7.8. Let X and Y be two nonempty compact and convex Hausdorff sets in two topological vector spaces, Z be a Hausdorff topological vector space and C be a convex cone of Z such that $intC \neq \emptyset$. Suppose that $F : X \times Y \to 2^Z$ is a set-valued mapping such that:

(i) the mapping $(x, y) \mapsto F(x, y)$ is upper semicontinuous with nonempty compact values and for each fixed $x \in X$, $y \mapsto F(x, y)$ is lower semicontinuous; and for each fixed $y \in Y$, $x \mapsto F(x, y)$ is lower semicontinuous; and

(ii) the mapping $x \longmapsto F(x, y)$ is convex and the mapping $y \longmapsto F(x, y)$ is concave.

Then there exists $(\overline{x}, \overline{y}) \in X \times Y$ such that

$$F(\overline{x}, \overline{y}) \cap \{(\mathrm{Min}_W \bigcup_{x\in X} \mathrm{Max}_W F(x, Y)) + C^0\} \neq \emptyset$$

and

$$F(\overline{x}, \overline{y}) \cap \{(\mathrm{Max}_W \bigcup_{y\in Y} \mathrm{Min}_W F(X, y)) - C^0\} \neq \emptyset.$$

Proof. By Theorem 7.5.24, F has a loose saddle point $(\overline{x}, \overline{y}) \in X \times Y$, so that $F(\overline{x}, \overline{y}) \cap \mathrm{Min}_W F(X, \overline{y}) \neq \emptyset$ and $F(\overline{x}, \overline{y}) \cap \mathrm{Max}_W F(\overline{x}, Y) \neq \emptyset$. By Theorem 7.7.5, it follows that

$$F(\overline{x}, \overline{y}) \cap \{(\mathrm{Min}_W \bigcup_{x\in X} \mathrm{Max}_W F(x, Y)) + C^0\} \neq \emptyset$$

and

$$F(\overline{x}, \overline{y}) \cap \{(\mathrm{Max}_W \bigcup_{y\in Y} \mathrm{Min}_W F(X, y)) - C^0\} \neq \emptyset$$

thus we complete the proof. □

Theorem 7.7.9. Let X and Y be two nonempty compact and convex sets in two Hausdorff topological vector spaces, Z be a Hausdorff topological vector space and C be a convex cone of Z such that $intC \neq \emptyset$. Suppose that $f : X \times Y \to Z$ is a single-valued mapping such that:

(i) the mapping $(x, y) \mapsto f(x, y)$ is continuous; and

(ii) the mapping $x \longmapsto f(x, y)$ is natural quasi-C-convex and the mapping $y \mapsto f(x, y)$ is natural quasi-C-concave.

Then there exists a weak C-saddle point (x^*, y^*) of f in $X \times Y$, and $z_1 \in \mathrm{Max}_W\mathrm{Min}_W f$ and $z_2 \in \mathrm{Min}_W\mathrm{Max}_W f$ such that $f(x^*, y^*) - z_1 \in -C^0$ and $f(x^*, y^*) - z_2 \in C^0$, i.e., we have

$$z_1 > f(x^*, y^*) > z_2.$$

Proof. By Theorem 7.5.25, f has a weak C-saddle point $(x^*, y^*) \in X \times Y$. By Theorem 7.7.5, we have

$$\mathrm{Max}_W f(x^*, Y) \subset \mathrm{Min}_W \bigcup_{u\in X} \mathrm{Max}_W f(u, Y) + C^0,$$

$$\mathrm{Min}_W f(X, y^*) \subset \mathrm{Max}_W \bigcup_{v\in Y} \mathrm{Min}_W f(X, v) - C^0.$$

Therefore there exists $X \times Y$ and $z_1 \in \mathrm{Max}_W\mathrm{Min}_W f$ and $z_2 \in \mathrm{Min}_W\mathrm{Max}_W f$ such that $f(x^*, y^*) - z_1 \in -C^0$ and $f(x^*, y^*) - z_2 \in C^0$, i.e., we have

$$z_1 > f(x^*, y^*) > z_2.$$

Thus we complete the proof. □

As an special case of Theorem 7.7.9, we have the following

Corollary 7.7.10. Let X and Y be two nonempty compact and convex sets in two Hausdorff topological vector spaces, Z be a Hausdorff topological vector space and C be a convex cone of Z such that $intC \neq \emptyset$. Suppose that $f : X \times Y \to Z$ is a single-valued mapping such that:

(i) the mapping $(x, y) \mapsto f(x, y)$ is continuous; and

(ii) the mapping $x \longmapsto f(x, y)$ is C-convex and the mapping $y \mapsto f(x, y)$ is C-concave.

Then there exists a weak C-saddle point (x^*, y^*) of f in $X \times Y$ and $z_1 \in \text{Max}_W\text{Min}_W f$ and $z_2 \in \text{Min}_W\text{Max}_W f$ such that $f(x^*, y^*) - z_1 \in -C$ and $f(x^*, y^*) - z_2 \in C$, i.e., we have

$$z_1 > f(x^*, y^*) > z_2.$$

If the solid-pointed convex cone C is assumed to be correct, as applications of Theorems 7.5.21, 7.5.24, 7.5.25 and Theorem 7.7.6, we have the following vector minimax inequalities for set-valued and single-valued mappings by the same proofs used in Theorems 7.7.7-7.7.9; thus we omit their proofs here.

Theorem 7.7.11. Let $(X, \{\Gamma_A\})$ and $(Y, \{\Gamma_B\})$ be two compact Hausdorff H-spaces, Z be a Hausdorff topological vector spaceand C be a solid-pointed convex and correct cone of Z such that $intC \neq \emptyset$. Suppose that $F : X \times Y \to 2^Z$ is a set-valued mapping and $\xi_1, \xi_2 : Z \to \mathbb{R}$ are strictly increasing continuous mappings such that:

(i) Suppose $F : X \times Y \to 2^Z$ is upper semicontinuous with nonempty compact-valued; and for each fixed $x \in X$, $y \mapsto F(x, y)$ is lower semicontinuous and for each fixed $y \in Y$, $x \mapsto F(x, y)$ is lower semicontinuous; and

(ii) the mapping $x \longmapsto \xi F(x, y)$ is H-quasiconvex and the set-valued mapping $y \longmapsto \xi F(x, y)$ is H-quasiconcave.

Then there exists $(\overline{x}, \overline{y}) \in X \times Y$ such that

$$F(\overline{x}, \overline{y}) \cap \{(\text{Min} \bigcup_{x \in X} \text{Max}_W F(x, Y)) + C\} \neq \emptyset$$

and

$$F(\overline{x}, \overline{y}) \cap \{(\text{Max} \bigcup_{y \in Y} \text{Min}_W F(X, y)) - C\} \neq \emptyset.$$

Moreover, if F is a single-valued mapping, then there exists $z_1 \in \text{Min} \bigcup_{x \in X} \text{Max}_W F(x, Y)$ and $z_2 \in \text{Max} \bigcup_{y \in Y} \text{Min}_W F(X, y)$ such that

$$z_2 \geq F(x^*, y^*) \geq z_1.$$

Proof. The conclusion follows by the same arguement as shown in the proof of Theorem 7.7.7 with the aid of Theorems 7.5.21 and 7.7.6. □

Theorem 7.7.12. Let X and Y be two nonempty compact and convex sets in two Hausdorff topological vector spaces, Z be a Hausdorff topological vector space and C be a solid-pointed convex and correct cone of Z such that $intC \neq \emptyset$. Suppose that $F : X \times Y \to 2^Z$ is a set-valued mapping such that:

(i) the mapping $(x, y) \mapsto F(x, y)$ is upper semicontinuous with nonempty compact values and for each fixed $x \in X$, $y \mapsto F(x, y)$ is lower semicontinuous; and for each fixed $y \in Y$, $x \mapsto F(x, y)$ is lower semicontinuous; and

(ii) the mapping $x \longmapsto F(x, y)$ is convex and the mapping $y \longmapsto F(x, y)$ is concave.

Then there exists $(\overline{x}, \overline{y}) \in X \times Y$ such that

$$F(\overline{x}, \overline{y}) \cap \{(\mathrm{Min} \bigcup_{x \in X} \mathrm{Max}_W F(x, Y)) + C\} \neq \emptyset$$

and

$$F(\overline{x}, \overline{y}) \cap \{(\mathrm{Max} \bigcup_{y \in Y} \mathrm{Min}_W F(X, y)) - C\} \neq \emptyset.$$

Proof. The conclusion follows from Theorems 7.5.24 and 7.7.6. □

Theorem 7.7.13. Let X and Y be two nonempty compact and convex sets in two Hausdorff topological vector spaces, Z be a Hausdorff topological vector space and C be a solid-pointed convex and correct cone of Z such that $intC \neq \emptyset$. Suppose that $f : X \times Y \to Z$ is a single-valued mapping such that:

(i) the mapping $(x, y) \mapsto f(x, y)$ is continuous; and

(ii) the mapping $x \longmapsto f(x, y)$ is natural quasi-C-convex and the mapping $y \mapsto f(x, y)$ is natural quasi-C-concave.

Then there exists a C-saddle point (x^*, y^*) of f in $X \times Y$, and $z_1 \in \mathrm{MaxMin}_W f$ and $z_2 \in \mathrm{MinMax}_W f$ such that $f(x^*, y^*) - z_1 \in -C$ and $f(x^*, y^*) - z_2 \in C$, i.e., we have

$$z_1 \geq f(x^*, y^*) \geq z_2.$$

Proof. From Theorems 7.5.25 and 7.7.6, the conclusion follows. □

As a special case of Theorem 7.7.13, we have the followingresult.

Corollary 7.7.14. Let X and Y be two nonempty compact and convex sets in two topological Hausdorff vector spaces, Z be a Hausdorff topological vector space and C be a solid-pointed convex and correct cone of Z such that $intC \neq \emptyset$. Suppose that $f : X \times Y \to Z$ is a single-valued mapping such that:

(i) the mapping $(x, y) \mapsto f(x, y)$ is continuous; and

(ii) the mapping $x \longmapsto f(x, y)$ is C-convex and the mapping $y \mapsto f(x, y)$ is C-concave.

Then there exists a point (x^*, y^*) of f in $X \times Y$ and $z_1 \in \text{MaxMin}_W f$ and $z_2 \in \text{MinMax}_W f$ such that $f(x^*, y^*) - z_1 \in -C$ and $f(x^*, y^*) - z_2 \in C$, i.e., we have

$$z_1 \geq f(x^*, y^*) \geq z_2.$$

Remark 7.7.15. Before we conclude this section, we note that Theorem 7.7.11 includes Theorem 4.1 and theorem 4.2 of Tanaka [524] as special cases. Theorem 7.7.13 also shows that Theorem 4.3 of Tanaka [524] (see also Theorem 5.2 of Tanaka [523], Theorem 3.1 of Tanaka [525]) still holds in topological vector spaces instead of locally convex spaces.

Chapter 8

Applications to Game Theory

8.1 Introduction

In this chapter, as applications of KKM theory in Chapter 2, we shall study multi-objective games by the Ky Fan minimax inequality method and fixed point approach. Then existence of Nash-equilibria of nonoperative N-person games have been established. As applications, we obtain the generic stability for three different kinds of models for nonoperative games. The stability results indicate that almost all nonoperative games are Pareto optima or, in other words, the solution of almost all games is essential in the sense of the Baire category.

Now we introduce some notes and definitions. Given the set $N = \{1, \cdots, n\}$ of players, for each player $i \in N$, denote by X_i the set of all his (resp., her) strategies and $f : X = \prod_{i=1}^{n} X_i \to R$ the payoffs for him (her). For each $i \in N$, denote $X_{\hat{i}} = \prod_{j \in N \setminus \{i\}} X_j$. A Nash equilibrium of a game is an n-tuple of strategies $x^* = \{x_1^*, cdots, x_n^*\} \in X$ such that for each $i \in N$,

$$f_i(x_i^*, x_{\hat{i}}^*) = \max_{u_i \in X_i} f_i(u_i, x_{\hat{i}}^*).$$

8.2 Weight Nash Equilibria of Multi-Objective Games

Recently, the study of the existence of Pareto equilibria in game theory with vector payoffs has attracted many authors, for example, see Bergstresser and Yu [41], Chose and Prasad [91], Szidarovszky et al. [500], Tanaka [523], Wang [571], Yu [601], Zeleny [615] and references therein. The motivation for the study of multicriteria models can be found in Bergstresser and Yu in [41], Szidarovszky et al. [500] and Zeleny [615]. As the existence of Pareto equilibria is one of the fundamental problem in the game theory. It is our purpose in this chaper to present some existence results of Pareto equilibria by both fixed point and the the Ky Fan minimax inequality approach. As the existence problem of Pareto equilibria can be formulated into the model in which solutions correspond to either fixed points of some kind of set-valued mappings or the Ky Fan type minimax inequality, thus both the fixed point method and the Ky Fan minimax inequality method provide an unified study for the existence of Pareto equilibria problems in multi-objective games under weak hypotheses. Our results in this section could be regarded as an unified improvement of the corresponding existence results of Pareto equilibria given in the literature.

Now we recall and introduce some notations and definitions. In this section, we shall consider a finite-players game with multicriterica in its strategic form (also called a normal from) $G := (X_i, F^i)_{i\in N}$, where $N := \{1, 2, \cdots, n\}$. For each $i \in N$, X_i is the set of strategies in $\mathbb{R}^{k_i}$ for the player i; and each F^i is a mapping from $X := \Pi_{i\in N} X_i$ into $\mathbb{R}^{k_i}$, which is called the payoff function (or say, multicriteria) of the i's player,here k_i is a positive integer. If a choose $x := (x_1, x_2, \cdots, x_n) \in X$ is played, each player i is trying to get his/her payoff function $F^i := (f_1^i(x), f_2^i(x), \cdots, f_{k_i}^i(x))$, which consists of non-commensurable outcomes. Each player i has a preference '$\succeq$' over the outcome space $\mathbb{R}^{k_i}$. For each player $i \in N$, its preference '$\succeq$' is given as follows:

$$z^1 \succeq_i z^2 \text{ if and only if } z_j^1 \geq z_j^2$$

for each $j = 1, 2, \cdots, k_i$, where $z^1 := (z_1^1, z_2^1, \cdots, z_{k_i}^1)$ and $z^2 = (z_1^2, z_2^2, \cdots, z_{k_i}^2)$ are any elements in $\mathbb{R}^{k_i}$. The players' preference relations induce the preferences on X, defined for each player i; and the chooses $x = (x_1, \cdots, x_n)$ and $y = (y_1, \cdots, y_n) \in X$ by

$$x \succeq_i y \text{ whenever } F^i(x) \succeq_i F^i(y).$$

Also we assume that that the model of a game in this section is a noncooperative game, i.e., there is no replay communicating between players and so players act as free agents, each player is trying to maximize his/her own payoff according to his/her preferences.

For the games with vector payoff functions (or say, multicriteria), as it is well known that in general there does not exists a strategy $\hat{x} \in X$ to minimize (or equivalently to say, maximize) all f_j^is for each player, for example, see the reference of Yu [602]. Hence we need to recall some solution concepts for multicriteria games. Throughout this section, for each given $m \in \mathbb{N}$, we shall denote by $\mathbb{R}_+^m$ the nonnegative orthant of $\mathbb{R}^m$, i.e.

$$\mathbb{R}_+^m := \{u := (u_1, \cdots, u_m) \in \mathbb{R}^m \text{ such that } u_j \geq 0 \text{ for } j := 1, \cdots, m\},$$

so that the nonnegative orthant $\mathbb{R}_+^m$ of $\mathbb{R}^m$ has a nonempty interior with the topology induced in terms of convergence of vectors with respect to the Euclidean metric. That is,

$$int\mathbb{R}_+^m := \{u := (u_1, \cdots, u_m) \in \mathbb{R}^m : u_j > 0 \text{ for all } j = 1, \cdots, m\}.$$

For each $i \in N$, denote $X_{\hat{i}} := \Pi_{j \in N \setminus \{i\}} X_j$. If $x = (x_1, \cdots, x_n) \in X$, we shall write $x_{\hat{i}} = (x_1, \cdots, x_{i-1}, x_{i+1}, \cdots x_n) \in X_{\hat{i}}$. If $x_i \in X_i$ and $x_{\hat{i}} \in X_{\hat{i}}$, we shall use $(x_{\hat{i}}, x_i)$ to denote $y = (y_1, \cdots, y_n) \in X$ such that $y_i = x_i$ and $y_{\hat{i}} = x_{\hat{i}}$. Let $\hat{x} := (\hat{x}_1, \cdots, \hat{x}_n) \in X$. Now we have the following definition.

Definition 8.2.1. A strategy $\hat{x}_i \in X_i$ of player i is said to be a *Pareto efficient strategy* (resp, a *weak Pareto efficient strategy*) with respect to $\hat{x}$ if there is no strategy $x_i \in X_i$ such that

$$F^i(\hat{x}) - F^i(\hat{x}_{\hat{i}}, x_i) \in \mathbb{R}_+^{k_i} \setminus \{0\} \text{ (resp, } F^i(\hat{x}) - F^i(\hat{x}_{\hat{i}}, x_i) \in int\mathbb{R}_+^{k_i} \text{).}$$

Definition 8.2.2. A strategy $\hat{x} \in X$ is said to be a *Pareto equilibrium* of a game $G := (X_i, F^i)_{i \in N}$ if, for each player i, $\hat{x}_i \in X_i$ is a Pareto efficient strategy against $\hat{x}$ (resp, a weak Pareto efficient strategy against $\hat{x}$).

¿From the above definition, it is clear that each Pareto equilibria is a weak Pareto equilibrium, but the converse is not always true. We also need the following definition which was first given by Wang in [571].

Definition 8.2.3. A strategy $\hat{x} \in X$ is said to be a *weight Nash-equilibrium* with respect to weight vector $W := (W^1, \cdots, W^n)$ of a game $G = (X_i, F^i)_{i \in N}$ if for each player $i \in N$, we have that

(1) $W^i \in \mathbb{R}_+^{k_i} \setminus \{0\}$; and

(2) $W^i \cdot F^i(\hat{x}) \leq W^i \cdot F^i(\hat{x}_{\hat{i}}, x_i)$ for each $x_i \in X_i$, where ' $\cdot$ ' denotes the *inner production.*

Remark 8.2.4. In particular, when $W^i \in T_+^{k_i}$ for all $i \in N$, the strategy $\hat{x} \in X$ is said to be a *normalized weight Nash-equilibrium* with respect to W, where $T_+^{k_i}$ is a simplex of $\mathbb{R}^{k_i}$, i.e.,

$$T_+^{k_i} := \{u := (u_1, \cdots, u_{k_i}) \in \mathbb{R}_+^{k_i} : \text{ such that } \Sigma_{j=1}^{k_i} u_j = 1\}.$$

Let $W^i \in \mathbb{R}^{k_i}_+ \setminus \{0\}$ be fixed for each $i \in N$. ¿From the definition above, it is not difficult to verify that a strategy $\hat{x} \in X$ is a weight Nash-equilibrium with respect to the weight vector $W = (W^1, \cdots, W^n)$ of a game $(X_i, F^i)_{i \in N}$ if and only if for each $i \in N$, $\hat{x}_i$ is an optimal solution of the following optimization problem:

$$Minimize_{x_i \in X_i} W^i \cdot F^i(\hat{x}_{\hat{i}}, x_i).$$

Now it is our time to study the existence of weight Nash-equilibria. In order to study the existence of Pareto equilibria as applications of either fixed point theorems or the Ky Fan minimax inequality in [164], we shall first study the existence of weight Nash-equilibria as Pareto equilibria problems can be reduced to the study of weight Nash-equilibria under certain conditions. Now we discuss the existence of weight Nash-equilibria as applications of fixed points.

8.2.1 Fixed Point Approach

For each $i = 1, 2, \cdots, n$, $W^i \in \mathbb{R}^{k_i}_+ \setminus \{0\}$ and $x = (x_1, \cdots, x_n)$, $y = (y_1, \cdots, y_n) \in X$. Define two mappings $S^W : X \times X \to \mathbb{R}$ and $M^W : X \to 2^X$ by

$$S^W(x, y) := \Sigma_{i=1}^n W^i \cdot F^i(x_{\hat{i}}, y_i)$$

and

$$M^W(x) := \{y^* \in X : S^W(x, y^*) = \min_{y \in X} S^W(x, y)\}$$

for each $(x, y) \in X \times X$. Then following simple fact which is Lemma 2.3 of Wang [571] shows that the existence of weight Nash-equilibria is equivalently to the existence of fixed points for the mapping M^W defined above, so that it enables us to study the weight Nash-equilibria by fixed point theorems (e.g., see Lemma 2.3 in [571]) and thus we will not give its proof here.

Lemma 8.2.5. Let $G = (X_i, F^i)_{i \in N}$ be a given multi-objective game. Suppose $W^i \in \mathbb{R}^{k_i}_+ \setminus \{0\}$ for each $i \in N$. A strategy $\hat{x} \in X$ is a weight Nash-equilibrium with the weight $W = (W^1, \cdots, W^n)$ of the game G is and only if $\hat{x}$ is a fixed point of the mapping M^W defined above.

For convenience of our discussion in this section, we now recall following two fixed point theorems; mainly, Fan-Glicksberg fixed point theorem (see Fan [156] or Glicksberg [196]) and the Browder-Fan fixed point theorem by Fan [157] and Browder [59] and we state them as follows.

Fan-Glicksberg Fixed Point Theorem. Let X be a nonempty compact and convex subset of a Hausdorff locally convex topological vector space E and $F : X \to 2^X$ is an upper semicontinuous set-valued mapping with nonempty closed and convex values. Then F has at least one fixed point, i.e., there exists $\hat{x} \in X$ such that $\hat{x} \in F(\hat{x})$.

Browder-Fan Fixed Point Theorem. Let X be a nonempty compact and convex subset of a Hausdorff topological vector space E. Suppose a set-valued mapping $F : X \to 2^X$ has the following properties:

(1) $F(x)$ is nonempty and convex for each $x \in X$; and

(2) F has open inverse valued, i.e., $F^{-1}(y) = \{x \in X : y \in F(x)\}$ for each $y \in X$ is open in X.

Then F has at least one fixed point.

We would like to note that the present form of Fan-browder fixed point theorem was first given by Browder [59]. However, the equivalent geometric form of the Browder-Fan fixed point theorem was also presented by Fan [164] as an application of the classical KKM principle [319]. Thus this fixed point is now shared by both of their names.

As an application of Fan-Glicksberg fixed point theorem, we first have the following:

Theorem 8.2.6. Let $G = (X_i, F^i)_{i\in N}$ be a given multi-objective game. For each $i \in N$, X_i is a nonempty compact and convex subset of a locally convex Hausdorff topological vector space E_i. If there is a weight vector $W := (W^1, \cdots, W^n)$ with $W^i \in \mathbb{R}^{k_i}_+ \setminus \{0\}$ such that the function S^W is jointly continuous on $X \times X$ and $S^W(x, \cdot)$ is quasi-convex on X for each fixed $x \in X$, then the game G has at least one weight Nash-equilibrium with respect to the weight vector $W = (W^1, \cdots, W^n)$.

Proof. In order to apply the Fan-Glicksberg fixed point theorem, it suffices to show that the mapping M^W defined above is upper semicontinuous with nonempty convex and closed values. Note that X is compact, $S^W : X \times X \to \mathbb{R}$ is jointly continuous, and $S^W(x, \cdot)$ is quasi-convex for each fixed $x \in X$. Thus the set $M^W(x)$ is nonempty, closed and convex for each fixed $x \in X$. By Proposition 23 of Aubin and Ekeland [15], it follows that the mapping $M^W : X \to 2^X$ defined by

$$M^W(x) := \{y^* \in X : S^W(x, y^*) = \min_{y\in X} S^W(x, y)\}$$

for each $x \in X$ is upper semicontinuous. Now by Fan-Glicksberg fixed point theorem, the mapping M^W has a fixed point $\hat{x} \in X$. ¿From Lemma 8.2.5, it follows that $\hat{x}$ is a weight Nash-equilibrium of the game $(X_i, F^i)_{i\in N}$ with respect to weight vector W, thus we complete the proof. □

Theorem 8.2.6 shows that the corresponding Theorem 3.1 of Wang [571] for the multi-objective game $G = (X_i, F^i)$ still holds when underlying space E_i is a locally convex Hausdorff topological vector space instead of being a normed space. Indeed, as an application of Theorem 8.2.B, we shall see that the hypotheses of Theorem 8.2.6 can be further weaken to guarantee the existence of weight Nash-equilibria even underlying space E_i is only a Hausdorff topological vector space without convex structure. Before we present this general result, we need the following fact.

Lemma 8.2.7. Let X and Y be two Hausdorff topological spaces and X be compact. Suppose $f : X \times Y \to \mathbb{R}$ is a real-valued function such that

(1) f is upper semicontinuous on $X \times Y$; and

(2) for each fixed $x \in X$, the mapping $y \mapsto f(x, y)$ is lower semicontinuous on Y.

Then the function $\psi : Y \to \mathbb{R}$ defined by $\psi(y) = \max_{u \in X} f(u, y)$ is continuous on Y.

Proof. As X is compact, ψ is upper semicontinuous on Y by (1) and Theorem 1 of Aubin [11]. By condition (2), ψ is also lower semicontinuous on Y. □

We now have the following existence theorem for the weight Nash-equilibria.

Theorem 8.2.8. Let $G = (X_i, F^i)_{i \in N}$ be a given multi-objective game. For each $i \in N$, X_i is a nonempty compact and convex subset of a Hausdorff topological vector space E_i. If there is a weight vector $W = (W^1, \cdots, W^n)$ with $W^i \in \mathbb{R}_+^{k_i} \setminus \{0\}$ such that the following are satisfied for each $i \in N$:

(1) the function $(x, y) \mapsto W^i \cdot F^i(x_{\hat{i}}, y_i)$ is jointly lower semicontinuous on $X \times X$;

(2) for each fixed $y \in X$, the mapping $x \mapsto W^i \cdot F^i(x_{\hat{i}}, y_i)$ is upper semicontinuous on X; and

(3) for each fixed $x \in X$, $W^i \cdot F^i(x_{\hat{i}}, y_i)$ is quasiconvex on X.

Then G has at least one weight Nash-equilibrium with respect to the weight vector W.

Proof. For each $k \in \mathbb{N}$, define a set-valued mapping $A_k : X \to 2^X$ by

$$A_k(x) := \Pi_{i=1}^n \{y_i \in X_i : W^i \cdot F^i(x_{\hat{i}}, y_i) < \min_{u_i \in X_i} W^i \cdot F^i(x^{\hat{i}}, u_i) + \frac{1}{k}\}$$

for each $x \in X$. Then $A_k(x)$ is nonempty and convex by the condition (3) for each $x \in X$. Note that $A_k^{-1}(y) := \cap_{i=1}^n \{x \in X : W^i \cdot F^i(x_{\hat{i}}, y_i) < \min_{u_i \in X_i} W^i \cdot F^i(x_{\hat{i}}, u_i) + \frac{1}{k}\}$, which is open in X by conditions (1) and (2) for each $y \in X$. Thus the Browder-Fan fixed point implies that the mapping A_k has a fixed point $x(k)$. ¿From the definition of A_k, it follows that,

$$W^i \cdot F^i(x_{\hat{i}}(k), x_i(k)) < \min_{u_i \in X_i} W^i \cdot F^i(x_{\hat{i}}(k), u_i) + \frac{1}{k}.$$

Without loss of the generality, we may assume that $x(k) \to \hat{x} \in X$ as X is compact. By condition (1) and Lemma 8.2.7, we have that

$$\begin{aligned} W^i \cdot F^i(\hat{x}_{\hat{i}},\ \hat{x}_i) &\leq \inf_{k\to\infty} W^i \cdot F^i(x_{\hat{i}}(k), x_i(k)) \\ &\leq \inf_{k\to\infty} \min_{u_i \in X_i} W^i \cdot F^i(x_{\hat{i}}(k), u_i) \\ &= \min_{u_i \in X_i} W^i \cdot F^i(\hat{x}_{\hat{i}}, u_i). \end{aligned}$$

Thus $\hat{x}$ is a weight Nash-equilibrium of G with respect to the weight vector W given above, and we complete the proof. □

It is clear that Theorem 8.2.8 includes Theorem 8.2.6 as an special case as the continuity and convexity of the mapping S^W in Theorem 8.2.6 automatically satisfy the corresponding conditions (1), (2) and (3) of Theorem 8.2.8, however the converse does not hold.

So far, we have given two existence results of weight Nash-equilibria as applications of fixed point theorems. In what follows, as an application of the Ky Fan minimax inequality, we shall give another existence result for weight Nash-equilibria which is different from both Theorems 8.2.1 and 8.2.2 above.

8.2.2 Ky Fan Minimax Inequality Method

In order to study the existence of Pareto equilibria as applications Ky Fan minimax inequality in [164], we first re-state the following non-compact version of Ky Fan minimax inequality again which is Theorem 2.5.3:

Ky Fan Type Minimax Principle. Let X be a nonempty convex set in a Hausdorff topological vector space and f a real-valued function defined on $X \times X$. Suppose there exists a non- empty compact and convex subset X_0 of X and a nonempty compact subset K of X such that

(1) for each fixed $x \in X$, $f(x, y)$ is a compactly lower semicontinuous function of y on X;

(2) for each $A \in \mathcal{F}(X)$ and $y \in co(A)$, we have $\min_{x\in A} f(x, y) \leq 0$ (in particular, condition (2) is satisfied if we assume that for each fixed $y \in X$, $f(x, y)$ is a quasi-concave function of x on X and $f(x, x) \leq 0$ for each $x \in X$); and

(3) for each $y \in X \setminus K$, there exists $x \in co(X_0 \cup \{y\})$ such that $f(x, y) > 0$.

Then there exists $y_0 \in K$ such that

$$\sup_{x\in X} f(x, y_0) \leq \sup_{x\in X} f(x, x).$$

Remark 8.2.9. Indeed, the famous Ky Fan minimax inequality above has been generalized in many ways. For more details, interested readers can find other generalizations of Ky Fan minimax inequalities and their applications in Tarafdar [536], Yuan [605] and many other references therein.

As an application of the Ky Fan minimax principe above, we have the following

Theorem 8.2.10. Let $G = (X_i, F^i)_{i\in N}$ be a given multi-objective game, where $F^i = (f_1^i, \cdots, f_{k_i}^i)$ for $i \in N$. For each $i \in N$, suppose X_i is a nonempty convex subset of a Hausdorff topological vector space E_i. Suppose X_i^0 and K_i are a nonempty compact and convex subset of X_i and a nonempty compact (not necessarily convex) subset of X_i, respectively for each $i \in I$. If there is a weighted vector $W = (W^1, \cdots, W^n)$ with $W^i \in \mathbb{R}_+^{k_i} \setminus \{0\}$ such that the followings are satisfied:

(1) for each fixed $y \in X$, the mapping $x \mapsto \Sigma_{i\in N} W^i \cdot F^i(x_{\hat{i}}, y_i)$ is upper semicontinuous on X;

(2) the mapping $(x, y) \mapsto \Sigma_{i\in N} W^i \cdot F^i(x_{\hat{i}}, y_i)$ is jointly lower semicontinuous on $X \times X$;

(3) for each fixed $x \in X$, the mapping $y \mapsto \Sigma_{i\in N} W^i \cdot F^i(x_{\hat{i}}, y_i)$ is quasi-convex; and

(4) for each $x \in X \setminus K$, there exists $y \in co(X_0 \cup \{x\})$ such that

$$\Sigma_{i=1}^n W^i \cdot F^i(x_{\hat{i}}, x_i) > \Sigma_{i=1}^n W^i \cdot F^i(x_{\hat{i}}, y_i)$$

where $X_0 := \Pi_{i=1}^n X_i^0$ and $K := \Pi_{i=1}^n K_i$.

Then G has at least one weighted Nash-equilibria.

Proof. Define a mapping $\phi : X \times X \to \mathbb{R}$ by

$$\phi(x, y) := \Sigma_{i=1}^n W^i \cdot (F^i(x_{\hat{i}}, x_i) - F^i(x_{\hat{i}}, y_i))$$

for each $(x, y) \in X \times X$. Then we have that:

(1) for each fixed $y \in X$, $x \mapsto \phi(x, y)$ is lower semicontinuous;
(2) for each fixed $x \in X$, the mapping $y \mapsto \phi(x, y)$ is quasi-concave;
(3) $\phi(x, x) \leq 0$ for all $x \in X$; and
(4) for each $x \in X \setminus K$, there exists $y \in co(X_0 \cup \{x\})$ such that $\phi(x, y) > 0$.

By the Ky Fan minimax inequality above, i.e., there exists $\hat{x} \in K$ such that for each $y \in X$,

$$\phi(\hat{x}, y) = \Sigma_{i=1}^n W^i \cdot (F^i(\hat{x}_{\hat{i}}, \hat{x}_i) - F^i(\hat{x}_{\hat{i}}, y_i)) \leq 0.$$

Let $y = (\hat{x}_{\hat{i}}, y_i) \in X$ for $i = 1, 2, \cdots, n$, then we have $W^i \cdot (F^i(\hat{x}_{\hat{i}}, \hat{x}_i) - F^i(\hat{x}_{\hat{i}}, y_i)) \leq 0$, which implies that

$$W^i \cdot F^i(\hat{x}_{\hat{i}}, \hat{x}_i) = \min_{u_i \in X_i} W^i \cdot F^i(\hat{x}_{\hat{i}}, u_i).$$

Thus $\hat{x}$ is a weighted Nash-equilibrium point of the game G with respect to the weighted vector W. □

Remark 8.2.11. We note that the both condition (3) of Theorem 8.2.8 and condition (3) of Theorem 8.2.10 are independent to each other, and thus Theorem 8.2.8 and Theorem 8.2.10 could not contain each other as a special case. Moreover, Theorem 8.2.10 does not only improve the corresponding existence results of weight Nash-equilibria for multi-objective games with multicriteria given by Wang [571, Theorem 3.1], but it also extends Theorem 1.11 of Szidarovszky et al. in [500] and Theorem 1 of Borm et al. [52] to Hausdorff topological vector spaces under weaker continuous hypotheses.

In this part, as applications of weight Nash-equilibria, we shall derive some existence of Pareto equilibria for multi-objective games. To this end, we first need the following Lemma which tells us that the existence problem for Pareto equilibria can be reduced to the existence of the weight Nash-equilibria under certain circumstance.

Lemma 8.2.12. Each normalized weight Nash-equilibrium $\hat{x} \in X$ with a weight $W = (W^1, \cdots, W^n) \in T_1^{k_1} \times \cdots \times T_+^{k_n}$ (resp., $W = (W^1, \cdots, W^n) \in intT_+^{k_1} \times \cdots \times intT_+^{k_n}$) is a weak Pareto equilibrium (resp., a Pareto equilibrium) of the game $(X_i, F^i)_{i \in N}$.

Proof. It is Lemma 2.1 of Wang [571, p.376-377]. □

Remark 8.2.13. We should note that the conclusion of above Lemma 3 still hold if $\hat{x}$ is a weight Nash-equilibrium with a weight $W = (W^1, \cdots, W^n)$ satisfying that $W^i \in \mathbb{R}_+^{k_i}$ (resp., $W^i \in int\mathbb{R}_+^{k_i}$) for each $i \in N$. Secondly, we note that a Pareto equilibria is not necessarily a weight Nash-equilibrium.

Now by combining Lemma 8.2.12 and Theorem 8.2.8, we have the following existence of Pareto equilibria for multi-objective games in Hausdorff topological vector spaces.

Theorem 8.2.4. Let $G = (X_i, F^i)_{i \in N}$ be a multi-objective game. For each $i \in N$, X_i is a nonempty compact and convex subset of a Hausdorff topological vector space E_i. If there is a weight combination $W = (W^1, \cdots, W^n)$ with $W^i \in \mathbb{R}_+^{k_i} \setminus \{0\}$ such that the following are satisfied for each $i \in N$:

(1) the function $(x, y) \mapsto W^i \cdot F^i(x_{\hat{i}}, y_i)$ is jointly lower semicontinuous on $X \times X$;

(2) for each fixed $y \in X$, the mapping $x \mapsto W^i \cdot F^i(x_{\hat{i}}, y_i)$ is upper semicontinuous on X; and

(3) for each fixed $x \in X$, $W^i \cdot F^i(x_{\hat{i}}, y_i)$ is quasiconvex on X.

Then G has at least one weak Pareto equilibrium. Furthermore, if $W^i \in intT_+^{k_i}$ for all $i = 1, \cdots, n$, then the game $G = (X_i, F^i)_{i=1}^n$ has at least one Pareto equilibrium.

Proof. By Theorem 8.2.8, G has at least one weight Nash-equilibrium $\hat{x}$ with respect to the weight vector W. Now Lemma 8.2.12 shows that $\hat{x}$ is also a weak Pareto equilibrium of G, and thus the conclusion follows. □

By the same arguement used in the proof of Theorem 8.2.14, as a consequence of Theorem 8.2.10 instead of Theorem 8.2.2, we have the following Pareto equilibrium theorem.

Theorem 8.2.15. Let $G = (X_i, F^i)_{i\in N}$ be a multi-objective game, where $F^i = (f_1^i, \cdots, f_{k_i}^i)$ for $i \in N$. For each $i \in N$, suppose X_i is a nonempty compact and convex subset of a Hausdorff topological vector space E_i. Suppose that for each $i \in N$, X_i^0 and K_i are a nonempty compact and convex subset of X_i and a nonempty compact (not necessarily convex) subset of X_i, respectively. If there is a weighted vector $W = (W^1, \cdots, W^n)$ with $W^i \in \mathbb{R}_+^{k_i} \setminus \{0\}$ such that the following are satisfied:

(1) for each fixed $y \in X$, the mapping $x \mapsto \Sigma_{i\in N} W^i \cdot F^i(x_{\hat{i}}, y_i)$ is upper semicontinuous on X;

(2) the mapping $(x, y) \mapsto \Sigma_{i\in N} W^i \cdot F^i(x_{\hat{i}}, y_i)$ is jointly lower semicontinuous on $X \times X$;

(3) for each fixed $x \in X$, the mapping $y \mapsto \Sigma_{i\in N} W^i \cdot F^i(x_{\hat{i}}, y_i)$ is quasi-convex; and

(4) for each $x \in X \setminus K$, there exists $y \in co(X_0 \cup \{x\})$ such that

$$\Sigma_{i=1}^n W^i \cdot F^i(x_{\hat{i}}, x_i) > \Sigma_{i=1}^n W^i \cdot F^i(x_{\hat{i}}, y_i)$$

where $X_0 := \Pi_{i=1}^n X_i^0$ and $K := \Pi_{i=1}^n K_i$.

Then G has at least one weak Pareto equilibrium in K. In addition, if $W^i \in intT_+^{k_i}$ for each $i = 1, \cdots, n$, then G has at least one Pareto equilibrium.

Proof. By Theorem 8.2.10, G has at least one weighted Nash-equilibrium $\hat{x}$ with respect to the weighted vector W. Now Lemma 8.2.12 shows that $\hat{x}$ is also a weak Pareto equilibrium of G, and thus the conclusion follows. □

As an immediate consequence of Theorem 8.2.15, we have the following

Theorem 8.2.16. Let $G = (X_i, F^i)_{i\in N}$ be a multi-objective game, where $F^i = (f_1^i, \cdots, f_{k_i}^i)$ for $i \in N$. For each $i \in N$, suppose X_i is a nonempty convex subset of a Hausdorff topological vector space E_i, and X_i^0 and K_i are a nonempty compact and convex subset of X_i and a nonempty compact (not necessarily convex) subset of X_i, respectively. Moreover we assume that the following conditions are satisfied for $j = 1, \cdots, k_i$:

(1) the f_j^i is jointly lower semicontinuous on X;

(2) for each fixed $y^i \in X_i$, the function $f_j^i(x_{\hat{i}}, y_i)$ is upper semicontinuous on $X_{\hat{i}}$;

(3) for each fixed $x_{\hat{i}} \in X_{\hat{i}}$, the function $f_j^i(x_{\hat{i}}, \cdot)$ is convex on X_i; and

(4) for each $x \in X \setminus K$, there exists $y \in co(X_0 \cup \{x\})$ such that for all $i = 1, \cdots, n$, we have $f_j^i(x_{\hat{i}}, x_i) > f_j^i(x_{\hat{i}}, y_i)$ where $X_0 := \Pi_{i=1}^n X_i^0$ and $K := \Pi_{i=1}^n K_i$.

Then the multi-objective game G has at least one Pareto equilibrium in K.

Proof. Taking any fixed weighted vector $W = (W^1, \cdots, W^n)$, where $W^i \in intT_+^{k_i}$ for each $i = 1, \cdots, n$. ¿From conditions (1)-(4), it follows that all hypotheses of Theorem 8.2.15 are satisfied. Thus G has at least one weighted Nash-equilibrium $\hat{x}$ with respect to the weighted vector W. As $W^i \in intT_+^{k_i}$ for each $i \in N$, $\hat{x}$ must be a Pareto equilibrium point of G and we complete the proof. □

Before we close this section, we state the following result which is a compact version of Theorem 8.2.16:

Corollary 8.2.17. Let $G = (X_i, F^i)_{i \in N}$ be a multi-objective game. For each $i \in N$, suppose X_i is a nonempty compact and convex subset of a Hausdorff topological vector space E_i and the following conditions are satisfied for $j = 1, \cdots, k_i$:

(1) the f_j^i is jointly lower semicontinuous on X;

(2) for each fixed $y_i \in X_i$, the function $f_j^i(x_{\hat{i}}, y_i)$ is upper semicontinuous on $X_{\hat{i}}$; and

(3) for each fixed $x_{\hat{i}} \in X_{\hat{i}}$, the function $f_j^i(x_{\hat{i}}, \cdot)$ is convex on X_i.

Then the multi-objective game G has at least one Pareto equilibrium.

Remark 8.2.18. In this section, we study the existence of weight Nash-equilibria and Pareto equilibria as applications of nonlinear analysis, mainly, fixed point theorems and the Ky Fan minimax inequalities. It is known that the fixed point method has wide applications in the study of economics, optimization and mathematical programming, for example, see Robinson [447] and other references cited therein. On the other hand, though the Ky Fan minimax inequality plays a very important role in the study of nonlinear analysis, and it is even regarded as one of the most important principles in nonlinear analysis, for instance, see Aubin [11], Aubin and Ekeland [15], Lin and Simons [345] et al. However, to the best knowledge of both authors, not enough attention has been put on the study of optimization, game theory and mathematical programming by using the Ky Fan minimax inequality. It is our purpose in this section to present how the Ky Fan type minimax inequality can be applied to study the existence of weight Nash-equilibria and Pareto equilibria. Hopefully the Ky Fan minimax inequality method will also play an important role in the study of optimization, in particular for the investigation of the existence of Pareto equilibria in the vector-valued optimization problems and multi-objective games. Indeed, some progress in this direction has been archeived in our joint section [549] and [600]. Also we shall consider the numerical aspects of weight Nash-equilibrium and Pareto equilibrium points for multi-objective games in our study by the computation method for fixed points, which has been established and used successfully in the computation of mathematical economics and optimization problems in last three decades, e.g., see Robinson [447] and references therein.

8.3 Nash-Equilibria of Non-Cooperative N-Person Games

In this section, as applications of the Browder-Fan fixed point theorem and Ky Fan minimax inequality, we first obtain an existence theorem of Nash equilbira for non-cooperative n-person games which generalizes the corresponding results of Nikaido and Isoda [389]. As applications, we give two new existence theorems of of ϵ-equilibrium points which generalize that of Tijs [558]. Finally, a saddle point theorem of Komiya [326] is deduced from one of our existence theorems of ϵ-equilibrium points.

Now we introduce some notions and definitions. Let $N = \{1, \cdots, n\}$ be a set of players. A non-cooperative n-person game is an ordered $2n$-tuple $(X_1, \cdots, X_n; f_1, \cdots, f_n)$ where for each player $i \in N$, the nonempty set X_i is his (pure) strategy space and $f_i : X = \Pi_{i=1}^n X_j \to \mathbb{R}$ is his payoff function. The set X is called an outcome space. For each $i \in N$, denote $X_{\hat{i}} = \Pi_{j \in N \setminus \{i\}} X_j$. If $x = (x_1, \cdots, x_n) \in X$, we shall write

$$x_{\hat{i}} = (x_i, \cdots, x_{i-1}, x_{i+1}, \cdots x_n) \in X_{\hat{i}}.$$

If $x_i \in X_i$ and $x_{\hat{i}} \in X_{\hat{i}}$, we shall use $(x_i, x_{\hat{i}})$ to denote $y = (y_1, \cdots, y_n) \in X$ such that $y_i = x_i$ and $y_{\hat{i}} = x_{\hat{i}}$. Let $\epsilon_1, \cdots, \epsilon_n$ be nonnegative real numbers. Then we shall call a point $(x^*, \cdots, x_n^*) \in X$ an $(\epsilon_1, \cdots, \epsilon_n)$-equilibrium point of the game Γ if for each $i \in N$, we have $f_i(x_i^*, x_{\hat{i}}^*) \geq \sup_{u_i \in X_i} f_i(u_i, x_{\hat{i}}^*) - \epsilon_i$. An $(0, \cdots, 0)$-equilibrium point of Γ is also called a Nash equilibrium point for the game Γ.

In 1955, Nikaido and Isoda [389, Theorem 3.2] (see also Aubin [15, Theorem 1, p.286]) proved the following existence theorem for Nash equilibrium points.

Theorem 8.3.1. Let $\Gamma = (X_1, \cdots, X_n; f_1, \cdots, f_n)$ be an N-person game with the following properties:

(1) for each $i \in N$, X_i is a nonempty compact convex subset of a Hausdorff topological vector space E_i;

(2) for each $i \in N$, f_i is continuous on X; and

(3) for each $i \in N$ and each fixed $x_{\hat{i}} \in X_{\hat{i}}$, the function $u_i \mapsto f_i(u_i, x_{\hat{i}})$ is concave on X_i.

Then Γ has a Nash equilibrium point.

Recall that a family $\mathcal{F}$ of real-valued functions on a set X is said to be upper bounded if there is $c \in \mathbb{R}$ such that $f(x) \leq c$ for all $x \in X$ and $f \in \mathcal{F}$. Tijs [558, Theorem 4.7] proved the following existence theorem for ϵ-equilibrium points:

Theorem 8.3.2. Let $\Gamma = (X_1, \cdots, X_n; f_1, \cdots, f_n)$ be an n-person game with the following properties:

(1) for each $i \in N$, X_i is a nonempty convex subset of a Hausdorff topological vector space E_i;

(2) for each $i \in N \setminus \{n\}$, X_i is compact;

(3) for each $i \in N$, f_i is continuous on X;

(4) for each $i \in N$ and each fixed $x_{\hat{i}} \in X_{\hat{i}}$, the function $u_i \mapsto f_i(u_i, x_{\hat{i}})$ is concave on X_i; and

(5) the family $\mathcal{F} = \{f_n(x_n, \cdot) : x_n \in X_n\}$ is an upper bounded and equicontinuous family of functions on $X_{\hat{n}}$.

Then Γ has an $(0, \cdots, 0, \epsilon)$-equilibrium point for each $\epsilon > 0$.

In this section, we first obtain existence theorems of Nash equilibria for noncooperative n-person games which generalize Theorem 8.3.1. As applications, we give two new existence theorems of ϵ-equilibrium points which generalize Theorem 8.3.B. Finally, a saddle point theorem of Komiya [326] is deduced from one of our existence theorems of ϵ-equilibrium points without using the coincidence theorem of set-valued mappings.

We shall first observe the following result.

Lemma 8.3.3. Let X and Y be two Hausdorff topological spaces and X be compact. Let f be a real-valued function defined on $X \times Y$ such that

(1) f is upper semicontinuous on $X \times Y$; and

(2) for each fixed $x \in X$, the function $y \mapsto f(x, y)$ is lower semicontinuous.

Then the function $\psi : Y \to \mathbb{R}$ defined by $\psi(y) = \max_{u \in X} f(u, y)$ is continuous on Y.

Proof. Since X is compact, ψ is upper semicontinuous on Y by (1) and Theorem 1 of Aubin [10, p.67]). By (2), ψ is also lower semicontinuous on Y. $\square$

Recall that a real-valued function f on a nonempty convex subset X of a vector space is said to be quasi-concave (respectively, quasi-convex) on X if the set $\{x \in X : f(x) > \lambda\}$ (respectively, $\{x \in X : f(x) < \lambda\}$) is convex for each $\lambda \in \mathbb{R}$.

Theorem 8.3.4. Let $\Gamma = (X_1, \cdots, X_n; f_1, \cdots, f_n)$ be an n-person game with the following properties:

(1) for each $i \in N$, X_i is a nonempty compact convex subset of a Hausdorff topological vector space E_i;

(2) for each $i \in N$, f_i is upper semicontinuous on $X = \Pi_{i\in N} X_j$;

(3) for each $i \in N$ and each fixed $x_i \in X_i$, the function $u_{\hat{i}} \mapsto f_i(x_i, u_{\hat{i}})$ is lower semicontinuous on $X_{\hat{i}}$; and

(4) for each $i \in N$ and each fixed $x_{\hat{i}} \in X_{\hat{i}}$, the function $u_i \mapsto f_i(u_i, x_{\hat{i}})$ is quasi-concave on X_i.

Then Γ has a Nash equilibrium point.

Proof. Fix a positive integer k. Define the set-valued mapping W_k from X to X by

$$W_k(x) = \Pi_{i=1}^n \{y_i \in X_i : f_i(y_i, x_{\hat{i}}) > \max_{u_i \in X_i} f_i(u_i, x_{\hat{i}}) - 1/k\}.$$

Then for each $x \in X$, $W_k(x)$ is nonempty and is convex by (4). For each $y \in X$, by Lemma 8.3.3 and (3), the function $x \mapsto f_i(y_i, x_{\hat{i}}) - \max_{u_i \in X_i} f_i(u_i, x_{\hat{i}})$ is lower semicontinuous and hence the set

$$W_k^{-1}(y) = \cap_{i=1}^n \{x \in X : f_i(y_i, x_{\hat{i}}) > \max_{u_i \in X_i} f_i(u_i, x_{\hat{i}}) - 1/k\}$$

is open in X. The Browder-Fan fixed point theorem (e.g., see Theorem 1 of Browder [59]) implies that there exists $x^k \in X$ such that $x^k \in W_k(x^k)$; that is, $f_i(x_i^k, x_{\hat{i}}^k) > \max_{u_i \in X_i} f_i(u_i, x_{\hat{i}}^k) - 1/k$ for each $i \in N$.

Since X is compact, there exists a subnet $\{z^\alpha\}$ of $\{x^k\}$ and $x^* \in X$ such that $z^\alpha \to x^*$. Let $z^\alpha = x^{k(\alpha)}$ where $k(\alpha) \to \infty$. By Lemma 8.3.3 and (2), for each $i \in N$,

$$f_i(x_i^*, x_{\hat{i}}^*) \geq \limsup_\alpha f_i(x_i^{k(\alpha)}, x_{\hat{i}}^{k(\alpha)}) \geq \lim_\alpha \max_{u_i \in X_i} f_i(u_i, x_{\hat{i}}^{k(\alpha)}) = \max_{u_i \in X_i} f_i(u_i, x_{\hat{i}}^*).$$

Hence x^* is a Nash equilibrium point of the game Γ. □

The following is a special case of Theorem 4.5 of Tan and Yu [505]. However, we shall give a direct and simple proof as follows.

Theorem 8.3.5. Let $\Gamma = (X_1, \cdots, X_n; f_1, \cdots, f_n)$ be an n-person game with the following properties:

(1) for each $i \in N$, X_i is a nonempty compact convex subset of a Hausdorff topological vector space E_i;

(2) $\Sigma_{i=1}^n f_i$ is upper semicontinuous on X;

(3) for each $i \in N$ and each fixed $x_i \in X_i$, the function $u_{\hat{i}} \mapsto f_i(x_i, u_{\hat{i}})$ is lower semicontinuous on $X_{\hat{i}}$; and

(4) for each $i \in N$ and each fixed $x_{\hat{i}} \in X_{\hat{i}}$, the function $u_i \mapsto f_i(u_i, x_{\hat{i}})$ is concave on X_i.

Then Γ has a Nash equilibrium point.

Proof. Define the function $\psi : X \times X \to \mathbb{R}$ by

$$\psi(x, y) = \Sigma_{i=1}^n [f_i(y_i, x_{\hat{i}}) - f_i(x_i, x_{\hat{i}})].$$

Then we have: (i) for each fixed $y \in X$, $x \mapsto \psi(x,y)$ is lower semicontinuous by (2) and (3); (ii) for each fixed $x \in X$, $y \mapsto \psi(x,y)$ is concave by (4) and (iii) for each $y \in X$, $\psi(y,y) = 0$. By the Ky Fan's minimax inequality (see Theorem 8.2.C above or Aubin [10, Theorem 3, p.214]), there exists $x^* \in X$ such that $\psi(x^*,y) \leq 0$ for all $y \in X$. For each $i \in N$ and each $u_i \in X_i$, set $y = (u_i, x^*_{\hat{i}}) \in X$, then $\psi(x^*,y) = f_i(u_i, x^*_{\hat{i}}) - f_i(x^*_i, x^*_{\hat{i}}) \leq 0$ so that $f_i(x^*_i, x^*_{\hat{i}}) = \max_{u_i \in X_i} f_i(u_i, x^*_{\hat{i}})$. Hence x^* is a Nash equilibrium point of the game Γ. $\square$

In order to obtain existence theorems of ϵ-equilibrium points, we need the concept of ϵ-domination introduced by Tijs [558]: Let $\epsilon > 0$ and $\mathcal{F}$ and $\mathcal{G}$ be two family of real-valued functions on a set K. We shall say that $\mathcal{G}$ ϵ-dominates $\mathcal{F}$ if for each $f \in \mathcal{F}$, there is a $g \in \mathcal{G}$ such that $f(x) \leq g(x) + \epsilon$ for all $x \in K$. Let $\Gamma = (X_1, \cdots, X_n; f_1 \cdots, f_n)$ be an n-person game and let $Y_1, \cdots, Y_n$ be nonempty subsets of $X_1, \cdots, X_n$, respectively. Then the game $\Gamma_1 = (Y_1, \cdots, Y_n; f_1, \cdots, f_n)$ in which the payoff functions are restrictions of the payoff functions in Γ to the set $\Pi_{i=1}^n Y_i$ is called a sub-game of Γ. Let $\epsilon = (\epsilon_1, \cdots, \epsilon_n)$, we shall say that Γ_1 ϵ-dominates Γ if for each $i \in N$ and each $x_i \in X_i$, there exists $y_i \in Y_i$ such that $f_i(x_i, u_{\hat{i}}) \leq f_i(y_i, u_{\hat{i}}) + \epsilon$ for all $u_{\hat{i}} \in X_{\hat{i}}$.

Theorem 8.3.6. Let $\Gamma = (X_1, \cdots, X_n; f_1, \cdots, f_n)$ be an n-person game with the following properties:

(1) for each $i \in N$, X_i is a nonempty convex subset of a Hausdorff topological vector space E_i;

(2) for each $i \in N \setminus \{n\}$, X_i is compact;

(3) for each $i \in N$, f_i is upper semicontinuous on X and for each fixed $x_i \in X_i$, the function $u_{\hat{i}} \mapsto f_i(x_i, u_{\hat{i}})$ is lower semicontinuous on $X_{\hat{i}}$;

(4) for each $i \in N$ and each fixed $x_{\hat{i}} \in X_{\hat{i}}$, the function $u_i \mapsto f_i(u_i, x_{\hat{i}})$ is concave on X_i; and

(5) the family $\mathcal{F} = \{f_n(x_n, \cdot) : x_n \in X_n\}$ is an upper bounded and equi-continuous family of functions on $X_{\hat{n}}$.

Then Γ has an $(0, \cdots, 0, \epsilon)$-Nash equilibrium point for each $\epsilon > 0$.

Proof. For any $\epsilon > 0$, since $X_{\hat{n}}$ is compact and (5) holds, by Proposition 4.2 of Tijs [558], there exists a finite subset Z_n of X_n such that the finite family $\mathcal{B} = \{f_n(x_n, \cdot) : x_n \in Z_n\}$ ϵ-dominates $\mathcal{F}$. This implies that the game $\Gamma_1 = (X_1, \cdots, X_{n-1}, Z_n; f_1, \cdots, f_n)$ ϵ-dominates Γ. Note that the game

$$\Gamma_2 = (X_1, \cdots, X_{n-1}, co(Z_n); f_1, \cdots, f_n)(0, \cdots, 0, \epsilon)$$

dominates Γ, where coZ_n is the convex hull of Z_n. Since $co(Z_n)$ is compact convex, the game Γ_2 satisfies all conditions of Theorem 8.2.6. Thus Γ_2 has at least one Nash equilibrium point and hence Γ has $(0, \cdots, 0, \epsilon)$-equilibrium points by Lemma 4.1 of Tijs [558]. $\square$

By the same arguement as in the proof of Theorem 8.3.6 where Theorem 8.3.2 is applied instead of Theorem 8.3.4, we have the following result.

Theorem 8.3.7. Let $\Gamma = (X_1, \cdots, X_n; f_1, \cdots, f_n)$ be an n-person game with the following properties:

(1) for each $i \in N$, X_i is a nonempty convex subset of a Hausdorff topological vector space E_i;

(2) for each $i \in N \setminus \{n\}$, X_i is compact;

(3) $\Sigma_{i=1}^n f_i$ is upper semicontinuous on X and for each fixed $i \in N$ and each fixed $x_i \in X_i$, the function $u_{\hat{i}} \mapsto f_i(x_i, u_{\hat{i}})$ is lower semicontinuous on $X_{\hat{i}}$;

(4) for each $i \in N$ and each fixed $x_{\hat{i}} \in X_{\hat{i}}$, the function $u_i \mapsto f_i(u_i, x_{\hat{i}})$ is concave on X_i; and

(5) the family $\mathcal{F} = \{f_n(x_n, \cdot) : x_n \in X_n\}$ is an upper bounded and equi-continuous family of functions on $X_{\hat{n}}$.

Then Γ has an $(0, \cdots, 0, \epsilon)$-Nash equilibrium point for each $\epsilon > 0$.

Both Theorems 8.3.6 and 8.3.7 also generalize Theorem 8.3.2 in several aspects.

As an application of Theorem 8.3.6, we have the following minimax inequality which is Theorem 3 of Komiya [326].

Theorem 8.3.8 (Komiya). Let X be a nonempty convex subset of a Hausdorff topological vector space E and let Y be a nonempty compact convex subset of a Hausdorff topological vector space F. Let f be a real-valued continuous function defined on $X \times Y$ which is quasi-concave in its first variable and quasi-convex in its second variable and satisfies $\sup_{x \in X} \min_{y \in Y} f(x,y) < \infty$. Let the family $\{f(x, \cdot) : x \in X\}$ of real-valued functions on Y be equi-continuous and closed in the Banach space $C(Y)$ of all real-valued continuous functions on Y with uniform norm. Then f has a saddle point $(x^*, y^*) \in X \times Y$; that is,

$$\max_{x \in X} f(x, y^*) = f(x^*, y^*) = \min_{y \in Y} f(x^*, y).$$

Proof. Let Γ be the game $(Y, X, -g, g)$, where $g : Y \times X \to \mathbb{R}$ is defined by $g(y, x) = f(x, y)$ for each $(y, x) \in Y \times X$. ¿From the proof of Theorem 3 of Komiya [326], f is upper bounded on $X \times Y$. Also it is easy to show that the game Γ satisfies all conditions of Theorem 8.3.6. Hence for $\epsilon_k = 1/k$, $k = 1, 2, \cdots$, the game Γ has an $(0, \epsilon_k)$-equilibrium point (y_k, x_k); that is, $f(x_k, y_k) = \min_{y \in Y} f(x_k, y)$ and $f(x_k, y_k) \geq \max_{x \in X} f(x, y_k) - 1/k$. Since $\{f(x_k, \cdot) : k = 1, 2, \cdots,\} \subset \mathcal{F}$ and $\mathcal{F}$ is equi-continuous and closed in $C(Y)$, by Arzela-Ascoli Theorem (see e.g., Rudin [449, p.369]), there exists a subsequence $\{x_{n_k}\}$ of $\{x_k\}$ and $x^* \in X$ such that $\{f(x_{n_k}, \cdot)\}$ uniformly converges to $f(x^*, \cdot)$ on Y. Since Y is compact, there exists a subnet $\{z_\alpha\}$ of $\{y_{n_k}\}$ and $y^* \in Y$ such that $z_\alpha \to y^*$. Let $z_\alpha = y_{n_{k(\alpha)}}$ where $k(\alpha) \to \infty$. Since

$|f(x_{n_{k(\alpha)}}, y_{n_{k(\alpha)}}) - f(x^*, y^*)| \leq |f(x_{n_{k(\alpha)}}, y_{n_{k(\alpha)}}) - f(x^*, y_{n_{k(\alpha)}})| + |f(x^*, y_{n_{k(\alpha)}}) - f(x^*, y^*)| \to 0$, we have $f(x_{n_{k(\alpha)}}, y_{n_{k(\alpha)}}) \to f(x^*, y^*)$. As $\{f(x_{n_{k(\alpha)}}, \cdot)\}$ uniformly converges to $f(x^*, \cdot)$ on Y, it is easy to show that $\min_{y \in Y} f(x_{n_{k(\alpha)}}, y) \to \min_{y \in Y} f(x^*, y)$; thus $f(x^*, y^*) = \min_{y \in Y} f(x^*, y)$. On the other hand, by Lemma 8.3.3, $\max_{x \in X} f(x, y_{n_{k(\alpha)}}) \to \max_{x \in X} f(x, y^*)$; it follows that $f(x^*, y^*) = \max_{x \in X} f(x, y^*)$. Hence $\max_{x \in X} f(x, y^*) = f(x^*, y^*) = \min_{y \in Y} f(x^*, y)$. □

8.4 Generic Stability of Nash Equilibria

In this section, we will study the generic stability of nash equilibria from a point of view of the Baire category theory. Our results show that almost all Nash-equilibria are stable in the sense of the secondary category; or equivalently, the majority of non-cooperative N-person game are Pareto optima.

To this end, we now recall some notions and introduce the following **Models A, B** and **C** (see also Tan et al. [520]).

Given the set $N = \{1, \cdots, n\}$ of players, for each player $i \in N$, denote by X_i the set of all his (resp., her) strategies and $f : X = \prod_{i=1}^{n} X_i \to R$ the payoffs for him (her). For each $i \in N$, denote $X^{\hat{i}} = \prod_{j \in N \setminus \{i\}} X_j$. A Nash equilibrium of a game is an n-tuple of strategies $x = \{x_1, \cdots, x_n\} \in X$ such that for each $i \in N$, $f_i(x_i, x^{\hat{i}}) = \max_{u_i \in X_i} f_i(u_i, x^{\hat{i}})$.

The following existence theorems of Nash equilibria have been proved in section 3 above (i.e., Theorems 8.3.2 and 8.3.1), for the convenience, we state them here as following:

Theorem 8.4.1. Let $X_1, \cdots, X_n$ be n nonempty compact convex subsets in Hausdorff topological vector spaces $E_1, \cdots, E_n$ respectively. Let $f_1, \cdots, f_n$ be n real-valued functions defined on $X = \prod_{i=1}^{n} X_i$ such that:

(1) $\sum_{i=1}^{n} f_i$ is upper semicontinuous on X;

(2) for each $i \in N$ and each fixed $x_i \in X_i$, $u_{\hat{i}} \to f_i(x_i, u_{\hat{i}})$ is lower semicontinuous on $X_{\hat{i}}$; and

(3) for each $i \in N$ and each fixed $x_{\hat{i}} \in X_{\hat{i}}$, $u_i \to f_i(u_i, X_{\hat{i}})$ is concave on X_i.

Then there is $x^* = (x_1^*, \cdots, x_n^*) \in X$ such that for each $i \in N$, $f_i(x_i^*, x_{\hat{i}}^*) = \max_{u_i \in X_i} f_i(u_i, x_{\hat{i}}^*)$.

Theorem 8.4.2. Let $X_1, \cdots, X_n$ be n nonempty compact convex subsets in Hausdorff topological vector spaces $E_1, \cdots, E_n$ respectively. Let $f_1, \cdots, f_n$ be

n real-valued functions defined on $X = \prod_{i=1}^{n} X_i$ such that:

(1) for each $i \in N$, f_i is upper semicontinuous on X;

(2) for each $i \in N$ and each fixed $x_i \in X_i$, $u_{\hat{i}} \to f_i(x_i, u_{\hat{i}})$ is lower semicontinuous on $X_{\hat{i}}$; and

(3) for each $i \in N$ and each fixed $x_{\hat{i}} \in X_{\hat{i}}$, $u_i \to f_i(u_i, x_{\hat{i}})$ is quasi-concave on X_i.

Then there is $x^* = (x_1^*, \cdots, x_n^*) \in X$ such that for each $i \in N$, $f_i(x_i^*, x_{\hat{i}}^*) = \max_{u_i \in X_i} f_i(u_i, x_{\hat{i}}^*)$.

Model (A): Let Y_1 (resp., Y_1') be the set of all $y = (f_1, \cdots, f_n)$ where $f_1, \cdots, f_n$ satisfy conditions (1), (2) and (3) (resp., (1)$'$, (2)$'$ and (3)$'$) in Theorem 8.4.1 (resp., Theorem 8.4.2) and $\sum_{i=1}^{n} \sup_{x \in X} |f_i(x)| < \infty$. For any $y_1 = (f_{11}, \cdots, f_{1n})$, $y_2 = (f_{21}, \cdots, f_{2n}) \in Y_1$ (resp., Y_1'), define the distance between y_1 and y_2 by

$$\rho(y_1, y_2) = \sum_{i=1}^{n} \sup_{x \in X} |f_{1i}(x) - f_{2i}(x)|.$$

Denoted by $F_1(y)$ all Nash-equilibria of the game $y \in Y_1$ (reps., $y \in Y_1'$), then $F_1(y) \neq \emptyset$ by Theorem 8.4.1 (resp., Theorem 8.4.2). It is easy to show the following results:

Lemma 8.4.3. Y_1 (resp., Y_1') is a complete metric space.

For each $y = (f_1, \cdots, f_n) \in Y_1$ (resp., Y_1'), it determines an n-person noncooperative game, where f_i is the payoffs for player $i \in N$. Denote by $F_1(y)$ (resp., $F_1'(y)$) all Nash equilibria of game y, then $F_1(y) \neq \phi$ (resp., $F_1'(y) \neq \phi$) by Theorem 8.4.1 (resp., Theorem 8.4.2).

Theorem 8.4.4. F_1 is an USCO mapping on Y_1.

Proof. We first prove that the $graph F_1 = \{(y, x) \in Y_1 \times X : y \in F_1(x)\}$ is closed. Let $\{y^\alpha, x^\alpha\}_{\alpha \in D}$ be a net in $graph F_1$ with $\{y^\alpha, x^\alpha\}_{\alpha \in D} \to (y, x) \in Y_1 \times X$. Then $y^\alpha \to y$ $x^\alpha \to x$ and $x^\alpha \in F_1(y^\alpha)$ for all $\alpha \in D$ and hence $f_i^\alpha(x_i^\alpha, x_{\hat{i}}^\alpha) = \max_{u_i \in X_i} f_i^\alpha(u_i, x_{\hat{i}}^\alpha)$ for each $i \in N$. Since

$$\begin{aligned}
\sum_{i=1}^{n} f_i(x_i, x_{\hat{i}}) \quad - \quad f_i^\alpha(x_i^\alpha, x_{\hat{i}}^\alpha)] &= \sum_{i=1}^{n} [f_i(x_i, x_{\hat{i}}) - f_i(x_i^\alpha, x_{\hat{i}}^\alpha) \\
&+ \sum_{i=1}^{n} f_i(x_i^\alpha, x_{\hat{i}}^\alpha) - f_i^\alpha(x_i^\alpha, x_{\hat{i}}^\alpha)] \\
&\geq \sum_{i=1}^{n} [f_i(x_i, x_{\hat{i}}) - f_i(x_i^\alpha, x_{\hat{i}}^\alpha) - \rho(y, y^\alpha),
\end{aligned}$$

we obtain that

$$\begin{aligned}\sum_{i=1}^{n} f_i(x_i, x_{\hat{i}}) &- \overline{\lim_{\alpha}}\sum_{i=1}^{n} f_i^{\alpha}(x_i^{\alpha}, x_{\hat{i}}^{\alpha}) = \underline{\lim_{\alpha}}\sum_{i=1}^{n} f_i(x_i, x_{\hat{i}}) - f_i^{\alpha}(x_i^{\alpha}, x_{\hat{i}}^{\alpha})\\ &\geq \underline{\lim_{\alpha}}\sum_{i=1}^{n} f_i(x_i, x_{\hat{i}}) - f_i(x_i^{\alpha}, x_{\hat{i}}^{\alpha})\\ &= \sum_{i=1}^{n} f_i(x_i, x_{\hat{i}}) - \overline{\lim_{\alpha}}\sum_{i=1}^{n} f_i(x_i^{\alpha}, x_{\hat{i}}^{\alpha}) \geq .\end{aligned}$$

For any $\varepsilon > 0$ and each $i \in N$, there exists $u_i^0 \in X_i$ such that $f_i(u_i^0, x_{\hat{i}}) > \sup_{u_i \in X_i} f_i(u_i, x_{\hat{i}}) - \varepsilon$. Since $\rho(y, y^{\alpha}) \to 0$ and $x_{\hat{i}}^{\alpha} \to x_{\hat{i}}$, there exists $\alpha_0 \in D$ such that $\rho(y, y^{\alpha}) < \varepsilon$ and $f_i(u_i^0, x_{\hat{i}}^{\alpha}) > f_i(u_i^0, x_{\hat{i}}) - \varepsilon$ for all $\alpha \geq \alpha_0$ and hence

$$\begin{aligned} f_i^{\alpha}(x_i^{\alpha}, x_{\hat{i}}^{\alpha}) &\geq f_i^{\alpha}(u_i^0, x_{\hat{i}}^{\alpha}) \geq f_i(u_i^0, x_{\hat{i}}^{\alpha}) - \rho(y, y^{\alpha})\\ &> f_i(u_i^0, x_{\hat{i}}^{\alpha}) - \varepsilon > f_i(u_i^0, x_{\hat{i}}) - 2\varepsilon > \sup_{u_i \in X_i} f_i(u_i, x_{\hat{i}}) - 3\varepsilon.\end{aligned}$$

Thus,

$$\begin{aligned}\sum_{i=1}^{n} f_i(x_i, x_{\hat{i}}) &\geq \overline{\lim_{\alpha}}\sum_{i=1}^{n} f_i^{\alpha}(x_i^{\alpha}, x_{\hat{i}}^{\alpha}) \geq \underline{\lim_{\alpha}}\sum_{i=1}^{n} f_i^{\alpha}(x_i^{\alpha}, x_{\hat{i}}^{\alpha})\\ &\geq \sum_{i=1}^{n} \underline{\lim_{\alpha}} f_i^{\alpha}(x_i^{\alpha}, x_{\hat{i}}^{\alpha}) \geq \sum_{i=1}^{n} \sup_{u_i \in X_i} f_i(u_i, x_{\hat{i}})\end{aligned}$$

and so that $f_i(x_i, x_{\hat{i}}) = \max_{u_i \in X_i} f_i(u_i, x_{\hat{i}})$ for each $i \in N$. Hence $x \in F_1(y)$ and the *graph*F_1 is closed.

Since *hraph*F_1 is closed, for each $y \in Y_1$, $F_1(y)$ is closed in X and must be compact since X is compact. By Proposition 3 of [11, p.72]. F_1 is upper semicontinuous on Y_1. Hence F_1 is an USCO mapping on Y_1. □

Remark 8.4.5. Note that we only need use the continuity conditions of $f_1, \cdots, f_n$ in the proof of Theorem 8.4.4. Since the continuity conditions of $f_1, \cdots, f_n$ in Theorem 8.4.2 are stronger than those in Theorem 8.4.1, F_1' is also an USCO mapping on Y_1'.

Model (B): Let $X_1, \cdots, X_n$ be n nonempty closed and convex subsets in Hausdorff topological vector spaces $E_1, \cdots, E_n$ respectively. For each $i \in N$, let $CK(X_i)$ be the space of all nonempty compact convex subsets of X_i endowed with the Vietoris topology. Let C be the space af all $f = (f_1, \cdots, f_n)$, where $f_1, \cdots, f_n$ be n real-valued continuous functions defined on $X = \prod_{i=1}^{n} X_i$ such that for each $i \in N$ and each fixed $x_{\hat{i}} \in X_{\hat{i}}$, $u_i \to f_i(u_i, x_{\hat{i}})$ is quasi-concave and $\sum_{i=1}^n \sup_{x\in X} |f_i(x)| < \infty$. For any $f_1 = (f_{11}, \cdots, f_{1n})$, $f_2 = (f_{21}, \cdots, f_{2n}) \in C$, define the distance between f_1 and f_2 by

$$\rho(f_1, f_2) = \sum_{i=1}^{n} \sup_{x\in X} |f_{1i}(x) - f_{2i}(x)|.$$

It is easy to see that C is complete. Define $Y = C \times \prod_{i=1}^{n} CK(X_i)$, for each $y = (f_1, \cdots, f_n, S_1, \cdots, S_n)$, it determines an n-person noncooperative game, where f_i is the payoffs for player $i \in N$ and S_i is the set of all strategies for player $i \in N$. A Nash equilibrium of games is an n-tuple of strategies $x^* = (x_1^*, \cdots, x_n^*) \in \prod_{i=1}^{n} S_i$ such that for each $i \in N$, $f_i(x_i^*, x_{\hat{i}}^*) = \max_{u_i\in S_i} f_i(u_i, x_{\hat{i}}^*)$. Denote by $F_2(y)$ all Nash equilibria of game y, then $F_2(y) \neq \phi$ by Theorem 8.4.2.

Theorem 8.4.6. F_2 is an USCO mapping on Y_2.

Proof. For any $y = (f_1, \cdots, f_n, S_1, \cdots, S_n) \in Y_2$, $x^\alpha \in F_2(y)$ for any $\alpha \in D$ and $x^\alpha \to x \in X$, we have $x \in \prod_{i=1}^{n} S_i$ and $f_i(x_i^\alpha, x_{\hat{i}}^\alpha) = \max_{u_i\in S_i} f_i(u_i, x_{\hat{i}}^\alpha)$ for any $i \in N$ and $\alpha \in D$. Since $f_i(x_i^\alpha, x_{\hat{i}}^\alpha) \to f_i(x_i, x_{\hat{i}})$ and by Theorem 8.3.2, $\max_{u_i\in S_i} f_i(u_i, x_{\hat{i}}^\alpha) \to \max_{u_i\in S_i} f_i(u_i, x_{\hat{i}})$, we obtain that $f_i(x_i, x_{\hat{i}}) = \max_{u_i\in S_i} f_i(u_i, x_{\hat{i}})$. Thus $x \in F_2(y)$ and $F_2(y)$ is closed. Since $F_2(y) \subset \prod_{i=1}^{n} S_i$ and $\prod_{i=1}^{n} S_i$ is compact, $F_2(y)$ is also compact. Suppose that F_2 were not upper semicontinuous at $y \in Y_2$, then there exists an open set G of X with $G \supset F_2(y)$ and a net $\{y^\alpha\}_{\alpha\in D}$ in Y_2 with $y^\alpha \to y$ such that for each $\alpha \in D$, there exists $x^\alpha \in F_2(y^\alpha)$ with $x^\alpha \notin G$. Denote $y^\alpha = (f_1^\alpha, \cdots, f_n^\alpha, S_1^\alpha, \cdots, S_n^\alpha)$ and $y = (f_1, \cdots, f_n, S_1, \cdots, S_n)$, then $\rho(f^\alpha, f) \to 0$ and $S_i^\alpha \to S_i$ for each $i \in N$. Since $\{\prod_{i=1}^{n} S_i^\alpha\}_{\alpha\in D}$ is a net in $\prod_{i=1}^{n} K(X_i)$ and for each $\alpha \in D$, $x^\alpha = (x_1^\alpha, \cdots, x_n^\alpha) \in \prod_{i=1}^{n} S_i^\alpha$, by Theorem 8.4.1(1), $\{x^\alpha\}$ has a cluster point $x = (x_1, \cdots, x_n)$ in $\prod_{i=1}^{n} S_i$. It is easy to see that $x \notin G$, hence $x \notin F_2(y)$ and there exists some $i \in N$ and $\varepsilon_0 > 0$ such that $f_i(x_i, x_{\hat{i}}) < \max_{u_i\in S_i} f_i(u_i, x_{\hat{i}}) - \varepsilon_0 = f_i(u_i^0, x_{\hat{i}}) - \varepsilon_0$ where $u_i^0 \in S_i$. Since f_i is continuous at $(x_i, x_{\hat{i}})$ and at $(u_i^0, x_{\hat{i}})$, there exists an open neighbourhood $O(x_i)$ of x_i in X_i, an open neighborhood $O(u_i^0)$ of u_i^0 in X_i and an open neighborhood $O(x_{\hat{i}})$ of $x_{\hat{i}}$ in $X_{\hat{i}}$ such that for any $x' \in O(x_i) \times O(x_{\hat{i}})$, $f_i(x_i', x_{\hat{i}}') - \frac{\varepsilon_0}{4} < f_i(x_i, x_{\hat{i}})$ and for any $x' \in O(u_i^0) \times O(x_{\hat{i}})$, $f_i(u_i^0, x_{\hat{i}}) < f_i(x_i', x_{\hat{i}}') + \frac{\varepsilon_0}{4}$. Since $f^\alpha \to f$, there exists $\alpha_0 \in D$ such that $\rho(f^\alpha, f) < \frac{\varepsilon_0}{4}$ for all $\alpha \geq \alpha_0$. Since x is a cluster point of $\{x^\alpha\}_{\alpha\in D}$, $O(u_i^0)\cap S_i\neq\phi$ and $S_1^\alpha\to S_i$, there exists $\alpha_1\geq\alpha_0$ such that $x^{\alpha_1}\in O(x_i)\times O(x_{\hat{i}})$ and $O(u_i^0)\cap S_i^{\alpha_1}\neq\phi$. Choose $u_i^{\alpha_1}\in O(u_i^0)\cap S_i^{\alpha_1}$, then

$$\begin{aligned} f_i^{\alpha_1}(x_i^{\alpha_1}, x_{\hat{i}}^{\alpha_1}) &< f_i(x_i^{\alpha_1}, x_{\hat{i}}^{\alpha_1})+\frac{\varepsilon_0}{4} < f_i(x_i, x_{\hat{i}})+\frac{\varepsilon_0}{2}\\ &< f_i(u_i^0, x_{\hat{i}})-\frac{\varepsilon_0}{2} < f_i(u_i^{\alpha_i}, x_{\hat{i}}^{\alpha_1})-\frac{\varepsilon_0}{4}\\ &< f_i^{\alpha_1}(u_i^{\alpha_1}, x_{\hat{i}}^{\alpha_1}) \end{aligned}$$

which contradicts the fact that $f_i^{\alpha_1}(x_i^{\alpha_1}, x_{\hat{i}}^{\alpha_1})\geq f_i^{\alpha_1}(u_i^{\alpha_1}, x_{\hat{i}}^{\alpha_1})$ since $u_i^{\alpha_1}\in S_i^{\alpha_1}$. Hence F_2 must be upper semicontinuous on Y_2. □

We now consider the following

Model (C): Let $X_1,\cdots,X_n$ be n nonempty compact convex subsets in normed spaces $E_1,\cdots,E_n$ respectively. Let C be the space of all $f=(f_1,\cdots,f_n)$, where $f_1,\cdots,f_n$ be n real-valued continuous functions defined on $X=\prod_{i=1}^n X_i$ such that for each $i\in N$ and each fixed $x_{\hat{i}}\in X_{\hat{i}}$, $u_i\to f_i(u_i, x_{\hat{i}})$ is quasi-concave. Let G be the space of all $G=(G_1,\cdots,G_n)$ such that for each $i\in N$, the set-valued map $G_i: X_{\hat{i}}\to K(X_i)$ is continuous with compact convex values. Denote $Y_3=C\times G$. For any $y_1=(f_{11},\cdots,f_{1n}, G_{11},\cdots,G_{1n})$, $y_2=(f_{21},\cdots,f_{2n}, G_{21},\cdots,G_{21})\in Y_3$, define the distance between y_1 and y_2 by

$$\rho(y_1, y_2)=\sum_{i=1}^n\sup_{x\in X}|f_{1i}(x)-f_{2i}(x)|+\sum_{i=1}^n\sup_{x_{\hat{i}}\in X_{\hat{i}}}H_i(G_{1i}(x_{\hat{i}}), G_{2i}(x_{\hat{i}}))$$

where for each $i\in N$, H_i is the Hausdorff distance induced by the norm on E_i. It is easy to prove the following result:

Lemma 8.4.7. Y_3 is a complete metric space.

For each $y=(f_1,\cdots,f_n, G_1,\cdots,G_n)\in Y_3$, it determines an n-person game, where f_i is the payoffs for player $i\in N$ and S_i is the feasible strategy correspondence of player $i\in N$. For every game $y\in Y_3$, replacing Kakutani fixed point theorem by Theorem 2.7.10 of Fan-Glicksberg fixed point (see also Theorem 9.2.3 of Smart [491]) in the proof of Theorem 4.3.1 of Ichiishi [245], there exists a Nash equilibrium $x^*=(x_1^*,\cdots,x_n^*)\in X$, that is $x_i^*\in G_i(x_{\hat{i}}^*)$ and $f_i(x_i^*, x_{\hat{i}}^*)=\max_{u_i\in G_i(x_{\hat{i}}^*)}f_i(u_i, x_{\hat{i}}^*)$ for each $i\in N$. Denote by $F_3(y)$ all Nash equilibria of the game y, then $F_3(y)\neq\phi$.

Proposition 8.4.8. Let X and Y be two Hausdorff topological spaces, $\{A_\alpha:\alpha\in D\}$ be a net in $K(X)$, $\{y^\alpha:\alpha\in D\}$ be a net in Y and $\{f^\alpha(x, y):\alpha\in D\}$ be a net of real-valued continuous functions defined on $X\times Y$. If $A_\alpha\to A\in K(X)$

in the Vietoris topology, $y^\alpha \to y \in Y$ and $\sup_{(x,y)\in X\times Y} |f^\alpha(x, y) - f(x, y)| \to 0$, where f is a real-valued continuous function defined on $X \times Y$. Then $\max_{u\in A_\alpha} f^\alpha(u, y^\alpha) \to \max_{u\in A} f(u, y)$.

Proof. By Theorem 3 of Aubin [11, p.70], we have $\max_{u\in A_\alpha} f(u, y^\alpha) \to \max_{u\in A} f(u, y)$. Hence it is sufficient to prove that

$$\max_{u\in A_\alpha} f^\alpha(u,y^\alpha) - \max_{u\in A_\alpha} f(u,y^\alpha) \to 0.$$

For each $\alpha \in D$, there exists $u_\alpha \in A_\alpha$ and $u'_\alpha \in A_\alpha$ such that

$$f^\alpha(u_\alpha, y^\alpha) = \max_{u\in A_\alpha} f^\alpha(u, y^\alpha) \quad \text{and} \quad f(u'_\alpha, y^\alpha) = \max_{u\in A_\alpha} f(u, y^\alpha).$$

Since

$$\begin{aligned} f^\alpha(u'_\alpha, g^\alpha) - f(u'_\alpha, y^\alpha) &\le \max_{u\in A_\alpha} f^\alpha(u, y^\alpha) - \max_{u\in A_\alpha} f(u, y^\alpha) \\ &\le f^\alpha(u_\alpha, y^\alpha) - f(u_\alpha, y^\alpha), \end{aligned}$$

we obtain that

$$\begin{aligned} |\max_{u\in A_\alpha} f^\alpha(u, y^\alpha) &- \max_{u\in A_\alpha} f(u, y^\alpha)| \\ &\le \max\{|f^\alpha(u'_\alpha, y^\alpha) - f(u'_\alpha, y^\alpha)|, |f^\alpha(u_\alpha, y^\alpha) - f(u_\alpha, y^\alpha)|\} \\ &\le \sup_{(x,y)\in X\times Y} |f^\alpha(x, y) - f(x, y)| \to 0 \end{aligned}$$

and thus the theorem is proved. □

Theorem 8.4.9. F_3 is an USCO mapping on Y_3.

Proof. We only need to prove that the graph of F_3 is closed. Indeed, if $\{(y^\alpha, x^\alpha)\}_{\alpha\in D}$ is a net is *graph*F_3 with $\{(y^\alpha, x^\alpha)\}_{\alpha\in D} \to (y, x) \in Y_3 \times X$, then $y^\alpha \to y$, $x^\alpha \to x$ and $x^\alpha \in F_3(y^\alpha)$ for all $\alpha \in D$ and hence $x_i^\alpha \in G_i^\alpha(x_{\hat{i}}^\alpha)$, $f_i^\alpha(x_i^\alpha, x_{\hat{i}}^\alpha) = \max_{u_i\in G_i^\alpha(x_{\hat{i}}^\alpha)} f_i^\alpha(u_i, x_{\hat{i}}^\alpha)$ for each $i \in N$. For each $i \in N$, since

$$\begin{aligned} d_i(x_i, G_i(x_{\hat{i}})) &\le d_i(x_i, x_i^\alpha)) \\ &+ d_i(x_i^\alpha, G_i^\alpha(x_{\hat{i}}^\alpha) + H_i(G_i^\alpha(x_{\hat{i}}^\alpha), G_i(x_{\hat{i}}^\alpha)) \\ &+ H_i(G_i(x_{\hat{i}}^\alpha), G_i(x_{\hat{i}})) \\ &\le d_i(x_i, x_i^\alpha) + \rho(y^\alpha, y) \\ &+ H_i(G_i(x_{\hat{i}}^\alpha), G_i(x_{\hat{i}})) \to 0 \end{aligned}$$

we obtain that $x_i \in G_i(x_{\hat{i}})$. Also as

$$\begin{aligned} |f_i^\alpha(x_i^\alpha, x_{\hat{i}}^\alpha) &- f_i(x_i, x_{\hat{i}})| \le |f_i^\alpha(x_i^\alpha, x_{\hat{i}}^\alpha) - f_i(x_i^\alpha, x_{\hat{i}}^\alpha)| \\ &+|f_i(x_i^\alpha, x_{\hat{i}}^\alpha) - f_i(x_i, x_{\hat{i}})| \le \rho(y^\alpha, y) \\ &+ |f_i(x_i^\alpha, x_{\hat{i}}^\alpha) - f_i(x_i, x_{\hat{i}})| \to 0 \end{aligned}$$

we have $f_i^\alpha(x_i^\alpha, x_{\hat{i}}^\alpha) \to f_i(x_i, x_{\hat{i}})$. Furthermore,

$$\begin{aligned} H_i(G_i^\alpha(x_{\hat{i}}^\alpha), G_i(x_{\hat{i}})) &\leq H_i(G_i^\alpha(x_{\hat{i}}^\alpha), G_i(x_{\hat{i}}^\alpha)) + H_i(G_i(x_{\hat{i}}^\alpha), G_i(x_{\hat{i}})) \\ &\leq \rho(y^\alpha, y) + H_i(G_i(x_{\hat{i}}^\alpha), G_i(x_{\hat{i}})) \to 0 \end{aligned}$$

it follows that $G_i^\alpha(x_{\hat{i}}^\alpha) \to G_i(x_{\hat{i}})$. By Proposition 8.4.8, $\max\limits_{u_i \in G_i^\alpha(x_{\hat{i}}^\alpha)} f_i^\alpha(u_i, x_{\hat{i}}^\alpha) \to \max\limits_{u_i \in G_i(x_{\hat{i}})} f_i(u_i, x_{\hat{i}})$. Therefore $f_i(x_i, x_{\hat{i}}) = \max\limits_{u_i \in G_i(x_{\hat{i}})} f_i(u_i, x_{\hat{i}})$ for each $i \in N$.

Hence $x \in F_3(y)$ and $graph F_3$ is closed. □

We are now ready to study the generic stability of Nash equilibria.

Let Y be any one of the spaces Y_1, Y_1', Y_2 and Y_3 of games and X be the spaces of solutions. For each game $y \in Y$, denote by $F(y)$ all Nash equilibria of y in X, then we know that $F(y) \neq \phi$ and the Nash equilibrium correspondence F is an USCO mapping on Y.

Definition 8.4.10. Let M be a nonempty and closed subset on Y and $y \in M$, then $x \in F(y)$ is said to be an *essential equilibrium* of the game y relative to M provided that for any open neighborhood $N(x)$ of x in X, there is an open neighborhood $O(y)$ of y in Y such that for any $y' \in M$ with $y' \in O(y)$, there exists $x' \in F(y')$ with $x' \in N(x)$. The game y is said to be *essential* relative to M if all its equilibria are essential relative to M. The game y is said to be weakly essential relative to M if it has at least one essential equilibrium relative to M.

It is easy to prove the following result.

Theorem 8.4.11. (1) The game $y \in M$ is essential relative to M if and only if the set-valued mapping $F : M \to K(X)$ is lower semicontinuous at y.

(2) The game $y \in M$ is weakly essential relative to M if and only if the set-valued mapping $F : M \to K(X)$ is almost lower semicontinuous at y.

Theorem 8.4.12. (1) Let M be a Baire space and X be a metric space. Then there exists a dense residual subset Q of M such that each game $y \in Q$ is essential relative to M.

(2) Let M be a Čech-complete space and X belong to the class $\mathcal{L}$. Then there exists a dense residual subset Q' of M such that each game $y \in Q'$ is weakly essential relative to M.

Before we obtain the stability of Nash eqilibria, we need the following result.

Proposition 8.4.13. Let X be a nonempty closed convex subset of a Hausdorff topological vector space E. Denote by $CK(X)$ the space of all nonempty compact convex subsets of X, endowed with the Vietoris topology. Then $CK(X)$ is closed in $K(X)$.

Proof. Let $\{A_\alpha : \alpha \in D\}$ be a net in $CK(X)$ such that $A_\alpha \to A \in K(X)$, suppose that A were not convex. Then there exists x_1 and $x_2 \in A$ and $\lambda \in (0, 1)$ such that $\lambda x_1 + (1-\lambda)x_2 \notin A$. Since A is compact, there exist an open set G in X which contains A and an open neighborhood $O(\lambda x_1 + (1-\lambda)x_2)$ of $\lambda x_1 + (1-\lambda)x_2$ in X such that $O(\lambda x_1 + (1-\lambda)x_2) \cap G = \phi$. Note that there exists an open neighborhood $O(x)$ of x_1 in X and an open neighborhood $O(x_2)$ of x_2 in X such that $\lambda O(x_1) + (1-\lambda)O(x_2) \subset O(\lambda x_1 + (1-\lambda)x_2)$. Since $O(x_1) \cap A \neq \phi$, $O(x_2) \cap A \neq \phi$, $G \supset A$ and $A_\alpha \to A$, there exists $\alpha_0 \in D$ such that for each $\alpha \geq \alpha_0$, $O(x_1) \cap A_\alpha \neq \phi$, $O(x_2) \cap A_\alpha \neq \phi$ and $G \supset A_\alpha$. Choose any $z_1 \in O(x_1) \cap A_{\alpha_0}$ and $z_2 \in O(x_2) \cap A_{\alpha_0}$. Since A_{α_0} is convex, $\lambda z_1 + (1-\lambda)z_2 \in A_{\alpha_0} \subset G$ which implies that $G \cap O(\lambda x_1 + (1-\lambda)x_2) \neq \phi$. This contradicts the assumption that $G \cap O(\lambda x_1 + (1-\lambda)x_2) = \phi$. Hence A must be convex. Therefore $CK(X)$ is closed in $K(X)$. □

Now we have the following result.

Theorem 8.4.14. If the game $y \in M$ is such that $F(y)$ is singleton, then game y is essential relative to M.

Theorem 8.4.15. For each model of (A), (B) and (C), we have the following:

(1) If M is a Baire space and X is a metric space, then there exists a dense residual subset Q of M such that each game $y \in Q$ is essential relative to M;

(2) If M is a Čech-complete space and X belongs to the class $\mathcal{L}$, then there exists a dense residual subset Q' of M such that each game $y \in Q'$ is weakly essential relative to M.

Proof. For the model (A): Since Y_1 (resp., Y_1') is a complete metric space and M, being a nonempty and closed subset of Y_1 (resp., Y_1') is itself a complete metric space. Note that M is Čech-complete. When E_i is a normed space for each $i \in N$, Theorem 8.4.12(1) is applicable. When $E_i \in \mathcal{L}$ for each $i \in N$, then $X = \prod_{i=1}^{n} X_i \in \mathcal{L}$ by Proposition 2(d) of Kenderov and Ribaraka [290], Theorem 8.4.12(2) is applicable.

For the model (B), when E_i is a Banach space for each $i \in N$, since $X_i \subset E_i$ is closed, X_i is a complete metric space and $K(X_i)$ is a complete metric space. As $CK(X_i) \subset K(X_i)$ is closed by Proposition 8.4.13, $CK(X_i)$ is a complete metric space and $Y_2 = C \times \prod_{i=1}^{n} CK(X_i)$ is complete metric space. Theorem 8.4.7 (1) is applicable. When X_i is a Čech-complete space and is in $\mathcal{L}$ for each $i \in N$, then $X = \prod_{i=1}^{n} X_i \in \mathcal{L}$ by Proposition 2(d) of Kenderov [288] and by Lemma 2.2 of Beer [27], $K(X_i)$ is Čech-complete. As $CK(X_i)$ is closed in $K(X_i)$, $CK(X_i)$ is Čech-complete by Theorem 3.9.6 of Engelking [154], it follows from Theorem

3.9.8 of Engelking [154] that $Y_2 := C \times \prod_{i=1}^{n} CK(X_i)$ is Čech-complete and hence M is also Čech-complete by Theorem 3.9.6 of Engelking [154]. Thus Theorem 8.4.12(2) is applicable.

For the Model (C), by applying Theorem 8.4.12(1) and (ii), the conclusions follow. □

Before we close this section, as an application of Theorem 2.5.31 for the existence of essential and connected components for KF points, we will study the existence of essential and connected components of Nash equilibria for Model (A). A similar or slightly modified approach could be used to study the existence of essential and connected components of Nash equilibria for both Models (B) and (C), hence we will not include their details here.

For any $f = (f_1, \cdots, f_n) \in Y_1$ in Model (A), by Theorem 8.4.1, there exists $x^* = (x_1^*, \cdots, x_n^*) \in X$, which is a Nash equilibrium point of the game f. For each $f, g \in Y_1$, the metric h between g and g is defined by $h(f,g) = \sup_{x\in X} \Sigma_{i=1}^{n} |f_i(x) - g_i(x)|$. Clearly (Y_1, h) is a complete metric space.

For each $f \in Y_1$, denote by $N(f)$ the set of all Nash equilibria of the game f. Then $N(f) \neq \emptyset$.

Lemma 8.4.16. For each $f = (f_1, \cdots, f_n) \in Y_1$ in Model (A), the mapping $T : (Y_1, h) \to (M_2, \rho)$ defined by $T(f) := \phi_f(x,y) := \Sigma_{i=1}^{n}[f_i(y_i, \hat{x}_i) - f_i(x_i, \hat{x}_i)]$ is continuous, where (M_2, ρ) is a space consisting of all functions ϕ defined on $X \times X$, which satisfy three conditions (i)-(iii) of Theorem 2.5.4 in and $\sup_{(x,y)\in X\times X} \phi(x,y) < \infty$ (e.g., see its definition in Section 5 of Chapter 2), and $\rho(f,g) = \sup_{(x,y)\in X\times X} |f(x,y) - g(x,y)|$ for each $f, g \in M_2$.

Proof. For each $f \in Y_1$, it is easy to check out that (1): for each fixed $y \in X$, $x \mapsto \phi_f(x,y)$ is lower semicontinuous; (2): for each fixed $x \in X$, $y \mapsto \phi_f(x,y)$ is concave; (3): $\phi_f(x,x) \leq 0$ for all $x \in X$ and (4): $\sup_{(x,y)\in X\times X} \phi_f(x,y) < \infty$. Therefore $T(f) = \phi_f \in M_2$. For any $f, g \in Y_1$, note that $\rho(\phi_f, \phi_g) < 2\epsilon$ if $h(f,g) < \epsilon$. This implies the mapping T is continuous from (Y_1, h) to (M_2, ρ). □

For each $f \in Y_1$, let $F(\phi_f)$ denote the set of all KY point for the function ϕ_f defined in Lemma 8.4.15, it is easy to see that $x^* \in N(f)$ if and only if As an application of Theorem 2.5.31, we are able to establish following general existence of essential and connected components of Nash equilibria for the game f in Model (A) which was first given by Yu and Xiang [598].

Theorem 8.4.17. For each game $f = (f_1, f_2, \cdots, f_n) \in Y_1$, there exists at least one essential and connected component of Nash equilibria.

Proof. For each $f \in Y_1$, it follows that $\phi_f \in M_1$. Since $N(f) = F(\phi_f)$. Suppose the decomposition of connected pairwise compact subsets of $F(\phi_f)$

is as $F(\phi_f) = \cup_{\alpha \in \Lambda} C_\alpha(\phi_f)$, where Λ is an index set and $C_\alpha(\phi_f)$ is a connected component of $F(\phi_f)$. By Theorem 2.5.31, there exists $\alpha_0 \in \Lambda$ such that $C_{\alpha_0}(\phi_f)$ is an essential and connected component of $F(\phi_f)$. For any open set $O \supset C_{\alpha_0}(\phi_f)$, there exists $\delta > 0$ such that for any $\psi \in M_2$ with $\rho(\psi, \phi_f) < 2\delta$, $F(\psi) \cap O \neq \emptyset$. By Theorem 8.4.15, for each $g \in Y_1$ with $h(f, g) < \delta$, we have that $\rho(\phi_f, \phi_g) < 2\delta$. Therefore $N(g) \cap O = F(\psi) \cap O \neq \emptyset$, that is, $C_{\alpha_0}(\phi_f)$ is an essential and connected component of $N(f)$. The proof is complete. □

For a given game $f \in Y_1$, we recall that a Nash equilibrium point $y \in N(f)$ is said to be essential (see Yu [596]) if for each open neighborhood O of x in X, there exists $\delta > 0$ such that $E(g) \cap O \neq \emptyset$ for each $g \in Y_1$ with $h(f, g) < \delta$. The game $f \in Y_1$ is said to be weakly essential if there exists $x \in N(f)$ which is essential;and f is said to be essential if its Nash equilibrium points $x \in N(f)$ are essential.

A proof analogous to that of Theorem 2.5.32 implies the following result and thus its proof is omitted.

Theorem 8.4.18. If the game $f \in Y_1$ is such that the Nash equilibria set $N(f)$ is totally disconnected, then f is a weakly essential game. In particular, if $N(f)$ is a singleton set, then f is an essential game.

Kohlberg and Mertens [323] show that for any finite n-person noncooperative game, its Nash equilibria set has at least one essential and connected component. Theorem 8.4.17 improves their results.

Chapter 9

Applications to Mathematical Economics

9.1 Introduction

In this chapter, as applications of KKM theory developed in Chapter 2, mainly the Browder-Fan fixed point theorems and fixed point theorems, in Section 2, we shall first study the generalization of the famous Debreu-Gale-Nikaido Lemma (e.g., see Debreu [117], Gale [188] and Nikaido [387] and see also Border [48, Theorem 18.1, p.81]) which is fundamental to proving the existence of a market equilibrium of an economy (e.g., see Border [48] and Debreu [120]).

In Section 3, we shall first study the existence of endpoints for set-valued dynamic systems which are upper or lower semicontinuous in metric spaces. Then the existence, uniqueness and algorithms of endpoints for set-valued dynamic systems are either generalized contractions (defined in metric spaces) or topological contractions (defined in topological spaces which are not necessarily metric spaces). These results are then applied to derive the existence of Pareto optima for mappings which take values in ordered Banach spaces. Then the continuous stability of (generalized) nucleolus sets is obtained in Section 4.

Finally in Section 5, by the concept of *essential equilibria* for production economies, we establishd the generic stability result for production economies which tell us that 'most' production economies in the sense of the Baire category all equilibria are essential.

9.2 Debreu-Gale-Nikaido Lemma

In this section, as applications of minimax inequalities, we shall follow Tan and Yu [506] to deduce a generalization of the Debreu-Gale-Nikaido Lemma to obtain its a new generalization, then a result of the corresponding generalization due to Grandmont is derived.

The Debreu-Gale-Nikaido Lemma (in short, DGN Lemma), see Debreu [117], Gale [189] and Nikaido [387] (see also Border [48, Theorem 18.1, p.81]) is fundamental to proving the existence of solutions for market equilibria of many problems in mathematical economics, e.g. see Aubin [13], Border [48] and Debreu [120]. Let $\Delta_n = \{x = (x_i) \in \mathbb{R}^{n+1}_+ : \Sigma_{i=1}^{n+1} x_i = 1\}$. The original version of the DGL Lemma can be stated as follows and it can be easily proved by the Ky Fan minimax inequality.

Lemma 9.2.1(DGN Lemma). Let $T : \Delta_n \to 2^{\mathbb{R}^{n+1}}$ be an upper semicontinuous set-valued mapping with nonempty values such that
(1) for each $x \in \Delta_n$, the set $T(x) - \mathbb{R}^{n+1}_+$ is convex closed; and
(2) for each $x \in \Delta_n$, $\sup_{y \in T(x)} \langle y, x\rangle \geq 0$ (Walras's law).

Then there exists $\hat{x} \in \Delta_n$ such that $T(\hat{x}) \cap \mathbb{R}^{n+1}_+ \neq \emptyset$ (the set $\{\hat{x} \in \Delta_n, T(\hat{x}) \cap \mathbb{R}^{n+1}_+ \neq \emptyset\}$ is also called the set of *free disposal equilibrium prices*).

Proof. Define a function $f : \Delta_n \times \Delta_n \to \mathbb{R} \cup \{-\infty, +\infty\}$ by

$$f(x, y) = - \sup_{p \in T(x)} \langle p, x\rangle$$

for each $(x, y) \in \Delta_n \times \Delta_n$. Then it is clear that $f(x, y)$ is concave in y for each fixed $x \in \Delta_n$ as $y \mapsto \sup_{p \in T(x)} \langle p, y\rangle$ is convex; and lower $f(x, y)$ is upper semicontinuous in x for each fixed $y \in \Delta_n$. Then the Ky Fan minimax inequality implies that there exists $\hat{x} \in \Delta_n$ such that $\sup_{y \in \Delta_n} f(\hat{x}, y) \leq 0$, i.e., we have $\sup_{p \in T(\hat{x}) - \mathbb{R}^{n+1}_+} \langle p, y\rangle \leq 0$ for each $y \in \mathbb{R}^{n+1}$. Note that $\sup_{p \in -\mathbb{R}^{n+1}_+} \langle p, y\rangle = 0$ if $y \in \mathbb{R}^n$ and $\sup_{p \in -\mathbb{R}^{n+1}_+} \langle p, y\rangle = +\infty$ when $y \notin \mathbb{R}^{n+1}_+$. We know that $T(\hat{x}) - \mathbb{R}^{n+1}_+$ is convex and closed, (2) implies that $0 \in T(\hat{x}) - \mathbb{R}^{n+1}_+$, so that $T(\hat{x}) \cap \mathbb{R}^{n+1}_+ \neq \emptyset$. $\square$.

Recently, there were many generalizations of this DGN Lemma, see Granas and Liu [209], Grandmont [211], Mehta and Tarafdar [370], Neuefeind [385] and Yannelis [585]. The objective of this section is two-fold: (1) we first give some minimax inequalities both in the compact case and in the non-compact case using the concept of escaping sequences introduced by Border [48]; (2) as applications of the minimax inequalities, (a) we deduce a generalization of the Debreu-Gale-Nikaido Lemma due to Mehta and Tarafdar [370] which in turn generalizes that of Yannelis [585] and (b) we obtain a new generalization of Debreu-Gale-Nikaido Lemma, from which the corresponding generalization due to Grandmont [211] is derived.

If A is a nonempty subset of $\mathbb{R}^m$ and $p \in \mathbb{R}^m$, dist (p, A) denotes the distance from p to A. If $p = (p_1, \cdots, p_m) \in \mathbb{R}^m$, then $p \geq 0$ (resp., $p > 0$, $p \leq 0$) if $p_i \geq 0$ (resp., $p_i > 0$, $p_i \leq 0$) for all $i = 1, \cdots, m$.

9.2.1 Compact Case

In order to establish the generalization of the DGN Lemma, we first have the followiny Ky Fan type minimax inequality as an application of the Browder-Fan fixed point theorem (e.g., see Theorem 2.3.19 in Chapter 2).

Theorem 9.2.2. Let X be a nonempty compact convex subset of a Hausdorff topological vector space E and Y be a nonempty convex subset of a Hausdorff topological vector space F. Suppose that the real-valued function $f : X \times Y \to \mathbb{R}$ and the set-valued mapping $T : X \to 2^Y$ satisfy the following conditions:

(i) for each fixed $y \in Y$, $x \to f(x, y)$ is upper semicontinuous and quasi-concave (resp., $x \to f(x, y)$ is concave);

(ii) for each fixed $x \in X$, $y \to f(x, y)$ is lower semicontinuous and quasi-convex (resp., $y \to f(x, y)$ is lower semicontinuous and convex);

(iii) for each $x \in X$, $T(x)$ is nonempty compact, convex and $\min_{y \in T(x)} f(x, y) \leq 0$; and

(iv) for each $x \in X$ with $\{u \in X : \min_{y \in T(x)} f(u, y) > 0\} \neq \emptyset$, there is $\overline{x} \in X$ such that

$$x \in \operatorname{int}\{v \in X : \min_{y \in T(v)} f(\overline{x}, y) > 0\}.$$

Then there exists $(x^*, y^*) \in X \times Y$ with $y^* \in T(x^*)$ such that $f(x, y^*) \leq 0$ for all $x \in X$.

Proof. Define the set-valued mapping $F : X \to 2^Y$ by $F(x) = \{u \in X : \min_{y \in T(x)} f(u, y) > 0\}$ for each $x \in X$. Fix an $x \in X$. For any $u_1 \in F(x)$ and $u_2 \in F(x)$, by (ii), (iii), there exists $y_1 \in T(x)$ and $y_2 \in T(x)$ such that $f(u_1, y_1) = \min_{y \in T(x)} f(u_1, y) > 0$ and $f(u_2, y_2) = \min_{y \in T(x)} f(u_2, y) > 0$. For any $\alpha \in [0, 1]$, there exists $y_\alpha \in T(x)$ such that $f(\alpha u_1 + (1 - \alpha) u_2, y_\alpha) = \min_{y \in T(x)} f(\alpha u_1 + (1 - \alpha) u_2, y)$.

By (i),

$$\begin{aligned} f(\alpha u_1 + (1-\alpha)u_2, y_\alpha) &\geq \min\{f(u_1, y_\alpha), f(u_2, y_\alpha)\} \\ &\geq \min\{f(u_1, y_1), f(u_2, y_2)\} > 0\} \end{aligned}$$

so that $\alpha u_1 + (1-\alpha)u_2 \in F(x)$. Thus $F(x)$ is convex for each $x \in X$.

By (iii), $x \notin F(x)$ for all $x \in X$.

For each $x' \in X$, $F^{-1}(x') = \{v \in X : \min_{y \in T(v)} f(x', y) > 0\}$. By (iv), if $F(x) \neq \emptyset$, then there is $\overline{x} \in X$ such that $x \in \text{int } F^{-1}(\overline{x})$. Then the Browder-Fan fixed point theorem implies that there exists $x^* \in X$ such that $F(x^*) = \emptyset$, i.e. $\min_{y \in T(x^*)} f(u, y) \leq 0$ for all $u \in X$. Now, since conditions (i), (ii) and (iii) hold, by Sion's minimax inequality in [490], we have

$$\min_{y \in T(x^*)} \max_{x \in X} f(x, y) = \max_{x \in X} \min_{y \in T(x^*)} f(x, y) \leq 0$$

(resp., $\min_{y \in T(x^*)} \sup_{x \in X} f(x, y) = \sup_{x \in X} \min_{y \in T(x^*)} f(x, y) \leq 0$). Hence there exists $y^* \in F(x^*)$ such that $\max_{x \in X} f(x, y^*) \leq 0$ (resp., $\sup_{x \in X} f(x, y^*) \leq 0$), i.e., $f(x, y^*) \leq 0$ for all $x \in X$. □

As an application of Theorem 9.2.2, we have the following generalization of Gale-Nikaido-Debreu Lemma due to Mehta and Tarafdar [370, Theorem 8]

Corollary 9.2.3. Let E be a real Hausdorff locally convex topological vector space, E^* be the topological dual of E equipped with the weak* topology, C be a closed convex cone of E having an interior point e, $C^* = \{p \in E^* : \langle p, y \rangle \leq 0$ for all $y \in C\} \neq \{0\}$ be the dual cone of C, and $\Delta = \{p \in C^* : \langle p, e \rangle = -1\}$. Suppose that the set-valued mapping $T : \Delta \to 2^E$ satisfies the following conditions:

(i) for each $p \in \Delta$, $T(p)$ is nonempty compact convex and $\min_{y \in T(p)} \langle p, y \rangle \leq 0$; and

(ii) for each $p \in \Delta$ with $\{q \in \Delta : \min_{y \in T(p)} \langle q, y \rangle > 0\} \neq \emptyset$, there is $\overline{p} \in \Delta$ such that

$$p \in \text{ int } \{\overline{q} \in \Delta : \min_{y \in T(\overline{q})} \langle \overline{p}, y \rangle > 0\}.$$

Then there exists $p^* \in \Delta$ such that $T(p^*) \cap C \neq \emptyset$.

Proof. Set $X = \Delta$, then X is convex. Since Δ is equi-continuous and w^*-closed, the Alaoglu Theorem of Jameson [265, Theorem 3.8, p.123] implies that it is w^*-compact. Set $Y = E$ and $f(p, y) = \langle p, y \rangle$, then conditions of Theorem 9.2.2 hold so that there exists $p^* \in \Delta$ and $y^* \in T(p^*)$ such that $\langle p, y^* \rangle \leq 0$ for all $p \in \Delta$. We shall prove that $y^* \in C$ and hence $y^* \in T(p^*) \cap C$. If $y^* \notin C$, since E is locally convex and C is closed convex, by Corollary 3.10 of Conway

[92, p.111], there exists $r \in E^*$ with $r \neq 0$ such that $\sup_{y \in C}\langle r, y\rangle < \langle r, y^*\rangle$. Since $0 \in C$ and C is a cone, we must have $\sup_{y \in C}\langle r, y\rangle = 0$. It follows that $r \in C^*$ and $\langle r, y^*\rangle > 0$. Since $e \in \text{ int } C$, we have $r(e) < 0$. Let $\overline{r} = -\frac{r}{r(e)}$, then $\overline{r} \in \Delta$. But $\langle \overline{r}, y^*\rangle > 0$ which contradicts the fact that $\langle p, y^*\rangle \leq 0$ for all $p \in \Delta$. Hence we must have $y^* \in C$. □

By Remark 1 of Mehta and Tarafdar [370], Corollary 9.2.1 is more general than Theorem 3.1 of Yannelis [585]. Hence Theorem 9.2.1 also generalizes Theorem 3.1 of Yannelis [585].

In what follows we deduce another minimax inequality from Theorem 9.2.1.

Theorem 9.2.4. Let X be a nonempty compact convex subset of a Hausdorff topological vector space E, Y be a nonempty convex subset of a Hausdorff topological vector space F. Suppose that the real-valued function $f : X \times Y \to \mathbb{R}$ and the set-valued mapping $T : X \to 2^Y$ satisfy the following conditions:

(i) for each fixed $y \in Y$, $x \to f(x, y)$ is upper semicontinuous and quasi-concave (resp., $x \to f(x, y)$ is concave);

(ii) for each fixed $x \in X$, $y \to f(x, y)$ is lower semicontinuous and quasi-convex (resp., $y \to f(x, y)$ is lower semicontinuous and convex);

(iii) for each $x \in X$, $T(x)$ is nonempty compact convex and $\min_{y \in T(x)} f(x, y) \leq 0$; and

(iv) for each $x \in X$, $\{u \in X : \min_{y \in T(u)} f(x, y) \leq 0\}$ is closed.

Then there exists $(x^*, y^*) \in X \times Y$ with $y^* \in T(x^*)$ such that $f(x, y^*) \leq 0$ for all $x \in X$.

Proof. We only need to prove that condition (iv) of Theorem 1 holds: Indeed, let $x \in X$ be such that $\{u \in X : \min_{y \in T(x)} f(u, y) > 0\} \neq \emptyset$ and take $\overline{x} \in \{u \in X : \min_{y \in T(x)} f(u, y) > 0\}$, i.e., $\min_{y \in T(x)} f(\overline{x}, y) > 0$. Thus $x \notin \{u \in X : \min_{y \in T(u)} f(\overline{x}, y) \leq 0\}$. Since $\{u \in X : \min_{y \in T(u)} f(\overline{x}, y) \leq 0\}$ is closed, so that $x \in \text{ int } \{u \in X : \min_{y \in T(u)} f(\overline{x}, y) > 0\}$. □

As an application of Theorem 9.2.4, we derive the following minimax inequality due to Granas and Liu [209, Theorem (13.1)].

Corollary 9.2.5. Let X be a nonempty compact convex subset of a Hausdorff topological vector space E, Y be a nonempty convex subset of a Hausdorff topological space F. Let $T : X \to 2^Y$ be upper semicontinuous with nonempty compact convex values and $g : X \times Y \to \mathbb{R}$ satisfying one of the following conditions:

$$(I) \quad \begin{cases} \text{For each fixed } y \in Y, \quad x \to g(x,y) \text{ is lower semicontinuous} \\ \text{and quasi-convex;} \\ \text{for each fixed } x \in X, \quad y \to g(x,y) \text{ is upper semicontinuous} \\ \text{and quasi-concave.} \end{cases}$$

$$(II) \quad \begin{cases} \text{For each fixed } y \in Y, \quad x \to g(x,y) \text{ is convex;} \\ \text{for each fixed } x \in X, \quad y \to g(x,y) \text{ is upper semicontinuous} \\ \text{and concave.} \end{cases}$$

Then for each $\lambda \in \mathbb{R}$, one of the following properties holds:

(A) there exists $\overline{x} \in X$ such that $\max_{y \in T(\overline{x})} g(\overline{x}, y) < \lambda$; or

(B) there exists $(x^*, y^*) \in X \times Y$ with $y^* \in T(x^*)$ such that $\min_{x \in X} g(x, y^*) \geq \lambda$.

Proof. Assume that (I) (resp., (II)) holds. If (A) is not true, then for each $x \in X$, $\max_{y \in T(x)}[g(x,y) - \lambda] \geq 0$. Define $f : X \times Y \to \mathbb{R}$ by $f(x,y) = \lambda - g(x,y)$ for each $(x,y) \in X \times Y$, then conditions (i), (ii) and (iii) of Theorem 9.2.2 hold. Now fix an $x \in X$. Define $W : X \times Y \to \mathbb{R}$ by $W(u,y) = g(x,y) - \lambda$ for each $(u,y) \in X \times Y$, then W is upper semicontinuous. Since T is upper semicontinuous such that for each $u \in X$, $T(u)$ is compact, by Theorem 5 of Aubin and Cellina [14, p.52], the mapping $V : X \to \mathbb{R}$ defined by $V(u) = \max_{y \in T(u)}[g(x,y) - \lambda]$ for each $u \in X$ is upper semicontinuous. It follows that the set $\{u \in X : \min_{y \in T(u)} f(x,y) \leq 0\} = \{u \in X : V(u) \geq 0\}$ is closed. Thus condition (iv) of Theorem 9.2.2 also holds. Hence there exists $(x^*, y^*) \in X \times Y$ with $y^* \in T(x^*)$such that $f(x, y^*) \leq 0$ for all $x \in X$, i.e., $\min_{x \in X} g(x,y) \geq \lambda$. □

The following simple example shows that Theorem 9.2.4 is a true generalization of Theorem (13.1) of Granas and Liu [209].

Example 9.2.6. Let $E = F = \mathbb{R}$, $X = [0, \frac{\pi}{2}]$ and $Y = \mathbb{R}$. Define $g : X \times Y \to \mathbb{R}$ by $g(x,y) = y - \sin x$ for all $(x,y) \in X \times Y$. Then for each fixed $y \in Y$, $x \to g(x,y)$ is continuous and convex and for each fixed $x \in X$, $y \to g(x,y)$ is continuous and concave. Define $T : X \to 2^Y$ by

$$T(x) = \begin{cases} [\frac{1}{2}\sin x, \sin x], & \text{if } x \neq 0, \\ \{1\}, & \text{if } x = 0 \end{cases}$$

for each $x \in X$. Then T has nonempty compact convex values, but T is not upper semicontinuous at $x = 1$ so that Theorem (13.1) of Granas and Liu [209] is not applicable. However, if we let $f = -g$, then we have $\min_{y \in T(x)} f(x,y) \leq 0$ for each $x \in X$. We shall show that for each $x \in X$, the set $\{u \in X : \min_{y \in T(u)} f(x,y) >$

$0\}$ is open in X. Indeed, let $x \in X$ be arbitrarily fixed. If $u \in X$ is such that $\min_{y\in T(u)} f(x,y) > 0$, we must have $0 < u < x$. Let $\delta > 0$ be such that $(u-\delta, u+\delta) \subset (0,x)$. It follows that for each $v \in (u-\delta, u+\delta)$, $\min_{y\in T(v)} f(x,y) = \sin x - \sin v > 0$. This shows that the set $\{u \in X : \min_{y\in T(u)} f(x,y) > 0\}$ is open in X so that the set $\{u \in X : \min_{y\in T(u)} f(x,y) \le 0\}$ is closed in X. Therefore Theorem 9.2.4 is applicable.

9.2.2 Non-Compact Case

We shall prove the following minimax inequality on a non-compact set.

Theorem 9.2.7. Let X be a nonempty subset of a Hausdorff topological vector space E such that $X = \cup_{n=1}^{\infty} C_n$ where $\{C_n\}_{n=1}^{\infty}$ is an increasing sequence of nonempty compact convex subsets of X, Y be a nonempty convex subset of a Hausdorff topological vector space F. Suppose that the real-valued function $f : X \times Y \to \mathbb{R}$ and the set-valued mapping $T : X \to 2^Y$ satisfy the following conditions:

(i) for each fixed $y \in Y$, $x \to f(x,y)$ is upper semicontinuous and quasi-concave (resp., $x \to f(x,y)$ is concave);

(ii) for each fixed $x \in X$, $y \to f(x,y)$ is lower semicontinuous and quasi-convex (resp., $y \to f(x,y)$ is lower semicontinuous and convex);

(iii) for each $x \in X$, $T(x)$ is nonempty compact convex and $\min_{y\in T(x)} f(x,y) \le 0$;

(iv) for each $n \in \mathbb{N}$ and each $x \in C_n$ with $\{u \in C_n : \min_{y\in T(x)} f(u,y) > 0\} \neq \emptyset$, there is $\overline{x} \in C_n$ such that $x \in \operatorname{int}\{v \in C_n : \min_{y\in T(v)} f(\overline{x},y) > 0\}$; and

(v) for each sequence $\{x_n\}_{n=1}^{\infty}$, where $x_n \in C_n$ for each $n = 1, 2, \cdots$, which is escaping from X relative to $\{C_n\}_{n=1}^{\infty}$ and each sequence $\{y_n\}_{n=1}^{\infty}$, where $y_n \in T(x_n)$ for each $n = 1, 2, \cdots$, there exists $n_0 \in \mathbb{N}$ and $x'_{n_0} \in C_{n_0}$ with $f(x'_{n_0}, y_{n_0}) > 0$.

Then there exists $(x^*, y^*) \in X \times Y$ with $y^* \in T(x^*)$ such that $f(x, y^*) \le 0$ for all $x \in X$.

Proof. For each $n \in \mathbb{N}$, by Theorem 9.2.1, there exists $(x_n, y_n) \in C_n \times Y$ with $y_n \in T(x_n)$ such that $f(x, y_n) \le 0$ for all $x \in C_n$. Suppose that the sequence $\{x_n\}_{n=1}^{\infty}$ were escaping from X relative to $\{C_n\}_{n=1}^{\infty}$. By (v), there exists $n_0 \in N$ and $x'_{n_0} \in C_{n_0}$ with $f(x'_{n_0}, y_{n_0}) > 0$ which is a contradiction. Therefore the sequence $\{x_n\}_{n=1}^{\infty}$ is not escaping from X relative to $\{C_n\}_{n=1}^{\infty}$, so that some subsequence of $\{x_n\}_{n=1}^{\infty}$ must lie entirely in some C_{n_1}. Since C_{n_1} is compact, there exists a subnet $\{z_\alpha\}_{\alpha\in\Gamma}$ of $\{x_n\}_{n=1}^{\infty}$ in C_{n_1} and $x^* \in C_{n_1}$ such that $z_\alpha \to x^*$. Let $z_\alpha = x_{n(\alpha)}$ where $n(\alpha) \to \infty$. If $\{u \in X : \min_{y\in T(x^*)} f(u,y) > 0\} \neq \emptyset$, there exists $n_2 \geq n_1$ such that $\{u \in C_{n_2} : \min_{y \in T(x^*)} f(u,y) > 0\} \neq \emptyset$. By (iv), there is $\overline{x} \in C_{n_2}$ such that $x^* \in int\{v \in C_{n_2} : \min_{y \in T(v)} f(\overline{x}, y) > 0\}$. Since $z_\alpha \to x^*$, there is α_0 such that $n(\alpha_0) \geq n_2$ and $\min_{y \in T(z_{\alpha_0})} f(\overline{x}, y) > 0$, hence $f(\overline{x}, y_{n(\alpha_0)}) > 0$, this contradicts the fact that $\overline{x} \in C_{n(\alpha_0)}$ and $f(\overline{x}, y_{n(\alpha_0)}) \leq 0$. Therefore $\{u \in X : \min_{y \in T(x^*)} f(u,y) > 0\} = \emptyset$, i.e. $\min_{y \in T(x^*)} f(u,y) \leq 0$ for all $u \in X$. By Corollary 3.5 of Sion [490], we have

$$\min_{y \in T(x^*)} \sup_{x \in X} f(x,y) = \sup_{x \in X} \min_{y \in T(x^*)} f(x,y) \leq 0.$$

Hence there exists $y^* \in T(x^*)$ such that $\sup_{x \in X} f(x, y^*) \leq 0$, i.e., $f(x,y^*) \leq 0$ for all $x \in X$. □

As an application of Theorem 9.2.7, we obtain the following generalization of the Debreu-Gale-Nikaido Lemma.

Theorem 9.2.8. Let $\Delta = \{p \in \mathbb{R}^m : p \geq 0, \sum_{i=1}^{m} p_i = 1\}$, $S = \{p \in \mathbb{R}^m : p > 0, \sum_{i=1}^{m} p_i = 1\}$ and $C_n = co\{p \in S : \text{dist } (p, \Delta \backslash S) \geq \frac{1}{n}\}$ for $n = 1, 2, \cdots$. Suppose that the set-valued map $T : S \to 2^{R^m}$ satisfies the following conditions:

(i) T is upper semicontinuous such that for each $p \in S$, $T(p)$ is nonempty compact convex;

(ii) for each $p \in S$ and each $y \in T(p)$, $\langle p, y \rangle = 0$; and

(iii) for each sequence $\{p_n\}_{n=1}^\infty$, where $p_n \in C_n$ for each $n = 1, 2, \cdots$, which is escaping from X relative to $\{C_n\}_{n=1}^\infty$ and for each sequence $\{y_n\}_{n=1}^\infty$, where $y_n \in T(p_n)$ for each $n = 1, 2, \cdots$, there exists $n_0 \in \mathbb{N}$ and $p'_{n_0} \in C_{n_0}$ with $\langle p'_{n_0}, y_{n_0} \rangle > 0$.

Then there exists $p^* \in S$ such that $0 \in T(p^*)$.

Proof. The conclusion clearly holds for $m = 1$. Now suppose that $m > 1$. We may assume that each C_n is nonempty. Note that each C_n is compact and convex and $S = \cup_{n=1}^\infty C_n$. Set $X = S$ and $Y = \mathbb{R}^m$. Let $f : X \times Y \to \mathbb{R}$ be defined by $f(p,y) = \langle p, y \rangle$ for each $(p,y) \in X \times Y$. Then similar to the proof of Corollary 9.2.5, we can prove that for each $n \in \mathbb{N}$ and each $p \in C_n$, the set $\{u \in C_n : \min_{y \in T(u)} f(p,y) \leq 0\}$ is closed. Also, similar to the proof of Theorem 9.2.4, condition (iv) of Theorem 9.2.7 holds. By Theorem 9.2.7, there exists $(p^*, y^*) \in S \times Y$ with $y^* \in T(p^*)$ such that $\langle p, y^* \rangle \leq 0$ for all $p \in S$.

If $y^* \leq 0$ does not hold, there is $i \in \{1, \cdots, m\}$ such that $y_i^* > 0$. We choose a with $0 < a < 1$ such that

$$\frac{1-a}{m-1} \sum_{j \neq i, 1 \leq j \leq m} y_j^* + a y_i^* > 0.$$

Let $p_i = a, p_j = \frac{1-a}{m-1}(j \neq i)$, then $p \in S$ and $\langle p, y^* \rangle > 0$, which is impossible. Thus we must have $y^* \leq 0$. On the other hand, since $p^* \in S$ and $\langle p^*, y^* \rangle = 0$, we obtain $y^* = 0$ and hence $0 \in T(p^*)$. □

Finally, we shall deduce the following generalization of Debreu-Gale-Nikaido Lemma due to Grandmont [211, Lemma 1]:

Corollary 9.2.9. Let $\Delta = \{p \in \mathbb{R}^m : p \geq 0, \sum_{i=1}^m p_i = 1\}$ and $S = \{p \in \mathbb{R}^m : p > 0, \sum_{i=1}^m p_i = 1\}$. Suppose that the set-valued mapping $T : S \to 2^{\mathbb{R}^m}$ satisfies the following conditions:

(i) T is upper semicontinuous such that for each $p \in S$, $T(p)$ is nonempty compact convex;

(ii) for each $p \in S$ and each $y \in T(p)$, $\langle p, y \rangle = 0$; and

(iii) for each sequence $\{p_n\}_{n=1}^\infty$ is S with $p_n \to p \in \Delta \backslash S$ and each sequence $\{y_n\}_{n=1}^\infty$, where $y_n \in T(p_n)$ for each $n = 1, 2, \cdots$, there is $\overline{p} \in S$ such that $\langle \overline{p}, y_n \rangle > 0$ for infinitely many n.

Then there exists $p^* \in S$ such that $0 \in T(p^*)$.

Proof. We shall show that condition (iii) of Theorem 9.2.8 holds. Indeed, let $\{p_n\}_{n=1}^\infty$ be a sequence, where $p_n \in C_n = co\{p \in S : \text{dist}\ (p, \Delta \backslash S) \geq \frac{1}{n}\}$ for $n = 1, 2, \cdots$, which is escaping from S relative to $\{C_n\}_{n=1}^\infty$ and $\{y_n\}_{n=1}^\infty$ be another sequence, where $y_n \in T(p_n)$ for $n = 1, 2, \cdots$. Since $\{p_n\}_{n=1}^\infty$ is a sequence in Δ and Δ is compact, without loss of generality, we may suppose that $p_n \to p^* \in \Delta \backslash S$. By (iii), there is $\overline{p} \in S$ such that $\langle \overline{p}, y_n \rangle > 0$ for infinitely many n. Since $S = \cup_{n=1}^\infty C_n$, there is $n_1 \in \mathbb{N}$ such that $\overline{p} \in C_n$ for all $n \geq n_1$. Choose any $n_0 \geq n_1$ such that $\langle \overline{p}, y_{n_0} \rangle > 0$, condition (iii) of Theorem 9.2.8 holds so that the conclusion follows. □

By Remark 18.15 of Boreder [48, p.86], the hypotheses of Lemma 1 of Grandmont [211] are weaker than the hypotheses of Lemma 2 of Neuefeind [385]. Hence Theorem 9.2.8 also generalizes Lemma 2 of Neuefeind [385].

9.3 Existence of Pareto Optima

In this section, we first study the existence of endpoints for set-valued dynamic systems, which are upper or lower semicontinuous in metric spaces. Then the existence, uniqueness and algorithms of endpoints for set-valued dynamic systems, which are either generalized contractions (defined in metric spaces) or topological contractions (defined in topological spaces which do not necessarily have any metric measures). These results are then applied to derive the existence of Pareto optima for mappings which take values in ordered Banach spaces. Finally the existence results of Pareto optima will be studied to establish the stability of (generalized) nucleolus sets in Section 4.

The study of set-valued dynamic systems has received much attention in recent years. For example, Aubin and Siegel [17] established some existence and stability of endpoint for set-valued dynamic systems (see also Aubin and Ekeland [15]) and Justman [273] provided some conditions that guarantee the existence of an endpoint and conditions that guarantee that each dynamic process converges and the limit of a dynamic process is an endpoint; while Maschler and Peleg [361] investigated the stability of (generalized) nucleolus sets for set-valued dynamic systems, and gave their applications to the study of game theory. This section will continue in the same stream. We shall first study the existence of endpoints for set-valued dissipative dynamic systems which are upper or lower semicontinuous in metric spaces. Then the existence, uniqueness and algorithms of endpoints for set-valued dynamic systems which are either generalized contractions in metric spaces or topological contractions in topological spaces are given. These existence theorems of endpoints for set-valued dynamic systems are then used to derive the existence of Pareto optima for mappings which take values in ordered Banach spaces. Finally the stability of (generalized) nucleolus sets is also established.

Now we recall some notions. If X is a nonempty set, we shall denote by 2^X the family of all subsets of X. The real line and the set of all positive integers are denoted by $\mathbb{R}$ and $\mathbb{N}$, respectively. We shall denote by $\mathbb{R}_+$ the non-negative part of $\mathbb{R}$. If $f : X \to \mathbb{R} \cup \{+\infty\}$ is an extended function, the domain of f is denoted by the set $Dom f := \{x \in X : f(x) \neq +\infty\}$. The function f is said to be *proper* if the domain of f is not empty, i.e., $Dom f \neq \emptyset$.

Definition 9.3.1. Let X be a nonempty subset of a topological space E and $F : X \to 2^E$ a set-valued mapping. A point $x \in X$ is said to be an *endpoint* of F if $F(x) = \{x\}$. We denote by $E(F)$ the set of all endpoints of F. If X is a subset of a topological space Y, we denote by $cl_Y X$ the closure of X in Y, or simply clX so there is no confusion.

9.3.1 Set-Valued Dynamic Systems

Definition 9.3.2. Let X be a metric space with the metric d. Following Aubin and Siegel [17], a *set-valued dynamic system* F on X is a set-valued mapping $F : X \to 2^X$ which takes nonempty values. Any infinite sequence $x^+ = x^0, x^1, \cdots, x^n, \cdots$, such that $x^{n+1} \in F(x^n)$ for each $n \in \mathbb{N}$ is called a *motion* of the system F at x_0. The set $\mathcal{F}(x^+) = \{x^n : n \in \mathbb{N} \text{ and } x^{n+1} \in F(x^n)\}$ is called the *trajectory of this motion* or *F-sequence* starting at x^0.

Definition 9.3.3. Let $U : X \to \mathbb{R}_+ \cup \{+\infty\}$ be a real-valued function with $DomU \neq \emptyset$. A dynamic system $F : X \to 2^X$ is said to be *dissipative* (with respect to U) if for each $x \in X$,

$$U(y) + d(x, y) \leq U(x) \text{ for all } y \in F(x). \tag{9.3.1}$$

We first have the following fixed point result.

Theorem 9.3.4. Let X be a complete metric space and $F : X \to 2^X$ a upper semicontinuous mapping with nonempty compact values such that the mapping $U'_F : X \to \mathbb{R}_+ \cup \{+\infty\}$ defined by

$$U'_F(x) = \inf\{\Sigma_{i=0}^{\infty} d(x^n, x^{n+1}) | (x^n)_{n=0}^{\infty} \text{ is any } F\text{-sequence with } x^0 = x\}$$

for each $x \in X$, is proper (i.e., $DomU'_F \neq \emptyset$). Then there exists a motion $\mathcal{F}(x^+)$ starting at some $x \in DomU'_F$ which converges to a limit $\hat{x}$, which is a fixed point of F, i.e., $\hat{x} \in F(\hat{x})$.

Proof. Since F is upper semicontinuous with nonempty compact values, then for each $x \in X$, there exists $y \in F(x)$ such that $U'_F(x) \geq U'_F(y) + d(x, y)$ by Proposition 4 of Aubin and Ekeland [15, p.242]. Note that the graph of F is closed and $DomU'_F \neq \emptyset$, by Proposition 2 of Aubin and Ekeland [15, p.240], there is a motion $\mathcal{F}(x^+)$ starting at some $x \in DomU'_F$ which converges to a limit $\hat{x}$, which is a fixed point of F, i.e., $\hat{x} \in F(\hat{x})$. □

In order to consider the existence of endpoints for set-valued dissipative dynamic systems which are lower semicontinuous, we need the following result.

Lemma 9.3.5. Let X be a complete metric space and $F : X \to 2^X$ a set-valued dynamic system. Suppose that F is dissipative with respect to the lower semicontinuous function $U : X \to \mathbb{R}_+ \cup \{+\infty\}$. Then F has an endpoint $\hat{x}$ in X, i.e., $F(\hat{x}) = \{\hat{x}\}$.

Proof. It is Proposition 1.3 of Aubin and Ekeland [15, p.248]. □

Lemma 9.3.6. Let X be a metric space and $F : X \to 2^X$ a set-valued dynamic system. Then the function $U_F : X \to \mathbb{R}_+ \cup \{+\infty\}$ defined by

$$U_F(x) = \sup\{\Sigma_{i=0}^{\infty} d(x^n, x^{n+1}) | (x^n)_{n=0}^{\infty} \text{ is any } F\text{-sequence with } x^0 = x\}$$

for each $x \in X$, satisfies the following inequality:

$$U_F(y) + d(x, y) \leq U_F(x) \text{ for all } y \in F(x)$$

for each $x \in X$. Moreover, if U is any other non-negative valued function which satisfies the inequality $(*)$, then $U_F(x) \leq U(x)$ for all $x \in X$.

Proof. It is clear that for each $x \in X$, $U_F(y) + d(x, y) \leq U_F(x)$ for all $y \in F(x)$ by the definition of U_F. Moreover it is trivial to verify that U_F is the smallest non-negative valued function which satisfies the property $(*)$.

Indeed, let U be any non-negative valued function satisfying $(*)$. Now for any F-sequence $(x^n)_{n=0}^{\infty}$ starting at $x^0 = x$. Note that $U(x^n) \geq d(x^n, x^{n+1}) + U(x^{n+1})$ and $U(x^n) \geq 0$ for all $n = 0, 1, \cdots, \infty$, we have that $U(x) \geq \Sigma_{n=0}^{\infty} d(x^n, x^{n+1})$. Thus $U_F(x) \leq U(x)$ by the definition of U_F. □

Lemma 9.3.7. Let X be a metric space and $F : X \to 2^X$ a lower semicontinuous set-valued mapping with nonempty values. Then the function $U_F : X \to \mathbb{R}_+ \cup \{+\infty\}$ defined by formula (b) in Lemma 9.3.6 is lower semicontinuous.

Proof. It is Lemma 2.1 of Maschler and Peleg [361, p.987]. □

Now we have the following result which is due to Mascheler and Peleg [361, Remark 2.8, p.988].

Theorem 9.3.8. Let X be a complete metric space and $F : X \to 2^X$ a lower semicontinuous mapping with nonempty values. Suppose that the function $U_F : X \to \mathbb{R}_+ \cup \{+\infty\}$ defined by the formula (b) is proper, i.e., $DomU_F \neq \emptyset$. Then F has an endpoint $\hat{x} \in X$, i.e., $F(\hat{x}) = \{\hat{x}\}$.

Proof. Note that F is lower semicontinuous, so that the function U_F is lower semicontinuous by Lemma 9.3.7. Then the conclusion follows from Lemma 9.3.5. □

For the convenience of our discussion, we shall call the function U_F defined in formula (b) of Lemma 9.3.3 the *function induced* by F in the remainder of this section.

As an immediate consequence of Theorem 9.3.8, we have the following result which is Corollary 10 of Aubin and Ekeland [15, p.246].

Corollary 9.3.8. Let X be a complete metric space and the dissipative dynamic system $F : X \to 2^X$ lower semicontinuous. Then F has an endpoint $\hat{x}$ in X, i.e., $F(\hat{x}) = \{\hat{x}\}$.

Proof. Since F is dissipative, there exists a function $U : X \to \mathbb{R}_+ \cup \{+\infty\}$ such that for each $x \in X$,

$$U(y) + d(y, x) \leq U(x) \text{ for all } y \in F(x)$$

and $DomU \neq \emptyset$. Now let $U_F : X \to \mathbb{R}_+ \cup \{+\infty\}$ be the induced function of F, i.e.,

$$U_F(x) = \sup\{\Sigma_{i=0}^{\infty} d(x^n, x^{n+1}) | (x^n)_{n=0}^{\infty} \text{ is any } F\text{-sequence with } x^0 = x\}$$

for each $x \in X$. By Lemma 9.3.6, $U_F(x) \leq U(x)$ for all $x \in X$. Hence $DomU_F \neq \emptyset$ because $DomU \neq \emptyset$. Note that F is lower semicontinuous with nonempty values, U_F is lower semicontinuous by Lemma 9.3.7. Therefore the conclusion follows from Theorem 9.3.8 and we complete the proof. □

In order to consider the existence, uniqueness and algorithms of endpoints for some special set-valued dynamic systems which are neither continuous in metric spaces nor defined in metric spaces, we recall the following definition which was given in [542].

Definition 9.3.9. Let (X, d) be a complete metric space. Then a set-valued mapping $F : X \to 2^X$ is said to be a *generalized contraction* if there exists a real number $\lambda \in (0,1)$ such that $\delta(F(A)) \leq \lambda\delta(A)$ for each nonempty closed bounded subset A of X with $F(A) \subset A$, where $F(A) := \cup_{x\in A}F(x)$ and $\delta(A)$ denotes the diameter of the subset A in X, i.e., $\delta(A) = \sup\{d(x,y) : x, y \in A\}$.

Suppose (X, d) is a metric space and $F : X \to 2^X$ is a set-valued mapping with nonempty values. For each $x \in X$, we define $F^0(x) := F(x) = F^1(x)$, $F^2(x) := F(F(x))$, $F^3(x) := F(F^2(x))$, and so on; i.e., $F^n(x) := \cup_{z\in F^{n-1}(x)}F(z)$ for each $n = 1, 2, \cdots$. A sequence $\{x^n\}_{n\in\mathbb{N}}$ with $x^n \in F(x^{n-1})$. (Here we do take $x^n \in F(x^{n-1})$ instead of $x^n \in F^{n-1}(x)$). Note that the set $F^{n-1}(x)$, in general is bigger than $F(x^{n-1})$ because F is a set-valued mapping) and $x^0 := x$ is said to be a *generalized sequence of iterations* with respect to x.

Note that a dynamic process or a trajectory of F starting at $x^0 \in X$ is a sequence $\{x^n\}_{n\in N}$ defined by $x^0 = x, x^n \in F(x^{n-1})$ for $n = 1, 2, \cdots$. We can easily see that each trajectory starting from $x^0 \in X$ is a generalized sequence of iterations with respect x^0, but the converse may not be true. Moreover it is clear that a Banach contraction mapping on a closed bounded set of a metric space is a generalized contraction mapping, however a generalized contraction mapping even in the single-valued case is not necessarily a Banach contraction. While a Banach contraction is always a continuous mapping, it is interesting to note that a generalized contraction even in the single-valued case may not be continuous. To substantiate our remarks, we provide an example after the following result.

Theorem 9.3.10. Let M be a nonempty closed bounded set of a complete metric space (X, d). Let F be a generalized contraction on M. Then F has a unique endpoint y, and for each $x \in X$, every generalized sequence of iterations of x converges to y.

Proof. We set $M_0 = M$ and define $M_n = clF(M_{n-1})$ for $n = 1, 2, \cdots$. Let $D = \cap_{i=1}^{\infty} M_i$. Then we can easily verify that the following holds:

(1) $M_0 \supset M_1 \supset M_2 \cdots \supset M_n, \cdots$ for $n \in \mathbb{N}$,
(2) $F(M_0) \subset M_n$ for $n = 1, 2, \cdots$, and
(3) $F(D) \subset D$.

It is clear that $D \neq \emptyset$. To see (3), let $u \in F(D)$ and k any positive integer. Since $u \in F(D)$, there is $a \in D$ such that $u \in F(a)$. Now note that $a \in M_k$, $F(a) \subset F(M_k)$ by (2). Hence $u \in M_k$. Since F is a generalized contraction and

(2) holds,

$$\begin{aligned}\delta(M_n) &= \delta(clF(M_{n-1})) = \delta(F(M_{n-1})) \leq \lambda\delta(M_{n-1}) \\ &= \lambda\delta(clF(M_{n-2})) = \lambda\delta(F(M_{n-2})) \leq \lambda^2\delta(M_{n-2}) \\ &\leq \cdots \leq \lambda^{n-1}\delta(M_1) \to 0\end{aligned}$$

as $n \to \infty$. In view of this and (1) we obtain by Cantor intersection theorem that D consists of single point, say, y. Hence it follows from (3) that $F(y) = \{y\}$.

We now prove the uniqueness of the endpoint y. Let $y' \in M$ such that $\{y'\} = F(y')$ and $y' \neq y$. Let $A = \{y', y\}$. Then $\delta(A) > 0$ and $F(A) = A$. Since F is generalized contraction so that $0 \neq \delta(A) = \delta(F(A)) \leq \lambda\delta(A)$ which is impossible, where $\lambda \in (0,1)$. Therefore $E(F)$ is a singleton.

Finally let $x \in M$ and $\{x^n\}_{n\in\mathbb{N}}$ be a generalized sequence of x. To show that $\{x^n\}$ converges to y we observe that

$$x^{n+1} \in F^n(x) \subset F^n(M_0) \subset M_{n-1}.$$

Now the convergence of $\{x^n\}_{n\in\mathbb{N}}$ to y follows from the already established fact that $\delta(M_n) \to 0$ as $n \to \infty$. This completes the proof. □

Example 9.3.11 (Tarafdar and Vyborny [542]). Let $F : [0,2] \to [0,2]$ by

$$F(t) = \begin{cases} t/2 + 1/2, & \text{if } t \in [0,1]; \\ t/2 - 1/2, & \text{if } t \in (1,2]. \end{cases}$$

In this case $F : [0,2] \to [0,1]$ and it is easy to see that F is a generalized contraction which is not continuous and, therefore not a Banach contraction. The point $t = 1$ is the unique fixed point of F.

So far, all existence, uniqueness and algorithms of endpoints for set-valued dynamic systems are given in metric spaces. In ordered to discuss these properties of endpoints in topological spaces without either metric or linear structures, we recall the following concept which was introduced by Tarafdar and Vyborny in [546]:

Definition 9.3.12. A set-valued mapping $F : X \to 2^X$ of a topological space X into itself is said to be a *set-valued topological contraction* if F is upper semicontinuous with nonempty closed values and if for each nonempty closed subset A of X with $F(A) = A$ implies that A is a single point, i.e., A is an endpoint of F.

Remark 9.3.13. If X is a bounded metric space (i.e., the diameter of $X = \delta(X)$ is finite), F is a set-valued self-mapping with nonempty closed values and is a generalized contraction with the contraction constant $\lambda \in (0,1)$, then clearly F is a topological contraction.

The following result was first proved in [546]; for the completeness, we shall include the outline of its proof here.

Theorem 9.3.14. Let X be a Hausdorff compact topological space and $F : X \to 2^X$ is a set-valued dynamic system which is a topological contraction. Then F has a unique endpoint $x^0 \in X$ and $\{x^0\} = \cap_{n=0}^{\infty} F^n(X)$, where $F^n(X) := F(F^{n-1}(X))$ for each $n \in \mathbb{N}$ and $F^0(X) := X$.

Proof. We shall first prove the existence and uniqueness of the endpoint for F. Let $\mathcal{F}$ be the family of all nonempty closed subsets S of X such that $F(S) \subset S$. Clearly $\mathcal{F}$ is nonempty as $X \in \mathcal{F}$. We order $\mathcal{F}$ partially with inclusion relation. Let C be a chain in $\mathcal{F}$. Then $M = \cap_{S \in C} F$ is a lower bound of C as $F(M) \subset M$, M is a closed set in X and $M \neq \emptyset$ by the finite intersection property. Hence by Zorn's lemma there is a minimal element $U \in \mathcal{F}$. Now we claim that $F(U) = U$. If not, let $D := F(U)$ be a proper subset of U. Clearly $D = F(U) \subset U$. As F is upper semicontinuous with nonempty compact values, S is a nonempty compact subset of X so that D is also closed. Hence $F(D) \subset F(U) = D$ which contradicts the minimality of U. Therefore $F(U) = U$. Since F is a set-valued topological contraction, we must have $U = \{x^0\}$ for some point $x^0 \in X$ and $F(x^0) = x^0$, i.e., x^0 is an endpoint of F. Let $\bar{S}$ be the set of all endpoints of F. Then $F(\bar{S}) = \bar{S}$ and $\bar{S}$ is a closed set by the upper semicontinuity of F (e.g., see Aubin and Frankowska [16] or Klein and Thompson [317, Theorem 7.3.8]). Hence $\bar{S} = \{x^0\}$ as F is a set-valued topological contraction and the proof is completed. □

Remark 9.3.15. We would like to point out that from the proof of Theorem 9.3.14, if topological contraction mapping F is single-valued with nonempty values, we have proved in fact that for each $x \in X$, the iteration $\{F^n(x) : n \in \mathbb{N}\}$ converges to the unique fixed point x_0, i.e., $\lim_{n\to\infty} F^n(x) = x_0$. For the convenience of the reader, we state it as follows (e.g., see Tarafdar and Yuan [546]).

Corollary 9.3.16. Let X be a Hausdorff topological compact space and $F : X \to X$ be a (single-valued) topological contraction self-mapping with nonempty values. Then there is a unique fixed point $x_0 \in X$. Furthermore, for each $x \in X$, the iteration $F^n(x)$ converges to the unique fixed point x_0 of F, i.e., $\lim_{n\to\infty} F^n(x) = x_0$.

Proof. It is Corollary 4 of Tarafdar and Yuan [546]. □

9.3.2 Applications to the Existence of Pareto Optima

In order to apply the existence results of a set-valued dynamical system and to establsih the existence of Pareto optimum in Banach spaces, we need to recall some notions and definitions.

A nonempty subset P of a real Banach space V is called a *cone* if $\overline{P} = P$, $P + P \subset P$, $\mathbb{R}_+ P \subset P$ and $P \cap (-P) = \{0\}$, where $\overline{P}$ is the closure of P and $\mathbb{R}_+ = [0, \infty)$. Each cone P induces in V an ordering '$\preceq$' defined by $x \preceq y$ if and only if $y - x \in P$. This relation '$\preceq$' is evidently reflexive, anti-symmetric and transitive. The pair (V, P) is called an *ordered Banach space* with the ordering '$\preceq$' induced by P (which is called the *positive cone* of V). The set $P^* = \{f \in V^*\ f(x) \geq 0 \text{ all } x \in P\}$ is called the *dual cone*, where V^* is the continuous dual of V, that is P^* is the set of order preserving continuous linear functional on V.

A cone P of V is said to be *normal* if and only if there exists a positive number e such that if $x, y \in P$ with $\|x\| \geq 1$ and $\|y\| \geq 1$, then $\|x + y\| \geq e$ (for other equivalent definitions, see Kelly and Namioka [287]). In what follows we always assume that V denotes an ordered Banach space and we write $x \prec y$ if $y - x \in P \setminus \{0\}$.

Definition 9.3.17. Let X be a nonempty convex subset of a Hausdorff real topological vector space. A mapping $F : X \to V$ is said to be *order* or *cone convex* if $F(\lambda x + \mu y) \preceq \lambda F(x) + \mu F(y)$ for all $x, y \in X$ and $\lambda \geq 0$, $\mu \geq 0$ with $\lambda + \mu = 1$, i.e. for such $x, y \in X$,

$$\lambda F(x) + \mu F(y) - F(\lambda x + \mu y) \in P.$$

In what follows, a Banach space V always means an ordered Banach space ordered by a normal cone C in the sense that $x \preceq y$ if and only if $y - x \in C$.

Lemma 9.3.18. Let X be a nonempty convex subset of a Haudorff real topological vector space and V an ordered Banach space. A mapping $F : X \to V$ is cone convex if and only if the function $f \cdot F$ is convex for all $f \in P^*$.

Proof. If F is cone convex, then $f \cdot F$ is convex for $f \in P^*$ as f is order preserving. Next, let $f \cdot F$ be convex for each $f \in V^*$. If possible, let for some $x, y \in X$ and $\lambda \geq 0$, $\mu \geq 0$ with $\lambda + \mu = 1$, $[\lambda F(x) + \mu F(y) - F(\lambda x + \mu y)] \notin P$. Then by a consequence of the separation theorem (e.g., see Shaefer [457, p.225]) there exists a $f \in P^*$ such that $f[\lambda F(x) + \mu F(y) - F(\lambda x + \mu y)] < 0$ which implies that $f \cdot F$ is not convex. □

Definition 9.3.19. Let X be a topological space. A mapping $F : X \to V$ is said to be *order* or *cone lower* (resp., *upper*) *semicontinuous* if $f \cdot F$ is lower (resp., upper) semicontinuous for each $f \in P^*$.

Lemma 9.3.20. If P is a normal cone of a real Banach space V and $x \in V$ with $x \succ 0$ (resp., $x \prec 0$), then there exists $f \in P^*$ such that $f(x) > 0$ (resp., $f(x) < 0$).

Proof. Let $x \succ 0$. Then $f(x) \geq 0$ for all $f \in P^*$. If possible, let $f(x) = 0$ for all $f \in P^*$. By the Hahn-Banach extension theorem there exists $p \in V^*$

such that $p(x) \neq 0$. Now since P is a normal cone, setting $p = g - h$, where $g, h \in P^*$ (e.g., see Theorem 23.5 of Kelly and Namioka [287, p.227]). Hence $p(x) = g(x) - h(x) = 0$ which is a contradiction. □

Definition 9.3.21. Let X be a nonempty set, V an ordered Banach space and $f : X \to V$ a (single-valued) mapping. Then (1) a point $a \in X$ is said to be a *Pareto maximal* (resp., *Parteo minimal*) *optimum* of f if $f(y) \succeq f(a)$ (resp., $f(y) \preceq f(a)$) for all $y \in X$ implies that $f(y) = f(a)$; and (2) if $a \in X$ is Pareto maximal (resp., minimal) optimum of f, then the set $N(f, a) = \{x \in X : f(a) = f(x)\}$ is said to be the (generalized) *nucleolus* or the *indifference set* of f with respect to a.

We denote by $P(f)$ the set of all Pareto maximal (resp., minimal) optima. It is easy to see that our definition for Pareto maximal (resp., minimal) optimum is equivalent to the corresponding definitions introduced in [537] and [538].

Also the Pareto optima always means *Pareto maximal optima* in the rest part of this section for simplicity unless otherwise specified.

Definition 9.3.22. Let X be a nonempty set, V an ordered Banach space and $f : X \to V$ a (single-valued) mapping and $\phi : X \to 2^X$ a set-valued mapping. Then f is said to be (1) *ϕ-monotone* if $x \in X$, $y \in \phi(x) \Rightarrow f(x) \preceq f(y)$; and (2) strictly *$\phi$-monotone* if $x \in X$, $y \in \phi(x)$ and $x \neq y \Rightarrow f(x) \prec f(y)$ (i.e., by Lemma 9.3.20 there exists $p \in P^*$ such that $pf(x) < pf(y)$ and $qf(x) \leq qf(y)$ for all $q \in P^*$).

It is clear that ϕ-monotone mapping generalizes ϕ-monotone vector function, introduced by Justman [273] which, in turn serves as the Lyapunov functions.

Remark 9.3.23. For a given nonempty set X and an ordered Banach space V and for each mapping $f : X \to V$, there exist two set-valued mappings $\phi_f, \hat{\phi}_f : X \to 2^X$ such that f is ϕ_f-monotone and f is $\hat{\phi}_f$-strictly monotone respectively, where $\phi_f, \hat{\phi}_f : X \to 2^X$ are defined by

$$\phi_f(x) = \{y \in X : f(x) \preceq f(y)\},$$

and

$$\hat{\phi}_f(x) = \{y \in X : f(x) \prec f(y)\}$$

for each $x \in X$.

Lemma 9.3.24. Let X be a nonempty set and V an ordered Banach space. Suppose $f : X \to V$ is a (single-valued) mapping and $\phi : X \to 2^X$ is a set-valued mapping. If f is strictly ϕ-monotone, then $N(f, a) \subset E(\phi)$ for each Pareto optimum a of f. Moreover if f is ϕ-monotone, then each $z \in E(\phi_f)$ is a Pareto optimum of f.

Proof. To see this, let $z \in N(f, a)$. If $z \notin E(\phi)$, then there exists $y \in X$ such that $y \neq z$ and $y \in \phi(z)$. The strict ϕ-monotone of f implies that $f(z) \preceq f(y)$

and $pf(z) < pf(y)$ for some $p \in P^*$ which contradicts the Pareto maximality of a of f. It is clear that if $f : X \to V$ is a mapping, then each $z \in E(\phi_f)$ is a Pareto optimum of f. In fact if $z \in E(\phi_f)$ and $z \notin P(f)$, then there exists $y \in X$ with $x \neq y$ and $f(z) \preceq f(y)$. Thus $y \in \phi_f(z)$ which is impossible since $\phi_f(z) = \{z\}$. □

Definition 9.3.25. Let X be a nonempty set and $\phi : X \to 2^X$ a set-valued mapping. Then a point $x_0 \in X$ is said to be a *maximal element* of ϕ if $\phi(x_0) = \emptyset$.

Lemma 9.3.26. Let X be a nonempty set and V be an ordered Banach space. Suppose $f : X \to V$ is a mapping. Then a point $x_0 \in X$ is a Pareto optimum of f if and only if $x_0 \in X$ is a maximal element of $\hat{\phi}_f$.

Proof. It is easy to verify from definitions. □

Now as applications of set-valued dynamic systems above, we have the following theorems on the existence of Pareto optima in metric spaces.

Theorem 9.3.27. Let X be a complete metric space, V an ordered Banach space and $f : X \to V$ a single-valued mapping. Suppose f is cone upper semicontinuous and the function $U_{\hat{\phi}_f} : X \to [0, \infty)$ which is defined from the function $\hat{\phi}_f$ by formula (b) of Lemma 9.3.6 is proper (i.e., $DomU_{\hat{\phi}_f} \neq \emptyset$), where $\hat{\phi}_f$ is a mapping defined by the formula (d) of Remark 9.3.23. Then $P(f) \neq \emptyset$.

Proof. In order to prove that $P(f) \neq \emptyset$, by Lemma 9.3.26 it is sufficient to show that the mapping $\hat{\phi}_f$ has a maximal element. If it were false, then $\hat{\phi}_f(x) \neq \emptyset$ for each $x \in X$. Note that f is cone upper semicontinuous, for each $y \in X$, the set $\hat{\phi}_f^{-1}(y) = \{x \in X : f(x) \prec f(y)\}$ is open in X, so that the mapping $\hat{\phi}_f$ is lower semicontinuous with nonempty values. By Theorem 9.3.8, there exists $x_0 \in X$ such that $x_0 = \hat{\phi}_f(x_0)$ which is impossible since $x \notin \hat{\phi}_f(x)$ for each $x \in X$. Thus there must exists some $x \in X$ such that $\hat{\phi}_f(x) = \emptyset$, so that $x \in P(f)$ and the proof is completed. □

Theorem 9.3.28. Let X be a closed bounded subset of a complete metric space, V an ordered Banach space and $f : X \to V$ a mapping. Suppose the induced set-valued mapping $\phi_f : X \to 2^X$ is a generalized contraction, where ϕ_f is a mapping defined by the formula (c) of Remark 9.3.23. Then $P(f) \neq \emptyset$.

Proof. By the assumption, the mapping ϕ_f is generalized contraction, so that there exists $x_0 \in X$ such that $x_0 = \phi_f(x_0)$ by Theorem 9.3.10. Then $x_0 \in P(f)$ follows from Lemma 9.3.24. □

Corresponding to Theorem 9.3.8, we also have the following result.

Theorem 9.3.29. Let X be a compact topological space, V an order Banach space and $f : X \to V$ a mapping. Suppose the set-valued mapping $\phi_f : X \to 2^X$ is a set-valued topological contraction, where ϕ_f is a mapping defined by the formula (c) of Remark 9.3.23. Then $P(f) \neq \emptyset$.

Proof. By the assumption, the mapping ϕ_f is topological contraction, so that there exists $x_0 \in X$ such that $x_0 = \phi_f(x_0)$ by Theorem 9.3.14. Then $x_0 \in P(f)$ follows from Lemma 9.3.24 and we complete the proof. □

Note that theorems 9.3.27, 9.3.28 and 9.3.29 give the existence of Pareto optima in metric spaces (which may not have linear structure), but they also need some other conditions such as continuous or generalized contraction hypotheses. In what follows, we shall study the existence of Pareto optimum solutions for f, defined in a nonempty convex subset of a topological vector space E under the assumption of compactness instead of continuity or generalized contraction conditions. In order to do so, we recall the following definition:

Definition 9.3.30. Let X be a convex subset of a vector space E and V a Banach space. A mapping $f : X \to V$ is said to be *quasi-cone concave* if for each $x \in X$, the set $\{y \in X : f(y) \succ f(x)\}$ is convex.

As an application of non-compact variant of the Browder-Fan fixed point theorem (e.g., see Theorem 4.3.4 in Chapter 4), we prove the existence of Pareto maximum for the mapping f in topological vector spaces as follows.

Theorem 9.3.31. Let X be a nonempty convex subset of a topological vector space E and V a Banach space. Suppose $f : X \to V$ is cone upper semicontinuous and cone quasi-concave. Furthermore there exists a nonempty compact and convex subset X_0 of X such that the set $\cap_{y \in X_0}\{x \in X : f(x) \not\prec f(y)\}$ is either empty or compact. Then $P(f) \neq \emptyset$.

Proof. We claim that the mapping $\hat{\phi}_f$ has a maximal element. Suppose it were false. Then $\hat{\phi}_f(x) \neq \emptyset$ for each $x \in X$ and $(\hat{\phi}_f)^{-1}(y) = \{x \in X : f(x) \prec f(y)\}$ is open since f is cone upper semicontinuous. Note that $\hat{\phi}_f(x)$ is convex since f is cone quasi-concave. Thus all hypotheses of Theorem 4.3.4 are satisfied and it implies that there exists $x_0 \in X$ such that $x_0 \in \hat{\phi}_f(x_0)$ which is impossible by the definition of $\hat{\phi}_f$. Thus there must exists $x \in X$ such that $\hat{\phi}_f(x) = \emptyset$. Thus $x \in P(f)$. □

As an immediate consequent of Theorem 9.3.31, we have the following result.

Corollary 9.3.32. Let X be a nonempty compact subset of a topological vector space E and V an ordered Banach space. Suppose $f : X \to V$ is cone upper semicontinuous and cone quasi-concave. Then $P(f) \neq \emptyset$.

9.4 Stability of Pareto Optima

In this section, as applications of the results in Section 3, we shall study the stability of Pareto optima for the mapping f which takes values in ordered Banach spaces.

Definition 9.4.1 (Maschler and Peleg [361]). Let Q be a nonempty subset of a topological space X. The set Q is called *stable* with respect to ϕ if for each open neighborhood U of Q, there exists a neighborhood V of Q such that if $x \in V$ and $\{x_n\}_{n\in\mathbb{N}}$ is a ϕ-sequence with $x_0 = x$, then $x_n \in U$ for $n = 1, 2, \cdots$. (or equivalently, $\phi^n(V) \subset U$ for $n = 1, 2, \cdots$).

We first have the following result.

Lemma 9.4.2. The set Q is stable w.r.t. ϕ if and only if for each open neighborhood U of Q, there exists another nonempty open subset $V \subset Q$ such that $\phi(V) \subset V$. Moreover if Q is closed and stable w.r.t. ϕ and X is regular, then $\phi(Q) \subset Q$.

Proof. If Q is stable and U is a open neighborhood of Q, then there is a neighborhood W of U such that $\phi^n(W) \subset U$ for $n \in \mathbb{N}$. Let $V = \cup_{n\in\mathbb{N}}\phi^n(W)$. Then V satisfies the stated properties. The converse is clear. Now suppose Q is stable and closed. Let $\{U_i : i \in I\}$ be the family of all nonempty open neighborhood of Q, then $\phi(Q) \subset \cap_{i\in I}U_i$. Note that Q is a closed subset of X, we claim that $\cap_{i\in I}U_i = Q$. It suffices to prove that $\cap_{i\in I}U_i \subset Q$. If not, there exists $w_0 \in \cap_{i\in I}U_i$ and $w_0 \notin Q$. As X is regular, and Q is closed, there exists a nonempty open neighborhood $O(w_0)$ of w_0 and a nonempty open neighborhood $\hat{U}$ of Q such that $O(w_0) \cap U = \emptyset$. But $x_0 \in \cap_{i\in I}U_i$, so that $x_0 \in \hat{U}$ as $\{U_i\}_{i\in I}$ is the family of all nonempty open neighborhoods of Q. Thus $\cap_{i\in I}U_i \subset Q$. Therefore we have $\phi(Q) \subset Q$. □

Theorem 9.4.3. Let $f : X \to V$ be a mapping from a compact metric space X into an ordered Banach space V. Let $N(f, a)$ be a generalized nucleolus of f. Suppose the following conditions hold:

(a) f is ϕ-monotone;

(b) f is cone upper semicontonuous on X; and

(c) f is continuous on $N(f, a)$.

Then $N(f, a)$ is stable with respect to ϕ.

Proof. First we note that by cone upper semicontinuity of f, the set $N(f, a) = \{x \in X : f(a) \preceq f(x)\}$ is a closed subset of X. Let $U \supset N(f, a)$ be an open subset of X. Then $S = X \setminus U$ is compact. Now for each $y \in S$, there exists by Lemma 9.3.20, a $p \in P^*$ and a positive integer $n \in \mathbb{N}$ such that

$pf(y) < pf(a) + 1/n$. Let

$$U_{p,n}(y) = \{x \in X : pf(x) < pf(a) + 1/n\}$$

for each $p \in P^*$ and $n \in \mathbb{N}$. Then by the upper semicontinuity of pf at $a \in X$, $U_{p,n}(y)$ is an open set containing y. Thus $\{U_{p,n}(y)\}_{p \in P^*, n \in \mathbb{N}}$ is an open covering of S. As S is compact, there is a finite subcover, say, $U_{p_1,n_1}(y_1), U_{p_2,n_2}(y_2), \cdots, U_{p_m,n_m}(y_m)$, i.e., $\cup_{i=1}^m U_{p_i,n_i}(y_i) \supset S$. For each $i = 1, 2, \cdots, m$, set

$$V_{p_i,n_i} = \{x \in X : p_i f(x) > p_i f(a) + 1/n_i\}$$

which is by the continuity of f on $N(f,a)$ an open set containing $N(f,a)$. Let $V = \cap_{i=1}^m V_{p_i,n_i}$. Then V is a neighborhood of $N(f,a)$ and is contained in U. To see this, let $y \in V$. If $y \notin U$, then $y \in S$ which would be a contradiction. Now if $x \in V$ and $\{x_n\}$ is ϕ-sequence with $x_0 = x$. Then by (a) we have for each $i = 1, 2, \cdots, m$,

$$p_i f(x_n) \geq p_i f(x_{n-1}) \geq \cdots \geq p_i f(x_0) > p_i f(a) + 1/n_i.$$

Hence $x_n \in V \subset U$, where $n = 1, 2, \cdots$. Thus $N(f,a)$ is stable. □

As an application of Theorem 9.4.3, we have the following result.

Theorem 9.4.4. Let X be a compact complete metric space, V an ordered Banach space and $f : X \to V$ a single-valued mapping which is continuous on $P(f)$. Suppose f is cone upper semicontinuous on X and the induced function $U_{\hat{\phi}_f}$ of $\hat{\phi}_f$ from X to $[0, \infty)$ is proper, i.e., $DomU_{\hat{\phi}_f} \neq \emptyset$, where $\hat{\phi}_f$ is a mapping defined by the formula (d) of Remark 9.3.23. Then $N(f,a)$ is stable with respect to $U_{\hat{\phi}_f}$ for each $a \in P(f)$.

Proof. By Theorem 9.3.27, $P(f) \neq \emptyset$. Note that X is compact, and then the conclusion follows from Theorem 9.4.3. □.

By employing Theorems 9.3.28, 9.3.29 and Corollary 9.3.32 and the same arguements used in the proof of Theorem 9.4.4, we have the following result.

Theorem 9.4.5. Let X be a compact subset of a complete metric space, V an ordered Banach space and $f : X \to V$ (single-valued) upper semicontinuous on X and continuous on $P(f)$. Suppose the set-valued mapping $\phi_f : X \to 2^X$ is a generalized contraction mapping, where ϕ_f is a mapping defined by the formula (d) of Remark 9.3.3. Then $N(f,a)$ is stable with respect to U_{ϕ_f} for each $a \in P(f)$.

Theorem 9.4.6. Let X be a compact subset of a metric space, V an ordered Banach space and let $f : X \to V$ be upper semicontinuous and continuous on $P(f)$. Suppose the set-valued mapping $\phi_f : X \to 2^X$ is a topological contraction mapping. Then $N(f,a)$ is stable with respect to U_{ϕ_f}, for each $a \in P(f)$.

Theorem 9.4.7. Let X be a nonempty compact subset of a topological vector space E and V be an ordered Banach space. Suppose $f : X \to V$ is cone upper semicontinuous and cone concave and f is continuous on $P(f)$. Then $N(f, a)$ is stable with respect to $\hat{\phi}_f$ for each $a \in P(f)$, where $\hat{\phi}_f$ is a mapping defined by the formula (d) of Remark 9.3.3.

In this section, as applications of endpoints for set-valued dynamic systems, we only studied existence of endpoints and stability of Pareto optima for mappings which take values in ordered Banach spaces. We conclude this section by noting that one possible application of our existence and stability results of dynamic systems is the study of stability of equilibria for economic models such as Arrow-Debreu economies which include private ownership economies and pure exchange economies et al. in mathematical economics as special cases.

9.5 Generic Stability of Production Economies

In this section, the concept of essential equilibria for production economies is given first. We then prove that in almost all production economies in the sense of the Baire category that all equilibria are essential.

In Section 8 of [128], Dierker introduced the concept of essential equilibria for pure exchange economies and proved that in '*most*' pure exchange economies (in the sense of the Baire category) all equilibria are essential.

The concept of essentiality for equilibria is a stability property. In [128], the stability of equilibria with perturbations on demand functions and with an initial endowment of each consumer were studied by Dierker [128].

In this section, the concept of essential equilibria for production economies is given first. We then study the stability of equilibria with perturbations on utility-maximizing consumptions and initial endowment of each consumer and on profit-maximizing productions of each producer. We also prove that in '*most* production economies (in the sense of the Baire category) all equilibria are essential.

Let (X, d) be a metric space and $K(X)$ be the space of all nonempty compact subsets of X equipped with the Hausdorff metric h which is induced by the metric d. For each $\epsilon > 0$ and $A \in K(X)$, let $U(\epsilon, A) = \{x \in X : d(u, x) < \epsilon$ for some $u \in A\}$. Let Y be a Hausdorff topological space and $F : Y \to K(X)$ be a set-valued mapping. Then (i) F is said to be upper semicontinuous (resp., lower semicontinuous) at $y \in Y$ if for each $\epsilon > 0$, there is an open neighborhood $O(y)$ of y in Y such that $F(y') \subset U(\epsilon, F(y))$ (resp., $F(y) \subset U(\epsilon, F(y')))$ for all $y' \in O(y)$; (2) F is said to be upper semicontinuous (resp., lower semicontinuous) on Y if it is upper semicontinuous (resp., lower semicontinuous) at each point $y \in Y$ and (3) F is said to be continuous at y if F is both upper semicontinuous and lower semicontinuous at $y \in Y$. Since F is compact-valued, if F is upper semicontinuous on Y, F is also called an USCO mapping. Recall that a subset $Q \subset Y$ is called a residual set in Y if it is a countable intersection of open dense subsets of Y.

The following Lemma 1 is due to Fort [178, Theorem 2] and we state it as:

Lemma 9.5.1. Let X be a metric space, Y be a Hausdorff topological space and $F : Y \to K(X)$ be an USCO mapping. Then the set of all points where F is lower semicontinuous is a residual set in Y.

Recall that a topological space X is a Baire space (e.g., see Dugundji [147, p.249]) if the intersection of each countable family of open dense sets in X is dense in X.

Lemma 9.5.2. Let X be a metric space, Y be a Baire space and F : $Y \to K(X)$ be an USCO mapping. Then the set of points where F is lower semicontinuous is a dense residual set in Y.

Proof. Since Y is a Baire space, a residual set in Y is dense, the result now follows from Lemma 9.5.1. □

In [408], Oxtoby introduced the notion of a pseudo-complete space as follows: A Hausdorff topological space Y is called quasi-regular if every nonempty open set in Y contains the closure of some nonempty open set in Y. A family $\mathbb{B}$ of nonempty open sets in Y is called a pseudo-base if every nonempty open set in Y contains at least one member of $\mathbb{B}$. The space Y is called pseudo-complete if Y is quasi-regular and if there exists a sequence $\{\mathbb{B}(k)\}_{k=1}^{\infty}$ of pseudo-bases in Y with the property that $\cap_{k=1}^{\infty} U_k \neq \emptyset$ whenever $U_k \in \mathbb{B}(k)$ and $U_k \supset \overline{U_{k+1}}$ for $k = 1, 2, \cdots$, where $\overline{U_{k+1}}$ denotes the closure of U_{k+1} in Y.

The mathematical model of a production economy which we shall study as follows (see also Debreu [120]):

Suppose that there are l such commodities. Let $P = \{x = (x_1, \cdots, x_l) \in \mathbb{R}^l : x_h > 0, h = 1, \cdots, l\}$ and $L = (0, \infty)$. The set $\Delta = \{x \in P : \Sigma_{h=1}^{l} x_h = 1\}$ is called the price simplex. Consider a production economy with m consumers and n producers. Suppose $e_i \in P$ is the initial endowment of the ith consumer, and for a given price vector $p \in \Delta$, the ith consumer chooses his utility-maximizing consumptions $\xi_i(p, p \cdot e_i) \subset \overline{P}$, $i = 1, 2, \cdots, m$ ($p \cdot e_i$ is the inner product of p and e_i) and the jth producer chooses his profit-maximizing productions $\eta_j(p) \subset \mathbb{R}^l$, $j = 1, \cdots, n$. The excess demand correspondence is defined by

$$\zeta(p, e) = \Sigma_{i=1}^{m} \xi_i(p, p \cdot e_i) - \Sigma_{j=1}^{n} \eta_j(p) - \Sigma_{i=1}^{m} e_i$$

where $e = (e_1, \cdots, e_m) \in P^m$ and $p \in \Delta$.

Lemma 9.5.3. Suppose that the following conditions hold:

(i) for each $p \in \Delta$ and each $w \in L$, $\xi_i(p, w)$ is nonempty compact convex and ξ_i is upper semicontinuous on $\Delta \times L$ for $i = 1, 2, \cdots, m$;

(ii) for each $p \in \Delta$, $\eta_j(p)$ is nonempty compact convex and η_j is upper semicontinuous and bounded from above on Δ for $j = 1, \cdots, n$;

(iii) for each $p \in \Delta$ and each $z \in \zeta(p, e)$, $p \cdot z = 0$ (Walrasian-Law); and

(iv) for each sequence $\{(p^k, w^k)\}_{k=1}^{\infty}$ in $\Delta \times L$ with $(p^k, w^k) \to (p, w) \in (\overline{\Delta} \setminus \Delta) \times L$, there is some i such that $d(0, \xi_i(p^k, w^k)) = \inf_{u \in \xi_i(p^k, w^k)} \|u\| \to \infty$.

Then there exists $p^* \in \Delta$ such that $0 \in \zeta(p^*, e)$.

Proof. Fix $e \in P^m$, by (i) and (ii), for each $p \in \Delta$, $T(p) = \zeta(p, e) = \Sigma_{i=1}^m \xi_i(p, p \cdot e) - \Sigma_{j=1}^n \eta_j(p) - \Sigma_{i=1}^m e_i$ is nonempty, compact convex and by Theorem 7.3.15 (ii) of Klein and Thompson [317], $T(p)$ is upper semicontinuous on Δ. Let $\{p^k\}_{k=1}^{\infty}$ be any sequence in Δ with $p^k \to p \in \overline{\Delta} \setminus \Delta$. If $z^k \in T(p^k)$ for $k = 1, 2, \cdots$, then since $w_i^k = p^k \cdot e_i \in L$ for $k = 1, 2, \cdots$, we have $w_i^k \to p \cdot e_i = w \in L$. By (i), (ii) and (iv), $d(0, T(p^k)) \to \infty$. Set $\hat{p} = (1/l, \cdots, 1/l) \in \Delta$, then there is k_1 such that $\hat{p} \cdot z^k > 0$ for all $k \geq k_1$. Thus all conditions of Corollary 9.2.9 (see also Lemma 1 of Grandmont [211] or Theorem 18.13 of Boreder [48]) are satisfied so that there exists $p^* \in \Delta$ such that $0 \in T(p^*) = \zeta(p^*, e)$. $\square$

Remark 9.5.4. The condition (iv) in Lemma 9.5.3 is a variant of the assumption (A) of Debreu [119, p.388] (see also Tarafdar and Thompson [541]). Condition (iv) in Lemma 9.5.3 expresses the idea that every commodity is demanded by some consumer.

Let C be the set of all $G = (\xi_1, \cdots, \xi_m; \eta_1, \cdots, \eta_n)$ which satisfy all conditions of Lemma 9.5.3. Let h be the Hausdorff metric on $K(\mathbb{R}^l)$ induced by the usual metric d on $\mathbb{R}^l$. Let $B = \{O(G, \epsilon) : G \in C \text{ and } 0 < \epsilon < 1\}$, where

$$\begin{aligned} O(G,\epsilon) &= \{\hat{G} = (\hat{\xi}_1, \cdots, \hat{\xi}_m; \hat{\eta}_1, \cdots, \hat{\eta}_n) \in C : \max_{1 \leq i \leq m} \sup_{(p,w) \in \Delta \times L} h(\hat{\xi}_i(p,w), \xi_i(p,w)) \\ &+ \max_{1 \leq j \leq n} \sup_{p \in \Delta} h(\hat{\eta}_j(p), \eta_j(p)) < \epsilon\}. \end{aligned}$$

Then B is a base for the topology of uniform convergence on C. If $\epsilon > 0$, denote

$$\begin{aligned} \overline{O}(G,\epsilon) &= \{\hat{G} = (\hat{\xi}_1, \cdots, \hat{\xi}_m; \hat{\eta}_1, \cdots, \hat{\eta}_n) \in C : \max_{1 \leq i \leq m} \sup_{(p,w) \in \Delta \times L} h(\hat{\xi}_i(p,w), \xi_i(p,w)) \\ &+ \max_{1 \leq j \leq n} \sup_{p \in \Delta} h(\hat{\eta}_j(p), \eta_j(p)) \leq \epsilon\}; \end{aligned}$$

it can be shown that the closure of $O(G, \epsilon)$ is $\overline{O}(G, \epsilon)$.

Lemma 9.5.5. C is a Baire space.

Proof. By the formulation (5.1) of Oxtoby [408], it is sufficient to prove that C is pseudo-complete. It is obvious that C is quasi-regular. Let

$$B(k) = \{O(G, a) : G \in C \text{ and } 0 < a < 1/k\}.$$

Then $B(k)$ is a pseudo-base for each $k = 1, 2, \cdots$. For each $k = 1, 2, \cdots$, let $U_k \in B(k)$ be such that $U_k \supset \overline{U_{k+1}}$. We need to prove that $\cap_{k=1}^{\infty} U_k \neq \emptyset$. Denote $U_k = O(G^k, a_k)$, then $0 < a_k < 1/k$, $k = 1, 2, \cdots$ and

$$O(G^1, a_1) \supset \overline{O}(G^2, a_2) \supset O(G^2, a_2) \supset \overline{O}(G^3, a_3) \supset \cdots$$

where $G^k = (\xi_1^k, \cdots, \xi_m^k; \eta_1^k, \cdots, \eta_n^k)$ for $k = 1, 2, \cdots$. Now for any $k = 1, 2, \cdots$ and any $q = 1, 2, \cdots$, since $G^{k+q} \in O(G^k, a_k)$, we have

$$\max_{1 \leq i \leq m} \sup_{(p,w) \in \Delta \times L} h(\xi_i^k(p, w), \xi_i^{k+q}(p, w)) < a_k < 1/k.$$

For any $i = 1, 2, \cdots, m$, $\{\xi_i^k(p, w)\}_{k=1}^{\infty}$ is a Cauchy sequence uniformly in $(p, w) \in \Delta \times L$, by Theorem 4.3.9 and Theorem 4.3.13 of Klein and Thompson [317], for each $(p, w) \in \Delta \times L$, there is a nonempty compact subset $\xi_i(p, w)$ of $\overline{P}$ such that $h(\xi_i^k(p, w), \xi_i(p, w)) \to 0$ as $k \to \infty$, uniformly in $(p, w) \in \Delta \times L$.

Similarly, for each $j = 1, 2, \cdots, n$ and each $p \in \Delta$, there is a nonempty compact convex subset $\eta_j(p)$ of $\mathbb{R}^l$ such that $h(\eta_j^k(p), \eta_j(p)) \to 0$ as $k \to \infty$, uniformly in $p \in \Delta$ and also for each $k = 1, 2, \cdots$, we have

$$\max_{1 \leq i \leq m} \sup_{(p,w) \in \Delta \times L} h(\xi_i^k(p, w), \xi_i(p, w))) + \max_{1 \leq j \leq n} \sup_{p \in \Delta} h(\eta_j^k(p), \eta_j(p)) \leq a_k.$$

Set $G = (\xi_1(p, w), \cdots, \xi_m(p, w); \eta_1(p), \cdots, \eta_n(p))$, we shall prove that $G \in C$ and $G \in \cap_{k=1}^{\infty} U_k$. For each $\epsilon > 0$, there is k_1 such that

$$\sup_{(p,w) \in \Delta \times L} h(\xi_i^k(p, w), \xi_i(p, w)) < \epsilon/3$$

for all $k \geq k_1$ and $i = 1, 2, \cdots, m$. For each $(p, w) \in \Delta \times L$, since $\xi_i^{k_1}$ is upper semicontinuous at $(p, w) \in \Delta \times L$, there is $\delta > 0$ such that $\xi_i^{k_1}(p', w') \subset U(\epsilon/3, \xi_i^{k_1}(p, w))$ whenever $p' \in \Delta$ with $\|p - p'\| < \delta$ and $w' \in L$ with $|w - w'| < \delta$ for all $i = 1, 2, \cdots, m$. Thus,

$$\xi_i(p', w') \subset U(\epsilon/3, \xi_i^{k_1}(p', w')) \subset U(2\epsilon/3, \xi_i^{k_1}(p, w)) \subset U(\epsilon, \xi_i(p, w))$$

and hence ξ_i is upper semicontinuous at $(p, w) \in \Delta \times L$ for all $i = 1, 2, \cdots, m$. By the same method, η_j is upper semicontinuous at $p \in \Delta$ for all $j = 1, \cdots, n$. Let $\{(p^q, w^q)\}_{q=1}^{\infty}$ be a sequence in $\Delta \times L$ with $(p^q, w^q) \to (p, w) \in (\overline{\Delta} \setminus \Delta) \times L$. Since $\{G^k\}_{k=1}^{\infty} \subset C$, without loss of generality, we may assume that $d(0, \xi_1^k(p^q, w^q)) \to \infty$ for all $k = 1, 2, \cdots$. Since $h(\xi_1^k(p^q, w^q), \xi_1(p^q, w^q)) < \epsilon/3$ for all $k \geq k_1$, we must also have $d(0, \xi_1(p^q, w^q)) \to \infty$ as $q \to \infty$. Now let $p \in \Delta$ and $z \in \zeta(p, e) = \Sigma_{i=1}^m \xi_i(p, p \cdot e_i) - \Sigma_{j=1}^n \eta_j(p) - \Sigma_{i=1}^m e_i$. We need to prove that $p \cdot z = 0$. Suppose the contrary that $p \cdot z \neq 0$. Since $\xi_i^k \to \xi_i$, $\eta_j^k \to \eta_j$ uniformly on $\Delta \times L$, there are k_2 and $x_i^{k_2} \in \xi_i^{k_2}(p, p \cdot e_i)$, $\eta_j^{k_2} \in \eta_j^{k_2}(p)$ for $i = 1, 2, \cdots, m$; $j = 1, 2, \cdots, n$ such that

$$p \cdot (\Sigma_{i=1}^m x_i^{k_2} - \Sigma_{j=1}^m y_j^{k_2} - \Sigma_{i=1}^m e_i) \neq 0.$$

This contradicts the Walrasian-Law that for each $p \in \Delta$ and each $z^{k_2} \in \zeta^{k_2}(p,e) = \Sigma_{i=1}^m \xi_i^{k_2}(p, p \cdot e_i) - \Sigma_{j=1}^n \eta_j^{k_2}(p) - \Sigma_{i=1}^m e_i$, $p \cdot z^{k_2} = 0$. Therefore $G \in C$.

By (*), $G \in \overline{O}(G^k, a_k)$ for all $k = 1, 2, \cdots$. Hence $G \in \cap_{k=2}^\infty \overline{O}(G^k, a_k) \subset \cap_{k=1}^\infty O(G^k, a_k) = \cap_{k=1}^\infty U_k$ so that $\cap_{k=1}^\infty U_k \neq \emptyset$. Thus C is pseudo-complete. □

Since P^m is a locally compact Hausdorff space, it is pseudo-complete by Oxtoby [408, p.164]. Let $Y = C \times P^m$, by Theorem 6 of [408], Y is pseudo-complete and by formulation (5.1) of Oxtoby [408] again, Y is a Baire space.

Definition 9.5.6. If $E = (\xi_1, \cdots, \xi_m; \eta_1, \cdots, \eta_n; e_1, \cdots, e_m) \in Y$, then $p \in \Delta$ is called an equilibrium point of the economy E if $0 \in \Sigma_{i=1}^m \xi_i(p, p \cdot e_i) - \Sigma_{j=1}^n \eta_j(p) - \Sigma_{i=1}^m e_i$.

Denote by $W(E)$ the set of all equilibria of the economy $E \in Y$; then $W(E) \neq \emptyset$ by Lemma 9.5.3. In the next part, we shall study the stability of $W(E)$.

Lemma 9.5.7. For each $E \in Y$, $W(E)$ is a compact set.

Proof. Since $W(E) \subset \overline{\Delta}$ and $\overline{\Delta}$ is compact, it is sufficient to prove that $W(E)$ is closed in $\overline{\Delta}$. Let $\{p^k\}_{k=1}^\infty$ be any sequence in $W(E)$ with $p^k \to p \in \overline{\Delta}$. Suppose $p \in \overline{\Delta} \setminus \Delta$. Since for each $i = 1, 2, \cdots, m$, $w_i^k = p^k \cdot e_i \in L$ so that $w_i^k \to p \cdot e_i = w_i \in L$, there is some $i \in \{1, \cdots, m\}$ such that

$$d(0, \xi_i(p^k, w^k)) \to \infty$$

as $k \to \infty$. Since each η_j is bounded from above, it follows from above that

$$d(0, \Sigma_{i=1}^m \xi_i(p^k, p^k \cdot e_i) - \Sigma_{j=1}^n \eta_j(p^k) - \Sigma_{i=1}^m e_i) \to \infty$$

as $k \to \infty$ which contradicts that

$$0 \in \Sigma_{i=1}^m \xi_i(p^k, p^k \cdot e_i) - \Sigma_{j=1}^n \eta_j(p^k) - \Sigma_{i=1}^m e_i$$

for $k = 1, 2, \cdots$. Hence $p \in \Delta$. If $0 \notin \Sigma_{i=1}^m \xi_i(p, p \cdot e_i) - \Sigma_{j=1}^n \eta_j(p) - \Sigma_{i=1}^m e_i$, since ξ_i is upper semicontinuous at $(p, p \cdot e_i)$, η_j is upper semicontinuous at p and $p^k \to p$ and $p^k \cdot e_i \to p \cdot e_i$, it is easy to show that there is k_1 such that $0 \notin \Sigma_{i=1}^m \xi_i(p^k, p^k \cdot e_i) - \Sigma_{j=1}^n \eta_j(p^k) - \Sigma_{i=1}^m e_i$ for all $k \geq k_1$. This contradicts that $p^k \in W(E)$ for $k = 1, 2, \cdots$. Therefore we must have $0 \in \Sigma_{i=1}^n \xi_i(p, p \cdot e_i) - \Sigma_{j=1}^n \eta_j(p) - \Sigma_{i=1}^n e_i$. Hence $p \in W(E)$ so that $W(E)$ is closed in $\overline{\Delta}$. □

By Lemma 9.5.7, the mapping $E \to W(E)$ indeed defines a set-valued mapping $W : Y \to K(\mathbb{R}^l)$.

Lemma 9.5.8. W is upper semicontinuous on Y.

Proof. Suppose that W were not upper semicontinuous at $E \in Y$, then there are $\epsilon_0 > 0$ and a sequence $\{E^k\}_{k=1}^\infty$ in Y with $E^k \to E$ such that for each $k = 1, 2, \cdots$, there exists $p^k \in W(E^k)$ with $p^k \notin U(\epsilon_0, W(E))$.

Since $\{p^k\}_{k=1}^{\infty} \subset \overline{\Delta}$ and $\overline{\Delta}$ is compact, we may assume that $p^k \to p \in \overline{\Delta}$. Note that we must have $p \notin U(\epsilon_0, W(E))$. If $p \in \overline{\Delta} \setminus \Delta$, then since $w_i^k = p^k \cdot e_i^k \in L$, $w_i^k \to p \cdot e_i \in L$, it follows that

$$d(0, \Sigma_{i=1}^m \xi_i(p^k, p^k \cdot e_i) - \Sigma_{j=1}^n \eta_j(p^k) - \Sigma_{i=1}^m e_i) \to \infty.$$

Since $\xi_i^k \to \xi_i$, $\eta_j^k \to \eta_j$ and $e_i^k \to e_i$ for $i = 1, 2, \cdots, m$ and $j = 1, \cdots, n$,

$$d(0, \Sigma_{i=1}^m \xi_i^k(p^k, p^k \cdot e_i^k) - \Sigma_{j=1}^n \eta_j^k(p^k) - \Sigma_{i=1}^m e_i^k) \to \infty.$$

This contradicts that $0 \in \Sigma_{i=1}^m \xi_i^k(p^k, p^k \cdot e_i^k) - \Sigma_{j=1}^n \eta_j^k(p^k) - \Sigma_{i=1}^m e_i^k$ for $k = 1, 2 \cdots$. Hence $p \in \Delta$.

If $0 \notin \Sigma_{i=1}^m \xi_i(p, p \cdot e_i) - \Sigma_{j=1}^n \eta_j(p) - \Sigma_{i=1}^m e_i$, since $\Sigma_{i=1}^m \xi_i(p, p \cdot e_i) - \Sigma_{j=1}^n \eta_j(p) - \Sigma_{i=1}^m e_i$ is compact, there is $\delta > 0$ such that

$$0 \notin U(\delta, \Sigma_{i=1}^m \xi_i(p, p \cdot e_i) - \Sigma_{j=1}^n \eta_j(p) - \Sigma_{i=1}^m e_i).$$

Since $\xi_i^k \to \xi_i$, $\eta_j^k \to \eta_j$ and $e_i^k \to e_i$, for $\epsilon = \delta/(2m + n)$, there is k_1 such that

$$h(\xi_i^k(p^k, p^k \cdot e_i^k), \xi_i(p^k, p^k \cdot e_i^k)) < \epsilon/2, \quad h(\eta_j^k(p^k), \eta_j(p^k)) < \epsilon/2, \quad \|e_i^k - e_i\| < \epsilon$$

for all $k \geq k_1$, $i = 1, 2, \cdots, m$ and $j = 1, 2, \cdots, n$. Since $p^k \to p$, $w_i^k = p^k \cdot e_i^k \in L$, $w_i^k \to p \cdot e_i \in L$, ξ_i is upper semicontunuous at $(p, p \cdot e)$ and η_j is upper semicontinuous at p, there is $k_2 \geq k_1$ such that $\xi_i(p^k, p^k \cdot e_i^k) \subset U(\epsilon/2, \xi_i(p, p \cdot e_i))$ and $\eta_j(p^k) \subset U(\epsilon/2, \eta_j(p))$ for all $k \geq k_2$, $i = 1, 2, \cdots, m$ and $j = 1, 2, \cdots, n$. Thus

$$\xi_i^k(p^k, p^k \cdot e_i^k) \subset U(\epsilon/2, \xi_i(p^k, p^k \cdot e_i^k)) \subset U(\epsilon, \xi_i(p, p \cdot e_i))$$

and

$$\eta_j^k(p^k) \subset U(\epsilon/2, \eta_j(p^k)) \subset U(\epsilon, \eta_j(p))$$

for all $k \geq k_2$, $i = 1, 2, \cdots, m$ and $j = 1, 2, \cdots, n$. It follows that

$$\Sigma_{i=1}^m \xi_i^k(p^k, p^k \cdot e_i^k) - \Sigma_{j=1}^m \eta_j^k(p^k) - \Sigma_{i=1}^m e_i^k \subset U(\delta, \Sigma_{i=1}^m \xi_i(p, p \cdot e_i) - \Sigma_{j=1}^m \eta_j(p) - \Sigma_{i=1}^m e_i)$$

for all $k \geq k_2$. This contradicts that

$$0 \in \Sigma_{i=1}^m \xi_i^k(p^k, p^k \cdot e_i^k) - \Sigma_{j=1}^n \eta_j^k(p^k) - \Sigma_{i=1}^m e_i^k$$

for all $k = 1, 2, \cdots$. Hence we must have

$$0 \in \Sigma_{i=1}^m \xi_i(p, p \cdot e_i) - \Sigma_{j=1}^n \eta_j(p) - \Sigma_{i=1}^m e_i$$

so that $p \in W(E)$ which again contradicts that $p \notin U(\epsilon_0, W(E))$. Therefore W must be upper semicontinuous at $E \in Y$. $\square$

Definition 9.5.9. If $E \in Y$, then $p \in W(E)$ is said to be an essential equilibrium point of the economy E provided that for each $\epsilon > 0$, there is $\delta > 0$

such that for each $E' \in Y$ with $E' \in O(E,\delta)$, there exists $p' \in W(E')$ with $\|p - p'\| < \epsilon$. The economy E is said to be essential if every $p \in W(E)$ is essential.

Theorem 9.5.10. W is lower semicontinuous at $E \in Y$ if and only if E is essential.

Proof. Suppose that W is lower semicontinuous at $E \in Y$. Then for each $\epsilon > 0$, there is $\delta > 0$ such that for each $E' \in Y$ with $E' \in O(E,\delta)$, we have $W(E) \subset U(\epsilon, W(E'))$. It follows that for each $p \in W(E)$, there is $p' \in W(E')$ with $\|p - p'\| < \epsilon$. Thus each $p \in W(E)$ is an essential equilibrium point and hence E is essential. Conversely, suppose that E is essential. If W were not lower semicontinuous at $E \in Y$, then there exists $\epsilon_0 > 0$ and a sequence $\{E^k\}_{k=1}^{\infty}$ in Y with $E^k \to E$ such that for each $k = 1, 2, \cdots$, there is $p^k \in W(E)$ with $p^k \notin U(\epsilon_0, W(E^k))$. Since $W(E)$ is compact, we may assume that $p^k \to p \in W(E)$. Since p is essential, $E^k \to E$ and $p^k \to p$, there is k_1 such that $\|p^k - p\| < \epsilon_0/2$ and $p \in U(\epsilon_0/2, W(E^k))$ for all $k \geq k_1$. Hence $p^k \in U(\epsilon_0, W(E^k))$ for all $k \geq k_1$ which contradicts that $p^k \notin U(\epsilon_0, W(E^k))$ for all $k = 1, 2, \cdots$. Hence W must be lower semicontinuous at E. □

Remark 9.5.11. By Lemma 9.5.8 and Theorem 9.5.1, we know that $E \in Y$ is essential if and only if W is continuous at $E \in Y$: for each $\epsilon > 0$, there is $\delta > 0$ such that $h(W(E'), W(E)) < \epsilon$ for each $E' \in Y$ and $E' \in O(E,\delta)$, i.e., the set $W(E)$ of equilibria is stable: $W(E')$ is close to $W(E)$ whenever E' is close to E.

Theorem 9.5.12. The set of essential economies in Y is a dense residual set in Y.

Proof. By Lemma 9.5.7 and Lemma 9.5.8, W is an USCO mapping. Since Y is a Baire space, by Lemma 9.5.1, the set of points where W is lower semicontinuous is a dense residual set in Y. By Theorem 9.5.1, the set of essential economies in Y is a dense residual set in Y. □

Remark 9.5.13. Thus, we prove that in '*most*' production economies (in the sense of the Baire category) all equilibria are essential.

Finally, we shall give a sufficient condition that $E \in Y$ is essential.

Theorem 9.5.14. If $E \in Y$ is such that $W(E)$ is singleton set, then E is essential.

Proof. Suppose $W(E) = \{p\}$. By lemma 9.5.6, W is upper semicontinuous at E. Thus for each $\epsilon > 0$, there is $\delta > 0$ such that for each $E' \in Y$, $E' \in O(E,\delta)$ implies $W(E') \subset U(\epsilon, W(E))$. But $W(E) = \{p\}$, so $W(E) = \{p\} \subset U(\epsilon, W(E'))$. This show that W is lower semicontinuous at E. By Theorem 9.5.10, E is essential. □

Remark 9.5.15. Before we conclude this section, we would like to point out the the study of generic stability and generic uniqueness have been adopted by many people. For example, by using the generic method, Yu [595] has established that almost optimization problems have a unique solutions (see also Beer [27], Kenderov [288], Kenderov and Ribaraka [290]) For more recent study on this topics, we refer interested readers to Kenderov and Revalsk [291], Lucchetti and Revalski [354], Tan et al. [520], and related references therein for more details.

References

1. P. Alexandrov, *Combinatorial Topology I*, Graylock Press, Baltimore, 1956.
2. C. Aliprantis and D. Brown, *Equilibria in markets with a Riesz space of commodities*, J. Math. Economics **11** (1983), 189-207.
3. C. Aliprantis, D. Brown and O. Burkinshaw, *Existence and Optimality of Competitive Equilibria*, Springer-Verlag, New York, 1989.
4. B. Allen, *General equilibrium with rational expectations*, Contribution to Mathematical Economics, in Honor of Gerard Debreu (W. Hildenbrand and A. Mas-Colell, eds.), North-Holland, New York, 1986, pp. 1-23.
5. G. Allen, *Variational inequalities, complementary problems and dual theorems*, J. Math. Anal. Appl. **58** (1977), 1-10.
6. R. M. Anderson and W. R. Zame, *Edgeworth's conjecture with infinitely many commodities*, 7th World Econometric Congress- August 22-29, 1995, Keio University, Tokyo, Japan.
7. N. Aronszajn and P. Panitchpakdi, *Extensions of uniformly continuous transformations and hyperconvex metric spaces*, Pacific J. Math. **6** (1956), 405-439.
8. K. J. Arrow and G. Debreu, *Existence of an equilibrium for a competitive economy*, Econometrica **22** (1954), 265-290.
9. K. J. Arrow and M. D. Intriligator, *Handbooks in Economics*, vol. 1-4, North-Holland, Amsterdam, 1983-1989.
10. J. P. Aubin, *Mathematical Methods of Game and Economic Theory (2nd Edition)*, North-Holland, Amsterdam, 1982.
11. J. P. Aubin, *Applied Functional Analysis*, Wiley-Interscience, New

York, 1979.

12. J. P. Aubin, *Viability Theory*, Birkhäuser, Boston, 1991.
13. J. P. Aubin, *Optima and Equilibria*, Springer-Verlag, Berlin, New York, 1993.
14. J. P. Aubin and A. Cellina, *Differential Inclusions*, Springer-Verlag, Berlin, New York, 1984.
15. J. P. Aubin and I. Ekeland, *Applied Nonlinear Analysis*, John Wiley & Sons, 1984.
16. J. P. Aubin and H. Frankowska, *Set-Valued Analysis*, Birkhäuser Boston, 1990.
17. J. P. Aubin and J. Siegel, *Fixed points and stationary points of dissipative multivalued maps*, Proc. Amer. Math. Soc. **78** (1980), 391-398.
18. R. J. Aumann, *Existence of competitive equilibria in markets with a continumm of agents*, Econometrica **34** (1966)), 1-17.
19. R. J. Aumann and S. Hart, *Handbook of Game Theory with Economic Applications*, vol. 1, North-Holland, Amsterdam, 1992.
20. J. S. Bae, W. K. Kim and K. K. Tan, *Another generalization of Ky Fan's minimax inequality and its applications*, Bull. Inst. Acad. Sinica **21** (1993), 229-244.
21. J. B. Baillon, *Nonexpansive mappings and hyperconvex spaces*, Fixed Point Theory and Its Applications (R. F. Brown, ed.), vol. 72, Contemp. Maths., Amer. Math. Soc., Providence, R. I., 1988, pp. 11-19.
22. C. Baiocchi and A. Capelo, *Variational and Quasivarational Inequalities, Applications to Free Boundary Problems*, John Wiley & Sons, New York, 1984.
23. R. Barbolla, *Equilibrium and quasi-equilibrium theorems for a general model without ordered preferences*, Economic Letters **19** (1985), 205-210.
24. C. Bardaro and R. Ceppitelli, *Some further generalizations of the Knaster -Kuratowski -Mazurkiewicz theorem and minimax inequalities*, J. Math. Anal. Appl. **132** (1988), 484-490.
25. C. Bardaro and R. Ceppitelli, *Fixed point theorems and vector-valued minimax theorems*, J. Math. Anal. Appl. **146** (1990), 363-373.
26. M. Baye, G. Tian and J. Zhou, *Characterizations of the existence of equilibria in games with discontinuous and nonquasiconcave payoffs*, Rev. Econom. Stud. **60** (1993), 935–948.
27. G. Beer, *On a generic optimization theorem of Kenderov*, Nonlinear Anal., TMA **12** (1988), 647-655.

28. G. Beer, *Topologies on Closed and Closed Convex Sets*, Kluwer Academic Publishers, Dordrecht, 1993.
29. J. C. Bellenger, *Existence of maximizable quasiconcave functions on paracompact convex spaces*, J. Math. Anal. Appl. **123** (1987), 333-338.
30. H. Ben-El-Mechaiekh, *Note on a class of set-valued maps having continuous selections*, Fixed Point Theory and Applications (M. A. Thera and J. B. Baillon, eds.), Longman Scientific Technical, New York, 1991, pp. 33-45.
31. H. Ben-El-Mechaiekh, S. Chebbi, M. Florenzano and J. V. Llinares, *Abstract convexity and fixed points*, J. Math. Anal. Appl. **222** (1998), 138-150.
32. H. Ben-El-Mechaiekh and P. Deguire, *General fixed point theorems for non-convex set-valued maps*, C. R. Acad. Sci. Paris **312** (1991), 433-438.
33. H. Ben-El-Mechaiekh and P. Deguire, *Approachability and fixed points for non-convex set-valued maps*, J. Math. Anal. Appl. **170** (1992), 477-450.
34. H. Ben-El-Mechaiekh and P. Deguire, *Equilibrium for abstract non-convex games*, C. R. Math. Rep. Acad. Sci. Canada **17** (1995), 1-6.
35. H. Ben-El-Mechaiekh, P. Deguire and A. Granas, *Une alternative non linéaire en analyse convexe et application. I*, C. R. Acad. Sci. Paris. I Math. **295** (1982), 257-259; II, C. R. Acad. Sci. Paris. I Math. **259** (1982), 381-384.
36. H. Ben-El-Mechaiekh, P. Deguire and A. Granas, *Point fixes et coïncidences pour les fonctions multivoques (applications de type M* et M) III*, C. R. Acad. Sci. Paris **305** (1987), 381-384.
37. A. Bensoussan, *Stochastic Control by Functional Analysis Method*, North-Holland, Amsterdam, 1982.
38. A. Bensoussan, M. Goursat and J. L. Lions, *Control impulsinnel et inéquations quasivariationelles stationaries* **276** (1973), C. R. Acad. Sci., 1279-1284.
39. A. Bensoussan and J. L. Lions, *Impulse Control and Quasivariational Inequalities*, Gauthiers-Villers, Bordas, Paris (English version Trans-Inter-Scientia), 1984.
40. C. Berge, *Espaces Topologiques, Fonctions Multivoques*, Dunod, Paris, 1959.
41. K. Bergstresser and P. L. Yu, *Domination structures and multicriteria problems in N-person games*, Theory and Decision **8** (1977), 5-47.

42. T. Bergstrom, *Maximal elements of acyclic preference relations on compact sets*, J. Economic Theory **10** (1975), 403-404.
43. T. R. Bergstrom, R. Parks and T. Rader, *Preferences which have open graphs*, J. Math. Economics **3** (1976), 265-268.
44. D. P. Bertsekas and J. N. Tsitsiklis, *Projection methods for variational inequalities with applications to the traffic assignment problem*, Math. Prog. Study **17** (1982), 139-159.
45. T. F. Bewley, *Existence of equilibria in economies with infinitely many commodities*, J. Econom. Theory **4** (1972), 514-540.
46. R. Bielawski, *Simplicial convexity and its applications*, J. Math. Anal. Appl. **127** (1987), 155-171.
47. D. Blackwell, *An analogy of the minimax theorem for vector payoffs*, Pacific J. Math. **6** (1956), 1-8.
48. K. C. Border, *Fixed Point Theorems with Applications to Economics and Game Theory*, Cambridge University Press, Cambridge, London, New York, Sydney, 1985.
49. Y. G. Borisovich, B. D. Gelman, A. D. Myskis and V. V. Obukhovskii, *Topological methods in the theory of fixed points of multivalued maps*, Uspehi Mat. Nauk **35** (1980), 52-126.
50. Y. G. Borisovich, B. D. Gelman, A. D. Myskis and V. V. Obukhovskii, *Multivalued mappings*, J. Soviet Math. **24** (1984), 719-791.
51. A. Borglin and H. Keiding, *Existence of equilibrium actions of equilibrium, "a note on the 'new' existence theorems"*, J. Math. Economic. **3** (1976), 313-316.
52. P. E. Borm, S. H. Tijs and J. van den Aarssen, *Pareto equilibrium in multiobjective games*, Methods Oper. Res. **60** (1990), 303-312.
53. J. M. Borwein, S. Fitzpatrick and P. Kenderov, *Minimal convex uscos and monotone operators on small sets*, Canad. J. Math. **43** (1991), 461–476.
54. H. Brezis, *Èquations et inèquations nonlinèaires dans les espaces vectorieis en dualitè*, Ann. Inst. Fourier (Grenoble) **18** (1968), 115-175.
55. H. Brezis, L. Nirenberg and G. Stampacchia, *A remark on Ky Fan's minimax principle*, Boll. Un. Mat. Ital. **6** (1972), 293-300.
56. L. E. J. Brouwer, *Über abbildungen vom mannigfaltigkeiten*, Math. Ann. **71** (1912), 97-115.
57. F. E. Browder, *Nonlinear mappings of nonexpensive and accretive type in Banach space*, Bull. Amer. Math. Soc. **73** (1967), 875-882.
58. F. E. Browder, *A new generalization of the Schauder fixed point theorem*, Math. Ann. **174** (1967), 285-290.

59. F. E. Browder, *The fixed point theory of multivalued mappings in topological vector spaces*, Math. Ann. **177** (1968), 283-301.
60. F. E. Browder, *Nonlinear Operators and Nonlinear Equations of Evolutions in Banach Spaces*, Proc. Symposia in Pure Mathematics, vol. 18, Part 2, Amer. Math. Soc., Prov. R. I., 1976.
61. F. E. Browder, *On a sharpened form of the Schauder fixed-point theorem*, Proc. Nat. Acad. Sci. U.S.A. **74** (1977), 4749-4751.
62. F. E. Browder, *Coincidence theorems, minimax theorems, and variational inequalities*, Contemporary Math., vol. 26, 1984, p. 67-80.
63. A.V. Bulavski, G. Isac and V. Kalashnikov, *Application of topological degree theory to complementarity problems*, Multilevel Optimization: Algorithms and Applications (A. Migdalas et al., eds.), Kluwer Academic Publishers, 1998, pp. 333-358.
64. D. Chan and J. S. Pang, *The generalized quasi-variational inequality problem*, Math. Operations Research **7** (1982), 211-222.
65. S. S. Chang, *Variational Inequalities and Complementarity Problem Theory with Applications*, Shanghai Scientific Literature Press (in Chinese), Shanghai, China, 1991.
66. S. S. Chang, X. Wu and S. W. Xiang, *A topological KKM theorem and minimax theorem*, J. Math. Anal. Appl. **182** (1994), 756-767.
67. S. S. Chang, Y. J. Cho, X. Wu and Y. Zhang, *The topological versions of KKM theorems and Fan's matching theorem with applications*, Top. Meth. Non. Anal. **1** (1993), 231-245.
68. S. S. Chang and N. J. Huang, *Generalized strongly nonlinear quasi-complementarity problems in Hilbert spaces*, J. Math. Anal. Appl. **158** (1991), 194-202.
69. S. S. Chang and N. J. Huang, *Generalized multi-valued implicit complementary problems in Hilbert spaces*, Math. Japonica **36** (1991), 1093-1100.
70. S. S. Chang and Y. H. Ma, *Generalized KKM theorem on H-space with applications*, J. Math. Anal. Appl. **163** (1992), 406-421.
71. S. S. Chang and J. W. Song, *Coincidence indices for set-valued compact mapping pairs*, J. Math. Anal. Appl. **148** (1990), 469-488.
72. S. S. Chang and G. S. Yang, *Some further generalizations of Ky Fan's minimax inequality and its applications*, Appl. Math. Mech. (English Ed.) **11** (1990), 1027-1034.
73. S. S. Chang, X. Z. Yuan, F. Wang and X. Long, *The study of algorithms and convergence for generalized multi-valued quasi-variational inclusions (in press)*, Topoloical Method in Nonlinear Analysis.
74. S. S. Chang, X. Z. Yuan, G. M. Lee and X. L. Zhang, *Saddle points and minimax theorems for vector-valued multifunctions on*

H-spaces, Applied Math. Letter **11** (1998), 101-107.

75. S. S. Chang, X. Z. Yuan, B. Thompson and E. Tarafdar, *The existence theorems of solutions for generalized vector-valued variational-like inequalities*, Reserach report, Department of Mathematics, The University of Queensland (1996).
76. S. S. Chang and Y. Zhang, *Generalized KKM theorem and variational inequalities*, J. Math. Anal. Appl. **159** (1991), 208-223.
77. S. Y. Chang, *On the Nash equilibrium*, Soochow J. Math. **16** (1990), 241-248.
78. S. Y. Chang, *A generalization of KKM principle and its applications*, Soochow J. Math. **15** (1989), 7-17.
79. J. Charatonik, K. Omiljanowski and B. Ricceri, *Graph intersection property*, Matematiche (Catania) **47** (1993), 103-115.
80. G. Y. Chen, *A generalized section theorem and a minimax inequality for a vector-valued mapping*, Optimization **22** (1991), 745-754.
81. G. Y. Chen, *Existence of solutions for a vector variational inequality: An extension of the Hartman-Stampacchia theorem*, J. Optim. Theory Appl. **74** (1992), 445-456.
82. G. Y. Chen and G. M. Cheng, *Vector variational inequality and vector optimization*, Lecture Notes in Economics and Mathematical System, vol. 285, pp. 408-416; Springer-Verlag, 1987.
83. G. Y. Chen and B. D. Craven, *A vector variational inequality and optimization over an efficient set*, Zeitscrift für Operations Research **3** (1990), 1–12.
84. G. Y. Chen and X. Q. Yang, *The vector complementarity problem and its equivalence with the weak minimal element in ordered sets*, J. Math., Anal. Appl. **153** (1990), 136–158.
85. C. Z. Cheng and B. L. Lin, *A two functions, noncompact topological minimax theorem*, Acta Math. Hungar **73** (1996), 65–69.
86. C. Z. Cheng and B. L. Lin, *Noncompact minimax theorems*, Bull. Inst. Math. Acad. Sinica **24** (1996), 23–32.
87. C. Z. Cheng and Y. H. Lin, *A generalization of minimax theorems of fixed-point type (Chinese)*, Acta Math. Sinica **34** (1991), 502–507.
88. G. Chichilnsky, *Limited arbitrage is necessary and sufficient for the existence of a competitive equilibrium* (1992), Work Paper, Columbia Universitty, U.S.A..
89. G. Chichilnsky, *Intersecting families of sets and the topology of cones in economics*, Bull. Amer. Math. Soc. (N.S.) **29** (1993), 189-207.

90. G. Chichilnsky and G. M. Heal, *Necessary and sufficient conditions for a resolution of the social choice paradox*, J. Econ. Theory **31** (1983), 68-87.

91. D. Chose and U. R. Prasad, *Concepts in two-person multicriteria games*, J. Optim. Theory Appl. **63** (1989), 167-189.

92. J.B. Conway, *A Course in Functional Analysis*, Springer-Verlag, Berlin, 1990.

93. J. P. Christensen, *Theorems of Namioka and Johnson type for upper semicontinuous and compact valued set-valued mappings*, Proc. Amer. Math. Soc. **86** (1982), 649-655.

94. H. P. R. Christensen and P. S. Kenderov, *Dense strong continuity of mappings and the Random-Nikodym property*, Math. Scand. **54** (1984), 70-78.

95. M. Coban, P. Kenderov, *Generic Gateaux differentiability of convex functionals in $C(T)$ and the topological properties of T*, Mathematics and Mathematical Education (Bulgarian), Bulgar. Acad. Sci., Sofia, 1986, pp. 141-149.

96. D. I. A. Cohen, *On the Sperner lemma*, J. Combin. Theory **2** (1967), 585-587.

97. H. W. Corley, *An existence result for maximizations with respect to cones*, J. Optim. Theory Appl. **31** (1980), 277-281.

98. R. W. Cottle, *Nonlinear programs with positive bounded Jacobians*, SIAM J. Appl. Math. **14** (1966), 147-158.

99. R. W. Cottle, *Complementarity and variational problems*, Symposia Mathematica, Vol. XIX (Convegno sulla Programmazione Matematica e sue Applicazioni, INDAM, Rome, 1974), Academic Press, London, 1976, pp. 177–208.

100. P. W. Cottle and G. B. Dantzig, *Complementarity pivot theory of mathematical programming*, Linear Algebra Appl. **1** (1968), 103-125.

101. R. W. Cottle, F. Giannessi and J. L. Lions, *Variational Inequalities and Complementarity Problems: Theory and Applications*, J. Wiley and Sons, New York, London, 1980.

102. R. W. Cottle, J. Kyparisis and J. S. Pang (eds), *Variational Inequality Problems*, vol. 48, Math. Prog. Series B, 1990.

103. R. W. Cottle and J. C. Yao, *Pseudo-monotone complementarity problems in Hilbert spaces*, J. Optim. Theory. Appl. **75: 2** (1992), 281–295.

104. J. Crank, *Free and Moving Boundary Problems*, Oxford Univ. Press, London, 1984.

105. P. Cubiotti, *Finite-dimensional quasi-variational inequalities associated with discontinuous functions*, J. Optim. Theory Appl. **72** (1992), 577-582.
106. P. Cubiotti, *Some remarks on the fixed points of lower semicontinuous multifunctions*, J. Math. Anal. Appl. **174** (1993), 407-412.
107. P. Cubiotti, *An existence theorem for generalized quasi-variational inequalities*, Set-Valued Anal. **1** (1993), 81-87.
108. P. Cubiotti, *An application of quasivariational inequalities to linear control systems*, J. Optim. Theory Appl. **89** (1996), 101–113.
109. P. Cubiotti, *A note on Chan and Pang's existence theorem for generalized quasi-variational inequalities*, Appl. Math. Lett. **9** (1996), 73–76.
110. P. Cubiotti, *A counter-example on a quasi-variational inequality without lower semicontinuity assumption*, Proc. Amer. Math. Soc. **123** (1995), 969–970.
111. P. Cubiotti and X. Z. Yuan, *A generalized quasi-variational inequality without upper semicontinuity*, Bull. Austal. Math. Soc. **54** (1996i), 247-254.
112. H. F. Cullen, *Introduction to General Topology*, D. C. Heath, Boston, 1968.
113. S. Dafermos, *Exchange price equilibria and variational inequalities*, Math. Prog. **46** (1990), 391-402.
114. M. M. Day, *Normed Linear Spaces*, Springer-Verlag, New York, 1973.
115. G. Debreu, *A social equilibrium existence theorem*, Proc. Nat. Acad. Sci. U.S.A. **38** (1952), 386-393.
116. G. Debreu, *A classical tax-subsidy problem*, Econometrica **22** (1954), 14–22.
117. G. Debreu, *Market equilibrium*, Proc. Nat. Acad. Sci. U.S.A. **42** (1956), 876–878.
118. G. Debreu, *Theory of Value: An axiomatic Analysis, New Haven and London*, Yale Univ. Press, 1959.
119. G. Debreu, *Economies with a finite set of equilibria*, Econometrica **38** (1970), 387-392.
120. G. Debreu, *Existence of competitive equilibrium,*, Handbook of Mathematical Economics, (K.J. Arrow and M.D. Intriligator, eds.), vol. II,, North-Holland, Amsterdam, 1982, pp. 697-743.
121. G. Debreu, *Mathematical Economics: Twenty papers of Gerard Debreu*, Cambride University Press, Cambridge, 1983.
122. P. Deguire, *Thèoremes de coïncidences, Théorie de minimax et applications*, Thèse de doctorat, Université de Montréal, 1987.

123. P. Deguire, *Browder-Fan fixed point theorem and related results*, Discuss. Math. Differential Incl. **15** (1995), 149-162.
124. P. Deguire and M. Lassonde, *Familles sélections*, Topol. Methods Nonlinear Anal. **5** (1995), 261-269.
125. P. Deguire, K. K. Tan and X. Z. Yuan, *Some maximal elements, fixed points and equilibria for L_S-majorizd mappings in product spaces and its applications (in press)*, Nonlinear Anal., T.M.A. (1998).
126. P. Deguire and X. Z. Yuan, *Maximal elements and coincident points for couple-majorized mappings in product topological vector spaces*, Z. Anal. Anwendungen ("Zeitschrift für Analysis und ihre Anwendungen (ZAA)) **14** (1995), 665-676.
127. K. Deimling, *Multivalued Differential Equations*, Walter de Gruyter & Co. Publ, 1992.
128. E. Dierker, *Topological Methods in Walrasian Economics*, Springer-Verlag, Berlin, 1974.
129. J. Diestel, *Geometry of Banach Spaces-Selected Topics*, vol. 485, Lectures Notes in Mathematics, Springer-Verlag, Berlin, 1975.
130. X. P. Ding, *Continuous selection theorems, coincidence theorems and intersection theorems concerning sets with H-convex sections*, J. Austral. Math. Soc.(Ser. A) **52** (1992), 11-25.
131. X. P. Ding, *Generalized strongly nonlinear quasi-variational inequalities*, J. Math. Anal. Appl. **173** (1993), 577-587.
132. X. P. Ding, *Existence of maximizable quasi-concave functions on H-Spaces*, Acta Math. Sinica (in Chinese) **36** (1993), 273-279.
133. X. P. Ding, *Quasi-variational inequalities and social equilibrium*, Appl. Math. Mech. (English Ed.) **12** (1991), 639–646.
134. X. P. Ding, *Borwder - Hartman - Stampacchia type variational inequalities for multi-valued quasi - monotone operator*, Journal of Sichuan Normal University **14, No. 2** (1991), 1-8.
135. X. P. Ding and K. K. Tan, *Generalized variational inequalities and generalized quasi - variational inequalities*, J. Math. Anal. Appl. **148** (1990), 497-508.
136. X. P. Ding and K. K. Tan, *A minimax inequality with applications to existence of equilibrium point and fixed point theorems*, Colloquium Math. **63** (1992), 233–247.
137. X. P. Ding and K. K. Tan, *Fixed point theorems and equilibria of non-compact generalized games*, Fixed Point Theory and Applications (Halifax, NS, 1991), World Sci. Publishing, River Edge, NJ, 1992, pp. 80-96.

138. X. P. Ding and K. K. Tan, *On equilibria of non-compact generalized games*, J. Math. Anal. Appl. **177** (1993), 226–238.
139. X. P. Ding and K. K. Tan, *A set-valued generalization of Fan's best approximation theorems*, Can. J. Maths. **44** (1992), 784-796.
140. X. P. Ding and K. K. Tan, *Covering properties of H-spaces and applications*, Applied. Math. Mech. (English Ed.) **14** (1993), 1079-1088.
141. X. P. Ding, W. K. Kim and K. K. Tan, *Equilibria of non-compact generalized games with L^*-majorized preference correspondences*, J. Math. Anal. Appl. **164** (1992), 508-517.
142. X. P. Ding, W. K. Kim and K. K. Tan, *A selection theorem and its applications*, Bull. Austral. Math. Soc. **46** (1992), 205-212.
143. X. P. Ding and E. Tarafdar, *Fixed point theorem and existence of equilibrium points of noncompact abstract economics*, Nonlinear World **1** (1994), 319-340.
144. X. P. Ding and E. Tarafdar, *Generalized nonlinear variational inequalities with non-monotone set-valued mappings*, Appl. Math. Letters **7** (1994), 5-11.
145. X. P. Ding and E. Tarafdar, *Monotone generalized variational inequalities and generalized complementarity problems*, J. Optim. Theory Appl. **88** (1996), 107-122.
146. S. Dolecki, *Tangency and differentiation: Some applications of convergence theory*, Ann. Mat. Pura. Appl. **130** (1982), 223-255.
147. J. Dugundji, *Topology*, Allyn and Bacon, Inc., Boston, 1966.
148. J. Dugundji and A. Granas, *KKM maps and variational inequalities*, Ann. Scuola Norm. Sup. Pisa **5** (1978), 679-682.
149. J. Dugundji and A. Granas, *Fixed Point Theory*, vol. I, Polish Scientific Publishers, Warszawa, 1982.
150. N. Dunford and J. T. Schwartz, *Linear Operators, Part I: General Theory*, Interscience Publisher Inc., New York, 1957.
151. G. Duvaut and J. L. Lions, *Inequalities in Mechanics and Physics*, Springer-Verlang, Berlin, 1976.
152. S. Eilenberg and D. Montgomery, *Fixed point theorems for multivalued transformations*, Amer. J. Math. **68** (1946), 214-222.
153. I. Ekeland and R. Temam, *Analyse Convexe et Problemes Variationnels*, Dunod, Paris, 1974.
154. R. Engelking, *General Topology, revised and completed edition*, Heldermann Verlag, Berlin, 1989.
155. R. S. Falk, *Error estimates for the approximation of a class of variational inequalities*, Math. Programm. **28** (1974), 963-972.

156. K. Fan, *Fixed points and minimax theorems in locally convex topological linear spaces*, Proc. Nat. Acad. Sci. U.S.A. **38** (1952), 131-136.
157. K. Fan, *A generalization of Tychonoff's fixed point theorem*, Math. Ann. **142** (1961), 305-310.
158. K. Fan, *Sur un théorème minimax*, C. R. Acad. Sci. Paris Ser. I. Math **259** (1964), 3925-3928.
159. K. Fan, *Applications of a theorem concerning sets with convex sections*, Math. Ann. **163** (1966), 189-203.
160. K. Fan, *Simplicial maps from an orientable n-pseudomanifold into S^m with the octahedral triangulation*, J. Comb. Theory **2** (1967), 588-602.
161. K. Fan, *A covering property of simplexes*, Math. Scand. **22** (1968), 17-20.
162. K. Fan, *Extensions of two fixed point theorems of F. E. Browder*, Math. Z. **112** (1969), 234-240.
163. K. Fan, *A combinatorial property of pseudoman folds and covering properties of simplexes*, J. Math. Anal. Appl. **31** (1970), 68-80.
164. K. Fan, *A minimax inequality and its applications*, Inequalities III (O. Shisha, ed.), Academic Press, 1972, pp. 103-113.
165. K. Fan, *Two applications of a consistency theorem for systems of linear inequalities*, Linear Algebra and Appl. **11** (1975), 171–180.
166. K. Fan, *Fixed-point and related theorems for noncompact convex sets*, Game Theory and Related Topics (Proc. Sem., Bonn and Hagen, 1978), North-Holland, Amsterdam-New York, 1979, pp. 151–156.
167. K. Fan, *Some properties of convex sets related to fixed point theorems*, Math. Ann. **266** (1984), 519-537.
168. K. Fan, *Sharpened forms of an inequality of von Neumann*, Math. Z. **194** (1987), 7–13.
169. K. Fan, *A survey of some results closely related to the Knaster - Kuratowski - Mazuriewicz theorem*, Game Theory and Applications, Econom. Theory Econometrics Math. Econom., Academic Press, 1990, p. 358-370.
170. S. C. Fang, *An iterative methods for generalized complementarity problems*, IEEE Trans. Auto. Control **25** (1980), 1225-1227.
171. S. C. Fang and E. L. Peterson, *Generalized variational inequalities*, J. Optim. Theory Appl. **38** (1982), 363-383.
172. F. Ferro, *A minimax theorem for vector-valued functions, Part 2*, J. Optim. Theory Appl. **68** (1991), 35-48.

173. G. Fichera, *Problemi elastostatic con vincoli unilateralil problema di singnorinicon ambigue condizioni al controno*, Atti. Accad. Naz. Lincei Mem. Cl. Sci. Fis. Mat. Natur. Sez. La. **7** (1963-64), 91-140.

174. P. M. Fitzpatick and W. V. Petryshyn, *Fixed point theorems for multivalued-valued non-compact inward maps*, J. Math. Anal. Appl. **46** (1974), 756-767.

175. P. M. Fitzpatick and W. V. Petryshyn, *Fixed point theorems for multivalued noncompact acyclic mappings*, Pacific J. Math. **54** (1974), 17-23.

176. S. D. Flam, *Abstract economy and games*, Soochow J. Math. **5** (1979), 155-162.

177. M. Florenzano, *On the existence of equilibria in economies with an infinitely dimensional commodity spaces*, J. Math. Econ. **12** (1983), 207-219.

178. M. K. Fort, Jr., *Points of continuity of semicontinuous functions*, Publ. Math. Debrecen **2** (1951), 100-102.

179. M. K. Fort, Jr., *Essential and nonessential fixed points*, Amer. J. Math. **72** (1952), 315-322.

180. D. H. Fremlin, *Topological Riesz Spaces and Measure Theory*, Cambridge Univ. Press, London, New York, 1974.

181. A. Friedman, *Stochastic Differential Equations and Applications*, vol. 1-2, Academic Press, New York, 1975-1976.

182. A. Friedman, *Optimal stopping problems in stochastic control*, SIAM Rev. **21** (1979), 71-80.

183. A. Friedman, *Variational Principle and Free Boundary Value Problems*, Wiley, New York, 1982.

184. M. Frigon and A. Granas, *Risultats du type de Leray-Schauder pour des contractions multivoques (French) [Leray-Schauder-type results for multivalued contractions]*, Topol. Methods Nonlinear Anal. **4** (1994), 197–208.

185. M. Frigon, A. Granas and Z. E. Guennoun, *Alternative non liniaire pour les applications contractantes. (French) [Nonlinear alternative for contracting mappings]*, Ann. Sci. Math. Quibec **19** (1995), 65–68.

186. J. Fu, *Implicit variational inequalities for multivalued mappings*, J. Math. Anal. Appl. **189** (1995), 801-814.

187. M. Furi and A. Vignoli, *A fixed point theorem in complete metric spaces*, Boll. Un. Mat. Ital. Ser. **4** (1969), 505-506.

188. D. Gale, *The law of supply and demand*, Math. Scand. **3** (1955), 155-169.

189. D. Gale, *The closed linear model of production, in Linear inequalities and Related Systems* **38** (1956), Annals of Mathematics Studies Princeton University Press, Princeton, N. J., 285–303.
190. D. Gale, *Equilibrium in a discrete exchange economy with money,* Int. J. Game Theory **13** (1984), 61-64.
191. D. Gale and A. Mas-Colell, *An equilibrium existence for a general model without ordered preferences*, J. Math. Economics. **2** (1975), 9-15.
192. D. Gale and A. Mas-Colell, *On the role of complete, translative preferences in equilibrium theory*, Equilibrium and Disequilibrium in Economics Theory (G. Schwödiauer, ed.), Reidel, Dordrecht, 1978, p. 7-14.
193. D. Gale and A. Mas-Colell, *Corrections to an equilibrium existence theorem for a general model without ordered preferences*, J. Math. Econ. **6** (1979), 297-298.
194. M. A. Geraghty and B. L. Lin, *Topological minimax theorems*, Proc. Amer. Math. Soc. **91** (1984), 377-380.
195. F. Giannessi, *Theorems of alternative, quadratic programs and complementarity problems, in Variational Inequalities and Complementarity Problems, (edited by R. W. Cottle, F. Giannessi and J. L. Lions)*, John Wiley and sons, Chichester, England (1980), 151–186.
196. I. L. Glicksberg, *A further generalization of the Kakutani fixed point theorem with applications to Nash equilibrium points*, Proc. Amer. Math. Soc. **3** (1952), 170-174.
197. R. Glowinski, J. Lions and R. Trmolieres, *Numerical Analysis of Variational Inequalities*, North-Holland, Amsterdam, 1982.
198. Goebel, K. and Kirk, W. A., *Topics in Metric Fixed Point Theory,* Cambridge University Press, Cambridge, 1990.
199. D. Goeleven, D. Motreanu and P. D. Panagiotopoulos, *Multiple solutions for a class of hemivariational inequalities involving periodic energy functionals*, Math. Methods Appl. Sci. **20** (1997), 547-568.
200. I. T. Gohberg, L. S. Goldenstein and A. S. Markus, *Investigations of some properties of bounded linear operators with their q-norms,* Uch. Zap. Kishinevsk. In-Ta. **29** (1957), 29-36.
201. L. Gorniewicz, *A Lefschetz-type fixed point theorem*, Fund. Math. **88** (1975), 103-115.
202. L. Gorniewicz and A. Granas, *Some general theorems in coincidence theory I*, J. Math. Pure Appl. **60** (1981), 361-373.
203. L. Gorniewicz, A. Granas and W. Kryszewski, *On the homotopy method in the fixed point index theory of multi-valued mappings of*

compact absolute neighborhood retracts, J. Math. Anal. Appl. **161** (1991), 457–473.

204. A. Granas, *KKM-maps and their applications to nonlinear problems*, The Scottish Book: Mathematic from the Scottish Cafe (R. Daniel Mauldin, ed.), Birkhäuser, Boston, 1982, p. 45-61.
205. A. Granas, *Continuation method for contractive maps*, Topol. Methods Nonlinear Anal. **3** (1994), 375–379.
206. A Granas, *On the Leray-Schauder alternative*, Topol. Methods Nonlinear Anal. **2** (1993), 225–231.
207. A. Granas and M. Lassonde, *Some elementary general principles of convex analysis*, Topol. Methods Nonlinear Anal. **5** (1995), 23–37.
208. A. Granas and F. C. Liu, *Théorèmes du minimax*, C. R. Acad. Sci. Paris **298** (1984), 329-332.
209. A. Granas and F. C. Liu, *Coincidences or set-valued maps and minimax inequalities*, J. Math. Pure. Appl. **65** (1986), 119-148.
210. A. Granas and F. C. Liu, *Théorèmes du minimax*, Nonlinear and Convex Analysis, Proceedings in Honor of Ky Fan (B. L. Lin and S. Simons, eds.), Marcel Dekker, 1987, pp. 115-117.
211. J.M. Grandmont, *Temporary general equilibrium theory*, Econometrica **45** (1977), 535–572.
212. J. Gwinner, *On fixed points and variational inequalities: A circular tour*, Nonlinear Analysis **5** (1981), 565-583.
213. C. W. Ha, *Minimax and fixed point theorems*, Math. Ann. **248 (1980),** (73-77).
214. C. W. Ha, *Extension of two fixed point theorems of Ky Fan*, Math. Z. **190** (1985), 13-16.
215. C. W. Ha, *On a minimax inequality of Ky Fan*, Proc. Amer. Math. Soc. **99** (1987), 680-682.
216. B. Halpern, *Fixed-Point Theorems for Outward Maps*, Doctoral Thesis, U.C. L.A. (1965).
217. B. Halpern, *Fixed point theorems for set-valued maps in infinite dimensional spaces*, Math. Ann. **189** (1970), 87-98.
218. B. Halpern and G. Bergman, *A fixed point theorem for inward and outward maps*, Trans. Amer. Math. Soc. **130** (1968), 353-358.
219. P. T. Harker, *Predicting Intercity Freight Flow*, VUN Science Press Utrecht, Holland, 1987.
220. P. T. Harker and J. S. Pang, *Finite-dimensional variational inequality and nonlinear complementarity problems: A survey of theory, algorithms, and applications*, Math. Prog. **48** (1990), 161-220.
221. P. Hartman and G. Stampacchia, *On some nonlinear elliptic differential functional equations*, Acta Math. **115** (1966), 271-310.

222. J. C. Harsanyi, *Equilibrium selection and the restricted game - A celebration of John F. Nash, Jr*, Duke Math. J. **81** (1996), 251–254.
223. J. C. Harsanyi, *A new theory of equilibrium selection for games with incomplete information*, Games Econom. Behav. **10** (1995), 318–332.
224. J. C. Harsanyi, *Rational Behavior and Bargaining Equilibrium in Games and Social Situations (Reprint of the 1986 edition)*, Cambridge University Press, Cambridge-New York, 1988.
225. J. C. Harsanyi, *Oddness of the number of equilibria points: a new proof*, Int. J. Game Theorey **2** (1973), 235-250.
226. J. C. Harsanyi and R. Selten, *A General Theory of Equilibrium Selection in Games*, MIT Press, Cambridge, 1988.
227. J. Hartley, *An extension of Sion's minimax theorem with an application to a method for constrained games*, Pacific J. Math. **103**, 401-408.
228. A. Hassouni and A. Moudafi, *A perturbed algorithm for algorithms for variational inclusions*, J. Math. Anal. Appl. **185** (1994), 706-712.
229. W. Hildenbrand, *Core and Equilibria of a Large Economy*, Princeton University Press, Princeton, New Jersey, 1974.
230. W. Hildebrand and H. Sonnenschein, *Handbook of Mathematical Economics*, vol. 1-4, North-Holland, Amsterdam, 1986-1991.
231. D. Hilbert, *Über das Dirichletsche Prinzip (lecture on the Dirichlet Principle)*, Math. Ann. (1904), 161-186.
232. D. Hilbert, *Mathematical Problems, Lecture delivered before the First International Congress of Mathematics at Paris*, Bull. Amer. Math. Soc. **8(1902)** (1900), 437-479 (Reprinted in Browder, E.F (ed) (1976), Vol. 1, pp.1-34).
233. C. J. Himmelberg, *Fixed points of compact multifunctions*, J. Math. Anal. Appl. **38** (1972), 205-207.
234. W. W. Hogan, *Point-to-set mappings in mathematical programming*, SIAM Rev. **15** (1973), 591-603.
235. R. B. Holmes, *Geometric Functional Analysis and Its Applications*, Springer-Verlag, New York, 1975.
236. C. Horvath, *Points fixes et coïncidences pour les applications multivoques sans convexite*, C. R. Acad. Sci. Paris **296** (1983), 403-406.
237. C. Horvath, *Some results on multivalued mappings and inequalities without convexity*, Nonlinear and Convex Analysis (B. L. Lin and S. Simons, eds.), Marcel Dekker, 1987, p. 96-106.
238. C. Horvath, *Quelques théorémes en théorie des mimi-max*, C. R. Acad. Sci. Paris **310** (1990), 269-272.

239. C. Horvath, *Contractiblity and generalized convexity*, J. Math. Anal. Appl. **156** (1991), 341-357.

240. C. Horvath, *Extension and selection theorems in topological spaces with a generalized convexity structure*, Ann. Fac. Sci. Toulouse Math. **2** (1993), 253–269.

241. C. Horvath and J. V. Llinares Ciscar, *Maximal elements and fixed points for binary relations on topological ordered spaces*, J. Math. Econom. **25** (1996), 291–306.

242. C. Horvath and M. Lassonde, *Intersection of sets with n-connected unions*, Proc. Amer. Math. Soc. **125** (1997), 1209-1214.

243. N. J. Huang and X. Q. Xu, *Generalized multi-valued nonlinear quasi-complementarity problems in Hilbert spaces*, J. Sichuan Univ. **31** (1994), 306-310.

244. D. H. Hyers, G. Isac and Th. M. Rassias, *Topics in Nonlinear Analysis and Applications*, World Scientific, Singapore, New Jersey, 1997.

245. T. Ichiishi, *Game Theory for Economic Analysis*, Academic Press, New York, 1983.

246. T. Ichiishi, *Alternative version of the Shapley's theorem on closed coverings of simples*, Proc. Amer. Math. Soc. **104** (1988), 759-763.

247. T. Ichiishi, *The Firm Theory and Game Theory*, Academic Press, New York, 1993.

248. T. Ichiishi and T. Idzik, *Theorems on closed coverings of a simplex and their applications to cooperative game theory*, J. Math. Anal. Appl. **146** (1990), 259-270.

249. A. Idzik, *Almost fixed point theorems*, Proc. Amer. Math. Soc. **104** (1988), 779–784.

250. A. Idzik and K. K. Tan, *Covering properties of simplexes*, J. Math. Anal. Appl. **176** (1993), 608-616.

251. I. S. Iohvidov, *On a lemma of Ky Fan generalizing the fixed point principle of A. N. Tihonov [Russian]*, Dokl. Akad. Nauk SSSR **159** (1964), 501-504 (English Trans., Soviet Math. Dokl., Vol. 5 (1964), p.1523-1526).

252. G. Isac A special variational inequality and the implicit complementarity problem, J. Fac. Sci. Univ. Tokyo **37** (1990), 109-127.

253. G. Isac, *Complementarity Problems, Lecture Notes in Mathematics*, vol. 1528, Springer-Verlag, 1992.

254. G. Isac, *Exceptional families of elements for k-set fields in Hilbert spaces and complementarity theory*, Proceedings of the International Conference ICOTA98, July 1-3, Perth, Australia, 1998.

255. I. Isac, A.V. Bulavski and V. Kalashnikov, *Exceptional families, topological degree and complementarity problems*, J. Global Optim. **10** (1997), 207-225.

256. G. Isac and W. T. Obuchowska, *Functions without exceptional families of elements and complementarity problems (in press)*, J. Optim. Theory Apppl. **98** (1998).

257. G. Isac and M. Thera, *Complementarity problem and the existence of the post-ceritial equilibrium state of a thin elastic plate*, J. Optim. Theory Appl. **58** (1988), 241-257.

258. G. Isac and X. Z. Yuan, *The essential components of coincident points for weakly inward and outward set-valued mappings (in press)*, App. Math. Letter (1998).

259. G. Isac and George X. Z. Yuan, *Remarks on the existence of maximal elements with respect to a binary relation in non-compact topological spaces*, J. Optimi. Theory Appl. (subject to the revised version) (1998).

260. G. Isac and X. Z. Yuan, *The existence of essentially connected components of solutions for variational inequalities*, Cadsmap 98-10 (1998).

261. G. Isac and X. Z. Yuan, *The generic stability and the existence of essentially connected components of solutions for nonlinear complementarity problems*, Cadsmap 98-9 (1998).

262. G. Isac and X. Z. Yuan, *The existence and generic stability of fixed points for uppet semicontinuous mappings in hyperconvex metric spaces*, Research Report, RMC, (1998).

263. J. R. Isbell, *Six theorems about injective metric spaces*, Comm. Math. Helvetici **39** (1964), 65-76.

264. V. I. Istratescu, *Fixed Point Theory*, P. Reidel Publishing Company, London, 1981.

265. G. Jameson, *Ordered Linear Spaces*, Springer-Verlag, Berlin, 1970.

266. J. Jaworowski, W. A. Kirk and S. Park, *Antipodal Points and Fixed Points*, vol. 28, Lecture Notes Series, Seoul National University, Korea, 1995, pp. 21-81.

267. J. H. Jiang, *Essential fixed points of the multivalued mappings*, Scientia Sinica **11** (1962), 293-298.

268. J. H. Jiang, *Essential fixed points of n-person non-cooperative games (II)*, Scientia Sinica **12** (1963), 651-671.

269. J. H. Jiang, *Fixed point theorems for paracompact convex sets, I and II,*, Acta Math. Sinica **4** (1988), 64-71 and p.234-241.

270. J. H. Jiang, *Fixed point theorems for convex sets*, Acta Math. Sinica **4** (1988), 356-363.

271. J. H. Jiang, *Coincidence theorems and minimax theorems*, Acta Math. Sinica **5** (1989), 307-320.
272. J. Jahn, *Scalarization in vector optimization*, Math. Programming **29** (1984), 203-218.
273. M. Justman, *Iterative processes with "nucleolar" restrictions*, Int. J. game Theory **6** (1978), 189-212.
274. I. Joó, *A simple proof for von Neumann's minimax theorems*, Acad. Sci. Math. Szeged **42** (1980), 91-94.
275. I. Joó, *On a fixed point theorem*, Publ. Math. Deb. **36** (1989), p.127-128.
276. I. Joó and L. L. Stachó, *A note on Ky Fan's minimax theorem*, Acta. Math. Hung. **39** (1982), 401-407.
277. I. Joó and A. P. Sovegjarto, *A fixed point theorem*, Ann. Univ. Sci. Budapest Math. **24** (1981), 9-11.
278. C. R. Jou and J. C. Yao, *Algorithms for generalized multi-valued variational inequalities in Hilbert spaces*, Comput. Math. Appl. **25** (1993), 7-16.
279. A. Kajii, *Note on equilibrium without ordered preferences in topological vector space*, Economics Letter **27** (1988), 1-4.
280. S. Kakutani, *A generalization of Brouwer's fixed point theorem*, Duke Math. J. **8** (1941), 457-459.
281. V. Kalashnikov, *Complementarity Problem and the Generalized Oligopoly Model*, Habilitation Thesis, CENI, Moscow, 1995.
282. S. Karamardian, *Generalized complementarity problem*, J. Optim. Theory Appl. **8** (1971), 161-168.
283. S. Karamardian and S. Schaible, *Seven kinds of monotone maps*, J. Optim. Theory Appl. **66** (1990), 37-46.
284. H. Keiding, *Existence of economic equilibria*, Lecture Notes in Economics and Math. System, vol. 22, 1984, pp. 223-243.
285. J. L. Kelley, *Banach spaces with the extension property*, Trans. Amer. Math. Soc. **72** (1952), 323-326.
286. J. L. Kelley, *General Topology*, Von-Nostand, 1955.
287. J. L. Kelley and I. Namioka, *Linear Topological Spaces*, Princeton, Van Nostrand, 1963.
288. P. S. Kenderov, *Most of the optimization problems have unique solutions*, Proceedings, Oberwolhfach Conference on Parametric Optimization (B. Brosowki and F. Deutsch, eds.), vol. 72, Birkhäuser, Basel., 1984, pp. 203-216.
289. P. Kenderov, *Monotone mappings are almost everywhere single-valued*, Studia Math. **56** (1976), 199–203.

290. P. Kenderov and Ribaraka, *Most of the two person zero-sum game have unique solution*, Workshop/Mini-Conference on Functional Analysis and Optimization, ANU, Canberra, Australia, 1988, pp. 73-82.
291. P. Kenderov and J. P. Revalski, *Generic well-posedness of optimization problems and the Banach-Mazur game*, Recent developments in well-posed variational Problems, Math. Appl., Kluwer Acad. Publ., Dordrecht, 1995, pp. 117–136.
292. M. A. Khamsi, *KKM and Ky Fan theorems in hyperconvex metric spaces*, J. Math. Anal. Appl. **204** (1996), 298-306.
293. M. A. Khan and R. Papageorgiou, *On Cournot-Nash equilibria in generalized quantitative games with an atomless measure space of agents*, Proc. Amer. Math. Soc. **100** (1987), 505-510.
294. M. A. Khan and R. Vohra, *Equilibrium in abstract economies without preferences and with a measure of agents*, J. Math. Economics **13** (1984), 133-142.
295. M. A. Khan and C. Yannelis (eds), *Equilibrium Theory in Infinite Dimensional Spaces*, Springer-Verlag, New York, 1991.
296. N. Kikuchi and J. T. Oden, *Contact Problems in Elasticity: Avariational Approach*, SIAM, Philadelphia, 1988.
297. W. K. Kim, *Some applications of the Kakutani fixed point theorem*, J. Math. Anal. Appl. **121** (1987), 119-122.
298. W. K. Kim, *Remarks on a generalized quasi-variational inequalities*, Proc. Amer. Math. Soc. **103** (1988), 667-668.
299. W. K. Kim, *Existence of maximal element and equilibrium for a nonparacompact N-person game*, Proc. Amer. Math. Soc. **116** (1992), 797-807.
300. W. Y. Kim, *A non-compact generalization of Horvath's intersection theorem*, Bull. Korean Math. Soc. **32** (1995), 153–162.
301. T. Kim and M. K. Richter, *Nontransitive-nontotal consumer theory*, J. Economic Theory **28** (1986), 324-363.
302. T. Kim, K. Prikry and N. C. Yannelis, *Equilibrium in abstract economies with a measure space and with an infinite dimensional strategy space*, J. Approx. Theory **56** (1989), 256-266.
303. W. K. Kim and K. K. Tan, *A variational inequality in non-compact sets and its applications*, Bull. Aust. Math. Soc. **46** (1992), 139-148.
304. D. Kinderlehrer and G. Stampacchia, *An Introduction to Variational Inequalities and Their Applications*, Academic Press, New York, 1980.
305. J. Kindler, *Topological intersection theorems*, Proc. Amer. Math. Soc. **117** (1993), 1003-1011.

306. J. Kindler, *Intersection theorems and minimax theorems based on connectedness*, J. Math. Anal. Appl. **178** (1993), 529-546.

307. J. Kindler, *Minimax theorems with applications to convex metric spaces*, Colloq. Math. **68** (1995), 179–186.

308. J. Kindler, *Intersecting sets in midset spaces I and II*, Arch. Math. **62** (1994), 49-57 and 168-176.

309. J Kindler, *Minimax theorems with one-sided randomization*, Acta Math. Hungar. **69** (1995), 337–345.

310. J. Kindler and R. Trost, *Minimax theorems for interval spaces*, Acta. Math. Hung. **54** (1989), 39-49.

311. S. Kinoshita, *On essential component of the set of fixed points*, Osaka J. Math. **4** (1952), 19–22.

312. W. A. Kirk, *Continuous mappings in compact hyperconvex metric spaces*, Numer. Funct. Anal. Optimiz. **17** (1996), 599-603.

313. W. A. Kirk and S. S. Shin, *Fixed point theorems in hyperconvex spaces*, Houston J. Math. **23** (1997), 175-187.

314. W. A. Kirk, B. Sims and X. Z. Yuan, *The KKM theory in hyperconvex metric spaces and some applications (in press)*, Nonlinear Analy., T.M.A (1998).

315. V. Klee, *On certain intersction properties for convex sets*, Canad. J. Math. **3** (1951), 272-275.

316. V. Klee, *Leray-Schauder theory without local convexity*, Math. Ann. **141** (1960), 286-296.

317. E. Klein and A. C. Thompson, *Theory of Correspondences: including applications to mathematical economics*, John Wiley & Sons, 1984.

318. H. Kneser, *Sur une théorème fondamental de la théorie des jeux*, C. R. Acad. Sci. Paris **234** (1952), 2418-2420.

319. H. Knaster, C. Kuratowski and S. Mazurkiewicz, *Ein Beweis des Fixpunktsatzes für n-dimensionale simplexe*, Fund. Math. **14** (1929), 132-137.

320. H. M. Ko and K. K. Tan, *A coincidence theorem with applications to minimax inequalities and fixed point theorems*, Tamkang J. Math. **17** (1986), 37-45.

321. H. M. Ko and K. K. Tan, *A note on a covering theorem of Ky Fan*, Tamkang J. Math. **22** (1991), 329-333.

322. H. M. Ko and K. K. Tan, *Coincidence theorems and matching theorems*, Tamkang J. Math. **23** (1992), 297-309.

323. E. Kohlberg and J. F. Mertens, *On the strategic stability of equilibia*, Econometrica **54** (1986), 1003-1037.

324. M. Kojima, *A unification of the existence theorems of the nonlinear complementarity problems*, Math. Programming **9** (1975), 257-277.
325. M. Kojima, A. Okada and S. Shindoh, *Strongly stable equilibrium points of N-person noncooperative games*, Math. Oper. Res. **10** (1985), 650–663.
326. H. Komiya, *Coincidence theorems and saddle point theorems*, Proc. Amer. Math. Soc. **196** (1986), 599-602.
327. H. Komiya, *A minimax theorem without linear structure*, Hiyoshi Review Nat. Sci. Keio. Univ. **8** (1990), 74-78.
328. V. Komornik, *Minimax theorems for upper semicontinuous functions*, Acta. Math. Sci. **40** (1982), 159-163.
329. H. König, *A general minimax theorems based on connectedness*, Arch. Math. **59** (1992), 55-64.
330. A. Kuica, *Personnal communication* (1997).
331. C. Kuratowski, *Sur les éspaces completès*, Fund. Math. **15** (1920), 301-309.
332. H. E. Lacey, *The Isometric theory of classical Babach Spaces*, vol. 208, Springer Verlag, New York, 1974.
333. K. Lan, *A generalization of H-space and some results on multivalued mappings without convexity*, J. Math. Anal. Appl. **194** (1995), 511–528.
334. M. Lassonde, *On the use of KKM multifunctions in fixed point theory and related topics*, J. Math. Anal. Appl. **97** (1983), 151-201.
335. M. Lassonde, *Sur le principle KKM*, C. R. Acad. Sci. Paris **310** (1990), 573-576.
336. M. Lassonde, *Fixed point for Kaktutani factorizable multifunctions*, J. Math. Anal. Appl. **152** (1990), 46-60.
337. M. Lassonde and C. Schenkel, *KKM principle, fixed points and Nash equilibria*, J. Math. Anal. Appl. **164** (1992), 542-548.
338. A. Lechicki and A. Spakowski, *A note on intersections of lower semicontinuous multifunctions*, Proc. Amer. Math. Soc. **95** (1985), 119-122.
339. G. M. Lee, D. S. Kim, B. S. Lee and S. J. Cho, *Generalized vector variational inequality and fuzzy extension*, Appl. Math. Lett. **6:6** (1993), 47–51.
340. C. M. Lee and K. K. Tan, *On Fan's extensions of Browder's fixed point theorems for multivalued inward mappings*, J. Austral. Math. Soc. (Ser A) **26** (1978), 169-174.
341. C. E. Lemke, *Bimatrix equilibrium points and mathematical programming*, Management Sci. **11** (1965), 681-689.

342. T. C. Lim, *A fixed point theorem for multivalued nonexpansive mappings in a uniformly convex Banach space*, Bull. Amer. Math. Soc. **80** (1974), 1123-1126.

343. B. L. Lin and X. C. Quan, *A symmetric minimax theorem without linear structure*, Arch. Math. (Basel) **52** (1989), 367–370.

344. B. L. Lin and X. C. Quan, *A non-compact topological minimax theorem*, J. Math. Anal. Appl. **161** (1991), 587-590.

345. B. L. Lin and S. Simons, *Nonlinear and Convex Analysis: Proceedings in Honor of Ky Fan*, Marcel Dekker, Inc., 1987.

346. M. Lin and R. C. Sine, *Retractions on the fixed point set of semigroup of nonexpansive maps in hyperconvex spaces*, Nonlinear Anal., T.M.A. **15** (1990), 943-954.

347. J. L. Lions, *Optimal Control of Systems Governed by Partial differential Equations*, Springer-Verlag, 1971.

348. J. V. Llinares, *Abstract Convexity, Fixed Points and Applications*, Ph.D Thesis, University of Alicante, Spain, 1994.

349. J. V. Llinares, *Unified treatment of the existence of maximal elements in binary relation: a characterization (in press)*, J. Math. Econom. (1998).

350. J. V. Llinares, *Existence of equilibrium in abstract economies with an abstract convexity*, Manuscript, CEPREMAP, France (1997).

351. D. T. Luc, *Theory of Vector Optimization, Lecture Notes in Economics and Mathematical Systems*, vol. 319, Springer-Verlag, Berlin, 1989.

352. D. T. Luc, *An existence theorem in vector optimization*, Math. Oper. Res. **14** (1989), 693–699.

353. D. T. Luc and C. Vargas, *A saddle point theorem for set-valued maps*, Nonlinear Anal., TMA **18** (1992), 1–7.

354. R. Lucchetti and J. Revalski (eds), *Recent developments in Well-Posed Variational Problems*, Kluwer Academic Publishers, 1995.

355. Q. Luo, X. Z. Yuan and T. X. Sun, *The existence of essential components for coincident points of set-valued mappings*, CADSMAP Research Report, 97-15, The University of Queensland, Australia (1997).

356. T. W. Ma, *On sets with convex sections*, J. Math. Anal. Appl. **27** (1969), 413-416.

357. Y. H. Ma and F. B. Zhang, *Generalized minimax inequalities and fixed point theorems*, Appl. Math. Mech. (English Ed.) **12** (1991), 465-472.

358. E. Marchi and J. E. Martinez-Legaz, *A generalization of Fan-Browder's fixed point theorems and its applications*, Topol. Method.

Nonlinear Anal. **2** (1993), 277-291.

359. E. Marchi and J. E. Martinez-Legaz, *Some results on approximate continuous selections, fixed points and minimax inequalities*, Fixed Point Theory and Applications (M. A. Thera and J. B. Baillon, eds.), Longman Scientific Technical, New York, 1991, p. 327-342.
360. S. A. Marano, *Controllability of partial differential inclusions depending on a parameter and distributed parameter control processes*, Le Matematiche **95** (1990), 283-300.
361. M. Maschler and B. Peleg, *Stable sets and stable points of set-valued dynamic systems with applications to game theory*, SIAM J. Control Optim. **14** (1976), 985-995.
362. A. Mas-Colell, *An equilibrium existence theorem without complete or transitive preferences*, J. Math. Economics **1** (1974), 237-246.
363. A. Mas-Colell, *The Theory of General Economic Equilibrium*, Cambridge Univ. Press, Cambridge, 1985.
364. A. Mas-Colell and S. Richard, *A new approach to the existence of equilibria in vector lattices*, J. Economic Theory **53** (1991), 1-11.
365. A. Mas-Colell and W. R. Zame, *Equilibrium Theory in Infinite Dimensional Spaces*, Handbook of Mathematical Economics (W. Hildenbrand and H. Sonnenschein, eds.), vol. IV, 1993, pp. 1835-1898.
366. L. McLinden, *Acyclic multifunctions without metrizability*, presented in the First International Conference on Fixed Point Theory and Applications, Marseille-Luminy (June, 1989).
367. G. Mehta, *Fixed points, equilibria and maximal elements in linear topological vector spaces*, Comm. Mathm. Univ. Carolinae **28** (1987), 377-385.
368. G. Mehta, *Maximal elements of condensing preference maps*, Appl. Math. Letter **3** (1990), 69-71.
369. G. Mehta and S. Sessa, *Coincidence theorems and maximal elements in topological vector spaces*, Math. Japonica **37** (1993), 839-845.
370. G. Mehta and E. Tarafdar, *Infinite-dimensional Gale-Debreu theorem and a fixed point theorem of Tarafdar*, J. Econ. Theory **41** (1987), 333-339.
371. E. Michael, *Continuous selections I*, Ann. Math. **63** (1956), 361-382.
372. G. Minity, *Monotone nonlinear equations in Hilbert spaces*, Duke J. Math. **29** (1962), 341-346.
373. M. D. P. Monteiro Marques, *Rafle par un convexe semi-continu inférieurement d'Intérieur non vide en dimension finie*, Séminaire

d' Analyse Convex, Montpellier, Exposé **No. 6** (1984).

374. J. J. Moreau, *Intersection of moving convex sets in a normed space*, Math. Scand. **36** (1975), 159-173.
375. U. Mosco, *Implicit problems and quasi-variational inequalities*, Lecture Notes in Mathematics, vol. 543, Springer-Verlag, 1976, p. 83-156.
376. U. Mosco and G. Strang, *One-side approximation and variational inequalities*, Bull. Amer. Math. Soc. **80** (1974), 308-312.
377. L. Nachbin, *A theorem of Hahn-Banach type for linear transformations*, Trans. Amer. Math. Soc. **68** (1960), 28-54.
378. S. B. Nalder, Jr, *Multi-valued contraction mappings*, Pacific J. Math. **30** (1969), 475-488.
379. Z. Naniewicz and P. D. Panagiotopoulos, *Mathematical theory of Hemivariational Inequalities and Applications*, Marcel Dekker, Inc., New York, 1995.
380. J. Nash, *Equilibrium points in N-person games*, Proc. Nat. Acad. Sci. U.S.A. **36** (1950), 48-49.
381. J. Nash, *The bargaining problem*, Econometrica **18** (1950), 155–162.
382. J. Nash, *Non-cooperative games*, Ann. Math. **54** (1951), 286-295.
383. J. Nash, *Two-person cooperative games*, Econometrica **21** (1953), 128–140.
384. O. N. Ricceri, *On the covering dimension of the fixed point set of certain multifunctions*, Comm. Math. Univ. Caroline **32** (1991), 281-286.
385. W. Neuefeind, *Notes on existence of equilibrium points and boundary behavior of supply*, Econometrica **48** (1980), 1831–1837.
386. J. W. Nieuwenhuis, *Some minimax theorems in vector-valued functions*, J. Optim Theory Appl. **40** (1983), 463-475.
387. H. Nikaido, *Note on the general economic equilibrium for nonlinear production functions*, Econometric **22** (1954), 49-53.
388. H. Nikaido, *Convex Structures and Economic Theory*, Academic Press, New York, 1967.
389. H. Nikaido, *Note on non-cooperative games*, Pacific. J. Math. **5** (1955), 807-815.
390. M. A. Noor, *Strongly nonlinear variational inequalities*, C. R. Math. Rep. Acad. Math. Sci. Canada **4** (1984), 213-218.
391. M. A. Noor, *An iterative scheme for a class of quasivariational inequalities*, J. Math. Anal. Appl. **110** (1985), 463-468.
392. M. A. Noor, *A Generalized quasimildly nonlinear variational inequalities*, Variational Methods in Engineering (C. A. Brebbia, ed.),

Springer-Verlag, Berlin, 1985, pp. 3.3-3.11.

393. M. A. Noor, *On the nonlinear complementarity problem*, J. Math. Anal. Appl. **123** (1987), 455-460.
394. M. A. Noor, *General variational inequalities*, Appl. Math. Letter **1** (1988), 119-122.
395. M. A. Noor, *Quasivariational inequalities*, Appl. Math. Letter **1** (1988), 367-370.
396. M. A. Noor, *Nonlinear variational inequalities in elastostatics*, Int. J. Eng. Sci. **26** (1988), 1043-1051.
397. M. A. Noor, *General algorithm for variational inequalities*, J. Optim. Theory Appl. **73** (1992), 409-413.
398. M. A. Noor, *An iterative algorithm for variational inequalities*, J. Math. Anal. Appl. **158** (1991), 448-455.
399. M. A. Noor, *General multivalued variational inequalities*, J. Natur. Geom. **6** (1994), 145–159.
400. M. A. Noor, *Some recent advances in variational inequalities*, New Developments in Approximation-Solvability of Nonlinear Equations - Abstract Differential and Integral Equations: Theory and Applications (Ram U. Verma, ed.), Marcel Dekker, Inc., New York, 1993.
401. M. A. Noor, *Some recent advances in variational inequalities: part I: basic concepts and part II: other concepts*, New Zealand J. Math. **26** (1997), 53-80; 229-255.
402. M. A. Noor, K. I. Noor and T. Rassias, *Invitation to variational inequalities*, Analysis, Geometry and Groups: a Riemann Legacy Volume, Hadronic Press Collect. Orig. Artic., Hadronic Press, Palm Harbor, FL., 1993, pp. 373–448.
403. M. A. Noor, K. I. Noor and Th. M. Rassias, *Some aspects of variational inequalities*, J. Comput. Appl. Math. **47** (1993).
404. M. A. Noor and S. Zarae, *Linear quasi-complementarity problems*, Utilitas Math. **27** (1987), 249-260.
405. R. D. Nussbaum,, *The fixed point index for local condensing maps*, Ann. Mat. Pura Appl. **81** (1971), 217-258.
406. V. V. Obukhovskii, Dissertation (1974), Voronezh.
407. J. M. Ortega and W. C. Rheinboldt, *Iterative Solutions of Nonlinear Equations in Several Variables*, Academic Press, New York, 1970.
408. J. C. Oxtoby, *Cartesian products of Baire spaces*, Fund. Math. **49** (1961), 157-166.
409. P. D. Panagiotopoulos, *Hemivariational Inequalities and Applications in Mechanics and Engineering*, Springer-Verlag, Berlin, 1993.

410. J. S. Pang, *On the convergence of a basic iterative method for the implicit complementarity problem*, J. Optim. Theory and Appl. **37** (1982), 149-162.
411. J. Parida and A. Sen, *A variational-like inequality for multifunctions with applications*, J. Math. Anal. Appl. **124** (1987), 73-81.
412. S. Park, *Eighty years of the Brouwer fixed point theorem*, Antipodal Points and Fixed Points, vol. 25, Lecture Notes Ser., Seoul Nat. Univ., Seoul, Korea, 1995, p. 55-97.
413. S. Park, *Generalized Leray-Schauder principles for compact admissible multifunctions*, Topol. Methods Nonlinear Anal. **5** (1995), 271–277.
414. S. Park, *Fixed points of approximable maps*, Proc. Amer. Math. Soc. **124** (1996), 3109-3114.
415. S. Park, *Acyclic maps, minimax inequalities and fixed points*, Nonlinear Anal., T.M.A **24** (1995), 1549-1554.
416. S. Park, *Best approximation theorems for composites of upper semicontinuous maps*, Bull. Austral. Math. Soc. **51** (1995), 263–272.
417. S. Park, *Foundations of the KKM theory via coincidences of composites of upper semicontinuous maps*, J. Korean Math. Soc. **31** (1994), 493–519.
418. S. Park, *Fixed point theorems in hyperconvex metric spaces (in press)*, Nonlinear Anal. T.M.A. (1998).
419. S. Park, *Fixed point theorems on compact convex sets in topological vector spaces*, Contemporary Mathematics, vol. 72, 1988, p. 183-191.
420. S. Park, *Generalization of Ky Fan's matching theorems and their application*, J. Math. Anal. Appl. **141** (1989), 164-176.
421. S. Park, *Some coincidence theorems on acyclic multifunctions and applications to KKM theory*, Fixed Point Theory and Applications (K. K. Tan, ed.), World Scientific, Singapore, 1992, pp. 248-278.
422. S. Park, *On minimax inequalities on spaces having certain contractible subsets*, Bull. Austral. Math. Soc. **47** (1993), 25-40.
423. S. Park and J. S. Bae, *Existence of maximizable quasiconcave functions on convex spaces*, J. Korean Math. Soc. **28** (1991), 285-292.
424. S. Park, J. S. Bae and H. K. Kang, *Geometric properties, minimax inequalities and fixed point theorems on convex spaces*, Proc. Amer. Math. Soc. **121** (1994), 429-439.
425. S. Park and M. P. Chen, *Generalized quasivariational inequalities*, Far East J. Math. Sci. **3** (1995), 185–190.
426. S. Park and K. S. Jeong, *A general coincidence theorem on contractible spaces*, Proc. Amer. Math. Soc. **124** (1996), 3203-3206.

427. S. Park and H. Kim, *Admissible classes of multifunctions on generalized convex spaces*, Proc. Coll. Nat. Sci. SNU **18** (1993), 1-21.
428. S. Park, S. P. Singh and B. Watson, *Remarks on best approximations and fixed points*, Indian J. Pure Appl. Math. **25** (1994), 459-462.
429. D. Pascali and S. Sburlan, *Nonlinear Mappings of Monotone Type*, Bucuresti, Romania, Sijthoff and Noordhoff International Publisher, 1976.
430. B. Peleg, *Equilibrium points for open acyclic relations*, Canadian J. Math. **19** (1967), 366-369.
431. A. Pitonyak, P. Shi and M. Shillor, *On an iterative method for variational inequalities*, Numer. Math. **58** (1990), 231-244.
432. M. Powers, *Lefschetz fixed point theorems for a new class of multi-valued maps*, Pacific. J. Math. **68** (1970), 619-630.
433. S. Reich, *Fixed points in locally convex spaces*, Math. Z. **125** (1972), 17-31.
434. S. Reich, *On fixed point theorems obtained from existence theorems for differential equations*, J. Math. Anal. Appl. **54** (1976), 26-36.
435. S. Reich, *Approximate selections, best approximations, fixed points, and invariant sets*, J. Math. Anal. Appl. **62** (1978), 104-113.
436. B. Ricceri, *Un théorème d'existence pour les inéquations*, C. R. Acad. Sci. Paris S'er I Math. **301** (1985), 885-888.
437. B. Ricceri, *Some topological mini-max theorems via an alternative principle for multifunctions*, Arch. Math. **60** (1993), 367-377.
438. B. Ricceri, *Recent uses of connectedness in functional analysis, presented*, Nonlinear Analysis and Convex Analysis (W. Takahashi, ed.), vol. RIMS Kokyuroku 939, Research Institute for Mathematical Science, Kyoto University, Kyoto, Japan, 1995, pp. 11-22.
439. B. Ricceri, *On some recent topological mini-max theorems*, Equilibrium Problems with Side Constraints, Lagrangean Theory and Duality (Acireale, 1994), Matematiche(Catania), vol. 49, 1995, p. 349–357.
440. B. Ricceri, *On the integrable selections of certain multifunctions*, Set-Valued Anal. **4** (1996), 91–99.
441. B. Ricceri, *Applications of a theorem concerning sets with connected sections*, Topol. Methods Nonlinear Anal. **5** (1995), 237–248.
442. B. Ricceri, *Existence of zeros via disconnectedness*, J. Convex Anal. **2** (1995), 287–290.
443. B. Ricceri, *A variational property of integral functionals on L_p-spaces of vector-valued functions. (French)*, C. R. Acad. Sci. Paris Sr. I Math. **318** (1994), 337–342.

444. P. T. Rockafellar, *Convex Analysis*, Princeton, New Jersey, Princeton University Press, 1970.
445. P. T. Rockafellar, *Lagrange multipliers and variational inequalities*, Variational Inequalities and Complementarity Problems, Theory and Applications (Cottle et al., eds.), New York, 1980, pp. 303-322.
446. P. T. Rockafellar and R. J.B Wets, *Variational Analysis*, Springer-Verlag, New York, 1998.
447. S. M. Robinson, *Analysis and Computation of Fixed Points*, Academic Press, New York, 1980.
448. J. F. Rodrigues, *Obstacle Problems in Mathematical Physics*, North-Holland, Amsterdam, 1987.
449. W. Rudin, *Functional Analysis*, McGraw-Hill, New York, 1973.
450. N. B. Sadovskii, *Ultimately compact and condensing mappings*, Uspehi Mat. Nauk **27** (1972), 81-146.
451. R. Saigal, *Extension of the generalized complementarity problem*, Math. Oper. Res. **1** (1976), 260-266.
452. H. Scarf, *The core of an n-person game*, Econometrica **35** (1967), 50-69.
453. H. H. Schaefer, *Topological Vector Spaces*, Springer-Verlag, New York, 1971.
454. M. Schechter, *A subgradient duality theorem*, J. Math. Anal. Appl. **61** (1977), 850-855.
455. V. M. Sehgal, S. P. Singh and R. E. Smithson, *Nearest points and some fixed point theorems for weakly compact sets*, J. Math. Anal.Appl. **128** (1987), 108-111.
456. S. Sessa and G. Mehta, *Coincidence theorems and maximal elements in topological vector spaces*, Mathematiea Japonica **37** (1993), 839-845.
457. H. H. Shaefer, *Topological Vector Spaces*, Springer-Verlag, New York, 1971.
458. W. Shafer, *Equilibrium in economies without ordered preferences or free disposal*, J. Math. Economics. **3** (1976), 135-137.
459. W. Shafer and H. Sonnenschein, *Equilibrium in abstract economies without ordered preferences*, J. Math. Econ. **2** (1975), 345-348.
460. L. S. Shapley, *On balanced games without side payments*, Mathematical programming (T. C. Hu and S. M. Robinson, eds.), New York, Academic Press, 1973, pp. 260-290.
461. L. Shapley and R. Vohra, *On Kakutani's fixed point theorem, the K-K-M-S theorem and the core of a balanced game*, Econom. Theory **1** (1991), 108–116.

462. Yy. A. Shashkin, *Fixed Points*, vol. 2, Amer. Math. Society and Math. Assoc. Amer., 1991.
463. P. Shi, *Equivalence of variational inequalities with Wiener-Hopf equations*, Proc. Amer. Math. Soc. **111** (1991), 339-346.
464. M. H. Shih and S. N. Lee, *A combinatorial Lefschetz fixed-point formula*, J. Combin. Theory, Ser. A **61** (1992), 123-129.
465. M. H. Shih and K. K. Tan, *The Ky Fan minimax principle, sets with convex sections, and variational inequalities, in: Differential Geometry, Calculus of Variations and Their Applications*, Lecture Notes in Pure Appl. Math. **100** (1985), Dekker, 471-481.
466. M. H. Shih and K. K. Tan, *Generalized quasi-variational inequalities in locally convex topological vector spaces*, J. Math. Anal. Appl. **108** (1985), 333-343.
467. M. H. Shih and K. K. Tan, *Minimax inequalities and applications*, Contemp. Math., vol. 54, 1986, pp. 45-63.
468. M. H. Shih and K. K. Tan, *Covering theorems of simplexes and systems of linear inequalities*, Linear and Multilinear Algebra **19** (1986), 309-320.
469. M. H. Shih and K. K. Tan, *Covering theorems of convex sets related to fixed point theorems*, Nonlinear and Convex Analysis, proceeding in Honor of Ky Fan (B. L. Lin and S. Simons, eds.), Marcel Dekker, Inc., New York, 1987, pp. 235-244.
470. M. H. Shih and K. K. Tan, *Shapley selections and covering theorems of simplexes*, Nonlinear and Convex Analysis (B. L. Lin and S. Simons, eds.), Marcel Dekker, Inc., 1987, pp. 245-251.
471. M. H. Shih and K. K. Tan, *The Browder-Hartman-Stampacchia Variational inequalities for multi-valued monotone operators*, J. Math. Anal. Appl. **134** (1988), 431-440.
472. M. H. Shih and K. K. Tan, *A minimax inequality and Browder-Hartman-Stampacchia variational inequalities for multi-valued monotone operators*, Proceedings of Fourth FRANCO-SEAMS Joint Conference, Chiang Mai, Thailand, 1988.
473. M. H. Shih and K. K. Tan, *A geometric property of convex sets with applications to minimax type inequalities and fixed point theorems*, J. Austral. Math. Soc. Ser. A **45** (1988), 169-183.
475. N. Shioji, *A further generalization of the Knaster-Kuratowski-Mazurkiewicz theorem*, Proc. Amer. Math. Soc. **111** (1991), 187-195.

475. N. Shioji, *A further generalization of the Knaster-Kuratowski-Mazurkiewicz theorem*, Proc. Amer. Math. Soc. **111** (1991), 187-195.

476. A. H. Siddiqi and Q. H. Ansari, *An integrative method for generalized variational inequalities*, Math. Japonica **34** (1989), 475-481.

477. A. H. Siddiqi and Q. H. Ansari, *Strong nonlinear quasivariational inequalities*, J. Math. Anal. Appl. **149** (1990), 444-450.

478. A. H. Siddiqi and Q. H. Ansari, *General strongly nonlinear variational inequalities*, J. Math. Anal. Appl. **166** (1992), 386-392.

479. A. H. Siddiqi, Q. H. Ansari and A. Khaliq, *On vector variational inequalities*, J. Optim. Theory Appl. **84** (1995), 171-180.

480. S. Simons, *Cyclical coincidences of multivalued maps*, J. Math. Soc. Japan **38** (1986), 515-525.

481. S. Simons, *Two-function minimax theorems and variational inequalities for functions on compact and noncompact sets, with some comments on fixed point theorems*, Proc. Sympos. Pure Math., vol. 45, Part 2, 1986, pp. 377-392.

482. S. Simons, *An existence theorem for quasiconcave functions with applications*, Nonlinear Analysis, TMA **10** (1986), 1133-1152.

483. S. Simons, *An upward-downward minimax theorem*, Arch. Math. **55** (1990), 275-279.

484. S. Simons, *A flexible minimax theorem*, Acta. Math. Hung. **63** (1994), 119-132.

485. S. Simons, *Minimax theorems and their proofs*, Minimax and Applications (D. Z. Du and P. Pardaalos, eds.), Kulwer Academic Publisher, Dordrecht, 1995, p. 1-23.

486. R. C. Sine, *On nonlinear contraction semigroups in Sup-norm spaces*, Nonlinear Anal., T. M. A. **3** (1979), 885-890.

487. R. Sine, *Hyperconvexity and approximate fixed points*, Nonlinear Analysis, T.M.A. **13** (1989), 863-869.

488. R. C. Sine, *Hyperconvexity and nonexpansive multifunctions*, Trans. Amer. Math. Soc. **315** (1989), 755-767.

489. S. P. Singh, B. Watson and P. Srivastava, *Fixed point theory and best approximation: the KKM-map principle*, Kluwer Academic Publisher, Dordrecht, 1997.

490. M. Sion, *On nonlinear variational inequalities*, Pacific J. Math. **8** (1958), 171-176.

491. D. R. Smart, *Fixed Point Theorems*, Cambridge Univ. Press, Cambridge, 1974.

492. S. Smale, *Exchange processes with price adjustment*, J. Math. Econom. **3** (1976), 211-226.

493. T. E. Smith, *A solution condition for complementarity problems with an application to spacial proce equilibrium*, Appl. Math. Comp. **15** (1984), 61-69.

494. P. M. Soardi, *Existence of fixed points of nonexpansinve mappings*

in certain Banach lattices, Proc. Amer. Math. Soc. **73** (1979), 25-29.

495. H. Sonnenschein, *Demand theory without transitive preference with applications to the theory of competitive equilibrium*, Preference, Utility and Demand (J. Chipman et al., eds.), Harcourt Brace Jovanovich, New York, 1971.
496. L. L. Stach, *A note on Ky Fan's minimax theorem*, Acta Math. Acad. Sci. Hungar. **39** (1982), 401–407.
497. G. Stampacchia, *Formes bilinéaires coercitives sur les ensembles convexes*, C. R. Acad. Sci. Paris **258** (1964), 4413-4416.
498. E. Sperner, *Neuer Beweis für die invarianz der dimensionszahl und des gebietes*, Abh. Math. Sem. Univ. Hamburg **6** (1928), 265-272.
499. L. Stachó, *Minimax theorems beyond topological vector spaces*, Acta Sci. Math. **42** (1980), 157-174.
500. F. Szidarovszky, M. E. Gershon and L. Duckstein, *Techniques for Multiobjective Decision Making in Systems Management*, Elsevier, Amsterdam, Holland, 1986.
501. W. Takahashi, *Fixed point, minimax and Hahn-Banach theorems*, Proc. Sympos. Pure Math., vol. 45, 1986, Part 2, pp. 419-427.
502. W. Takahashi, *Nonlinear Analysis and Mathmeatical Economics*, vol. 861, RIMS, Kyoto University, Japan, 1994.
503. W. Takahashi, E. Tarafdar and X. Z. Yuan, *The existence and stability of saddle points and minimax theorems for vector-valued optimization theory*, Research Reports, Department of Mathematics, The University of Queensland **Feb.-1** (1996).
504. K. K. Tan, *Comparison theorems on minimax inequalities, variational inequalities, and fixed point theorems*, J. London Math. Soc. **28** (1983), 555-562.
505. K. K. Tan and J. Yu, *A new minimax inequality with applications to existence theorems of equilibrium points*, J. Optim. Theory. Appl. **82** (1994), 105-120.
506. K. K. Tan and J. Yu, *Minimax inequalities and generalisations of the Gale-Nikaido-Debreu lemma*, Bull. Austral. Math. Soc. **49** (1994), 267–275.
507. K. K. Tan and X. Z. Yuan, *A minimax inequality with applications to existence of equilibrium points*, Bull. Austral. Math. Soc. **47** (1993), 483-503.
508. K. K. Tan and X. Z. Yuan, *Equilibria of generalized games with $\mathcal{U}$-majorized preference correspondences*, Econom. Lett. **41** (1993), 379-383.

509. K. K. Tan and X. Z. Yuan, *Lower semicontinuity of multivalued mapping and equilibrium points*, Proceedings of the First World Congress of Nonlinear Analysis -1992 (Tampa, FL, 1992), vol. II, Walter de Gruyter Publishers, 1996, pp. 1849-1860.
510. K. K. Tan and X. Z. Yuan, *Variational inequalities on reflexive Banach spaces and applications*, J. Natural Geometry **5** (1994), 43-58.
511. K. K. Tan and X. Z. Yuan, *Existence of equilibrium for abstract economies*, J. Math. Econmics **23** (1994), 243-251.
512. K. K. Tan and X. Z. Yuan, *Approximation methods and equilibria of generalized games*, Proc. Amer. Math. Soc. **122** (1994), 503-510.
513. K. K. Tan and X. Z. Yuan, *Equilibria of generalized games*, Nonlinear Studies (former named 'Nonlinear Time and Digest') **1** (1994), 61-70.
514. K. K. Tan, J. Yu and X. Z. Yuan, *The stability of Coincidence points and fixed points for multivalued mappings*, Nonlinear Anal. T. M. A. **25** (1995), 163-168.
515. K. K. Tan, J. Yu and X.Z. Yuan, *The stability of Ky Fan's Points*, Proc. Amer. Math. Soc. **123** (1995), 1511-1519.
516. K.K. Tan, J. Yu and X.Z. Yuan, *The uniqueness of saddle points*, Bull. Polish. Acad. Sci. Math. **43** (1995), 119-129.
517. K.K. Tan, J. Yu and X.Z. Yuan, *Existence theorems of Nash equilibria for noncooperative N-person games*, Int. J. Game Theory **24** (1995), 217-222.
518. K. K. Tan, J. Yu and X. Z. Yuan, *Some new minimax inequalities and applications to generalized games in H-spaces*, Nonlinear Anal. T. M. A. **24** (1995), 1457-1470.
519. K. K. Tan, J. Yu and X. Z. Yuan, *The stability of generalized quasi-variational inequalities jour Panamer Math. J.* **5** (1995), 51-61.
520. K. K. Tan, J. Yu and X. Z. Yuan, *The Generic Stability of Some Nonlinear Probelms*, vol. October- 1, Research Report, Department of Mathematics, Statistics and Computing Scienece, Dalhousie University, 1995.
521. K. K. Tan, J. Yu and X. Z. Yuan, *Existence theorems for saddle points of vector-valued maps*, J. Optim. Theory Appl. **89** (1996), 731–747.
522. K. K. Tan, J. Yu and X.Z. Yuan, *The Stability of production economies*, J. Austra. Math. Soc., Ser. A **61** (1996), 162-170.
523. T. Tanaka, *Generalized quasiconvexities, cone saddle points, and a minimax theorem fro vector-valued functions*, J. Optim. Theory Appl. **81** (1994), 355-377.

524. T. Tanaka, *Minimax theorems of vector-valued functions*, Nonlinear Analysis and Mathematical Economics (W. Takahashi, ed.), vol. 861, Research Institute for Mathematical Sciences, Kyoto University, Japan, 1994, pp. 180-196.
525. T. Tanaka, *Two types of minimax theorems for vector-valued functions*, J. Optim Theory Appl. **68** (1991), 321-334.
526. T. Tanaka, *A characterization of generalized saddle points for vector-valued functions via scalarization*, Nihonkai Math. J. **1** (1990), 209-227.
527. T. Tanaka, *Some minimax problems of vector-valued functions*, J. Optim. Theory Appl. **59** (1988), 505-524.
528. T. Tanaka, *Existence theorems for cone saddle points of vector-valued function in infinite-dimensional spaces*, J. Optim. Theory Appl. **62** (1989), 127–138.
529. T. Tanaka, *Generalized quasiconvexities, cone saddle points and a minimax theorem for vector valued functions*, J. Optim. Theory Appl. **81** (1994), 355–377.
530. T. Tanaka and D. Kuroiwa, *The convexity of A and B assures* $intA + B = int(A + B)$, Appl. Math. Letter **6** (1993), 83-86.
531. E. Tarafdar, *On nonlinear variational inequalities*, Proc. Amer. Math. Soc. **67** (1977), 95-98.
532. E. Tarafdar, *Variational problems via a fixed point theorem*, Indian J. Math. **28** (1986), 229-240.
533. E. Tarafdar, *A fixed point theorem equivalent to Fan - Knaster - Kuratowski - Mazurkiewicz theorem*, J. Math. Anal. Appl. **128** (1987), 475–479.
534. E. Tarafdar, *A fixed point theorem in H-space and related results*, Bull. Austral. Math. Soc. **42** (1990), 133-140.
535. E. Tarafdar, *A fixed point theorem and equilibrium point of an abstract economy*, J. Math. Econom. **20** (1991), 211-218.
536. E. Tarafdar, *A fixed point theorem on H-spaces and equilibrium points of abstract economies*, J. Aust. Math. Soc. (ser. A) **53** (1992), 252-260.
537. E. Tarafdar, *Pareto solutions of cone inequality and Pareto optimality of a mapping*, Proceedings of the First World Congress of Nonlinear Analysts-1992, Florida, U.S.A., vol. III, Walter de Gruyter Publisher, 1995, pp. 2431-2439.
538. E. Tarafdar, *Applications of Pareto optimality of a mapping to mathematical economics*, Proceedings of the First World Congress of Nonlinear Analysts-1992, Florida, U.S.A., vol. III, Walter de Gruyter Publisher, 1995, pp. 2511-2519.

539. E. Tarafdar, *Fixed point theorems in locally H-convex uniform spaces*, Nonlinear. Anal., T. M. A. **29** (1997), 971-978.
540. E. Tarafdar and G. Mehta, *A generalized version of the Gale-Nikaio-Debreu theorem*, Comm. Math. Univ. Carolinae **28** (1987), 655-659.
541. E. Tarafdar and B. Thompson, *On the existence of the price equilibrium by different methods*, Comm. Math. Univ. Carolinae **34** (1993), 413-417.
542. E. Tarafdar and R. Vyborny, *A generalized (multivalued) contraction mapping principle*, Research report of Pure Mathematics, No.54, The University of Queensland, Australia (1976).
543. E. Tarafdar and P. Watson, *Coincidence and the Fan-Glicksberg fixed point theorem in locally H-convex uniform spaces*, Research Report, The University of Queensland (1997).
544. E. Tarafdar and X. Z. Yuan, *Equilibria for non-compact generalized games*, Bull. Polish Acad. Sci. Math. **41** (1993), 229-239.
545. E. Tarafdar and X. Z. Yuan, *Non-compact generalized quasi - variational inequalities in locally convex topological vector spaces*, Nonlinear World **1** (1994), 373-383.
546. E. Tarafdar and X. Z. Yuan, *Set-valued contraction mapping principle, Applied Math. Letter* **8** (1995), 79-81.
547. E. Tarafdar and X. Z. Yuan, *Existence of equilibria of generalized games without compactness and paracompactness*, Nonlinear Anal., TMA **26** (1996), 893-902.
548. E. Tarafdar and X. Z. Yuan, *Equilibria of non-compact generalized games and non-compact quasi-variational inequalities*, Indian J. Pure Appl. Math. **25** (1994), 1207–1222.
549. E. Tarafdar and X. Z. Yuan, *Non-compact Pareto equilibria for multiobjective games*, J. Math. Anal. Appl. **204** (1996), 156-163.
550. F. Terkelson, *Some minimax theorems*, Math. Scand **31** (1972), 405-413.
551. E. Thompson and X. Z. Yuan, *Topological intersection theorems for two set-valued mappings and applications to minimax inequalities*, Numer. Funct. Anal. Optim. **17** (1996), 437–452.
552. G. Tian, *Equilibrium in abstract economies with a non-compact infinite dimensional strategy space an infinite number of agents and without ordered preferences*, Economic Letters **33** (1990), 203-206.
553. G. Tian, *Existence of equilibrium in abstract economies with discontinuous payoffs and noncompact choice spaces*, J. Math. Econom. **21** (1992), 379–388.

554. G. Tian, *Generalizations of the FKKM theorem and the Ky Fan minimax inequality, with applications to maximal elements, price equilibrium, and complementarity*, J. Math. Anal. Appl. **170** (1992), 457-471.
555. G. Tian and J. Zhou, *Transfer continuities, generalizations of the Weierstrass and maximum theorems: a full characterization*, J. Math. Econom. **24** (1995), 281-303.
556. G. Tian and J. Zhou, *Quasi-variational inequalities with non-compact sets*, J. Math. Anal. Appl. **160** (1991), 583-595.
557. J. V. Tiel, *Convex Analysis*, John Wiley and Sons Ltd., 1984.
558. S. H. Tijs, *Nash equilibria for noncooperative n-person game in normal form*, SIAM Review **23** (1981), 225-237.
559. R. Toscano and A. Maceri, *On the elastic stability of beams under unilateral constraints*, Meccanica, March (1980), 31-38.
560. S. Toussaint, *On the existence of equilibria in economies with infinitely many commodities and without ordered preferences*, J. Econ. Theory **33** (1984), 98-115.
561. C. I. Tulcea, *On the equilibriums of generalized games*, The Center for Mathematical Economics Studies in Economics and Managements Sciences, University of Maryland (1986).
562. C. I. Tulcea, *On the approximation of upper-semicontinuous correspondences and the equilibriums of generalized games*, J. Math. Anal. Appl. **136** (1988), 267-289.
563. H. Tuy, *On a general minimax theorem*, Soviet. Math. Dokl. **15** (1974), 1689-1693.
564. H. Tuy, *On the general minimax theorem*, Colloq. Math. **33** (1975), 145-158.
565. A. Tychonoff, *Ein Fixpunktsatz*, Math. Ann. **111** (1935), 767-776.
566. E. van Damme, *Stability and Prefection of Nash Equilibrian*, Springer-Verlag, Berlin, 1987.
567. J. von Neumann, *Zur theorie der gesellschaftsspiele*, Math. Ann. **100** (1928), 295-320.
568. J. von Neumann, *Uber ein okonomsiches Gleichungssystem und eine Verallgemeinerung des Browerschen Fixpunktsatzes*, Ergebnisse eines Mathematischen Kolloquiums **8** (1937), 73-83; English transl. *A model of general economic equilibrium (G. Morgenstern, Trans.)*, Review Econom. Stud. **13** (1945-1946), 1-9.
569. J. von. Neumann and O. Morgenstern, *Theory of Games and Economic Behavior*, Princeton University Press, Princeton, N.J., 1944.
570. M. Walker, *On the existence of maximal elements*, J. Economic Theory **16** (1977), 470-474.

571. S. Y. Wang, *Existence of a Pareto equilibrium*, J. Optim. Theory Appl. **79** (1993), 373-384.

572. C. W. Waters, *Some Fixed Point Theorems for Radial Contractions, Nonexpansive and Set-valued Mappings*, Ph.D Thesis, Univ. of Wyoming, 1984.

573. A. Wilansky, *Topology for Analysis*, Waltham Massachusetts: Ginn, 1972.

574. Y. C. Wong, *Introductory Theory of Topological Vector Spaces*, Marcel Dekker, Inc., 1992.

575. X. Wu, *Kakutani-Fan-Glicksberg fixed point theory with applications in H-spaces (in press)*, J. Math. Anal. Appl..

576. X. Wu, *A new fixed point theorem and its applications*, Proc. Amer. Math. Soc. **125** (1997), 1779-1783.

577. X. Wu and S. K. Shen, *A further generalization of Yannelis- Prabhakar's continuous selection theorem and its applications*, Journal of Mathematical Analysis and Applications **197** (1996), 61-74.

578. X. Wu and X. Z. Yuan, *Approximate selection, fixed Point, almost fixed point of multivalued mappings and generalized quasi-variational inequality in H-spaces (in press)*, Nonlinear Analysis, T.M.A. (1998).

579. W. T. Wu, *A remark on the fundamental theorem in the theory of games*, Science Record (New Ser.) **3** (1959), 229-233.

580. W. T. Wu and J. H. Jiang, *Essential equilibrium points of n-person non-cooperative games*, Sci. Sinina **11** (1962), 1307-1322.

581. X. Q. Yang, *Vector variational inequality and its duality*, Nonlinear Anal. TMA **21** (1993), 869–877.

582. X. Q. Yang, *Generalized convex functions and vector variational inequalities*, J. Optim. Theory Appl. **79** (1993), 563–580.

583. G. S. Yang and X. Z. Yuan, *A necessary and sufficient condition for upper hemicontinuous set-valued mappings without Compact-Values being upper demicontinuous*, Proc. Amer. Math. Soc. **126** (1998), 3539-3544.

584. G. S. Yang and X. Z. Yuan, *A characteristic of strong separation property of convex sets without compactness in locally convex spaces and some applications*, CADSMAP 98-7, The University of Queensland (1998).

585. N. C. Yannelis, *Maximal elements over non-compact subsets of linear topological spaces*, Economics Letters **17** (1985), 133-136.

586. N. C. Yannelis, *Existence of maxima elements and equilibria in linear topological spaces*, J. Math. Econ. **12** (1987), 96-111.

587. N. C. Yannelis and N. D. Prabhakar, *Existence of maximal elements and equilibria in linear topological spaces*, J. Math. Econ. **12** (1983), 233-245.
588. J. C. Yao, *eneralized-quasi-variational inequality problems with discontinuous mappings*, Math. Oper. Res. **20** (1995), 465–478.
589. J. C. Yao, *On the generalized complementarity problem*, J. Austral. Math. Soc. (Ser. B) **35** (1994), 420-428.
590. J. C. Yao and J. S. Guo, *Variational and generalized variational inequalities with discontinuous mappings*, J. Math. Anal. Appl. **182** (1994), 371–392.
591. C. L. Yen, *A minimax inequality and its applications to variational inequalities*, Pacific J. Math. **97** (1981), 477-481.
592. N. D. Yen, *On an existence theorem for generalzied quasi-variational inequalities*, Set-valued Analysis **3**((1995), 1-10.
593. J. Yu, *A generalization of Ky Fan's Lemma and its applications (in Chinese)*, Acta Math. Appl. Sinica **11** (1988), 423-428.
594. J. Yu, *On Nash equilibrium in N-person games over reflexive Banach spaces*, J. Optim. Theory and Appl. **73** (1992), 211-214.
595. J. Yu, *Essential weak efficient solution in multiobjective optimization problems*, J. Math. Anal. Appl. **166** (1992), 230-235.
596. J. Yu, *Essential equilibria of n-person non-cooperative games (in press)*, J. Math. Economics (1998).
597. J. Yu and Q. Luo, *On essential components of the solution set of generalized games*, Research Report, Guizhou University of Technology (1997).
598. J. Yu and S. W. Xiang, *On essential components of Nash equilibrium points*, submitted to Nonlinear Analy., T.M.A. (1998).
599. X. T. Yu, *The Geometry of Banach Spaces*, East China Normal University Press, Shanghai, China, 1986.
600. J. Yu and X. Z. Yuan, *An saddle point beyond topological vector spaces*, J. Optim. Theory Appl. **98** (1998), 211-219.
601. P. L. Yu, *Cone convexity, cone extreme points, and nondominated solutions in decision problems with multiobjectives*, J. Optim. Theory Appl. **14** (1974), 319–377.
602. P. L. Yu, *Second-order game problems: Decision dynamicsin gaming phenomena*, J. Optim. theory Appl. **27** (1979), 147-166.
603. X. Z. Yuan, *Remarks of random equilibria*, Panamer. Math. J. **3** (1993), 23-35.
604. X. Z. Yuan, *Contributions to Nonlinear Analysis and Applications*, Department of Matgematics, Statistcis and Computing Science, Dalhousie University, Canada, 1993.

605. X. Z. Yuan, *A note to equilibria for $\mathcal{U}$-majorized preferences*, C. R. Math. Acad. Sci. Canada **XV** (1994), 243-248.
606. X. Z. Yuan, *The KKM principle, Ky Fan minimax inequalitiesand fixed point theorems*, Nonlinear World **2** (1995), 131-169.
607. X. Z. Yuan, *The study of KKM theory and applications to abstract economics*, Nonlinear Analysis and Convex Analysis (W. Takahashi, ed.), vol. RIMS Kokyuroku 939, Research Institute for Mathematical Science, Kyoto University, Kyoto, Japan, 1995, pp. 30-48.
608. X. Z. Yuan, *Extensions of Ky Fan section theorems and minimax inequality theorems*, Acta. Math. Hungrica **71** (1996).
609. X. Z. Yuan, *The Study of Minimax and Applications to Economics and Variational Inequalities*, vol. 132, Mem. Amer. Math. Soc., 1998, pp. 140.
610. George X. Z. Yuan, *Fixed point theorems for set-valued mappings in abstract Convexity-MC spaces*, Cadsmap, 97-17, The University of Queensland (1998).
611. George X. Z. Yuan, *Fixed point theorems for upper semicontinuous set-valued mappings in locally G-convex spaces (in press)*, Bull. Austral. Math. Soc. (1998).
612. W. R. Zame, *Competitive equilibria in production economies with an infinite - dimensional commodity spaces*, Econometrica **55** (1987), 1075-1108.
613. E. Zeidler, *Nonlinear Functional Analysis and its Applications I: Fixed Point Theorems*, Springer-Verlag, Berlin, New York, 1985.
614. E. Zeidler, *Nonlinear Functional Analysis and its Applications II/B: Nonlinear Monotone Operators*, Springer-Verlag, Berlin, New York, 1985.
615. M. Zeleny, *Game with multiple payoffs*, Int. Game Theory **4** (1976), 179-191.
616. L. C. Zeng, *Iterative algorithms for finding approximate solutions of completely generalized strongly nonlinear quasi-variational inequalities*, Appl. Math. Mech. **15** (1994), 1069-1080.
617. Y. B. Zhao, *Exceptional families anf finite-dimensional variational inequalities over polyhedral convex sets*, Appl. Math. Comput. **87** (1997), 111-126.
618. Y. B. Zhao, *Existence Theory and Algorithms for Finite-Dimensional Variational Inequalities and Complementarity Problems*, Ph.D Dissertation, Institute of Applied Mathematics, Chinese Academy of Sciences, Beijing, 1998, pp. 1-100.
619. X. Y. Zheng, *Approximate selection theorems and their applications*, Journal of Mathematical Analysis and Applications **212** (1997),

88-97.

620. J. Zhou, *Necessary and sufficient conditions for KKM property and variational inequalities for preference relations*, Mime., Department of Mathematics, Texas A & M University, U.S.A. (1992).
621. J. Zhou, *On the existence of equilibrium for abstract economies*, J. Math. Anal. Appl. **193** (1995), 839–858.
622. J. Zhou and G. Chen, *Diagonal convexity conditions for problems in convex analysis and quasi - variational inequalities*, J. Math. Anal. Appl. **132** (1988), 213-225.
623. J. Zhou and G. Tian, *Transfer method for characterizing the existence of maximal elements of binary relations on compact or noncompact sets*, SIAM J. Optim. **2** (1992), 360-375.
624. Y. G. Zhu and George X. Z. Yuan, *The existence of nontrivial olutions for variational inqualities by using solution index approach*, Cadsmap 97-11, The University of Queensland (1997).

Index